Multiples and Prefixes for Metric Units*

Multiple	Prefix (and Abbreviation)	Pronunciation
10^{24}	yotta- (Y)	yot'ta (*a* as in *a*bout)
10^{21}	zetta- (Z)	zet'ta (*a* as in *a*bout)
10^{18}	exa- (E)	ex'a (*a* as in *a*bout)
10^{15}	peta- (P)	pet'a (as in *pet*al)
10^{12}	tera- (T)	ter'a (as in *terra*ce)
10^{9}	giga- (G)	ji'ga (*ji* as in *ji*ggle, *a* as in *a*bout)
10^{6}	mega- (M)	meg'a (as in *mega*phone)
10^{3}	kilo- (k)	kil'o (as in *kilo*watt)
10^{2}	hecto- (h)	hek'to (*heck-toe*)
10	deka- (da)	dek'a (*deck* plus *a* as in *a*bout)
10^{-1}	deci- (d)	des'i (as in *deci*mal)
10^{-2}	centi- (c)	sen'ti (as in *senti*mental)
10^{-3}	milli- (m)	mil'li (as in *mili*tary)
10^{-6}	micro- (μ)	mi'kro (as in *micro*phone)
10^{-9}	nano- (n)	nan'oh (*an* as in *an*nual)
10^{-12}	pico- (p)	pe'ko (*peek-oh*)
10^{-15}	femto- (f)	fem'toe (*fem* as in *fem*inine)
10^{-18}	atto- (a)	at'toe (as in *at*omy)
10^{-21}	zepto- (z)	zep'toe (as in *zep*pelin)
10^{-24}	yocto- (y)	yock'toe (as in *sock*)

*For example, 1 gram (g) multiplied by 1000 (10^{3}) is 1 kilogram (kg); 1 gram multiplied by 1/1000 (10^{-3}) is 1 milligram (mg).

SI Base Units

Physical Quantity	Name of Unit	Symbol
Length	meter	m
Mass	kilogram	kg
Time	second	s
Electric current	ampere	A
Temperature	kelvin	K
Amount of substance	mole	mol
Luminous intensity	candela	cd

Some SI Derived Units

Physical Quantity	Name of Unit	Symbol	SI Unit
Frequency	hertz	Hz	s^{-1}
Energy	joule	J	$kg \cdot m^2/s^2$
Force	newton	N	$kg \cdot m/s^2$
Pressure	pascal	Pa	$kg/(m \cdot s^2)$
Power	watt	W	$kg \cdot m^2/s^3$
Electric charge	coulomb	C	$A \cdot s$
Electric potential	volt	V	$kg \cdot m^2/(A \cdot s^3)$
Electric resistance	ohm	Ω	$kg \cdot m^2/(A^2 \cdot s^3)$
Capacitance	farad	F	$A^2 \cdot s^4/(kg \cdot m^2)$
Inductance	henry	H	$kg \cdot m^2/(A^2 \cdot s^2)$
Magnetic field	tesla	T	$kg/(A \cdot s^2)$

SI Units of Some Other Physical Quantities

Physical Quantity	SI Unit
Speed	m/s
Acceleration	m/s^2
Angular speed	rad/s
Angular acceleration	rad/s^2
Torque	$kg \cdot m^2/s^2$ or $m \cdot N$
Entropy	J/K or $kg \cdot m^2/(K \cdot s^2)$ or $N \cdot m/K$
Thermal conductivity	$W/(m \cdot K)$
Electric field	N/C or V/m

Conversion Factors

Mass
- 1 g = 10^{-3} kg
- 1 kg = 10^3 g
- 1 u = 1.66×10^{-24} g = 1.66×10^{-27} kg
- 1 metric ton = 1000 kg

Length
- 1 nm = 10^{-9} m
- 1 cm = 10^{-2} m = 0.394 in.
- 1 m = 10^{-3} km = 3.28 ft = 39.4 in.
- 1 km = 10^3 m = 0.621 mi
- 1 in. = 2.54 cm = 2.54×10^{-2} m
- 1 ft = 0.305 m = 30.5 cm
- 1 mi = 5280 ft = 1609 m = 1.609 km

Area
- 1 cm^2 = 10^{-4} m^2 = 0.1550 in^2
  = 1.08×10^{-3} ft^2
- 1 m^2 = 10^4 cm^2 = 10.76 ft^2 = 1550 in^2
- 1 in^2 = 6.94×10^{-3} ft^2 = 6.45 cm^2
  = 6.45×10^{-4} m^2
- 1 ft^2 = 144 in^2 = 9.29×10^{-2} m^2 = 929 cm^2

Volume
- 1 cm^3 = 10^{-6} m^3 = 3.35×10^{-5} ft^3
  = 6.10×10^{-2} in^3
- 1 m^3 = 10^6 cm^3 = 10^3 L = 35.3 ft^3
  = 6.10×10^4 in^3 = 264 gal
- 1 liter = 10^3 cm^3 = 10^{-3} m^3 = 1.056 qt
  = 0.264 gal
- 1 in^3 = 5.79×10^{-4} ft^3 = 16.4 cm^3
  = 1.64×10^{-5} m^3
- 1 ft^3 = 1728 in^3 = 7.48 gal = 0.0283 m^3
  = 28.3 L
- 1 qt = 2 pt = 946 cm^3 = 0.946 L
- 1 gal = 4 qt = 231 in^3 = 0.134 ft^3 = 3.785 L

Time
- 1 h = 60 min = 3600 s
- 1 day = 24 h = 1440 min = 8.64×10^4 s
- 1 y = 365 days = 8.76×10^3 h
  = 5.26×10^5 min = 3.16×10^7 s

Angle
- 1 rad = 57.3°
  - 1° = 0.0175 rad
  - 15° = $\pi/12$ rad
  - 30° = $\pi/6$ rad
  - 45° = $\pi/4$ rad
  - 60° = $\pi/3$ rad
  - 90° = $\pi/2$ rad
  - 180° = π rad
  - 360° = 2π rad
- 1 rev/min = $\pi/30$ rad/s = 0.1047 rad/s

Speed
- 1 m/s = 3.60 km/h = 3.28 ft/s
  = 2.24 mi/h
- 1 km/h = 0.278 m/s = 0.621 mi/h
  = 0.911 ft/s
- 1 ft/s = 0.682 mi/h = 0.305 m/s
  = 1.10 km/h
- 1 mi/h = 1.467 ft/s = 1.609 km/h
  = 0.447 m/s
- 60 mi/h = 88 ft/s

Force
- 1 N = 0.225 lb
- 1 lb = 4.45 N
- Equivalent weight of a mass of 1 kg on Earth's surface = 2.2 lb = 9.8 N

Pressure
- 1 Pa (N/m^2) = 1.45×10^{-4} lb/in^2
  = 7.5×10^{-3} torr (mm Hg)
- 1 torr (mm Hg) = 133 Pa (N/m^2)
  = 0.02 lb/in^2
- 1 atm = 14.7 lb/in^2 = 1.013×10^5 N/m^2
  = 30 in. Hg = 76 cm Hg
- 1 lb/in^2 = 6.90×10^5 Pa (N/m^2)
- 1 bar = 10^5 Pa
- 1 millibar = 10^2 Pa

Energy
- 1 J = 0.738 ft·lb = 0.239 cal
  = 9.48×10^{-4} Btu = 6.24×10^{18} eV
- 1 kcal = 4186 J = 3.968 Btu
- 1 Btu = 1055 J = 778 ft·lb = 0.252 kcal
- 1 cal = 4.186 J = 3.97×10^{-3} Btu
  = 3.09 ft·lb
- 1 ft·lb = 1.36 J = 1.29×10^{-3} Btu
- 1 eV = 1.60×10^{-19} J
- 1 kWh = 3.6×10^6 J

Power
- 1 W = 0.738 ft·lb/s = 1.34×10^{-3} hp
  = 3.41 Btu/h
- 1 ft·lb/s = 1.36 W = 1.82×10^{-3} hp
- 1 hp = 550 ft·lb/s = 745.7 W
  = 2545 Btu/h

Mass–Energy Equivalents
- 1 u = 1.66×10^{-27} kg ↔ 931.5 MeV
- 1 electron mass = 9.11×10^{-31} kg
  = 5.49×10^{-4} u ↔ 0.511 MeV
- 1 proton mass = 1.673×10^{-27} kg
  = 1.007 267 u ↔ 938.28 MeV
- 1 neutron mass = 1.675×10^{-27} kg
  = 1.008 665 u ↔ 939.57 MeV

Temperature
- $T_F = \frac{9}{5} T_C + 32$
- $T_C = \frac{5}{9}(T_F - 32)$
- $T_K = T_C + 273$

cgs Force
- 1 dyne = 10^{-5} N = 2.25×10^{-6} lb

cgs Energy
- 1 erg = 10^{-7} J = 7.38×10^{-6} ft·lb

College Physics

College Physics

Fourth Edition
Volume 1

Jerry D. Wilson
Lander University

Anthony J. Buffa
California Polytechnic State University
San Luis Obispo

PRENTICE HALL
Upper Saddle River, NJ 07458

Library of Congress Cataloging-in-Publication Data

Wilson, Jerry D.
   College physics.—4th ed. / Jerry D. Wilson, Anthony J. Buffa.
     p. cm.
   Includes bibliographical references and index.
   ISBN 0-13-084152-8 (Vol. 1)
   1. Physics.   I. Buffa, Anthony J.   II. Title.
QC21.2.W548   2000
530—dc21   99–12353
                                                             CIP

Executive Editor: Alison Reeves
Senior Development Editor: Karen Karlin
Editor in Chief: Paul F. Corey
Assistant Vice President of Production and Manufacturing: David W. Riccardi
Executive Managing Editor: Kathleen Schiaparelli
Assistant Managing Editor: Lisa Kinne
Editor in Chief–Development: Carol Trueheart
Senior Marketing Manager: Danny Hoyt
Project Manager: Elizabeth Kell
Director of Marketing: John Tweeddale
Manufacturing Manager: Trudy Pisciotti
Director of Creative Services: Paula Maylahn
Associate Creative Director: Amy Rosen
Art Director: Amy Rosen/Joseph Sengotta
Art Manager: Gus Vibal
Art Editor: Karen Branson
Art Studio: Academy Artworks
Interior Design: Rosemarie Votta
Cover Design: Joseph Sengotta
Assistant to Art Director: John M. Christiana
Cover Photographs: U.S. skater Christine Witty from West Allis, Wisconsin, skates during the women's 1500-meter speed skating competition at the Winter Olympics in Nagano on Monday, February 16, 1998. Witty won the bronze medal. Lionel Cironneau, AP/Wide World Photos; Homer Levi Dodge (1887–1983), the first President of the American Association of Physics Teachers, and Mrs. Margaret Dodge skating on the Iowa River, ca. January 1921. American Institute of Physics/Niels Bohr Library; Bont Sonic Clap Skate. Bont Skates Pty Ltd.
Photo Editor: Melinda Reo
Photo Researcher: Beaura K. Ringrose
Editorial Assistant: Gillian Buonanno
Assistant Editor: Wendy Rivers/Amanda Griffith
Senior Media Editor: Laura Pople
Production Supervision/Composition: WestWords, Inc.

© 2000, 1997, 1994, 1990 by Prentice-Hall, Inc.
Upper Saddle River, New Jersey 07458

All rights reserved. No part of this book may be
reproduced, in any form or by any means,
without permission in writing from the publisher.

Printed in the United States of America
10 9 8 7 6 5 4 3 2 1

ISBN 0-13-084152-8

Prentice-Hall International (UK) Limited, *London*
Prentice-Hall of Australia Pty. Limited, *Sydney*
Prentice-Hall Canada Inc., *Toronto*
Prentice-Hall Hispanoamericana, S.A., *Mexico*
Prentice-Hall of India Private Limited, *New Delhi*
Prentice-Hall of Japan, Inc., *Tokyo*
Prentice-Hall (*Singapore*) Pte Ltd
Editora Prentice-Hall do Brasil, Ltda., *Rio de Janeiro*

Brief Contents

Preface *xvii*

Part One: Mechanics

1 Units and Problem Solving *1*
2 Kinematics: Description of Motion *32*
3 Motion in Two Dimensions *64*
4 Force and Motion *99*
5 Work and Energy *137*
6 Momentum and Collisions *172*
7 Circular Motion and Gravitation *210*
8 Rotational Motion and Equilibrium *251*
9 Solids and Fluids *294*

Part Two: Thermodynamics

10 Temperature *333*
11 Heat *360*
12 Thermodynamics *388*

Part Three: Oscillations and Wave Motion

13 Vibrations and Waves *418*
14 Sound *452*

Appendices

I Mathematical Relationships *A-1*
II Kinetic Theory of Gases *A-2*
III Planetary Data *A-3*
IV Alphabetical Listing of the Chemical Elements *A-3*
V Properties of Selected Isotopes *A-4*
Answers to Follow-up Exercises *A-7*
Answers to Odd-numbered Exercises *A-15*
Photo Credits *P-1*
Index *I-1*

About the Authors

Jerry D. Wilson Jerry Wilson, a native of Ohio, is now Emeritus Professor of Physics and former Chair of the Division of Biological and Physical Sciences at Lander University in Greenwood, South Carolina. He received his B.S. degree from Ohio University, M.S. degree from Union College, and in 1970, a Ph.D. from Ohio University. He earned his M.S. degree while employed as a Materials Behavior Physicist by the General Electric Co.

As a doctoral graduate student, Professor Wilson held the faculty rank of Instructor and began teaching physical science courses. During this time, he co-authored a physical science text that is now in its eighth edition. In conjunction with his teaching career, Professor Wilson continued his writing and has authored or co-authored six titles. Having retired from full-time teaching, he continues to write, including *The Curiosity Corner*, a weekly column for local newspapers, which now can also be found on the Internet.

With several competitive books available, one may wonder why another algebra-based physics text was written. Having taught introductory physics many times, I was well aware of the needs of students and the difficulties they have in mastering the subject. I decided to write a text that presents the basic physics principles in a clear and concise manner, with illustrative examples that help resolve the major difficulty in learning physics: problem solving. Also, I wanted to write a text that is relevant so as to show students how physics applies in their everyday world—how things work and why things happen. Once the basics are learned, these follow naturally.

—*Jerry Wilson*

Anthony J. Buffa Anthony Buffa received his B.S. degree in physics from Rensselaer Polytechnic Institute and both his M.S. and Ph.D. degrees in physics from the University of Illinois, Urbana–Champaign. In 1970, Professor Buffa joined the faculty at California Polytechnic State University, San Luis Obispo, where he is currently Professor of Physics, and has been a research associate with the Department of Physics Radioanalytical Facility since 1980.

Professor Buffa's main interest continues to be teaching. He has taught courses at Cal Poly ranging from introductory physical science to quantum mechanics, has developed and revised many laboratory experiments, and has taught elementary physics to local teachers in an NSF-sponsored workshop. Combining physics with his interests in art and architecture, Dr. Buffa develops his own artwork and sketches, which he uses to increase his effectiveness in teaching physics.

I try to teach my students the crucial role physics plays in understanding all aspects of the world around them—whether it be technology, biology, astronomy, or any other field. In that regard, I emphasize conceptual understanding before number crunching. To this end, I rely heavily on visual methods. I hope the artwork and other pedagogical features in this book assist you in achieving your own teaching goals for your students.

—*Tony Buffa*

Contents

Preface *xiii*

1 Units and Problem Solving 1

1.1 Why and How We Measure 1
1.2 SI Units of Length, Mass, and Time 2
Insight: A 2000-Year-Old Calculation: The Earth's Circumference 3
1.3 More about the Metric System 6
Insight: A Bit of History: Why 5280 Feet in a Mile? 8
1.4 Dimensional Analysis and Unit Analysis 9
1.5 Unit Conversions 13
1.6 Significant Figures 17
1.7 Problem Solving 21
Chapter Review 26
Exercises 26

2 Kinematics: Description of Motion 32

2.1 Distance and Speed: Scalar Quantities 33
2.2 One-Dimensional Displacement and Velocity: Vector Quantities 35
Learn by Drawing: Cartesian Coordinates and One-dimensional Displacement 35
2.3 Acceleration 42
Learn by Drawing: Signs of Velocity and Acceleration 42
2.4 Kinematic Equations (Constant Acceleration) 46
2.5 Free Fall 50
Insight: Galileo Galilei and the Leaning Tower of Pisa 52
Chapter Review 56
Exercises 57

3 Motion in Two Dimensions 64

3.1 Components of Motion 65
3.2 Vector Addition and Subtraction 69
3.3 Relative Velocity 78
3.4 Projectile Motion 83
Chapter Review 90
Exercises 91

4 Force and Motion 99

4.1 The Concepts of Force and Net Force 100
4.2 Inertia and Newton's First Law of Motion 101
4.3 Newton's Second Law of Motion 102
Demonstration 1: Newton's First Law and Inertia 103
4.4 Newton's Third Law of Motion 109
4.5 More on Newton's Laws: Free-body Diagrams and Translational Equilibrium 111
Insight: Action and Reaction: The Harrier "Jump" Jet 112
Demonstration 2: Tension in a String: Action and Reaction Forces 113
Learn by Drawing: Drawing a Free-body Diagram 116
4.6 Friction 121
Chapter Review 129
Exercises 130

5 Work and Energy 137

5.1 Work Done by a Constant Force 138
Learn by Drawing: Determining the Sign of Work 140
5.2 Work Done by a Variable Force 142
5.3 The Work–Energy Theorem: Kinetic Energy 145
5.4 Potential Energy 149
5.5 The Conservation of Energy 153
Learn by Drawing: Energy Exchanges: A Falling Ball 158
5.6 Power 160
Insight: More Broken Records: The Clap Skate 163
Chapter Review 164
Exercises 165

6 Momentum and Collisions 172

6.1 Linear Momentum *173*
6.2 Impulse *178*
Insight: The Automobile Air Bag *180*
6.3 The Conservation of Linear Momentum *182*
Demonstration 3: Internal Forces (and Conservation of Momentum?) *184*
6.4 Elastic and Inelastic Collisions *187*
6.5 Center of Mass *194*
6.6 Jet Propulsion and Rockets *200*
Demonstration 4: Jet Propulsion *202*
Chapter Review *202*
Exercises *203*

7 Circular Motion and Gravitation 210

7.1 Angular Measure *211*
Learn by Drawing: The Small-Angle Approximation *213*
7.2 Angular Speed and Velocity *214*
Demonstration 5: Constant Angular Velocity, but Not Tangential Velocity *215*
7.3 Uniform Circular Motion and Centripetal Acceleration *217*
Insight: The Centrifuge: Separating Blood Components *221*
7.4 Angular Acceleration *224*
7.5 Newton's Law of Gravitation *227*
Insight: Space Exploration: Gravity Assists *234*
7.6 Kepler's Laws and Earth Satellites *234*
Insight: Space Colonies and Artificial Gravity *242*
Chapter Review *243*
Exercises *244*

8 Rotational Motion and Equilibrium 251

8.1 Rigid Bodies, Translations, and Rotations *252*
8.2 Torque, Equilibrium, and Stability *254*
Demonstration 6: Stability and Center of Gravity *263*
Insight: The Leaning Tower of Pisa: Stability in Question *264*
8.3 Rotational Dynamics *265*
Insight: Stability in Action *266*
8.4 Rotational Work and Kinetic Energy *274*
8.5 Angular Momentum *277*
Insight: Slide or Roll to a Stop?—Antilock Brakes *278*
Chapter Review *284*
Exercises *285*

9 Solids and Fluids 294

9.1 Solids and Elastic Moduli *295*
Insight: Feat of Strength or Knowledge of Materials? *300*
9.2 Fluids: Pressure and Pascal's Principle *300*
Insight: Blood Pressure *308*
9.3 Buoyancy and Archimedes' Principle *309*
Demonstration 7: Buoyancy and Density *312*
9.4 Fluid Dynamics and Bernoulli's Equation *314*
Demonstration 8: Bernoulli-trapped Ball *318*
Insight: Throwing a Curveball *320*
*9.5 Surface Tension; Viscosity and Poiseuille's Law *321*
Chapter Review *325*
Exercises *326*

10 Temperature 333

10.1 Temperature and Heat *334*
10.2 The Celsius and Fahrenheit Temperature Scales *335*
Insight: Human Body Temperature *337*
10.3 Gas Laws and Absolute Temperature *338*
Demonstration 9: Boyle's Shaving Cream *339*
10.4 Thermal Expansion *343*
Learn by Drawing: Thermal Area Expansion *345*
10.5 The Kinetic Theory of Gases *348*
Insight: Physiological Diffusion *351*
*10.6 Kinetic Theory, Diatomic Gases, and the Equipartition Theorem *351*
Chapter Review *354*
Exercises *355*

11 Heat 360

11.1 Units of Heat 361
11.2 Specific Heat 363
11.3 Phase Changes and Latent Heat 367
Insight: Phase Changes and Ice Skating 372
11.4 Heat Transfer 373
Insight: The Greenhouse Effect 379
Insight: Heat Transfer by Radiation:
  The Microwave Oven 380
Chapter Review 382
Exercises 383

12 Thermodynamics 388

12.1 Thermodynamic Systems, States,
  and Processes 389
12.2 The First Law of Thermodynamics 390
Learn by Drawing: Leaning on Isotherms 394
12.3 The Second Law of Thermodynamics
  and Entropy 396
12.4 Heat Engines and Thermal Pumps 400
Insight: Life, Order, and the Second Law 401
Insight: The Otto Cycle 404
12.5 The Carnot Cycle and Ideal Heat Engines 408
Chapter Review 411
Exercises 412

13 Vibrations and Waves 418

13.1 Simple Harmonic Motion 419
Learn by Drawing: Oscillating in a Parabolic
  Potential Well 422
13.2 Equations of Motion 424
Demonstration 10: Simple Harmonic Motion (SHM)
  and Sinusoidal Oscillation 431
13.3 Wave Motion 432
Insight: Earthquakes, Seismic Waves, and Seismology 436
13.4 Wave Properties 437

13.5 Standing Waves and Resonance 440
Demonstration 11: Flame Standing Wave Pattern 442
Insight: Desirable and Undesirable Resonances 445
Chapter Review 446
Exercises 447

14 Sound 452

14.1 Sound Waves 453
14.2 The Speed of Sound 455
Insight: Speech and Hearing 458
14.3 Sound Intensity and Sound Intensity Level 458
14.4 Sound Phenomena 463
14.5 The Doppler Effect 467
Insight: Doppler Radar 472
14.6 Musical Instruments and Sound
  Characteristics 473
Chapter Review 477
Exercises 478

Appendix I Mathematical Relationships A-1
Appendix II Kinetic Theory of Gases A-2
Appendix III Planetary Data A-3
Appendix IV Alphabetical Listing of the
  Chemical Elements A-3
Appendix V Properties of Selected Isotopes A-4
Answers to Follow-up Exercises A-7
Answers to Odd-numbered Exercises A-15
Photo Credits P-1
Index I-1

Learn by Drawing

Cartesian Coordinates and One-dimensional Displacement 35
Signs of Velocity and Acceleration 42
Drawing a Free-body Diagram 116
Determining the Sign of Work 140
Energy Exchanges: A Falling Ball 158
The Small-Angle Approximation 213
Thermal Area Expansion 345
Leaning on Isotherms 394
Oscillating in a Parabolic Potential Well 422

Demonstrations

Demonstration 1: Newton's First Law and Inertia 103
Demonstration 2: Tension in a String: Action and Reaction Forces 113
Demonstration 3: Internal Forces (and Conservation of Momentum?) 184
Demonstration 4: Jet Propulsion 202
Demonstration 5: Constant Angular Velocity, but Not Tangential Velocity 215
Demonstration 6: Stability and Center of Gravity 263
Demonstration 7: Buoyancy and Density 312
Demonstration 8: Bernoulli-trapped Ball 318
Demonstration 9: Boyle's Shaving Cream 339
Demonstration 10: Simple Harmonic Motion (SHM) and Sinusoidal Oscillation 431
Demonstration 11: Flame Standing Wave Pattern 442

Applications (Insights in **boldface**)

Chapter 1
Nutrition labels 30

Chapter 2
Reaction time 53
Transverse and longitudinal seismic waves 58

Chapter 3
Tacking a sailboat 78
Air resistance and range 89

Chapter 4
Harrier jump jet 112
Air foils 128
Sky diving and terminal velocity 128
Down force and race cars 135

Chapter 5
Power rating of motors 161
The clap skate for speed skating 163
Mechanical efficiency of automobiles 164

Chapter 6
How to catch a fast hard ball 179
Impulse and automobile air bags 180
Following through in sports 180
Colliding-balls toy and conservation laws 194
Center of mass of the human body 200
Recoil of a rifle 200
Rocket thrust 200
Propulsion of fan boats 205
Ballistic pendulum 206
Tightrope walking 207

Chapter 7
Measuring angular distance 212
Merry-go-rounds and rotational speeds 216
Satellites and centripetal acceleration 219
Centrifuge and separating blood components 221
Driving on a curved road 221
Banked roads 223, 246
Compact discs and angular acceleration 225
Cooking evenly in a microwave oven 226
Gravity assists in space exploration 234
Earth satellites and escape speed 237
Retrorockets, conservation of energy, and *Mir* 240
Weightlessness and apparent weightlessness 241
Space colonies and artificial gravity 242
Amusement park bottomless room 247
Ocean tides 249

Chapter 8
Muscle torque 255
Center of gravity of human body 262
Stabilizing the Leaning Tower of Pisa 264
Bicycle stability 266
Yo-yo torque 273
Antilock brakes 278
Tornadoes and hurricanes 280
Throwing a football spiral 280
Helicopters 283
Russell traction 286

Chapter 9
Karate chops 300
Hydraulic brakes, lifts, and jacks 303
Manometers, tire gauges, and barometers 305
Blood pressure and its measurement 308
Rate of blood flow 316
Chimneys, smokestacks, and the Bernoulli effect 317
Airplane lift 319
Throwing a curveball 320
Motor oils and viscosity 323
IV blood transfusions 324
Water dispenser for pets 327

Plimsoll mark for loading depth *328*
Zeppelins *329*
Indy cars and the Venturi tunnel *329*

Chapter 10
Thermometers and thermostats *335*
Human body temperature *337*
Expansion gaps *346*
Physiological diffusion *351*
Osmosis and kidneys *351*
Gas leaks *358*
Atmospheric lapse rate *359*

Chapter 11
Counting (kilo)calories *361*
Specific heat and burning your mouth *363*
Liquid crystal displays (LCDs) *367*
Cooking at Pike's Peak *371*
Phase changes and ice skating *372*
Evaporation and overheating of the body *372*
Copper-bottomed pots and plastic foams *374*
Day–night atmospheric convection cycles *376*
Forced convection in refrigerators, heating and cooling systems, and the body *377*
Polymer foam insulation *377*
Greenhouse effect *378, 379*
Thermography *378*
Microwave ovens *380*
Solar panels *381*
Saving fruit trees from frost *382*
Dressing for the desert *382*
Thermos bottle *385*
R-values *386*
Solar collectors for heating *387*

Chapter 12
Exhalation *395*
Perpetual motion machines *396*

Cell processes and the second law *401*
Geothermal energy *402*
Drinking birds and heat engines *402*
Thermal efficiency of engines *403*
Internal combustion engines and the Otto cycle *404*
Refrigerators as thermal pumps *406*
Heating/cooling with heat pumps *407*
The ocean as a heat engine *416*

Chapter 13
Damping—bathroom scales, shock absorbers, and earthquake protection *432*
Earthquakes, seismic waves, and seismology *436*
Stringed musical instruments *442*
Tuning a violin *443*
Tuning a guitar *444*
Pushing a swing *444*
Tuning a radio *445*
Desirable and undesirable resonances *445*
Echolocation *450*

Chapter 14
Ultrasonic whistles *453*
Sonar *454*
Medical ultrasound *454*
Speech and hearing *458*
Noise reduction *465*
Beats *466*
Radar speed guns *471*
Sonic booms *471*
Doppler radar in weather and airplane safety *472*
Stringed instruments *473*
Pipe organs *474*
Wind and brass instruments *475*
The Richter scale *478*

Preface

We continue to believe that there are two goals any introductory physics course must accomplish, regardless of the approach, emphasis, or pedagogical techniques: (1) to impart an understanding of the basic physics principles and (2) to enable students to solve a variety of reasonable problems in topics presented in the text material.

These goals are linked. An understanding of physical principles is of limited use if it does not enable students to solve problems. Physics is a problem-solving science—and in the real world, students will be evaluated on their ability to produce correct answers on final exams or on the MCAT. Yet learning to solve problems by rote is not the same thing as learning physics. Knowing and doing—insight and skill—must go hand in hand.

Any deficiency in meeting the first goal is likely to be obvious. Test scores quickly get the attention of both test takers and test graders. Low grades demoralize instructors while discouraging students who, understandably, conclude that physics is "too hard" for any but the phenomenally gifted. Deficiencies in meeting the second goal tend to be more subtle. Research in physics education has shown that a surprising number of students who learn to solve typical problems well enough to pass examinations do so without ever arriving at a real understanding of the most elementary physical concepts. Such students often get high marks on exams, yet when asked to answer simple, qualitative questions designed to test their grasp of basic principles, they betray a surprising lack of insight. Simply put, they can solve quantitative problems and get the right answer, but they do not know why it is right.

Achieving Our Goals— Features of the Fourth Edition

Most of the specific features of *College Physics* can be understood in light of these goals.

Conceptual Basis. We believe that giving students a secure grasp of physical principles will almost invariably enhance their problem-solving abilities. Central to this belief is an approach to the development of problem-solving skills that stresses the understanding of basic concepts, rather than the mechanical and rote use of formulas, as the essential foundation. Throughout the writing of *College Physics*, we have organized discussions and incorporated pedagogical tools to ensure that conceptual insight drives the development of practical skills.

Concise Coverage. To maintain a sharp focus on essential concepts, a book should emphasize the basics and minimize superfluous material. Topics of marginal interest have been avoided, as have those that present formal or mathematical difficulties for students. Similarly, we have not wasted space deriving relationships when they shed no additional light on the principle involved. It is usually more important for students in a course such as this to understand what a relationship means and how it can be used rather than the mathematical or analytical techniques employed to derive it.

Applications. *College Physics* has always been known for the strong mix of applications related to technology, science, architecture, medicine, and everyday life in its Insight boxes and text narrative. While the Fourth Edition continues to have a wider range of applications than do most texts, we have also increased the number of biological applications in recognition of the high percentage of pre-med and allied health majors who take this course. Some examples are the new Insight boxes Human Body Temperature, Electric Potential and Nerve Signal Transmission, and Magnetism in Nature. Overall, one-third of our Insight boxes are new. The Fourth Edition also contains many new applications within the text narrative and has an increased emphasis on real-world and applied topics in the worked Examples and end-of-chapter Exercises. A list of the most important applications with page references is found on p. x.

Insight
Human Body Temperature

We commonly take "normal" human body temperature to be 98.6°F (or 37.0°C). The source of this value is a study of human temperature readings done in 1868—more than 130 years ago! A more recent study, done in 1992, notes that the 1868 study used thermometers that were not as accurate as modern electronic thermometers. The new study has some interesting results.

The "normal" human body temperature from oral measurements varies with individuals over a range of about 96°F to 101°F, with an average temperature of 98.2°F. After strenuous exercise, the oral temperature can rise as high as 103°F. When the body is exposed to cold, oral temperatures can fall below 96°F. A rapid drop in temperature of 2 to 3 F° produces uncontrollable shivering. There is not only a contraction of the skeletal muscles, but also of the tiny muscles attached to the hair follicles. The result is "goose bumps."

Your body temperature is typically lower in the morning, after you have slept and your digestive processes are at a low point. "Normal" body temperature generally rises during the day to a peak and then recedes. The study also indicated that women have a slightly higher average temperature than men do (98.4°F versus 98.1°F).

What about the extremes? A fever temperature is typically between 102°F and 104°F. A body temperature above 106°F is extremely dangerous. The enzymes for certain chemical reactions in the body begin to be inactivated, and a total breakdown of body chemistry can result.

On the cold side, decreasing body temperature results in memory lapse and slurred speech, muscular rigidity, erratic heartbeats, loss of consciousness, and, below 78°F, death due to heart failure. However, mild hypothermia (lower-than-normal body temperature) can be beneficial. A decrease in body temperature slows down the body's chemical reactions, and cells use less oxygen than they normally do. This fact is applied in some surgeries (Fig. 1). A patient's body temperature may be lowered significantly to avoid damage to the heart, which must be stopped during such procedures, and to the brain.

FIGURE 1 Lower than normal During some surgeries, the patient's body temperature is lowered to slow down the body's chemical reactions and to reduce the need for blood to supply oxygen to the tissues.

Visualization: "Learn by Drawing." Visualization is one of the most important problem-solving tools in physics. In many cases, if students can make a sketch of a problem, they can solve it. "Learn by Drawing" features offer students specific help on making certain types of sketches and graphs that will provide key insights into a variety of physical situations. The Fourth Edition has three new Learn by Drawing features, on the following topics: Cartesian coordinates and one-dimensional motion, oscillation in a parabolic potential well, and the independence of potential difference on reference point.

LEARN BY DRAWING
Graphical Relationship between Electric Field Lines and Equipotentials

Since it takes no work to move a charge along an equipotential, equipotential surfaces and electric field lines must be perpendicular to one another. Also, the electric field at any location points in the direction in which the potential decreases most rapidly and has a magnitude equal to the change in potential per meter. We can use these facts to construct equipotentials for a given pattern of electric field lines. The reverse is also true: Given the equipotentials, we can construct the electric field lines. If we know the potential value (in volts) associated with each equipotential, we can estimate the strength and direction of the field.

A couple of examples should provide graphical insight into the close connection between equipotentials and fields. Consider Fig. 1, in which you are given the electric field lines and want to determine the shape of the equipotentials. Pick any point, such as A, and begin moving at right angles to the electric field (mapped by the field lines) at that point. Keep moving, always maintaining this same perpendicular orientation to the field lines. Between lines you may have to approximate, but plan ahead to the next field line so that you can cross it at a right angle. To find another equipotential, start at another point, such as B, and proceed the same way. Sketch as many equipotentials as you need to map the area of interest. Figure 1 shows the result of sketching four such equipotentials, from A (at the highest potential value—can you tell why?) to D (at the lowest potential).

Suppose that you are given the equipotentials instead of the field lines (Fig. 2). The electric field lines point in the direction of decreasing V and are perpendicular to the equipotential surfaces. Thus, to map the field, start at any point and move in such a way that your path intersects each equipotential surface at a right angle. The resulting field line starting at point A in Fig. 2 is shown. Starting at points B, C, and D gives rise to additional field lines that suggest the complete electric field pattern; you need only add the arrows in the direction of decreasing potential.

Suppose that you want to estimate the magnitude of **E** at some point P, knowing the values of the equipotentials 1.0 cm on either side of it (Fig. 3). From this information you can easily tell that the field points in the direction from A to B (why?). Its approximate magnitude is

$$|E| = \frac{\Delta V}{\Delta x} = \frac{(1000\text{ V} - 950\text{ V})}{2.0 \times 10^{-2}\text{ m}}$$

$$= 2.5 \times 10^3 \text{ V/m}$$

FIGURE 1 Sketching equipotentials from electric field lines Given the electric field pattern, pick a point in the region of interest and move so that your path is always perpendicular to the next field line. Keep your path as smooth as possible, since kinks and loops in the equipotentials are not allowed. (Why?) To map a higher potential value, move opposite the direction of the electric field and repeat the process. Here $V_A > V_B$ and so on.

FIGURE 2 Mapping the electric field from equipotentials Start at a convenient point and trace a line that crosses each equipotential at a right angle. Repeat the process as often as needed to reveal the field pattern, adding arrows to indicate the direction of the field lines from high to low potential.

FIGURE 3 Estimating the magnitude of the electric field The magnitude of the potential change per meter at any point gives the strength of the electric field at that point.

Demonstrations. Photo sequences of 16 physics demonstrations bring physical principles to life, helping students understand that the information and equations on the page describe real-world phenomena.

Integrated Learning Objectives. Specific learning objectives, located at the beginning of each chapter section, help students structure their reading and facilitate review.

Suggested Problem-Solving Procedure. An extensive section (Section 1.7) provides a framework for thinking about problem solving. This section includes:
- An overview of problem-solving strategies;
- A seven-step procedure that is general enough to apply to most problems in physics but is easily used in specific situations;
- Three Examples that illustrate the problem-solving process, showing how the general procedure is applied in practice.

Problem-Solving Strategies and Hints. The initial treatment of problem solving is followed up throughout *College Physics* with an abundance of suggestions, tips, cautions, shortcuts, and useful techniques for solving specific kinds of problems. These strategies and hints help students apply general principles to specific contexts as well as avoid common pitfalls and misunderstandings.

Conceptual Examples. College Physics was among the first physics text to include examples that are conceptual in nature in addition to quantitative ones. Our Conceptual Examples ask students to think about a physical situation and choose the correct prediction on the basis of an understanding of relevant principles. The discussion that follows (Reasoning and Answer) explains clearly how the correct answer can be identified as well as why the other answers are wrong.

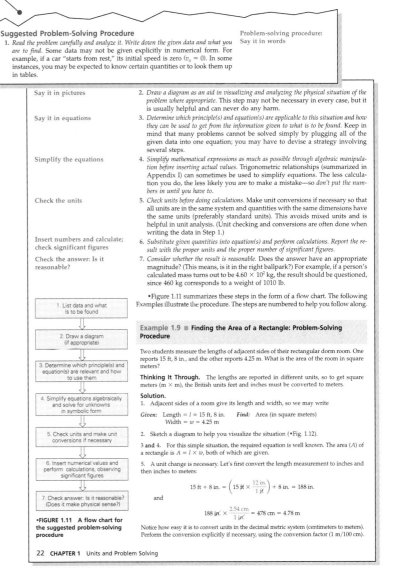

More Explanation in Examples. Too many solutions to worked examples in other texts rely on formulas such as "From Eq. 6.7 we have. . . ." We have tried to make the solutions to in-text Examples as clear, patient, and detailed as possible. The aim is not merely to show students which equations to use but to explain the strategy being employed and the role of each step in the overall plan. Students are encouraged to learn the "why" of each step along with the "how." This technique will make it easier for students to apply the demonstrated techniques to other problems that are not identical in structure.

Thinking It Through. New to the Fourth Edition, every worked Example now includes a "Thinking It Through" section after the problem statement and before the Solution to focus students on the critical thinking and analysis they should do before beginning to use equations.

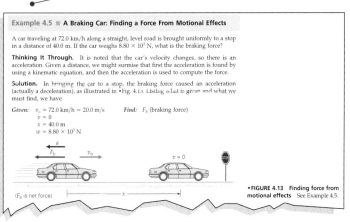

Follow-up Exercises. Follow-up Exercises at the end of each Conceptual Example and each regular worked Example further reinforce the importance of conceptual understanding and offer additional practice. (Answers to Follow-up Exercises are given at the back of the book.)

Chapter Review. Each Chapter Review is made up of three parts:
1. *Important Terms:* A listing, with page references, of the key terms introduced in the chapter that students should be able to define and explain.
2. *Important Concepts:* A summary of the key principles of each chapter.
3. *Important Equations:* A listing, cross-referenced to the equations in the chapter, of the major laws and mathematical relationships introduced. Specific applicability and limiting conditions are clearly stated for each expression.

Exercises. Each chapter ends with a wealth of Exercises, organized by chapter section and ranked by general level of difficulty. In addition, the Exercises offer the following special features to help students refine both their conceptual understanding and their problem-solving skills:

- *Integration of Conceptual and Quantitative Exercises.* To help break down the artificial and ultimately counterproductive barrier between conceptual questions and quantitative problems, we do not distinguish between these categories in the end-of-chapter Exercises. Instead, each section begins with a series of multiple-choice and short-answer questions that provide content review, test conceptual understanding, and ask students to reason from principles. The aim is to show students that the same kind of conceptual insight is required regardless of whether the desired answer involves words, equations, or numbers. The conceptual or "thought" questions are marked by a bold **TQ** in the *Annotated Instructor's Edition* of the text for easy reference when assigning questions. Unlike most other texts, *College Physics* offers short answers to all odd-numbered conceptual questions (as well as to all odd-numbered quantitative problems) in the back of the text so that students can check their understanding. About 35% of all Thought Questions and Exercises in the Fourth Edition are new.

- *Interactive Exercises.* New to the Fourth Edition, many of the end-of-chapter Exercises are keyed to simulations on Prentice Hall's multimedia study guide, the *Interactive Journey through Physics*. Exercises that have a corresponding simulation are indicated with a CD-ROM icon. The *College Physics Media Pack* (ISBN 0-13-085346-1), a specially discounted package consisting of the text and *Interactive Journey through Physics* CD-ROM, includes a cross-reference/location guide to allow you to match Exercise numbers to corresponding simulations.

- *Paired and Trio Exercises.* Most numbered sections include at least one set of paired Exercises and, new to the Fourth Edition, one set of trio Exercises, that deal with similar situations. The first problem in a pair or trio is solved in the Study Guide; the second problem, exploring a similar situation, has only an answer at the back of the book, thereby encouraging students to work out the problem on their own. The third problem in a trio is answered in the *Student Study Guide and Solutions Manual*.

- *Additional Exercises.* Each chapter includes a supplemental section of Additional Exercises drawn from all sections of the chapter to ensure that students can synthesize concepts.

The Absolutely Zero Tolerance for Errors Club (The AZTECs). This team approach to accuracy checking worked quite well in the third edition, so we did it again. Bo Lou of Ferris State University, the author of our *Instructor's Solutions Manual*, headed the AZTEC team and was supported by the text authors and two additional accuracy checkers, Bill McCorkle of West Liberty State University and Dave Curott of the University of North Alabama. Each member of the team individually and independently worked all end-of-chapter Exercises. The results were then collected, and any discrepancies were resolved by a "team" discussion. All data in the chapters, as well as the answers at the back of the book, were checked and rechecked in first and second page proofs. In addition, two other physics teachers—J. Erik Hendrickson and K. W. Nicholson—read all first pages in detail, checking for errors in the chapter narrative and text art. Although it is probably not humanly possible to produce a physics text with absolutely no errors, that was our goal; we worked very hard to make the book as error-free as we could.

New Multimedia Explorations of Physics

New to the Fourth Edition are a state-of-the art Website and a CD-ROM media package.

Companion Website. Our Website (at http://www.prenhall.com/wilson), with contributions from leaders in physics education research, provides students with a variety of interactive explorations of each chapter's topics, easily accommodating differences in learning styles. Student tools include Warm-Ups, Puzzles, and "What Is Physics Good For?" applications by Gregor Novak and Andy Gavrin (Indiana University–Purdue University, Indianapolis); award-winning Java-based Physlet problems by Wolfgang Christian (Davidson College); algorithmically generated numerical Practice Problems, multiple-choice Practice Questions, on-line Destinations, and Net Search key words by Carl Adler (East Carolina University); Ranking Task Exercises edited by Tom O'Kuma (Lee College), David Maloney (Indiana University–Purdue University, Fort Wayne) and Curtis Hieggelke (Joliet Junior College); downloadable PDF files for a Mechanics Problem-Solving Workbook by Dan Smith (South Carolina State University); and MCAT Questions by Glen Terrell (University of Texas at Arlington) and from ARCO's MCAT Supercourse. Using the Preferences module at the opening of the site or the tool in the "Results reporter" part of each module, students can, at a professor's request, have the results of their work on the Companion Website e-mailed to the professor or teaching assistant. Instructor tools include on-line grading capabilities and a Syllabus Manager. See pp. xxiv–xxv for further information about the modules in this site.

Media Pack for College Physics, Fourth Edition (0-13-085346-1). *College Physics, Fourth Edition* can be purchased in a specially discounted package called the Media Pack, which includes the student text, the dual-platform *Interactive Journey through Physics* (*IJTP*) CD-ROM by Cindy Schwarz (Vassar College) and Logal, Inc., *Science on the Internet: A Student's Guide, 1999* by Andrew Stull, and a cross-reference/location guide to correlate Exercises in the text marked with a CD-ROM icon and corresponding simulations on the *IJTP* CD-ROM. This CD-ROM is a multimedia study guide for physics, with simulations, animations, videos, hyperlinked topic reviews, MCAT review questions and problems (including Ranking Task Exercises, context-rich problems, and video problems) for mechanics, thermodynamics, electricity and magnetism, and light and optics. It also includes a built-in scientific calculator with a library of key physics constants, a glossary, and pertinent tables and equations. See pp. xxvi–xxvii for a complete description of the types of materials on the *IJTP* CD-ROM.

Additional Supplements

The pedagogical value of *College Physics* is enhanced by a variety of supplements developed to address the needs of both students and instructors.

For the Instructor

Annotated Instructor's Edition (0-13-084167-6). The margins of the *Annotated Instructor's Edition* (*AIE*) contain an abundance of suggestions for classroom demonstrations and activities, along with teaching tips (points to emphasize, discussion suggestions, and common misunderstandings to avoid). In addition, the *AIE* contains:
- Icons that identify each illustration reproduced as a transparency in the *Transparency Pack*.
- Answers to end-of-chapter Exercises (following each Exercise).
- References to applicable video demonstrations from the *Physics You Can See* videotape.

Instructor's Solutions Manual (0-13-084168-4). Prepared by Bo Lou of Ferris State University, the *Instructor's Solutions Manual* supplies answers with complete, worked-out solutions to all end-of-chapter Exercises. Each solution has been checked for accuracy by a minimum of five instructors. This manual is also available electronically on both Windows and Macintosh platforms.

Test Item File (0-13-084160-9). Fully revised by Dave Curott of the University of North Alabama, the *Test Item File* now offers more than 2300 questions—approximately 30% of them new to this edition—and includes several new conceptual questions per chapter. The questions are now organized and referenced by type and by section.

Prentice Hall Custom Test (Windows: 0-13-084171-4; Macintosh: 0-13-084172-2). Based on the powerful testing technology developed by Engineering Software Associates, Inc. (ESA), the *Prentice Hall Custom Test* allows instructors to create and tailor exams to their own needs. With the On-line Testing Program, exams can also be administered on-line and data can then be automatically transferred for evaluation. A comprehensive desk reference guide is included, along with on-line assistance.

Transparency Pack (0-13-084175-7). The *Transparency Pack* contains more than 300 full-color acetates of text illustrations useful for class lectures. It is available upon adoption of the text.

Physics You Can See Video Demonstrations (0-205-12393-7). Each segment, 2–5 minutes long, demonstrates a classical physics experiment. Eleven segments are included, such as "Coin & Feather" (acceleration due to gravity), "Monkey & Gun" (rate of vertical free fall), "Swivel Hips" (force pairs), and "Collapse a Can" (atmospheric pressure).

Presentation Manager CD-ROM (0-13-084174-9). This new CD-ROM contains all the text art and videos from the *Physics You Can See* videotape as well as additional lab and demonstration videos and animations from the *Interactive Journey through Physics* CD-ROM, which is also available from Prentice Hall (see below).

Just-in-Time Teaching: Blending Active Learning with Web Technology (0-13-085034-9). Just-in-Time Teaching (JiTT) is an exciting new teaching and learning methodology designed to engage students. Using feedback from pre-class Web assignments, instructors can adjust classroom lessons so that students receive rapid response to the specific questions and problems they are having—instead of more generic lectures that may or may not address topics with which students actually need help. Many teachers have found that this process makes students become active and interested learners. In this resource book for educators, authors Gregor Novak (Indiana University–Purdue University, Indianapolis), Evelyn Patterson (United States Air Force Academy), Andrew Gavrin (Indiana University–Purdue University, Indianapolis), and Wolfgang Christian (Davidson College) more fully explain what Just-in-Time Teaching is, its underlying goals and philosophies, and how to implement it. They also provide an extensive section of tested resource materials that can be used in introductory physics courses with the JiTT approach.

For the Student

Student Study Guide and Solutions Manual (0-13-084365-2). Significantly revised by Bo Lou of Ferris State University, the *Student Study Guide and Solutions Manual* presents chapter-by chapter reviews, chapter summaries, key terms, additional worked problems, and solutions to selected problems.

The New York Times "Themes of the Times" Program. This innovative program, made possible through an exclusive partnership between Prentice Hall and *The New York Times*, brings current and relevant applications into the classroom. Through this program, adopters of *College Physics, Fourth Edition* are eligible to receive our free, unique "mini-newspapers," which bring together a collection of the latest and best physics articles from the highly respected pages of *The New York Times*. They are updated annually and are free to qualified adopters up to the quantity of texts purchased. Contact your local representative for ordering.

MCAT Physics Study Guide (0-13-627951-1). This study resource by Joseph Boone of California Polytechnic State University–San Luis Obispo references all of the physics topics on the MCAT to the appropriate sections in the text. Since most MCAT questions require more thought and reasoning than simply plugging numbers into an equation, this study guide is designed to refresh students' memory about the topics they've covered in class. Additional review, practice problems, and review questions are included.

Science on the Internet: A Student's Guide, 1999 (0-13-021308-X). This guide helps students gain a greater understanding of the Internet and the ways in which they can access information on the Web relating to their study of physics.

Other Related Multimedia Materials

Interactive Physics II Player Workbook (Windows: 0-13-667312-0; Macintosh: 0-13-477670-4). Written by Cindy Schwarz of Vassar College, this highly interactive workbook/software package contains simulation projects of varying difficulty. Each includes a physics review, simulation details, hints, an explanation of results, math help, and a self-test.

Interactive Journey through Physics CD-ROM (0-13-254103-3). This highly interactive CD-ROM can be used as a stand-alone supplement for any introductory physics course or as a general reference tool. Through simulation, animation, video, and interactive problem solving, students can visualize difficult physics concepts in ways not available through the traditional lecture, lab, and text. The innovative concept checks and extension exercises within the simulations facilitate the reinforcement of important physical concepts. The content of this CD-ROM is organized according to the main topics in physics—mechanics, electricity and magnetism, thermodynamics, and light and optics. The numerous analysis tools are easily navigated through a user-friendly interface. See the previous description under Media Pack and pp. xxvi–xxvii for more detail about the *IJTP CD-ROM*, and see ordering information for the discounted price when purchased with *College Physics, Fourth Edition*, in the Media Pack.

Acknowledgments

We would like to acknowledge the generous assistance we received from many people during the preparation of the Fourth Edition. First, our sincere thanks go to Bo Lou of Ferris State University for his vital contributions and meticulous, conscientious help with checking problem solutions and answers, preparing the *Instructor's Solutions Manual* as well as the answer keys for the back of the book, and preparing a major revision of the *Student Study Guide and Solutions Manual*. We are similarly grateful to Dave Curott of the University of North Alabama for preparing the *Test Item File* as well as for his participation as an accuracy checker for all solutions to end-of-chapter Exercises.

Indeed, all the members of AZTEC—Bo Lou, Dave Curott, and Bill McCorkle (West Liberty State University)—as well as the reviewers of first and second page proofs—Xiaochun He (Georgia State University), J. Erik Hendrickson (University of Wisconsin–Eau Claire), K. W. Nicholson (Central Alabama Community College), Peter Shull (Oklahoma State University), and Lorin Vant-Hull (University of Houston)—deserve more than a special thanks for their tireless, timely, and extremely thorough review of all materials in the book for scientific accuracy.

Dozens of other colleagues, listed next helped us with reviews of the Third Edition to help us plan the Fourth Edition as well as with reviews of manuscript as it was developed. We are indebted to them, as their thoughtful and constructive suggestions benefited the book greatly.

At Prentice Hall, the editorial staff continued to be particularly helpful. We are especially grateful to Karen Karlin, Senior Development Editor, for everything—her cheerful competence, experienced hand, insight, creativity, extreme attention to detail, and even her queries, all of which have helped make this book one of the most carefully crafted introductory physics texts available. Pat McCutcheon, Project Manager for the book, and Lisa Kinne, Assistant Managing Editor, kept the whole complex endeavor moving forward, while designers Joseph Sengotta and Rosemarie Votta made sure the ultimate physical presentation would be both visually engaging and clean and easy to use. We also thank Danny Hoyt, Senior Marketing Manager, for his cheerful enthusiasm; Wendy Rivers for her work on the supplements; Alison Reeves, Executive Editor, and Gillian Buonanno, Editorial Assistant, for their help in coordinating all of these facets; and Paul Corey, Editor in Chief, for his support and encouragement.

In addition, I (Tony Buffa) once again extend my thanks to my coauthor, Jerry Wilson, for his helpfulness and professional approach to the work at hand during this revision. As always, several colleagues of mine at Cal Poly gave unselfishly of their time for fruitful discussions. Among them are Professors Joseph Boone, Ronald Brown, Theodore Foster, and Richard Frankel. My family—my wife, Connie, and daughters, Jeanne and Julie—was a continuous source of support and inspiration that helped me through the hectic parts of the schedule. I also acknowledge my recently deceased mother, Florence, for her help and advice over the years.

Finally, both of us would like to urge anyone using the book—student or instructor—to pass on to us any suggestions that you have for its improvement. We look forward to hearing from you.

—*Jerry D. Wilson*
—*Anthony J. Buffa*

Reviewers of the Fourth Edition

Anand Batra
Howard University

Jeffrey Braun
University of Evansville

Aaron Chesir
Lucent Technologies

Lattie Collins
East Tennessee State University

James Cook
Middle Tennessee State University

James Ellingson
College of DuPage

John Flaherty
Yuba College

Rex Gandy
Auburn University

Xiaochun He
Georgia State University

J. Erik Hendrickson
University of Wisconsin–Eau Claire

Jacob W. Huang
Towson University

John Kenny
Bradley University

Dana Klinck
Hillsborough Community College

Chantana Lane
University of Tennessee–Chattanooga

R. Gary Layton
Northern Arizona University

Mark Lindsay
University of Louisville

Dan MacIsaac
Northern Arizona University

Trecia Markes
University of Nebraska–Kearney

Aaron McAlexander
Central Piedmont Community College

William McCorkle
West Liberty State University

Michael McGie
California State University–Chico

Gerhard Muller
University of Rhode Island

K. W. Nicholson
Central Alabama Community College

William Pollard
Valdosta State University

David Rafaelle
Glendale Community College

Robert Ross
University of Detroit–Mercy

Gerald Royce
Mary Washington College

Peter Shull
Oklahoma State University

Larry Silva
Appalachian State University

Christopher Sirola
Tri-County Technical College

Soren P. Sorensen
University of Tennessee–Knoxville

Frederick Thomas
Sinclair Community College

Jacqueline Thornton
St. Petersburg Junior College

Anthony Trippe
ITT Technical Institute–San Diego

Lorin Vant-Hull
University of Houston

Kevin Williams
ITT Technical Institute–Earth City

Linda Winkler
Appalachian State University

John Zelinsky
Southern Illinois University

Reviewers of Previous Editions

William Achor
West Maryland College

Arthur Alt
College of Great Falls

Zaven Altounian
McGill University

Frederick Anderson
University of Vermont

Charles Bacon
Ferris State College

Ali Badakhshan
University of Northern Iowa

William Berres
Wayne State University

Hugo Borja
Macomb Community College

Bennet Brabson
Indiana University

Michael Browne
University of Idaho

David Bushnell
Northern Illinois University

Lyle Campbell
Oklahoma Christian University

Lowell Christensen
American River College

Philip A. Chute
University of Wisconsin–Eau Claire

Lawrence Coleman
University of California–Davis

Lattie F. Collins
East Tennessee State University

David M. Cordes
Belleville Area Community College

James R. Crawford
Southwest Texas State University

William Dabby
Edison Community College

J. P. Davidson
University of Kansas

Donald Day
Montgomery College

Richard Delaney
College of Aeronautics

Arnold Feldman
University of Hawaii

Rober J. Foley
University of Wisconsin–Stout

Donald Foster
Wichita State University

Donald R. Franceschetti
Memphis State University

Frank Gaev
ITT Technical Institute–Ft. Lauderdale

Simon George
California State–Long Beach

Barry Gilbert
Rhode Island College

Richard Grahm
Ricks College

Tom J. Gray
University of Nebraska

Douglas Al Harrington
Northeastern State University

Al Hilgendorf
University of Wisconsin–Stout

Joseph M. Hoffman
Frostburg State University

Omar Ahmad Karim
University of North Carolina–Wilmington

S.D. Kaviani
El Camino College

Victor Keh
ITT Technical Institute–Norwalk, California

James Kettler
Ohio University, Eastern Campus

Phillip Laroe
Carroll College

Rubin Laudan
Oregon State University

Bruce A. Layton
Mississippi Gulf Coast Community College

Federic Liebrand
Walla Walla College

Bryan Long
Columbia State Community College

Michael LoPresto
Henry Ford Community College

Robert March
University of Wisconsin

John D. McCullen
University of Arizona

Gary Motta
Lassen College

J. Ronald Mowrey
Harrisburg Area Community College

R. Daryl Pedigo
Austin Community College

T. A. K. Pillai
University of Wisconsin–La Crosse

Darden Powers
Baylor University

Donald S. Presel
University of Massachusetts–Dartmouth

E. W. Prohofsky
Purdue University

Dan R. Quisenberry
Mercer University

W. Steve Quon
Ventura College

George Rainey
California State Polytechnic University

Michael Ram
SUNY–Buffalo

William Riley
Ohio State University

William Rolnick
Wayne State University

Roy Rubins
University of Texas, Arlington

R. S. Rubins
The University of Texas at Arlington

Sid Rudolph
University of Utah

Anne Schmiedekamp
Pennsylvania State University–Ogontz

Ray Sears
University of North Texas

Mark Semon
Bates College

Thomas Sills
Wilbur Wright College

Michael Simon
Housatonic Community Technical College

Dennis W. Suchecki
San Diego Mesa College

Frederick J. Thomas
Sinclair Community College

Pieter B. Visscher
University of Alabama

Arthur J. Ward
Nashville State Technical Institute

John C. Wells
Tennessee Technical University

Arthur Wiggins
Oakland Community College

Dean Zollman
Kansas State University

New Multimedia Explorations of Physics

Companion Website   www.prenhall.com/wilson

This exciting text-specific interactive Website for introductory physics provides students and instructors with a wealth of innovative on-line materials for use with *College Physics, Fourth Edition*.

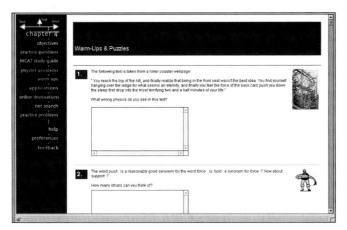

Warm-Ups & Puzzles by Gregor Novak and Andrew Gavrin (Indiana University–Purdue University, Indianapolis)

Warm-Up and Puzzle questions are real-world short-answer questions based on important concepts in the text chapters. Both types of questions get students' attention, often refer to current events, and are good discussion starters. The **Warm-Ups** are designed to help introduce a topic, whereas the **Puzzles** are more complex and often require the integration of more than one concept. Thus, professors can assign Warm-Up questions after students have read a chapter but before the class lecture on that topic, and Puzzle questions as follow-up assignments submitted after class.

Applications by Gregor Novak and Andrew Gavrin (Indiana University–Purdue University, Indianapolis)

The **Applications** modules answer the question "What is physics good for?" by connecting physics concepts to real-world phenomena and new developments in science and technology. These illustrated essays include embedded Web links to related sites, one for each chapter. Each essay is followed by short-answer/essay questions, which professors can assign for extra credit.

Physlet Problems by Wolfgang Christian (Davidson College)

Physlet Problems are multimedia-focused problems based on Wolfgang Christian's award-winning Java applets for physics, called *Physlets*. With these problems, students use multimedia elements to help solve a problem by observing, applying appropriate physics concepts, and making measurements of parameters they deem important. No numbers are given, so students are required to consider a problem qualitatively instead of plugging in formulas.

Practice Questions by Carl Adler (East Carolina University)
Twenty-five to 30 multiple-choice **Practice Questions** are available for review and drill with each chapter.

Practice Problems by Carl Adler (East Carolina University)
Ten algorithmically generated numerical **Practice Problems** per chapter allow students to get multiple iterations of each problem set for practice.

Mechanics Problem-Solving Workbook
by Dan Smith (South Carolina State University)
The **Mechanics Problem-Solving Workbook** is a self-contained workbook designed for student self-study and is available from the Wilson/Buffa Website in downloadable PDF files. It helps students explore issues of central importance in mechanics by beginning with conceptual solutions to problems and then building slowly to final results. Each topic has between 4 and 10 problems that students work out in successive stages, beginning with a conceptual description of the problem and ending with algebraic analyses. Students must complete each stage before proceeding to the next stage, thereby ensuring that they understand the physics before using the algebra.

Ranking Task Exercises
edited by Tom O'Kuma (Lee College), David Maloney (Indiana University–Purdue University, Fort Wayne), and Curtis Hieggelke (Joliet Junior College)
Available for most text chapters as PDF files downloadable from the Wilson/Buffa Website, **Ranking Tasks** are conceptual exercises that require students to rank a number of situations or variations of a situation. Engaging in this process of making comparative judgments helps students reason about physical situations and often gives them new insights into relationships among various concepts and principles.

MCAT Study Guide by Glen Terrell (University of Texas at Arlington) and ARCO's MCAT Supercourse
For all relevant chapters, the MCAT Study Guide module provides students with an average of 25 multiple-choice questions on topics and concepts covered on the MCAT exam. As with all multiple-choice modules, the computer automatically grades and scores student responses and provides cross-references to corresponding text sections.

Destinations
Destinations are links to relevant Websites for each chapter, either about the physics topic in the chapter or about related applications.

Net Search
The **Net Search** feature automatically configures key words according to the key word search conventions of the top three Internet search engines.

Syllabus Manager
Wilson/Buffa's **Syllabus Manager** provides instructors with an easy, step-by-step process for creating and revising a class syllabus with direct links to the text's Companion Website and other on-line content. Through this on-line syllabus, instructors can add assignments and send announcements to the class with the click of a button. The completed syllabus is hosted on Prentice Hall's servers, allowing the syllabus to be updated from any computer with Internet access.

On-line Grading
Scoring for all objective questions and problems, as well as responses to essay questions, can be e-mailed back to the instructor.

Media Pack for *College Physics, Fourth Edition*

To help your students succeed in physics, the Fourth Edition includes end-of-chapter Exercises that have corresponding simulations on the *Interactive Journey through Physics* CD-ROM. These problems are marked with a CD-ROM icon, and a key in the front of the textbook gives the location of the relevant simulation on the CD-ROM. The **Media Pack** version of *College Physics, Fourth Edition* (ISBN 0-13-085346-1) includes the textbook, the *Interactive Journey through Physics* CD-ROM, and *Science on the Internet: A Student's Guide, 1999* by Andrew Stull. Encourage your students to explore the world of physics further with the interactive learning tools found on the CD-ROM:

Interactive Journey through Physics

Whether your students are exploring concepts, improving their grades, or reviewing for a test such as the MCAT, this interactive CD-ROM augments the traditional learning experiences of lecture, lab, and text.

Four dynamic components—simulation, animation, video, and interactive problem solving—enable students to visualize and interact with concepts in ways not possible in traditional learning programs. The content of the CD-ROM is organized according to the main topics in physics—**Mechanics, Electricity and Magnetism, Thermodynamics,** and **Light and Optics.** Although the CD is designed to be interactive, instructors can also use any of the simulations, animations, or videos to add multimedia to their lectures.

Topic Coverage

I. Mechanics. *Overview:* **1D-2D Motion**—Horizontal motion. Free fall. Projectile motion. Boats. Airplanes. Inclined planes. Frames of reference. **Forces**—Inclined planes. Collisions. Friction. Vector components. Kinds of forces. Elevators. **Energy**—Kinetic energy. Potential energy. Work. Work/energy connections. Friction. **Rotation**—Circular motion. Planets. Rotation and translation. Angular momentum. Circular motion II. **Vibrations and Waves**—Simple harmonic motion. Pendulums. Standing waves. Sound waves. Doppler effect. **II. Thermodynamics.** *Overview:* **Gas Laws**—PV plots. PT Plots. TV plots. $PV = nRT$. **Temperature Changes**—Temperature scales. Calorimetry. Effects on solids. Kinetic theory. **First Law**—First law. Special processes. Cycles. Carnot cycle. **III. Electricity & Magnetism.** *Overview:* **Circuits**—Ohm's law. Resistors. Capacitors. RC circuits. **Fixed charges**—Coloumb's law. Electric field. Electric potential. **Moving Charges**—Constant E field. Constant B field. Through a potential. E and B fields. Induced EMF. **IV. Light & Optics.** *Overview:* **Reflection**—Law of reflection. Concave mirror. Convex mirror. Mystery mirrors. Mirror equation. **Refraction**—Snell's law. Critical angle. Index of refraction. Converging lenses. Diverging lenses. Prisms. Lens equation. **Diffraction**—Principle. Single slit. Gratings. **Interference**—Double slit. Thin films.

SIMULATE

Students can ask questions and alter the parameters of experiments to discover the answers. More than 120 simulations of physical systems are provided, with Concept Check questions, built-in spreadsheets, and Extensions for further exploration.

SOLVE

A collection of 231 interactive problems (including "content-rich," ranking-task, and video problems) helps students build critical problem-solving skills. Also featured are 149 MCAT review questions that are based on ARCO's bestselling MCAT review guides and offer pre-meds review both during and after their physics course.

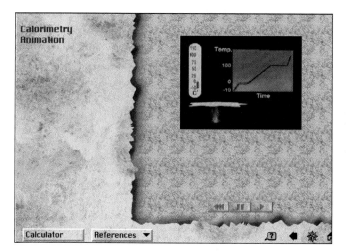

ANIMATE
Thirty-nine narrated animations illustrate physics principles at work, guiding students through explorations of different physical phenomena. This CD-ROM also includes animations that allow students to observe a simulation in an automated mode with audio descriptions to clarify the content.

VISUALIZE
Sixty-two narrated video segments allow students to view actual laboratory experiments and demonstrations. Students can use a special video analysis tool to conduct "video-based labs," a research-proven way to help students make mental links between a physical event and its graphical representation.

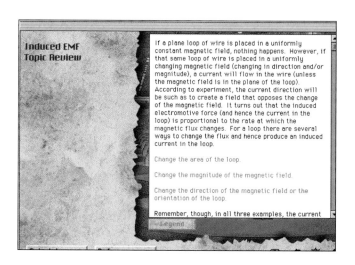

REVIEW
For a brief overview of concepts, students can access a written review of the subtopics covered on the CD-ROM. Hyperlinks let students check definitions, reference equations, or run a related animation or video in the context of the written review.

College Physics

Units and Problem Solving

CHAPTER 1

1.1 Why and How We Measure
1.2 SI Units of Length, Mass, and Time
1.3 More about the Metric System
1.4 Dimensional Analysis and Unit Analysis
1.5 Unit Conversions
1.6 Significant Figures
1.7 Problem Solving

Insights
- A 2000-Year-Old Calculation
- A Bit of History

Not everyone has had the experience of laying out a course on a marine chart, as in the photo above. But most people have done something quite similar with a road map—plotting the shortest distance between two cities, measuring it, and estimating how long the trip will take. You might be planning a 500-mile trip; traveling at an average speed of 50 mi/h, the trip would take about 10 hours. Knowing your car's fuel economy, you might also estimate the amount of gas you expect to need and what it is likely to cost. (At 25 mpg and $1.25 per gallon, how much should you be prepared to spend?)

Measurement and problem solving are not confined to science—they are part of our lives. But they play a particularly central role in our attempts to describe and understand the physical world, as we shall see in this chapter.

1.1 Why and How We Measure

Objective: To distinguish standard units and systems of units.

Imagine that someone is giving you directions to her house. Would you find it helpful to be told, "Drive along Elm Street for a little while and turn right at one of the lights, then keep going for quite a long way"? Or would you want to deal with a bank that sent you a statement at the end of the month saying, "You still have some money left in your account. Not a great deal, though."?

Measurement is important to all of us. It is one of the concrete ways in which we deal with our world. This is particularly true in physics. *Physics is concerned with the description and understanding of nature,* and measurement is one of its most important tools.

There are certainly ways of describing the physical world that do not involve measurement. For instance, we might talk about the color of a flower or a dress. But color perception is subjective: It may vary from one person to another. Indeed, many people are color blind and cannot tell certain colors apart. Light can also be described in terms of wavelengths and frequencies. Different wavelengths are associated with different colors because of the physiological response of our eyes to light. But unlike the sensations or perceptions of color, the wavelengths can be measured. They are the same for everyone. In other words, they are *objective. Physics attempts to describe nature in an objective way through measurements.*

Standard Units

Measurements are expressed in unit values, or units. As you are probably aware, a large variety of units are used to express measured values. Some of the earliest units of measurement, such as the foot, were originally referenced to parts of the human body. (Even today, the hand is still used as a unit to measure the height of horses.) If a unit becomes officially accepted, it is called a **standard unit**. Traditionally, a government or international body establishes standard units.

A group of standard units and their combinations is called a **system of units**. Two major systems of units are in use today—the metric system and the British system. The latter is still widely used in the United States but has virtually disappeared in the rest of the world, having been replaced by the metric system.

Different units in the same system or units from different systems can be used to describe the same thing. For example, your height can be expressed in inches, feet, centimeters, meters—or even miles, for that matter (although this would not be a very *convenient* unit). It is always possible to convert from one unit to another, however, and this is sometimes necessary.

You may be surprised at how easy it is to determine very basic facts about our world through simple measurements and calculations that almost anyone can perform. Some 2200 years ago, a Greek scientist, using only some basic facts of geometry, was able to calculate the size of the Earth with considerable accuracy (see the Insight on p. 3).

Modern science may utilize more sophisticated mathematics and technology, but the approach is still the same. Much of our knowledge rests on a foundation of ingenious measurement and simple calculation. In a sense, mathematics is the language of physics. Principles can be expressed in words but are stated most clearly—and usually most concisely—in mathematical terms. One equation is worth a thousand words, so to speak. But, in order to use equations to do calculations, you must have the proper background and understanding. You will begin your study of physics with a look at some basic quantities and units and a general approach to problem solving. A firm understanding of these will benefit you throughout this entire course of study.

1.2 SI Units of Length, Mass, and Time

Objectives: To (a) describe the SI and (b) specify the references for the three main base quantities of this system.

Length, mass, and time are fundamental physical quantities that describe a great many objects and phenomena. In fact, the topics of mechanics (the study of motion and force) covered in the first part of this book require *only* these physical

Insight

A 2000-Year-Old Calculation: The Earth's Circumference

One of the first measurements of the Earth's circumference was done sometime before 200 B.C. by the Greek scientist Eratosthenes (276–196 B.C.). His method, although quite simple, shows a great deal of insight and ingenuity. The basis for Eratosthenes' technique lies in the understanding that because the distance between the Earth and the Sun is so great compared to our planet's size, the Sun's rays arriving at Earth are virtually parallel. The Sun would have to be at an infinite distance for the rays to be exactly parallel, but the divergence is so small that it is insignificant for practical purposes.

Eratosthenes lived in Alexandria, Egypt. He had learned that on the first day of summer at the town of Syene (now Aswan), some 5000 stadia to the south of Alexandria, the noon Sun was directly overhead. (A *stadium*—plural, *stadia*—was a Greek unit of length believed to be equal to about $\frac{1}{6}$ km.) The Sun's position overhead was apparent because deep wells at Syene were lighted all the way to the bottom, and a vertical stick would cast no shadow at noon on that day.

However, on the first day of summer at Alexandria, some 800 km to the north, Eratosthenes observed that a vertical stick *did* cast a noon-day shadow with a length $\frac{1}{8}$ that of the stick. This corresponds to an angle of 7.1° between the Sun's rays and the vertical (see Fig. 1 insert). Since lines extending through the well at Syene and through the vertical stick at Alexandria intersect at the Earth's center, the angular distance between these locations is also 7.1° (Fig. 1).

Then Eratosthenes reasoned that the 800-kilometer arc length was to the 7.1° angle as the Earth's total circumference was to 360°, the number of degrees in a complete circle. That is, in ratio form,

$$\frac{\text{circumference}}{360°} = \frac{800 \text{ km}}{7.1°}$$

and

$$\text{circumference} = \frac{360°}{7.1°} \times 800 \text{ km} = 40\,600 \text{ km}$$

The Earth's radius (R) can easily be calculated from the circumference (c), since $c = 2\pi R$:

$$R = \frac{c}{2\pi} = \frac{40\,600 \text{ km}}{2\pi} = 6460 \text{ km}$$

This is quite close to the modern value of 6378 km for Earth's equatorial radius (Appendix I). We cannot be certain of the accuracy of Eratosthenes' result because several different Greek stadia of various lengths were in use at that time. (You can see why it is important to have *standard* units.) Even so, his method was sound, and the circumference of the Earth was measured more than 2000 years ago.

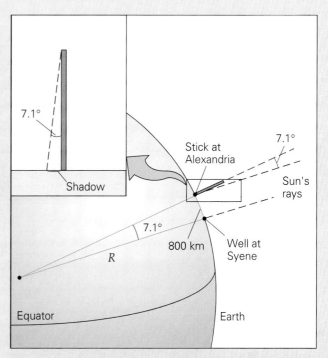

FIGURE 1 A 2000-year-old calculation Observations on the incidence of sunlight allowed the Greek scientist Eratosthenes to calculate the circumference and radius of the Earth sometime before 200 B.C.

---

quantities. The system of units used by scientists to represent these and other quantities is based on the metric system.

Historically, the metric system was the outgrowth of proposals for a more uniform system of weights and measures in France during the seventeenth and eighteenth centuries. The modernized version of the metric system is called the **International System of Units**, officially abbreviated as **SI** (from the French *Système International des Unités*).

The SI

The SI includes *base quantities* and *derived quantities*, which are described by base units and derived units, respectively. **Base units,** such as the meter and the kilogram, are represented by standards. Other quantities that may be expressed in terms of combinations of base units are called **derived units**. (Think of how we commonly measure the length of a trip in miles and the time in hours. To express how fast we travel, we use the derived unit of miles per hour, which represents distance per unit of time, or length per time.)

One of the refinements of the SI was the adoption of new standard references for some base units, including those of length and of time.

Length

Length is the base quantity used to measure distances or dimensions in space. We commonly say that length is the distance between two points. But the distance between any two points depends on how the space between them is traversed, which may be in a straight or a curved line.

The SI unit of length is the **meter (m)**. The meter was originally defined as 1/10 000 000 of the distance from the North Pole to the Equator along a meridian running through Paris (•Fig. 1.1a).* A portion of this meridian between Dunkirk, France, and Barcelona, Spain, was surveyed to establish the standard length, which was assigned the name *metre* (from the Greek word *metron*, meaning "a measure"; the American spelling is *meter*). A meter is 39.37 inches—slightly longer than a yard.

The length of the meter was initially preserved in the form of a material standard: the distance between two marks on a metal bar (made of a platinum–iridium alloy) that was stored under controlled conditions and called the Meter of the Archives. However, it is not desirable to have a reference standard that changes with external conditions, such as temperature. Research provided more accurate standards. In 1983, the meter was redefined in terms of an unvarying property of light: the length of the path traveled by light in a vacuum during an interval of 1/299 792 458 of a second (Fig. 1.1b). In other words, light travels 299 792 458 meters in a second, and the speed of light in a vacuum is defined to be 299 792 458 meters per second. Note that the length standard is referenced to time, which can be measured with great accuracy.

Mass

Mass is the base quantity used to describe amounts of matter. The more massive an object, the more matter it contains. (We will encounter more precise definitions of mass in Chapters 4 and 7.)

•**FIGURE 1.1 The SI length standard: the meter** (a) The meter was originally defined as 1/10 000 000 (or 10^{-7}) of the distance from the North Pole to the Equator along a meridian running through Paris, of which a portion was measured between Dunkirk and Barcelona. A metal bar (called the Meter of the Archives) was constructed as a standard. (b) The meter is currently defined in terms of the speed of light.

*Note that this book and most physicists have adopted the practice of writing large numbers with a thin space for three-digit groups—for example, 10 000 000 (not 10,000,000). The reason for this practice is to avoid confusion with the European practice of using a comma as a decimal point. For instance, 3.141 in the United States would be written as 3,141 in Europe. Large decimal numbers, such as 0.537 84, may also be separated for consistency. Spaces are generally used for numbers with more than four digits on either side of the decimal point.

The SI unit of mass is the **kilogram** (**kg**). The kilogram was originally defined in terms of a specific volume of water but is now referenced to a specific material standard: the mass of a prototype platinum-iridium cylinder kept at the International Bureau of Weights and Measures in Sèvres, France. The United States has a duplicate of the prototype cylinder (•Fig. 1.2). (The kilogram may eventually be referenced to something less subject to time and chance than a piece of metal.)

You may have noticed that the phrase "weights and measures" is generally used instead of "masses and measures." In the SI, mass is a base quantity, but in the familiar British system, weight is used instead to describe amounts of mass—for example, weight in pounds instead of mass in kilograms. The weight of an object is the gravitational attraction that Earth exerts on an object. For example, when you weigh yourself on a scale, your weight is a measure of the downward gravitational force exerted on you by Earth. We can use weight in this way because near Earth's surface, mass and weight are directly proportional to each other: They differ only by a particular constant.

But treating weight as a base quantity creates some problems. A base quantity is naturally most useful if its value is the same everywhere. This is the case with mass—an object has the same mass, or amount of matter, regardless of its location. *But this is not true of its weight.* For example, the weight of an object on the Moon is less than its weight on Earth. This is because the Moon is less massive than Earth, and the gravitational attraction (weight) for an object on the Moon is less than that on Earth. That is, an object with a given amount of mass has a particular weight on Earth, but on the Moon, the same amount of mass will weigh only about one-sixth as much. Similarly, the weight of an object would vary on the surfaces of different planets.

For now, keep in mind that *weight is related to mass, but they are not the same.* Since the weight of an object of a certain mass can vary with location, it is much more useful to take mass as the base quantity, as the SI does. Base quantities should remain the same regardless of where they are measured, under normal or standard conditions. The distinction between mass and weight will be more fully explained in a later chapter. Our discussion until then will be chiefly concerned with mass.

Time

Time is a difficult concept to define. (Try it.) A common definition is that time is the forward flow of events. This is not so much a definition as an observation that time has never been known to run backward (as it might appear to do when you view a film run backward in a projector). Time is sometimes said to be a fourth dimension, accompanying the three dimensions of space. Thus, if something exists in space, it also exists in time. In any case, events can be used to mark time measurements. The events are analogous to the marks on a meterstick used for length measurements.

The SI unit of time is the **second** (**s**). The solar "clock" was originally used to define the second. A solar day is the interval of time that elapses between two successive crossings of the same longitude line (meridian) by the Sun, at its highest point in the sky at that longitude (•Fig. 1.3a). A second was fixed as $1/86\,400$ of this apparent solar day (1 day = 24 h = 1440 min = 86 400 s). However, the elliptical path of the Earth's motion around the Sun causes apparent solar days to vary in length.

As a more precise standard, an average, or mean, solar day was computed from the lengths of the apparent solar days during a solar year. In 1956, the second was referenced to this mean solar day. But the mean solar day is not exactly the same for each yearly period because of minor variations in the Earth's motions and a steady slowing of its rate of rotation due to tidal friction. So scientists kept looking for something better.

MASS: KILOGRAM

(a)

(b)

•**FIGURE 1.2 The SI mass standard: the kilogram** (a) The kilogram was originally defined in terms of a specific volume of water, that of a cube 0.10 m on a side, thereby associating the mass standard with the length standard. The standard kilogram is now defined by a metal cylinder. (b) Prototype Kilogram No. 20 is the national standard of mass for the United States. It is kept under a double bell jar in a vault at the National Institute of Standards and Technology (NIST, formerly the National Bureau of Standards) in Washington, D.C.

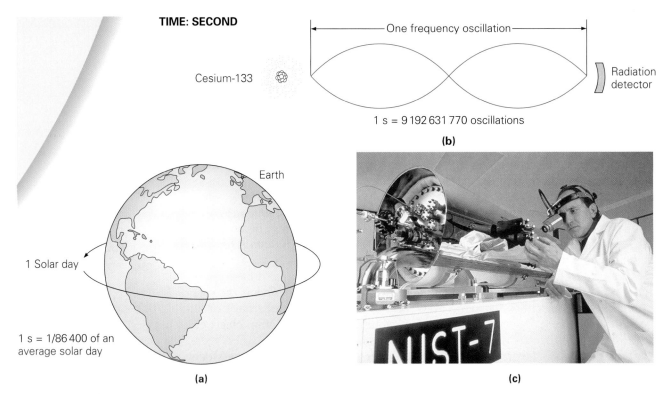

•**FIGURE 1.3 The SI time standard: the second** (a) The second was once defined in terms of the average solar day. (b) It is currently defined in terms of the frequency of the radiation associated with an atomic transition. (c) This atomic "clock" is the primary frequency standard at the National Institute of Standards and Technology. The device keeps time with an accuracy of about three millionths of a second per year.

In 1967, an atomic standard was adopted as a more precise reference. Currently, the second is defined by the radiation frequency of the cesium-133 atom. This particular cesium atom is used in an atomic "clock" that maintains our time standard. The atomic clock doesn't look much like an ordinary clock (Fig. 1.3b,c).

SI Base Units

The complete SI has seven base quantities and base units. In addition to the meter, kilogram, and second for (1) length, (2) mass, and (3) time, respectively, we measure (4) electric current (charge) in amperes (A), (5) temperature in kelvins (K), (6) amount of substance in moles (mol), and (7) luminous intensity in candelas (cd). See Table 1.1.

This is thought to be the smallest number of base quantities needed for a full description of everything observed or measured in nature. There are also two supplemental units for angular measure: the two-dimensional plane angle and the three-dimensional solid angle. There has not been general agreement about whether these geometric units are base or derived.

1.3 More about the Metric System

Objectives: To use common (a) metric prefixes and (b) nonstandard metric units.

The metric system involving the standard units of length, mass, and time now incorporated in the SI was once called the **mks system** (for *m*eter-*k*ilogram-*s*econd). Another metric system that has been used in dealing with relatively small quantities is the **cgs system** (for *c*entimeter-*g*ram-*s*econd). In the United States, the system still generally in use is the British (or English) engineering system, in which the standard units of length, mass, and time are foot, slug, and second, respectively. You may not have heard of the slug, because, as we mentioned earlier, gravitational force (weight) is commonly used instead of mass—pounds instead of

TABLE 1.1 The Seven Base Units of the SI

Name of Unit (abbreviation)	Property Measured
meter (m)	length
kilogram (kg)	mass
second (s)	time
ampere (A)	electric current
kelvin (K)	temperature
mole (mol)	amount of substance
candela (cd)	luminous intensity

slugs—to describe quantities of matter. As a result, the British system is sometimes called the **fps system** (*foot-pound-second*).

The metric system is predominant throughout the world and is coming into increasing use in the United States. Primarily because it is simple mathematically, it is the preferred system of units for science and technology. SI units are used throughout most of this book. All quantities can be expressed in SI units. However, some units from other systems are accepted for limited use as a matter of practicality—for example, the time unit of hour and the temperature unit of degree Celsius. British units will sometimes be used in the early chapters for comparison purposes, since these units are still employed in everyday activities and for many practical applications.

The increasing worldwide use of the metric system means that you should be familiar with it. One of the greatest advantages of the metric system is that it is a decimal, or base-10, system. This means that larger or smaller units are obtained by multiplying or dividing a base unit by powers of 10. A list of the various multiples and corresponding prefixes is given in Table 1.2. In decimal measurements, the prefixes milli-, centi-, and kilo- are the ones most commonly used. The decimal characteristic makes it convenient to change measurements from one size metric unit to another. With the familiar British system, different factors must be used, such as 16 for converting pounds to ounces and 12 for converting feet to inches. The British system developed historically and not very scientifically (see the Insight on p. 8).

You are already familiar with one base-10 system—U.S. currency. Just as a meter can be divided into 10 decimeters, 100 centimeters, or 1000 millimeters, the "base unit" of the dollar can be broken down into 10 "decidollars" (dimes), 100 "centidollars" (cents), or 1000 "millidollars" (tenths of a cent, or mills, used in figuring property taxes and bond levies). Since all the metric prefixes are powers of 10, there are no metric analogues for quarters or nickels.

TABLE 1.2 Multiples and Prefixes for Metric Units*

Multiple[†]	Prefix (and Abbreviation)	Pronunciation	Multiple[†]	Prefix (and Abbreviation)	Pronunciation
10^{24}	yotta- (Y)	yot'ta (*a* as in *a*bout)	10^{-1}	deci- (d)	des'i (as in *deci*mal)
10^{21}	zetta- (Z)	zet'ta (*a* as in *a*bout)	10^{-2}	centi- (c)	sen'ti (as in *senti*mental)
10^{18}	exa- (E)	ex'a (*a* as in *a*bout)	10^{-3}	milli- (m)	mil'li (as in *mili*tary)
10^{15}	peta- (P)	pet'a (as in *peta*l)	10^{-6}	micro- (μ)	mi'kro (as in *micro*phone)
10^{12}	tera- (T)	ter'a (as in *terra*ce)	10^{-9}	nano- (n)	nan'oh (*an* as in *ann*ual)
10^{9}	giga- (G)	ji'ga (*ji* as in *jiggle*, *a* as in *a*bout)	10^{-12}	pico- (p)	pe'ko (*peek-oh*)
10^{6}	mega- (M)	meg'a (as in *mega*phone)	10^{-15}	femto- (f)	fem'toe (*fem* as in *femi*nine)
10^{3}	kilo- (k)	kil'o (as in *kilo*watt)	10^{-18}	atto- (a)	at'toe (as in *ana*tomy)
10^{2}	hecto- (h)	hek'to (*heck-toe*)	10^{-21}	zepto- (z)	zep'toe (as in *zep*pelin)
10	deka- (da)	dek'a (*deck* plus *a* as in *a*bout)	10^{-24}	yocto- (y)	yock'toe (as in *sock*)

*For example, 1 gram (g) multiplied by 1000 (10^3) is 1 kilogram (kg); 1 gram multiplied by 1/1000 (10^{-3}) is 1 milligram (mg).

[†]The most commonly used prefixes are printed in color. Note that the abbreviations for the multiples 10^6 and greater are capitalized, whereas the abbreviations for the smaller multiples are lowercased.

Insight

A Bit of History: Why 5280 Feet in a Mile?

Have you ever wondered why there are 5280 feet in a mile? It is certainly a strange number, and 1000 meters in a kilometer is clearly more convenient. It all has to do with the historical development of British units.

The word "mile" come from the Latin *mille passus*, literally meaning one thousand (*mille*, as in millennium) paces. This unit was introduced to Britain during the Roman occupation, circa A.D. 50. Each *passus* or pace contained 5 *pes*, the Roman foot unit; thus a mi*lle passus* was 5000 *pes* or feet, even though the Roman foot was slightly shorter than our current foot. (The length of the foot has varied with different cultures, as you might expect when a length is referenced to an anatomical unit.)

A common length in early England was the furlong, which is said to have originated as the average length of a furrow a yoke of oxen could plow before needing a rest—a "furrow long." The furlong was used in measuring land acreage and is still used today in horse racing. About A.D. 1500, the "Old London" mile was defined to be 8 furlongs. At the time, the furlong, in terms of a longer (German) foot, contained 625 feet. So, by this definition, the mile again equaled (8 × 625 ft =) 5000 ft.

The other 280 feet were added in 1593 when Queen Elizabeth I decreed (by statute) that the furlong would be defined in terms of the shorter British foot, which made the furlong 660 feet. Hence, the 8-furlong mile became 5280 ft (= 8 × 660 ft).

The 5280-foot mile is called a *statute* mile, because it was established by the queen's statute. We also have a *nautical* mile, which is generally used at sea (and in air transportation), as the name implies. How does it differ from a regular, or statute, mile? The nautical mile's definition is a bit more scientific and somewhat like that of the meter. In principle, the nautical mile is equal to the length of one minute of arc (latitude) on a meridian, or a great circle. An international committee in 1929 established this length to be 6076 ft, or 1.852 km. The knot, a unit of speed, is one nautical mile per hour.

With unusual numbers such as these, it is easy to see the advantages of the metric decimal system.

The official metric prefixes can help eliminate confusion. In the United States, a billion is a thousand million (10^9); in Great Britain, a billion is a million million (10^{12}). The use of metric prefixes eliminates any confusion, since giga- indicates 10^9 and tera- stands for 10^{12}.

Volume

In the SI, the standard unit of volume is the cubic meter (m^3)—the three-dimensional derived unit of the meter base unit. Since this is a rather large unit, it is often more convenient to use the nonstandard unit of volume (or capacity) of a cube 10 cm (centimeters) on a side (•Fig. 1.4a). This volume was given the

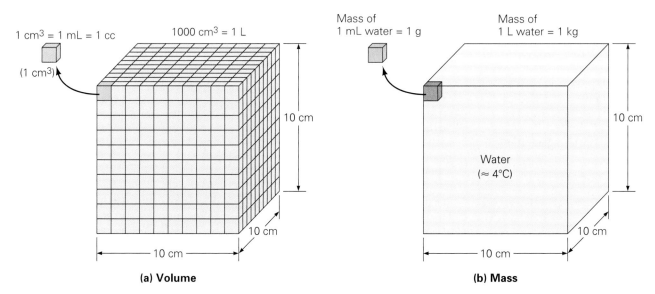

•**FIGURE 1.4 The liter and the kilogram** Other metric units are derived from the meter. **(a)** A unit of volume (capacity) was taken to be the volume of a cube 1.0 dm (10 cm or 0.10 m) on a side and was given the name liter. **(b)** The mass of a liter of water (at its maximum density) was defined to be 1 kg. Note that the decimeter cube contains 1000 cm^3, or 1000 mL. Thus, 1 cm^3, or 1 mL, of water has a mass of 1 g.

name *litre*, which is spelled **liter** (**L**) in the United States. The volume of a liter is 1000 cm³ (10 cm × 10 cm × 10 cm). Since 1 L = 1000 mL (milliliters), it follows that 1 mL = 1 cm³. (The cubic centimeter is sometimes abbreviated cc, particularly in chemistry and biology.)

Recall that the standard unit of mass, the kilogram, was originally defined to be the mass of a cubic volume of water 10 cm or 0.10 m on a side, or the volume of one *liter* of water (at its maximum density, at about 4°C). The density of water does not vary greatly with temperature, so we can say with fairly good accuracy that *1 L of water has a mass of 1 kg*, an approximation that we will use throughout this book (Fig. 1.4b). Also, since 1 kg = 1000 g and 1 L = 1000 cm³, 1 cm³ (or 1 mL) of water has a mass of 1 g.

You are probably more familiar with the liter than you think. The use of the liter is becoming quite common in some instances in the United States, as •Fig. 1.5 indicates.

Note: Liter is sometimes abbreviated as a lowercase "ell" (l), but a capital "ell" (L) is preferred in the United States so that the abbreviation is less likely to be confused with the numeral one. (Isn't 1 L clearer than 1 l?)

Example 1.1 ■ The Metric Ton (or Tonne): Another Unit of Mass

As we have seen, the metric unit of mass was originally related to the length standard, with a liter (1000 cm³) of water having a mass of 1 kg. The standard metric unit of volume is the cubic meter (m³), and this volume of water was used to define a larger mass unit called the *metric ton* (or *tonne*, as it is sometimes spelled). A metric ton is equivalent to how many kilograms?

Thinking It Through. A cubic meter is a relatively large volume and holds a large amount of water (more than a cubic yard; why?). The key is to find how many cubic volumes measuring 10 cm on a side (liters) are in a cubic meter. We expect, therefore, a large number.

Solution. Each liter of water has a mass of 1 kg, so we must find how many liters there are in 1 m³. Since there are 100 cm in a meter, we can visualize a cubic meter as simply a cube with sides 100 cm in length. Therefore, a cubic meter (1 m³) has a volume of 10^2 cm × 10^2 cm × 10^2 cm = 10^6 cm³. Since 1 L = 10^3 cm³, there must be 10^6 cm³/(10^3 cm³/L) = 1000 L in 1 m³. Thus, 1 metric ton is equivalent to 1000 kg.

Note that this entire line of reasoning can be expressed very concisely in a single calculation:

$$\frac{1 \text{ m}^3}{1 \text{ L}} = \frac{100 \text{ cm} \times 100 \text{ cm} \times 100 \text{ cm}}{10 \text{ cm} \times 10 \text{ cm} \times 10 \text{ cm}} = 1000$$

Follow-up Exercise. One kilogram has an equivalent weight of 2.2 lb. Using this relationship, how does the metric ton compare with the British ton? (*See the Answers to Follow-up Exercises section.*)*

•**FIGURE 1.5 Two, three, one, and one-half liters** The liter is now a common volume unit for soft drinks.

Because the metric system is coming into increasing use in the United States, you may find it helpful to have an idea of how metric and British units compare. The relative sizes of some units are illustrated in •Fig. 1.6. The mathematical conversion from one unit to another will be discussed shortly.

1.4 Dimensional Analysis and Unit Analysis

Objectives: To explain the advantages of and apply (a) dimensional analysis and (b) unit analysis.

The fundamental, or base, quantities used in physical descriptions are called *dimensions*. For example, length, mass, and time are dimensions. You could measure the distance between two points and express it in units of meters, centimeters, or feet, but the quantity would still have the dimension of length.

---

*The Answers to Follow-up Exercises section after the appendices contains the answers—and, for Conceptual Exercises, the reasoning—for all Follow-up Exercises in this book. We'll remind you here in Chapter 1 to refer to that section, but then you're on your own.

•**FIGURE 1.6 Comparison of some SI and British units** The bars illustrate the relative magnitudes of each pair of units. (The comparison scales are different in each case.)

Definition of:
Dimensional analysis

It is common to express dimensional quantities by bracketed symbols, such as [L], [M], and [T] for length, mass, and time, respectively. Derived quantities are combinations of dimensions; for example, velocity (v) has the dimensions [L]/[T] (think of miles/hour or meters/second), and volume (V) has the dimensions [L] × [L] × [L], or [L^3]. Addition and subtraction can be done only with quantities that have the same dimensions, for example, 10 s + 20 s = 30 s, or [T] + [T] = [T].

Dimensional analysis is a procedure by which the dimensional consistency of any equation may be checked. You have used equations and know that an equation is a mathematical equality. Since physical quantities used in equations have dimensions, *the two sides of an equation must be equal not only in numerical value, but also in dimensions.* For example, suppose you had the length quantities a = 3.0 m and b = 4.0 m. Inserting them into the equation $a \times b = c$ gives 3.0 m × 4.0 m = 12 m^2. Both sides of the equation are numerically equal (3 × 4 = 12), and both sides are dimensionally equal ([L] × [L] = [L^2]). Notice that we are not algebraically manipulating the dimension symbol [L] in the same way you normally would a variable, such as L. (Note that variables are italicized.) The expression [L] × [L] = [L^2] is read, "when you take a length dimension and multiply it by a length dimension, the result has the dimension of length squared." A second example will help clarify the point: If a and b are inserted into the equation $a + b = c$, we get 3.0 m + 4.0 m = 7.0 m. Comparing both sides of the equation dimensionally, we would say [L] + [L] = [L], or a length plus a length gives another length, but *not* [L] + [L] = 2[L].

One of the major advantages of dimensional analysis is its usefulness for checking whether an equation that has been derived or is being used in solving a problem has the correct form. For example, suppose that you think you can solve a problem about the distance an object travels by using the equation $x = at$, where x represents distance, or length, and has the dimension [L]; a represents acceleration, which, as we will see in Chapter 2, has the dimensions [L]/[T^2]; and t represents time and has the dimension [T]. (The dimensions and units of some common quantities are given in Table 1.3.) First, before trying to use numbers with the equation, you can check to see if it is dimensionally correct:

TABLE 1.3 Dimensions and Units

Quantity	Dimension	Unit
mass	[M]	kg
time	[T]	s
length	[L]	m
area	[L^2]	m^2
volume	[L^3]	m^3
velocity (v)	[L]/[T]	m/s
acceleration (a or g)	[L]/[T^2]	m/s^2

$$x = at$$

is expressed dimensionally as

$$[L] = \frac{[L]}{[T^2]} \times [T] \quad \text{or} \quad [L] = \frac{[L]}{[T]}$$

which is not true. Thus, $x = at$ *cannot* be a correct equation.

Dimensional analysis will tell you if an equation is incorrect, but a dimensionally consistent equation may *not* correctly express the real relationship of quantities. For example, in terms of dimensions,

$$x = at^2$$

is

$$[L] = \frac{[L]}{[T^2]} \times [T^2] \quad \text{or} \quad [L] = [L]$$

This equation is dimensionally correct. But, as you will see in Chapter 2, it is not physically correct. The correct form of the equation—both dimensionally and physically—is $x = \frac{1}{2}at^2$. The fraction $\frac{1}{2}$ has no dimensions; it is a dimensionless number. Thus, you must be sure you have the correct form of an equation before you use it to find solutions to Exercises.

Doing dimensional analysis with the symbols [L], [T], and [M] is fine, but in practice it is often more convenient to use the actual unit abbreviations, such as m, s, and kg. Units can also be treated as algebraic quantities, and like symbols can be canceled. Using units instead of symbols in dimensional analysis is called **unit analysis**. If an equation is correct by unit analysis, it must be dimensionally correct. The following Example demonstrates the use of unit analysis.

Definition of:
Unit analysis

Example 1.2 ■ Checking Dimensions: Unit Analysis

A professor puts two equations on the board: (a) $v = v_o + at$ and (b) $x = v/2a$, where x is distance in meters (m), v and v_o are velocities in meters/second (m/s), a is acceleration in (meters/second)/second or meters/second^2 (m/s^2), and t is time in seconds (s). Are the equations dimensionally correct? Use unit analysis to find out.

Thinking It Through. Simply insert the units for the quantities in each equation, cancel, and check the units on both sides.

Solution.
(a) The equation is

$$v = v_o + at$$

Inserting units for the physical quantities gives

$$\frac{m}{s} = \frac{m}{s} + \left(\frac{m}{s^2} \times s\right) \quad \text{or} \quad \frac{m}{s} = \frac{m}{s} + \left(\frac{m}{s \times \cancel{s}} \times \cancel{s}\right)$$

Notice that units cancel like numbers in a fraction. Then, we have

$$\frac{m}{s} = \frac{m}{s} + \frac{m}{s} \quad \begin{bmatrix} \text{Dimensionally} \\ \text{correct} \end{bmatrix}$$

The equation is dimensionally correct, since the units on each side are meters per second. (The equation is also a correct relationship, as we shall see in Chapter 2.)

1.4 Dimensional Analysis and Unit Analysis 11

(b) By unit analysis, the equation

$$x = \frac{v}{2a}$$

is

$$m = \frac{\left(\frac{m}{s}\right)}{\left(\frac{m}{s^2}\right)} = \frac{\cancel{m}}{\cancel{s}} \times \frac{s^{\cancel{2}}}{\cancel{m}} \quad \text{or} \quad m = s \quad \begin{bmatrix} \text{Not dimensionally} \\ \text{correct} \end{bmatrix}$$

Meters (m) cannot equal seconds (s), so in this case, the equation is dimensionally incorrect and therefore not physically correct.

Follow-up Exercise. Is the equation $x = v^2/a$ dimensionally correct? *(See the Answers to Follow-up Exercises section.)*

Mixed Units

Unit analysis also allows you to check for mixed units. In general, in working problems, you should always use the same unit for a given dimension throughout a problem. For example, suppose that you wanted to buy a new carpet to fit a rectangular room. To compute the floor area, you would measure the lengths of two adjacent sides of the room and multiply them. Would you measure the sides in different units such as 10 ft × 3.0 m and express the area as 30 ft·m? Probably not. Such mixed units are not usually very useful.

Let's look at mixed units in an equation. Suppose that you used centimeters as the unit for x in the equation

$$v^2 = v_o^2 + 2ax$$

and the units for the other quantities as in Example 1.2. In terms of units, this would give

$$\left(\frac{m}{s}\right)^2 = \left(\frac{m}{s}\right)^2 + \left(\frac{m}{s^2} \times cm\right)$$

or

$$\frac{m^2}{s^2} = \frac{m^2}{s^2} + \frac{m \times cm}{s^2}$$

which is dimensionally correct. That is, it is dimensionally equivalent to

$$\frac{[L^2]}{[T^2]} = \frac{[L^2]}{[T^2]} + \frac{[L^2]}{[T^2]}$$

But the units are mixed (m and cm). The terms on the right-hand side should not be added together without centimeters first being changed to meters.

Determining the Units of Quantities

Another aspect of unit analysis that is very important in physics is the determination of the units of quantities from defining equations. For example, **density** (the Greek letter rho, ρ) is defined by the equation

$$\rho = \frac{m}{V} \tag{1.1}$$

where m is mass and V is volume. (Density is the mass of an object or substance per unit volume and is a measure of the compactness of the mass of that object or substance.) What are the units of density? In SI units, mass is measured in kilograms and volume in cubic meters. Hence the defining equation

$$\rho = \frac{m}{V} \quad \left(\frac{\text{kg}}{\text{m}^3}\right)$$

gives the derived SI unit for density as kilograms per cubic meter (kg/m³).

What are the units of π? The relationship between the circumference (c) and the diameter (d) of a circle is given by the equation $c = \pi d$, so $\pi = c/d$. If length is measured in meters, then

$$\pi = \frac{c}{d} \left(\frac{\cancel{m}}{\cancel{m}}\right)$$

Thus, the constant π has no units, because they cancel out. It is unitless or a dimensionless constant.

1.5 Unit Conversions

Objectives: To (a) explain conversion factor relationships, and (b) apply them in converting units within a system or from one system of units to another.

Because units in different systems, or even different units in the same system, can express the same quantity, it is sometimes necessary to convert the units of a quantity from one unit to another. For example, we may need to convert feet to yards or inches to centimeters. You already know how to do many unit conversions. If a room is 12 ft long, what is its length in yards? Your immediate answer is 4 yd.

How did you do this? Well, you must have known a relationship between the units of foot and yard. That is, you know that 1 yd = 3 ft. This is what we call an *equivalence statement*. As we saw in Section 1.4, the numerical values and units on both sides of an equation must be the same. In equivalence statements, we commonly use an equal sign to indicate that 1 yd and 3 ft stand for the *same* or *equivalent length*. The numbers are different because they stand for different *units* of length.

Mathematically to change units, we use **conversion factors**, which are simply equivalence statements expressed in the form of ratios—for example, 1 yd/3 ft or 3 ft/1 yd. (The 1 is often omitted in the denominators of such ratios for convenience; for example, 3 ft/yd.) To understand why such ratios are useful, note that dividing the expression 1 yd = 3 ft by 3 ft (or 3 ft = 1 yd by 1 yd) on both sides gives

$$\frac{(1\text{ yd} = 3\text{ ft})}{3\text{ ft}} = \frac{1\text{ yd}}{3\text{ ft}} = \frac{3\text{ ft}}{3\text{ ft}} = 1 \quad \text{or} \quad \frac{(3\text{ ft} = 1\text{ yd})}{1\text{ yd}} = \frac{3\text{ ft}}{1\text{ yd}} = \frac{1\text{ yd}}{1\text{ yd}} = 1$$

As you can see from these examples, a conversion factor always has an actual value of 1—and you can multiply any quantity by 1 without changing its value, or size. Thus, a *conversion factor simply lets you express a quantity in terms of other units without changing its physical value or size.*

What is done in converting 12 feet to yards may be expressed mathematically as follows:

$$12\ \cancel{\text{ft}} \times \frac{1\text{ yd}}{3\ \cancel{\text{ft}}} = 4\text{ yd} \quad (\textit{feet cancel})$$

Using the appropriate conversion factor form, the feet cancel, as shown by the slash marks, giving the correct unit analysis, yd = yd.

Suppose you are asked to convert 12.0 inches to centimeters. You may not know the conversion factor in this case, but you can get it from a table (such as the one that appears inside the front cover of this book) that gives the needed relationships: 1 in. = 2.54 cm or 1 cm = 0.394 in. It makes no difference which of these equivalence statements is used. The question, once you have expressed the equivalence statement as a conversion factor, is whether to divide or multiply by that factor to make the conversion. *In doing unit conversions, take advantage of unit analysis*—that is, let the units determine the appropriate form of conversion factor, so to speak.

Note that the equivalence statement 1 in. = 2.54 cm can give rise to two forms of the conversion factor: 1 in./2.54 cm or 2.54 cm/1 in. Unit analysis tells you that the second form is the appropriate one to multiply by in this case:

$$12.0 \text{ in.} \times \frac{2.54 \text{ cm}}{1 \text{ in.}} = 30.5 \text{ cm}$$

What would be the result if you used the other form of the first conversion factor, 1 in./2.54 cm? Let's see:

$$12.0 \text{ in.} \times \frac{1 \text{ in.}}{2.54 \text{ cm}} = 4.72 \frac{\text{in}^2}{\text{cm}}$$

This result is dimensionally correct (why?), but it is not a conversion to centimeters. The units in^2/cm for length are nonstandard and confusing. To use this ratio properly, you must *divide* a quantity in inches by the conversion factor:

$$\frac{12.0 \text{ in.}}{\left(\frac{1 \text{ in.}}{2.54 \text{ cm}}\right)} = 12.0 \left(\frac{2.54 \text{ cm}}{1}\right) = 30.5 \text{ cm}$$

(Remember to invert the fraction in the denominator.) Compare this procedure with the first conversion of 12.0 in. above. Can you see why they are equivalent?

Alternatively, you could multiply by a conversion factor derived from the equivalence statement 1 cm = 0.394 in.:

$$12.0 \text{ in.} \times \frac{1 \text{ cm}}{0.394 \text{ in.}} = 30.5 \text{ cm}$$

Or you could divide by the reciprocal form of this conversion factor:

$$\frac{12.0 \text{ in.}}{\left(\frac{0.394 \text{ in.}}{1 \text{ cm}}\right)} = 30.5 \text{ cm}$$

Note: 1 kg of mass has an equivalent weight of 2.2 lb near the surface of the Earth.

A few commonly used equivalence statements are not dimensionally correct; for example, consider 1 kg = 2.2 lb, which is used for conversions. The kilogram is a unit of mass, and the pound is a unit of weight. This means that 1 kilogram is *equivalent* to 2.2 pounds; that is, a 1-kilogram *mass* has a *weight* of 2.2 lb, but only on the Earth's surface. Since mass and weight are directly proportional, differing only by a particular constant, we can use the dimensionally incorrect conversion factor 1 kg/2.2 lb (near the Earth's surface).

Example 1.3 ■ Converting Units: Use of Conversion Factors

(a) A basketball player is 6.5 ft tall. What is the player's height in meters? (b) How many seconds are there in a 30-day month? (c) What is 50 mi/h in meters per second?

Thinking It Through. If we use the correct conversion factors, the rest is basic mathematics.

Solution.
(a) From the conversion table, 1 ft = 0.3048 m, so

$$6.5 \text{ ft} \times \frac{0.3048 \text{ m}}{1 \text{ ft}} = 2.0 \text{ m}$$

Another feet–meter conversion is shown in •Fig. 1.7 (top). Is it correct? You should be able to do this one in your head.

14 CHAPTER 1 Units and Problem Solving

(b) The conversion factor for days and seconds is available from the table (1 day = 86 400 s), but you may not always have a table handy. You can always use several better-known conversion factors to get the result:

$$30 \text{ days} \times \frac{24 \text{ h}}{\text{day}} \times \frac{60 \text{ min}}{\text{h}} \times \frac{60 \text{ s}}{\text{min}} = 2.6 \times 10^6 \text{ s}$$

Note how unit analysis checks the conversion factors for you. The rest is simple arithmetic.

(c) In this case, from the conversion table, 1 mi = 1609 m and 1 h = 3600 s. (The latter is easily computed.) These ratios are used to cancel the units that are to be changed, leaving behind the ones that are wanted:

$$\frac{50 \text{ mi}}{1 \text{ h}} \times \frac{1609 \text{ m}}{1 \text{ mi}} \times \frac{1 \text{ h}}{3600 \text{ s}} = 22 \text{ m/s}$$

Follow-up Exercise. (a) Convert 50 mi/h directly to meters per second by using a single conversion factor, and (b) show that this single conversion factor can be derived from those in part c of this Example. *(See the Answers to Follow-up Exercises section.)*

•**FIGURE 1.7 Converting units** Road signs show the elevation in both feet and meters, and distances in both miles and kilometers.

Example 1.4 ■ More Conversions: A Really Long Capillary System

Capillaries, the smallest blood vessels of the body, connect the arterial system with the venous system and supply our tissues with oxygen and nutrients (•Fig. 1.8). It is estimated that if all of the capillaries of an average adult were unwound and spread out end to end, they would extend to a length of about 64 000 km. (a) How many miles is this? (b) Compare this length to the circumference of the Earth.

Thinking It Through. (a) This is a simple conversion—just use the right conversion factor. (b) How do we calculate the circumference of a circle or sphere? There is an equation, and the radius or diameter of the Earth must be known. (If you do not remember it, see the table inside the back cover.)

Solution.
(a) We see in the conversion table that 1 km = 0.621 mi, so

$$64\,000 \text{ km} \times \frac{0.621 \text{ mi}}{1 \text{ km}} = 40\,000 \text{ mi} \quad \text{(rounded off)}$$

•**FIGURE 1.8 Capillary system** Capillaries connect the arterial and venous systems in our bodies. They are the smallest blood vessels, but their total combined length is impressive.

(b) A length of 40 000 mi is substantial. To see how this length compares to the circumference (c) of the Earth, recall that the radius of the Earth is about 4000 mi, or a diameter (d) of 8000 mi. The circumference of a circle is given by $c = \pi d$, as we saw in Section 1.4. Then,

$$c = \pi d \approx 3 \times 8000 \text{ mi} = 24\,000 \text{ mi}$$

[To make a general comparison, we round π ($= 3.14\ldots$) off to 3.] So,

$$\frac{\text{capillary length}}{\text{Earth's circumference}} = \frac{40\,000 \text{ mi}}{24\,000 \text{ mi}} = 1.7$$

The capillaries of your body have a total length that would extend 1.7 times around the world. Wow!

Follow-up Exercise. Taking the average distance between the east and west coasts of the continental United States to be 4800 km, how many times would the capillary length of your body cross the country? *(See the Answers to Follow-up Exercises section.)*

Example 1.5 ■ Converting Units of Area: Choosing the Correct Conversion Factor

A hall bulletin board has an area of 2.5 m². What is this area in square centimeters (cm²)?

Thinking It Through. This is a conversion of area units, and we know that 1 m = 100 cm. So, some squaring must be done to get square meters and square centimeters.

Solution. A common error in such conversions is the use of incorrect conversion factors. Because 1 m = 100 cm, it is sometimes assumed that 1 m² = 100 cm², which is *wrong*. The correct area conversion factor may be obtained directly from the correct linear conversion factor, 100 cm/1 m or 10^2 cm/1 m, by squaring it:

$$\left(\frac{10^2 \text{ cm}}{1 \text{ m}}\right)^2 = \frac{10^4 \text{ cm}^2}{1 \text{ m}^2}$$

Hence, 1 m² = 10^4 cm² (= 10 000 cm²). We can therefore write:

$$2.5 \text{ m}^2 \times \left(\frac{10^2 \text{ cm}}{1 \text{ m}}\right)^2 = 2.5 \text{ m}^2 \times \frac{10^4 \text{ cm}^2}{1 \text{ m}^2} = 2.5 \times 10^4 \text{ cm}^2$$

Follow-up Exercise. How many cubic centimeters are there in one cubic meter? *(See the Answers to Follow-up Exercises section.)*

Conceptual Example 1.6

Selecting Units: Units and Magnitudes

You are selling a car and want to express its fuel economy so that it will sound the most impressive. Which set of units would give the largest numerical value? (a) miles per gallon, (b) kilometers per gallon, (c) miles per liter, (d) kilometers per liter. (Note: 1 mi = 1.61 km.) *Clearly establish the reasoning used in determining your answer before you check it below. That is, **why** did you select your answer?*

Reasoning and Answer. To obtain the largest numerical value, we require the larger numerical value distance unit and the smaller numerical value volume unit. Since 1 mi is equivalent to 1.61 km, 1.61 is the larger numerical value, (b) would have a larger numerical value than (a), and (d) would have a larger value than (c). That is, 1 mi/gal = 1.61 km/gal and 1 mi/L = 1.61 km/L. Hence, the length unit would be kilometers, and we must decide which volume unit would give the larger apparent value—kilometers per gallon or kilometers per liter. Since a liter is roughly a quart, a gallon is much larger than a liter (about 4 times larger). A gallon of gas would therefore run a car many more kilometers than would a liter of gas. That is, 1 gal ≈ 4 L; so, for example, 40 km/gal ≈ 10 km/L. Thus the answer is (b).

Follow-up Exercise. (a) Which set of units would give the *smallest* numerical value? (b) Given that 1 L = 0.264 gal, justify your answer mathematically. *(See the Answers to Follow-up Exercises section.)*

1.6 Significant Figures

Objectives: To (a) determine the number of significant figures in a numerical value and (b) report the proper number of significant figures after performing simple calculations.

Most of the time, you will be given numerical data when asked to solve a problem. In general, such data are either exact numbers or measured numbers (quantities). **Exact numbers** are those without any uncertainty or error. These include numbers such as the 100 used to calculate a percentage and the 2 in the equation $r = d/2$ relating the radius and diameter of a circle. **Measured numbers** are those obtained from measurement processes and so generally have some degree of uncertainty or error.

When calculations are done with measured numbers, the error of measurement is propagated, or carried along, by the mathematical operations. A question of how to report a result arises. For example, suppose that you are asked to find time (t) from the formula $x = vt$ and are given that $x = 5.3$ m and $v = 1.67$ m/s. Then

$$t = \frac{x}{v} = \frac{5.3 \text{ m}}{1.67 \text{ m/s}} = ?$$

Doing the division operation on a calculator yields a result such as 3.173 652 695 (•Fig. 1.9). How many figures, or digits, should you report in the answer?

The error or uncertainty of the result of a mathematical operation may be computed by statistical methods. A simpler, widely used procedure for estimating this uncertainty involves the use of **significant figures** (sf), sometimes called significant digits. The degree of accuracy of a measured quantity depends on how finely divided the measuring scale of the instrument is. For example, you might measure the length of an object as 2.5 cm with one instrument and 2.54 cm with another; the second instrument provides more significant figures and a greater degree of accuracy.

In general, *the number of significant figures of a numerical quantity is the number of reliably known digits it contains*. For a measured quantity, this is usually defined as all of the digits that can be read directly from the instrument used in making the measurement plus one uncertain digit that is obtained by estimating the fraction of the smallest division of the instrument's scale.

The quantities 2.5 cm and 2.54 cm have two and three significant figures, respectively. This is rather evident. However, some confusion may arise when a quantity contains one or more zeros. For example, how many significant figures does the quantity 0.0254 m have? What about 104.6 m? 2705.0 m? In such cases, we will use the following rules:

1. Zeros at the beginning of a number are not significant. They merely locate the decimal point.

    0.0254 m   three significant figures   (2, 5, 4)

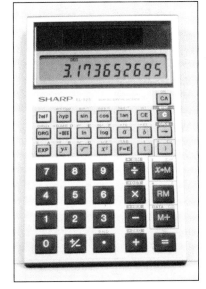

•FIGURE 1.9 **Significant figures and insignificant figures** For the division operation of 5.3/1.67, a calculator with a floating decimal point gives many digits. A calculated quantity can be no more accurate than the least accurate quantity involved in the calculation, so this result should be rounded off to two significant figures—that is, 3.2.

2. Zeros within a number are significant.

$$104.6 \text{ m} \quad \text{four significant figures} \quad (1, 0, 4, 6)$$

Rules for determining the number of significant figures in quantities with zeros

3. Zeros at the end of a number after the decimal point are significant.

$$2705.0 \text{ m} \quad \text{five significant figures} \quad (2, 7, 0, 5, 0)$$

4. In whole numbers without a decimal point that end in one or more zeros (trailing zeros)—for example, 500 kg—the zeros may or may not be significant. In such cases, it is not clear which zeros serve only to locate the decimal point and which are actually part of the measurement. That is, if the first zero from the left (5$\underline{0}$0 kg) was the estimated digit in the measurement, then only two digits are reliably known and there are only two significant figures. Similarly, if the last zero was the estimated digit (50$\underline{0}$ kg), then there are three significant figures. This ambiguity may be removed by using scientific (powers-of-ten) notation:

$$5.0 \times 10^2 \text{ kg} \quad \text{two significant figures}$$

$$5.00 \times 10^2 \text{ kg} \quad \text{three significant figures}$$

This notation is helpful in expressing the results of calculations with the proper numbers of significant figures, as we shall see shortly. (Appendix I includes a review of scientific notation.)

Note: To avoid confusion with numbers having trailing zeroes used as given quantities in text examples and exercises, we will consider the trailing zeroes to be significant. For example, assume that a time of 20 s has two significant figures, even if it is not written out as 2.0×10^1 s.

It is important to report the results of mathematical operations with the proper number of significant figures. The following general rule tells how to determine the number of significant figures in the result of a multiplication or division:

Multiplication and division with significant figures

The final result of a multiplication or division operation should have the same number of significant figures as the quantity with the least number of significant figures that was used in the calculation.

What this means is that the result of a calculation can be no more accurate than the least accurate quantity used; that is, you cannot gain accuracy in performing mathematical operations. Thus, the result that should be reported for the division operation discussed at the beginning of this section is

$$\frac{\overset{(2\,sf)}{5.3 \text{ m}}}{\underset{(3\,sf)}{1.67 \text{ m/s}}} = 3.2 \text{ s} \quad (2\,sf)$$

The result is rounded off to two significant figures (see Fig. 1.9).

For multiple operations, rounding off to the proper number of significant figures should not be done at each step, because rounding errors may accumulate. It is usually suggested that one or two insignificant figures be carried along, or if a calculator is being used, rounding off may be done on the final result of the multiple calculations.

Rounding rules

The rules for rounding off numbers are as follows:

1. If the next digit after the last significant figure is 5 or greater, round up: Increase the last significant figure by 1. (For example, 2.136 becomes 2.14 rounded to 3 significant figures.)

2. If the next digit after the last significant figure is less than 5, round down: Do not change the last significant figure. (For example, 2.132 becomes 2.13 rounded to 3 significant figures.)

Example 1.7 ■ Using Significant Figures in Multiplication and Division: Rounding Applications

The following operations are performed and the results rounded off to the proper number of significant figures (sf).

Multiplication

$$2.4 \text{ m} \times 3.65 \text{ m} = 8.76 \text{ m}^2 = 8.8 \text{ m}^2 \quad \text{(rounded to 2 sf)}$$
(2 sf) (3 sf)

Division

$$\frac{725.0 \text{ m} \,(4\,sf)}{0.125 \text{ s} \,(3\,sf)} = 5800 \text{ m/s} = 5.80 \times 10^3 \text{ m/s} \quad \text{(represented with 3 sf; why?)}$$

Follow-up Exercise. Perform the following operations with answers expressed in the standard powers-of-ten notation (one digit to the left of the decimal point) with the proper number of significant figures. (a) $(2.0 \times 10^5 \text{ kg})(0.035 \times 10^2 \text{ kg})$ and (b) $(148 \times 10^{-6} \text{ m})/(0.4906 \times 10^{-6} \text{ m})$ *(See the Answers to Follow-up Exercises section.)*

There is also a general rule for determining the number of significant figures for the results of addition and subtraction.

Addition and subtraction with significant figures

The final result of the addition or subtraction of numbers should have the same number of decimal places as the quantity with the least number of decimal places that was used in the calculation.

This rule is applied by rounding off numbers to the least number of decimal places *before* adding or subtracting. The rounding off ensures that a result will have no more reliability than the quantity read from the measuring instrument with the least fine scale. Another way of looking at this is that the rounding off is determined by the quantity with the first doubtful or estimated figure from the left. (Why?)

Example 1.8 ■ Using Significant Figures in Addition and Subtraction: Application of Rules

The following operations are performed by finding the number that has the first doubtful figure in a column (counting from the left with the decimal points aligned) and rounding the other numbers to this column. (Units have been omitted for convenience.)

Addition
In the numbers to be added, note that 23.1 has the least number of decimal places, or has the first doubtful figure in a column counting from the left, the 1 (evaluating from the left with the decimal points of the numbers aligned). The other numbers are rounded to this column and the addition performed:

```
    23.1                  23.1
     0.546   ─────→        0.5
     1.45   (rounding off) 1.5
    ─────                 ─────
                          25.1
```

Subtraction

The same rounding procedure is used. Here, the 157 has the least number of decimal places (none). Rounding the 5.5 and subtracting, we get

$$\begin{array}{r} 157 \\ -\ 5.5 \end{array} \text{ (rounding off)} \quad \begin{array}{r} 157 \\ -\ 6 \\ \hline 151 \end{array}$$

(The decimal point, although omitted, is understood in aligning the numbers.)

Follow-up Exercise. Given the numbers 23.15, 0.546, and 1.058, (a) add the first two and (b) subtract the last number from the first. *(See the Answers to Follow-up Exercises section.)*

Suppose you must deal with mixed operations—multiplication and/or division *and* addition and/or subtraction. What do you do in this case? Just follow your regular rules for order of algebraic operations, and observe significant figures as you go.

The number of digits reported in a result depends on the number of digits in the given data. The rules for rounding will generally be observed for examples in this book. However, there will be exceptions that may make a difference, as explained in the following hint.

Problem-Solving Hint: The "Correct" Answer

When working problems, you naturally strive to get the correct answer and will probably want to check your answers against those listed in the Answers to Odd-numbered Exercises section in the back of the book. However, on occasion you may find that your answer differs slightly from that given even though you have solved the problem correctly. There are several reasons why this could happen.

As stated previously, it is best to round off only the final result of a multi-part calculation, but this is not always convenient in elaborate calculations. Sometimes the results of intermediate steps are important in themselves and need to be rounded off to the appropriate number of digits as if each were a final answer. Similarly, text Examples in this book are often worked in steps for pedagogical reasons, to show the stages in the *reasoning* of the solution. The results obtained when intermediate steps are rounded off may differ slightly from those obtained when only the final answer is rounded.

Rounding differences may also occur when using conversion factors. For example, in changing 5.0 mi to kilometers with the conversion factors listed in the front of the book,

$$5.0 \text{ mi} \left(\frac{1.609 \text{ km}}{1 \text{ mi}} \right) = (8.045 \text{ km}) = 8.0 \text{ km} \quad \text{(2 significant figures)}$$

and

$$5.0 \text{ mi} \left(\frac{1 \text{ km}}{0.621 \text{ mi}} \right) = (8.051 \text{ km}) = 8.1 \text{ km} \quad \text{(2 significant figures)}$$

The difference arises because of rounding of the conversion factors. Actually, 1 km = 0.6214 mi, so 1 mi = (1/0.6214) km = 1.609 269 km = 1.609 km. (Try repeating these conversions with the unrounded factors and see what you get.) To avoid rounding differences in conversions, we will generally use the multiplication form of a conversion factor, as in the first of the equations above, unless there is a convenient exact factor, such as 1 min/60 s.

Finally, slight differences in answers may occur when different methods are used to solve a problem, even though both may be legitimate. Keep in mind that when solving a problem (a general procedure for which is given in Section 1.7), *if your answer differs from that in the text in only the last digit, the disparity is most likely the result of rounding difference or of an alternative method of solution being used.*

1.7 Problem Solving

Objectives: To (a) establish a problem-solving procedure and (b) apply it to typical problems.

An important aspect of physics is problem solving. In general, problem solving involves the application of physical principles and equations to data from a particular situation in order to find some unknown or wanted quantity. There is no universal method for approaching a problem that will automatically produce a solution. (•Fig 1.10).

A few general points, though, are worth keeping in mind:

- *Make sure you understand the problem.* The foundation of successful problem solving is having a thorough grasp of the problem. All too often a student will start working on a problem without really understanding the situation described or knowing what is to be found. (If you don't know exactly what you want, it is very difficult to find it!) Before starting to work a problem, make it a habit to list everything that is given or known (as we will do for most of the Examples in this text). Otherwise, it's all too easy to overlook a critical piece of information.

- To proceed from the given data to the ultimate solution, you may have to devise a strategy or plan. Many problems cannot be solved merely by finding one equation and "plugging in" the given quantities to get the solution. Often you will need to perform intermediate steps, each of which will bring you closer to the final answer. The process is a lot like crossing a stream by stepping from stone to stone—you need to keep in mind where you are and where you're going, and as you plan each step you must think ahead to the next one.

- Remember that equations are expressions of physical principles. Solving problems involves translating principles into the "language" of equations, but you should also be able to look at an equation and say what physical relationship it embodies. If you use an equation without knowing what it means, you *may* be lucky and get the right answer, but you will not be learning much physics—and you probably won't be able to repeat your success when confronted with a slightly different problem.

- A common cause of failure in problem solving is the *misapplication* of physical principles or equations. Principles and equations are generally subject to certain conditions or are limited in their applicability to physical situations.

- Many problems can be solved by more than one method. As with many things in life, however, there is often a hard way and an easy way. If you understand the problem and the relevant principles well, you can probably figure out how to solve the problem in the fastest and easiest way.

Although there is no magic formula for problem solving, there are some sound practices that can be very useful. The steps in the following procedure are intended to provide you with a framework that can be applied to solving most of the problems that you will encounter in this text. We will generally use these steps in dealing with the Example problems throughout this text. Additional helpful problem-solving hints will be given where appropriate in subsequent chapters.

•**FIGURE 1.10 Problem solving** There is no set approach in working problems. However, you may find the suggested procedure given here to be helpful.

Suggested Problem-Solving Procedure

1. *Read the problem carefully and analyze it. Write down the given data and what you are to find.* Some data may not be given explicitly in numerical form. For example, if a car "starts from rest," its initial speed is zero ($v_o = 0$). In some instances, you may be expected to know certain quantities or to look them up in tables.

Problem-solving procedure: Say it in words

Say it in pictures

2. *Draw a diagram as an aid in visualizing and analyzing the physical situation of the problem where appropriate.* This step may not be necessary in every case, but it is usually helpful and can never do any harm.

Say it in equations

3. *Determine which principle(s) and equation(s) are applicable to this situation and how they can be used to get from the information given to what is to be found.* Keep in mind that many problems cannot be solved simply by plugging all of the given data into one equation; you may have to devise a strategy involving several steps.

Simplify the equations

4. *Simplify mathematical expressions as much as possible through algebraic manipulation before inserting actual values.* Trigonometric relationships (summarized in Appendix I) can sometimes be used to simplify equations. The less calculation you do, the less likely you are to make a mistake—so *don't put the numbers in until you have to.*

Check the units

5. *Check units before doing calculations.* Make unit conversions if necessary so that all units are in the same system and quantities with the same dimensions have the same units (preferably standard units). This avoids mixed units and is helpful in unit analysis. (Unit checking and conversions are often done when writing the data in Step 1.)

Insert numbers and calculate; check significant figures

6. *Substitute given quantities into equation(s) and perform calculations. Report the result with the proper units and the proper number of significant figures.*

Check the answer: Is it reasonable?

7. *Consider whether the result is reasonable.* Does the answer have an appropriate magnitude? (This means, is it in the right ballpark?) For example, if a person's calculated mass turns out to be 4.60×10^2 kg, the result should be questioned, since 460 kg corresponds to a weight of 1010 lb.

•Figure 1.11 summarizes these steps in the form of a flow chart. The following Examples illustrate the procedure. The steps are numbered to help you follow along.

Example 1.9 ■ Finding the Area of a Rectangle: Problem-Solving Procedure

Two students measure the lengths of adjacent sides of their rectangular dorm room. One reports 15 ft, 8 in., and the other reports 4.25 m. What is the area of the room in square meters?

Thinking It Through. The lengths are reported in different units, so to get square meters (m × m), the British units feet and inches must be converted to meters.

Solution.

1. Adjacent sides of a room give its length and width, so we may write

   Given: Length = l = 15 ft, 8 in.   *Find:* Area (in square meters)
   Width = w = 4.25 m

2. Sketch a diagram to help you visualize the situation (•Fig. 1.12).

3 **and** 4. For this simple situation, the required equation is well known. The area (A) of a rectangle is $A = l \times w$, both of which are given.

5. A unit change is necessary. Let's first convert the length measurement to inches and then inches to meters:

$$15 \text{ ft} + 8 \text{ in.} = \left(15 \text{ ft} \times \frac{12 \text{ in.}}{1 \text{ ft}}\right) + 8 \text{ in.} = 188 \text{ in.}$$

and

$$188 \text{ in.} \times \frac{2.54 \text{ cm}}{1 \text{ in.}} = 478 \text{ cm} = 4.78 \text{ m}$$

Notice how easy it is to convert units in the decimal metric system (centimeters to meters). Perform the conversion explicitly if necessary, using the conversion factor (1 m/100 cm).

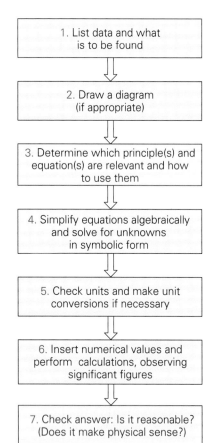

•**FIGURE 1.11 A flow chart for the suggested problem-solving procedure**

6. Now perform the calculation:

$$A = l \times w = 4.78 \text{ m} \times 4.25 \text{ m}$$
$$= 20.315 \text{ m}^2 = 20.3 \text{ m}^2 \quad \textit{(computed value rounded to 3 sf—why?)}$$

7. The answer appears reasonable. Since 1 m ≈ 3 ft, the dorm room would be about 13 ft by 14 ft, which is about right (but, as always, too small). Suppose you had inadvertently input 47.8 instead of 4.78 on your calculator. The result would be $A = 47.8 \text{ m} \times 4.25 \text{ m} = 203 \text{ m}^2$. A room with an area of about 200 m² would have dimensions of about 10 m by 20 m, which is roughly 30 ft by 60 ft. Since this is not the size of a typical dorm room, the magnitude of the result should make you suspect there may be an error.

Follow-up Exercise. The dimensions of a textbook are 0.22 m × 0.26 m × 4.0 cm. What volume in a backpack would the book take up? Give the answer in both cubic meters and cubic centimeters. *(See the Answers to Follow-up Exercises section.)*

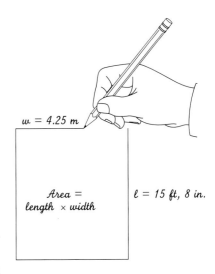

•**FIGURE 1.12 A helpful step in problem solving** Drawing a diagram helps you visualize and better understand the situation. See Example 1.9.

Many problems will involve basic trigonometric functions. The most common functions are given in the marginal note on this page; for other trigonometric relations, see Appendix I or the tables inside the back cover.

Example 1.10 ■ Finding the Length of One Side of a Triangle: Trigonometry Application

A flower bed is laid out in the form of a triangle as shown in •Fig. 1.13. What is the length of the side of the bed that runs along the flagstone walkway?

Thinking It Through. The length of one side of a right triangle and an angle are given. Basic trigonometry applies.

Solution.
1 and 2. In some problems we are given diagrams with data. Here we have

Given: $x = 5.4$ m   Find: r (long side of triangle)
$\theta = 40°$

3 and 4. Noting in the figure that the flower bed is a right triangle (as indicated by the right angle symbol at the large angle), we can use a trigonometric function to find the hypotenuse r. Recalling that x and r are associated with the cosine for a standard right triangle, we can write

$$\cos\theta = \frac{x}{r} \quad \text{or} \quad r = \frac{x}{\cos\theta}$$

5 and 6. The units are OK. (Since x is in meters and $\cos\theta$ is dimensionless, r will come out in meters.) So put in the data:

$$r = \frac{x}{\cos\theta} = \frac{5.4 \text{ m}}{\cos 40°} = \frac{5.4 \text{ m}}{0.766} = 7.0 \text{ m}$$

7. The magnitude of r seems reasonable compared to the value of x.

Follow-up Exercise. To keep animals out of the flower bed in Fig. 1.13, the owner decides to fence in the perimeter of the bed. What will be the total length of the fence? *(See the Answers to Follow-up Exercises section.)*

Trigonometric functions:

$$\cos\theta = \frac{x}{r} \left(\frac{\text{side adjacent}}{\text{hypotenuse}}\right)$$

$$\sin\theta = \frac{y}{r} \left(\frac{\text{side opposite}}{\text{hypotenuse}}\right)$$

$$\tan\theta = \frac{\sin\theta}{\cos\theta} = \frac{y}{x} \left(\frac{\text{side opposite}}{\text{side adjacent}}\right)$$

•**FIGURE 1.13 The length of a triangle's side** See Example 1.10.

It is understood that you do not need to write out the steps of problem-solving procedure each time. However, it is a good practice to run through them mentally when working a problem. In the examples in the following chapters, the problem-solving steps will not always be listed as was done here for illustration. Even so, you should be able to see the general pattern outlined above. In some instances, especially when introducing the application of a new principle or concept, problem-solving steps or hints will be listed to promote better understanding.

The main point of this section is that in solving problems, you should have some systematic procedure for analyzing the situation and extracting the information wanted. The suggested problem-solving procedure given here is one option. Let's review it again.

Example 1.11 ■ Finding the Capacity of a Cylinder: Problem Solving in Three Dimensions

A cylindrical container with a height of 28.5 cm and an inside diameter of 10.4 cm is filled with water. What is the mass of the water in kilograms?

Thinking It Through. The dimensions of the cylinder are given, so we can find its volume, or the volume of water it would hold. How is volume related to mass? By density ($\rho = m/V$), and the density of water is known.

Solution.
1. *Read the problem carefully and analyze it.*
   From the problem, we have

   Given: $h = 28.5$ cm (height) *Find:* m (mass in kg)
   $d = 10.4$ cm (diameter)

2. *Draw a diagram as an aid in visualizing and analyzing the situation.*
   Sketch the cylinder, showing its dimensions, and add any general information you might know about a cylinder. For example, the circular base has an area of $A = \pi r^2$, where r is the radius and $r = d/2$ (•Fig. 1.14).

3. *Determine which principle(s) and equation(s) are applicable and how to use them.*
   We are given the height and diameter of the cylinder, so we can find its volume, which is the volume of the water it contains. But how is volume related to mass, which is what we want? The link is density (ρ), which we saw earlier (Eq. 1.1) is mass per unit volume:

   $$\rho = \frac{m}{V} \quad \text{or} \quad m = \rho V$$

   The density of water is not given. However, you can look that up in a table, or you may remember from the definition of the kilogram that $\rho_{H_2O} = 1.00$ g/cm³ (at maximum density, which will be assumed here). So, knowing the volume and the density of water, we can find the mass.

4. *Simplify expressions algebraically if possible.*
   Should you not remember the formula for the volume of a cylinder, your sketch can be helpful. The area of the circular end of the cylinder is $A = \pi r^2 = \pi d^2/4$ (since $r = d/2$), and the volume is this area times the height of the cylinder, or $V = Ah$. (Formulas for areas and volumes of common shapes are given in Table 5 of Appendix I.)

   We can therefore write an equation for the mass entirely in terms of known quantities:

   $$m = \rho V = \rho A h = \rho \left(\frac{\pi d^2}{4}\right) h$$

5. *Check units before doing calculations.*
   The cylinder dimensions are in centimeters and the density in grams per cubic centimeter. Hence, the mass will come out in grams, which we can convert to kilograms as requested in the problem. (Another option would be to convert to meter and kilogram units first, but this would require more steps.)
   Note: If the given data have mixed units, it is a good practice to convert to consistent units when writing down what is given. This avoids mistakes that might be made later if this step were overlooked.

6. *Substitute the given quantities into the equation and perform calculations, observing significant figures.*

•**FIGURE 1.14 The capacity of a cylinder** Drawing a diagram helps in problem solving. See Example 1.11.

Using the equation derived above and inserting the known quantities, we have

$$m = \rho V = \rho A h = \rho\left(\frac{\pi d^2}{4}\right)h$$

$$= \left(\frac{1.00 \text{ g}}{\text{cm}^3}\right)\left[\frac{\pi(10.4 \text{ cm})^2}{4}\right](28.5 \text{ cm}) = 2.42 \times 10^3 \text{ g} = 2.42 \text{ kg}$$

7. *Consider whether the result is reasonable.*
It may be difficult to determine whether a result is reasonable because of unfamiliarity with units. Here, we might calculate that about 2.5 kg of water has a weight of about 5.0 lb (since 1 kg is equivalent to 2.2 lb or $\approx$ 2.0 lb). This seems to be a reasonable weight for this volume of water. (The dimensions are approximately those of a 2-liter soda bottle.)

Follow-up Exercise. If the outside diameter of the cylinder in Fig. 1.14 is 10.5 cm, what is the total outside surface area of the sides and base of the cylinder? *(See the Answers to Follow-up Exercises section.)*

Conceptual Example 1.12

Doubling: Big Time

A wise woman (early physicist?) did a favor for a king. The king wanted to repay her and asked what she would like. The wise woman noticed a chessboard in the throne room and told the king she had a humble request, only a single grain of wheat. But, for each of the next 64 days, this grain of wheat would be doubled on each square of the chessboard (64 total squares, 8 squares by 8 squares, as on a checkerboard). She would then take only what was on the last square. The king was delighted and granted the wish. However, the king and his kingdom went bankrupt. Why? *Clearly establish the reasoning used in determining your answer before you check it below. That is,* **why** *did you select your answer?*

Reasoning and Answer. To use something more familiar for our analysis, let's use a penny instead of a grain of wheat. The king would still think it's a bargain, right? (Be careful.)

This doubling builds faster than you think—something like compounding interest rates, but a bit faster. So, on the first day of doubling, there would be 2^1 (two) pennies on the first square; on the second day, 2^2 ($2 \times 2 = 4$) pennies on the next square; on the third day, 2^3 ($2 \times 2 \times 2 = 8$) pennies on the next; on the fourth day, 2^4 ($2 \times 2 \times 2 \times 2 = 16$) pennies on that square, and so on. The doubling would continue for exactly 64 days.

Rather than doing 2×2 this many times, we simply want to know the value of 2^{64}. On your scientific calculator, you should have a y^x function: Enter ($y =$) 2, push the y^x key, and enter ($x =$) 64. You should get 1.8×10^{19} pennies! Slipping a couple digits (dividing by 100) to convert to dollars, you have 1.8×10^{17} dollars.

A trillion is 10^{12}, so this is $18\,000$ trillion dollars—which beats the U.S. national debt of $5–6 trillion dollars. A very wise, and rich, woman.

Follow-up Exercise. Back to the king and the wheat. Let's say an average grain of wheat has mass 0.010 g. How many metric tons of wheat would the king owe the wise woman? (As a comparison, the annual output of wheat in the United States is on the order of 60 million metric tons.) *(See the Answers to Follow-up Exercises section.)*

Chapter Review

Important Terms

You should be able to define and explain these chapter terms clearly.

standard unit 2
system of units 2
International System
  of Units (SI) 3
SI base units 3
SI derived units 3
meter (m) 4

kilogram (kg) 5
second (s) 5
mks system 6
cgs system 6
fps system 7
liter (L) 9
dimensional analysis 10

unit analysis 11
density (ρ) 12
conversion factor 13
exact number 17
measured number 17
significant figures (sf) 17

Important Concepts

- The SI units of length, mass, and time are the meter (m), kilogram (kg), and second (s), respectively.
- To a good approximation, a liter (L) of water has a mass of 1 kg.
- Dimensional analysis and/or unit analysis can be used to determine whether an equation is correct, and unit analysis can be used to find the unit of a quantity.
- The number of significant figures of a numerical quantity is the number of reliably known digits it contains.
- Exercises should be worked with a consistent problem solving procedure.

Important Equations

This equation is a mathematical statement of a concept or principle presented in the chapter. It will be needed in working some Exercises, so a good understanding is important. You should be able to identify the symbols and to explain their relationships before proceeding. (The in-text equation reference number is given for convenience.) Most chapters will have more than one important equation.

Density: $$\rho = \frac{m}{V} \quad \left(\frac{\text{mass}}{\text{volume}}\right) \tag{1.1}$$

Exercises

Note: Exercises with a CD-ROM icon have corresponding simulations on the Interactive Journey through Physics CD-ROM; a complete cross-reference of text problems to CD-ROM activities appears in the Media Pack.

Throughout the text, many exercise sections will include "paired" and "trio" exercises. These exercise pairs and trios, identified with colored numbers, are intended to assist you in problem solving and learning. In a pair, the first exercise (even numbered) is worked out in the Study Guide so that you can consult it should you need assistance in solving it. The second exercise (odd numbered) is similar in nature, and its answer is given at the back of the book. In a trio, the first exercise (even numbered) is worked out in the Study Guide. The second exercise (odd numbered) is similar in nature, and its answer is given at the back of the book. The third exercise (even numbered) is based on the same physical principle but is to be solved in a different manner. Its answer is given in the Study Guide.

1.2 Units

1. The only SI standard currently represented by an artifact is the (a) meter, (b) kilogram, (c) second, (d) electric charge.

2. Which of the following is *not* an SI base quantity? (a) length, (b) mass, (c) weight, (d) time.

3. Which of the following is the SI unit of mass? (a) pound, (b) gram, (c) kilogram, (d) ton.

4. Are the following reasonable statements? (Justify your answers.) (a) It took 300 L of gasoline to fill up the car's tank. (b) The center on the basketball team is 225 cm tall. (c) The area of a dorm room is 120 m².

5. ■ The metric system is a decimal (base-10) system, and the British system is in part a duodecimal (base-12) system. Discuss the ramifications if the monetary system of the United States used a duodecimal base. What would be the possible values of our coins if this were the case?

6. ■ (a) In the British system, 16 oz = 1 pt and 16 oz = 1 lb. Is there something wrong here? Explain. (b) Here's an old one: a pound of feathers weighs more than a pound of gold. How can that be? (Hint: Look up *ounce* in the dictionary.)

7. ■■ A sailor told you that if his ship is traveling at 25 knots (nautical miles per hour), it is faster than the 25 mi/h your car travels. How can that be?

1.4 Dimensional Analysis*

8. Both sides of an equation are equal in (a) numerical value, (b) units, (c) dimensions, (d) all of the preceding.

9. Unit analysis of an equation cannot tell you if (a) it is dimensionally correct, (b) the equation is correct, (c) the mathematics is correct, (d) both (b) and (c).

10. Can dimensional analysis tell you whether you have used the correct equation in solving a problem? Explain.

11. If an equation has the same dimensions or units on both sides, the equation is (a) dimensionally correct, (b) correct, (c) may be correct, (d) both (a) and (c).

12. ■ Show that the equation $x = x_o + vt$, where v is velocity and x and x_o are lengths, is dimensionally correct.

13. ■ If x refers to distance, v_o and v to velocities, a to acceleration, and t to time, which of the following equations is dimensionally correct? (a) $x = v_o t + at^3$, (b) $v^2 = v_o^2 + 2at$, (c) $x = at + vt^2$, (d) $v^2 = v_o^2 + 2ax$.

14. ■ Show that $v^2 = v_o^2 + 2ax$ is dimensionally correct. (Here a is acceleration, v and v_o are velocities, and x is length.)

15. ■■ Use SI unit analysis to show that the equation $A = 4\pi r^2$, where A is the area and r is the radius of a sphere, is dimensionally correct.

16. ■■ You are told that the volume of a sphere is given by $V = \pi d^3/4$, where V is the volume and d is the diameter of a sphere. Is this equation dimensionally correct? (Use SI unit analysis to find out.)

17. ■■ The correct equation for the volume of a sphere is $V = 4\pi r^3/3$, where r is the radius of the sphere. Is the equation in Exercise 16 correct?

18. ■■ If $x = gt^2/2$, where x is length and t is time, is dimensionally correct, what are the SI units of the constant g?

19. ■■ Is the equation $v = v_o \sin\theta - gt$ dimensionally correct? Show by using SI unit analysis. (Here v and v_o are velocities, θ is an angle, t is time, and g is acceleration.)

20. ■■ Using unit analysis, determine whether the following equations are dimensionally correct (t is time, x is length, a is acceleration, and v is velocity): (a) $t = \sqrt{2x/a}$ and (b) $v = \frac{1}{2}(v_o + a)t$. (See Section 1.4 for units.)

21. ■■ Is the equation for the area of a trapezoid, $A = \frac{1}{2}a(b_1 + b_2)$, where a is the height and b_2 and b_1 are the bases, dimensionally correct (•Fig. 1.15)? If not, how should it be changed to correct it?

•FIGURE 1.15
The area of a trapezoid
See Exercise 21.

22. ■■ One student, using unit analysis, says that the equation $v = \sqrt{2ax}$ is dimensionally correct. Another says it isn't. With whom do you agree?

23. ■■ The equation for the frequency (f) of oscillation of a simple pendulum is $f = \frac{1}{2\pi}\sqrt{\frac{g}{L}}$, where L is the length of the pendulum string and g is the acceleration due to gravity. Frequency is commonly expressed in units of hertz (Hz). What is a hertz unit in terms of SI base units?

24. ■■■ Newton's second law of motion (Chapter 4) is expressed by the equation $F = ma$, where F represents force, m is mass, and a is acceleration. (a) The SI unit of force is, appropriately, called the newton (N). What are the units of the newton in terms of SI base quantities? (b) Another equation for force associated with uniform circular motion (Chapter 7) is $F = mv^2/r$, where v is speed and r is the radius of the circular path. Does this equation give the same units for the newton?

25. ■■■ Einstein's famous mass–energy equivalence is expressed by the equation $E = mc^2$, where E is energy, m is mass, and c is the speed of light. (a) What are the SI base units of energy? (b) Another equation for energy is $E = mgh$, where m is mass, g is the acceleration due to gravity, and h is height. Does this equation give the same units as in (a)?

1.5 Unit Conversions*

26. A good way to ensure proper unit conversion is to (a) use another measurement instrument, (b) always work in one system of units, (c) use unit analysis, (d) use dimensional analysis.

*Dimensions and/or units of velocity and acceleration are given in the chapter.

*Conversion factors are listed inside the front cover.

27. A conversion factor written as 1 mi = 1609 m means that (a) 1 mi is equivalent to 1609 m, (b) this is a true equation, (c) 1 m = 1609 mi, (d) none of the preceding.

28. ■ Figure 1.7 (top) shows the elevation of a location in both feet and meters. If a town is 130 ft above sea level, what is the elevation in meters?

29. ■ If you wanted to express your height as a larger number, which unit in each of the following pairs would you use: (a) meter or yard; (b) decimeter or foot; (c) centimeter or inch?

30. ■ What is the length in feet of (a) a 100-meter dash and (b) a 2.4-meter high jump?

31. ■ If the capillaries of an average adult were unwound and spread out end to end, they would extend to a length of more than 40 000 mi (Fig. 1.8). If you are 1.75 m tall, how many times would the capillary length equal your height?

32. ■ Standing at 1454 ft, the Sears Tower in Chicago is one of the tallest buildings in the world. What is its height in meters?

33. ■ A Boeing 777 jet has a length of 209 ft, 1 in.; a wingspan of 199 ft, 11 in.; and a fuselage diameter of 20 ft, 4 in. What are these dimensions in meters, and how big around is the fuselage?

34. ■ (a) Which holds more soda, and how much more: a half-gallon bottle or a 2-liter bottle? (b) Suppose that a 16-ounce soda and a 500-milliliter soda sell for the same price. Which would give you the most for your money, and how much more (in milliliters) would you get? [Hint: 1 pint = 16 oz.]

35. ■ Jeff regularly buys 18 gal of gas, but the gas station has installed new pumps that are measured in liters. How many liters of gas (rounded off to a whole number) should he ask for?

36. ■ (a) A football field is 300 ft long and 160 ft wide. What are the field's dimensions in meters? (b) A football is 11.0 to $11\frac{1}{4}$ in. long. What is its length in centimeters?

37. ■ Suppose that when the United States goes completely metric, the dimensions of a football field are established as 100 m by 54 m. Which would be larger, the metric football field or a current football field (see Exercise 36a), and what would be the difference in the areas?

38. ■■ A car travels at a constant speed of 15 m/s. How many miles does it travel in 1 h?

39. ■■ Royal Air Force pilot Andy Green broke the sound barrier on land for the first time and achieved a record land speed of more than 763 mi/h in Black Rock Desert, NV, Oct. 15, 1997 (•Fig. 1.16). (a) What is this speed expressed in meters per second? (b) How long would it take the jet-powered car to travel the length of a 300-foot football field at this speed?

•Figure 1.16 Record run See Exercise 39.

40. ■■ Which one of the following quantities represents the greatest speed? (a) 1 m/s, (b) 1 km/h, (c) 1 ft/s, (d) 1 mi/h.

41. ■■ An automobile speedometer is shown in •Fig. 1.17. (a) What would the equivalent scale readings (for each empty box) be in kilometers per hour? (b) What would the 65 mi/h speed limit be in kilometers per hour?

•FIGURE 1.17 Speedometer readings See Exercise 41.

42. ■■ A student has a car that gets on average 25.0 mi/gal of gasoline. She plans to spend a year in Europe and take her car with her. (a) What should she expect the car's average gas mileage to be in kilometers per liter? (b) During the year there, she drove 6000 km. Assuming gas costs $5.00/gal in Europe, how much did she spend on fuel? (Compute to the nearest dollar.)

43. ■■ Some common product labels are shown in •Fig. 1.18. From the units on the labels, find (a) the number of milliliters in 2.0 fl. oz and (b) the number of ounces in 100 g.

•FIGURE 1.18 Conversion factors See Exercise 43.

44. ■■ Another set of product labels is shown in •Fig. 1.19, but there is an inconsistency. What is the problem, and which is correct? [Hint: 1 L = 1.057 618 qt.]

•FIGURE 1.19 An extra milliliter? See Exercise 44.

45. ■■ A student was 18 in. long when she was born. She is now 5 ft 6 in. and 20 years old. On average, how many centimeters a year did she grow?

46. ■■ In a "19-inch TV" the measurement indicates the diagonal length of the TV tube. Assuming that the face of the tube is flat and rectangular and that the diagonal makes a 37° angle with the base of the tube, what is the area of the tube face in (a) square inches and (b) square centimeters? (Hint: A diagram as suggested in the problem-solving procedure is helpful here.)

47. ■■ The width and length of a room are 3.2 yd and 4.0 yd, respectively. If the height of the room is 8.0 ft, what is the volume of the room in (a) cubic meters and (b) cubic feet?

48. ■■■ The density of metal mercury is 13.6 g/cm³. (a) What is this density expressed in kilograms per cubic meter? (b) How many kilograms of mercury would be required to fill a 0.250-liter container?

49. ■■■ (a) What is the density, in kilograms per cubic meter, of a sphere with radius 10 cm and a mass of 10 kg? (b) A second sphere of the same density has a radius of 20 cm. What is its mass?

50. ■■■ In the Bible, Noah is instructed to build an ark 300 cubits long, 50.0 cubits wide, and 30.0 cubits high (•Fig. 1.20). Historical records indicate a cubit is equal to half of a yard. (a) What would the dimensions of the ark be in meters? (b) What would its volume be in cubic meters? To approximate, assume that the ark was to be rectangular.

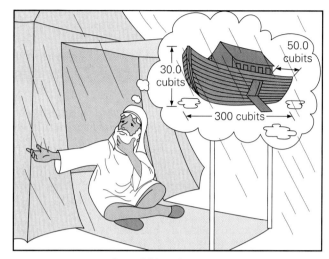

•FIGURE 1.20 Noah and his ark See Exercise 50.

1.6 Significant Figures

51. Which of the following has the greatest number of significant figures? (a) 103.07, (b) 124.5, (c) 0.099 16, (d) 5.408×10^5

52. In a multiplication and/or division operation involving the numbers 15 437; 201.08; and 408.0×10^5, the result should be rounded to how many significant figures? (a) 3, (b) 4, (c) 5, (d) any number.

53. ■ Express the length 50 500 μm (micrometers) in centimeters, decimeters, and meters to three significant figures.

54. ■ Using a meterstick, a student measures a length and reports it to be 0.8755 m. What is the smallest division on the meterstick scale?

55. ■ If a measured length is reported as 25.483 cm, could this length have been measured with an ordinary meterstick whose smallest division is millimeters? Discuss in terms of significant figures.

56. ■ Determine the number of significant figures in the following measured numbers: (a) 1.007 m, (b) 8.03 cm, (c) 16.272 kg, (d) 0.015 μs (microseconds).

57. ■ Express each of the numbers in Exercise 56 with two significant figures.

58. ■ Which of the following quantities has three significant figures: (a) 305.0 cm, (b) 0.0500 mm, (c) 1.000 81 kg, (d) 8.06×10^4 m²?

59. ■ Express each of the following numbers to only three significant figures: (a) 10.072 m, (b) 775.4 km, (c) 0.002 549 kg, (d) 93 000 000 mi.

60. ■■ A rectangle is 3.7 m long and 2.37 m wide. What is its area?

61. ■■ What is the area of a circle of radius 6.37×10^6 m?

62. ■■ The side of a cube is measured, and its volume is reported to be 2.5×10^2 cm^3. What was the measured length of the side of the cube?

63. ■■ The outside dimensions of a cylindrical soda can are reported as 12.559 cm for the diameter and 5.62 cm for the height. What is the total outside area of the can?

64. ■■■ In doing a problem, a student adds 46.9 m and 5.72 m and then subtracts 38 m from the result. What is the final answer?

65. ■■■ Work the following Exercise by two procedures as directed, commenting on and explaining any difference in the answers. Use your calculator for the calculations. Compute $p = mv$, where $v = x/t$. (a) First compute v and then p. (b) Compute $p = mx/t$ without an intermediate step. Given: $x = 8.5$ m, $t = 2.7$ s, and $m = 0.66$ kg.

1.7 Problem Solving

66. An important step in problem solving before mathematically solving an equation is (a) checking units, (b) checking significant figures, (c) consulting with a friend, (d) checking to see if the result is reasonable.

67. An important final step in problem solving before reporting an answer is (a) reading the problem again, (b) saving your calculations, (c) seeing if the answer is reasonable, (d) checking your results with another student.

68. ■ Earth has a mass of 6.0×10^{24} kg and a volume of 1.1×10^{21} m^3. What is Earth's average density?

69. ■ The density of sea water is 1.03 g/cm^3. What is the mass of 1 L of sea water?

70. ■ Does each reported measurement have an appropriate magnitude? (a) A subcompact car has a mass of 1000 kg. (b) A person is 283 cm tall. (c) A runner runs a mile in 3.00×10^8 μs. (d) A normal highway driving speed is 25 m/s. (e) A dachshund stands 10 dm tall. (f) A professor has a waist of 40 cm.

71. ■■ Nutrition Facts labels now appear on most foods. An abbreviated label concerned with fat is shown in •Fig. 1.21. When burned in the body, each gram of fat supplies 9 Calories. (A food Calorie is really a kilocalorie, as we shall see in Chapter 11.) (a) What percentage of the Calories in one serving is supplied by fat? (b) You may notice that your answer doesn't agree with the listed Total Fat percentage in Fig. 1.21. This is because the given Percent Daily Values are the percentages of the maximum recommended amounts of nutrients (in grams) in a 2000-Calorie diet. What are the maximum recommended amounts of total fat and saturated fat for a 2000-Calorie diet?

Nutrition Facts
Serving Size: 1 can
Calories: 310

Amount Per Serving	% Daily Value*
Total Fat 18 g	28%
Saturated Fat 7g	35%

* Percent Daily Values are based on a 2,000 Calorie diet.

•FIGURE 1.21 **Nutrition Facts** See Exercise 71.

72. ■■ The thickness of the numbered pages of a textbook is measured to be 3.75 cm. If the last page of the book is numbered 860, what is the average thickness of a page?

73. ■■ A light-year is a unit of distance corresponding to the distance light travels in a vacuum in 1 y. If the speed of light is 3.00×10^8 m/s, what is the length of a light-year in meters?

74. ■■ To go to a football game, you first drive 1000 m north and then 500 m west. What is the straight-line distance from your home to the football field?

75. ■■ Two equally long chains of length 1.0 m are used to support a lamp as shown in •Fig. 1.22. The distance between the two chains is 1.0 m along the ceiling. What is the vertical distance from the lamp to the ceiling?

•FIGURE 1.22 **Support the lamp** See Exercise 75.

76. ■■ Tony's Pizza Palace sells a 9.0-inch (diameter) pizza for $7.95 and a 12-inch pizza for $13.50. Which is the better buy?

77. ■■ In •Fig. 1.23, which black region has greater area, the center circle or the outer ring?

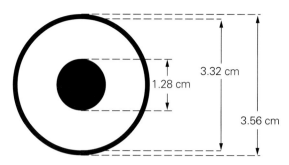

•FIGURE 1.23 Which black area is greater? See Exercise 77.

78. ■■ The distance between the bases in a baseball field is 90 ft. What is the straight-line distance in meters between first base and third base?

79. ■■ The Channel Tunnel, or "Chunnel," that runs under the English Channel between Great Britain and France is 31 mi long. (There are actually three separate tunnels.) A shuttle train that carries passengers through the tunnel travels with an average speed of 75 mi/h. On the average, how long in minutes does it take the shuttle to make a one-way trip through the Chunnel?

80. ■■■ When trains travel through the Chunnel (see Exercise 79), friction heats the air. Left unchecked, the air temperature in the tunnel would be unbearable, so a giant air-conditioning system was installed. In the cooling process, water is pumped through a 300-mile network of 24-inch-diameter pipes. How many metric tons of water does the network contain?

81. ■■■ A student wants to determine the distance of a small island from the lake shore (•Fig. 1.24). He first draws a 50-meter line parallel to the shore. Then, he goes to the ends of the line and measures the angles of the lines of sight from the island relative to the line he has drawn. The angles are 30° and 40°. How far is the island from the shore?

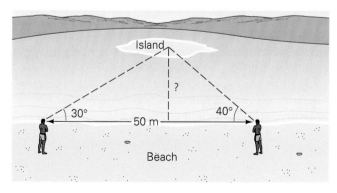

•FIGURE 1.24 Measuring with lines of sight See Exercise 81.

Additional Exercises

82. Express the following calculations to the proper number of significant figures: (a) $12.634 + 2.1$, (b) $13.5 - 2.134$, (c) $\pi (0.25 \text{ m})^2$, (d) $\sqrt{2.37/3.5}$

83. Suppose you are paying $1.20 for 1 gal of gas. Then the United States switches to SI units, and you find that gas costs $0.32/L. Are you paying more after the switch? If so, how much would you be paying?

84. On the average, the human heart beats 70 times a minute (called the pulse rate). On the average, how many times does the heart beat in a 70-year lifetime?

85. The base and the height of a right triangle are 11.2 cm long and 7.5 cm high, respectively. What is the area of the triangle?

86. The general equation for a parabola is $y = ax^2 + bx + c$, where a, b, and c are constants. What are the units of each constant if y and x are in meters?

87. A rectangular block has the dimensions 4.85 cm, 6.5 cm, and 15.51 cm. What is the volume of the block in cubic centimeters?

88. The radius of a solid sphere is 12 cm. What is the surface area of the sphere in (a) square centimeters and (b) square meters? (c) If the mass of the sphere is 9.0 kg, what is the sphere's density in kilograms per cubic meter?

89. A cylindrical drinking glass has an inside diameter of 8.0 cm and a depth of 12 cm. If a person drinks a completely full glass of water, how much (in liters) will be consumed?

90. The top of a rectangular table measures 1.245 m by 0.760 m. (a) What is the smallest division on the scale of the measurement instrument? (b) What is the area of the tabletop?

91. When computing the average speed of a cross-country runner, a student gets 25 m/s. Is this a reasonable result? Justify your answer.

92. The average density of the Moon is 3.36 g/cm³, and the Moon's diameter is 2160 mi. What is the total mass of the Moon in kilograms? (Compare your answer to the value given inside the back cover.)

93. The inner and outer diameters of a thick-walled spherical metal shell are 18.5 cm and 24.6 cm, respectively. What is the volume occupied by the shell itself?

94. An airplane flies 100 mi south from city A to city B, 200 mi east from city B to city C, and then 300 mi north from city C to city D. (a) What is the straight-line distance from A to D? (b) What is the direction of city D relative to city A?

CHAPTER 2

Kinematics: Description of Motion

2.1 Distance and Speed: Scalar Quantities

2.2 One-Dimensional Displacement and Velocity: Vector Quantities

2.3 Acceleration

2.4 Kinematic Equations (Constant Acceleration)

2.5 Free Fall

Insight

- Galileo Galilei and the Leaning Tower of Pisa

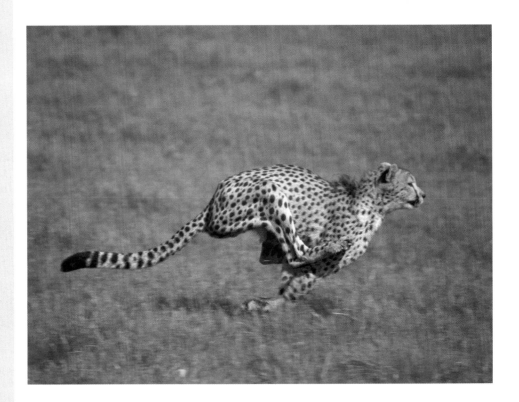

The cheetah is running at full stride. This fastest of all land animals is capable of attaining speeds up to 113 km/h, or 70 mi/h. The sense of motion in this photograph is so strong you can almost feel the air rushing by you. And yet, this is an illusion. Motion takes place in time, but the photo can only "freeze" a single instant. You'll find that, without the dimension of time, you can hardly describe motion at all.

The description of motion involves the representation of a restless world. Nothing is ever perfectly still. You may sit, apparently at rest, but your blood flows, and air moves into and out of your lungs. The air is composed of gas molecules moving at different speeds and in different directions. And, while you experience stillness, you, your chair, the building you are in, and the air you breathe are all revolving through space with the Earth, part of a solar system in a spiraling galaxy in an expanding universe.

The branch of physics concerned with the study of motion and what produces and affects motion is called **mechanics**. The roots of mechanics and of human interest in motion go back to early civilizations. The study of the motions of heavenly bodies, or celestial mechanics, grew out of the need to measure time and location. Several early Greek scientists, notably Aristotle, put forth theories of motion that were useful descriptions but were later proved to be incorrect. Our currently accepted concepts of motion were formulated in large part by Galileo (1564–1642) and Isaac Newton (1642–1727).

Mechanics is usually divided into two parts: (1) kinematics and (2) dynamics. **Kinematics** deals with the description of the motion of objects without consideration of what causes the motion. **Dynamics** analyzes the causes of motion. This chapter covers kinematics and reduces the description of motion to its simplest terms by considering the simple case of motion in a straight line, or linear motion, which is motion in one dimension (of space). We'll learn to analyze changes in motion—speeding up, slowing down, stopping. Along the way, we'll deal with a particularly interesting case of accelerated motion: free fall under the influence only of gravity. Chapter 3 focuses on motion in two dimensions (which can easily be extended to three dimensions). Chapter 4 investigates dynamics to show what causes changes in motion.

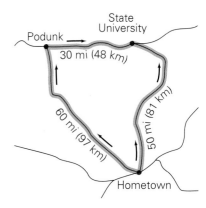

•FIGURE 2.1 Distance—total path length In driving to State University from Hometown, one student may take the shortest route and travel a distance of 50 mi (81 km). Another student takes a longer route in order to visit a friend in Podunk before returning to school. The longer trip is in two segments; the distance traveled is the total path length, 60 mi + 30 mi = 90 mi (145 km).

2.1 Distance and Speed: Scalar Quantities

Objectives: To (a) define distance and calculate speed and (b) explain what is meant by a scalar quantity.

Distance

We can observe motion in many instances around us. What is motion? This seems a simple question, but you might have some difficulty giving an immediate answer (and it's not fair to use the verb "moving" to describe motion). After a little thought, you should be able to conclude that **motion** (or moving) involves the changing of position. Motion can be described in part by specifying *how far* something travels in changing position—that is, the distance it travels. **Distance** is simply the *total path length* traversed in moving from one location to another. For example, you may drive to school from your hometown and express the distance traveled in miles or kilometers. In general, the distance between two points depends on the path traveled (•Fig. 2.1).

Distance: How far

Along with many other quantities in physics, we say that distance is a scalar quantity. A **scalar quantity** is one with only magnitude, or size. That is, a *scalar* (short for scalar quantity) has only a numerical value, such as 160 km or 100 mi. (Note that the magnitude includes units.) Distance is a scalar quantity; it tells you the magnitude only—how far, but not how far in any physical direction. Other examples of scalars are quantities such as 10 s (time), 3.0 kg (mass), and −20°C (temperature). As we will learn in Section 2.2, there is another type of quantity (a *vector*) that has direction associated with it.

A scalar quantity has magnitude but no direction.

Speed

When something is in motion, its position changes with time. That is, it moves a certain distance in a given time. Both length and time are therefore important quantities in describing motion. For example, imagine a car and a pedestrian moving down a street and traveling a distance (length) of one block. You would expect the car to travel faster, to cover the distance in a shorter time, than the person does. This relation can be expressed by using length and time to give the rate at which distance is traveled, or **speed**, for each.

Speed: How fast

Average speed is the distance d traveled, the actual path length, divided by the total time t elapsed in traveling that distance:

2.1 Distance and Speed: Scalar Quantities   **33**

Definition of: Average speed

$$\text{average speed} = \frac{\text{distance traveled}}{\text{total time to travel that distance}}$$

$$avg.\ sp. = \frac{d}{t} \quad (2.1)$$

SI unit of speed: meters per second (m/s)

The SI standard unit of speed is meters per second (length/time), although kilometers per hour is used in many everyday applications. The British standard unit is feet per second, but we often use miles per hour.

Since distance is a scalar quantity (as is time), speed is also a scalar. The distance does not have to be in a straight line (see Fig. 2.1). For example, you probably have computed the average speed of an automobile trip by using the distance obtained from the starting and ending odometer readings. Suppose these readings were 17 455 km and 17 775 km, respectively, for a 4-hour trip. (We'll assume that you have a car with odometer readings in kilometers.) Subtracting the readings gives a traveled distance d of 320 km, so the average speed of the trip is $d/t = 320$ km/4.0 h $= 80$ km/h (or about 50 mi/h).

Average speed gives a general description of motion over a time interval t. In the case of the auto trip with an average speed of 80 km/h, the car's speed wasn't *always* 80 km/h. With various stops and starts on the trip, the car must have been moving more slowly than the average speed part of the time. It therefore had to be moving more rapidly than the average speed another part of the time. With an average speed, you really don't know how fast the car was moving at any particular time during the trip. Similarly, the average test score of a class doesn't tell you the score of any particular student.

If the time interval t considered becomes smaller and smaller and approaches zero, the speed calculation gives an **instantaneous speed**. This is how fast something is moving *at a particular instant of time*. The speedometer of a car gives an approximate instantaneous speed. For example, the speedometer shown in •Fig. 2.2 indicates a speed of about 44 mi/h, or 70 km/h. If the car travels with constant speed (so the speedometer reading does not change), then the average and instantaneous speeds will be equal. (Do you agree? Think of the average test-score analogy.)

•**FIGURE 2.2 Instantaneous speed** The speedometer of a car gives the speed over a very short interval of time, so its reading approaches the instantaneous speed. Add the direction of the car's motion, and you have a good approximation of the instantaneous velocity (covered in Section 2.2).

Example 2.1 ■ Slow Motion: *Sojourner* Moves Along

On July 4, 1997, the *Pathfinder Lander* touched down on the surface of Mars. Out rolled the rover named *Sojourner* (•Fig. 2.3)

Sojourner could move at a maximum average speed of about 0.600 m/min. At this speed, how long would it take the rover to travel 3.00 m to get to another rock to analyze?

Thinking It Through. Knowing the average speed and distance, we can compute the time to travel the distance from the defining equation of average speed.

Solution. We list the data and what is to be found in symbol form:

Given: $avg.\ sp. = 0.600 \dfrac{\text{m}}{\text{min}} \left(\dfrac{1 \text{ min}}{60 \text{ s}} \right)$ **Find:** t (time to travel distance d)

$\qquad\qquad\quad = 0.0100$ m/s

$\qquad d = 3.0$ m

where the unit meters per minute was converted to the standard meters per second.

From Eq. 2.1, we have

$$avg.\ sp. = \frac{d}{t}$$

•**FIGURE 2.3 Away we go!** *Sojourner* speeding at 0.600 m/min (0.0100 m/s) along the Martian surface. See Example 2.1.

Rearranging, we get

$$t = \frac{d}{avg.\ sp.} = \frac{3.00\ \text{m}}{0.0100\ \text{m/s}} = 300\ \text{s} = 5.00\ \text{min}$$

Note that this value is at *Sojourner's maximum* average speed. The trip would probably take longer, as the rover usually had to maneuver around or over small rocks.

Follow-up Exercise. (a) Was it necessary to convert meters per minute to meters per second? Explain. (b) Suppose *Sojourner* took 15.0 min to travel the 3.00 m. What would be the rover's average speed in this case?

2.2 One-Dimensional Displacement and Velocity: Vector Quantities

Objectives: To (a) define displacement and calculate velocity, and (b) explain the difference between scalar and vector quantities.

Displacement

As we have seen, distance is a scalar quantity with only magnitude (and units). However, often when we describe motion, more information can be given by adding a *direction*. This is particularly convenient for a change of position in a straight line. We define **displacement** as the straight-line distance between two points, along with the *direction* from the starting point to the final position.

As such, displacement is a **vector quantity**. A vector (short for vector quantity) has both magnitude *and* direction. For example, when we describe the displacement of an airplane as 25 km north, we are giving a *vector* description (magnitude and direction). Other vector quantities include velocity and acceleration, as we will see shortly.

Graphically, vectors are represented by arrows, with the direction of the arrowhead giving the direction of the vector (•Fig. 2.4). The length of the arrow can be drawn to scale—that is, made proportional to the magnitude or numerical value of the vector. Note in the figure that one vector is twice as long as the other.

For straight-line, or linear, motion, it is convenient to specify position by using the familiar two-dimensional Cartesian coordinate system with *x* and *y* axes at right angles. A straight-line path can be in any direction, but for convenience we usually orient the coordinate axes so that the motion is along one of them. (See Learn by Drawing.)

Algebra applies to vectors, but we have to know how to specify and deal with the direction part of the vector. This process is relatively simple in one dimension. Vectors are discussed more fully in Chapter 3, but let's take a sneak preview here for displacements used in straight-line, or linear, motion.

To illustrate this, suppose a student moves in a straight line from the lockers toward the physics lab as shown in •Fig. 2.5. The scalar distance traveled (along the *x* axis) is determined to be 8.0 m (Fig. 2.5a). But, in using vectors to specify the student's displacement, we first use vectors $\mathbf{x}_1$, and $\mathbf{x}_2$ to locate the positions x_1 and x_2 relative to the origin (Fig. 2.5b). It is a common practice to use boldface symbols to represent vectors. So, when you see such symbols, keep in mind that they have both magnitude *and* direction.

To specify the student's displacement between x_1 and x_2, we subtract the position vectors,

$$\Delta \mathbf{x} = \mathbf{x}_2 - \mathbf{x}_1 \quad (2.2)$$

where the Greek letter Δ (delta) is used to represent a change or difference in a quantity. Here, it means the difference in *position* vectors along the *x* axis.

Equation 2.2 essentially means to subtract the length of $\mathbf{x}_1$ from that of $\mathbf{x}_2$. As a result, we have a new vector $\Delta \mathbf{x}$ that represents the displacement from position x_1 to x_2 (Fig. 2.5c). (Note that we have indicated a direction.)

•**FIGURE 2.4 Vector representation** A vector quantity is represented graphically by an arrow. The vector direction is indicated by the arrow direction, and the length of the arrow is proportional to the magnitude of the quantity.

Displacement: How far and in what direction

LEARN BY DRAWING

Cartesian Coordinates and One-dimensional Displacement

(a) A two-dimensional Cartesian coordinate system. A displacement vector **d** locates a point (*x*, *y*).

(b) For one-dimensional, or straight-line, motion, it is convenient to orient the motion along one of the coordinate axes.

•**FIGURE 2.5 Distance (scalar) and displacement (vector)**
(a) The distance (straight-line path length) between the lockers and the physics lab is 8.0 m and is a scalar quantity. (b) To indicate a displacement, x_1 and x_2 locate the specific initial and final positions. (c) The displacement is then $\Delta x = x_2 - x_1 = 9.0 \text{ m} - 1.0 \text{ m} = +8.0$ m—that is, 8.0 m in the $+x$ direction.

(a) Distance (magnitude or numerical value)

(b) Initial and final position vectors

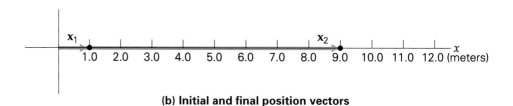

$\Delta \mathbf{x} = \mathbf{x}_2 - \mathbf{x}_1 = 9.0 \text{ m} - 1.0 \text{ m} = +8.0 \text{ m}$

(c) Displacement (magnitude and direction)

Note: In the classroom, many instructors may use arrow notation for vectors: $\vec{A}$. This notation should not be confused with the short line that denotes an average value. Using precise notation is well worth the extra time it takes. Poor or confusing notation can cause problems.

In one dimension, or along a coordinate axis, it is convenient to indicate the direction of a displacement by a plus sign (+) or minus sign (−). The boldface vector notation has the direction implied. But by using magnitudes and directional signs explicitly for the student's displacement in Fig. 2.5, we have

$$\Delta x = x_2 - x_1 = 9.0 \text{ m} - 1.0 \text{ m} = +8.0 \text{ m}$$

Hence, his displacement (magnitude and direction) is 8.0 m in the positive x direction, as indicated by the positive (+) result. (As in "regular" mathematics, the plus sign is often omitted as being understood.)

Suppose the other student in Fig. 2.5 walks from the physics lab (initial, $x_1 = 9.0$ m) to the end of the lockers (final, $x_2 = 1.0$ m). Her displacement would be

$$\Delta \mathbf{x} = \mathbf{x}_2 - \mathbf{x}_1 = 1.0 \text{ m} - 9.0 \text{ m} = -8.0 \text{ m}$$

The minus sign indicates that the direction of the displacement was in the negative x direction, and a representative vector arrow would point that way. We say that the two students' displacements are equal and opposite.

Velocity

As we have seen, speed, like the distance it incorporates, is a scalar quantity—it has magnitude only. Another quantity used to describe motion is *velocity*. Speed and velocity are often used synonymously in everyday conversation, but the terms have different meanings in physics. Speed is a scalar, and velocity is a vector—it has both magnitude and direction.

Velocity tells you how fast something is moving *and* in what direction. And just as we can speak of average and instantaneous speeds, we have average and instantaneous velocities, involving vector displacements. The **average velocity** is the displacement divided by the total travel time:

Velocity: How fast and in what direction

$$\text{average velocity} = \frac{\text{displacement}}{\text{total travel time}}$$

$$\bar{\mathbf{v}} = \frac{\Delta \mathbf{x}}{\Delta t} = \frac{\mathbf{x} - \mathbf{x}_o}{t - t_o} \quad (2.3)$$

Definition of:
Average velocity

SI unit of velocity: meters per second (m/s)

where the boldfacing indicates vector quantities. (A bar above a symbol, such as $\bar{\mathbf{v}}$, is a common way of indicating an average value.)

Note the general notation that is more convenient with fewer subscripts. In the previous discussion on displacement, $\mathbf{x}_1$ and $\mathbf{x}_2$ were used specifically to refer to the initial and final displacements for clarity. But we can generalize. The subscripts on the $\mathbf{x}_o$ and the t_o refer to the *original*, or initial, conditions—where you start the measurements. The $\mathbf{x}$ and t are the displacement and time, respectively, for some later condition—after a change has occurred. These general variables fit whatever the described condition may be. No numerical subscript is needed.

It is common to take $\mathbf{x}_o = 0$ and $t_o = 0$, since we generally (but not always) start a measurement at the zero end of the measurement scale. Keeping this in mind, we have $\Delta \mathbf{x} = \mathbf{x}$ and $\Delta t = t$ in Eq. 2.3, which give the simpler form of

$$\bar{\mathbf{v}} = \frac{\mathbf{x}}{t} \quad \text{or} \quad \mathbf{x} = \bar{\mathbf{v}}t \quad (2.4)$$

Note: In Eq. 2.4, t means Δt and x means Δx.

As this equation indicates, the standard units of velocity are the same as those for speed: meters per second (m/s) or feet per second (ft/s).

Also, as discussed for motion in one dimension, it is convenient to use plus and minus signs to indicate the directions of displacements and velocities along the positive and negative axes—for example, $+x$ and $+v$ (or simply x and v) and $-x$ and $-v$. In the case of more than one displacement, the average velocity is equal to the total or net displacement divided by the total time. The total displacement is found by adding the displacements algebraically according to the directional signs.

You might be wondering whether there is a relationship between average speed and average velocity. A quick look at Fig. 2.5 will show you that *if* all the motion is in one direction, the distance is equal to the magnitude of the displacement, and the average speed is the magnitude of the average velocity. However, be careful. This is not true if there is motion in both directions, as the following Example shows.

Example 2.2 ■ There and Back: Average Speed versus Average Velocity

A jogger jogs from one end to the other of a straight 300-meter track (from point A to point B in •Fig. 2.6) in 2.50 min and then turns around and jogs 100 m back toward the starting point (to point C) in another 1.00 min. What are the jogger's average speeds and average velocities in going (a) from A to B and (b) from A to C after turning around?

Thinking It Through. The average speed and average velocity are computed by using the defining equations. But keep in mind that velocity and displacement are vectors, so their directions must be specified.

•**FIGURE 2.6 Average speed and velocity** (a) The jogger starts at A, jogs to B, then back to C. What are the average speeds and velocities of both segments of the lap? See Example 2.2. (b) If the jogger returns to the starting point (A), the total or net displacement, $x_C + x_B + x_A$, is zero.

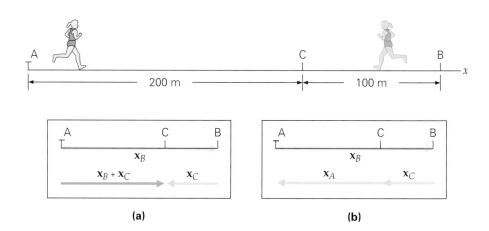

Solution. From the problem we have

Given: $x_B = 300\text{ m} - 0 = +300\text{ m}$ (from A to B)
$\quad\quad\;\; x_C = 200\text{ m} - 300\text{ m} = -100\text{ m}$ (from B to C)
$\quad\quad\;\; t_B = (2.50\text{ min})(60\text{ s/min}) = 150\text{ s}$ ⎫ (conversion to
$\quad\quad\;\; t_C = (1.00\text{ min})(60\text{ s/min}) = 60.0\text{ s}$ ⎭ standard units)

Find: Average speeds and average velocities

(a) The jogger's average speed in going from A to B is computed as follows:

$$\text{avg. sp.}_B = \frac{d}{t} = \frac{300\text{ m}}{150\text{ s}} = 2.00\text{ m/s} \quad \text{(a scalar)}$$

The average velocity in going from A to B with $x_1 = 0$ and $t_1 = 0$ is also easily found, but direction must be indicated. Using Eq. 2.4 and indicating directions by signs,

$$\bar{v} = \frac{x_B}{t_B} = \frac{+300\text{ m}}{150\text{ s}} = +2.00\text{ m/s} \quad \text{(a vector)}$$

The positive direction is to the right in Fig. 2.6. Note that the average speed is equal to the magnitude of the average velocity in this instance.

(b) The average speed in going from A to C involves the total *distance* traveled, so

$$\text{avg. sp.}_C = \frac{d}{t} = \frac{300\text{ m} + 100\text{ m}}{150\text{ s} + 60.0\text{ s}} = 1.90\text{ m/s}$$

where there are no directional signs. (Why?)

The average velocity, as noted above, involves the *sum* of the vector displacements (vector addition):

$$\bar{v}_C = \frac{x_B + x_C}{t_B + t_C} = \frac{+300\text{ m} + (-100\text{ m})}{150\text{ s} + 60.0\text{ s}} = +0.952\text{ m/s}$$

Note that direction makes a difference; the average speed is not equal to the magnitude of the average velocity in this case. This is because the displacements add vectorially. Notice in Fig. 2.6a, which illustrates this simple case of vector addition, that adding x_B and x_C gives an effective, or resultant, displacement from the initial starting point (A) to the stopping point (C), or +200 m.

Suppose that next the jogger sprints from point C back to the starting point (A) in 0.500 min. What are the average speed and average velocity for the whole lap? The additional data are $x_A = 0 - 200\text{ m} = -200\text{ m}$ (Fig. 2.6b) and $t_A = 30.0\text{ s}$. Then, for the round trip, the average speed is

$$\text{avg. sp.}_A = \frac{d}{t} = \frac{300\text{ m} + 100\text{ m} + 200\text{ m}}{150\text{ s} + 60.0\text{ s} + 30.0\text{ s}} = 2.50\text{ m/s}$$

The average velocity is

$$\bar{v}_A = \frac{x_B + x_C + x_A}{t_B + t_C + t_A} = \frac{+300 \text{ m} + (-100 \text{ m}) + (-200 \text{ m})}{240 \text{ s}} = 0$$

The average velocity in this case is zero!

The total displacement is measured from the initial starting point to the final stopping point. When the jogger comes back to the starting point, the displacement is zero and so, therefore, is the average velocity. This is true for any round trip (•Fig. 2.7). Notice in Fig. 2.6b that the (vector) addition of $x_C + x_A$ gives a combined, or resultant, vector that is equal and opposite (in direction) to x_B. The sum of all three vectors is zero—they cancel each other, so to speak.

Follow-up Exercise. (a) Example 2.2 shows that it is possible to have a zero average velocity. Is it possible for the average speed to be zero? Explain. (b) It took 6 y and 2.3 billion mi for the *Galileo* spacecraft to reach Jupiter in late 1995, where it went into orbit and sent a probe into the Jovian atmosphere. What was *Galileo's* average speed in miles per hour for the trip to Jupiter?

As Example 2.2 shows, average velocity provides only a very general description of motion. One way to obtain a closer look at motion is to take smaller time segments, that is, to let the time of observation (Δt) become smaller. As with speed, when Δt approaches zero, we obtain the **instantaneous velocity,** which describes how fast something is moving and in what direction at a particular instant of time.

In one dimension, uniform motion means motion with a constant velocity (constant magnitude *and* constant direction). For example, the car in •Fig. 2.8 has a uniform velocity (as well as a uniform speed). It travels the same distance in equal time intervals (50 km each hour), and the direction of its motion does not change.

Graphical Analysis

Graphical analysis is often helpful in understanding motion and its related quantities. For example, the motion of the car in Fig. 2.8a may be represented on a plot of position versus time, or x versus t. As can be seen from Fig. 2.8b, a straight line is obtained for a uniform, or constant, velocity on such a graph.

Recall from Cartesian graphs of y versus x that the slope of a straight line is given by $\Delta y/\Delta x$. Here, with a plot of x versus t, the slope of the line, $\Delta x/\Delta t$, is equal to the average velocity $\bar{v} = \Delta x/\Delta t$). For uniform motion, this is equal to the instantaneous velocity. That is, $\bar{v} = v$. (Why?) The numerical value of the slope is the magnitude of the velocity and the sign of the slope gives direction. A positive slope indicates that x increases with time, so the motion is in the positive x direction. (The plus sign is often omitted as being understood.)

Suppose that a plot of position versus time for a car's motion was a straight line with a negative slope, as in •Fig. 2.9. What does this slope indicate? As the figure shows, the position (x) values get smaller with time at a constant rate, indicating that the car was traveling in uniform motion in the negative x direction.

In most instances the motion of an object is nonuniform, meaning that different distances are covered in equal intervals of time. An x versus t plot for such motion in one dimension is a curved line, as illustrated in •Fig. 2.10. The average velocity of a particular interval of time is the slope of a straight line between the two points on the curve that correspond to the starting and ending times of the interval. In the figure, the average velocity of the total trip is the slope of the straight line joining the beginning and ending points of the curve (t_1 and t_2).

•**FIGURE 2.7 Back home again!** Despite having covered nearly 110 m on the base paths at the moment the runner slides through the batter's box (his original position) into home plate, his displacement is zero—at least, if he is a right-handed hitter. No matter how fast he ran the bases, his average velocity for the round trip is also zero. (What would his displacement be if he bats lefty? his average velocity?)

Instantaneous velocity: How fast and in what direction *right now*

The word *uniform* means constant.

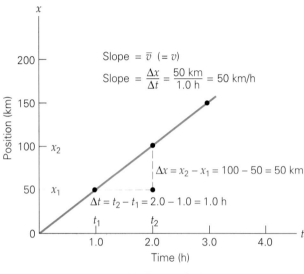

•FIGURE 2.8 **Uniform linear motion—constant velocity** In uniform linear motion, an object travels at a constant velocity, covering the same distance in equal time intervals. **(a)** Here, a car travels exactly 50 km each hour. **(b)** An x-versus-t plot is a straight line, since equal displacements are covered in equal times. The numerical value of the slope of the line is equal to the magnitude of the velocity, and the sign of the slope gives its direction. (Average velocity equals instantaneous velocity in this case. Why?)

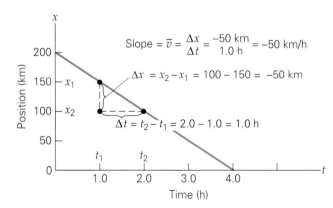

•FIGURE 2.9 **Position-versus-time graph for an object in uniform motion in the $-x$ direction** A straight line on an x-versus-t plot with a negative slope indicates uniform motion in the $-x$ direction. Note that the object's location changes at a constant rate. At $t = 4.0$ h, the object is at $x = 0$. How does the graph look if the motion continues for $t > 4.0$ h?

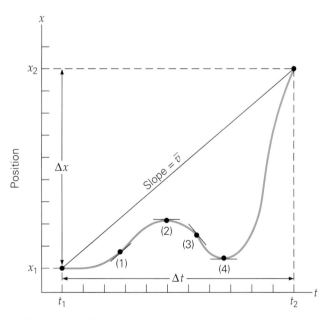

•FIGURE 2.10 **Position-versus-time graph for an object in nonuniform linear motion** For a nonuniform velocity, an x-versus-t plot is a curved line. The slope of the line between two positions is the average velocity between those positions, and the instantaneous velocity is the slope of a line tangent to the curve at any point. Four tangent lines are shown. Can you describe the object's motion in words?

The instantaneous velocity is equal to the slope of a straight line tangent to the curve at a specific point. Four tangent lines are shown in Fig. 2.10. At (1) the slope is positive, and the motion is in the positive x direction. At (2) the slope of a horizontal tangent line is zero, so there is no motion. That is, the object has stopped ($v = 0$). At (3) the slope is negative, so the object is moving in the negative x direction. Thus, it stopped and changed direction at point (2). What is happening at point (4)?

Drawing various tangent lines along the curve, we see that their slopes vary, indicating that the instantaneous velocity is changing with time. An object in nonuniform motion speeds up and/or slows down. How we describe motion with changing velocity is the topic of Section 2.3.

But, before we speed things up (or slow them down), consider the following Example.

Conceptual Example 2.3

A Round Trip: Using Definitions

In driving the usual route to school, a student computes the average speed for the trip to be 20 km/h. On the return trip along the same route, there is less traffic and the average speed is 30 km/h. What is the average speed for the total trip: (a) less than 25 km/h, (b) 25 km/h, (c) greater than 25 km/h? *Clearly establish the reasoning used in determining your answer before checking it below. That is, **why** did you select your answer?*

Reasoning and Answer. It is tempting to choose 25 km/h as the correct answer, as this is the average of the two average speeds. However, this is not the case. The average of the average speeds would be 25 km/h only if the student spent the *same amount of time* traveling at each speed. For example, suppose you spent an hour traveling at each of the given speeds. Then it is easy to see that the average speed for the total trip would be 25 km/h: avg. sp. = (20 km + 30 km)/2.0 h = 25 km/h.

But in this case, the *distances traveled* rather than the times traveled are the same. To travel the same distance, more time must be spent traveling at 20 km/h than at 30 km/h. Hence the total average is weighted toward the slower average speed, and the answer is (a).

You can confirm this by finding the actual average speed, using the definition of average speed: the total distance divided by the time to travel this distance. We are not given the distance, but let's say that it is d one way, for a total round-trip distance of $2d$. The times for each part of the trip are given by $t_1 = d/\text{avg. sp.}_1$ and $t_2 = d/\text{avg. sp.}_2$. Thus, the average speed of the total trip, by definition, is

$$\text{avg. sp.} = \frac{\text{total distance}}{\text{time to travel distance}}$$

$$= \frac{2d}{t_1 + t_2} = \frac{2d}{d/\text{avg. sp.}_1 + d/\text{avg. sp.}_2} = \frac{2}{1/\text{avg. sp.}_1 + 1/\text{avg. sp.}_2}$$

where the d in the denominator was factored out and canceled. So,

$$\text{avg. sp.} = \frac{2}{1/\text{avg. sp.}_1 + 1/\text{avg. sp.}_2}$$

$$= \frac{2}{1/20 + 1/30} = \frac{2}{3/60 + 2/60} = \frac{2}{5/60} = 120/5 = 24 \text{ km/h}$$

where the units were omitted in the calculation for convenience.

Follow-up Exercise. What average speed on the return trip would give an average of 25 km/h for the total trip?

2.3 Acceleration

Objectives: To (a) explain the relationship between velocity and acceleration and (b) perform graphical analyses of acceleration.

The basic description of motion involves the time rate of change of position, which we call velocity. Going one step further, we can consider how this *rate of change* changes. Suppose that something is moving at a constant velocity and then the velocity changes. Such a change in velocity is called an *acceleration*. The gas pedal on an automobile is commonly called the accelerator. When you press down on the accelerator, the car speeds up; when you let up on the accelerator, the car slows down. In either case, there is a change in velocity with time. We define **acceleration** as the time rate of change of velocity.

Analogous to average velocity is the **average acceleration**, or the change in velocity divided by the time taken to make the change:

$$\text{average acceleration} = \frac{\text{change in velocity}}{\text{time to make the change}}$$

$$\bar{\mathbf{a}} = \frac{\Delta \mathbf{v}}{\Delta t} = \frac{\mathbf{v} - \mathbf{v}_o}{t - t_o} \qquad (2.5)$$

SI unit of acceleration: meters per second squared (m/s²)

In Eq. 2.5, we use the boldface vector notation because, in general, the velocities may be in different directions. Since velocity is a vector quantity, so is acceleration. Analogous to instantaneous velocity is **instantaneous acceleration**, which is the acceleration at a particular instant of time.

The dimensions of acceleration are $([L]/[T])/[T]$, from $\Delta v/\Delta t$. The SI units of acceleration are therefore meters per second per second, that is, (m/s)/s or m/s·s, commonly expressed as meters per second squared (m/s²). In the British system, the units are feet per second squared (ft/s²).

Since velocity is a vector quantity, having both magnitude and direction, a change in velocity may involve either or both of these factors. An acceleration, therefore, may result from a change in *speed* (magnitude), a change in *direction*, or a change in *both*, as illustrated in •Fig. 2.11.

For the special case of straight-line, or linear, motion, plus and minus signs can be used to indicate velocity directions (so we can do away with the boldface), as was done for linear displacements. Then, Eq. 2.5 can be written as

$$\bar{a} = \frac{v - v_o}{t} \qquad (2.6)$$

where t_o is taken to be zero (v_o may not be zero, so it cannot generally be omitted).

Example 2.4 ■ Slowing It Down: Average Acceleration

A couple in their sport utility vehicle (SUV) is traveling 90 km/h down a straight highway. They see an accident in the distance, so the driver slows down to 40 km/h in 5.0 s. What is the average acceleration of the SUV?

Thinking It Through. To find the average acceleration, the variables as defined in Eq. 2.6 must be given, and they are.

Solution. From the problem, we have the following data. [With the motion in a straight line, the instantaneous velocities are assumed to be in the positive direction, and conversions to standard units (kilometers per second to meters per second) are

---

Acceleration: The time rate of change of velocity

Definition of: Average acceleration

Instantaneous acceleration: a change in velocity *right now*

Note: In compound units, multiplication is indicated by a dot.

LEARN BY DRAWING

Signs of Velocity and Acceleration

•FIGURE 2.11 Acceleration—the time rate of change of velocity Since velocity is a vector quantity, with magnitude and direction, an acceleration can occur when there is **(a)** a change in magnitude but not direction, **(b)** a change in direction but not magnitude, or **(c)** a change in both magnitude and direction.

made right away since it is noted that the time is given in seconds. In general, we always work with acceleration in standard units.]

Given: $v_o = (90 \text{ km/h})\left(\dfrac{0.278 \text{ m/s}}{1 \text{ km/h}}\right)$ **Find:** $\bar{a}$ (average acceleration)

$= 25 \text{ m/s}$

$v = (40 \text{ km/h})\left(\dfrac{0.278 \text{ m/s}}{1 \text{ km/h}}\right)$

$= 11 \text{ m/s}$

$t = 5.0 \text{ s}$

Given the initial and final velocities and the time interval, the average acceleration can be found by using Eq. 2.6:

$$\bar{a} = \dfrac{v - v_o}{t} = \dfrac{11 \text{ m/s} - 25 \text{ m/s}}{5.0 \text{ s}} = -2.8 \text{ m/s}^2$$

The minus sign indicates the direction of the (vector) acceleration. In this case, the acceleration is opposite to the direction of the motion ($+v$), and the car slows. Such an acceleration is sometimes called a *deceleration*, since the car is slowing.

Follow-up Exercise. Does a negative acceleration necessarily mean that a moving object is slowing down (decelerating) or its speed is decreasing?

Although acceleration can vary with time, our study of motion in this chapter will be restricted to constant accelerations for simplicity. (An important constant acceleration is the acceleration due to gravity near the Earth's surface, which will be considered in the next section.) Since for a constant acceleration the average is equal to the constant value ($\bar{a} = a$), the bar over the acceleration in Eq. 2.6 may be omitted. Thus, for a constant acceleration, the equation relating velocity, acceleration, and time is commonly written (rearranging Eq. 2.6) as follows:

$$v = v_o + at \quad \text{(constant acceleration only)} \quad (2.7)$$

(Note that the term at represents the *change* in velocity that occurs, since $at = v - v_o = \Delta v$.)

Example 2.5 ■ Fast Start, Slow Stop: Motion with Constant Acceleration

A drag racer starting from rest accelerates in a straight line at a constant rate of 5.5 m/s² for 6.0 s. (a) What is the racer's velocity at the end of this time? (b) If a parachute deployed at this time causes the racer to slow down uniformly at a rate of 2.4 m/s², how long will it take the racer to come to a stop?

Thinking It Through. It is noted that the racer first speeds up and then slows down, so close attention must be given to the directional signs of the vector quantities. The answers can then be found by using the appropriate equations.

Solution. Taking the initial motion to be in the positive direction, we have

Given: (a) $v_o = 0$ (at rest) *Find:* v (final velocity)
 $a = 5.5 \text{ m/s}^2$
 $t = 6.0 \text{ s}$
 (b) $v_o = v$ [from part (a)] *Find:* t (time)
 $v = 0$ (comes to stop)
 $a = -2.4 \text{ m/s}^2$ (opposite direction of v_o)

The data have been listed in two parts. This helps avoid confusion with symbols. Note that the final velocity (v) that is to be found in part (a) becomes the initial velocity (v_o) for part (b).

(a) To find v, we use Eq. 2.7 directly:

$$v = v_o + at = 0 + (5.5 \text{ m/s}^2)(6.0 \text{ s}) = 33 \text{ m/s}$$

(b) Here we want time, so solving Eq. 2.6 for t and using $v_o = 33$ m/s from part (a), we have

$$t = \frac{v - v_o}{a} = \frac{0 - 33 \text{ m/s}}{-2.4 \text{ m/s}^2} = 14 \text{ s}$$

Note that the time comes out positive, as it should. We start implicitly at zero time (taking $t_o = 0$ when the parachute is deployed), and time goes forward or in a "positive direction."

Follow-up Exercise. What is the racer's instantaneous velocity 10 seconds after the parachute is deployed?

Motions with constant accelerations are easy to represent graphically by using instantaneous velocity versus time. A v versus t plot is a straight line whose slope is equal to the acceleration, as illustrated in •Fig. 2.12. Note that Eq. 2.7 can be written $v = at + v_o$, which, as you may recognize, has the form of an equation of a straight line, $y = mx + b$ (slope m and intercept b). In Fig. 2.12a, the motion is in the positive direction and the acceleration adds to the velocity for a time t, as illustrated by the

•FIGURE 2.12 Velocity-versus-time graphs for motions with constant accelerations The slope of a v versus t plot is the acceleration. **(a)** A positive slope indicates an increase in the velocity in the positive direction. The vertical arrows to the right indicate how the acceleration adds velocity to the initial velocity v_o. **(b)** A negative slope indicates a decrease in the initial velocity v_o, or a deceleration. **(c)** Here a negative slope indicates a negative acceleration, but the initial velocity is in the negative direction, $-v_o$, so the speed of the object increases in that direction. **(d)** The situation here is initially similar to that in (b) but ends by resembling that in (c). Can you explain what happened at time t_1?

vertical arrows at the right of the graph. Here the slope is positive, $+a$. In Fig. 2.12b, the negative slope ($-a$) indicates a negative acceleration that produces a slowing down, or deceleration. However, Fig. 2.12c illustrates how a negative acceleration can speed things up (for motion in the negative direction). The situation in Fig. 2.12d is slightly more complex. Can you explain what is happening there?

When an object moves at a constant acceleration, its velocity changes by the same amount in each time unit. For example, if the acceleration is 10 m/s² in the same direction as the initial velocity, the object's velocity increases by 10 m/s in each second. Suppose that the object has an initial velocity (v_o) of 20 m/s at $t_o = 0$. Then, for $t = 0, 1.0, 2.0, 3.0,$ and 4.0 s, the velocities are 20, 30, 40, 50, and 60 m/s, respectively. The average velocity over the 4-second interval is $\bar{v} = 40$ m/s.

This average velocity may be computed in the regular manner, or you may immediately recognize that the uniformly increasing series of numbers 20, 30, 40, 50, and 60 has an average value of 40 (the midway value of the series). Note that the average of the extreme (initial and final) values also gives the average of the series—that is, $(20 + 60)/2 = 40$. This is true in general: When the velocity changes at a uniform rate because of a constant acceleration, $\bar{v}$ is the average of the initial and final velocities:

$$\bar{v} = \frac{v + v_o}{2} \quad \text{(constant acceleration only)} \quad (2.8)$$

Example 2.6 ■ On the Water: Using Multiple Equations

A motorboat starting from rest on a lake accelerates in a straight line at a constant rate of 3.0 m/s² for 8.0 s. How far does the boat travel during this time?

Thinking It Through. Since we have only one equation for distance (Eq. 2.4, $x = \bar{v}t$), this equation cannot be used directly. The average velocity must first be found, so multiple equations are involved.

Solution. Reading the problem and summarizing the given data and what is to be found, we have

Given: $v_o = 0$ Find: x (distance)
$\quad\quad\; a = 3.0$ m/s²
$\quad\quad\; t = 8.0$ s

(It is noted that all of the units are standard.)

In analyzing the problem, we might reason as follows: To find x, we will need to use Eq. 2.4, $x = \bar{v}t$. (The average velocity $\bar{v}$ must be used because the velocity is changing and not constant.) With t given, the solution to the problem then involves finding $\bar{v}$. By Eq. 2.8, $\bar{v} = (v + v_o)/2$, so with $v_o = 0$, we need find only the final velocity v to solve the problem. Equation 2.7, $v = v_o + at$, enables us to calculate v from the given data. So, we have the following:

The velocity of the boat at the end of 8.0 s is

$$v = v_o + at = 0 + (3.0 \text{ m/s}^2)(8.0 \text{ s}) = 24 \text{ m/s}$$

The average velocity over that time interval is

$$\bar{v} = \frac{v + v_o}{2} = \frac{24 \text{ m/s} + 0}{2} = 12 \text{ m/s}$$

Finally, the magnitude of the displacement, which in this case is the same as the distance traveled, is

$$x = \bar{v}t = (12 \text{ m/s})(8.0 \text{ s}) = 96 \text{ m}$$

Follow-up Exercise. (Sneak preview.) In Section 2.4, the following equation will be derived: $x = v_o t + \frac{1}{2}at^2$. Use the data in this Example to see if this equation gives the distance traveled.

2.4 Kinematic Equations (Constant Acceleration)

Objectives: To (a) explain the constant acceleration kinematic equations and
(b) apply them to physical situations.

The description of motion in one dimension with constant acceleration requires only three basic equations. From previous sections, these are

$$x = \bar{v}t \quad\quad\quad (2.4)$$

$$\bar{v} = \frac{v + v_o}{2} \quad \text{(constant acceleration only)} \quad\quad (2.8)$$

$$v = v_o + at \quad \text{(constant acceleration only)} \quad\quad (2.7)$$

(Keep in mind that the first equation is general and is not limited to situations in which there is constant acceleration, as the latter two are.)

However, as Example 2.6 showed, the description of motion in some instances requires multiple applications of these equations, which may not be obvious at first. It would be helpful if there were a way to reduce the number of operations in solving kinematic problems, and there is—combining equations algebraically.

For instance, suppose we want an expression that gives displacement (x) in terms of time and acceleration rather than in terms of time and average velocity (as in Eq. 2.4). We can eliminate $\bar{v}$ from Eq. 2.4 by substituting for $\bar{v}$ from Eq. 2.8 into Eq. 2.4:

$$x = \bar{v}t = \left(\frac{v + v_o}{2}\right)t$$

Then substituting for v from Eq. 2.7 gives

$$x = \left(\frac{v + v_o}{2}\right)t = \left[\frac{(v_o + at) + v_o}{2}\right]t$$

Simplifying gives

$$x = v_o t + \tfrac{1}{2} at^2 \quad \text{(constant acceleration only)} \quad (2.9)$$

Essentially, this series of steps was done in Example 2.6. This combined equation allows the distance traveled by the motorboat in that Example to be computed directly:

$$x = v_o t + \tfrac{1}{2} at^2 = 0 + \tfrac{1}{2}(3.0 \text{ m/s}^2)(8.0 \text{ s})^2 = 96 \text{ m}$$

Much easier, isn't it?

Perhaps we want an expression that gives velocity as a function of position (x) rather than time (as in Eq. 2.7). We can eliminate t from Eq. 2.7 by using Eq. 2.4 in the form $t = x/\bar{v}$:

$$v = v_o + a\left(\frac{x}{\bar{v}}\right)$$

Then we replace $\bar{v}$, using Eq. 2.8:

$$v = v_o + \frac{ax}{\frac{v + v_o}{2}} \quad \text{or} \quad v - v_o = \frac{2ax}{v + v_o}$$

Simplify by using the algebraic relationship $(v + v_o)(v - v_o) = v^2 - v_o^2$, we have

$$v^2 = v_o^2 + 2ax \quad \text{(constant acceleration only)} \quad (2.10)$$

Problem-Solving Hint

Students in introductory physics courses are sometimes overwhelmed by the various kinematic equations—and this is only Chapter 2! Keep in mind that equations and mathematics are the "tools" of physics; as any mechanic or carpenter will tell you, tools make your work easier as long as you are familiar with them and know how to use them. The same is true with our physics "tools."

Basically, just this set of equations

$$v = v_o + at \quad (2.7)$$

$$x = v_o t + \tfrac{1}{2}at^2 \quad (2.9)$$

$$v^2 = v_o^2 + 2ax \quad (2.10)$$

is used to solve the majority of kinematic problems. (Occasionally we are interested in average speed or velocity, but, as pointed out earlier, averages don't tell you a great deal.) Now, note that each of these three equations has four variables. Three of these must be known in a particular equation to solve for the fourth, unknown variable. Always try to understand and visualize a problem, but listing the data as in the Suggested Problem-Solving Procedure in Chapter 1 may help you decide which equation to use, since you generally need to know three variables. Remember this as you go over the remaining Examples in the chapter.

Also, don't overlook any *implied data*. For example, what does "dropped" in the following Example imply?

Example 2.7 ■ A Dropped Ball: Using the Kinematic Equations

A ball is dropped out of a window near the top of a building. If the ball accelerates toward the ground at a rate of 9.80 m/s², what is its velocity when it has fallen 4.00 m?

Thinking It Through. Usually, one of the kinematic equations will provide a solution. A good way to determine which equation to use is to look at what is given and what is to be found.

Solution. Here, we list the given data and make a sketch to help visualize the situation (•Fig. 2.13).

Given: $v_o = 0$ (dropped) *Find:* v (velocity)
$a = -9.80$ m/s²
$x = -4.00$ m

Note that three variables are given: The statement that the ball is dropped means that the initial velocity is zero, $v_o = 0$. Upward is usually taken as the positive direction and downward as the negative direction. The magnitudes of **a** and **x** are then written with minus signs to indicate their directions.

No directional sign is explicitly assigned to the velocity we wish to find. In this case you should realize that the final velocity will be negative, or in the downward direction.

Since time is not directly involved, the velocity may be obtained directly from Eq. 2.10:

$$v^2 = v_o^2 + 2ax = 0 + 2(-9.80 \text{ m/s}^2)(-4.00 \text{ m}) = 78.4 \text{ m}^2/\text{s}^2$$

and

$$v = \pm\sqrt{78.4 \text{ m}^2/\text{s}^2} = -8.85 \text{ m/s}$$

where the negative root is chosen because the motion is downward (that is, in the negative direction) in this physical situation. (Keep in mind that when you take the square root of a quantity you always obtain both positive and negative roots.)

Using the combined forms of the equations can often save you a lot of steps and calculations.

Follow-up Exercise. In this Example, how long does it take for the ball to fall the 4.00 m?

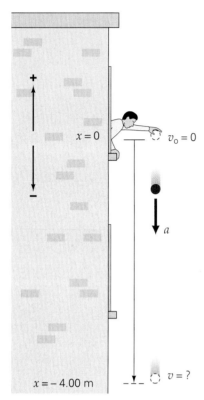

•**FIGURE 2.13 A dropped ball** A sketch to help visualize Example 2.7.

For your convenience, the equations for linear motion with constant acceleration are summarized in Table 2.1. Note that only the first three are basic equations; the last two are convenient combinations. Also, they are quite versatile.

Conceptual Example 2.8

Two Race Cars: The Effect of Squared Quantities

During some time trials, race car A, starting from rest, accelerates uniformly along a straight, level track for a particular time interval. Car B, also starting from rest, accelerates at the same rate but for twice the time. At the ends of their respective acceleration periods, which of these statement is true: (a) Car A has traveled a greater distance, (b) car B has traveled twice as far as car A, (c) car B has traveled four times

TABLE 2.1 Equations for Linear Motion with Constant Acceleration*

$x = \bar{v}t$	(2.4)†
$\bar{v} = \dfrac{v + v_o}{2}$	(2.8)
$v = v_o + at$	(2.7)
$x = v_o t + \tfrac{1}{2}at^2$	(2.9)
$v^2 = v_o^2 + 2ax$	(2.10)

*It is assumed that $x_o = 0$ and that the velocity is v_o at $t_o = 0$. The initial position and time, x_o and t_o, may be included for general cases, for example, $x - x_o = \bar{v}(t - t_o)$.

†Note that Eq. 2.4 is not limited to constant acceleration but applies generally.

as far as car A, (d) both cars have traveled the same distance? *Clearly establish the reasoning and physical principle(s) used in determining your answer before checking it below. That is, **why** did you select your answer?*

Reasoning and Answer. It is given that $v_o = 0$ and that the acceleration is the same for both cars. To find the distances traveled in a time t, you would use Eq. 2.9, which, with $v_o = 0$, becomes $x = \frac{1}{2}at^2$. The important thing to note here is that the distance increases as t^2. That is, if you double the time, the distance quadruples (increases by a factor of four).

In this example, car B accelerates twice as long as car A, or $t_B = 2t_A$, so car B travels four times as far as car A and the answer is (c). Expressed mathematically,

$$x_A = \tfrac{1}{2}at_A^2 \quad \text{and} \quad x_B = \tfrac{1}{2}at_B^2 = \tfrac{1}{2}a(2t_A)^2 = \tfrac{1}{2}a(4t_A^2) = 4(\tfrac{1}{2}at_A^2) = 4x_A$$

How do their distances compare if $t_B = 3t_A$?

Follow-up Exercise. In this Example, how do the speeds of the cars compare at the ends of the acceleration periods?

Example 2.9 ■ Putting on the Brakes: Vehicle Stopping Distance

The stopping distance of a vehicle after the brakes have been applied is an important factor in road safety. This distance depends on the initial speed $(+v_o)$ and the braking capacity, or deceleration, $-a$, which is assumed to be constant. (Recall that the minus sign indicates that the acceleration is in the negative direction. In this case the sign of acceleration is opposite that of the velocity, which is taken to be positive. Thus the car slows to a stop.) Express the stopping distance x in terms of these quantities.

Thinking It Through. Again, a kinematic equation is required, and the appropriate one is determined by what is given and what is to be found. Notice the distance x is wanted, and time is not involved.

Solution. Here, we are working with variables, so we can represent quantities only in symbolic form.

Given: v_o (positive direction)   *Find:* x (in terms of the given variables)
        $-a$ (opposite direction of v_o)
        $v = 0$ (car comes to stop)

Again, it is helpful to make a sketch of the situation, particularly when directional vector quantities are involved (•Fig. 2.14). Since Eq. 2.10 has the variables we want, it should allow us to find the stopping distance. Expressing the negative acceleration explicitly gives

$$v^2 = v_o^2 + 2(-a)x = v_o^2 - 2ax$$

Since the vehicle comes to a stop ($v = 0$), we can solve for x:

$$x = \frac{v_o^2}{2a}$$

This gives us x expressed in terms of the vehicle's initial speed and stopping acceleration.

•**FIGURE 2.14 Vehicle stopping distance** A sketch to help visualize Example 2.9.

Notice that the stopping distance x is proportional to the *square* of the initial speed. Doubling the initial speed therefore increases the stopping distance by a factor of 4 (for the same deceleration). That is, if the stopping distance is x_1 for an initial speed of v_1, then for a two-fold initial speed ($v_2 = 2v_1$), the stopping distance would increase four-fold:

$$x_1 = \frac{v_1^2}{2a}$$

$$x_2 = \frac{v_2^2}{2a} = \frac{(2v_1)^2}{2a} = 4\left(\frac{v_1^2}{2a}\right) = 4x_1$$

We can get the same result by directly using ratios:

$$\frac{x_2}{x_1} = \frac{v_2^2}{v_1^2} = \left(\frac{v_2}{v_1}\right)^2 = 2^2 = 4.$$

Do you think this is an important consideration in setting speed limits, for example, in school zones? (The driver's reaction time should also be considered. A method for approximating a person's reaction time is given in Section 2.5.)

Follow-up Exercise. A driver traveling at 25 mi/h in a school zone can brake to an emergency stop in 8.0 ft. Under similar conditions, what would be the braking distance if the car were traveling at 40 mi/h?

2.5 Free Fall

Objective: To use the kinematic equations to analyze free fall.

One of the more common cases of constant acceleration is the acceleration due to gravity near the Earth's surface. When an object is dropped, its initial velocity is zero (at the instant it is released). At a later time while falling, it has a nonzero velocity. There has been a change in velocity and so, by definition, an acceleration. This **acceleration due to gravity** (g) has an approximate magnitude of

Definition of: Acceleration due to gravity

$$g = 9.80 \text{ m/s}^2 \quad \textit{acceleration due to gravity}$$

or 980 cm/s² and is directed downward (toward the center of the Earth). In British units, the value of g is about 32.2 ft/s².

The values given here for g are only approximate because the acceleration due to gravity varies slightly at different locations as a result of differences in elevation and regional average mass density of the Earth. These small variations will be ignored in this book unless otherwise noted. (Gravitation is studied in more detail in Chapter 7.) Air resistance is another factor that affects the acceleration of a falling object. But for relatively dense objects and over the short distances of fall commonly encountered, air resistance produces only a small effect, which will also be ignored here for simplicity. (The frictional effect of air resistance will be considered in Chapter 4.)

Definition of: Free fall

*Objects in motion solely under the influence of gravity are said to be in **free fall**.* The words "free fall" bring to mind dropped objects that are moving downward under the influence of gravity ($g = 9.80$ m/s² in the absence of air resistance). However, the term can be applied in general to any vertical motion under the sole influence of gravity. Objects released from rest or thrown upward or downward are in free fall once they are released. That is, after $t = 0$ (the time of release), only gravity is acting and influencing the motion. (Even when an object projected upward is traveling upward, *it is still accelerating downward*.) Thus, the set of equations for motion in one dimension (in Table 2.1) can be used to describe generalized free fall.

The acceleration due to gravity g is the *constant* acceleration for all free-falling objects, regardless of their mass or weight. It was once thought that heavier bodies fall faster than lighter bodies. This was part of Aristotle's theory of motion. You can easily observe that a coin falls faster than a sheet of paper when dropped simultaneously from the same height. But in this case air resistance plays a noticeable role. If the paper is crumpled into a compact ball, it gives the coin a better race. Similarly, a feather "floats" down much more slowly than a coin falls. However, in a near-vacuum, where there is negligible air resistance, the feather and the coin fall with the same acceleration—the acceleration due to gravity (•Fig. 2.15).

Astronaut David Scott performed a similar experiment on the Moon in 1971 by simultaneously dropping a feather and a hammer from the same height. He did not need a vacuum pump: The Moon has no atmosphere and therefore no air resistance. The hammer and the feather reached the lunar surface together, but both fell at a slower rate than on Earth. The acceleration due to gravity near the Moon's surface is about one-sixth of that near the Earth's surface ($g_M \approx g/6$).

Currently accepted ideas about the motion of falling bodies are due in large part to Galileo. He challenged Aristotle's theory and experimentally investigated the motion of objects. Legend has it that Galileo studied the accelerations of falling bodies by dropping objects of different weights from the top of the Leaning Tower of Pisa (see the Insight on p. 52).

It is customary to use y to represent the vertical direction and to take upward as positive (as with the vertical y axis of Cartesian coordinates). Since the acceleration due to gravity is always downward, it is in the negative direction. This negative acceleration, $a = -g = -9.80 \text{ m/s}^2$, should be substituted into the equations of motion. However, the relationship $a = -g$ may be expressed explicitly in the equations for linear motion (see Table 2.1):

$$y = \bar{v}t \qquad (2.4')$$

$$\bar{v} = \frac{v + v_o}{2} \qquad (2.8')$$

Free-fall equations with

$$v = v_o - gt \qquad (2.7')$$

$-g$ expressed explicitly

$$y = v_o t - \tfrac{1}{2}gt^2 \qquad (2.9')$$

$$v^2 = v_o^2 - 2gy \qquad (2.10')$$

Note: While traveling upward ($+v$), a projected object is accelerating downward, or has an acceleration in the downward direction ($-g$).

Note: Here g refers to the magnitude of acceleration due to gravity.

(a) (b)

•**FIGURE 2.15 Free fall and air resistance** (a) When dropped simultaneously from the same height, a feather falls more slowly than a coin because of air resistance. But when both objects are dropped in an evacuated container with a good partial vacuum, where air resistance is negligible, the feather and the coin fall together with a constant acceleration. (b) An actual demonstration with multiflash photography. An apple and a feather are released simultaneously through a trap door into a large vacuum chamber, and they fall together—almost. Although the chamber has a partial vacuum, there is still some air resistance. (How can you tell?)

Insight

Galileo Galilei and the Leaning Tower of Pisa

Galileo Galilei (Fig. 1) was born in Pisa, Italy, in 1564 during the Renaissance. Today, he is known throughout the world by his first name and often referred to as the father of modern science or the father of modern mechanics and experimental physics, which attests to the magnitude of his scientific contributions.

One of Galileo's greatest contributions to science was the establishment of the scientific method—that is, investigation through experiment. In contrast, Aristotle's approach was based on logical deduction. By the scientific method, for a theory to be valid it must predict or agree with experimental results. If it doesn't, it is invalid or requires modification. Galileo said, "I think that in the discussion of natural problems we ought not to begin at the authority of places of Scripture, but at sensible experiments and necessary demonstrations."*

Probably the most popular and well-known legend about Galileo is that he performed experiments with falling bodies by dropping objects from the Leaning Tower of Pisa (Fig. 2). There is some doubt as to whether Galileo actually did this, but there is little doubt that he questioned Aristotle's view on the motion of falling objects. In 1638, Galileo wrote:

> Aristotle says that an iron ball of one hundred pounds falling from a height of one hundred cubits reaches the ground before a one-pound ball has fallen a single cubit. I say that they arrive at the same time. You find, on making the experiment, that the larger outstrips the smaller by two finger-breadths, that is, when the larger has reached the ground, the other is short of it by two finger-breadths; now you would not hide behind these two fingers the ninety-nine cubits of Aristotle.†

This and other writings show that Galileo was aware of the effect of air resistance.

The experiments at the Tower of Pisa were supposed to have taken place around 1590. In his writings of about that time, Galileo mentions dropping objects from a high tower, but never specifically names the Tower of Pisa. A letter written to Galileo from another scientist in 1641 describes the dropping of a cannon ball and a musket ball from the Tower of Pisa. The first account of Galileo doing a similar experiment was written a dozen years after his death by Vincenzo Viviani, his last pupil and first biographer. It is not known whether Galileo told this story to Viviani in his declining years or Viviani created this picture of his former teacher.

The important point is that Galileo recognized (and probably experimentally showed) that free-falling objects fall with the same acceleration regardless of their mass or weight (see Fig. 2.15). Galileo gave no reason why all objects in free fall have the same acceleration, but Newton did, as you will learn in a later chapter.

*From *Growth of Biological Thought: Diversity, Evolution & Inheritance*, by F. Meyr (Cambridge, MA: Harvard University Press, 1982).

†From *Aristotle, Galileo, and the Tower of Pisa*, by L. Cooper (Ithaca, NY: Cornell University Press, 1935).

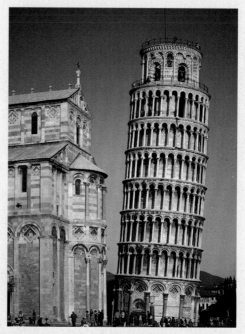

FIGURE 2 The Leaning Tower of Pisa The tower, constructed as a belfry for the nearby cathedral, was built on shifting subsoil. Construction began in 1173, and the tower started to shift one way and then the other, before inclining in its present direction. Today, the tower leans about 5 m (16 ft) from the vertical at the top. It has been closed since 1990, and efforts are being made to stabilize the leaning.

FIGURE 1 Galileo Galileo is alleged to have performed free-fall experiments by dropping objects off the Leaning Tower of Pisa.

The origin ($y = 0$) of the reference frame is usually taken to be at the initial position of the object. Since upward is generally taken to be in the positive direction (+y axis on a graph), writing $-g$ explicitly in the equations reminds you of directional differences, and the value of g is inserted as 9.80 m/s^2. However, the choice is arbitrary. The equations can be written with $a = g$, for example, $v = v_o + gt$, with the directional minus sign associated directly with g. In this case, a value of -9.80 m/s^2 must be substituted for g each time.

Note that you have to be explicit about the directions of vector quantities. The displacement y and the velocities v and v_o may be positive or negative, depending on the direction of motion. The use of these equations and the sign convention are illustrated in the following Examples.

Example 2.10 ■ A Stone Thrown Downward: The Kinematic Equations Revisited

A boy on a bridge throws a stone vertically downward with an initial velocity of 14.7 m/s toward the river below. If the stone hits the water 2.00 s later, what is the height of the bridge above the water?

Thinking It Through. A free-fall problem, but note that the initial velocity is downward, or negative. It is important to express this explicitly.

Solution. As usual, we first write down what is given and what is to be found.

Given: $v_o = -14.7 \text{ m/s}$ (downward taken as   Find: y (height)
$t = 2.00 \text{ s}$ the negative direction)
$g\ (= 9.80 \text{ m/s}^2)$

Notice that g is now a positive number, since the directional minus sign has already been put into the previous equations of motion. After a while, you will probably just write down the symbol g, since you will be familiar with its numerical value. This time, draw a sketch of your own to help you analyze the situation.

Which equation(s) will provide the solution using the given data? It should be evident that the distance the stone travels in a time t is given directly by Eq. 2.9':

$$y = v_o t - \tfrac{1}{2}gt^2 = (-14.7 \text{ m/s})(2.00 \text{ s}) - \tfrac{1}{2}(9.80 \text{ m/s}^2)(2.00 \text{ s})^2$$
$$= -29.4 \text{ m} - 19.6 \text{ m} = -49.0 \text{ m}$$

The minus sign indicates that the displacement is downward, which agrees with what you know from the statement of the problem.

Follow-up Exercise. How much longer would it take for the stone to reach the river if the boy in this Example had dropped the ball rather than thrown it?

Example 2.11 ■ Measuring Reaction Time: Free Fall

Reaction time is the time it takes a person to notice, think, and act in response to a situation—for example, the time between first observing and then responding to an obstruction on the road ahead while you are driving an automobile. Reaction time varies with the complexity of the situation (and with the individual). In general, the largest part of a person's reaction time is spent thinking, but practice in dealing with a given situation can reduce this time.

A person's reaction time can be measured by having another person drop a ruler (without warning) from above and through the thumb and forefinger as shown in •Fig. 2.16. The first person grasps the falling ruler as quickly as possible, and the length of the ruler below the top of the finger is noted. If on the average the ruler descends 18.0 cm before it is caught, what is the person's average reaction time?

•FIGURE 2.16 **Reaction time** A person's reaction time can be measured by having the person grasp a dropped ruler. See Example 2.11.

Thinking It Through. Both distance and time are involved. This indicates which kinematic equation should be used.

Solution. Notice that only the distance of fall is given. However, we know a couple of other things, such as v_o and g, so

Given: $y = -18.0 \text{ cm} = -0.180 \text{ m}$ Find: t (reaction time)
$v_o = 0$
$g\, (= 9.80 \text{ m/s}^2)$

(Note that the distance y has been converted directly to meters.) We can see that Eq. 2.9' applies here:

$$y = v_o t - \tfrac{1}{2} g t^2$$

or

$$y = -\tfrac{1}{2} g t^2 \quad (\text{with } v_o = 0)$$

Solving for t gives

$$t = \sqrt{\frac{2y}{-g}} = \sqrt{\frac{2(-0.180 \text{ m})}{-9.80 \text{ m/s}^2}} = 0.192 \text{ s}$$

Try this with a fellow student and measure your reaction time. Why do you think another person should drop the ruler rather than yourself?

Follow-up Exercise. A popular trick is to substitute a crisp dollar bill lengthwise for the ruler in Fig. 2.16, telling the person that he or she can have the dollar if caught. Is this a good deal? (The length of a dollar is 15.7 cm.)

Example 2.12 ■ Free Fall Up and Down: Using Implicit Data

A worker on a scaffold on a billboard throws a ball straight up. It has an initial velocity of 11.2 m/s when it leaves his hand at the top of the billboard (•Fig. 2.17). (a) What is the maximum height the ball reaches relative to the top of the billboard? (b) How long does it take to reach this height? (c) What is the position of the ball at $t = 2.00$ s?

Thinking It Through. In (a), only the upward part of the motion has to be considered. Note that the ball stops (zero velocity) at the maximum height, which allows this height to be found. (b) Knowing the maximum height, we can determine the upward time of flight. In (c), the distance–time equation (Eq. 2.9') applies at any time and gives the position (y) of the ball relative to the launch point.

Solution. It might appear that all that is given in the general problem is the initial velocity v_o. However, a couple of other things are implied because they are understood. One is the acceleration g, and the other is the velocity at the maximum height where the ball stops. Here, in changing direction, the velocity of the ball is momentarily zero, so we have

Given: $v_o = 11.2$ m/s   Find: (a) $y_{\max}$ (maximum height)
$g\, (= 9.80 \text{ m/s}^2)$ (b) t_u (time upward)
$v = 0$ (at $y_{\max}$)   (c) y (at $t = 2.00$ s)
$t = 2.00$ s (for part c)

(a) Notice that we reference the height ($y = 0$) to the top of the billboard. For this part of the problem, we need be concerned with only the upward motion—a ball is thrown upward and stops at its maximum height $y_{\max}$. With $v = 0$ at this height, $y_{\max}$ may be found directly from Eq. 2.10':

$$v^2 = 0 = v_o^2 - 2g y_{\max}$$

and

•**FIGURE 2.17 Free fall up and down** Note the lengths of the velocity and acceleration vectors at different times. (The upward and downward paths of the ball are horizontally displaced for illustration.) See Example 2.12.

$$y_{max} = \frac{v_o^2}{2g} = \frac{(11.2 \text{ m/s})^2}{2(9.80 \text{ m/s}^2)} = 6.40 \text{ m}$$

relative to the top of billboard ($y = 0$, see Fig. 2.17).

(b) The time the ball travels upward is designated t_u. This is the time it takes to reach y_{max}, where $v = 0$. Then, we know v_o and v, so the time t_u can be found directly from Eq. 2.7′:

$$v = 0 = v_o - gt_u$$

and

$$t_u = \frac{v_o}{g} = \frac{11.2 \text{ m/s}}{9.80 \text{ m/s}^2} = 1.14 \text{ s}$$

(c) The height of the ball at $t = 2.00$ s is given directly by Eq. 2.9′:

$$y = v_o t - \tfrac{1}{2}gt^2$$
$$= (11.2 \text{ m/s})(2.00 \text{ s}) - \tfrac{1}{2}(9.80 \text{ m/s}^2)(2.00 \text{ s})^2 = 22.4 \text{ m} - 19.6 \text{ m} = 2.8 \text{ m}$$

Note that this is 2.8 m above, or measured upward from, the reference point ($y = 0$). The ball has reached its maximum height and is on its way back down.

Considered from another reference point, this situation is like dropping a ball from a height of y_{max} above the top of the billboard with $v_o = 0$ and asking how far it falls in a time $t = 2.00$ s $- t_u = 2.00$ s $- 1.14$ s $= 0.86$ s. The answer is

$$y = v_o t - \tfrac{1}{2}gt^2 = 0 - \tfrac{1}{2}(9.80 \text{ m/s}^2)(0.86 \text{ s})^2 = -3.6 \text{ m}$$

This is the same as the position found above but is measured with respect to the maximum height as the reference point; that is,

$$y_{max} - 3.6 \text{ m} = 6.4 \text{ m} - 3.6 \text{ m} = 2.8 \text{ m}$$

Follow-up Exercise. At what height does the ball in this Example have a speed of 5.00 m/s? (Hint: This occurs twice — once on the way up, and once on the way down.)

Here are a couple of interesting facts about the vertical projectile motion of an object thrown upward in the absence of significant air resistance. First, the times of flight upward and downward are the same. That is, the time it takes the object to reach its maximum height is the same as the time it takes the object to fall from the maximum height back to the initial starting point. Note that at the very top of the trajectory, the object's velocity is zero for an instant, but the acceleration (even at the top) remains a constant 9.8 m/s² downward. If the acceleration went to zero, it would remain there and gravity would be turned off!

Second, the object returns to the starting point with the same speed as it was launched. (The velocities have the same magnitude but are opposite in direction.)

These facts can be shown mathematically. See Exercises 68 and 81.

Problem-Solving Hint

When working vertical projection problems involving motions up and down, it is often convenient to divide the problem into two parts and consider each separately. As seen in Example 2.12, for the upward part of the motion, the velocity is zero at the maximum height. A zero quantity simplifies the calculations. Similarly, the downward part of the motion is analogous to that of an object dropped from a height where the initial velocity is zero.

However, as Example 2.12 shows, the appropriate equations may be used directly at any position or time of the motion. For instance, note in part (c) that the height was found directly for a time *after* the ball had reached the maximum height. The velocity of the ball at that time could also have been found directly from Eq. 2.7', $v = v_o - gt$.

Chapter Review

Important Terms

You should be able to define and explain these chapter terms clearly.

mechanics   32
kinematics   33
dynamics   33
motion   33
distance   33
scalar (quantity)   33
speed   33

average speed   33
instantaneous speed   34
displacement   35
vector (quantity)   35
velocity   37
average velocity   37
instantaneous velocity   39

acceleration   42
average acceleration   42
instantaneous acceleration   42
acceleration due to gravity   50
free fall   50

Important Concepts

- Motion involves a change of position; it can be described in terms of the distance moved (a scalar) or the displacement (a vector).
- A scalar quantity has magnitude (value and units) only; a vector quantity has magnitude *and* direction.
- Speed (a scalar) is the time rate of change of position, and average velocity (a vector) is the time rate of change of displacement.

- Acceleration is the time rate of change of velocity, and hence it is a vector quantity.
- Most kinematic equations are limited to constant accelerations only.
- An object in free fall has a constant acceleration of magnitude $g = 9.80$ m/s² (acceleration due to gravity) near the surface of the Earth.

Important Equations

These equations are mathematical statements of the concepts and principles presented in the chapter. They will be needed in working problem Exercises, so a good understanding is important. You should be able to identify the symbols and to explain the relationships and limitations before proceeding. (In-text equation reference numbers are given for convenience.)

Average speed: $\quad\text{avg. sp.} = \dfrac{d}{t}$ (2.2)

Kinematic Equations for Linear Motion with Constant Acceleration

(with $x_o = 0$): $\quad x = \bar{v}t$ (2.4)

(general, not limited to constant acceleration):

$$\left.\begin{aligned}\bar{v} &= \dfrac{v + v_o}{2} \\ v &= v_o + at \\ x &= v_o t + \tfrac{1}{2}at^2 \\ v^2 &= v_o^2 + 2ax\end{aligned}\right\}\text{ constant acceleration only}$$

(2.8)
(2.7)
(2.9)
(2.10)

Acceleration Due to Gravity:

$g = 9.80 \text{ m/s}^2 = 980 \text{ cm/s}^2 \; (\approx 32.2 \text{ ft/s}^2)$

Kinematic Equations Applied to Free Fall

(with downward taken as the negative direction, $y_o = 0$, and $-g$ expressed explicitly):

$$\left.\begin{aligned}y &= \bar{v}t \\ \bar{v} &= \dfrac{v + v_o}{2} \\ v &= v_o - gt \\ y &= v_o t - \tfrac{1}{2}gt^2 \\ v^2 &= v_o^2 - 2gy\end{aligned}\right\}\; g \text{ constant}$$

(2.4')
(2.8')
(2.7')
(2.9')
(2.10')

Exercises

Note: Exercises with a CD-ROM icon have corresponding simulations on the Interactive Journey through Physics CD-ROM; a complete cross-reference of text problems to CD-ROM activities appears in the Media Pack.

2.1 Distance and Speed: Scalar Quantities
and
2.2 Displacment and Velocity: Vector Quantities

1. Distance is always (a) equal to the magnitude of the corresponding displacement, (b) less than or equal to the magnitude of the corresponding displacement, (c) greater than or equal to the magnitude of the corresponding displacement.

2. Two people choose different reference points to specify an object's position. Does this difference affect the coordinates of the object? How about its displacement? Explain.

3. A scalar quantity has (a) only magnitude, (b) only direction, (c) both direction and magnitude.

4. Can a displacement from one point to another be zero, yet the distance involved in moving between these points be nonzero? How about the reverse situation? Explain.

5. You are told that a person walks 500 m. What can you safely say about the person's final position relative to the starting point?

6. An object travels at a constant velocity. What is the relationship of the object's speed to its velocity?

7. Speed is the magnitude of velocity. Is average speed the magnitude of average velocity? Explain.

8. What is the magnitude of the displacement of a car that travels half a lap along a circle of radius of 150 m?

9. A student throws a rock straight upward from shoulder level, which is 1.65 m above the ground. What is the displacement of the rock when it hits the ground?

10. A small airplane flies in a straight line at a speed of 150 km/h. How long does it take the plane to fly 250 km?

11. A bus travels at an average speed of 90 km/h. On the average, how far does the bus travel in 20 min? Would this be the actual travel distance? Explain.

12. A motorist drives 150 km from one city to another in 2.5 h but makes the return trip in only 2.0 h. What are the average speeds for (a) each half of the round trip and (b) the total trip?

13. (a) A senior citizen walks 0.30 km in 10 min going around a shopping mall. What is her average speed in meters per second? (b) If she wants to increase her average speed by 20% in walking a second lap, what would her travel time in minutes have to be?

14. A race car travels a complete lap on a circular track of radius 500 m in 50 s. (a) What is the average speed of the race car? (b) What is the average velocity of the race car?

15. A boy runs 30 m east, 40 m north, 50 m west. What is his net displacement?

16. ■■ A student throws a ball vertically upward such that it travels 10 m to its maximum height. If the ball is caught at the initial height 2.4 s after being thrown, (a) what is the ball's average speed? (b) Its average velocity?

17. ■■ An insect crawls along the edge of a rectangular swimming pool of length 27 m and width 21 m (•Fig. 2.18). If it crawls from corner A to corner B in 30 min, (a) what is its average speed in centimeters per second? (b) What is the magnitude of its average velocity?

•FIGURE 2.18 **Speed versus velocity** See Exercise 17. (Not drawn to scale; insect is displaced for clarity.)

18. ■■ The distance of one lap around an oval dirt bike track is 1.50 km. If a rider going at a constant speed makes one lap in 1.10 min, what is the speed of the bike and rider in meters per second? Is the velocity of the bike also constant? Explain.

19. ■■ (a) Given that the speed of sound is 340 m/s and the speed of light is 3.00×10^8 m/s (186 000 mi/s), how much time will elapse between a lightning flash and the resulting thunder if the lightning strikes 2.50 km away from the observer? (b) Does your answer change if you assume light travels with an infinite speed (that is, the lightning flash is seen instantaneously)?

20. ■■ ◉ A plot of position versus time is in •Fig. 2.19 for an object in linear motion. (a) What are the average velocities for the segments AB, BC, CD, DE, EF, FG, and BG? (b) State whether the motion is uniform or nonuniform in each case. (c) What is the instantaneous velocity at point D?

21. ■■ ◉ In demonstrating a dance step, a person moves in one dimension as shown in •Fig. 2.20. What are the (a) average speed and (b) average velocity for each phase of the motion? (c) What are the instantaneous velocities at $t = 1.0$ s, 2.5 s, 4.5 s, and 6.0 s? (d) What is the average velocity for the interval between $t = 4.5$ s and $t = 9.0$ s? [Hint: Recall that the overall displacement is the displacement between the starting point and the ending point.]

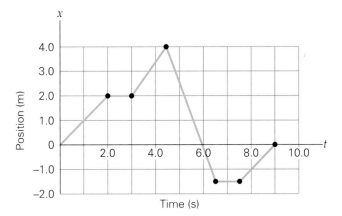

•FIGURE 2.20 **Position versus time** See Exercise 21.

22. ■■ You can determine the speed of a car by measuring the time it takes to travel between the mile markers. (a) How many seconds should elapse between mile markers if the car's average speed is 65 mi/h? (b) What is the average speed in miles per hour if it takes 65 s to travel between mile markers?

23. ■■ An earthquake releases two types of traveling seismic waves, called *transverse* and *longitudinal* waves. The average speeds of transverse and longitudinal waves in rock are 8.9 km/s and 5.1 km/s, respectively. A seismograph records the arrival of the transverse waves 73 s before that of the longitudinal waves. Assuming the waves travel in straight lines, how far away is the center of the earthquake?

24. ■■ The Indianapolis 500, a 500-mile auto race, was first run in 1911 in a time of 6 h, 42 min, and 8 s. In 1990, the race was run in a record time of 2 h, 41 min, and 18 s. (a) What were the average speeds for the Indy 500 for these years in miles per hour (to 0.1 mi/h)? (b) What was the percentage change in the average speed from 1911 to 1990?

25. ■■■ A student driving home for the holidays starts at 8:00 A.M. to make the 675-kilometer trip, practically all of which is on non-urban interstate highway. If she wants to arrive home no later than 3:00 P.M., what must be her minimum average speed? Will she have to exceed the 65 mi/h speed limit?

26. ■■■ Two runners approaching each other on a straight track have constant speeds of +4.50 m/s and −3.50 m/s, respectively, when they are 100 m apart (•Fig. 2.21). How

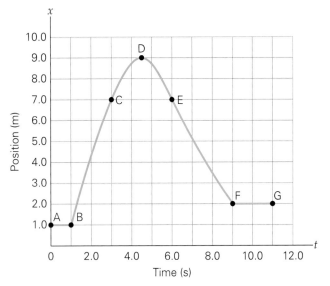

•FIGURE 2.19 **Position versus time** See Exercise 20.

Important Equations

These equations are mathematical statements of the concepts and principles presented in the chapter. They will be needed in working problem Exercises, so a good understanding is important. You should be able to identify the symbols and to explain the relationships and limitations before proceeding. (In-text equation reference numbers are given for convenience.)

Average speed: $\quad \text{avg. sp.} = \dfrac{d}{t}$ (2.2)

Kinematic Equations for Linear Motion with Constant Acceleration

(with $x_o = 0$): $\qquad x = \bar{v}t$ (2.4)

(general, not limited to constant acceleration):

$$\left. \begin{array}{l} \bar{v} = \dfrac{v + v_o}{2} \\ v = v_o + at \\ x = v_o t + \tfrac{1}{2}at^2 \\ v^2 = v_o^2 + 2ax \end{array} \right\} \text{constant acceleration only}$$

(2.8)
(2.7)
(2.9)
(2.10)

Acceleration Due to Gravity:

$g = 9.80 \text{ m/s}^2 = 980 \text{ cm/s}^2 \ (\approx 32.2 \text{ ft/s}^2)$

Kinematic Equations Applied to Free Fall
(with downward taken as the negative direction, $y_o = 0$, and $-g$ expressed explicitly):

$$\left. \begin{array}{l} y = \bar{v}t \\ \bar{v} = \dfrac{v + v_o}{2} \\ v = v_o - gt \\ y = v_o t - \tfrac{1}{2}gt^2 \\ v^2 = v_o^2 - 2gy \end{array} \right\} g \text{ constant}$$

(2.4')
(2.8')
(2.7')
(2.9')
(2.10')

Exercises

Note: Exercises with a CD-ROM icon have corresponding simulations on the Interactive Journey through Physics CD-ROM; a complete cross-reference of text problems to CD-ROM activities appears in the Media Pack.

2.1 Distance and Speed: Scalar Quantities and
2.2 Displacment and Velocity: Vector Quantities

1. Distance is always (a) equal to the magnitude of the corresponding displacement, (b) less than or equal to the magnitude of the corresponding displacement, (c) greater than or equal to the magnitude of the corresponding displacement.

2. Two people choose different reference points to specify an object's position. Does this difference affect the coordinates of the object? How about its displacement? Explain.

3. A scalar quantity has (a) only magnitude, (b) only direction, (c) both direction and magnitude.

4. Can a displacement from one point to another be zero, yet the distance involved in moving between these points be nonzero? How about the reverse situation? Explain.

5. You are told that a person walks 500 m. What can you safely say about the person's final position relative to the starting point?

6. An object travels at a constant velocity. What is the relationship of the object's speed to its velocity?

7. Speed is the magnitude of velocity. Is average speed the magnitude of average velocity? Explain.

8. ■ What is the magnitude of the displacement of a car that travels half a lap along a circle of radius of 150 m?

9. ■ A student throws a rock straight upward from shoulder level, which is 1.65 m above the ground. What is the displacement of the rock when it hits the ground?

10. ■ A small airplane flies in a straight line at a speed of 150 km/h. How long does it take the plane to fly 250 km?

11. ■ A bus travels at an average speed of 90 km/h. On the average, how far does the bus travel in 20 min? Would this be the actual travel distance? Explain.

12. ■ A motorist drives 150 km from one city to another in 2.5 h but makes the return trip in only 2.0 h. What are the average speeds for (a) each half of the round trip and (b) the total trip?

13. ■ (a) A senior citizen walks 0.30 km in 10 min going around a shopping mall. What is her average speed in meters per second? (b) If she wants to increase her average speed by 20% in walking a second lap, what would her travel time in minutes have to be?

14. ■ A race car travels a complete lap on a circular track of radius 500 m in 50 s. (a) What is the average speed of the race car? (b) What is the average velocity of the race car?

15. ■■ A boy runs 30 m east, 40 m north, 50 m west. What is his net displacement?

16. ■■ A student throws a ball vertically upward such that it travels 10 m to its maximum height. If the ball is caught at the initial height 2.4 s after being thrown, (a) what is the ball's average speed? (b) Its average velocity?

17. ■■ An insect crawls along the edge of a rectangular swimming pool of length 27 m and width 21 m (•Fig. 2.18). If it crawls from corner A to corner B in 30 min, (a) what is its average speed in centimeters per second? (b) What is the magnitude of its average velocity?

•FIGURE 2.18 Speed versus velocity See Exercise 17. (Not drawn to scale; insect is displaced for clarity.)

18. ■■ The distance of one lap around an oval dirt bike track is 1.50 km. If a rider going at a constant speed makes one lap in 1.10 min, what is the speed of the bike and rider in meters per second? Is the velocity of the bike also constant? Explain.

19. ■■ (a) Given that the speed of sound is 340 m/s and the speed of light is 3.00×10^8 m/s (186 000 mi/s), how much time will elapse between a lightning flash and the resulting thunder if the lightning strikes 2.50 km away from the observer? (b) Does your answer change if you assume light travels with an infinite speed (that is, the lightning flash is seen instantaneously)?

20. ■■ ● A plot of position versus time is in •Fig. 2.19 for an object in linear motion. (a) What are the average velocities for the segments AB, BC, CD, DE, EF, FG, and BG? (b) State whether the motion is uniform or nonuniform in each case. (c) What is the instantaneous velocity at point D?

21. ■■ ● In demonstrating a dance step, a person moves in one dimension as shown in •Fig. 2.20. What are the (a) average speed and (b) average velocity for each phase of the motion? (c) What are the instantaneous velocities at $t = 1.0$ s, 2.5 s, 4.5 s, and 6.0 s? (d) What is the average velocity for the interval between $t = 4.5$ s and $t = 9.0$ s? [Hint: Recall that the overall displacement is the displacement between the starting point and the ending point.]

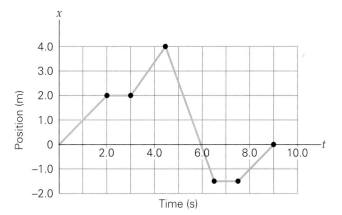

•FIGURE 2.20 Position versus time See Exercise 21.

22. ■■ You can determine the speed of a car by measuring the time it takes to travel between the mile markers. (a) How many seconds should elapse between mile markers if the car's average speed is 65 mi/h? (b) What is the average speed in miles per hour if it takes 65 s to travel between mile markers?

23. ■■ An earthquake releases two types of traveling seismic waves, called *transverse* and *longitudinal* waves. The average speeds of transverse and longitudinal waves in rock are 8.9 km/s and 5.1 km/s, respectively. A seismograph records the arrival of the transverse waves 73 s before that of the longitudinal waves. Assuming the waves travel in straight lines, how far away is the center of the earthquake?

24. ■■ The Indianapolis 500, a 500-mile auto race, was first run in 1911 in a time of 6 h, 42 min, and 8 s. In 1990, the race was run in a record time of 2 h, 41 min, and 18 s. (a) What were the average speeds for the Indy 500 for these years in miles per hour (to 0.1 mi/h)? (b) What was the percentage change in the average speed from 1911 to 1990?

25. ■■■ A student driving home for the holidays starts at 8:00 A.M. to make the 675-kilometer trip, practically all of which is on non-urban interstate highway. If she wants to arrive home no later than 3:00 P.M., what must be her minimum average speed? Will she have to exceed the 65 mi/h speed limit?

26. ■■■ Two runners approaching each other on a straight track have constant speeds of +4.50 m/s and −3.50 m/s, respectively, when they are 100 m apart (•Fig. 2.21). How

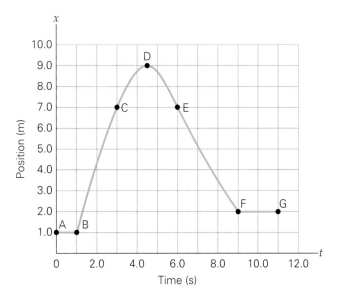

•FIGURE 2.19 Position versus time See Exercise 20.

58 CHAPTER 2 Kinematics: Description of Motion

long will it take for the runners to meet, and at what position will this occur if they maintain these speeds?

•FIGURE 2.21 **When and where do they meet?** See Exercise 26.

27. ■■■ In driving the usual route to school, a student computes his average speed to be 30 km/h. In a hurry to get home that afternoon, he wants to average 60 km/h for the round trip. What would his average speed for the return trip over the same route have to be in order to do this? (Hint: See Example 2.3, and be ready for a surprise.)

2.3 Acceleration

28. The gas pedal of an automobile is commonly referred to as the accelerator. Which of the following might also be called an accelerator: (a) the brakes, (b) the steering wheel, (c) the gear shift, (d) all of the preceding?

29. A car is traveling at a constant speed of 55 mi/h on a circular track. Is the car accelerating? Explain.

30. 💿 On a velocity-versus-time plot for an initially moving object that has a constant acceleration, (a) the graph is a curved line, (b) the y-intercept is nonzero, (c) the slope of the line is necessarily positive, (d) the slope of the line is necessarily negative.

31. An object traveling at a constant velocity v_o experiences a constant acceleration in the same direction for a time t. Then, an acceleration of equal magnitude is experienced in the opposite direction of v_o for the same time t. What is the velocity after all of this?

32. Describe the motions of the two objects that have the velocity-versus-time plots shown in •Fig. 2.22.

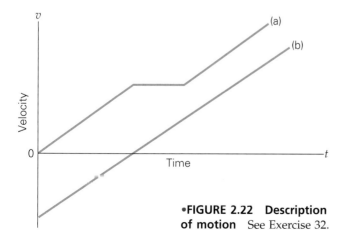

•FIGURE 2.22 **Description of motion** See Exercise 32.

33. ■ An automobile traveling at 25.0 km/h along a straight road accelerates to 65.0 km/h in 6.00 s. What is the magnitude of the average acceleration?

34. ■ A sports car can accelerate from zero to 60 mi/h in 3.9 s. What is the magnitude of the average acceleration of the car in meters per second squared?

35. ■ If the sports car in Exercise 34 can accelerate at a rate of 7.2 m/s^2, how long does it take for the car to accelerate from zero to 60 mi/h?

36. ■ If an automobile moving with a velocity of 40 km/h along a straight road brakes uniformly to rest in 5.0 s, by how much must the automobile's velocity change each second?

37. ■ A skier with speed of 5.00 m/s crests a hill. On the downslope, she accelerates at 3.00 m/s^2 for 3.50 s. What is her final speed?

38. ■■ An object moves in the positive x direction at an initial speed of 3.5 m/s. If the object experiences an acceleration of 0.50 m/s^2 in the negative x direction, after how long a time will the object have a negative velocity?

39. ■■ What is the acceleration for each graph segment in •Fig. 2.23? Describe the motion of the object over the total time interval.

•FIGURE 2.23 **Velocity versus time** See Exercises 39 and 61.

40. ■■ •Figure 2.24 shows a plot of velocity versus time for an object in linear motion. (a) Compute the acceleration for each phase of motion. (b) Describe how the object moves during the last time segment.

41. ■■■ A train normally travels at a uniform speed of 72 km/h on a long stretch of straight, level track. On a particular day, the train must make a 2.0-minute stop at a station along this track. If the train decelerates at a uniform rate of 1.0 m/s^2 and accelerates at a rate of 0.50 m/s^2, how much time is lost in stopping at the station?

Exercises 59

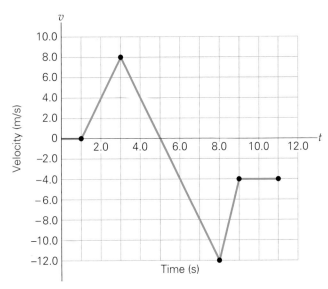

•FIGURE 2.24 Velocity versus time   See Exercises 40 and 62.

2.4 Kinematic Equations (Constant Acceleration)

42. For a constant linear acceleration, the velocity-versus-time graph would be (a) a horizontal line, (b) a vertical line, (c) a non-horizontal and non-vertical straight line, (d) a curve.

43. An object accelerates uniformly from rest for t seconds. The average speed for this time interval is (a) $\frac{1}{2}at$, (b) $\frac{1}{2}at^2$ (c) $2at$, (d) $2at^2$.

44. ● A classmate states that a negative acceleration always means that a moving object is decelerating. Is this true? Explain.

45. ■ At a sports car rally, a car starting from rest accelerates uniformly at a rate of 9.0 m/s² over a straight-line distance of 100 m. The time to beat is 4.5 s. Does the driver do it? If not, what must the acceleration be to do so?

46. ■ A car accelerates from rest at a constant rate of 2.0 m/s² for 5.0 s. (a) What is the speed of the car at that time? (b) How far does the car travel in this time?

47. ■ A car traveling at 35 mi/h is to stop on a 35-meter-long shoulder of the road. What minimum deceleration is required?

48. ■ A motorboat traveling on a straight course slows down uniformly from 75 km/h to 40 km/h in a distance of 50 m. What is the required acceleration?

49. ■■ The driver of a pickup truck going 100 km/h applies the brakes, giving the truck a uniform deceleration of 6.50 m/s² while it travels 20.0 m. (a) What is the speed of the truck in kilometers per hour at the end of this distance? (b) How much time has elapsed?

50. ■■ An experimental rocket car starting from rest reaches a speed of 560 km/h after a straight 400-meter run on a level salt flat. Assuming the acceleration to be constant, what was the time of the run?

51. ■■ A rocket car is traveling at a constant speed of 250 km/h on a salt flat. The driver gives the car a reverse thrust, and it experiences a continuous and constant deceleration of 8.25 m/s². How much time elapses until the car is 175 m from the point where the reverse thrust is applied?

52. ■■ ● Two identical cars capable of accelerating at 3.00 m/s² are racing on a straight track and cross the starting line together with running starts. Car A has an initial speed of 2.50 m/s; car B starts with an initial speed of 5.00 m/s. (a) What is the separation of the two cars after 10 s? (b) Which car is moving faster after 10 s?

53. ■■ An object moves in the positive x direction at a speed of 40 m/s. As it passes through the origin, it starts to experience a constant acceleration of 3.5 m/s² in the negative x direction. How much time elapses before the object returns to the origin?

54. ■■ A rifle bullet with a muzzle speed of 330 m/s is fired directly into a special, dense material that stops the bullet in 30 cm. Assuming the deceleration to be constant, what is its magnitude?

55. ■■ A bullet traveling horizontally at a speed of 350 m/s hits a board perpendicular to the surface, passes through it, and emerges on the other side at a speed of 210 m/s. If the board is 4.00 cm thick, how long does the bullet take to pass through it?

56. ■■ A rock hits the ground at a speed of 10 m/s and leaves a hole 25 cm deep. What is the magnitude of the (assumed) uniform deceleration of the rock?

57. ■■ A jet aircraft being launched from an aircraft carrier is accelerated from rest along a 94-meter track for 2.5 s. What is the launch speed of the jet?

58. ■■ The speed limit in a school zone is 40 km/h (about 25 mi/h). A driver traveling at this speed sees a child run into the road 13 m ahead of his car. He applies the brakes, and the car decelerates at a uniform rate of 8.0 m/s². If the driver's reaction time is 0.25 s, will the car stop before hitting the child?

59. ■■ Assuming a reaction time of 0.50 s for the driver in Exercise 58, will the car stop before hitting the child?

60. ■■ Show that the area under the curve of a velocity-versus-time plot for a constant acceleration is equal to the displacement. Do this for the cases in which (a) $a = 0$, (b) $v_o = 0$, and (c) $v_o \neq 0$. [Hint: The area of a triangle is $ab/2$, or half the altitude times the base.]

61. ■■ Compute the distance traveled for the motion represented by Fig. 2.23. [Hint: See Exercise 60.]

62. ■■■ Figure 2.24 shows velocity versus time for an object in linear motion. (a) What are the instantaneous velocities at $t = 8.0$ s and $t = 11.0$ s? (b) Compute the total displacement of the object. (c) Compute the total distance the object travels.

63. ■■■ In our kinematic equations, it was assumed for simplicity and convenience that $x_o = 0$ and $t_o = 0$. Rewrite the equations with x_o and t_o expressed explicitly. What is the difference between the x in Eq. 2.9 and the x in your rewritten equation?

64. ■■■ An object moves in the positive x direction with a constant acceleration. At $x = 5.0$ m, its speed is 10 m/s; 2.5 s later, the object is at $x = 65$ m. What is its acceleration?

2.5 Free Fall
(Neglect air resistance throughout.)

65. ● An object is thrown vertically upward. Which one of the following is true? (a) Its velocity changes nonuniformly, (b) the maximum height is independent of the initial velocity, (c) the travel time upward is slightly greater than its travel time downward, (d) the speed on returning to its starting point is the same as its initial speed.

66. A dropped object in free fall (a) falls 9.8 m each second, (b) falls 9.8 m during the first second, (c) has an increase in speed of 9.8 m/s each second, (d) has an increase in acceleration of 9.8 m/s each second.

67. An object is thrown straight upward. When it reaches the highest point in the air, what is its acceleration? Explain.

68. Given the data for an object projected vertically upward, you are asked to find the time the object is at height y. The equation $y = v_o t - \frac{1}{2}gt^2$ is solved by using the quadratic formula, and two roots are obtained. Does this mean there are two answers? Explain. [Hint: Do the roots of a quadratic equation necessarily have to be both positive and negative?]

69. ■ If a dropped object falls 19.6 m in 2.00 s, how far will it fall in 4.00 s?

70. ■ ● For the motion of a dropped object in free fall, sketch the general forms of the graphs of (a) v versus t and (b) y versus t.

71. ■ ● Sketch the general forms of the graphs of (a) v versus t and (b) y versus t for an object projected vertically upward.

72. ■ A student drops a ball from the top of a tall building; it takes 2.8 s to reach the ground. (a) What was its speed just before hitting the ground? (b) What was the height of the tall building?

73. ■ An object dropped from the top of a cliff takes 1.80 s to hit the water in the lake below. What is the cliff's height above the water?

74. ■ A boy throws a stone straight upward with an initial speed of 15 m/s on the surface of the Earth. What maximum height above the starting point will the stone reach before falling back down?

75. ■ In Exercise 74, what would be the maximum height if the boy and the stone were on the surface of the Moon, where the acceleration due to gravity is only 1.67 m/s^2?

76. ■■ A spring-loaded gun shoots a 0.0050-kilogram bullet vertically upward with an initial velocity of 21 m/s. (a) What is the height of the bullet 3.0 s after firing? (b) At what times is the bullet 12 m above the muzzle of the gun?

77. ■■ The ceiling of a classroom is 3.75 m above the floor. A student tosses an apple vertically upward, releasing it 0.50 m above the floor. What is the maximum initial speed that can be given to the apple if it is not to touch the ceiling?

78. ■■ (a) The World Trade Center and the Empire State Building in New York City have heights of about 417 m and 381 m, respectively. If objects were dropped from the top of each, what would be the difference in time in their reaching the ground? (b) What would be the time of fall for Chicago's Sears Tower, at 443 m?

79. ■■ A stone is thrown vertically downward at an initial speed of 14 m/s from a height of 65 m above the ground. (a) How far does the stone travel in 2.0 s? (b) What is its velocity just before it hits the ground?

80. ■■ A ball is projected vertically downward at a speed of 4.00 m/s. (a) How far does the ball travel in 1.80 s? (b) What is the velocity of the ball at that time?

81. ■■ Referring to the thrown ball in Example 2.12 and Fig. 2.17, (a) compare the travel time upward (t_u) with the travel time downward (t_d) required for the ball to return to its starting point. (b) Compare the initial velocity of the ball with the velocity it has when back at the starting point.

82. ■■ ● Draw graphs of y versus t and v versus t for the following: (a) an object dropped from rest from a height of 30 m above the ground and (b) an object that is projected vertically upward with an initial velocity of 34.3 m/s and that returns to the same point. (c) Draw plots of a versus t for these motions.

83. ■■ A superball is dropped from a height of 4.00 m. Assuming the ball rebounds with 95% of its impact speed, how high will the ball go? (Would it be 95.0% of the initial height?)

84. ■■ Two balls are thrown vertically, both at the initial speed of 10.0 m/s from a height of 60.0 m above the ground. Ball A is thrown upward, while ball B is projected downward. (a) Which ball will hit the ground first? (b) What is the time

lapse between the two balls hitting the ground? (c) Do the masses of the balls matter?

85. ■■■ In •Fig. 2.25, a student at a window on the top floor of a dorm sees his math professor walking beside the building. He drops a water balloon from 18.0 m above the ground when the prof is 1.00 m from the point directly beneath the window. If the prof is 170 cm tall and walks at a rate of 0.450 m/s, does the balloon hit her? If not, how close does it come?

•FIGURE 2.25 **Hit the prof.** See Exercise 85. (Not drawn to scale.)

86. ■■■ A photographer in a helicopter ascending vertically at a constant rate of 12.5 m/s accidentally drops a camera out the window when the helicopter is 60.0 m above the ground. (a) How long will it take the camera to reach the ground? (b) What will its speed be when it hits?

87. ■■■ The acceleration due to gravity on the Moon is about one-sixth of that on Earth. (a) If an object were dropped from the same height on the Moon and on Earth, how much longer (by what factor) would it take it to hit the surface of the Moon? (b) For a projectile with an initial velocity of 18.0 m/s upward, what would be the maximum height and total time of flight on the Moon and on Earth?

88. ■■■ It takes 0.210 s for a dropped object to pass a window that is 1.35 m tall (•Fig. 2.26). From what height above the top of the window was the object released?

Additional Exercises

89. In throwing an object vertically upward at a speed of 7.25 m/s from the top of a tall building, Jennifer leans over the edge so that the object will not strike the building on the return trip. (a) What is the velocity of the object when it has traveled a total distance of 25.0 m? (b) How long does it take to travel this distance?

90. On landing, a jet plane on a straight runway comes to rest at an average velocity of −35.0 km/h. If this takes 7.00 s, what is the plane's acceleration?

•FIGURE 2.26 **From where did it come?** See Exercise 88.

91. A vertically moving projectile reaches a maximum height of 23 m above its starting position. (a) What was the projectile's initial speed? (b) What is its height above the starting point at $t = 1.3$ s?

92. A drag racer traveling at a speed of 200 km/h on a straight track ejects a parachute and slows uniformly to a speed of 20 km/h in 12 s. (a) What is the racer's acceleration? (b) How far does it travel in the 12-second interval?

93. On a cross-country trip, a couple drives 500 mi in 10 h on the first day, 380 mi in 8.0 h on the second day, and 600 mi in 15 h on the third day. What was their average speed for the whole trip?

94. A car traveling at 25 mi/h has 1.5 s to come to a stop approaching a red light. The constant deceleration of the car is 7.0 m/s^2. Will the car stop safely before it reaches the red light?

95. What is the magnitude of the displacement if a car travels three quarters of a lap on a circular track of radius 50 m?

96. A circular track has a diameter of 0.50 km. A go-cart with a constant speed of 10.0 m/s makes two complete laps around the track. How long does it take the cart to complete the two laps?

97. At what speed must an object be projected vertically upward for the object to reach a maximum height of 14.0 m above its starting point?

98. A train on a straight, level track has an initial speed of 45.0 km/h. A uniform acceleration of 1.50 m/s^2 is applied while the train travels 200 m. (a) What is the speed of the train at the end of this distance? (b) How long did it take for the train to travel the 200 m?

99. A car going 85 km/h on a straight road is brought uniformly to a stop in 10 s. How far does the car travel during that time?

100. A car and a motorcycle start from rest at the same time on a straight track, but the motorcycle is 25.0 m behind the car (•Fig. 2.27). The car accelerates at a uniform rate of 3.70 m/s^2, and the motorcycle at a uniform rate of 4.40 m/s^2. (a) How much time elapses before the motorcycle overtakes the car? (b) How far will each have traveled during that time? (c) How far ahead of the car will the motorcycle be 2.00 s later? (Both vehicles are still accelerating.)

•FIGURE 2.27 A tie race   See Exercise 100. (Not drawn to scale.)

101. A person throws a stone straight upward at an initial speed of 15 m/s on a bridge that is 25 m above the surface of the water. If the stone just misses the bridge on the way down, (a) what is the speed of the stone just before it hits the water? (b) What is the total time the stone is in the air?

102. An object initially at rest experiences an acceleration of 1.5 m/s^2 for 6.0 s and then travels at that constant velocity for another 8.0 s. What is the magnitude of the object's average velocity over the 14 s interval?

103. To find the depth of the water surface in a well, a person drops a stone from the top of the well and simultaneously starts a stopwatch. The watch is stopped when the splash is heard, giving a reading of 3.65 s. The speed of sound is 340 m/s. Find the depth of the water surface below the top of the well. Take the person's reaction time for stopping the watch to be 0.250 s.

104. An arrow shot from a bow vertically upward has an initial speed of 50.0 m/s. (a) What is the maximum height of the arrow above its launch point? (b) How long does it take for the arrow to return to its launch point? (Assume that the arrow travels in a straight line.)

105. An object starting from rest and traveling in only one direction has velocities of 5.0 m/s, 10 m/s, 15 m/s, 20 m/s, and 25 m/s at the ends of the first, second, third, fourth, and fifth seconds, respectively. (a) What is the magnitude of the object's acceleration? (b) What is the object's average velocity for the 5-second interval? (c) Sketch a graph of v versus t. (d) Sketch a graph of x versus t.

106. On a walk in the country, a couple goes 1.80 km east along a straight road in 20.0 min and then 2.40 km directly north in 35.0 min. Answer the following in units of kilometers per hour. (a) What is the average velocity for each segment of their walk? (b) What is the average speed for the total trip? (c) If they take a straight-line path back to their original starting place in 25.0 min, what are the average speed and average velocity for the total trip?

CHAPTER 3
Motion in Two Dimensions

3.1 Components of Motion
3.2 Vector Addition and Subtraction
3.3 Relative Velocity
3.4 Projectile Motion

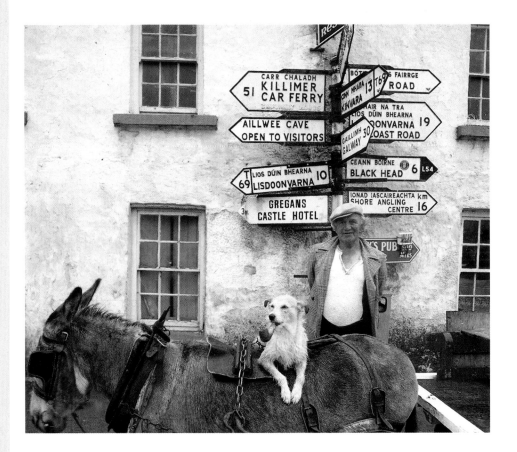

You *can* get there from here! It's just a matter of knowing which way to head at the crossroads. But did you ever wonder why so many roads meet at right angles? There's a good reason. Living on Earth's surface, we are used to describing locations in two dimensions, and one of the easiest ways to do this is by referring to a pair of mutually perpendicular axes. When you want to tell someone how to get to a particular place in the city, you might say, "Go uptown 4 blocks and then crosstown for three more." In the country, it's "Go south for 5 miles and then another half mile east." Either way, though, you need to know how far to go in each of *two* directions that are 90° apart.

You can use the same approach to describe motion—and the motion doesn't have to be in a straight line. As you will shortly see, we can use vectors, introduced in Chapter 2, to describe motion in curved paths as well. Such analysis of *curvilinear* motion will eventually allow you to analyze the behavior of batted balls, planets circling the Sun, and even electrons in atoms.

Curvilinear motion can be analyzed by using rectangular components of motion. Essentially, you break down, or *resolve*, the curved motion into rectangular (x and y) components and look at the motion in both dimensions simultaneously. To those components you can apply the kinematic equations introduced in Chap-

99. A car going 85 km/h on a straight road is brought uniformly to a stop in 10 s. How far does the car travel during that time?

100. A car and a motorcycle start from rest at the same time on a straight track, but the motorcycle is 25.0 m behind the car (•Fig. 2.27). The car accelerates at a uniform rate of 3.70 m/s^2, and the motorcycle at a uniform rate of 4.40 m/s^2. (a) How much time elapses before the motorcycle overtakes the car? (b) How far will each have traveled during that time? (c) How far ahead of the car will the motorcycle be 2.00 s later? (Both vehicles are still accelerating.)

•FIGURE 2.27 A tie race See Exercise 100. (Not drawn to scale.)

101. A person throws a stone straight upward at an initial speed of 15 m/s on a bridge that is 25 m above the surface of the water. If the stone just misses the bridge on the way down, (a) what is the speed of the stone just before it hits the water? (b) What is the total time the stone is in the air?

102. An object initially at rest experiences an acceleration of 1.5 m/s^2 for 6.0 s and then travels at that constant velocity for another 8.0 s. What is the magnitude of the object's average velocity over the 14 s interval?

103. To find the depth of the water surface in a well, a person drops a stone from the top of the well and simultaneously starts a stopwatch. The watch is stopped when the splash is heard, giving a reading of 3.65 s. The speed of sound is 340 m/s. Find the depth of the water surface below the top of the well. Take the person's reaction time for stopping the watch to be 0.250 s.

104. An arrow shot from a bow vertically upward has an initial speed of 50.0 m/s. (a) What is the maximum height of the arrow above its launch point? (b) How long does it take for the arrow to return to its launch point? (Assume that the arrow travels in a straight line.)

105. An object starting from rest and traveling in only one direction has velocities of 5.0 m/s, 10 m/s, 15 m/s, 20 m/s, and 25 m/s at the ends of the first, second, third, fourth, and fifth seconds, respectively. (a) What is the magnitude of the object's acceleration? (b) What is the object's average velocity for the 5-second interval? (c) Sketch a graph of v versus t. (d) Sketch a graph of x versus t.

106. On a walk in the country, a couple goes 1.80 km east along a straight road in 20.0 min and then 2.40 km directly north in 35.0 min. Answer the following in units of kilometers per hour. (a) What is the average velocity for each segment of their walk? (b) What is the average speed for the total trip? (c) If they take a straight-line path back to their original starting place in 25.0 min, what are the average speed and average velocity for the total trip?

CHAPTER 3

Motion in Two Dimensions

3.1 Components of Motion
3.2 Vector Addition and Subtraction
3.3 Relative Velocity
3.4 Projectile Motion

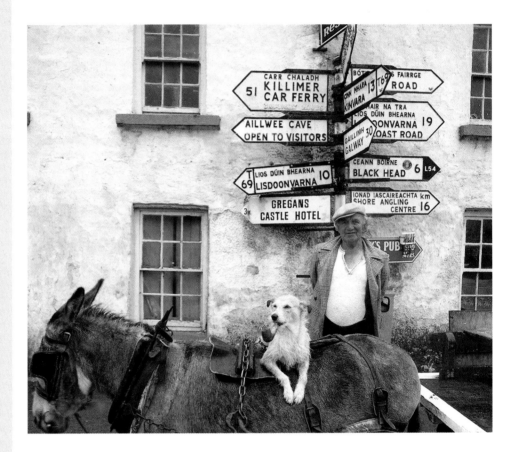

You *can* get there from here! It's just a matter of knowing which way to head at the crossroads. But did you ever wonder why so many roads meet at right angles? There's a good reason. Living on Earth's surface, we are used to describing locations in two dimensions, and one of the easiest ways to do this is by referring to a pair of mutually perpendicular axes. When you want to tell someone how to get to a particular place in the city, you might say, "Go uptown 4 blocks and then crosstown for three more." In the country, it's "Go south for 5 miles and then another half mile east." Either way, though, you need to know how far to go in each of *two* directions that are 90° apart.

You can use the same approach to describe motion—and the motion doesn't have to be in a straight line. As you will shortly see, we can use vectors, introduced in Chapter 2, to describe motion in curved paths as well. Such analysis of *curvilinear* motion will eventually allow you to analyze the behavior of batted balls, planets circling the Sun, and even electrons in atoms.

Curvilinear motion can be analyzed by using rectangular components of motion. Essentially, you break down, or *resolve*, the curved motion into rectangular (*x* and *y*) components and look at the motion in both dimensions simultaneously. To those components you can apply the kinematic equations introduced in Chapter 2. For an object moving in a curved path, for example, the x and y coordinates of the motion at any time give the object's position as the point (x, y).

3.1 Components of Motion

Objectives: To (a) analyze motion in terms of its components and (b) apply the kinematic equations to components of motion.

An object moving in a straight line was considered in Chapter 2 to be moving along one of the Cartesian axes (x or y). But what if the motion is not along an axis? For example, consider the situation illustrated in •Fig. 3.1. Here, the balls are moving uniformly across a tabletop. The ball rolling in a straight line along the side of the table designated as the x direction is moving in one dimension. That is, its motion can be described with a single coordinate, x, as was done for motions in Chapter 2. Similarly, the motion of the ball rolling in the y direction can be described by a single y coordinate. However, both x and y coordinates are needed to describe the motion of the ball rolling diagonally across the table. We call this motion in two dimensions.

You might observe that if the diagonally moving ball were the only object you had to consider, the x axis could be chosen in that direction and the motion reduced to one dimension. This is true, but once the coordinate axes are fixed, motions not along the axes must be described with two coordinates (x, y) or in two

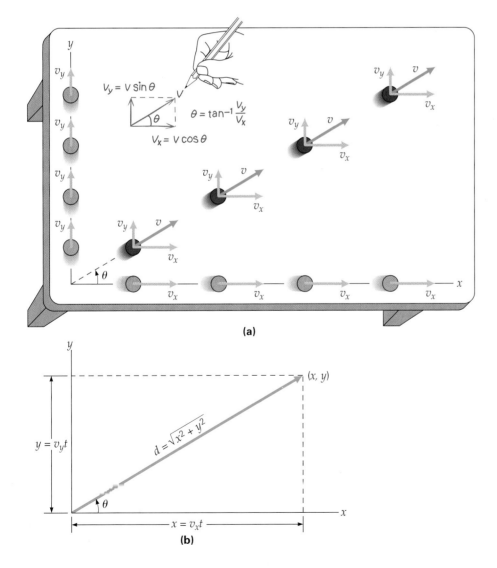

•**FIGURE 3.1 Components of motion** (a) The velocity (and displacement) for uniform straight-line motion—the dark purple ball—may have x and y components (v_x and v_y as shown in the pencil drawing) because of the chosen orientation of the coordinate axes. Note that the velocity and displacement of the ball in the x direction are exactly the same as those that a ball rolling along the x axis with a uniform velocity of v_x would have. A comparable relationship holds true for the ball's motion in the y direction. Since the motion is uniform, the ratio v_y/v_x (and therefore θ) is constant. (b) The coordinates (x, y) of the ball's position and the distance d it has traveled from the origin can be found for any time t.

dimensions. Also, keep in mind that not all motions in a plane (two dimensions) are in straight lines. Think about the path of a ball you toss to another person.

In considering the motion of the ball moving diagonally across the table in Fig. 3.1a, we can think of the ball moving in the x and y directions simultaneously. That is, it has a velocity in the x direction (v_x) *and* a velocity in the y direction (v_y) at the same time. The combined velocity components describe the actual motion of the ball. If the ball has a constant velocity (**v**) in a direction at an angle θ relative to the x axis, then the velocities in the x and y directions are obtained by resolving, or breaking down, the velocity vector into **components of motion** in these directions, as shown in the pencil drawing in Fig. 3.1a. As this drawing shows, the v_x and v_y components have magnitudes of

Magnitudes of velocity components in x and y directions

$$v_x = v \cos \theta \quad (3.1a)$$

$$v_y = v \sin \theta \quad (3.1b)$$

(Notice that $v = \sqrt{v_x^2 + v_y^2}$, so v is a combination of the velocities in the x and y directions.)

You are familiar with the use of two-dimensional length components in finding the x and y coordinates in a Cartesian system. For the ball rolling on the table, its position, (x, y), or the distance traveled from the origin in each of the component directions at time t, is given by

$$x = v_x t \quad \text{\textit{Magnitude of displacement}} \quad (3.2a)$$
$$\quad\quad\quad \text{\textit{components (constant velocity---}}$$
$$y = v_y t \quad \text{\textit{zero acceleration---only)}} \quad (3.2b)$$

just as it is for motion in the x and y directions separately. (Here, as usual, we assume that $x_o = y_o = 0$ at $t = 0$.) The ball's straight-line distance from the origin is then $d = \sqrt{x^2 + y^2}$ (Fig. 3.1b).

Note that $\tan \theta = v_y / v_x$, so the direction of the motion relative to the x axis is given by $\theta = \tan^{-1}(v_y / v_x)$. (See the hand-drawn sketch in Fig. 3.1a.) Also, $\theta = \tan^{-1}(y/x)$. Why?

Example 3.1 ■ On a Roll: Using Components of Motion

If the diagonally moving ball in Fig. 3.1a has a velocity of 0.50 m/s at an angle of 37° relative to the x axis, find how far it travels in 3.0 s by using x and y components.

Thinking It Through. Given the magnitude and direction (angle) of the velocity of the ball, the x and y components of the velocity can be found. Then, the distance in each direction can be computed. Since x and y are at right angles, the Pythagorean theorem gives the straight-line path distance of the ball. (Note the procedure: Separate the motion into components, calculate what is needed in each direction, and recombine if necessary.)

Solution. Organizing the data, we have

Given: $v = 0.50$ m/s   *Find:* d (distance)
$\quad\quad\theta = 37°$
$\quad\quad t = 3.0$ s

The distance traveled by the ball in terms of its x and y components is given by $d = \sqrt{x^2 + y^2}$. To find x and y as given by Eq. 3.2, we must first compute the velocity components v_x and v_y (Eq. 3.1):

$$v_x = v \cos 37° = (0.50 \text{ m/s})(0.80) = 0.40 \text{ m/s}$$

$$v_y = v \sin 37° = (0.50 \text{ m/s})(0.60) = 0.30 \text{ m/s}$$

Then, the component distances are

$$x = v_x t = (0.40 \text{ m/s})(3.0 \text{ s}) = 1.2 \text{ m}$$

$$y = v_y t = (0.30 \text{ m/s})(3.0 \text{ s}) = 0.90 \text{ m}$$

and the actual path distance is

$$d = \sqrt{x^2 + y^2} = \sqrt{(1.2 \text{ m})^2 + (0.90 \text{ m})^2} = 1.5 \text{ m}$$

Follow-up Exercise. Suppose that a ball were rolling diagonally across a table with the same speed as in Example 3.1 *but* from the lower right corner, which is taken as the origin of the coordinate system, toward the upper left corner at an angle of 37° relative to the $-x$ axis. What would be the velocity components in this case?

Problem-Solving Hint

Note that for this simple case, the distance can also be obtained directly from $d = vt = (0.50 \text{ m/s})(3.0 \text{ s}) = 1.5 \text{ m}$. However, we have solved this Example in a more general way to illustrate the use of components of motion. The direct solution would have been evident if the equations had been combined algebraically before calculation, that is,

$$x = v_x t = (v \cos \theta) t$$

$$y = v_y t = (v \sin \theta) t$$

and

$$d = \sqrt{x^2 + y^2} = \sqrt{(v \cos \theta)^2 t^2 + (v \sin \theta)^2 t^2} = \sqrt{v^2 t^2 (\cos^2 \theta + \sin^2 \theta)} = vt$$

Before embarking on the first solution strategy that occurs to you, pause for a moment to see whether there might be an easier or more direct way of approaching the problem.

Kinematic Equations for Components of Motion

Example 3.1 involved two-dimensional motion in a plane. With a constant velocity (constant components v_x and v_y), the motion is in a straight line. The motion may also be accelerated. For motion in a plane *with a constant acceleration* having components a_x and a_y, the displacement (from the origin) and velocity components are given by the kinematic equations of Chapter 2 for the x and y directions:

$$x = v_{x_o} t + \tfrac{1}{2} a_x t^2 \quad (3.3a)$$

$$y = v_{y_o} t + \tfrac{1}{2} a_y t^2 \quad (3.3b)$$

(constant acceleration only)

$$v_x = v_{x_o} + a_x t \quad (3.3c)$$

$$v_y = v_{y_o} + a_y t \quad (3.3d)$$

Kinematic equations for displacement and velocity components

If an object is initially moving with a constant velocity and suddenly experiences an acceleration in the direction of the velocity (0°) or opposite to it (180°), it would continue in a straight-line path, either speeding up or slowing down, respectively.

If, however, the acceleration vector is at some angle other than 0° or 180° to the velocity vector, the motion is along a curved path. For the motion of an object to be *curvilinear*—that is, to vary from a straight-line path—an acceleration is required. For a curved path, the ratio of the velocity components varies with time. That is, the direction of the motion, $\theta = \tan^{-1}(v_y/v_x)$, varies with time because one or both of the velocity components do, so the motion is not in a straight line.

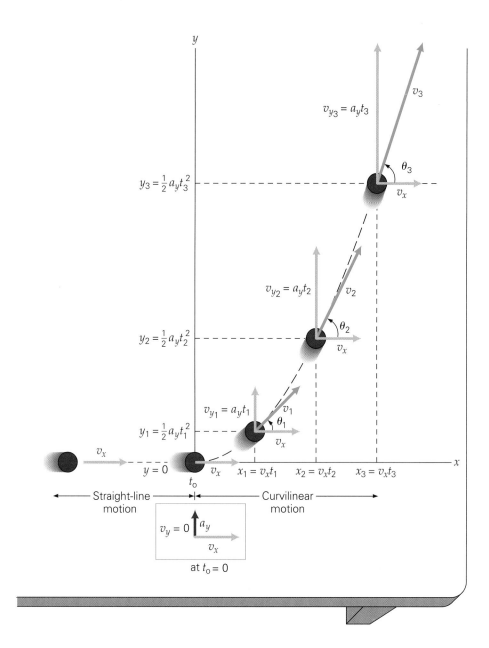

•FIGURE 3.2 Curvilinear motion An acceleration not parallel to the instantaneous velocity produces a curved path. Here an acceleration ($\mathbf{a}_y$) is applied at $t_o = 0$ to a ball initially moving with a constant velocity ($\mathbf{v}_x$). The result is a curved path with the velocity components as shown. Notice how $\mathbf{v}_y$ increases with time.

Consider a ball initially moving along the x axis, as illustrated in •Fig. 3.2. Assume that starting at a time $t_o = 0$ it receives a constant acceleration a_y in the y direction. The magnitude of the x component of the ball's displacement is given by $x = v_x t$; the $\frac{1}{2}a_x t^2$ term of Eq. 3.3a drops out since there is no acceleration in the x direction. Prior to t_o, the motion is in a straight line along the x axis. But at any time after t_o, there is a y displacement with a magnitude of $y = \frac{1}{2}a_y t^2$ (as given by Eq. 3.3b with $v_{y_o} = 0$). The result is a curved path for the ball.

Note that the length (magnitude) of the velocity component v_y changes with time. The total velocity vector ($\mathbf{v}$) *at any time* is tangent to the curved path of the ball. It is at an angle θ relative to the x axis, given by $\theta = \tan^{-1}(v_y/v_x)$, which now changes with time as we see in Fig. 3.2 and in the following Example.

Example 3.2 ■ A Curving Path: Vector Components

Suppose that the ball in Fig. 3.2 has an initial velocity of 1.50 m/s along the x axis and, starting at $t_o = 0$, receives an acceleration of 2.80 m/s² in the y direction. (a) What is the position of the ball 3.00 s after t_o? (b) What is the velocity of the ball at that time?

Thinking It Through. Keep in mind that the motions in the x and y directions can be analyzed independently—the connecting factor is time. That is, both component motions occur at the same time. In (a), simply compute the x and y positions at the given time, taking into account that there is an acceleration in the y direction. For (b), find the component velocities and combine to get the velocity.

Solution. Referring to Fig. 3.2, we have the following:

Given: $v_{x_o} = v_x = 1.50$ m/s
$v_{y_o} = 0$
$a_x = 0$
$a_y = 2.80$ m/s^2
$t = 3.00$ s

Find: (a) (x, y) (position coordinates)
(b) $\mathbf{v}$ (velocity)

(a) At 3.00 s after $t_o = 0$, Eqs. 3.3a and 3.3b tell us that the ball has traveled the following distances from the origin in the x and y directions:

$$x = v_{x_o}t + \tfrac{1}{2}a_x t^2 = (1.50 \text{ m/s})(3.00 \text{ s}) + 0 = 4.50 \text{ m}$$

$$y = v_{y_o}t + \tfrac{1}{2}a_y t^2 = 0 + \tfrac{1}{2}(2.80 \text{ m/s}^2)(3.00 \text{ s})^2 = 12.6 \text{ m}$$

Thus, its position is $(x, y) = (4.50 \text{ m}, 12.6 \text{ m})$. If you computed the distance $d = \sqrt{x^2 + y^2}$, what would this be? (Note that it is not the actual distance the ball has traveled in 3.00 s but the magnitude of the *displacement*, or straight-line distance, from the origin at $t = 3.00$ s.)

(b) The x component of the velocity is given by Eq. 3.3c:

$$v_x = v_{x_o} + a_x t = 1.50 \text{ m/s} + 0 = 1.50 \text{ m/s}$$

(This is constant since there is no acceleration in the x direction.) Similarly, the y component of the velocity is given by Eq. 3.3d:

$$v_y = v_{y_o} + a_y t = 0 + (2.80 \text{ m/s}^2)(3.00 \text{ s}) = 8.40 \text{ m/s}$$

The velocity therefore has a magnitude of

$$v = \sqrt{v_x^2 + v_y^2} = \sqrt{(1.50 \text{ m/s})^2 + (8.40 \text{ m/s})^2} = 8.53 \text{ m/s}$$

and its direction relative to the x axis is

$$\theta = \tan^{-1}\left(\frac{v_y}{v_x}\right) = \tan^{-1}\left(\frac{8.40 \text{ m/s}}{1.50 \text{ m/s}}\right) = 79.9°$$

Note: Don't confuse the direction of the velocity with the direction of the displacement from the origin. The direction of the velocity is always tangent to the path.

Follow-up Exercise. Suppose that the ball in Example 3.2 also received an acceleration of 1.00 m/s^2 in the $+x$ direction at t_o. What would be the position of the ball 3.00 s after t_o in this case?

3.2 Vector Addition and Subtraction

Objectives: To add and subtract vectors (a) graphically and (b) analytically.

Many physical quantities have a direction associated with them—that is, they are vectors. You have worked with a few such quantities related to motion (displacement, velocity, and acceleration) and will encounter more during this course of study. A very important technique in the analysis of many physical situations is the addition (and subtraction) of vectors. By adding or combining such quantities (**vector addition**), you can obtain the overall, or net, effect that occurs—the *resultant*, as we call the vector sum.

You have already been adding vectors. In Chapter 2, displacements were added to get the net displacement. In this chapter, vector components of motion were added to give net effects. Notice in the preceding Example 3.2 that the velocity components v_x and v_y were added to give the resultant velocity.

Note: One thing that scalar addition and vector addition have in common is that the quantities being added must have the same units. You cannot meaningfully add a displacement vector to a velocity vector, just as you cannot add the scalars of distance and speed and obtain a result with any physical meaning.

In this section, we will look at vector addition and subtraction in general, along with common vector notation. As you will learn, these operations are not the same as scalar or numerical addition and subtraction, with which you are already familiar. Vectors have magnitudes *and* directions, so different rules apply.

In general, there are geometric (graphical) methods and analytical (computational) methods of vector addition. The geometric methods are useful in helping you visualize the concepts of vector addition. Analytical methods are more commonly used, however, because they are faster and more precise.

Vector Addition: Geometric Methods

Triangle Method. To add two vectors—say, to add **B** to **A** (that is, to find **A** + **B**) by the **triangle method**—you first draw **A** on a sheet of graph paper to some scale (•Fig. 3.3a). For example, if **A** is a displacement in meters, a convenient scale is 1 cm : 1 m, or 1 cm of vector length on the graph corresponds to 1 m of displacement. As shown in Fig. 3.3b, the direction of the **A** vector is specified as being at an angle (θ_A) relative to a coordinate axis, usually the x axis.

Next draw **B** with its tail starting at the tip of **A**. (Thus, this method is also called the *tip-to-tail method*.) The vector from the tail of **A** to the tip of **B** is then the vector sum, or the resultant of the two vectors: **R** = **A** + **B**.

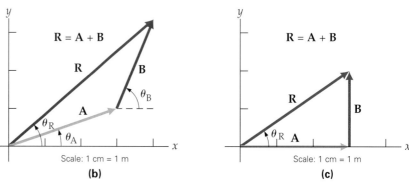

•**FIGURE 3.3 Triangle method of vector addition** (a) The vectors **A** and **B** are placed tip to tail. The vector that extends from the tail of **A** to the tip of **B**, forming the third side of the triangle, is the resultant **R** = **A** + **B**. (b) When drawn to scale, the magnitude of **R** can be found by measuring its length and using the scale conversion, and the direction angle θ_R can be measured with a protractor. Analytical methods can also be used. For a non-right triangle as in (b), the laws of sines and cosines would be used. (c) If the vector triangle is a right triangle, **R** is easily obtained via the Pythagorean theorem and the direction angle is given by an inverse trigonometric function.

If drawn to scale, the magnitude of **R** can be found by measuring its length and using the scale conversion. In such a graphical approach, the direction angle (θ_R) is measured with a protractor. Knowing the magnitudes and directions (angles θ) of **A** and **B**, the magnitude and direction of **R** can also be found analytically by using trigonometric methods. For the non-right triangle in Fig. 3.3b, the laws of sines and cosines would be used (see Appendix I).

The resultant of the vector right triangle in Fig. 3.3c would be much easier to find by using the Pythagorean theorem for the magnitude and an inverse trig function to find the direction angle. [Notice how **R** in this case is made up of x and y components (**A** and **B**).]

Parallelogram Method. Another graphical method of vector addition similar to the triangle method is the **parallelogram method**. In •Fig. 3.4, **A** and **B** are drawn tail to tail, and a parallelogram is formed as shown. The resultant **R** lies along the diagonal of the parallelogram. Drawing the diagram to scale with proper orientations, the magnitude and direction of **R** can be measured directly from the diagram as in the triangle method.

Notice that **B** could be moved to the other side of the parallelogram, forming the **A** + **B** triangle (and demonstrating why the triangle and parallelogram methods are equivalent). In general, a vector (arrow) can be moved around in vector addition methods. As long as you don't change its length (magnitude) or direction, you don't change the vector. In Fig. 3.4, this shifting of vector arrows shows that **A** + **B** = **B** + **A**—that is, the vectors can be added in either order.

Polygon Method. The triangle method can be extended to include the addition of any number of vectors. The method is then called the **polygon method** because the resulting graphical figure is a polygon. This is illustrated for four vectors in •Fig. 3.5, where **R** = **A** + **B** + **C** + **D**. Note that this addition is essentially three applications of the triangle method. The length and direction of the resultant could be found analytically by successive applications of the laws of sines and cosines, but an easier analytical method, the component method, will be described shortly. As in the parallelogram method, the four vectors (or any number of vectors) can be added in any order.

Vector Subtraction. Vector subtraction is a special case of vector addition:

$$\mathbf{A} - \mathbf{B} = \mathbf{A} + (-\mathbf{B})$$

That is, to subtract **B** from **A**, a *negative* **B** is added to **A**. In Chapter 2, you learned that a minus sign simply means that the direction of a vector is opposite that of one with a plus sign (for example, $+x$ and $-x$). The same is true of a vector. The

Note: A common *error* is to draw **R** from the tip of **A** to the tip of **B**.

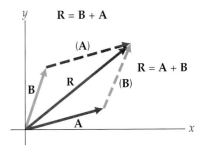

•**FIGURE 3.4 Parallelogram method of vector addition** The diagonal of the parallelogram formed after **A** and **B** are drawn with their tails at the origin is the resultant **R** = **A** + **B**. Drawn to scale, the magnitude and direction of **R** can be measured directly from the diagram as in the triangle method. The parallelogram method is equivalent to the triangle method. Shifting **B** to the right forms the same **A** + **B** vector triangle shown in Fig. 3.3b. Shifting **A** up has the same effect, with **R** = **B** + **A**.

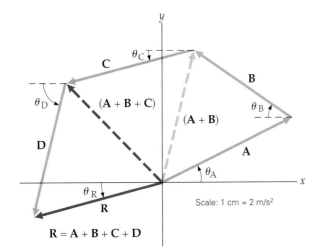

•**FIGURE 3.5 Polygon method of vector addition** The vectors to be added are placed tip to tail. The resultant **R** is the vector from the tail of the first vector (**A**) to the tip of the last vector (**D**), which completes the polygon. This method is essentially multiple applications of the triangle method: (**A** + **B**) + **C**, and so on.

vector $-\mathbf{B}$ has the same magnitude as the vector $\mathbf{B}$ but is in the opposite direction (•Fig. 3.6). The vector diagram in the figure provides a graphical representation of $\mathbf{A} - \mathbf{B}$.

Example 3.3 ■ Vector Subtraction: A Special Case of Vector Addition

Consider yourself in a car (A) traveling along a straight, level highway with a speed $v_A = 75$ km/h. Another car (B) travels at a speed $v_B = 90$ km/h. Find the differences in the velocities $\mathbf{v}_{BA} = \mathbf{v}_B - \mathbf{v}_A$ when (a) the other car travels in the same direction in front of you and (b) the other car is approaching you traveling in the opposite direction.

Thinking It Through. This is a case of simple vector addition (or subtraction). The important things to note are the directions of the velocities so that you will use the appropriate signs.

Solution. Taking the direction in which you are traveling as positive, we have, where the signs indicate vector directions,

Given: $v_A = +75$ km/h
(a) $v_B = +90$ km/h (same direction)
(b) $v_B = -90$ km/h (opposite direction)

Find: $\mathbf{v}_{BA} = \mathbf{v}_B - \mathbf{v}_A$

(a) With both cars traveling in the same (+) direction,

$$\mathbf{v}_{BA} = \mathbf{v}_B - \mathbf{v}_A = 90 \text{ km/h} - (+75 \text{ km/h}) = +15 \text{ km/h}$$

(b) When traveling in opposite directions,

$$\mathbf{v}_{BA} = \mathbf{v}_B - \mathbf{v}_A = -90 \text{ km/h} - (+75 \text{ km/h}) = -165 \text{ km/h}$$

Here, $\mathbf{v}_{BA}$ is what is called a *relative* velocity—the velocity of B relative to A. In part (a), relative to you in car A, car B appears to travel at a speed of 15 km/h. However, when approaching you, car B appears to travel at 165 km/h. It's as though you consider yourself or your car's reference frame to be at rest. (We will further explore the subject of relative velocity in Section 3.3.)

Follow-up Exercise. What is the reference frame for the given velocities in Example 3.3?

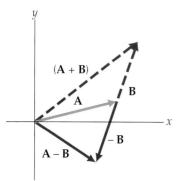

•FIGURE 3.6 Vector subtraction Vector subtraction is a special case of vector addition; that is, $\mathbf{A} - \mathbf{B} = \mathbf{A} + (-\mathbf{B})$, where $-\mathbf{B}$ has the same magnitude as $\mathbf{B}$ but is in the opposite direction (see sketch). Thus $\mathbf{A} + \mathbf{B}$ is *not* the same as $\mathbf{A} - \mathbf{B}$. Can you show that $\mathbf{B} - \mathbf{A} = -(\mathbf{A} - \mathbf{B})$ geometrically?

Vector Components and the Analytical Component Method

Probably the most widely used analytical method for adding multiple vectors is the **component method**. It will be used again and again throughout the course of our study, so a basic understanding of the method is *essential*. Learn this section well.

Adding Rectangular Vector Components. By *rectangular components*, we mean those at right (90°) angles to each other and usually taken in the rectangular coordinate x and y directions. You have already had an introduction to the addition of such components in the discussion of the velocity components of motion. For the general case, suppose that $\mathbf{A}$ and $\mathbf{B}$, two vectors at right angles, are added, as illustrated in •Fig. 3.7a. The right angle makes the math easy. The magnitude of $\mathbf{C}$ is given by the Pythagorean theorem:

$$C = \sqrt{A^2 + B^2} \tag{3.4a}$$

The orientation of $\mathbf{C}$ relative to the x axis is given by the angle

$$\theta = \tan^{-1}\left(\frac{B}{A}\right) \tag{3.4b}$$

Note: The notation $\tan^{-1} x$ stands for arctangent x, or the angle whose tangent is x.

Magnitude-angle form of a vector

This is how a resultant is expressed in **magnitude-angle form**.

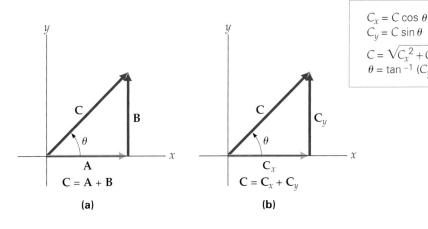

•FIGURE 3.7 **Vector components** (a) The vectors **A** and **B** along the x and y axes, respectively, add to give **C**. (b) A vector **C** may be resolved into rectangular components $\mathbf{C}_x$ and $\mathbf{C}_y$.

Resolving a Vector into Rectangular Components; Unit Vectors. Resolving a vector into rectangular components is essentially the reverse of adding the components. Given a vector **C**, Fig. 3.7b illustrates how it may be resolved into x and y vector components, $\mathbf{C}_x$ and $\mathbf{C}_y$. You simply complete the vector triangle with x and y components. As the diagram shows, the magnitudes or vector lengths of these components are given by

$$C_x = C \cos \theta \quad \text{magnitudes of} \quad (3.5a)$$
$$C_y = C \sin \theta \quad \text{components} \quad (3.5b)$$

(similar to $v_x = v \cos \theta$ and $v_y = v \sin \theta$ in Example 3.1).* The angle of direction of **C** can also be expressed in terms of the components, since $\tan \theta = C_y/C_x$, or

$$\theta = \tan^{-1}\left(\frac{C_y}{C_x}\right) \quad \begin{array}{l}\text{direction of vector from}\\ \text{magnitudes of components}\end{array} \quad (3.6)$$

A general notation for expressing the magnitude and direction of a vector involves the use of unit vectors. For example, as illustrated in •Fig. 3.8, a vector **A** can be written as $\mathbf{A} = A\hat{\mathbf{a}}$. The numerical magnitude is represented by A, and $\hat{\mathbf{a}}$ is called a **unit vector**. That is, it has a magnitude of unity, or one, but no units and so simply indicates the vector's direction. For example, a velocity along the x axis might be written $\mathbf{v} = (4.0 \text{ m/s}) \hat{\mathbf{x}}$ (that is, 4.0 m/s magnitude in the $+x$ direction).

Note in Fig. 3.8 how $-\mathbf{A}$ would be represented in this notation. Although the minus sign is sometimes put in front of the numerical magnitude, this is an absolute number; the minus actually goes with the unit vector; $-\mathbf{A} = -A\hat{\mathbf{a}} = A(-\hat{\mathbf{a}})$.† That is, the unit vector is in the $-\hat{\mathbf{a}}$ direction (opposite $\hat{\mathbf{a}}$). A velocity of $\mathbf{v} = (-4.0 \text{ m/s}) \hat{\mathbf{x}}$ has a magnitude of 4.0 m/s in the $-x$ direction; that is, $\mathbf{v} = (4.0 \text{ m/s})(-\hat{\mathbf{x}})$.

This notation can be used to express explicitly the rectangular components of a vector. For example, the ball's displacement from the origin in Example 3.2 could be written $\mathbf{d} = (4.50 \text{ m}) \hat{\mathbf{x}} + (12.6 \text{ m}) \hat{\mathbf{y}}$, where $\hat{\mathbf{x}}$ and $\hat{\mathbf{y}}$ are unit vectors in the x and y directions, respectively. In some instances it may be more convenient to express a general vector in this unit vector **component form**:

$$\mathbf{C} = C_x \hat{\mathbf{x}} + C_y \hat{\mathbf{y}} \quad (3.7)$$

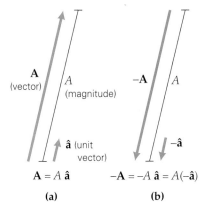

•FIGURE 3.8 **Unit vectors** (a) A unit vector $\hat{\mathbf{a}}$ has a magnitude of unity, or one, and so simply indicates a vector's direction. Written with the magnitude A, it represents the vector **A**, and $\mathbf{A} = A\hat{\mathbf{a}}$. (b) For the vector $-\mathbf{A}$, the unit vector is $-\hat{\mathbf{a}}$, and $-\mathbf{A} = -A\hat{\mathbf{a}} = A(-\hat{\mathbf{a}})$. The magnitude A is a positive number—that is, $|A|$.

Component form of a vector

---

*Only a vector in the first quadrant is illustrated in Fig. 3.7b, but the equations hold for all quadrants when vectors are referenced to the nearest x axis. The directions of the components will be indicated by $+$ and $-$ signs, as will be shown shortly.

†The notation is sometimes written $\mathbf{A} = |A|\hat{\mathbf{a}}$ or $-\mathbf{A} = -|A|\hat{\mathbf{a}}$ so as to express the absolute value of the magnitude of A.

Vector Addition Using Components. The **analytical component method** of vector addition involves resolving the vectors into rectangular components and adding the components for each axis independently. This method is illustrated graphically in •Fig. 3.9 for two vectors, $\mathbf{F}_1$ and $\mathbf{F}_2$. The sums of the x and y components of the vectors being added are then equal to the corresponding components of the resultant vector.

The same principle applies if you are given three (or more) vectors to add. You could find the resultant by applying the graphical tip-to-tail method as illustrated in •Fig. 3.10a. However, this technique involves drawing the vectors to scale and using a protractor to measure angles, which is time-consuming. Note that the magnitude of the directional angle θ is not obvious in the figure and would have to be measured (or computed trigonometrically).

However, if you use the component method, you do not have to draw the vectors tip to tail. In fact, it is usually more convenient to put all of the tails together at the origin, as shown in Fig. 3.10b. Also, the vectors do not have to be drawn to scale, since the sketch is just a visual aid in applying the analytical method.

In the component method, you simply resolve the vectors to be added into their x and y components, add the respective components, and recombine to find the resultant. This procedure is illustrated in Fig. 3.10c. Notice that it is just linear vector addition in the component directions and that components may add or subtract. A vector in the x or y direction has a "component" only in that direction (for example, $\mathbf{v}_2$ in the y direction in the figure).

Notice also in Fig. 3.10c that the directional angle θ of the resultant is referenced to the x axis, as are the individual vectors in Fig. 3.10b. *In adding vectors by the component method, we will reference all vectors to the nearest x axis—that is, the +x axis or −x axis.* This policy eliminates angles greater than 90° (as occurs when we customarily measure angles counterclockwise from the $+x$ axis) and the use of

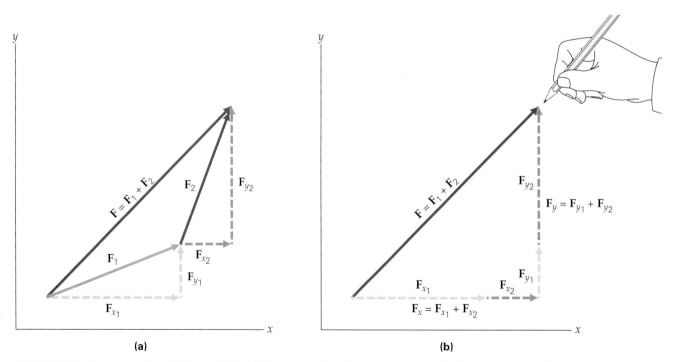

•**FIGURE 3.9 Component addition** (a) In adding vectors by the component method, each vector is first resolved into its x and y components. (b) The sums of the x and y components of $\mathbf{F}_1$ and $\mathbf{F}_2$ are the x and y components of the resultant $\mathbf{F}$; that is, $\mathbf{F}_x = \mathbf{F}_{x_1} + \mathbf{F}_{x_2}$ and $\mathbf{F}_y = \mathbf{F}_{y_1} + \mathbf{F}_{y_2}$.

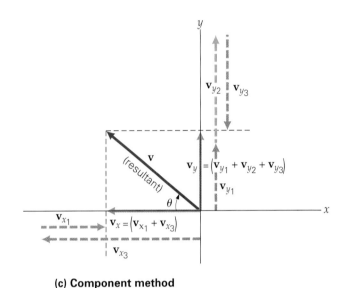

•**FIGURE 3.10 Component method of vector addition** (a) Several vectors may be added graphically to find the resultant **v**, but this technique is time-consuming. (b) In the analytical component method all the vectors to be added are first placed with their tails at the origin so that they may be easily resolved into rectangular components. (c) The respective summations of all the x components and all the y components are then added to give the components of the resultant **v**. Notice how the negative components subtract. See Example 3.4.

double angle formulas, such as $\cos(\theta + 90°)$. This restriction greatly simplifies calculations. The recommended procedures for adding vectors analytically by the component method can be summarized as follows:

Procedures for Adding Vectors by the Component Method

1. Resolve the vectors to be added into their x and y components. Use the acute angles (those less than 90°) between the vectors and the x axis, and indicate the directions of the components by plus and minus signs (•Fig. 3.11).
2. Add all of the x components together and all of the y components together vectorially to obtain the x and y components of the resultant, or vector sum. (This is done algebraically by using plus and minus signs.)
3. Express the resultant vector, using:
   (a) the component form—for example, $\mathbf{C} = C_x \hat{\mathbf{x}} + C_y \hat{\mathbf{y}}$ —or
   (b) the magnitude-angle form.

For the latter, find the magnitude of the resultant by using the summed x and y components and the Pythagorean theorem:

$$C = \sqrt{C_x^2 + C_y^2}$$

Find the angle of direction (relative to the x axis) by taking the inverse tangent

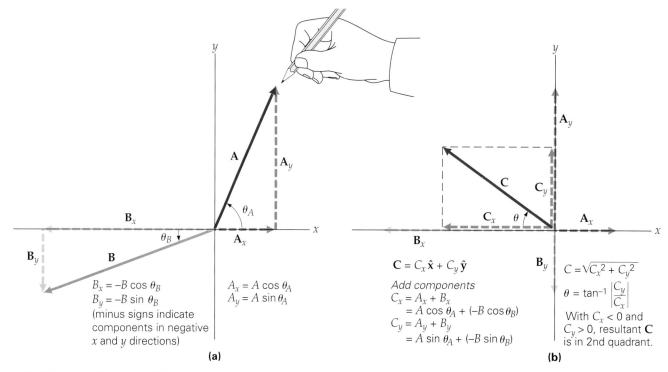

•**FIGURE 3.11 Vector addition by the analytical component method** (a) Resolve the vectors into their x and y components. (b) Add all of the x components and all of the y components together vectorially to obtain the x and y components (C_x and C_y) of the resultant. Express the resultant in either component form or magnitude-angle form. All angles are referenced to the $+x$ or $-x$ axis to keep them less than 90°.

Note: The absolute value indicates that minus signs are ignored (for example, $|-3| = 3$). This is done to avoid negative values and angles greater than 90°.

($\tan^{-1}$) of the *absolute value* of the ratio of the y and x components (that is, the positive value, ignoring, any minus signs):

$$\theta = \tan^{-1}\left|\frac{C_y}{C_x}\right|$$

Designate the quadrant in which the resultant lies. This information is obtained from the signs of the summed components or from a sketch of their addition via the triangle (or parallelogram) method (see Fig. 3.11). The angle θ is the angle between the resultant and the x axis in that quadrant.

Example 3.4 ■ Applying the Analytical Component Method: Separating and Combining x and y Components

Let's apply the procedural steps of the component method to the addition of the vectors in Fig. 3.10b. The vectors with units of meters per second represent velocities.

Thinking It Through. Follow and learn the steps given on page 75. Basically, you resolve the vectors into components and add the respective components to get the components of the resultant, which then may be expressed in component form or magnitude-angle form.

Solution.

1. The rectangular components of the vectors are shown in Fig. 3.10b.
2. Summing these components gives

$$\mathbf{v} = v_x\hat{\mathbf{x}} + v_y\hat{\mathbf{y}} = (v_{x_1} + v_{x_2} + v_{x_3})\hat{\mathbf{x}} + (v_{y_1} + v_{y_2} + v_{y_3})\hat{\mathbf{y}}$$

where

$$v_x = v_{x_1} + v_{x_2} + v_{x_3} = v_1\cos 45° + v_2\cos 90° - v_3\cos 30°$$
$$= (4.5 \text{ m/s})(0.707) + (5.0 \text{ m/s})(0) - (9.0 \text{ m/s})(0.866) = -4.6 \text{ m/s}$$

$$v_y = v_{y_1} + v_{y_2} + v_{y_3} = v_1 \sin 45° + v_2 \sin 90° - v_3 \sin 30°$$

$$= (4.5 \text{ m/s})(0.707) + (5.0 \text{ m/s})(1) - (9.0 \text{ m/s})(0.50) = 3.7 \text{ m/s}$$

In tabular form, the components are:

	x components		y components
v_{x_1}	$v_1 \cos 45° = +3.2$ m/s	v_{y_1}	$v_1 \sin 45° = +3.2$ m/s
v_{x_2}	$v_2 \cos 90° = 0$ m/s	v_{y_2}	$v_2 \sin 0° = +5.0$ m/s
v_{x_3}	$v_3 \cos 30° = -7.8$ m/s	v_{y_3}	$v_3 \sin 30° = -4.5$ m/s
Sums:	$v_x = -4.6$ m/s		$v_y = +3.7$ m/s

The directions of the components are indicated by signs (the + sign is usually omitted as being understood). In this case, v_2 has no x component. Note that in general for the analytical component method, the x components are cosine functions and the y components are sine functions, as long as we reference to the nearest x axis.

3. In component form, the resultant vector is

$$\mathbf{v} = (-4.6 \text{ m/s})\,\hat{\mathbf{x}} + (3.7 \text{ m/s})\,\hat{\mathbf{y}}$$

In magnitude-angle form, the resultant velocity has a magnitude of

$$v = \sqrt{v_x^2 + v_y^2} = \sqrt{(-4.6 \text{ m/s})^2 + (3.7 \text{ m/s})^2} = 5.9 \text{ m/s}$$

Since the x component is negative and the y component is positive, the resultant lies in the second quadrant at an angle of

$$\theta = \tan^{-1}\left|\frac{v_y}{v_x}\right| = \tan^{-1}\left(\frac{3.7 \text{ m/s}}{4.6 \text{ m/s}}\right) = 39°$$

above the negative x axis (see Fig. 3.10c).

Follow-up Exercise. Suppose in Example 3.4 there were an additional $\mathbf{v}_4 = (+4.6 \text{ m/s})\,\hat{\mathbf{x}}$ velocity vector. What would be the resultant of all four vectors in this case?

Although this discussion is limited to motion in two dimensions (in a plane), the component method is easily extended to three dimensions. For a velocity in three dimensions, the vector has x, y, and z components: $\mathbf{v} = v_x\,\hat{\mathbf{x}} + v_y\,\hat{\mathbf{y}} + v_z\,\hat{\mathbf{z}}$.

Conceptual Example 3.5

Vector Components in Action: Sailing "Into" the Wind

A sailboat (or wind surfer) on a lake travels in the direction the wind is blowing and then returns. Traveling "into" the wind, how does the boat get back home? *Clearly establish the reasoning and physical principle(s) used in determining your answer before checking it below. That is, **why** did you select your answer?*

Reasoning and Answer. Sailing into the wind is called *tacking*. This is not a direct mode of sailing but a wise use of vector components. Here, we will consider components of force, which is a vector quantity. As you know from experience, a single force (or force component), such as a push, gives rise to motion in the direction in which the force is applied. (We shall learn much more about forces in Chapter 4.) For simplicity, let's assume that the wind fills the sail and exerts a force perpendicular to the sail (F_s), as illustrated in •Fig. 3.12a. (Ignore friction and water current effects.)

Note that this force is resolved into components. The force parallel ($F_\parallel$) to the boat is at an acute angle (<90°) to the direction of the wind and is an "upwind" force on the boat.

•FIGURE 3.12 Let's go tacking. (a) The wind filling the sail exerts a force perpendicular to the sail (F_s). We can resolve this force vector into components. The one parallel to the motion of the boat ($F_\parallel$) has an upwind component, and there is motion in the upwind direction. (b) By changing the direction of the sail, the sailor can "tack" upwind. See Conceptual Example 3.5.

But, before heading too far to the northeast (as in Fig. 3.12a), the sail is turned so that the direction of $F_\parallel$ is changed by about 90° (Fig. 3.12b). The boat then comes back more in line with the desired upwind course. Using this zig-zag process, the boat sails "into" the wind and eventually gets home.

Follow-up Exercise. The zig-zag, straight-line path in Fig. 3.12b was drawn for simplicity. Would this be the actual path of the boat? Explain. [Hint: Consider $F_\perp$.]

3.3 Relative Velocity

Objective: **To determine relative velocities through vector addition and subtraction.**

Measurements must be made with respect to some reference. This reference is usually taken to be the origin of a coordinate system. The point you designate as the origin of a set of coordinate axes is arbitrary and entirely a matter of choice. For example, you may "attach" the coordinate system to the road or the ground and then measure the displacement or velocity of a car relative to these axes.

We can analyze a situation from any frame of reference. For example, the origin of the coordinate axes may be attached to a car moving along a highway. In analyzing motion from another reference frame, you do not change the physical situation or what is taking place, only the point of view from which you describe it. Hence, we say that motion is *relative* (to some reference frame), and we refer to **relative velocity**. Since velocity is a vector, vector addition and subtraction are helpful in determining relative velocities.

Relative Velocities in One Dimension

When the velocities are linear (along a straight line) in the same or opposite directions and all have the same reference (such as the ground), we can find relative velocities by vector subtraction. As an illustration, consider cars moving with

•**FIGURE 3.13 Relative velocity** The observed velocity of a car depends on, or is relative to, the frame of reference. The velocities shown in (a) are relative to the ground or to the parked car A. In (b) the frame of reference is with respect to car B, and the velocities are those that the driver of car B would observe. (c) These aircraft, performing air-to-air refueling, are normally described as traveling at hundreds of kilometers per hour. To what frame of reference do their velocities refer? What is their velocity relative to each other?

constant velocities along a straight, level highway, as in •Fig. 3.13. The velocities shown in the figure are *relative to the Earth or the ground*, as indicated by the reference set of coordinate axes in Fig. 3.13a. They are also relative to the stationary observers standing by the highway and sitting in the parked car A. That is, these observers see the cars as moving with velocities $\mathbf{v}_B = +90$ km/h and $\mathbf{v}_C = -60$ km/h. The relative velocity of two objects is given by the velocity (vector) difference between them. For example, the velocity of car B *relative to car A* is given by

$$\mathbf{v}_{BA} = \mathbf{v}_B - \mathbf{v}_A = +90 \text{ km/h} - 0 = +90 \text{ km/h}$$

Note: Use subscripts carefully! v_{AB} = velocity *of A relative to B*.

Thus, a person sitting in car A would see car B move away (in the positive x direction) with a speed of 90 km/h. For this linear case, the directions of the velocities are indicated by plus and minus signs (in addition to the minus sign in the formula).

Similarly, the velocity of car C relative to an observer in car A is

$$\mathbf{v}_{CA} = \mathbf{v}_C - \mathbf{v}_A = -60 \text{ km/h} - 0 = -60 \text{ km/h}$$

The person in car A would see car C approaching (in the negative x direction) with a speed of 60 km/h.

But suppose that you want to know the velocities of the other cars *relative to car B* (that is, from the point of view of an observer in car B) or relative to a set of coordinate axes with the origin fixed to car B (Fig. 3.13b). Relative to those axes, car B is not moving and acts as the fixed reference point. The other cars are moving relative to car B. The velocity of car C relative to car B is

$$\mathbf{v}_{CB} = \mathbf{v}_C - \mathbf{v}_B = -60 \text{ km/h} - (+90 \text{ km/h}) = -150 \text{ km/h}$$

Similarly, car A has a velocity relative to car B of

$$\mathbf{v}_{AB} = \mathbf{v}_A - \mathbf{v}_B = 0 - (+90 \text{ km/h}) = -90 \text{ km/h}$$

Notice that relative to B, the other cars are both moving in the negative x direction. That is, C is approaching B with a velocity of 150 km/h in the $-x$ direction and A appears to be receding from B with a velocity of 90 km/h in the $-x$ direction. (Imagine yourself in car B, and take that position as stationary. Car C would appear to be coming toward you at a high rate of speed, and car A would be getting farther and farther away, as though it were moving backward relative to you.) Note that, in general,

$$\mathbf{v}_{AB} = -\mathbf{v}_{BA}$$

What about the velocities relative to car C? From the point of view (or reference point) of car C, cars A and B would both appear to be approaching or moving in the positive x direction. For B relative to C,

$$\mathbf{v}_{BC} = \mathbf{v}_B - \mathbf{v}_C = 90 \text{ km/h} - (-60 \text{ km/h}) = +150 \text{ km/h}$$

Can you show that $\mathbf{v}_{AC} = +60$ km/h? Also note the situation in Fig. 3.13c.

In some instances, we may need to work with velocities that do not all have the same reference point. In such cases, relative velocities can be found by means of vector addition. To solve problems of this kind, *it is essential to identify the velocity references with care.*

Let's look first at a one-dimensional (linear) example. Suppose that a straight moving walkway in a major airport moves with a velocity of $\mathbf{v}_{wg} = +1.0$ m/s, where the subscripts indicate the velocity of the walkway (w) with respect to, or relative to, the ground (g). A passenger (p) on the walkway trying to make a flight connection walks with a velocity of $\mathbf{v}_{pw} = +2.0$ m/s relative to the walkway. What is the passenger's velocity relative to an observer standing next to the walkway (that is, relative to the ground)?

The velocity we are seeking, $\mathbf{v}_{pg}$, is given by

$$\mathbf{v}_{pg} = \mathbf{v}_{pw} + \mathbf{v}_{wg} = +2.0 \text{ m/s} + 1.0 \text{ m/s} = +3.0 \text{ m/s}$$

Thus the stationary observer sees the passenger as traveling with speed of 3.0 m/s down the walkway. (Make a sketch and show how the vectors add.)

Problem-Solving Hint

Notice the pattern of the subscripts in this example. On the right side of the equation, the two inner subscripts are the same (w). The outer subscripts (p and g) are sequentially the same as those for the relative velocity on the left side of the equation. When adding relative velocities, always check to make sure that the subscripts have this relationship—it indicates that you have set up the equation correctly.

What if a passenger got on the walkway going in the opposite direction and walked with the same speed as that of the walkway? Now it is essential to indicate the direction in which the passenger is walking by means of a minus sign: $\mathbf{v}_{pw} = -1.0$ m/s. In this case, relative to the stationary observer,

$$\mathbf{v}_{pg} = \mathbf{v}_{pw} + \mathbf{v}_{wg} = -1.0 \text{ m/s} + 1.0 \text{ m/s} = 0$$

so the passenger is stationary with respect to the ground and the walkway acts as a treadmill. (Excellent physical exercise!)

Relative Velocities in Two Dimensions

Of course, velocities are not always in the same or opposite directions. However, knowing how to use rectangular components to add or subtract vectors, we can solve problems involving relative velocities in two dimensions, as Examples 3.6 and 3.7 show.

Example 3.6 ■ Across and Down the River: Relative Velocity and Components of Motion

The current of a 500-meter-wide straight river has a flow rate of 2.55 km/h. A motorboat that travels with a constant speed of 8.00 km/h in still water crosses the river (●Fig. 3.14). (a) If the boat's bow points directly across the river toward the opposite shore, what is the velocity of the boat relative to the stationary observer sitting at the corner of the bridge? (b) How far downstream will the boat's landing point be from the point directly opposite its starting point? (c) What is the distance traveled by the boat in crossing the river? (Assume that the boat comes instantaneously to rest on a grassy shore.)

Thinking It Through. Careful designation of the given quantities is very important—the velocity of what, relative to what. Once this is done, part (a) should be straightforward. (See the previous Problem-Solving Hint.) For parts (b) and (c), we use kinematics, where the time it takes the boat to cross the river is the key.

Solution. As indicated in Fig. 3.14, we take the river's flow velocity ($\mathbf{v}_{rs}$, river to shore) to be in the x direction and the boat's velocity ($\mathbf{v}_{br}$, boat to river) to be in the y direction. Note that the river's flow velocity is *relative to the shore* and that the boat's velocity is *relative to the river*, as indicated by the subscripts. Listing the data, we have

Given: $y_{max} = 500$ m (river width)
$\mathbf{v}_{rs} = (2.55 \text{ km/h}) \, \hat{\mathbf{x}}$
$= (0.709 \text{ m/s}) \, \hat{\mathbf{x}}$
(velocity of river relative to shore)
$\mathbf{v}_{br} = (8.00 \text{ km/h}) \, \hat{\mathbf{y}}$
$= (2.22 \text{ m/s}) \, \hat{\mathbf{y}}$
(velocity of boat relative to river)

Find: (a) $\mathbf{v}_{bs}$ (velocity of boat relative to shore)
(b) x (distance downstream)
(c) d (distance traveled by boat)

●**FIGURE 3.14 Relative velocity and components of motion** As the boat moves across the river, it is carried downstream by the current. See Example 3.6.

Notice that as the boat moves toward the opposite shore, it is also carried downstream by the current. These velocity components would be clearly apparent relative to the jogger crossing the bridge and to the person sauntering downstream in Fig. 3.14. If both observers stay even with the boat, the velocity of each will match one of the components of the boat's velocity. Since the velocity components are constant, the boat travels in a straight line diagonally across the river (much like the ball rolling across the table in Example 3.1).

(a) The velocity of the boat relative to the shore ($\mathbf{v}_{bs}$) is given by vector addition. In this case, we have

$$\mathbf{v}_{bs} = \mathbf{v}_{br} + \mathbf{v}_{rs}$$

Since the velocities are not along one axis, their magnitudes cannot be added directly. Notice in Fig. 3.14 that the vectors form a right triangle, so we can apply the Pythagorean theorem to find the magnitude of $\mathbf{v}_{bs}$:

$$v_{bs} = \sqrt{(v_{br})^2 + (v_{rs})^2} = \sqrt{(2.22 \text{ m/s})^2 + (0.709 \text{ m/s})^2}$$
$$= 2.33 \text{ m/s}$$

The direction of this velocity is defined by

$$\theta = \tan^{-1}\left(\frac{v_{rs}}{v_{br}}\right) = \tan^{-1}\left(\frac{0.709 \text{ m/s}}{2.22 \text{ m/s}}\right) = 17.7°$$

(b) To find the distance x that the current carries the boat downstream, we use components. Note that in the y direction, $y_{max} = v_{br}t$, and

$$t = \frac{y_{max}}{v_{br}} = \frac{500 \text{ m}}{2.22 \text{ m/s}} = 225 \text{ s}$$

which is the time it takes the boat to cross the river.

During this time, the boat is carried downstream by the current a distance of

$$x = v_{rs}t = (0.709 \text{ m/s})(225 \text{ s}) = 160 \text{ m}$$

(c) We could find the distance d the boat travels by using x and y_{max} with the Pythagorean theorem again, but let's use the magnitude of the relative velocity and time:

$$d = v_{bs}t = (2.33 \text{ m/s})(225 \text{ s}) = 524 \text{ m}$$

Follow-up Exercise. In part (c), find the distance d by using the Pythagorean theorem. (The answer may be slightly different. Why?)

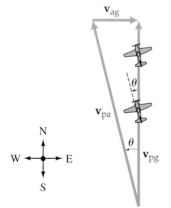

•**FIGURE 3.15 Flying into the wind** To fly directly north, the plane's heading (θ direction) must be west of north. See Example 3.7.

Example 3.7 ■ Flying Into the Wind: Relative Velocity

An airplane with an air speed of 200 km/h (its speed in still air) flies in a direction such that with a west wind of 50.0 km/h blowing, it travels in a straight line northward. (Wind direction is specified by the direction *from* which the wind blows, so a west wind blows from west to east.) To maintain its course due north, the plane must fly at an angle as illustrated in •Fig. 3.15. What is the speed of the plane along its northward path?

Thinking It Through. Here again, the velocity designations are important, but Fig. 3.15 shows that the velocity vectors form a right triangle, and the magnitude of the unknown velocity can be found by using the Pythagorean theorem.

Solution. As always, it is important to identify the reference frame to which the given velocities are relative.

Given: $\mathbf{v}_{pa} = 200$ km/h at angle θ *Find:* v_{pg} (ground speed of plane)
(velocity of plane with
respect to still air = air speed)

$v_{ag} = 50.0$ km/h east (velocity of air with respect to Earth or ground = wind speed)

plane flies due north with velocity $\mathbf{v}_{pg}$

The speed of the plane with respect to the Earth or the ground, v_{pg}, is called its ground speed; v_{pa} is its air speed. Vectorially, the respective velocities are related by

$$\mathbf{v}_{pg} = \mathbf{v}_{pa} + \mathbf{v}_{ag}$$

If there were no wind blowing ($v_{ag} = 0$), the ground speed and air speed would be equal. However, a head wind (one blowing directly toward the plane) would cause a slower ground speed; and a tail wind, a faster ground speed. The situation is analogous to that of a boat going upstream versus downstream.

Here $\mathbf{v}_{pg}$ is the resultant of the other two vectors, which can be added by the triangle method. We use the Pythagorean theorem to find v_{pg}, noting that v_{pa} is the hypotenuse of the triangle:

$$v_{pg} = \sqrt{v_{pa}^2 - v_{ag}^2} = \sqrt{(200 \text{ km/h})^2 - (50.0 \text{ km/h})^2} = 194 \text{ km/h}$$

(Note that it was convenient to use the units of kilometers per hour, since the calculation did not involve any other units.)

Follow-up Exercise. What must be the plane's heading (θ direction) in Example 3.7 for the plane to fly directly north?

3.4 Projectile Motion

Objectives: To analyze projectile motion to find (a) position, (b) time of flight, and (c) range.

A familiar example of two-dimensional, curvilinear motion is the motion of objects that are thrown or projected by some means. The motion of a stone thrown across a stream or a golf ball driven off a tee is **projectile motion**. A special case of projectile motion in one dimension occurs when an object is projected vertically upward. This case was treated in Chapter 2 in terms of free fall with air resistance neglected. We will generally neglect air resistance here also so that the only acceleration acting on a projectile is that due to gravity.

Note: Review Section 2.5 (free fall).

We can use vector components to analyze projectile motion. You simply break up the motion into its x and y components and treat them separately.

Horizontal Projections

It is worthwhile to analyze first the special case of the motion of an object projected horizontally, or parallel to a level surface. Suppose that you throw an object horizontally with an initial velocity v_{x_o} (•Fig. 3.16a). Projectile motion is analyzed beginning at the instant of release ($t = 0$). Once the object is released, there is no longer a horizontal acceleration ($a_x = 0$), so throughout the object's path the horizontal velocity remains constant: $v_x = v_{x_o}$

According to the equation $x = v_x t$ (Eq. 3.2a), the projected object would continue to travel in the horizontal direction indefinitely. However, you know that this is not what happens. As soon as the object is projected, it is in free fall in the vertical direction, with $v_{y_o} = 0$ (as though it were dropped) and $a_y = -g$. In other words, the projected object travels at a uniform velocity in the horizontal direction while *at the same time* undergoing acceleration in the downward direction under the influence of gravity. The result is a curved path, as illustrated in Fig. 3.16. (Compare the motions in Fig. 3.16 and Fig. 3.2. Do you see any similarities?) If there were no horizontal motion, the object would simply drop to the ground in

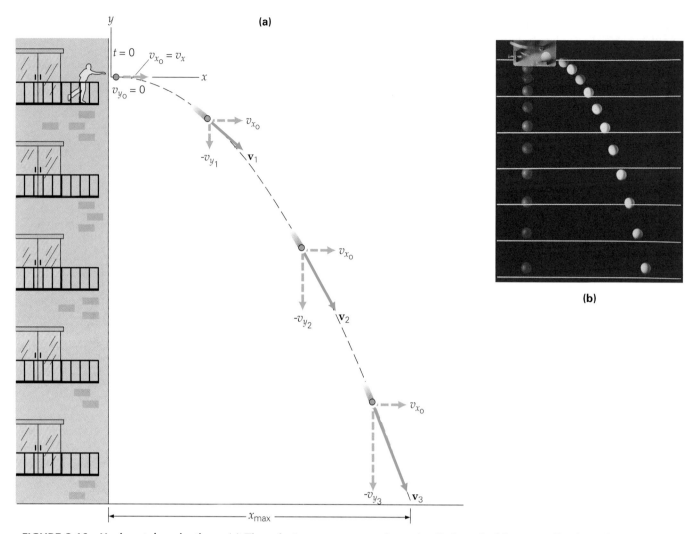

•**FIGURE 3.16 Horizontal projection** (a) The velocity components of a projectile launched horizontally show that it travels to the right as it falls downward. (b) A multiflash photograph shows the paths of two golf balls. One was projected horizontally at the same time that the other was dropped straight down. The horizontal lines are 15 cm apart, and the time interval between flashes was 1/30 s. The vertical motions of the balls are the same. Why? Can you describe the horizontal motion of the yellow ball?

a straight line. In fact, the time of flight of the projected object is *exactly the same as if it were falling vertically*.

Note the components of the velocity vector in Fig. 3.16a. The length of the horizontal component of the velocity vector remains the same, but the length of the vertical component increases with time. What is the instantaneous velocity at any point along the path? (Think in terms of vector addition, covered in Section 3.3.) The photo in Fig. 3.16b shows the actual motions of a horizontally projected golf ball and one that is simultaneously dropped from rest. The horizontal reference lines show that the balls fall vertically at the same rate. The only difference is that the horizontally projected ball also travels to the right as it falls.

Example 3.8 ■ Starting at the Top: Horizontal Projection

Suppose that the ball in Fig. 3.16a is projected from a height of 25.0 m above the ground and is thrown with an initial horizontal velocity of 8.25 m/s. (a) How long is the ball in flight before striking the ground? (b) How far from the building does the ball strike the ground?

Thinking It Through. In looking at the components of motion, part (a) involves the time it take the ball to fall vertically, analogous to a ball dropped from that height. This is also the time the ball travels in the horizontal direction. The horizontal speed is constant, so we can find the horizontal distance, part (b).

Solution. Writing the data with the origin chosen as the point from which the ball is thrown and downward taken as the negative direction, we have

Given: $y = -25.0$ m   Find: (a) t (time of flight)
$v_{x_o} = 8.25$ m/s   (b) x (horizontal distance)
$a_x = 0$
$v_{y_o} = 0$
$a_y = -g$

(a) As noted previously, the time of flight is the same as the time it takes for the ball to fall vertically to the ground. To find this time, we can use the equation $y = v_{y_o}t - \frac{1}{2}gt^2$, in which the negative direction of g is expressed explicitly as was done in Chapter 2. With $v_{y_o} = 0$,

$$y = -\tfrac{1}{2}gt^2$$

and

$$t = \sqrt{\frac{2y}{-g}} = \sqrt{\frac{2(-25.0 \text{ m})}{-9.80 \text{ m/s}^2}} = 2.26 \text{ s}$$

(b) The ball travels in the x direction for the same amount of time it travels in the y direction (that is, 2.26 s). Since there is no acceleration in the horizontal direction, it does so with a uniform velocity. Thus, with $a_x = 0$,

$$x = v_{x_o}t = (8.25 \text{ m/s})(2.26 \text{ s}) = 18.6 \text{ m}$$

Follow-up Exercise. What is the velocity (in component form) of the ball just before it strikes the ground?

Projections at Arbitrary Angles

The general case of projectile motion involves an object projected at an arbitrary angle θ relative to the horizontal; for example, a golf ball struck by a club (•Fig. 3.17). During projectile motion, the object travels up and down while traveling horizontally with a constant velocity.

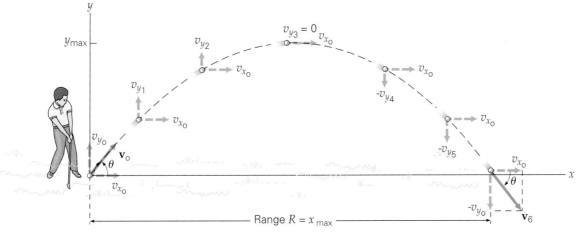

•**FIGURE 3.17 Projection at an angle** The velocity components of the ball are shown for various times. Note that $v_y = 0$ at the top of the arc, or at y_{max}. The range (R) is the maximum horizontal distance, or x_{max}.

This motion is also analyzed by using its components. As before, upward is taken as the positive direction and downward as the negative direction. The initial velocity v_o is first resolved into rectangular components:

$$v_{x_o} = v_o \cos\theta \qquad \text{(3.8a)}$$
$$v_{y_o} = v_o \sin\theta \qquad \text{(3.8b)}$$

initial velocity components $(t_o = 0)$

Since there is no horizontal acceleration and gravity acts in the negative y direction, the x component of the velocity is constant and the y component varies with time (see Eq. 3.3):

$$v_x = v_{x_o} = v_o \cos\theta \qquad \text{(3.9a)}$$
$$v_y = v_{y_o} - gt = (v_o \sin\theta) - gt \qquad \text{(3.9b)}$$

projectile motion velocity components

The components of the instantaneous velocity at various times are illustrated in Fig. 3.17. The instantaneous velocity is the sum of these components and is tangent to the curved path of the ball at any point. Notice that the ball strikes the ground at the same speed as it was launched (but with $-v_{y_o}$) and at the same angle below the horizontal.

Similarly, the displacement components are given by

$$x = v_{x_o}t = (v_o \cos\theta)t \qquad \text{(3.10a)}$$
$$y = v_{y_o}t - \tfrac{1}{2}gt^2 = (v_o \sin\theta)t - \tfrac{1}{2}gt^2 \qquad \text{(3.10b)}$$

projectile motion displacement components

The curve produced by these equations, or the path of motion of the projectile, is called a **parabola**. The path of projectile motion is often referred to as a parabolic arc. Such arcs are commonly observed (•Fig. 3.18).

Note as in the case of horizontal projection, *time is the common feature shared by the components of motion.* Aspects of projectile motion that may be of interest in various situations include the time of flight, the maximum height reached, and the **range** (R), which is the maximum horizontal distance traveled.

(a)

(b)

•**FIGURE 3.18 Parabolic arcs** (a) Fireworks and (b) streams of water move in parabolic arcs, though air resistance may distort the trajectories to some extent.

Example 3.9 ■ Teeing Off: Projection at an Angle

Suppose a golf ball is hit off the tee with an initial velocity of 30.0 m/s at an angle of 35° to the horizontal as in Fig. 3.17. (a) What is the maximum height reached by the ball? (b) What is its range?

Thinking It Through. The maximum height involves the y component; finding it is like finding the maximum height of a ball projected vertically upward. The ball travels in the x direction for the same amount of time it would take the ball to go up and down.

Solution.

Given: $v_o = 30.0$ m/s   Find: (a) y_{max}
$\theta = 35°$ (b) $R = x_{max}$
$a_y = -g$

Let us compute v_{x_o} and v_{y_o} explicitly to be able to use simplified kinematic equations:

$$v_{x_o} = v_o \cos 35° = (30.0 \text{ m/s})(0.819) = 24.6 \text{ m/s}$$
$$v_{y_o} = v_o \sin 35° = (30.0 \text{ m/s})(0.574) = 17.2 \text{ m/s}$$

(a) Just as for an object thrown vertically upward, $v_y = 0$ at the maximum height (y_{max}). Thus, we can find the time to reach the maximum height (t_{max}) by using Eq. 3.9b with v_y set equal to zero:

$$v_y = 0 = v_{y_o} - gt_{max}$$

$$t_{max} = \frac{v_{y_o}}{g} = \frac{17.2 \text{ m/s}}{9.80 \text{ m/s}^2} = 1.76 \text{ s}$$

As in the case of vertical projection, the time in going up is equal to the time in coming down, so the total time of flight is $t = 2t_{max}$ (to return to the elevation from which the object was projected).

The maximum height (y_{max}) is then obtained by substituting t_{max} into Eq. 3.10b:

$$y_{max} = v_{y_o}t_{max} - \tfrac{1}{2}gt_{max}^2 = (17.2 \text{ m/s})(1.76 \text{ s}) - \tfrac{1}{2}(9.80 \text{ m/s}^2)(1.76 \text{ s})^2 = 15.1 \text{ m}$$

The maximum height could also be obtained directly from Eq. 2.10′, $v_y^2 = v_{y_o}^2 - 2gy$, with $y = y_{max}$ and $v_y = 0$. However, the method of solution used here illustrates how the time of flight is obtained.

(b) The range (R) is equal to the horizontal distance traveled (x_{max}), which is easily found by substituting the total time of flight $t = 2t_{max} = 2(1.76 \text{ s}) = 3.52 \text{ s}$ into Eq. 3.10a:

$$R = x_{max} = v_x t = v_{x_o}(2t_{max}) = (24.6 \text{ m/s})(3.52 \text{ s}) = 86.6 \text{ m}$$

Follow-up Exercise. What is the location of the golf ball in Example 3.9 at $t = 3.00$ s?

Note: A projectile launched at an angle to the horizontal follows a parabolic path.

The range of a projectile is an important consideration in various applications. This is particularly true in those sports in which a maximum range is desired, such as in driving a golf ball or throwing a javelin.

In general, what is the range of a projectile launched with velocity v_o at an angle θ? To answer this question requires us to consider the equation used in Example 3.9 to calculate the range, $R = v_x t$. First let's look at the expressions for v_x and t. Since there is no acceleration in the horizontal direction, we know that

$$v_x = v_{x_o} = v_o \cos \theta$$

and $t = 2t_{max}$, where (as we found in Example 3.9)

$$t_{max} = \frac{v_{y_o}}{g} = \frac{v_o \sin \theta}{g}$$

Then,

$$R = v_x t = (v_o \cos \theta)(2t_{max}) = (v_o \cos \theta)\left(\frac{2v_o \sin \theta}{g}\right) = \frac{2v_o^2 \sin \theta \cos \theta}{g}$$

Using the trigonometric identity $\sin 2\theta = 2 \cos \theta \sin \theta$ (see Appendix I), we have

$$R = \frac{v_o^2 \sin 2\theta}{g} \quad \begin{array}{l} \text{projectile range, } x_{max} \\ (\text{only for } y_{initial} = y_{final}) \end{array} \quad (3.11)$$

Projectile range—the maximum horizontal distance traveled

Projectile range—for $y_{initial} = y_{final}$ only

Note that the range depends on the magnitude of the initial velocity (or speed, v_o) and the angle of projection (θ); g is assumed to be constant. Keep in mind that this equation applies only to the *special*, but common, case of $y_{initial} = y_{final}$—that is, the landing point is at the same height as the launch point.

Example 3.10 ■ The Last Shot: A Win or a Loss?

The college basketball game is close. The home team trails by two points, and 1.00 s remains in the game. The hometown guard takes a jump shot (in other words, his feet leave the floor). The ball is launched at the angle of 60° relative to the horizontal, with a speed of 10.0 m/s, and an initial height of 3.05 m (10 ft) above the floor, the same height as the hoop (•Fig. 3.19). The ball goes in, and the crowd goes wild!

(a) Baskets made from outside a semicircle of radius 6.02 m (19 ft, 9 in.) from beneath the center of the hoop are given 3 points, whereas those made from inside the semicircle are worth 2 points (see Fig 3.19). Did the home team win, or did the game go into overtime in a tie?

(b) Suppose the home guard stops at a distance of 7.00 m from a point directly below the point nearest to him on the hoop and takes a set shot (that is, his feet are on the floor) with the same launch speed as in part (a) but at an angle of 30° and from an initial height of 1.80 m above the floor. Is there a possibility of him making the shot?

•**FIGURE 3.19 A close game!** A home player takes a final shot, and the basket is good. But is it a win or a tie? See Example 3.10.

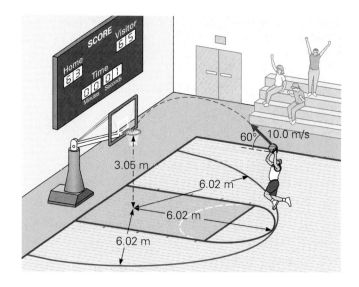

Thinking It Through. Part (a) involves the range of the ball. Note that the launch point is the same height as the basket. In part (b), there is a question of the height of the ball at the hoop. Was there enough arc for the ball to get there and go through? The vertical height of the ball at the basket must be determined, but *note* that the range equation (Eq. 3.11) *cannot* be used. Why?

Solution.

Given: (a) $v_o = 10.0$ m/s *Find:* (a) if $R > 6.02$ m
$\theta = 60°$ (b) if $y_{final} > 3.05$ m
$t = 1.00$ s   (note $y_{initial} = y_{final}$)
(b) $x = 7.00$ m
$\theta = 30°$
$v_o = 10.0$ m/s   (note $y_{initial} \neq y_{final}$)

(a) We must determine whether the ball was shot at a distance greater than 6.02 m—that is, whether the range was greater than 6.02 m. Since the launch point and the basket are at the same height, Eq. 3.11 applies, and

$$R = \frac{v_o^2 \sin 2\theta}{g} = \frac{(10.0 \text{ m/s})^2 \sin 2(60°)}{9.80 \text{ m/s}^2} = 8.84 \text{ m}$$

The shot was made from outside the semicircle, and the home team wins!

(b) Here, we want to know where the ball is vertically when it gets to the basket—that is, its y position. Knowing the horizontal distance and the initial velocity, we have

$$v_x = v_o \cos 30° = (10.0 \text{ m/s})(0.866) = 8.66 \text{ m/s}$$

and

$$v_y = v_o \sin 30° = (10.0 \text{ m/s})(0.500) = 5.00 \text{ m/s}$$

The time of flight is found from the x direction kinematic equation:

$$t = \frac{x}{v_x} = \frac{7.00 \text{ m}}{8.66 \text{ m/s}} = 0.808 \text{ s}$$

And at this time, the vertical height of the ball is

$$y = v_{y_o}t - \tfrac{1}{2}gt^2 = (5.00 \text{ m/s})(0.808 \text{ s}) - \tfrac{1}{2}(9.80 \text{ m/s}^2)(0.808 \text{ s})^2 = 0.84 \text{ m}$$

Note: This is the height of the ball *above* the height of the launch point. (Why?) Hence, the ball is at a height of 1.80 m + 0.84 m = 2.64 m from the floor—well under the 3.02 m of the basket height, so the shot would not be good.

Follow-up Exercise. In part (a) of Example 3.10, did time run out before the ball went through the hoop? (Even if it did, the basket counts.)

Equation 3.11 allows us to compute the range for a particular projection angle and initial velocity. However, we are sometimes interested in the maximum range for a given initial velocity—for example, the maximum range of an artillery piece that fires a projectile with a particular muzzle velocity. Is there an optimum angle? Under ideal conditions, the answer is yes.

For a particular v_o, the range is a maximum (R_{max}) when $\sin 2\theta = 1$, since this is the maximum value of the sine function (which varies from 0 to 1). Thus

$$R_{max} = \frac{v_o^2}{g} \tag{3.12}$$

Because this maximum range is obtained when $\sin 2\theta = 1$ and because $\sin 90° = 1$, we have

$$2\theta = 90° \quad \text{or} \quad \theta = 45°$$

for the maximum range for a given initial speed when the projectile returns to the elevation from which it was projected. At a greater or smaller angle, for a projectile with the same initial speed, the range will be less, as illustrated in •Fig. 3.20. Also, the range is the same for angles equally above and below 45°, such as 30° and 60°. (See Exercise 88.)

Thus, to get the maximum range, a projectile should *ideally* be projected at an angle of 45°. However, up to now we have neglected air resistance. In actual situations, such as when a ball or object is thrown or hit hard, this factor may have a significant effect. Air resistance reduces the speed of the projectile, thereby reducing the range. As a result, when air resistance is a factor the angle of projection for maximum range is less than 45°, which gives a greater initial horizontal velocity (•Fig. 3.21). Other factors, such as spin and wind, may affect the range of a projectile. For example, backspin on a driven golf ball provides lift, and the projection angle for the maximum range may be considerably less than 45°.

Keep in mind that the 45° maximum range requires that the initial velocity components be equal: $\tan^{-1}(v_{y_o}/v_{x_o}) = 45°$ and $\tan 45° = 1$, so $v_{y_o} = v_{x_o}$. However, this condition may not always be physically possible, as the following Example shows.

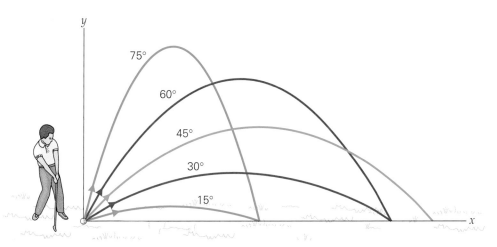

•**FIGURE 3.20 Range** For a projectile with a given initial speed, the maximum range is ideally attained with a projection angle of 45°. For projection angles above and below 45°, the range is shorter, and it is equal for angles equally different from 45° (for example, 30° and 60°).

•FIGURE 3.21 **Air resistance and range** When air resistance is a factor, the angle of projection for maximum range is less than 45°.

•FIGURE 3.22 **Hanging in there** Basketball players seem to "hang" in the air at the peak of their jump. Why is this? See the Follow-up Exercise, Conceptual Example 3.11.

Conceptual Example 3.11

The Longest Jump: Theory and Practice

In a long jump event, the jumper normally has a launch angle of (a) less than 45°, (b) exactly 45°, (c) greater than 45°? *Clearly establish the reasoning and physical principle(s) used in determining your answer before checking it below. That is, **why** did you select your answer?*

Reasoning and Answer. Air resistance is not a major factor here (although wind speed is taken into account for record settings in track and field events). Therefore, it would seem that, to achieve maximum range, the jumper would take off at an angle of 45°. But there is another physical consideration. Let's look more closely at the jumper's initial velocity components.

To maximize a long jump, the jumper runs as fast as possible and then pushes upward as strongly as possible to maximize the velocity components. The initial vertical velocity component (v_{y_o}) depends on the upward push of the jumper's legs, whereas the initial horizontal component (v_{x_o}) depends mostly on the running speed toward the jump point. In general, a greater velocity can be achieved by running than by jumping, so $v_{x_o} > v_{y_o}$. Then, since $\theta = \tan^{-1}(v_{y_o}/v_{x_o})$, we have $\theta < 45°$, where $v_{y_o}/v_{x_o} < 1$ in this case. Hence, the answer is (a) and certainly could not be (c). A typical launch angle for a long jump is 20° to 25°. (If a jumper increased her launch angle closer to the ideal 45°, her running speed would have to decrease, resulting in a decrease in range.)

Follow-up Exercise. When jumping to score, basketball players seem to be suspended momentarily or to "hang" in the air (•Fig. 3.22). Explain the physics of this effect.

Chapter Review

Important Terms

components of motion 66
vector addition (subtraction) 69
triangle method 70
parallelogram method 71
polygon method 71
component method 72
magnitude-angle form 72
unit vector 73
component form 73
analytical component method 74
relative velocity 78
projectile motion 83
parabola 86
range 86

Important Concepts

- Motion in two dimensions is analyzed by considering linear component motion in each dimension individually.
- Vector subtraction is a special case of vector addition.
- Of the various methods of vector addition, the component method is the most powerful. A resultant vector can be expressed in the magnitude-angle form or the unit-vector component form.

- Relative velocity is a velocity expressed *relative* to a particular reference frame.
- Projectile motion is analyzed by considering horizontal and vertical components separately—constant velocity in the horizontal direction and an acceleration due to gravity, g, in the downward vertical direction.
- To get the maximum range, a projectile should ideally be projected at an angle of 45°. However, when air resistance is a factor, the angle of projection for the maximum range is less than 45°.

Important Equations

Components of Initial Velocity:
$$v_{x_o} = v_o \cos\theta \quad (3.1a)$$
$$v_{y_o} = v_o \sin\theta \quad (3.1b)$$

Components of Displacement (constant acceleration only):
$$x = v_{x_o} t + \tfrac{1}{2} a_x t^2 \quad (3.3a)$$
$$y = v_{y_o} t + \tfrac{1}{2} a_y t^2 \quad (3.3b)$$

Component of Velocity (constant acceleration only):
$$v_x = v_{x_o} + a_x t \quad (3.3c)$$
$$v_y = v_{y_o} + a_y t \quad (3.3d)$$

Vector Representation:
$$C = \sqrt{C_x^2 + C_y^2} \quad (3.4a)$$
$$\theta = \tan^{-1}\left|\frac{C_y}{C_x}\right| \quad (3.4b)$$
magnitude-angle form

$$\mathbf{C} = C_x \hat{\mathbf{x}} + C_y \hat{\mathbf{y}} \quad \text{component form} \quad (3.7)$$

Exercises

Note: Exercises with a CD-ROM icon have corresponding simulations on the Interactive Journey through Physics CD-ROM; a complete cross-reference of text problems to CD-ROM activities appears in the Media Pack.

3.1 Components of Motion

1. On Cartesian axes, the x component of a vector is generally associated with a (a) cosine, (b) sine, (c) tangent, (d) none of these.

2. For an object in curvilinear motion, (a) its velocity components are constant, (b) the y-velocity component is necessarily greater than the x-velocity component, (c) there is an acceleration nonparallel to the object's path, (d) the velocity and acceleration vectors must be at right angles (90°).

3. Describe the motion of an object that is initially traveling with a constant velocity and then receives an acceleration of constant magnitude (a) in a direction parallel to the initial velocity, (b) in a direction perpendicular to the initial velocity, (c) that is always perpendicular to the instantaneous velocity or direction of motion.

4. ■ A golf ball is hit with an initial speed of 35 m/s at an angle 37° above the horizontal. What are the initial horizontal and vertical velocity components?

5. ■ The x and y components of an acceleration vector are 3.0 m/s² and 4.0 m/s², respectively. What are the magnitude and direction of the acceleration vector?

6. ■ If the magnitude of a velocity vector is 7.0 m/s and the x component is +3.0 m/s, what are the possible values of its y component?

7. ■■ The x component of a velocity vector that has an angle of 37° to the $+x$ axis has a magnitude of 4.8 m/s. (a) What is the magnitude of the velocity? (b) What is the magnitude of the y component of the velocity?

8. ■■ A student walks 100 m west and 50 m south. What displacement will bring him back to the starting point?

9. ■■ A student strolls diagonally across a level rectangular campus plaza, covering the 50-meter distance in 1.0 min (•Fig. 3.23). (a) If the diagonal route makes a 37° angle with the long side of the plaza, what would be the distance if she walked half way around the outside of the

•FIGURE 3.23 Which way? See Exercise 9.

plaza instead of along the diagonal route? (b) If she walked the outside route in 1.0 min at a constant speed, how much time in minutes would she spend on each leg?

10. ■■ The displacement vector of a moving object initially at the origin has a magnitude of 12.5 cm and is at an angle of 30° below the $-x$ axis at a particular instant. What are the coordinates of the object at that instant?

11. ■■ A ball rolls with a constant velocity of 1.50 m/s at an angle of 45° below the $+x$ axis in the 4th quadrant. Taking the ball to be at the origin at $t = 0$, what are its (x, y) coordinates 1.65 s later?

12. ■■ A ball rolling on a table has velocity with rectangular components $v_x = 0.60$ m/s and $v_y = 0.80$ m/s. What is the displacement of the ball in an interval of 2.5 s?

13. ■■ A ball has an initial velocity of 1.30 m/s along the $+y$ axis and, starting at t_o, receives an acceleration of 2.10 m/s² in the $+x$ direction. (a) What is the position of the ball 2.50 s after t_o? (b) What is the velocity of the ball at that time?

14. ■■ A small plane takes off with a constant velocity of 150 km/h at an angle of 37°. In 3.00 s, (a) how high is the plane above the ground, and (b) what horizontal distance has it traveled from the lift-off point?

15. ■■ A ball rolls diagonally across a table top from corner to corner with a constant speed of 0.75 m/s. The table top is 3.0 m wide and 4.0 m long. Another ball, starting at the same time as the first one, rolls along the long edge of the table. What constant speed must the second ball have in order to reach the same corner at the same time as the ball traveling diagonally?

16. ■■ A particle moves with a speed of 2.5 m/s in the $+x$ direction. Upon reaching the origin, it receives a continuous constant acceleration of 0.75 m/s² in the $-y$ direction. What is the position of the particle 4.0 s later?

17. ■■■ An automobile travels 800 m with a constant speed of 60 km/h along a straight highway that is inclined 15° to the horizontal. An observer notes only the vertical motion of the car. What is (a) the magnitude of the car's vertical velocity and (b) the vertical travel distance?

3.2 Vector Addition and Subtraction

18. Two vectors, of magnitudes 3 and 4, respectively, are added. The magnitude of the resultant vector is (a) 1, (b) 7, (c) between 1 and 7.

19. In Exercise 18, under what condition will the magnitude of the resultant equal 1? How about 7?

20. The resultant of $\mathbf{A} - \mathbf{B}$ is the same as that of (a) $\mathbf{B} - \mathbf{A}$, (b) $-\mathbf{A} + \mathbf{B}$, (c) $-(\mathbf{A} + \mathbf{B})$, (d) $-(\mathbf{B} - \mathbf{A})$.

21. ■ Find the rectangular components of a velocity vector that has a magnitude of 10.0 m/s and is oriented at an angle of 30° above the $+x$ axis.

22. ■ Via the triangle method, show graphically that (a) $\mathbf{A} + \mathbf{B} = \mathbf{B} + \mathbf{A}$ and that (b) if $\mathbf{A} - \mathbf{B} = \mathbf{C}$, then $\mathbf{A} = \mathbf{B} + \mathbf{C}$.

23. ■ Is vector addition associative; that is, does $(\mathbf{A} + \mathbf{B}) + \mathbf{C} = \mathbf{A} + (\mathbf{B} + \mathbf{C})$? Justify your answer.

24. ■ A vector has an x component of -2.5 m and a y component of 4.2 m. What is the vector expressed in magnitude-angle form?

25. ■ In Chapter 2, the displacement between two points on the x axis was written $\Delta x = x_2 - x_1$. Show what this means graphically in terms of vector subtraction.

26. ■ For the two vectors $\mathbf{x}_1 = (20 \text{ m})\,\hat{\mathbf{x}}$ and $\mathbf{x}_2 = (15 \text{ m})\,\hat{\mathbf{x}}$, compute and show graphically (a) $\mathbf{x}_1 + \mathbf{x}_2$, (b) $\mathbf{x}_1 - \mathbf{x}_2$, and (c) $\mathbf{x}_2 - \mathbf{x}_1$.

27. ■ During a takeoff with no wind, an airplane moves with a speed of 120 mi/h at an angle of 25° above the ground. What is the ground speed of the plane?

28. ■■ A dog walks northeast for 15 m and then east for 25 m. Find the resultant (or sum) displacement vector by (a) the graphical method and (b) the component method.

29. ■■ An airplane flies northwest for 250 mi and then west for 150 mi. Find the resultant (or sum) displacement vector by (a) the graphical method and (b) the component method.

30. ■■ Two boys are pulling a box across a horizontal floor as shown in •Fig 3.24. If $F_1 = 50.0$ N and $F_2 = 100$ N, find the resultant (or sum) force by (a) the graphical method and (b) the component method.

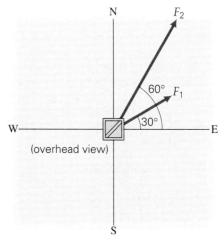

•**FIGURE 3.24 Adding force vectors** See Exercises 30 and 47.

31. ■■ For each of the following vectors, give another vector that, when added to the first, yields a *null vector* (a vector with a magnitude of zero). Express the vector in the alternate component or magnitude-angle form: (a) **A** = 4.5 cm at an angle of 40° above the +x axis, (b) **B** = (2.0 cm) $\hat{x}$ − (4.0 cm) $\hat{y}$, (c) **C** = 8.0 cm at an angle of 60° above the −x axis.

32. ■■ Find the resultant (or sum) of the three vectors **A** = (−4.0 m) $\hat{x}$ + (2.0 m) $\hat{y}$, **B** = (6.0 m) $\hat{x}$ + (3.5 m) $\hat{y}$, and **C** = (−5.5 m) $\hat{y}$.

33. ■■ (a) Find the resultant (or sum) of the force vectors in •Fig. 3.25. (b) If **F**$_1$ in the figure were at an angle of 27° instead of 37° with the +x axis, what would be the resultant (or sum) of **A** and **B**?

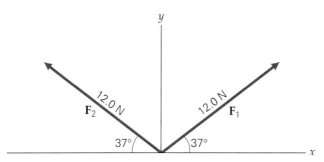

•**FIGURE 3.25 Vector addition** See Exercises 33 and 34.

34. ■■ For the vectors in Fig. 3.25, determine (a) **F** = **F**$_1$ − **F**$_2$ and (b) **F** = **F**$_2$ − **F**$_1$. Do you think that the resultants will be the same? That is, does **F**$_1$ − **F**$_2$ = **F**$_2$ − **F**$_1$? (c) Is the following a true relationship: **F**$_1$ − **F**$_2$ = −(**F**$_2$ − **F**$_1$)?

35. ■■ For the vectors shown in •Fig. 3.26, determine **A** + **B** + **C**.

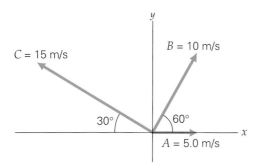

•**FIGURE 3.26 Adding velocity vectors** See Exercises 35, 36, and 37.

36. ■■ For the vectors shown in Fig. 3.26, determine **A** − **B** − **C**.

37. ■■ For the vectors shown in Fig. 3.26, determine **A** + **B** − **C**.

38. ■■ Two force vectors, **F**$_1$ = (3.0 N) $\hat{x}$ − (3.0 N) $\hat{y}$ and **F**$_2$ = (−6.0 N) $\hat{x}$ + (4.5 N) $\hat{y}$, are applied to a particle. What third force **F**$_3$ would make the net or resultant force on the particle zero?

39. ■■ Two force vectors, **F**$_1$ = 8.0 N at an angle of 60° above the +x axis and **F**$_2$ = 5.5 N at an angle of 45° below the +x axis, are applied to a particle at the origin. What third force **F**$_3$ would make the net or resultant force on the particle zero?

40. ■■ A student works three problems involving the addition of different vectors **F**$_1$ and **F**$_2$. He states that the magnitudes of the three resultants are given by (a) $F_1 + F_2$, (b) $F_1 − F_2$, (c) $\sqrt{F_1^2 + F_2^2}$. Is this possible? If so, describe the vectors in each case.

41. ■■ A block weighing 50 N rests on an inclined plane. Its weight is a force directed vertically downward, as illustrated in •Fig. 3.27. Find the components of the force along the surface of the plane and perpendicular to it.

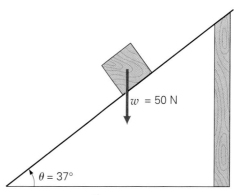

•**FIGURE 3.27 Block on an inclined plane** See Exercise 41.

42. ■■■ An amusement park roller coaster starts out on a level track 50.0 m long and then goes up a 25.0-meter incline at an angle of 30° to the horizontal. It then goes down a 15.0-meter ramp with an incline of 40° to the horizontal. When the roller coaster has reached the bottom of the ramp, what is its displacement from its starting point?

43. ■■■ A person walks from point A to point B as shown in •Fig. 3.28. What is the person's displacement relative to A?

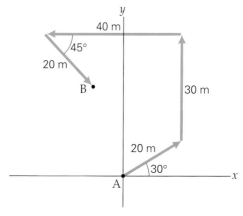

•**FIGURE 3.28 Adding displacement vectors** See Exercise 43.

44. ■■■ A meteorologist tracks the movement of a cloud with Doppler radar. At 8:00 P.M., the cloud was 60 mi northeast of her station. At 10:00 P.M., the cloud is at 75 mi north. What is the average velocity of the cloud?

45. ■■■ A flight controller determines that an airplane is 20.0 mi south of him. Half an hour later the same plane is 35.0 mi northwest of him. If the plane is flying with a constant velocity, what is its velocity during this time?

46. ■■■ A ship was seen on a radar screen to be 10 km east of a particular location. Some time later, the ship is at 15 km north of west. What is the displacement of the ship?

47. ■■■ Two students pull on a box at rest as shown in Fig. 3.24. If $F_1 = 100$ N and $F_2 = 150$ N and a third student wants to stop the box from moving initially, what force should he apply?

3.3 Relative Velocity

48. A student walks on a treadmill moving at 4.0 m/s and remains at the same place in the gym. (a) What is the student's velocity relative to the gym floor? (b) What is the student's speed relative to the treadmill?

49. You are walking along a straight sidewalk to your dorm in the rain. (Your umbrella is in your room, of course.) A wind is blowing horizontally in the direction of your motion. Which part of your body (face or back) will stay dry if you walk (a) slower than the wind? (b) as fast as the wind? (c) faster than the wind?

50. ■ While you are traveling in a car on a straight, level interstate highway at 90 km/h, another car passes you in the same direction; its speedometer reads 120 km/h. (a) What is your velocity relative to the other driver? (b) What is the other car's velocity relative to you?

51. ■ A shopper is in a hurry to catch a bargain in a department store. She walks up the escalator, rather than letting it carry her, with a speed of 1.0 m/s relative to the escalator. If the escalator is 20 m long and moves with a speed of 0.50 m/s, how long does it take the shopper to get to the next floor?

52. ■ A person riding in the back of a pickup truck traveling at 70 km/h on a straight, level road throws a ball with a speed of 15 km/h relative to the truck in the direction opposite its motion. What is the velocity of the ball (a) relative to a stationary observer by the side of the road? (b) relative to the driver of a car moving in the same direction as the truck at a speed of 90 km/h?

53. ■ In Exercise 52, what are the relative velocities if the ball is thrown in the direction of the truck?

54. ■ A boat can travel with a speed of 5.0 m/s relative to still water. If the boat is heading upstream in a river with a current of 3.0 m/s, what distance does the boat travel in 30 s?

55. ■■ In a 500-meter stretch of a river, the speed of the current is a steady 5.0 m/s. How long does it take in minutes for a boat to finish one round trip (upstream and downstream) if the speed of the boat is 7.5 m/s relative to still water?

56. ■■ A passenger on a train traveling 90 km/h on a straight, level track notes that it takes another train (180 m long and moving with constant velocity) 4.0 s to pass by. What is the velocity of that passing train relative to the ground?

57. ■■ A moving walkway in an airport is 75 m long and moves with a speed of 0.30 m/s. A passenger, after traveling 25 m while standing on the walkway, starts to walk with a speed of 0.50 m/s relative to the walkway surface. How long does it take her to travel the total walkway distance?

58. ■■ A swimmer swims north across a river that flows at 0.20 m/s from west to east. If the speed of the swimmer is 0.15 m/s relative to still water, what is the swimmer's velocity relative to the river bank?

59. ■■ A swimmer maintains a speed of 0.15 m/s relative to the water when he swims directly toward the opposite shore of a river. The river is straight and has a current that flows at 0.75 m/s. (a) How far downstream is he carried in 1.5 min? (b) What is his velocity relative to an observer on shore?

60. ■■ Swimming at 0.15 m/s relative to still water, a swimmer heads directly across a river of width 100 m. He arrives 50 m downstream from the point directly across the river from his starting point. (a) What is the speed of the current in the river? (b) What direction should he head so as to arrive at a point directly opposite his starting point?

61. ■■ It is raining and there is no wind. (a) When you are sitting in a stationary car, the rain falls straight down relative to the car and the ground. But when you're driving, the rain appears to hit the windshield at an angle, which increases as the velocity of the car increases. Explain this effect in terms of velocity components. [Hint: Consider the velocity of the rain relative to that of the car.] (b) If the raindrops fall straight down with a speed of 10 m/s but appear to make an angle of 25° to the vertical, what is the speed of the car?

62. ■■ A boat that travels with a speed of 6.75 m/s in still water is to go directly across a river and back (•Fig. 3.29). The current flows at 0.50 m/s. (a) At what angle(s) must the boat be steered? (b) How long does it take to make a round trip? (Assume that the boat's speed is constant at all times, and neglect turnaround time.)

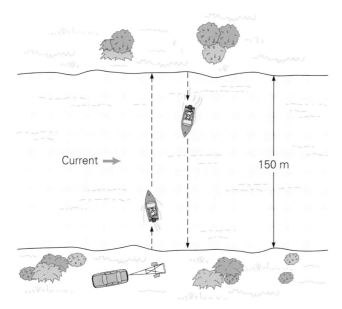

• **FIGURE 3.29 A boat race** See Exercise 62. (Not drawn to scale.)

63. ■■■ Two cars travel on level streets that intersect each other at right angles. Car A approaches the intersection from the east with a speed of 30 km/h. Car B has just passed through the intersection and is traveling 25 km/h northward. (a) What is the velocity of car B relative to car A? (b) the velocity of car A relative to car B?

3.4 Projectile Motion*

64. If air resistance is neglected, the motion of an object projected at an angle consists of a uniform downward acceleration combined with (a) an equal horizontal acceleration, (b) a uniform horizontal velocity, (c) a constant upward velocity, (d) an acceleration always perpendicular to the path of motion.

65. Using a sketch of components, show that the velocity of a projectile at the same elevation as the launch point ($y_{\text{final}} = y_{\text{initial}}$) has the same magnitude as the initial velocity and the same size directional angle but is directed below the horizontal.

66. Figure 3.16b shows the multiflash photographs of one ball dropping from rest and at the same time, another ball projected horizontally at the same height. The two balls hit the ground at the same time. Why? Explain.

67. In •Fig. 3.30, a spring-loaded "cannon" on a wheeled car can fire a metal ball vertically. The car is given a push and set in motion horizontally with constant velocity. A pin is pulled with a string to launch the ball, which travels upward and then falls back into the moving cannon every time. Why? Explain.

*Assume angles to be exact for significant figures.

• **FIGURE 3.30 A ballistics car** See Exercises 67 and 77.

68. ■ A ball with a horizontal speed of 1.5 m/s rolls off a bench 2.0 m high. (a) How long will it take the ball to reach the floor? (b) How far from a point on the floor directly below the edge of the bench will the ball land?

69. ■ An electron is ejected horizontally, with a speed of 1.5×10^6 m/s, from the electron gun of a TV set. If the viewing screen is 35 cm away from the end of the gun, how far will the electron fall or travel in the vertical direction before hitting the screen?

70. ■ A ball is thrown horizontally, with a speed of 15 m/s, from the top of a 6.0-meter-tall hill. How far from the point on the ground directly below the launch point does the ball strike the ground?

71. ■ If Exercise 70 were taking place on the surface of the Moon, where the acceleration due to gravity is only 1.67 m/s², what would be the answer?

72. ■ A ball rolls horizontally off the edge of a 6.5-meter-tall platform. If the ball lands 8.7 m from the point on the ground directly below the edge of the platform, what is the speed of the ball when it rolls off the platform?

73. ■ A golf ball is hit with a speed of 30 m/s at an angle of 30° above the horizontal. What are the horizontal and vertical components of the ball's velocity?

74. ■■ A pitcher throws a fastball horizontally with a speed of 140 km/h toward home plate 18.4 m away. (a) If the batter's combined reaction and swing times total 0.350 s, how long before swinging can the batter watch the ball after it has left the pitcher's hand? (A major reason why some players illegally use hollowed-out bats with cork or superball filling is to make the bat lighter, which decreases the swing time and allows more time to follow the ball before deciding to swing.) (b) In traveling to the plate, how far does the ball drop from its original horizontal line?

75. ■■ One ball rolls with a constant speed of 0.25 m/s on a table 0.95 m above a floor, and another ball rolls on the floor directly under the first ball and with the same speed and direction. Will the balls collide after the first ball rolls off the table? If so, how far from the point directly below

the edge of the table will they be when they strike each other?

76. ■■ An airplane must drop a package of supplies so that it hits the ground at a designated spot near some campers. The airplane, moving horizontally with a constant velocity of 140 km/h, approaches the spot at an altitude of 0.500 km above level ground. Having the designated point in sight, the pilot prepares to drop the package. (a) What should the angle between the horizontal and the pilot's line of sight be when the package is released? (b) What is the location of the plane when the package hits the ground?

77. ■■ A wheeled car with a spring-loaded cannon fires a metal ball vertically (Fig. 3.30). If the initial vertical speed of the ball is 5.0 m/s as the cannon moves horizontally with a speed of 0.75 m/s, (a) how far from the launch point does the ball fall back into the cannon? (b) What would happen if the cannon were accelerating?

78. ■■ ● A soccer player kicks a stationary ball, giving it a speed of 20.0 m/s at an angle of 15.0° to the horizontal. (a) What is the maximum height reached by the ball? (b) What is its range? (c) How could the range be increased?

79. ■■ ● A rifle fires a bullet with a speed of 250 m/s at an angle of 37° above the horizontal. (a) Relative to the launch height, what height does the bullet reach? (b) How long is the bullet in the air? (c) What is its horizontal range?

80. ■■ A golf ball, hit off a tee, has an initial speed of 30.0 m/s at an angle of 37° to the horizontal. At what other time would the golf ball have the same vertical height as it has at $t = 3.00$ s? [Hint: See Example 3.9.]

81. ■■ An arrow has an initial launch speed of 18 m/s. If it must strike a target 31 m away at the same elevation, what should be the projection angle?

82. ■■ An astronaut on the Moon fires a projectile from a launcher on a level surface so as to get the maximum range. If the launcher gives the projectile a muzzle velocity of 25 m/s, what is the range of the projectile? [Hint: Don't forget where the launcher is and gravitational effects. The acceleration due to gravity on the Moon is only one-sixth of that on the Earth.]

83. ■■ A stone thrown off a bridge 20 m above a river has an initial velocity of 12 m/s at an angle of 45° above the horizontal (•Fig. 3.31). (a) What is the range of the stone? (b) At what velocity does the stone strike the water?

84. ■■ William Tell is said to have shot an apple off his son's head with an arrow. If the arrow was shot with an initial speed of 55 m/s and the boy was 15 m away, at what launch angle did Bill aim the arrow? (Assume that the arrow and apple are initially at the same height above the ground.)

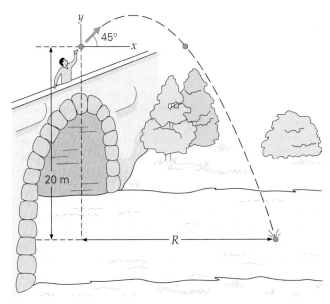

•FIGURE 3.31 **A view from the bridge** See Exercise 83.

85. ■■■ This time William Tell is shooting at an apple that hangs on a tree (•Fig. 3.32). The apple is a horizontal distance of 20.0 m away and at a height of 4.00 m above the ground. If the arrow is released from a height of 1.00 m above the ground and hits the apple 0.500 s later, what is its initial velocity?

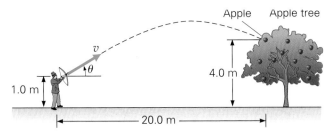

•FIGURE 3.32 **Hit the apple** See Exercise 85. (Not drawn to scale.)

86. ■■■ A ditch 2.5 m wide crosses a trailbike path (•Fig. 3.33). An upward incline of 15° has been built up on the approach so that the top of the incline is level with the top of the other side of the ditch. With what minimum speed must a trailbike be moving to clear the ditch? (Add 1.4 m to the range for the back of the bike to clear the ditch safely.)

•FIGURE 3.33 **Clear the ditch?** See Exercise 86. (Not drawn to scale.)

87. ■■■ A quarterback passes a football with a speed of 50 ft/s at an angle of 40° to the horizontal toward an intended receiver 30 yd downfield. The pass is released 5.0 ft above the ground. Assume that the receiver is stationary and that he will catch the ball if it comes to him. Will the pass be completed? If not, will the throw be long or short?

88. ■■■ Three identical balls are thrown off a cliff at a height (h) above a horizontal plain. One ball is projected at an angle of 45° above the horizontal, the second is thrown horizontally, and the third is thrown at an angle of 30° below the horizontal (downward). If all of the balls have the same initial speed, which one will have the greatest speed when it hits the ground? (Justify your answer mathematically.)

89. ■■■ The hole on a level, elevated golf green is a horizontal distance of 150 m from the tee and at an elevation of 12.0 m above the tee. A golfer hits a ball at an angle 10.0° greater than that of the hole above the tee and makes a hole-in-one! (a) What was the initial speed of the ball? (b) Suppose the next golfer hits her ball toward the hole with the same speed but at an angle of 10.5° greater than that of the hole above the tee. Does the ball go in the hole, or is the shot too long or too short?

Additional Exercises

90. The apparatus for a popular lecture demonstration is shown in •Fig. 3.34. The gun is aimed directly at the can, which is simultaneously released when the gun is fired. This gun won't miss as long as the initial speed of the bullet is sufficient to reach the falling target before the target hits the floor. Verify this, using the figure. [Hint: Note that $y_o = x \tan \theta$.]

•FIGURE 3.34 **A sure shot** See Exercise 90. (Not drawn to scale.)

91. What is the resultant velocity vector of the following three velocity vectors: $\mathbf{v}_1 = (3.0 \text{ m/s})\hat{\mathbf{x}} + (4.0 \text{ m/s})\hat{\mathbf{y}}$; $\mathbf{v}_2 = (-4.0 \text{ m/s})\hat{\mathbf{x}} + (5.0 \text{ m/s})\hat{\mathbf{y}}$; $\mathbf{v}_3 = (5.0 \text{ m/s})\hat{\mathbf{x}} - (7.0 \text{ m/s})\hat{\mathbf{y}}$? Express your result in both component and magnitude-angle forms.

92. ● A motorboat travels with a speed of 40.0 km/h in a straight path on a still lake. Suddenly, a strong, steady wind pushes the boat with a speed of 15.0 km/h perpendicularly to its straight-line path for 5.00 s. Relative to its position just when the wind started to blow, where is the boat located at the end of this time?

93. A golf ball lies 2.00 m directly south of the hole on a level green. On the first putt, the ball travels 3.00 m along a straight-line path at an angle of 5° east of north; on the second putt, it travels a straight-line distance of 1.20 m at an angle of 6° south of west. What would be the displacement of a third putt that would put the ball in the hole?

94. If the flow rate of the current in a straight river is greater than the speed of a boat in the water, the boat cannot make a trip *directly across* the river. Prove this statement.

95. A student throws a softball horizontally from a dorm window 15.0 m above the ground. Another student standing 10.0 m away from the dorm catches the ball at a height of 1.50 m above the ground. What is the initial speed of the ball?

96. ● A javelin is thrown at angles of 35° and 60°, respectively, to the horizontal from the same height and with the same speed in each case. For which throw does the javelin go farther, and how many times farther? (Assume that the landing place is at the same height as the launching.)

97. A firefighter holds the nozzle of a hose a horizontal distance of 25.0 m from a flaming building. If the speed of the water coming from the nozzle is 20.0 m/s, (a) show that the stream will not reach a third-story window 11.0 m above the nozzle level for a projection angle of 45°. (b) A greater height can be achieved by directing the nozzle at a greater angle. Could the window be reached with a stream projection angle of 50°?

98. In a movie, a monster climbs to the top of a building 30 m above the ground and hurls a boulder downward with a speed of 25 m/s at an angle of 45° below the horizontal. How far from the base of the building does the boulder land?

99. The shells fired from an artillery piece have a muzzle speed of 150 m/s, and the target is at a horizontal distance of 2.00 km. (a) At what angle relative to the horizontal should the gun be aimed? (b) Could the gun hit a target 3.00 km away?

100. A ball moving with a speed of 2.5 m/s along the $+y$ axis experiences an acceleration of 0.45 m/s² in the $+x$ direction when the ball is at $(x, y) = (0, 1.0 \text{ m})$. What are (a) the velocity components and (b) the position of the ball 4.0 s after the acceleration is applied?

101. A ball is thrown horizontally from the top of a building at a height of 32.5 m above the ground and hits the level ground 56.0 m from the base of the building. (a) What is the initial speed of the ball? (b) What is the velocity of the ball just before it hits the ground?

102. A field goal is attempted when the football is at the center of the field 40 yd from the goalposts. If the kicker gives the ball a velocity of 70 ft/s toward the goalposts at an angle of 45° to the horizontal, will the kick be good? (The crossbar of the goalposts is 10 ft above the ground, and the ball must be higher than this when it reaches the goalposts for the field goal to be good.)

103. (a) An object moving with a speed of 10 m/s in a straight line has a velocity with an x component of $+6.0$ m/s. In what direction is the object moving? (b) What is the y component of the velocity?

104. At a track and field meet, the best long jump is measured as 8.20 m. The jumper took off at an angle of 37° to the horizontal. (a) What was the jumper's initial speed? (b) If there were another meet on the Moon and the same jumper could attain only half of the initial speed he had on the Earth, what would be the maximum jump there? (Air resistance does not have to be neglected in part (b). Why?)

105. An airplane with a speed of 150 km/h heads directly north while a west wind blows with a constant speed of 30 km/h. How far does the plane travel during the 4.0 h of flight?

Force and Motion

CHAPTER 4

4.1 The Concepts of Force and Net Force
4.2 Inertia and Newton's First Law of Motion
4.3 Newton's Second Law of Motion
4.4 Newton's Third Law of Motion
4.5 More on Newton's Laws: Free-body Diagrams and Translational Equilibrium
4.6 Friction

Insight

- Action and Reaction: The Harrier "Jump" Jet

You don't have to understand any physics to know what's needed to get the car (or anything else) moving: a push or a pull. If the frustrated motorist (or the tow truck that he will soon call) can apply enough *force*, the car will move. But what's keeping the car stuck in the snow? A car's engine can generate plenty of force—so why doesn't he just put the car into reverse and back out? For a car to move, another force is needed besides that exerted by the engine: *friction*. Here, the problem is most likely that there is not enough friction between the tires and the snow.

In Chapters 2 and 3 we learned how to analyze motion in terms of kinematics. Now we need to know more about the *dynamics* of motion—that is, what causes motion and changes in motion. This inquiry leads us to the concept of force.

The study of force and motion occupied many early scientists. It was the English scientist Isaac Newton (1642–1727) (•Fig. 4.1) who summarized the various relationships and principles of those early scientists into three statements, or laws, which not surprisingly are known as Newton's laws of motion. These laws sum up the concepts of dynamics. In this chapter, you'll learn what Newton had to say about force and motion.

99

4.1 The Concepts of Force and Net Force

Objectives: To (a) relate force and motion and (b) explain what is meant by a net or unbalanced force.

Let's first take a closer look at the meaning of force. It is easy to give examples of forces, but how would you generally define this concept? An operational definition of force is based on observed effects. That is, a force is described in terms of what it does. From your own experience, you know that *forces can produce changes in motion*. A force can set a stationary object into motion. It can also speed up or slow down a moving object, or change the direction of its motion. In other words, a force can produce a *change* in velocity (speed and/or direction)—that is, an acceleration. Therefore, an observed change in motion, including motion starting from rest, is evidence of a force. This leads to a common definition of **force**:

A force is something capable of changing an object's state of motion (its velocity).

The word "capable" is very significant here. It takes into account the fact that a force may be acting on an object, but its capability to produce a change in motion may be balanced or canceled by one or more other forces. The net effect is then zero. Thus, a force may not *necessarily* produce a change in motion. However, it follows that if a force acts *alone*, the object on which it acts *will* accelerate.

Since a force can produce an acceleration—a vector quantity—force must itself be a vector quantity, with both magnitude and direction. When several forces act on an object, you will often be interested in their combined effect—the unbalanced or net force. The **unbalanced** or **net force** is the vector sum, $\Sigma_i \mathbf{F}_i$, or resultant, of all the forces acting on an object or system. Consider the opposite forces illustrated in •Fig. 4.2a. The net force is zero when forces of equal magnitude act in opposite directions (Fig. 4.2b). Such forces are said to be *balanced forces*. A nonzero net force is referred to as an *unbalanced force* (Fig. 4.2c). In this case, *the situation can be analyzed as though only one force equal to the net force were acting*. An unbalanced or net force produces an acceleration. In some instances, an applied

•**FIGURE 4.1 Isaac Newton**
Newton (1642–1727), one of the greatest scientific minds of all time, made fundamental contributions to mathematics, astronomy, and several branches of physics, including optics and mechanics. He formulated the laws of motion and universal gravitation (Chapter 7) and was one of the inventors of calculus. He did some of his most profound work when he was in his middle twenties.

Note: In the notation $\Sigma_i \mathbf{F}_i$, the Greek letter sigma means the "sum of" the individual forces, as indicated by the *i* subscripts: $\Sigma_i \mathbf{F}_i = \mathbf{F}_1 + \mathbf{F}_2 + \mathbf{F}_3 + \cdots$— a vector sum. The *i* subscripts are sometimes omitted as being understood, and we write $\Sigma \mathbf{F}$.

(a)

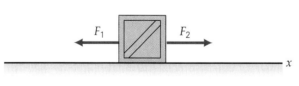

(b) Zero net force (balanced forces)

(c) Nonzero net force (unbalanced forces)

$F_{net} = F_2 - F_1 = 0$

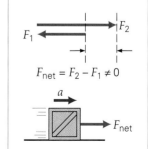

$F_{net} = F_2 - F_1 \neq 0$

•**FIGURE 4.2 Net force**
(a) Opposite forces are applied to a crate. (b) If the forces are of equal magnitude, the vector resultant, or the net force acting on the crate in the *x* direction, is zero. The forces acting on the crate are said to be balanced. (c) If the forces are unequal in magnitude, the resultant is not zero. A nonzero net force, or an unbalanced force, then acts on the crate, producing an acceleration (for example, setting it into motion if it was initially at rest).

unbalanced force may also deform an object, that is, change its size and/or shape (as we shall see in Chapter 9). A deformation involves a change in motion for some part of an object, hence there is an acceleration.

Forces are sometimes divided into two types or classes. The more familiar of these are *contact forces*. Such forces arise because of physical contact between objects. For example, when you push on a door to open it or throw or kick a ball, you exert a contact force on the door or ball.

The other class of forces is called *action-at-a-distance forces*. Examples of these forces are gravity, the electrical force between two charges, and the magnetic force between two magnets. The Moon is attracted to Earth and maintained in orbit by a gravitational force, but there seems to be nothing physically transmitting that force. In Chapter 30, you will learn the modern view of how such action-at-a-distance forces are thought to be transmitted.

Now, with our better understanding of the concept of force, let's see how force and motion are related through Newton's laws.

4.2 Inertia and Newton's First Law of Motion

Objectives: To (a) state and explain Newton's first law of motion and (b) describe inertia and its relationship to mass.

The groundwork for Newton's first law of motion was laid by Galileo. In his experimental investigations, Galileo dropped objects to observe motion under the influence of gravity. However, the relatively large acceleration due to gravity causes dropped objects to move quite fast and quite far in a short time. From the kinematic equations in Chapter 2, you can see that 3.0 s after being dropped (if we neglect air resistance), an object in free fall has a speed of about 29 m/s (64 mi/h) and has fallen a distance of 44 m (about 48 yd, or almost half the length of a football field). Thus, experimental measurements of free-fall distance versus time were particularly difficult to make with the instrumentation available in Galileo's time.

To slow things down so that he could study motion, Galileo used balls rolling on inclined planes. He allowed a ball to roll down one inclined plane and then up another with a different degree of incline (•Fig. 4.3). Galileo noted that the ball rolled to approximately the same height in every case, but it rolled farther in the horizontal direction when the angle of this incline was smaller. When allowed to roll onto a horizontal surface, the ball traveled a considerable distance and went even farther when the surface was made smoother. Galileo wondered how far the ball would travel if the horizontal surface could be made perfectly smooth (frictionless). Although this situation was impossible to attain experimentally, Galileo reasoned that in this ideal case with an infinitely long surface, the ball would continue to travel (slide) indefinitely with straight-line, uniform motion, since there would be nothing (no force) to cause its motion to change.

According to Aristotle's theory of motion, which had been accepted for about 1500 years prior to Galileo's time, the normal state of a body was to be at rest (with the exception of celestial bodies, which were naturally in motion). Aristotle probably observed that objects moving on a surface tend to slow down and come to rest, so this conclusion would have seemed logical to him. However, from his experiments, Galileo concluded that bodies in motion exhibit the behavior of

•FIGURE 4.3 **Galileo's experiment** A ball rolls farther along the upward incline as the angle of incline is decreased. On a smooth, horizontal surface, the ball rolls a greater distance before coming to rest. How far would the ball travel on an ideal, perfectly smooth surface?

maintaining that motion, and that if an object is initially at rest, it will remain so, unless something causes it to move.

Galileo called this tendency of an object to maintain its initial state of motion **inertia**. That is,

Definition of: Inertia

Inertia is the natural tendency of an object to maintain a state of rest or to remain in uniform motion in a straight line (constant velocity).

For example, if you've ever tried to stop a slowly rolling automobile by pushing on it, you felt its resistance to a change in motion, to slowing down. Physicists describe the property of inertia in terms of observed behavior, as they do for all physical phenomena. A comparative example of inertia is illustrated in •Fig. 4.4. If two punching bags have the same density (mass per unit volume; see Chapter 1), the large one will have more mass and more inertia, as you would quickly notice when you tried to punch them.

Newton related the concept of inertia to mass. Originally, he called mass a quantity of matter, but he later redefined it as follows:

Relationship of mass and inertia

Mass is a measure of inertia.

That is, a massive object has more inertia, or more resistance to a change in motion, than a less massive object does. For example, a car has more inertia than a bicycle.

Newton's first law—the law of inertia

Newton's first law of motion, sometimes called the *law of inertia*, summarizes these observations:

In the absence of an unbalanced applied force, a body at rest remains at rest and a body already in motion remains in motion with a constant velocity (constant speed and direction).

That is, if the net force acting on an object is zero, then its acceleration is zero. (See Demonstration 1.)

4.3 Newton's Second Law of Motion

Objectives: To (a) state and explain Newton's second law of motion, (b) apply it to physical situations, and (c) distinguish between weight and mass.

A change in motion, or an acceleration (change in speed and/or direction), is evidence of a force. All experiments indicate that an acceleration is directly proportional to, and in the direction of, the applied net force; that is,

$$\mathbf{a} \propto \mathbf{F}_{net}$$

where the boldface symbols indicate vector quantities. For example, if you hit a ball twice as hard (applied twice as much force), you would expect the acceleration of the ball to be twice as great (but still in the direction of the force).

However, as Newton recognized, the inertia or mass of the object also plays a role. For a given force, the more massive the object, the less its acceleration will be. That is, the acceleration and mass (m) of an object are inversely proportional:

$$a \propto \frac{1}{m}$$

For example, if you hit two different balls with the same force, the less massive ball would acquire a greater acceleration.

Then combining these relationships, we have,

$$\mathbf{a} \propto \frac{\mathbf{F}_{net}}{m}$$

or in words,

•FIGURE 4.4 **A difference in inertia** The larger punching bag has more mass and hence more inertia or resistance to a change in motion.

Demonstration 1 ■ Newton's First Law and Inertia

According to Newton's first law, an object remains at rest or in motion with a constant velocity unless acted on by an unbalanced force.

(a) A pen is at rest on an embroidery hoop on top of a bottle.

(b) The hoop is struck sharply and accelerates horizontally. Because the friction between the pen and hoop is small and acts only for a short time, the pen does not move appreciably in the horizontal direction. However, there is now an unbalanced force acting vertically on it—gravity.

(c) The pen falls into the bottle.

The acceleration of an object is directly proportional to the *net* force acting on it and inversely proportional to its mass. The direction of the acceleration is in the direction of the applied net force.

Newton's second law—cause and effect

• Figure 4.5 presents some illustrations of this principle.

The SI unit of force is, appropriately, called the **newton** (N). With $\mathbf{F}_{net} \propto m\mathbf{a}$, in equation form **Newton's second law of motion** is commonly expressed as follows:

The newton (N), unit of force

(a) A net force accelerates the crate: $a \propto F/m$

(b) If the net force is doubled, the acceleration is doubled.

(c) If the mass is doubled, the acceleration is halved.

• **FIGURE 4.5 Newton's second law** The relationships among force, acceleration, and mass shown here are expressed by Newton's second law of motion (assuming no friction).

$$\mathbf{F} = m\mathbf{a} \quad \text{Newton's second law} \quad (4.1)$$

SI unit of force: newton (N) or
kilogram-meter per second squared (kg · m/s²)

*(Note: The net force is commonly written as just **F** for convenience, with the understanding that this is the vector sum $\mathbf{F} = \Sigma_i \mathbf{F}_i = \mathbf{F}_{net}$. This simpler notation will generally be used hereafter; however, you may wish to write $\mathbf{F}_{net}$, or $\Sigma_i \mathbf{F}_i$, explicitly as a reminder.)*

Equation 4.1 shows that (by unit analysis) a newton in base units is defined as $1\text{ N} \equiv 1$ kg·m/s². That is, a net force of 1 N gives a mass of 1 kg an acceleration of 1 m/s² (•Fig. 4.6). The British-system unit of force is the pound (lb). One pound is equivalent to about 4.5 N (actually, 4.448 N). An average apple weighs about 1 N.

Thus, if the net force acting on an object is zero its acceleration is zero and it remains at rest or in uniform motion, which is consistent with the first law. For a nonzero net force (an unbalanced force), the resulting acceleration is in the same direction as the force.*

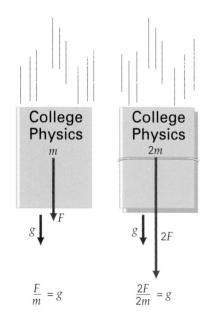

•**FIGURE 4.6 The newton (N)** A net force of 1.0 N acting on a mass of 1.0 kg produces an acceleration of 1.0 m/s² (on a frictionless surface).

Weight

Equation 4.1 can be used to relate mass and weight. Recall from Chapter 1 that weight is the gravitational force of attraction that a celestial body exerts on an object. For us, this is the gravitational attraction of Earth. Its effects are easily demonstrated: When you drop an object, it falls (accelerates) toward Earth. Since there is only one force acting on the object, its **weight** (**w**) is the net force **F**, and the acceleration due to gravity (**g**) can be substituted for a in Eq. 4.1. We can therefore write, in terms of magnitude,

$$w = mg \quad (4.2)$$

$$(F = ma)$$

The weight of 1.0 kg of mass is then $w = mg = (1.0\text{ kg})(9.8\text{ m/s}^2) = 9.8$ N.

Thus, 1.0 kg of mass has a weight of approximately 9.8 N, or 2.2 lb. But, although weight and mass are simply related through Eq. 4.2, keep in mind that *mass is the fundamental property*. Mass doesn't depend on the value of g, but weight does. As pointed out previously, the acceleration due to gravity on the Moon is about $\frac{1}{6}$ that on Earth. The weight of an object on the Moon would thus be one-sixth of its weight on Earth; but its mass, which reflects the quantity of matter it contains, would be the same in both places.

Newton's second law also shows why all objects in free fall have the same acceleration. Consider, for example, two falling objects, one with twice the mass of the other. The object with twice as much mass would have twice as much weight, or two times as much gravitational force acting on it. But the more massive object also has twice the inertia, so twice as much force is needed to give it the same acceleration. Expressing this relationship mathematically, for the smaller mass (m) we can write $F/m = g$, and for the larger mass ($2m$), we have the same acceleration: $2F/2m = g$ (•Fig. 4.7).

Newton's second law allows you to analyze dynamic situations. In using this equation, you should keep in mind that F is the *magnitude of the net force* and m

•**FIGURE 4.7 Newton's second law and free fall** In free fall, all objects fall with the same constant acceleration g. An object with twice the mass of another has twice as much gravitational force acting on it. But with twice the mass, it also has twice as much inertia, so twice as much force is needed to give it the same acceleration.

---

*It may appear that Newton's first law is a special case of the second law, but that is not so. The first law *defines* what is called an *inertial reference system* (as we shall see in Chapter 26): one on which there is no net force or that is not accelerating, or one in which an isolated object is stationary or moves with a constant velocity. If Newton's first law holds, then the second law in the form $\mathbf{F} = m\mathbf{a}$ applies to the system.

is the *total mass of the system*. The boundaries defining a system may be real or imaginary. For example, a system might consist of all the gas molecules in a particular sealed vessel. But you might also define a system to be all the gas molecules in an arbitrary cubic meter of air. In studying dynamics we often have occasion to work with systems made up of one or more discrete masses: the Earth and Moon, for instance, or a series of blocks on a table top, or a tractor and wagon, as will be seen in the following Example.

In $F = ma$, m is the *total* mass of the system.

Example 4.1 ■ Newton's Second Law: Finding Acceleration

A tractor pulls a loaded wagon with a constant force of 440 N (•Fig. 4.8). If the total mass of the wagon and its contents is 275 kg, what is the wagon's acceleration? (Ignore any frictional forces.)

Thinking It Through. This is a direct application of Newton's second law. Note that the *total* mass is given; we treat the two separate masses as one.

Solution. Listing the data, we have

Given: $F = 440$ N   Find: a (acceleration)
$m = 275$ kg

In this case, F is the net force and the acceleration is given by Eq. 4.1, $F = ma$. Solving for a,

$$a = \frac{F}{m} = \frac{440 \text{ N}}{275 \text{ kg}} = 1.60 \text{ m/s}^2$$

in the direction that the tractor is pulling.

Note that m was the *total* mass of the wagon and its contents. If the masses of wagon and contents had been given separately—say, $m_1 = 75$ kg and $m_2 = 200$ kg, respectively—they would be added together in Newton's law, $F = ma = (m_1 + m_2)a$.

In reality there would be an opposing force of friction. Suppose there were an effective frictional force of $f = 140$ N. In this case, the net force would be the vector sum of the force exerted by the tractor and the frictional force, and the acceleration would be

$$a = \frac{F - f}{m} = \frac{440 \text{ N} - 140 \text{ N}}{275 \text{ kg}} = 1.09 \text{ m/s}^2$$

in the direction that the tractor is pulling.

With a constant net force, the acceleration is also constant, so the kinematic equations of Chapter 2 can be applied. Suppose the wagon started from rest ($v_o = 0$). Could you find how far it traveled in 4.00 s? Using the appropriate kinematic equation,

$$x = v_o t + \tfrac{1}{2}at^2 = 0 + \tfrac{1}{2}(1.09 \text{ m/s}^2)(4.00 \text{ s})^2 = 8.72 \text{ m}$$

Follow-up Exercise. Suppose the applied force on the wagon is 550 N. With the same frictional force, what would be the wagon's velocity 4.0 s after starting from rest?

•FIGURE 4.8 **Force and acceleration** See Example 4.1.

4.3 Newton's Second Law of Motion

Example 4.2 ■ Newton's Second Law: Finding Mass

A student weighs 588 N. What is her mass?

Thinking It Through. Newton's second law allows us to determine a mass if we know the weight (force), since, g is known.

Solution.

Given: $w = 588$ N   *Find:* m (mass)

Recall that weight is a (gravitational) force and that Newton's second law, $F = ma$, can be written in the form $w = mg$ (Eq. 4.2), where g is the acceleration due to gravity (9.80 m/s²). Rearranging the equation, we have

$$m = \frac{w}{g} = \frac{588 \text{ N}}{9.80 \text{ m/s}^2} = 60.0 \text{ kg}$$

On the surface of Earth, this is equivalent to 60.0 kg (2.2 lb/kg) = 132 lb. In countries using the metric system, the kilogram unit of mass, rather than a force unit, is used to express "weight." It would be said that the student weighs 60.0 "kilos."

Follow-up Exercise. (a) A person in Europe is a bit overweight and would like to lose 5.0 "kilos." What would be the equivalent pound loss? (b) What is your "weight" in kilos?

A dynamic system may consist of more than one discrete mass. In applications of Newton's second law, it is often advantageous, and sometimes necessary, to isolate a given mass within a system. This is possible because any part of a system can be treated as a discrete system to which the second law can be applied, as the following Example shows.

Example 4.3 ■ Newton's Second Law: All or Part of the System

Two blocks with masses, $m_1 = 2.5$ kg and $m_2 = 3.5$ kg, rest on a frictionless surface and are connected by a light string (•Fig. 4.9). A horizontal force (F) of 12.0 N is applied to m_1 as shown in the figure. (a) What is the magnitude of the acceleration of the masses (that is, of the system)? (b) What is the magnitude of the force (T) in the string? [When a rope or string is stretched taut, it is said to be under *tension*, which is represented by

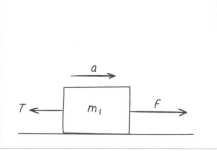

•**FIGURE 4.9 An accelerated system** See Example 4.3.

Isolating the masses

the magnitude of the force acting at any point. For a very light string, the force at the right end of the string has the same magnitude (T) as the force at the left end.]

Thinking It Through. It is important to remember that the second law may be applied to a total system or any part of it (a subsystem, so to speak). This makes for the analysis of a particular component of a system if desired. Identification of the acting forces is critical, as this Example shows. We then apply $F = ma$ to each subsystem or component.

Solution. Carefully listing the data and what we want to find, we have

Given: $m_1 = 2.5$ kg Find: (a) a (acceleration)
$m_2 = 3.5$ kg (b) T (tension, a force)
$F = 12.0$ N

[Since the surface is ideally frictionless, the weight forces of the blocks, which ordinarily affect the frictional forces (Section 4.6), can be ignored.]

Given an applied force that produces motion, the acceleration of the masses can be found from Newton's second law. In using the second law, it is important to keep in mind that it applies to the total system *or to any part of it*—that is, to the total mass $(m_1 + m_2)$, to m_1 individually, or to m_2 individually. However, *we must be sure to identify correctly the appropriate force or forces in each case*. The net force acting on the combined masses, for example, is not the same as the net force acting on m_2 considered separately.

(a) Drawing an imaginary line around m_1 and m_2, we see that the net force acting on this system is F. Representing the total mass simply as m, we can thus write

$$a = \frac{F_{net}}{m} = \frac{F}{m_1 + m_2} = \frac{12.0 \text{ N}}{2.5 \text{ kg} + 3.5 \text{ kg}} = 2.0 \text{ m/s}^2$$

The acceleration is in the direction of the applied force, as the figure indicates. Note that m is the *total* mass of the system, or all the mass that is accelerated. (The mass of the light string is small enough to be ignored.)

Note: When an object is described as "light," you can ignore its mass in analyzing the problem situation. That is, the mass is negligible relative to the other masses.

(b) Under tension a force is exerted *on* an object by flexible strings (or ropes or wires) and is directed along the string. Note in the figure that we are assuming the tension to be transmitted *undiminished* through the string. That is, the tension is the same everywhere in the string. Thus, the magnitude of **T** acting on m_2 is the same as that acting on m_1. This is actually true only if the string has zero mass. Only such idealized *light* (negligible mass) strings or ropes will be considered in this book.

So, there is a force of magnitude T on each of the masses because of tension in the connecting string. The T forces on the masses are equal in magnitude and opposite in direction. Note that the tension forces did not appear in part (a), where the total system of both masses was considered. In this case, the *internal* equal and opposite T forces cancel each other.

However, each mass may also be considered a separate system to which Newton's second law applies. In these systems, the tension comes into play explicitly. Looking at the sketch of the isolated m_2 in Fig. 4.9, we see that the only force acting to accelerate this mass is T. From the values of m_2 and a, the magnitude of this force is given directly by

$$F_{net} = T = m_2 a = (3.5 \text{ kg})(2.0 \text{ m/s}^2) = 7.0 \text{ N}$$

An isolated sketch of m_1 is also shown in Fig. 4.9, and the second law can equally well be applied to this mass to find T. We must add the forces vectorially to get the net force on m_1 that produces its acceleration. That is,

$$F_{net} = F - T = m_1 a \quad \text{(direction of } F \text{ taken as positive)}$$

Then, solving for T,

$$T = F - m_1 a$$
$$= 12.0 \text{ N} - (2.5 \text{ kg})(2.0 \text{ m/s}^2) = 12.0 \text{ N} - 5.0 \text{ N} = 7.0 \text{ N}$$

Follow-up Exercise. Suppose there were an additional horizontal force to the left of 3.0 N applied to m_2 in Fig. 4.9. What would be the tension in the connecting string in this case?

The Second Law in Component Form

Not only does Newton's second law hold for any part of a system, it also applies to the components of motion. A force may be expressed in component notation in two dimensions as follows:

$$\mathbf{F} = m\mathbf{a}$$

and

$$F_x \hat{\mathbf{x}} + F_y \hat{\mathbf{y}} = m(a_x \hat{\mathbf{x}} + a_y \hat{\mathbf{y}}) = ma_x \hat{\mathbf{x}} + ma_y \hat{\mathbf{y}} \quad (4.3a)$$

Hence we have the x and y components

$$F_x = ma_x$$
$$F_y = ma_y \quad (4.3b)$$

and Newton's second law applies separately to each component of motion. Note that *both* equations must be true. An example of how the second law is applied to components follows.

Example 4.4 ■ Newton's Second Law: Components of Force

A block of mass 0.50 kg travels with a speed of 2.0 m/s in the x direction on a flat, frictionless surface. On passing through the origin, it experiences a constant force of 3.0 N at an angle of 60° relative to the x-axis for 1.5 s (•Fig. 4.10). What is the velocity of the block at the end of this time?

Thinking It Through. With the force at an angle to the initial motion, it would appear that the solution is complicated. But note in the Fig. 4.10 insert that the force can be resolved into components. The motion can then be analyzed in each component direction.

Solution. First we write down the given data and what is to be found.

Given: $m = 0.50$ kg *Find:* $\mathbf{v}$ (velocity at the end of 1.5 s)
$v_o = v_{x_o} = 2.0$ m/s
$v_{y_o} = 0$
$F = 3.0$ N, $\theta = 60°$
$t = 1.5$ s

Let's find the magnitudes of the forces in the component (x and y) directions:

$$F_x = F\cos 60° = (3.0 \text{ N})(0.500) = 1.5 \text{ N}$$
$$F_y = F\sin 60° = (3.0 \text{ N})(0.866) = 2.6 \text{ N}$$

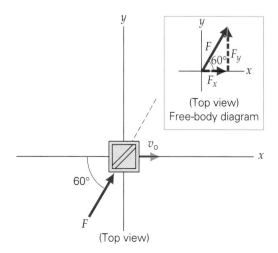

•**FIGURE 4.10 Off the straight and narrow** A force is applied to a moving block when it reaches the origin, and the block is deviated from its straight-line path. See Example 4.4.

Then, applying Newton's second law to each direction to find the acceleration components, we get

$$a_x = \frac{F_x}{m} = \frac{1.5 \text{ N}}{0.50 \text{ kg}} = 3.0 \text{ m/s}^2$$

$$a_y = \frac{F_y}{m} = \frac{2.6 \text{ N}}{0.50 \text{ kg}} = 5.2 \text{ m/s}^2$$

Next, from the kinematic equation relating velocity and acceleration, the velocity of the block is

$$v_x = v_{x_o} + a_x t = 2.0 \text{ m/s} + (3.0 \text{ m/s}^2)(1.5 \text{ s}) = 6.5 \text{ m/s}$$

$$v_y = v_{y_o} + a_y t = 0 + (5.2 \text{ m/s}^2)(1.5 \text{ s}) = 7.8 \text{ m/s}$$

And, at the end of the 1.5 s, the velocity of the block is

$$\mathbf{v} = v_x \hat{\mathbf{x}} + v_y \hat{\mathbf{y}} = (6.5 \text{ m/s}) \hat{\mathbf{x}} + (7.8 \text{ m/s}) \hat{\mathbf{y}}$$

Follow-up Exercise. (a) What is the direction of motion at the end of the 1.5 s? (b) If the force were applied at an angle of 30° (rather than 60°) relative to the x axis, how would this change the results of Example 4.4?

4.4 Newton's Third Law of Motion

Objectives: To (a) state and explain Newton's third law of motion and (b) identify action–reaction force pairs.

Newton formulated a third law that is as far-reaching in its physical significance as the first two. For a simple introduction to the third law, consider the forces involved in seat belt safety. When the brakes are suddenly applied in a moving car, you continue to move forward (the frictional force on the seat of your pants is not enough to stop you). In doing so, you exert forces on the seat belt and shoulder strap. The belt and strap exert corresponding reaction forces on you, causing you to slow down with the car. If you haven't buckled up, you may keep on going (Newton's first law) until another applied force, such as that applied by the dashboard or windshield, slows you down.

We commonly think of single forces. However, Newton recognized that it is impossible to have a single force. He observed that in any application of force, there is always a mutual interaction, and forces always occur in pairs. An example given by Newton was this: If you press on a stone with a finger, then the finger is also pressed by, or receives a force from, the stone.

Newton termed the paired forces *action* and *reaction*, and **Newton's third law of motion** is as follows:

For every force (action), there is an equal and opposite force (reaction).

Newton's third law—action and reaction

In symbol notation,

$$\mathbf{F}_{12} = -\mathbf{F}_{21}$$

That is, $\mathbf{F}_{12}$ is the force exerted *on* object 1 *by* object 2, and $-\mathbf{F}_{21}$ is the equal and opposite force exerted *on* object 2 *by* object 1. (The minus sign indicates the opposite direction.) *Which force is considered the action or the reaction is arbitrary*; $\mathbf{F}_{21}$ may be the reaction to $\mathbf{F}_{12}$ or vice versa.

Newton's third law may seem to contradict Newton's second law. If there are always equal and opposite forces, how can there be a nonzero net force? An important thing to remember about a force pair of the third law is that *the action–reaction forces do not act on the same object*. The second law is concerned with force(s) acting on a particular object (or system). The opposing forces of the third law act on different objects. Hence, the forces cannot cancel each other out or have a vector sum of zero when we apply the second law to the objects.

To illustrate this distinction, consider the situations shown in •Fig. 4.11. We often tend to forget the reaction force. For example, in Fig. 4.11a (left), the obvious force that acts on a block sitting on a table is the Earth's gravitational attraction, which is expressed by the weight mg. But, there *has to be another force* acting on it. In fact, the other force must be equal and opposite in weight. Otherwise, by Newton's second law, the block would be accelerating downward.

The *block* exerts a downward force of magnitude mg on the table, and in turn the *table* exerts an equal but opposite force upward on the block (N), as illustrated in Fig. 4.11a (right). The net force *on the block* is then zero, and the block doesn't accelerate; thus, $F_{net_y} = N - mg = 0$. You could easily demonstrate that this upward force on the block is there by placing the block on your hand and holding it stationary—you would exert an upward force on the block. (And you would feel a reaction force of $-N$ on your hand.)

We call the force a surface exerts on an object a *normal* force and use the symbol N to denote this. *Normal* means *perpendicular*. The **normal force** a surface exerts on an object is always perpendicular to the surface. In Fig. 4.11a, the normal force is equal and opposite to the weight of the block.

The normal force is not always equal and opposite to an object's weight, however. The normal force is a "reaction" force; it reacts to the situation. If a force is applied to the block downward at an angle, as shown in Fig. 4.11b, then a downward component (F_y) acts on the block in addition to the block's weight, and the normal force must balance both forces ($F_{net_y} = N - mg - F_y = 0$.) and $N = mg + F_y$.

In Fig. 4.11b, the block could accelerate to the right (why?). However, this component might be balanced by a frictional force, as it often is. We'll bring friction into the picture in a later section; the normal force will be quite important there.

For the situation in Fig. 4.11c, the normal force is reduced. Can you see why? Note that the F_y component of the applied force is opposite in direction to the weight. Hence, in magnitude, $N = mg - F_y$.

Inclined planes are important in physics; for a block on an inclined plane (Fig. 4.11d), the normal force on the block is *perpendicular to the plane surface*. This force is the reaction to the block's force on the plane surface, and since $a_y = 0$ (no motion in the y direction), we have $F_{net_y} = N - F_y = 0$ and $N = F_y = mg \cos \theta$. The weight component (F_x) accelerates the block down the plane in the absence of an equal opposing friction force between the block and the plane surface.

Finally, Fig. 4.11e shows a case in which there is acceleration in the direction of the normal force. Here, a person is accelerating the block vertically, and $N - mg = ma_y$. Hence in this case, $N > mg$.

As another example, consider the situation in •Fig. 4.12a. Two third-law force pairs are acting when the person is holding the briefcase. There is a pair of contact forces: The person's hand exerts an upward force on the handle, and the handle exerts an equal downward force on the hand. That is, $F_1 = -F_1$. This is an action–reaction pair, with the forces acting on different objects. The other third-law force pair consists of action-at-a-distance forces associated with gravitational attraction: The Earth attracts the briefcase (its weight), and the briefcase attracts the Earth, or $F_2 = -F_2$.

Concentrating on the briefcase in isolation, we can see that only two of the four forces in Fig. 4.12a act on it—the upward force on the handle (F_1) and the downward force of the briefcase's weight ($-F_2$). These do *not* form a third-law force pair, however, because they act on the *same* object. Since the briefcase does not accelerate, these forces must be equal and opposite. Thus, the net force on the isolated stationary briefcase is zero, as required by Newton's first and second laws.

•**FIGURE 4.11 Second and third law distinctions** Newton's second law deals with the forces acting on a particular object (or system). The third law deals with the force pair that acts on different objects. This distinction is illustrated for various situations.

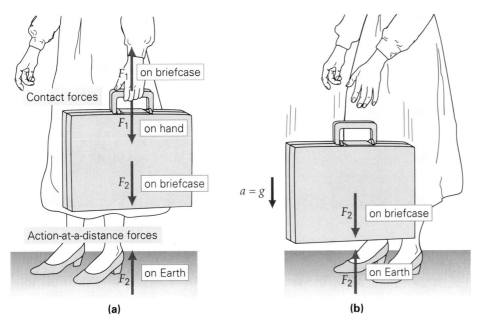

•FIGURE 4.12 Force pairs of Newton's third law (a) When the person holds the briefcase, there are two force pairs: a contact pair (F_1 and $-F_1$) and an action-at-a-distance (gravity) pair (F_2 and $-F_2$). The net force acting on the briefcase is zero: The upward contact force balances the downward weight force. Note, however, that the upward contact force and downward weight force do *not* form a third-law pair. (b) When the briefcase is released, there is an unbalanced force acting on the case (its weight force), and it accelerates downward (at g for free fall).

When the person drops the briefcase (Fig. 4.12b), there is then a nonzero net force (the unbalanced force of gravity) on the case, and it accelerates toward the Earth. But there is still a pair of third law forces *acting on different objects*—the briefcase and the Earth. (The Earth actually accelerates upward toward the briefcase as well, but the amount of the acceleration is infinitesimal. Can you explain why?)

Jet propulsion is yet another example of Newton's third law in action. In the case of a rocket, the rocket and exhaust gases exert equal and opposite forces on each other. As a result, the exhaust gases are accelerated away from the rocket, and the rocket is accelerated in the opposite direction. When big rockets "blast off" as in a Space Shuttle launch, there is a fiery exhaust from the rocket. A common misconception is that the exhaust gases "push" against the launch pad to accelerate the rocket. If this misconception were true, there would be no space travel, since there is nothing to "push" against in space. The correct explanation is one of action (gases exerting a force on the rocket) and reaction (rocket exerting an opposite force on the gases).

Another action–reaction pair is given in the Insight on page 112, and Demonstration 2 shows action–reaction forces in terms of the tension in a string.

4.5 More on Newton's Laws: Free-body Diagrams and Translational Equilibrium

Objectives: To (a) apply Newton's laws in analyzing various situations, using free-body diagrams, and (b) understand the concept of translational equilibrium.

Now that you have been introduced to Newton's laws and some applications in analyzing motion, the importance of these laws should be evident. They are so simply stated, yet so far-reaching. The second law is probably the most often applied, because of its mathematical relationship. However, the first and third laws are often used in qualitative analysis, as our study of physics will reveal.

The simple relationship expressed by Newton's second law, $\mathbf{F} = m\mathbf{a}$, allows the quantitative analysis of force and motion. We can think of it as a cause-and-effect relationship, with force being the cause and acceleration being the motional effect.

Note: Remember that **F** means the *net* force, or $\Sigma_i \mathbf{F}_i$.

In general, you will be concerned with applications involving constant forces. Constant forces result in constant accelerations and allow us to use the kinematic equations from Chapter 2 in analyzing the motion. When there is a variable force, Newton's second law holds for the *instantaneous* force and acceleration, but the

Insight

Action and Reaction: The Harrier "Jump" Jet

The action–reaction principle of Newton's third law is used to power jet aircraft (and propeller-driven planes), usually for propulsion rather than directly for lift. However, there is one type of aircraft that uses jet propulsion to take off vertically—no runway needed, like a helicopter: the Harrier "Jump" jet originally developed by the British (Fig. 1). This plane has four engine nozzles that can be swiveled to point at any angle from vertical to horizontal. These engines provide the power for both vertical and horizontal flight.

A Harrier takeoff is illustrated in Fig. 2. (There are other small jets on the plane to help control and stabilize the aircraft that are not shown.) As shown in Fig. 2a, the nozzles are directed downward for takeoff, and the reaction forces raise the plane. Achieving a proper altitude, the nozzles are swiveled and a forward component of the force gives the plane a forward acceleration (Fig. 2b). Finally, with the jets horizontal, the planes flies conventionally (Fig. 2c).

But wait! Did you notice something wrong with Fig. 2c? The plane has weight, or a downward force acting on it, that is not shown. What is the counterbalancing force to this weight that keeps the plane from falling? As you will learn in Chapter 9, this *lift* force results from the curved wing surface and fluid (air) dynamics.

FIGURE 1 **"Jump" jet** A Harrier "Jump" jet lifting off vertically.

FIGURE 2 **Up, up, and away** (a) The reaction force to the downward exhaust force from the engines raises the plane. Only one nozzle is shown for simplicity. (b) At the appropriate altitude, the engine nozzles are swiveled and a forward force component gives the plane a forward acceleration (while still gaining altitude—why?). (c) With the nozzles swiveled 90° from their position in (a), the plane flies conventionally.

Demonstration 2 ■ Tension in a String: Action and Reaction Forces

(a) Two suspended 2-kilogram masses are attached to opposite sides of a scale (calibrated in newtons). The total suspended weight is $w = mg = (4.00 \text{ kg})(9.80 \text{ m/s}^2) = 39.2$ N, yet the scale reads about 20 N. Is something wrong with the scale?

(b) No, think of it in this manner. The effect of the weight of the mass on the right is replaced by fixing the end of the string. The other mass stretches the scale spring, giving a reading of about 20 N [or $w = mg = (2.00 \text{ kg})(9.80 \text{ m/s}^2) = 19.6$ N].

(c) Similarly, the other end of the scale can be fixed. (A fixed pulley merely changes the direction of the force, and the scale can be hung vertically with the same effect.) In all cases, the tension in the string is 19.6 N, as the scale shows.

acceleration will vary with time. We will generally limit ourselves to constant accelerations and forces.

This section presents several examples of applications of Newton's second law so that you can become familiar with its use. This small but powerful equation will be used again and again throughout this text.

Newton's second law gives the acceleration resulting from an applied force, but it can also be used to find the force from motional effects, as the following Example shows.

Example 4.5 ■ A Braking Car: Finding a Force From Motional Effects

A car traveling at 72.0 km/h along a straight, level road is brought uniformly to a stop in a distance of 40.0 m. If the car weighs 8.80×10^3 N, what is the braking force?

Thinking It Through. It is noted that the car's velocity changes, so there is an acceleration. Given a distance, we might surmise that first the acceleration is found by using a kinematic equation, and then the acceleration is used to compute the force.

Solution. In bringing the car to a stop, the braking force caused an acceleration (actually a deceleration), as illustrated in ●Fig. 4.13. Listing what is given and what we must find, we have

Given: $v_o = 72.0$ km/h $= 20.0$ m/s *Find:* F_b (braking force)
$v = 0$
$x = 40.0$ m
$w = 8.80 \times 10^3$ N

●FIGURE 4.13 Finding force from motional effects See Example 4.5.

We know that $F = ma$, so we can easily calculate F_b if we can find m and a. The car's mass m can be obtained from its given weight. The other given quantities should remind you of a kinematic equation from Chapter 2 from which the acceleration a can be found. Since the car is brought uniformly to a stop, the acceleration is constant and we may use Eq. 2.10, $v^2 = v_o^2 + 2ax$, to find a:

$$a = \frac{v^2 - v_o^2}{2x} = \frac{0 - (20.0 \text{ m/s})^2}{2(40.0 \text{ m})} = -5.00 \text{ m/s}^2$$

The minus sign indicates that the acceleration is opposite to v_o, as expected for a braking force, which slows the car.

The mass of the car is obtained from the weight: $w = mg$, or $m = w/g$. Using this expression along with the acceleration obtained previously gives a braking force of

$$F_b = ma = \left(\frac{w}{g}\right)a = \left(\frac{8.80 \times 10^3 \text{ N}}{9.80 \text{ m/s}^2}\right)(-5.00 \text{ m/s}^2) = -4.49 \times 10^3 \text{ N}$$

Follow-up Exercise. A 1000-kilogram car starting from rest travels in a straight line and uniformly reaches a speed of 10 m/s in 5.0 s. What is the magnitude of the net force that accelerates the car?

Example 4.6 ■ The Atwood Machine: Finding the Acceleration of a System

The *Atwood machine* consists of two masses suspended from a fixed pulley, as shown in •Fig. 4.14. If $m_1 = 0.55$ kg and $m_2 = 0.80$ kg, (a) what is the acceleration of the system? (b) What is the magnitude of the tension in the string? (Consider the pulley to be frictionless and the masses of the string and the pulley to be negligible.)

Thinking It Through. The system consists of the two masses that are acted on by forces, so this appears to be a straightforward application of the second law. To find the tension in the string, the masses will have to be isolated as separate systems so as to have T appear in the second-law equations.

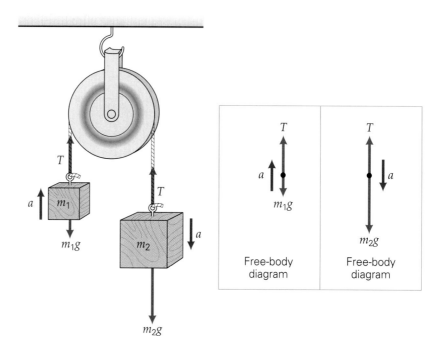

•**FIGURE 4.14 The Atwood machine** A single, fixed pulley is simply a direction changer. Free-body diagrams are shown for the forces in the box. See Example 4.6.

Solution. Listing the two given quantities, we have

Given: $m_1 = 0.55$ kg Find: (a) a (acceleration)
$m_2 = 0.80$ kg   (b) T (tension)

(a) Since m_2 is greater than m_1, if released from rest, m_2 will fall and m_1 will rise with accelerations of the same magnitude in opposite directions. Looking at the two masses as separate systems, we are free to assign directional plus and minus signs arbitrarily. Here, *the positive direction is taken to be the direction of acceleration for both masses.* That is, for m_2, downward is taken to be the positive direction; for m_1, upward is positive. The pulley is simply a direction changer. By analogy, you might think of the system as two masses that are connected by a horizontal string and are moving to the right (as in Fig. 4.9).

Then, applying Newton's second law to the system as a whole (and thus ignoring the T forces, which are internal) gives

$$F_2 - F_1 = m_2 g - m_1 g$$
$$= (m_1 + m_2)a$$

net force = total mass × acceleration

Solving for a gives

$$a = \frac{(m_2 - m_1)g}{m_1 + m_2}$$
$$= \frac{(0.80 \text{ kg} - 0.55 \text{ kg})(9.8 \text{ m/s}^2)}{0.55 \text{ kg} + 0.80 \text{ kg}} = 1.8 \text{ m/s}^2$$

(b) To find the magnitude of the tension, the second law must be applied to an isolated mass on which the tension acts. This will give an equation with T. Note the boxed sketches in Fig. 4.14, which show all the forces of the system. These sketches are **free-body diagrams**, which represent an object as a "particle" or point-sized mass and shows all the forces acting on that particle. Free-body diagrams, discussed in the following Problem-Solving Strategy, are very helpful in analyzing situations involving several bodies and multiple forces.

Applying the second law to each of the masses, we have

$$T - m_1 g = m_1 a$$
$$m_2 g - T = m_2 a$$

where the direction of the acceleration for each mass (upward for m_1 and downward for m_2) is taken as positive so as to avoid a minus sign for a. The magnitudes of the accelerations of the masses are equal. Eliminating T from the two equations gives the equation derived above:

$$m_2 g - m_1 g = m_2 a + m_1 a = (m_1 + m_2)a$$

Once the acceleration is found, use it in either of the previous two equations, solved for T. Using the first equation gives

$$T = m_1 a + m_1 g = m_1(a + g)$$
$$= (0.55 \text{ kg})(1.8 \text{ m/s}^2 + 9.8 \text{ m/s}^2) = 6.4 \text{ N}$$

Follow-up Exercise. Suppose there were a constant retarding force of friction, f, in the pulley of an Atwood machine when the masses were in motion. What would be the general (symbol) form of the acceleration in this case?

Incidentally, the Atwood machine is named after the British scientist George Atwood (1746–1807), who used the arrangement to study motion and measure the value of g. The masses can be chosen to minimize the acceleration, making it easier to measure the time of fall.

LEARN BY DRAWING

Drawing a Free-body Diagram

1 Space diagram

2

3

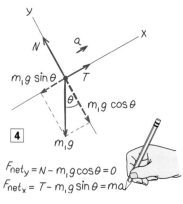

4

$F_{net_y} = N - m_1 g \cos\theta = 0$
$F_{net_x} = T - m_1 g \sin\theta = m_1 a$

Problem-Solving Strategy: Free-body Diagrams

When working with problems in which two or more forces or components of force act on a body, it is convenient and instructive to draw a free-body diagram of the forces, as was done in Example 4.6. In such a diagram, we show all the forces acting on the body or object. If several bodies are involved, we may make a diagram for each body *separately*, showing all of the forces acting on each individual body.

In illustrations of physical situations, sometimes called *space diagrams* (such as the Atwood machine in Fig. 4.14), force vectors may be drawn at different locations to indicate their points of application. However, since we are concerned only with linear motions, vectors in free-body diagrams (such as those boxed in Fig. 4.14) may be shown emanating from a common point, which is chosen as the origin of x–y axes. One of the axes is generally chosen along the direction of the net force acting on a body, since that is the direction in which the body will accelerate. Also, it is often important to resolve force vectors into components, and properly chosen x–y axes simplify this.

In a free-body diagram, the vector arrows do not have to be exactly to scale. However, it should be made apparent if there is a net force, and whether forces balance each other in a particular direction. When the forces aren't balanced, we know from Newton's second law that there must be an acceleration.

In summary, the general steps in constructing and using free-body diagrams are as follows (refer to the accompanying Learn by Drawing as you read):

1. Sketch a space diagram (if one is not already available) and identify the forces acting on each body of the system.

2. Isolate the body for which the free-body diagram is to be constructed. Draw a set of Cartesian axes with the origin at a point through which the forces act and with one of the axes along the direction of the body's acceleration. (This will be in the direction of the net force if there is one.)

3. Draw properly oriented force vectors on the diagram emanating from the origin of the axes. If there is an unbalanced force, assume a direction of acceleration and indicate it with an acceleration vector. Be sure to include only those forces that act on the isolated body.

4. Resolve any forces that are not directed along the x or y axes into x or y components. Use the free-body diagram to analyze the forces in terms of Newton's second law of motion. (Note: If the acceleration is in the direction opposite that selected, indicate this by an acceleration with an opposite sign in the solution. For example, if the motion is taken to be in the positive direction and it is actually in the opposite direction, the acceleration will be negative.)

Free-body diagrams are a particularly useful way of following one of the Suggested Problem-Solving Procedures in Chapter 1: Draw a diagram as an aid in visualizing and analyzing the physical situation of the problem. *Make it a practice to draw free-body diagrams for force problems, as is done in the following Examples.*

Example 4.7 ■ Components of Force and Free-body Diagrams

A force of 15.0 N is applied at an angle of 30° to the horizontal on a 0.750-kilogram block at rest on a frictionless surface as illustrated in •Fig. 4.15. (a) What is the magnitude of the resulting acceleration of the block? (b) What is the magnitude of the normal force?

Thinking It Through. The applied force may be resolved into components. The horizontal component accelerates the block. The vertical component affects the normal force (review Fig. 4.11).

Solution. First we write down the given data and what is to be found.

Given: $F = 15.0$ N *Find:* (a) a (acceleration)
$m = 0.750$ kg   (b) N (normal force)
$\theta = 30°$
$v_o = 0$

Then we draw a free-body diagram for the block as in Fig. 4.15. Note that the forces act in the same directions as in the space diagram.

(a) The acceleration of the block is given by Newton's second law. We choose our axes so that a is in the $+x$ direction. As the free-body diagram shows, only a component (F_x) of the applied force F acts in this direction. The component of F in the direction of motion is $F_x = F \cos \theta$. We apply Newton's law in the $+x$ direction,

$$F_x = F \cos 30° = ma_x$$

$$a_x = \frac{F \cos 30°}{m} = \frac{(15.0 \text{ N})(0.866)}{0.750 \text{ kg}} = 17.3 \text{ m/s}^2$$

(b) This is the total acceleration of the block, since the block moves only in the x direction (it does not accelerate in the y direction). With $a_y = 0$, the sum of the forces in the y direction must then be zero. That is, the downward component of F acting on the block, F_y, and its downward weight force w must be balanced by the upward normal force N that the surface exerts on the block. If this were not the case, then there would be a net force and an acceleration in the y direction.

We sum the forces in the y direction with upward as positive:

or
$$N - F_y - w = 0$$

$$N - F \sin 30° - mg = 0$$

and

$$N = F \sin 30° + mg = (15.0 \text{ N})(0.500) + (0.750 \text{ kg})(9.80 \text{ m/s}^2) = 14.85 \text{ N}$$

The surface then exerts a force of 14.85 N upward on the block, which balances the downward forces acting on it.

Follow-up Exercise. (a) Suppose the applied force on the block is applied for only a short time. What is the magnitude of the normal force after the applied force is removed? (b) If the block slides off the edge of the table, what would be the net force on the block just after leaving the table (with the applied force removed)?

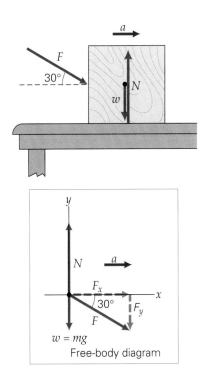

• **FIGURE 4.15** Newton's second law and components of force See Example 4.7.

Problem-Solving Hint

There is no single fixed way to go about solving a problem. However, there are general strategies or procedures that are helpful in solving problems involving Newton's second law. Using our Suggested Problem-Solving Procedures introduced in Chapter 1, you might include the following in force applications:

Note: Review pp. 21–24.

- Draw a free-body diagram for each individual body, showing all of the forces acting on that body.
- Depending on what is to be found, Newton's second law may be applied to the system as a whole (in which case internal forces cancel) or to a part of the system. Basically, *you want to obtain an equation containing the quantity for which you want to solve.* Review Example 4.3. (If there are two unknown quantities, application of Newton's law to two parts of the system may give you two equations and two unknowns. See Example 4.8.)
- Keep in mind that Newton's second law may be applied to components of motion and that forces may be resolved into components to do this. Review Example 4.7.

Example 4.8 ■ The Atwood Machine Revisited: Motion on a Frictionless Inclined Plane

Two masses are connected by a light string running over a light pulley of negligible friction as illustrated in •Fig. 4.16. One mass (m_1 = 5.0 kg) is on a frictionless 20° inclined plane, and the other (m_2 = 1.5 kg) is freely suspended. What is the acceleration of the masses?

Thinking It Through. Apply the preceding Problem-Solving Hint.

Solution. Following our usual procedure, we write

Given: m_1 = 5.0 kg   Find: **a** (acceleration)
m_2 = 1.5 kg
θ = 20°

To help us visualize the forces involved, we isolate m_1 and m_2 and draw free-body diagrams for each mass. For mass m_1, there are three concurrent forces (forces acting through a common point). These are T, $m_1 g$, and N, where T is the tension force in the string and N is the normal force of the plane on the block. The forces are shown emanating from their common point of action. (Recall that a vector arrow can be moved as long as its direction is not changed.)

We will start by assuming that m_1 accelerates up the plane, which is taken to be in the x direction. (It makes no difference whether it is assumed that m_1 accelerates up or

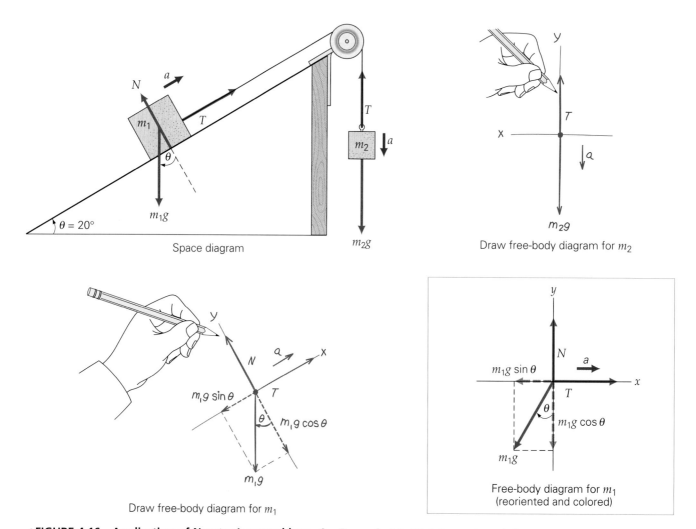

•**FIGURE 4.16 Application of Newton's second law**   See Example 4.8. (Not drawn to scale.)

down the plane, as we shall see shortly.) Notice that m_1g (the weight) is broken down into components. The x component is in the assumed direction of acceleration, and the y component acts perpendicularly to the plane and is balanced by the normal force N. (There is no acceleration in the y direction, so there is no net force in this direction.)

Then, applying Newton's second law in component form to m_1, we have

$$F_{x_1} = T - m_1 g \sin\theta = m_1 a$$

$$F_{y_1} = N - m_1 g \cos\theta = 0 \quad \text{(no net force, forces cancel)}$$

And for m_2,

$$F_{y_2} = m_2 g - T = m_2 a$$

where the masses of the string and pulley have been neglected.

Adding the first and last equations to eliminate T, we have

$$m_2 g - m_1 g \sin\theta = (m_1 + m_2)a$$

(net force = *total* mass × acceleration)

(Note that this is the equation that would be obtained by applying the second law to the system as a whole, in which the T forces cancel.)

Then, we solve for a:

$$a = \frac{m_2 g - m_1 g \sin 20°}{m_1 + m_2}$$

$$= \frac{(1.5 \text{ kg})(9.8 \text{ m/s}^2) - (5.0 \text{ kg})(9.8 \text{ m/s}^2)(0.342)}{5.0 \text{ kg} + 1.5 \text{ kg}}$$

$$= -0.32 \text{ m/s}^2$$

The minus sign indicates that the acceleration is opposite to the assumed direction. That is, m_1 actually accelerates down the plane and m_2 accelerates upward. As this example shows, if you assume the acceleration to be in the wrong direction, the sign on the result will give you the correct direction anyway.

Could you find the tension force T in the string if you were asked to do so? How this could be done should be quite evident from the free-body diagram.

Follow-up Exercise. (a) In Example 4.8, what mass m_2 would cause m_1 to accelerate up the plane? (b) Keeping the masses the same as in the Example, how should the angle of incline be adjusted so that m_1 would accelerate up the plane?

Translational Equilibrium

Forces may act on an object without producing an acceleration. In such a case, with $a = 0$, we know from Newton's second law that

$$\Sigma \mathbf{F} = 0 \qquad (4.4)$$

That is, the vector sum of the forces, or the net force, is zero, so the object either remains at rest (as in •Fig. 4.17) or moves with a constant velocity. In such cases, objects are said to be in **translational equilibrium**. When remaining at rest, an object is said to be in *static translational equilibrium*.

It follows that the sums of the rectangular components of the forces for an object in translational equilibrium are also zero. (Why?)

$$\Sigma F_x = 0 \qquad \text{(translational} \qquad (4.5)$$
$$\Sigma F_y = 0 \qquad \text{equilibrium only)}$$

For three-dimensional problems, we would add $\Sigma F_z = 0$. However, we will restrict our discussion of forces to two dimensions.

(a)

(b)

•**FIGURE 4.17 Many forces, no acceleration** (a) At least five different external forces act on this physics professor. (Here **f** is the force of friction.) Nevertheless, she experiences no acceleration. Why? (b) Adding the force vectors by the polygon method reveals that the vector sum of the forces is zero. The professor is in static translational equilibrium. (She is also in static rotational equilibrium; we'll see why in Chapter 8.)

These equations give what is often referred to as the **condition for translational equilibrium**. (A condition for rotational considerations will be given in Chapter 8.) Let's apply this translational equilibrium condition to a static equilibrium case.

Example 4.9 ■ A Hanging Sign: Static Translational Equilibrium

A 3.0-kilogram sign hangs in a hall in the Physics Department as shown in •Fig. 4.18a. What is the minimum tensile strength necessary for the cord that is used to hang the sign?

Thinking It Through. The sign hangs motionless on the wall, so it is in static translational equilibrium. Hence, the summation of the forces is zero, and here we see that there will be components in two dimensions.

Solution.

Given: $m = 3.0$ kg Find: T_1 and T_2
$\theta_1 = \theta_2 = 45°$ (tensions in the cord)

The minimum *tensile strength* of the cord is the amount of tension the cord must be able to support without breaking. With a cord of sufficient strength, the sign will hang in static equilibrium. We want to find the values of T_1 and T_2 that will *just* support the weight (mg) of the sign.

The rectangular components of the tensions are shown in the free-body diagram (Fig. 4.18b). Applying the component conditions for static translational equilibrium (Eq. 4.5), we have

$$\Sigma F_x = T_{1_x} - T_{2_x} = 0$$

or

$$T_{1_x} = T_{2_x}$$

Hence, the magnitudes of the x components of the tensions are equal, as you might have expected. (They are the only forces in the x direction, and they balance each other.) However, this gives no direct information about the magnitude of the tensions.

Since angles θ_1 and θ_2 are equal, we also know that $T_{1_y} = T_{2_y}$. (Why?) We apply the other component condition:

$$\Sigma F_y = T_{1_y} + T_{2_y} - mg = 2T_{1_y} - mg$$
$$= 2(T_1 \sin 45°) - mg = 0$$

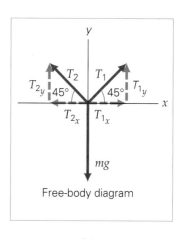

•**FIGURE 4.18 Static translational equilibrium** See Example 4.9.

(a) (b)

Solving for T_1,

$$T_1 = \frac{mg}{2 \sin 45°} = \frac{(3.0 \text{ kg})(9.8 \text{ m/s}^2)}{2(0.707)} = 21 \text{ N}$$

With the x and y tension components equal, $T_1 = T_2 = 21$ N, so the sign should be hung from a cord with a tensile strength of at least 21 N.

Follow-up Exercise. How could a heavier sign be hung in Example 4.9 from a cord with a tensile strength of less than 21 N?

4.6 Friction

Objectives: To explain (a) the causes of friction and (b) how it is described by using coefficients of friction.

Friction refers to the ever-present resistance to motion that occurs whenever two materials, or media, are in contact with each other. This resistance occurs for all types of media—solids, liquids, and gases—and is characterized as the **force of friction** (f). We have up to now generally ignored all kinds of friction (including air resistance) in examples and problems for simplicity. Now that you know how to describe motion, we are ready to consider situations that are more realistic, in that the effects of friction are included.

In some real situations, we want to increase friction—for example, by putting sand on an icy road or sidewalk to improve traction. This might seem contradictory, since an increase in friction presumably would increase the resistance to motion. However, consider the forces involved in walking as illustrated in •Fig. 4.19. Without friction, the foot would slip backward. (Think about walking on a slippery surface.) The force of friction prevents this and sometimes needs to be increased on slippery surfaces (•Fig. 4.20a). In other situations, we try to reduce friction (Fig. 4.20b). For instance, we lubricate moving machine parts to allow them to move more freely, lessen wear, and reduce expenditure of energy. Automobiles would not run without friction-reducing oils and greases.

This section is concerned chiefly with friction between solid surfaces. All surfaces are microscopically rough, no matter how smooth they appear or feel. It was originally thought that friction was primarily due to the mechanical interlocking of surface irregularities, or *asperities* (high spots). However, research has shown that the friction between the contacting surfaces of ordinary solids (metals in particular) is mostly due to local adhesion. When surfaces are pressed together, local welding or bonding occurs in a few small patches where the largest asperities make contact. To overcome this local adhesion, a force great enough to pull apart the bonded regions must be applied. Once contacting surfaces are in relative motion, another form of friction may result when the asperities of a harder material dig into a softer material, with a "plowing" effect.

Friction between solids is generally classified into three types: static, sliding (kinetic), and rolling. **Static friction** includes all cases in which the frictional force is sufficient to prevent relative motion between surfaces. Suppose you want to move a large desk. You push it, but the desk doesn't move. The force of static friction between the desk's legs and the floor opposes and equals the horizontal force you are applying, so there is no motion—a static condition.

Sliding friction, or **kinetic friction**, occurs when there is relative (sliding) motion at the interface of the surfaces in contact. In pushing on the desk, you eventually get it sliding, but there is still a great deal of resistance between the desk's legs and the floor—kinetic friction.

Rolling friction occurs when one surface rotates as it moves over another surface but does not slip or slide at the point or area of contact. Rolling friction, such as occurs between a train wheel and a rail, is attributed to small local deformations in the contact region. This type of friction is somewhat difficult to analyze.

•**FIGURE 4.19 Friction and walking** The force of friction is shown in the direction of the walking motion. This may seem wrong at first glance, but it's not. The force of friction prevents the foot from slipping backward while the other foot is brought forward. If you walk on a deep-pile rug, F is evident in that the pile will be bent backward.

(a)

(b)

• FIGURE 4.20 **Increasing and decreasing friction** (a) To get a fast start, drag racers need to make sure that their wheels don't slip when the light goes on and they floor the accelerator. They therefore try to maximize the friction between their tires and the track by "burning in" the tires just before the start of the race. This is done by spinning the wheels with the brakes on until the tires are extremely hot. The rubber becomes so sticky that it almost welds itself to the road surface. (b) Water serves as a good lubricant to reduce friction in rides such as this one.

Frictional Forces and Coefficients of Friction

Here, we will consider the forces of friction on stationary and sliding objects. These are called the force of static friction and the force of kinetic (or sliding) friction, respectively. Experimentally, it is found that the force of friction depends on both the nature of the two surfaces and on the *load*, or the force that presses the surfaces together. For an object on a horizontal surface, this force is equal in magnitude to the object's weight. However, as was shown in Fig. 4.16, on an inclined plane only a component of the weight force contributes to the load.

Thus, to avoid confusion, remember that the force of friction is proportional to the normal force N in magnitude; that is, $f \propto N$. As we learned earlier, the *normal force* always acts perpendicular to, and away from, the surface. (It is the force exerted *by* the surface *on* the object.) In the absence of other perpendicular forces, the normal force is equal in magnitude to the component of the weight force acting perpendicular to the surface by the third law.

The force of static friction (f_s) between parallel surfaces in contact acts in the direction that opposes the initiation of relative motion between the surfaces. The magnitude has different values such that

$$f_s \leq \mu_s N \qquad \text{(static conditions)} \qquad (4.6)$$

where μ_s is the **coefficient of static friction**. (Note that it is a dimensionless constant. Why?)

The less-than-or-equal-to sign ($\leq$) indicates that the force of static friction may have different values or magnitudes up to some maximum. To understand this, look at • Fig. 4.21. In Fig. 4.21a, one person pushes on a file cabinet and it doesn't move. With no acceleration, the net force on the cabinet is zero, and $F - f_s = 0$, or $F = f_s$. Suppose that a second person also pushes and the file cabinet still doesn't budge (Fig. 4.21b). Then f_s must now be larger, since the applied force has been increased. Finally, if the applied force is made large enough to overcome the static friction, motion occurs (Fig. 4.21c). The greatest, or maximum, force of static friction is exerted just before the cabinet starts to slide (Fig. 4.21b), and for this case Eq. 4.6 can be written with an equal sign:

$$f_{s_{max}} = \mu_s N \qquad (4.7)$$

Once an object is sliding, there is a force of kinetic friction (f_k) acting on it. That force acts in the direction opposite to the direction of motion and has a magnitude of

$$f_k = \mu_k N \qquad \text{(sliding conditions)} \qquad (4.8)$$

where μ_k is the **coefficient of kinetic friction** (sometimes called the coefficient of sliding friction). Generally, the coefficient of kinetic friction is less than the coefficient of static friction ($\mu_k < \mu_s$) for two surfaces, which means that the force of kinetic friction is less than $f_{s_{max}}$ as illustrated in Fig. 4.21. The coefficients of friction between some common materials are listed in Table 4.1.

Note that the force of static friction (f_s) exists in response to an applied force. The magnitude of f_s and its direction depend on the magnitude and direction of the applied force. Up to its maximum value, the force of static friction is equal in magnitude and opposite to the applied force (F), since there is no acceleration ($F - f_s = 0 = ma$). Thus, if the person in Fig. 4.21a pushed on the cabinet in the opposite direction, f_s would also be in the opposite direction. If there were no applied force F, then f_s would be zero. When F exceeds $f_{s_{max}}$, the cabinet slides and kinetic friction comes into effect, with $f_k = \mu_k N$. If F is equal to f_k, the cabinet will slide with a constant velocity; if F is greater than f_k, the cabinet will accelerate.

It has been experimentally determined that the coefficients of friction (and therefore the forces of friction) are nearly independent of the size of the contact area between metal surfaces. This means that the force of friction between a

•**FIGURE 4.21 Force of friction versus applied force** (a) In the static region of the graph, as the applied force F increases, so does f_s: That is, $f_s = F$ and $f_s < \mu_s N$. (b) When the applied force F exceeds $f_{s_{max}} = \mu_s N$, the heavy file cabinet is set into motion. (c) Once the cabinet is moving, the frictional force is decreased, since kinetic friction is less than static friction ($f_k < f_{s_{max}}$). Thus, if the applied force is maintained at $F = f_{s_{max}}$ there is a net force, and the cabinet is accelerated. For the cabinet to move with constant velocity, the applied force must be reduced to equal the kinetic friction force, $f_k = \mu_k N$.

TABLE 4.1 Approximate Values for Coefficients of Static and Kinetic Friction between Certain Surfaces

Friction between Materials	μ_s	μ_k
aluminum on aluminum	1.90	1.40
glass on glass	0.94	0.35
rubber on concrete		
dry	1.20	0.85
wet	0.80	0.60
steel on aluminum	0.61	0.47
steel on steel		
dry	0.75	0.48
lubricated	0.12	0.07
Teflon on steel	0.04	0.04
Teflon on Teflon	0.04	0.04
waxed wood on snow	0.05	0.03
wood on wood	0.58	0.40
lubricated ball bearings	<0.01	<0.01
synovial joints (at the ends of most long bones—for example, elbows and hips)	0.01	0.01

4.6 Friction

brick-shaped metal block and a metal surface is the same regardless of whether the block is lying on a larger side or a smaller side. The observation is not generally valid for other surfaces, such as wood, and does not apply to plastic or polymer surfaces.

Finally, keep in mind that although the equation $f = \mu N$ holds in general for frictional forces, friction may not be linear over a wide range. That is, μ is not always constant. For example, the coefficient of kinetic friction varies somewhat with the relative speed of the surfaces. However, for speeds up to several meters per second the coefficients are relatively constant. Thus, this discussion will neglect any variations due to speed (or area), and the forces of static and kinetic friction will be assumed to depend only on the load (N) and the nature of the materials as expressed by the given coefficients of friction.

Example 4.10 ■ Pulling a Crate: Static and Kinetic Forces of Friction

(a) If the coefficient of static friction between the 40.0-kilogram crate in •Fig. 4.22 and the floor is 0.650, with what horizontal force must the worker pull to move the crate? (b) If the worker maintains that force once the crate starts to move and the coefficient of kinetic friction between the surfaces is 0.500, what is the magnitude of the acceleration of the crate?

Thinking It Through. This scenario involves applications of the forces of friction. In (a), the maximum force of static friction must be calculated. In (b), if the worker maintains an applied force of this magnitude after the crate is in motion, there will be an acceleration, since $f_k < f_{s_{max}}$.

Solution. Listing the given data and what we want to find, we have

Given: $m = 40.0$ kg *Find:* (a) F (force necessary to move crate)
$\mu_s = 0.650$ (b) a (acceleration)
$\mu_k = 0.500$

(a) The crate will not move until the applied force F slightly exceeds the maximum static frictional force $f_{s_{max}}$. So we must find $f_{s_{max}}$ to see what force the worker must apply. The weight of the crate and the normal force are equal in magnitude in this case (see the free-body diagram in Fig. 4.22), so the maximum force of static friction is

•**FIGURE 4.22 Forces of static and kinetic friction** See Example 4.10.

$$f_{s_{max}} = \mu_s N = \mu_s(mg)$$
$$= (0.650)(40.0 \text{ kg})(9.80 \text{ m/s}^2) = 255 \text{ N} \quad \text{(about 57 lb)}$$

The crate moves if the applied force exceeds this value: $F > 255$ N.

(b) Now the crate is in motion and the worker maintains a constant applied force $F = f_{s_{max}} = 255$ N. The force of kinetic friction f_k acts on the crate, but this is smaller than F because $\mu_k < \mu_s$. Hence, there is a net force, and the acceleration of the crate may be found by using Newton's second law:

$$F - f_k = F - \mu_k N = ma$$

or

$$a = \frac{F - \mu_k N}{m} = \frac{F - \mu_k(mg)}{m}$$
$$= \frac{255 \text{ N} - (0.500)(40.0 \text{ kg})(9.80 \text{ m/s}^2)}{40.0 \text{ kg}} = 1.5 \text{ m/s}^2$$

Follow-up Exercise. On the average, by what factor does μ_s exceed μ_k for nonlubricated, metal-on-metal surfaces? (See Table 4.1.)

Let's look at the worker and the crate again, but this time assume that he applies the force at an angle (•Fig. 4.23).

Example 4.11 ■ Pulling at an Angle: A Closer Look at the Normal Force

A worker pulling a crate applies a force at an angle of 30° to the horizontal as shown in Fig. 4.23. How large a force must he apply to move the crate? (Before looking at the solution, would you expect that the force needed in this case would be greater or lesser than in Example 4.10?)

Thinking It Through. Here we see from the figure that the applied force is at an angle to the horizontal surface, so the vertical component will affect the normal force (see Fig. 4.11). This will in turn affect the maximum force of static friction.

Solution. The data are the same as in Example 4.10, except that the force is applied at an angle.

Given: $\theta = 30°$ *Find:* F (force necessary to move crate)

 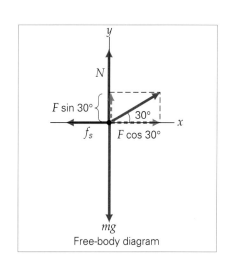

•**FIGURE 4.23 Normal force** See Example 4.11.

In this case, the crate will move when the *horizontal component* of the applied force, $F \cos 30°$, slightly exceeds the maximum static friction force. So, we may write for the maximum friction:

$$F \cos 30° = f_{s_{max}} = \mu_s N$$

However, the magnitude of the normal force is *not* equal to that of the weight of the crate here because of the upward component of the applied force (see the free-body diagram in the figure). By the second law, since $a_y = 0$, we have

$$N + F \sin 30° - mg = 0$$

or

$$N = mg - F \sin 30°$$

In effect, the applied force partially supports the weight of the crate. Substituting this expression for N into the first equation gives

$$F \cos 30° = \mu_s(mg - F \sin 30°)$$

Solving for F gives

$$F = \frac{mg}{(\cos 30°/\mu_s) + \sin 30°}$$

$$= \frac{(40.0 \text{ kg})(9.80 \text{ m/s}^2)}{(0.866/0.650) + 0.500} = 214 \text{ N (about 48 lb)}$$

Thus, less applied force is needed in this case, reflecting the fact that the frictional force is less because of the reduced normal force.

Follow-up Exercise. Note that in this Example, applying the force at an angle produces two effects. As the angle between the applied force and the horizontal increases, the horizontal component of the applied force is reduced. However, the normal force also gets smaller, resulting in a lower $f_{s_{max}}$. Does one effect always outweigh the other? That is, does the applied force F necessary to move the crate *always decrease* with increasing angle? (*Hint*: Investigate F for different angles. For example, compute F for 20° and 50°. You already have a value for 30°. What do the results tell you?)

Now let's look at ways to determine coefficients of friction experimentally.

Example 4.12 ■ A Sliding Block: Determining the Coefficient of Kinetic Friction

A block slides with a constant velocity down a plane inclined at 37° to the horizontal (•Fig. 4.24). What is the coefficient of kinetic friction between the block and the plane?

Thinking It Through. If the block slides down the plane with a constant velocity (zero acceleration), then the weight component down the plane is balanced by the force of kinetic friction up the incline. Hence, they may be equated, which may provide a way to find μ_k.

Solution.

Given: $a = 0$ (because v is constant) *Find:* μ_k (coefficient of kinetic friction)
$\theta = 37°$

Note that we are not given the mass of the block. We will not need it. Since the acceleration is zero, there is no net force on the block in either the x or y direction. So, noting the forces in the free-body diagram and using Newton's second law,

$$F_x = 0 = mg \sin \theta - f_k$$
$$F_y = 0 = N - mg \cos \theta$$

$$f_{s_{max}} = \mu_s N = \mu_s(mg)$$
$$= (0.650)(40.0 \text{ kg})(9.80 \text{ m/s}^2) = 255 \text{ N} \quad \text{(about 57 lb)}$$

The crate moves if the applied force exceeds this value: $F > 255$ N.

(b) Now the crate is in motion and the worker maintains a constant applied force $F = f_{s_{max}} = 255$ N. The force of kinetic friction f_k acts on the crate, but this is smaller than F because $\mu_k < \mu_s$. Hence, there is a net force, and the acceleration of the crate may be found by using Newton's second law:

$$F - f_k = F - \mu_k N = ma$$

or

$$a = \frac{F - \mu_k N}{m} = \frac{F - \mu_k(mg)}{m}$$
$$= \frac{255 \text{ N} - (0.500)(40.0 \text{ kg})(9.80 \text{ m/s}^2)}{40.0 \text{ kg}} = 1.5 \text{ m/s}^2$$

Follow-up Exercise. On the average, by what factor does μ_s exceed μ_k for nonlubricated, metal-on-metal surfaces? (See Table 4.1.)

Let's look at the worker and the crate again, but this time assume that he applies the force at an angle (•Fig. 4.23).

Example 4.11 ■ Pulling at an Angle: A Closer Look at the Normal Force

A worker pulling a crate applies a force at an angle of 30° to the horizontal as shown in Fig. 4.23. How large a force must he apply to move the crate? (Before looking at the solution, would you expect that the force needed in this case would be greater or lesser than in Example 4.10?)

Thinking It Through. Here we see from the figure that the applied force is at an angle to the horizontal surface, so the vertical component will affect the normal force (see Fig. 4.11). This will in turn affect the maximum force of static friction.

Solution. The data are the same as in Example 4.10, except that the force is applied at an angle.

Given: $\theta = 30°$ *Find:* F (force necessary to move crate)

•**FIGURE 4.23 Normal force** See Example 4.11.

In this case, the crate will move when the *horizontal component* of the applied force, $F \cos 30°$, slightly exceeds the maximum static friction force. So, we may write for the maximum friction:

$$F \cos 30° = f_{s_{max}} = \mu_s N$$

However, the magnitude of the normal force is *not* equal to that of the weight of the crate here because of the upward component of the applied force (see the free-body diagram in the figure). By the second law, since $a_y = 0$, we have

$$N + F \sin 30° - mg = 0$$

or

$$N = mg - F \sin 30°$$

In effect, the applied force partially supports the weight of the crate. Substituting this expression for N into the first equation gives

$$F \cos 30° = \mu_s(mg - F \sin 30°)$$

Solving for F gives

$$F = \frac{mg}{(\cos 30°/\mu_s) + \sin 30°}$$

$$= \frac{(40.0 \text{ kg})(9.80 \text{ m/s}^2)}{(0.866/0.650) + 0.500} = 214 \text{ N (about 48 lb)}$$

Thus, less applied force is needed in this case, reflecting the fact that the frictional force is less because of the reduced normal force.

Follow-up Exercise. Note that in this Example, applying the force at an angle produces two effects. As the angle between the applied force and the horizontal increases, the horizontal component of the applied force is reduced. However, the normal force also gets smaller, resulting in a lower $f_{s_{max}}$. Does one effect always outweigh the other? That is, does the applied force F necessary to move the crate *always decrease* with increasing angle? (*Hint*: Investigate F for different angles. For example, compute F for 20° and 50°. You already have a value for 30°. What do the results tell you?)

Now let's look at ways to determine coefficients of friction experimentally.

Example 4.12 ■ A Sliding Block: Determining the Coefficient of Kinetic Friction

A block slides with a constant velocity down a plane inclined at 37° to the horizontal (•Fig. 4.24). What is the coefficient of kinetic friction between the block and the plane?

Thinking It Through. If the block slides down the plane with a constant velocity (zero acceleration), then the weight component down the plane is balanced by the force of kinetic friction up the incline. Hence, they may be equated, which may provide a way to find μ_k.

Solution.

Given: $a = 0$ (because v is constant) *Find:* μ_k (coefficient of kinetic friction)
$\theta = 37°$

Note that we are not given the mass of the block. We will not need it. Since the acceleration is zero, there is no net force on the block in either the x or y direction. So, noting the forces in the free-body diagram and using Newton's second law,

$$F_x = 0 = mg \sin \theta - f_k$$
$$F_y = 0 = N - mg \cos \theta$$

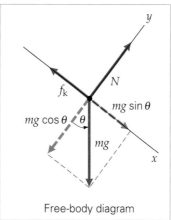

• **FIGURE 4.24 Coefficient of kinetic friction** See Example 4.12.

Rearranging the equations, we have

$$f_k = mg \sin \theta$$

$$N = mg \cos \theta$$

Then, since $f_k = \mu_k N$,

$$\mu_k = \frac{f_k}{N} = \frac{mg \sin \theta}{mg \cos \theta} = \tan \theta = \tan 37° = 0.75$$

Thus, adjusting the angle of incline until the velocity of the block sliding down the plane is constant allows μ_k to be determined experimentally from the angle of incline. From the preceding general result, we can write

$$\mu_k = \tan \theta$$

Suppose that μ_s between the plane and the block is 0.90. Can you determine the angle of incline at which the block will start to move down the plane? It will be an angle just slightly greater than that at which the component of the weight force down the plane equals the maximum force of static friction. That is,

$$mg \sin \theta = f_{s_{max}} = \mu_s N = \mu_s (mg \cos \theta)$$

and

$$\frac{\sin \theta}{\cos \theta} = \tan \theta = \mu_s$$

Thus,

$$\theta = \tan^{-1} \mu_s = \tan^{-1}(0.90) = 42°$$

Therefore, the block will move if the angle of incline exceeds 42°. Adjusting the angle of incline until the block just starts to slide down the plane is an experimental way of approximating μ_s. (This critical angle is called the *angle of repose*.)

Notice how the difference in angles in this Example points out the difference between static friction and kinetic friction. You must tilt the incline to 42° to start the block moving but must reduce it to 37° to keep the block's velocity constant.

Follow-up Exercise. What would be the dynamic situation if the plane in Example 4.12 were inclined at an angle greater than 37°? Would this be a convenient way to determine μ_k?

• **FIGURE 4.25 Air foil** The air foil at the top of the truck's cab makes the truck streamlined and therefore reduces air resistance.

(a) As v increases, so does f.

(b) When $f = mg$, the object falls with a constant (terminal) velocity.

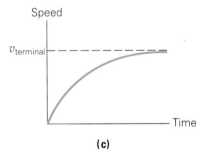

(c)

• **FIGURE 4.26 Air resistance and terminal velocity** **(a)** As the speed of a falling object increases, so does the frictional force of air resistance. **(b)** When this force of friction equals the weight of the object, the net force is zero and the object falls with a constant (terminal) velocity. **(c)** A plot of speed versus time, showing these relationships.

Air Resistance

Air resistance refers to the resistance force acting on an object as it moves through air. In other words, air resistance is a frictional force. In analyses of free fall, you can generally ignore the effect of air resistance and still get valid approximations for objects falling relatively short distances. However, for longer distances, air resistance cannot be ignored.

Air resistance arises because a moving object collides with air molecules. Therefore, air resistance depends on the object's shape and size (which determine the area exposed to collisions) as well as its speed. The larger the object and the faster it moves, the more collisions there will be with air molecules. (Air density is also a factor, but this can be assumed to be constant near the Earth's surface.) To reduce air resistance (and fuel consumption), automobiles are made more "streamlined" and air foils are used on trucks and campers (•Fig. 4.25).

Consider now a falling object. Since air resistance depends on speed, as a falling object accelerates under the influence of gravity, the retarding force of air resistance increases (•Fig. 4.26a). Eventually, the magnitude of the retarding force equals that of the object's weight force (Fig. 4.26b), so the net force on the object is zero. The object then falls with a maximum constant velocity, which is called the **terminal velocity**.

This can be easily seen from Newton's second law. For the falling object, we have

$$F_{net} = ma$$

$$mg - f = ma$$

where downward has been taken as positive for convenience. Solving for a,

$$a = g - \frac{f}{m}$$

where a is the magnitude of the *instantaneous* acceleration.

Notice that the acceleration for a falling object when air resistance is included is less than g; that is, $a < g$ or $a < 9.8 \text{ m/s}^2$. As the object continues to fall, the force of air resistance f increases until $a = 0$ when $f = mg$ and $f - mg = 0$. The object then falls at its constant terminal velocity.

For a skydiver with an unopened parachute, the terminal velocity is about 200 km/h (about 125 mi/h). To reduce the terminal velocity, so that it can be reached sooner and the time of fall be extended, a skydiver will try to increase exposed body area to a maximum by assuming a spread-eagle position (•Fig. 4.27). Doing this takes advantage of the dependence of air resistance on the size and shape of the falling object. Once the parachute is open (giving a larger exposed area and a shape that catches the air), the additional air resistance slows the diver down to about 40 km/h (or 25 mi/h), which is preferable for landing.

Conceptual Example 4.13

Race You Down: Air Resistance and Terminal Velocity

From a high altitude, a balloonist simultaneously drops two balls of identical size but appreciably different weight. Assuming that both balls reach terminal velocity during the fall, (a) the heavier ball reaches terminal velocity first, (b) the balls reach terminal velocity at the same time, (c) the heavier ball hits the ground first, (d) the balls hit the ground at the same time. *Clearly establish the reasoning and physical principle(s) used in determining your answer before checking it below. That is, **why** did you select your answer?*

Reasoning and Answer. Terminal velocity is reached when the weight of a ball is balanced by the frictional air resistance. Both balls start to fall with the same acceleration, g, and their speeds and the retarding forces of air resistance increase at the same rate. The weight of the lighter ball will be balanced first, so (a) and (b) are incorrect. The lighter ball reaches terminal velocity ($a = 0$), but the heavier ball continues to accelerate and pulls ahead of the lighter ball. Hence, the heavier ball hits the ground first and the answer is (c), and (d) does not apply.

Follow-up Exercise. Suppose the heavier ball were much larger than the lighter ball. How might this difference affect the outcome?

•**FIGURE 4.27 Terminal velocity** Sky divers assume a spread-eagle position to maximize air resistance. This causes them to reach terminal velocity more quickly and prolongs the time of fall.

Chapter Review

Important Terms

force *100*
net (unbalanced) force *100*
inertia *102*
Newton's first law of motion
  (law of inertia) *102*
newton (unit) *104*
Newton's second law of
  motion *104*
weight *104*
Newton's third law of
  motion *109*
normal force *110*
free-body diagram *115*
translational equilibrium *119*
condition for translational
  equilibrium *120*
force of friction *121*
static friction *121*
kinetic (sliding) friction *121*
rolling friction *121*
coefficient of static friction *122*
coefficient of kinetic friction *122*
air resistance *128*
terminal velocity *128*

Important Concepts

- A force is something capable of changing an object's state of motion. To produce a change in motion, there must be a net or unbalanced force.
- Newton's first law of motion is also called the law of inertia, where inertia is the natural tendency of an object to maintain its state of motion. It states that in the absence of a net applied force, a body at rest remains at rest and a body in motion remains in motion with constant velocity.
- Newton's second law relates the net force acting on an object or system to the (total) mass and the resulting acceleration. It defines the cause-and-effect relationship between force and acceleration ($\mathbf{F} \propto m\mathbf{a}$).
- Newton's third law states that for every force there is an equal and opposite reaction force. The opposing forces of a third-law pair always act on different objects.
- Friction is the resistance to motion that occurs between contacting surfaces.
- The frictional force between surfaces is characterized by coefficients of friction (μ), one for the static case and one for the kinetic (moving) case. In many cases, $f = \mu N$, where N is the normal force—the force perpendicular to the surface (the force exerted *by* the surface *on* the object). As a ratio of forces, μ is unitless.
- The force of air resistance on a falling object increases with increasing speed until the object falls at a constant rate called the terminal velocity.

Important Equations

Newton's Second Law:
$$\mathbf{F} = m\mathbf{a} \quad (4.1)$$
where $\mathbf{F} = \sum_i \mathbf{F}_i = \mathbf{F}_{net}$

Weight:
$$w = mg \quad (4.2)$$

Component Form of Newton's Second Law:
$$F_x \hat{\mathbf{x}} + F_y \hat{\mathbf{y}} = ma_x \hat{\mathbf{x}} + ma_y \hat{\mathbf{y}} \quad (4.3a)$$

Condition for Translational Equilibrium:
$$\Sigma \mathbf{F} = 0 \quad (4.4)$$
or
$$\Sigma F_x = 0 \quad \text{and} \quad \Sigma F_y = 0 \quad (4.5)$$

Force of Static Friction:
$$f_s \leq \mu_s N \quad (4.6)$$
$$f_{s_{max}} = \mu_s N \quad (4.7)$$

Force of Kinetic (Sliding) Friction:
$$f_k = \mu_k N \quad (4.8)$$

Exercises

Note: Exercises with a CD-ROM icon have corresponding simulations on the Interactive Journey through physics CD-ROM; a complete cross-reference of text problems to CD-ROM activities appears in the Media Pack.

Note: Unless otherwise stated, all objects are located near the Earth's surface, where $g = 9.80 \text{ m/s}^2$.

4.1 The Concepts of Force and Net Force
and
4.2 Inertia and Newton's First Law of Motion

1. If an object remains at rest, there must be no force acting on it. Is this statement correct? Explain.

2. The tendency of an object to maintain its state of motion is called (a) Newton's first law, (b) Galileo's principle, (c) inertia.

3. If an object is moving with constant velocity, (a) there must be a force in the direction of the velocity, (b) there must be no force in the direction of the velocity, (c) there must be no net force.

4. If the net force on an object is zero, the object must (a) be at rest, (b) be in motion with a constant velocity, (c) have zero acceleration, (d) none of these.

5. An object weighs 300 N on Earth and 50 N on the Moon. Does the object also have less inertia on the Moon?

6. Consider an air-bubble level that is sitting on a horizontal surface (•Fig. 4.28). Initially, the air bubble is in the middle of the horizontal glass tube. (a) If the level is pushed and a force is applied to accelerate it, which way would the bubble move? Which way would the bubble move if the force is then removed and the level slows down due to friction? (b) Such a level is sometimes used as an "accelerometer" to indicate the direction of the acceleration. Explain the principle involved. [Hint: Think about pushing a pan of water.]

7. As a follow-up to Exercise 6, consider the situation of a child holding a helium balloon in a closed car at rest. What would the child observe when the car (a) accelerates from rest and (b) brakes to a stop? (The balloon does not touch the roof of the car.)

8. Objects have no weight in outer space. How can you then distinguish their masses?

9. If a tablecloth is yanked out very quickly, the dishes on top of it will barely move (•Fig. 4.29). Why?

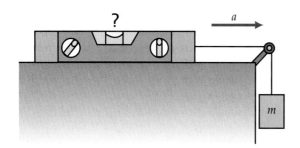

•FIGURE 4.28 An air-bubble level/accelerometer See Exercise 6.

•FIGURE 4.29 No more dinner? See Exercise 9.

10. If your professor's car is traveling west with a constant velocity of 55 mi/h west on a straight highway, what is the net force acting on it?

11. ■ Which has more inertia, 20 cm³ of water or 10 cm³ aluminum, and how many times more? (See Table 9.2.)

12. ■■ 🔘 You are told that an object has zero acceleration. Two forces on the object are F_1 = 3.6 N at 74° below the $+x$ axis and F_2 = 3.6 N at 34° above the $-x$ axis. Is there a third force on the object? If yes, what is it? Can you tell whether the object is at rest or in motion?

13. ■■ 🔘 A 5.0-kilogram block at rest on a frictionless surface is acted on by forces F_1 = 5.5 N and F_2 = 3.5 N, as illustrated in •Fig. 4.30. What additional horizontal force will keep the block at rest?

• FIGURE 4.30 **Two applied forces** See Exercises 13 and 86.

14. ■■■ 🔘 A 1.5-kilogram object moves up the y axis with a constant speed. When it reaches the origin, the forces F_1 = 5.0 N at 37° above the $+x$ axis, F_2 = 2.5 N in the $+x$ direction, F_3 = 3.5 N at 45° below the $-x$ axis, and F_4 = 1.5 N in the $-y$ direction are applied to it. (a) Will the object continue to move along the y axis? (b) If not, what simultaneously applied force will keep it moving along the y axis with a constant speed?

4.3 Newton's Second Law of Motion

15. The newton unit of force is equivalent to (a) kg·m/s, (b) kg·m/s², (c) kg·m²/s, (d) none of these.

16. In general, this chapter considered forces that were applied to objects of constant mass. What would be the situation if mass were added to or lost from a system while a force was being applied to the system? Give examples of situations in which this might happen.

17. Good football wide receivers usually have "soft" hands for catching balls (•Fig. 4.31). How would you interpret this description on the basis of Newton's second law?

18. ■ A 3.0-newton net force is applied to a 1.5-kilogram mass. What is its acceleration?

19. ■ What is the mass of an object that accelerates at 3.0 m/s² under the influence of a 5.0-newton net force?

20. ■ A worker pushes on a crate, which experiences a net force of 75 N. If the crate also experiences an acceleration of 0.50 m/s², what is its weight?

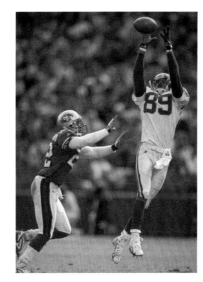

• FIGURE 4.31 **Soft hands** See Exercise 17.

21. ■ An ocean tanker has a gross mass of 7.0 × 10⁷ kg. What constant net force would give the tanker an acceleration of 0.10 m/s²?

22. ■ A 6.0-kilogram object is brought to the Moon, where the acceleration due to gravity is only one-sixth of that on the Earth. What is the mass of the object on the Moon? (a) 1.0 kg (b) 6.0 kg (c) 59 kg (d) 9.8 kg

23. ■ What is the mass of a person weighing 740 N?

24. ■ What is the net force acting on a 1.0-kilogram object in free fall?

25. ■ What is the weight of an 8.0-kilogram mass in newtons? How about in pounds?

26. ■■ What is the weight of a 150-pound person in newtons? What is his mass in kilograms?

27. ■■ (a) Is the product label shown in •Fig. 4.32 correct on the Earth? (b) Is it correct on the Moon, where the acceleration due to gravity is only about one-sixth of that on Earth? If not, what should the label be on the Moon?

• FIGURE 4.32 **Correct label?** See Exercise 27.

28. ■■ In a college orientation competition, 18 students lift a sports car. While holding the car off the ground, each student exerts an upward force of 600 N. (a) What is the mass of the car in kilograms? (b) What is its weight in pounds?

29. ■■ A horizontal force of 12 N acts on an object that rests on a level, frictionless surface on the Earth, where the object has a weight of 98 N. (a) What is the magnitude of the acceleration of the object? (b) What would be the acceleration of the same object in a similar situation on the Moon?

30. ■■ The engine of a 1.0-kilogram toy plane exerts a 15-newton forward force on the plane. If the air exerts an 8.0-newton resistive force on the plane, what is the magnitude of the acceleration of the plane?

31. ■■ When a horizontal force of 300 N is applied to a 75.0-kilogram box, the box slides on a level floor, opposed by a force of kinetic friction of 120 N. What is the magnitude of the acceleration of the box?

32. ■■ An ocean-going oil tanker that has a mass of 6.4×10^7 kg and that travels with a speed of 15 knots takes 5.0 km to come to a stop. What is the magnitude of the constant force required to do this? [1 knot = 1.15 (statute) mi/h]

33. ■■ A stalled 1500-kilogram automobile is pushed toward a gas station by a man and a woman on a level road. The applied horizontal forces are 200 N for the woman and 300 N for the man. (a) If there is an effective force of friction of 300 N on the car as it moves, what is the car's acceleration? (b) Once the car is moving appreciably, what would be an appropriate combined applied force, and why?

34. ■■ In an emergency stop to avoid an accident, a shoulder-strap seat belt holds a 60-kilogram passenger firmly in place. If the car was initially traveling at 90 km/h and came to a stop in 5.5 s along a straight, level road, what was the average force applied to the passenger by the seat belt?

35. ■■ A jet catapult on an aircraft carrier accelerates a 2000-kilogram plane uniformly from rest to a launch speed of 320 km/h in 2.0 s. What is the magnitude of the net force on the plane?

36. ■■ In serving, a tennis player accelerates a 56-gram tennis ball horizontally from rest to a speed of 35 m/s. Assuming the acceleration is uniform when the racquet applies a force over a distance of 0.50 m, what is the magnitude of the force exerted on the ball?

4.4 Newton's Third Law of Motion

37. A brick hits a glass window. The brick breaks the glass, so (a) the magnitude of the force of the brick on the glass is greater than the magnitude of the force of the glass on the brick, (b) the magnitude of the force of the brick on the glass is smaller than the magnitude of the force of the glass on the brick, (c) the magnitude of the force of the brick on the glass is equal to the magnitude of the force of the glass on the brick, (d) none of the above.

38. The force pair of Newton's third law (a) consists of forces that are always opposite but not always equal, (b) always cancel each other when the second law is applied to a body, (c) always act on the same object, (d) consists of forces that are equal and opposite but act on different objects.

39. Is there something wrong with the following statement? A baseball bat hits a baseball, and there are equal and opposite forces on the baseball. The forces then cancel, and there is no motion.

40. Draw a free-body diagram of a book sitting on a horizontal surface. Identify the reaction forces to all the forces in the diagram.

41. By using the right technique, a karate master can exert huge forces on objects. If a fist hits a brick with a force of 800 N, what else do we know according to Newton's third law? Is this something you can try at home?

42. ■ A person pushes on a block of wood that has been placed against a wall. Draw a free-body diagram and identify the reaction forces to all the forces on the block.

43. ■■ In the pairs competition of Olympic figure skating, a male skater with a mass of 60 kg pushes a female skater who has a mass 45 kg, causing her to accelerate at a rate of 2.0 m/s². At what rate will the male skater accelerate? What is the direction of his acceleration?

44. ■■ Jane and John, with masses of 50 kg and 60 kg, respectively, stand on a frictionless surface 10 m apart. John pulls on a connecting rope, giving Jane an acceleration of 0.92 m/s² toward him. (a) What is John's acceleration? (b) If the pulling force is applied constantly, where will they meet?

4.5 More on Newton's Laws: Free-body Diagrams and Translational Equilibrium

Note: In Exercises with strings and pulleys, "ideal conditions" means that the masses of the string(s) and pulley(s), as well as the friction of the pulley, should be neglected.

45. ■■ Is it possible for a 500-kilogram object on a scale to appear to weigh only 4000 N in a moving elevator? If so, explain and describe the motion.

46. ■■ 🌐 A 75.0-kilogram person is standing on a scale in an elevator. What is the reading of the scale in newtons if the elevator is (a) at rest, (b) moving up with a constant velocity of 2.00 m/s, (c) accelerating up at 2.00 m/s²?

47. ■■ 🌐 In Exercise 46, what if the elevator is accelerating downward at the same rate?

48. ■■ Two boats pull a 75.0-kilogram water skier as illustrated in •Fig. 4.33. (a) If each boat pulls with a force of 600 N and the skier travels with a constant velocity, what is the magnitude of the retarding force between the water and the skis? (b) Assuming the retarding force remains constant, if each boat pulls with a force of 700 N, what is the magnitude of the acceleration of the skier?

•FIGURE 4.33 Double tow   See Exercise 48.

49. ■■ (a) A 65-kilogram water skier is pulled due east by a boat with a horizontal force of 400 N with a water drag of 300 N. A sudden gust of wind supplies another horizontal force of 50 N on the skier at an angle of 60° north of east. What is the skier's acceleration at that instant? (b) What would be the acceleration if the wind force were in the opposite direction of that in part (a)?

50. ■■ A boy pulls a box of mass 30 kg with a force of 25 N in the direction shown in •Fig. 4.34. (a) What is the acceleration of the box? (b) What is the normal force on the box by the ground? Ignore friction.

•FIGURE 4.34 Pulling a box   See Exercise 50.

51. ■■ A girl pushes a 25-kilogram lawn mower as shown in •Fig. 4.35. If $F = 30$ N and $\theta = 37°$, (a) what is the acceleration of the mower? (b) What is the normal force exerted on the mower by the lawn? Ignore friction.

52. ■■ ⊙ An Olympic skier coasts down a slope with an angle of inclination of 37°. (a) Neglecting friction, what is the acceleration of the skier? (b) If the skier has a speed

•FIGURE 4.35 Mowing the lawn   See Exercise 51.

of 5.0 m/s at the top of the slope, what is his speed when he reaches the bottom of the 35-meter-long slope?

53. ■■ A car is coasting (engine off) up a 30° grade. If the speed of the car is 25 m/s at the bottom of the grade, what is the distance traveled by the car before coming to rest?

54. ■■ A horizontal force of 40 N acting on a block on a frictionless level surface produces an acceleration of 2.5 m/s². A second block with a mass of 4.0 kg is dropped onto the first. What is the magnitude of the acceleration of the combination if the same force continues to act? (Assume that the second block does not slide on the first block.)

55. ■■ Two blocks, weighing 80 N and 50 N, respectively, sit side by side in contact with each other on a frictionless horizontal surface. (a) If a constant horizontal force of 40 N is applied to one of the blocks in the direction of the other block, what is the magnitude of the resulting acceleration? (b) If the force is applied instead to the block downward at an angle of 25° to the horizontal, what would be the acceleration?

56. ■■ A 50-kilogram gymnast hangs vertically from a pair of parallel rings. (a) If the ropes supporting the rings are attached to the ceiling directly above, what is the tension in each rope? (b) If the ropes are supported so that they make an angle of 45° with the ceiling, what is the tension in each rope?

57. ■■ A truck of mass 3000 kg tows a 1500-kilogram car by a chain. If the net forward force on the truck by the ground is 3200 N, (a) what is the acceleration of the car? (b) What is the tension in the connecting chain?

58. ■■ ⊙ Three blocks are pulled along a frictionless surface by a horizontal force as shown in •Fig. 4.36. (a) What is the acceleration of the system? (b) What are the tension forces in the light strings? [Hint: Can T_1 equal T_2? Investigate by drawing free-body diagrams of each block separately.]

•FIGURE 4.36 Three-block system   See Exercises 58 and 87.

59. ■■ Assume ideal conditions for the apparatus illustrated in •Fig. 4.37. What is the acceleration of the system if (a) $m_1 = 0.25$ kg, $m_2 = 0.50$ kg, and $m_3 = 0.25$ kg, and (b) $m_1 = 0.35$ kg, $m_2 = 0.15$ kg, and $m_3 = 0.50$ kg?

•**FIGURE 4.37 Which way will they accelerate?** See Exercises 59, 89, and 90.

60. ■■ An Atwood machine (Fig. 4.14) has suspended masses of 0.25 kg and 0.20 kg. Under ideal conditions, what will be the acceleration of the smaller mass?

61. ■■ Consider an ideal Atwood machine (Fig. 4.14) in which m_2 (0.15 kg) is released 1.0 m above the floor and m_1 (0.14 kg) rests on the floor. If it takes m_2 2.4 s to reach the floor, what is the experimental value of g?

62. ■■■ One mass, $m_1 = 0.215$ kg, of an ideal Atwood machine rests on the floor 1.10 m below the other mass, $m_2 = 0.255$ kg. (a) If the masses are released from rest, how long does it take m_2 to reach the floor? (b) How high will mass m_1 ascend from the floor? [Hint: When m_2 hits the floor, m_1 continues to move upward.]

63. ■■■ A 0.20-kilogram ball is released from a height of 10 m above the beach; the impression the ball makes in the sand is 5.0 cm deep while the ball stops uniformly. What is the force acting on the ball by the sand?

64. ■■■ In the ideal apparatus shown in •Fig. 4.38, $m_1 = 2.0$ kg. What is m_2 if both masses are at rest? How about if both are moving at a constant velocity?

•**FIGURE 4.38 Inclined Atwood machine** See Exercises 64, 65, and 91.

65. ■■■ In the ideal setup shown in Fig. 4.38, $m_1 = 3.0$ kg and $m_2 = 2.5$ kg. (a) What is the acceleration of the masses? (b) What is the tension in the string?

4.6 Friction

Note: Neglect air resistance unless otherwise stated.

66. In general, the frictional force (a) is greater for smooth surfaces, (b) depends on sliding speeds, (c) is proportional to the normal force, (d) depends significantly on the surface area.

67. Identify the direction of the friction force in the following cases: (a) a book sitting on a table, (b) a box sliding on a horizontal surface, (c) a car making a turn on a flat road, (d) the initial motion of a machine part delivered on a conveyor belt in an assembly line.

68. The coefficient of kinetic friction μ_k (a) is usually greater than the coefficient of static friction μ_s, (b) usually equals μ_s, (c) is usually smaller than μ_s, (d) equals the applied force that exceeds the maximum static force.

69. The purpose of anti-lock brakes is to avoid the lock-up of the wheels to keep the car rolling rather than sliding. Why would rolling decrease the stopping distance?

70. Is it easier to push or pull a lawnmower at an angle? Explain.

71. •Figure 4.39 shows the front and rear wings of an Indy car. These wings generate *down force*, the vertical downward force by the air moving over the car. Why is such a down force desired? Will the additions of the wings slow down the car? An Indy car can create a down force equal to twice its weight. Why not simply make the cars heavier?

•**FIGURE 4.39 Down force** See Exercise 71.

72. (a) We commonly say that friction opposes motion. Yet, when we walk, the frictional force is in the direction of our motion (Fig. 4.19). Is there an inconsistency in terms of Newton's second law? Explain. (b) What effects would wind have on air resistance? [Hint: The wind can blow in different directions.]

73. Why are racing tires wide and smooth whereas regular tires are narrower and have tread (•Fig. 4.40)? Are there frictional and/or safety considerations?

• **FIGURE 4.40 Racing tires versus passenger tires: safety or speed?** See Exercise 73.

74. ■ ⚙ In moving a 35.0-kilogram desk from one side of a classroom to the other, a professor finds that a horizontal force of 275 N is necessary to set the desk in motion and a force of 195 N is necessary to keep it in motion with a constant speed. What are the coefficients of (a) static and (b) kinetic friction between the desk and the floor?

75. ■ ⚙ A 40-kilogram crate is at rest on a level surface. If the coefficient of static friction between the crate and the surface is 0.69, what horizontal force is required to get the crate moving?

76. ■ A 20-kilogram box sits on a horizontal surface. The forces required to set it in motion and then to keep it in motion at constant velocity are 150 N and 120 N, respectively. What are the coefficients of static and kinetic friction?

77. ■ The coefficients of static and kinetic friction between a 50-kilogram box and a horizontal surface are 0.60 and 0.40, respectively. (a) What is the acceleration of the object if a 250-newton horizontal force is applied to the box? (b) What if the applied force is 350 N?

78. ■■ ⚙ A 1500-kilogram automobile travels at a speed of 90 km/h along a straight concrete highway. Faced with an emergency situation, the driver jams on the brakes, and the car skids to a stop. What will the stopping distance be for (a) dry pavement and (b) wet pavement?

79. ■■ ⚙ A hockey player hits a puck with her stick, giving the puck an initial speed of 5.0 m/s. If the puck slows uniformly and comes to rest in a distance of 20 m, what is the coefficient of kinetic friction between the ice and the puck?

80. ■■ ⚙ A packing crate is placed on a 20° inclined plane. If the coefficient of static friction between the crate and the plane is 0.65, will the crate slide down the plane? (Justify your answer.)

81. ■■ A wood block with a weight of 3.5 N is placed on a wood board, measuring 0.75 m × 0.75 m, that is elevated at one end to act as an inclined plane. What is the elevation of the higher end of the board when the block, once it is in motion, slides down with a constant velocity?

82. ■■ ⚙ Suppose the slope conditions for the skier shown in •Fig. 4.41 are such that he travels with a constant velocity. From the photo, could you find the coefficient of kinetic friction between the snowy surface and the skis? If so, describe how this would be done.

• **FIGURE 4.41 A downslope run** See Exercise 82.

83. ■■ ⚙ A 5.0-kilogram wooden block is placed on an adjustable wooden inclined plane. (a) What is the angle of incline above which the block will *start* to slide down the plane? (b) At what angle of incline will the block then slide down the plane with a constant speed?

84. ■■ A block that has a mass of 2.0 kg and is 10 cm wide on each side just slides down an inclined plane with a 30° angle of incline (•Fig. 4.42). Another block of the same height and same material has base dimensions of 20 cm × 10 cm and thus a mass of 4.0 kg. (a) At what critical angle will the more massive block start to slide down the plane? (b) Estimate the coefficient of static friction between the block and the plane.

• **FIGURE 4.42 At what angle will it begin to slide?** See Exercise 84.

85. ■■ ⚙ While being unloaded from a moving truck, a 10-kilogram suitcase is placed on a flat ramp inclined at 37°. When released from rest, the suitcase accelerates down the ramp at 0.15 m/s². What is the coefficient of kinetic friction between the suitcase and the ramp?

86. ■■ For the situation shown in Fig. 4.30, what is the minimum coefficient of static friction between the block and the surface that will keep the block from moving? ($F_1 = 5.0$ N, $F_2 = 4.0$ N, and $m = 5.0$ kg.)

87. ■■ ⊙ For the system illustrated in Fig. 4.36, if $\mu_s = 0.45$ and $\mu_k = 0.35$ between the blocks and the surface, what applied forces will (a) set the blocks in motion and (b) move the blocks with a constant velocity?

88. ■■ ⊙ In the apparatus shown in •Fig. 4.43, $m_1 = 10$ kg and the coefficients of static and kinetic friction between m_1 and the table are 0.60 and 0.40, respectively. (a) What mass of m_2 will set the system in motion? (b) With mass m_2 as in part (a), after the system moves, what is the acceleration?

•FIGURE 4.43 Friction and motion. See Exercise 88.

89. ■■■ For the apparatus in Fig. 4.37, what is the minimum value of the coefficient of static friction between the block (m_3) and the table that would keep the system at rest if $m_1 = 0.25$ kg, $m_2 = 0.50$ kg, and $m_3 = 0.75$ kg? (Assume ideal conditions for the string and pulleys.)

90. ■■■ If the coefficient of kinetic friction between the block and the table in Fig. 4.37 is 0.560, and $m_1 = 0.150$ kg and $m_2 = 0.250$ kg, (a) what should m_3 be if the system is to move with a constant speed? (b) If $m_3 = 0.100$ kg, what is the acceleration of the system? (Assume ideal conditions for the string and pulleys.)

91. ■■■ In the apparatus shown in Fig. 4.38, $m_1 = 2.0$ kg and the coefficients of static and kinetic friction between m_1 and the inclined plane are 0.30 and 0.20, respectively. (a) What is m_2 if both masses are to remain at rest? (b) What is m_2 if both masses are moving with constant velocity? Neglect all friction.

Additional Exercises

92. A force acts on a mass. If the force is halved and the mass is doubled, how is the acceleration affected? (Give a factor of change—for example, 2 times as great or 1/2 as great.)

93. A rifle weighs 50.0 N, and its barrel is 0.750 m long. It shoots a 25.0-gram bullet, which leaves the barrel at a speed of 300 m/s (muzzle velocity) after being uniformly accelerated. What is the magnitude of the force exerted on the rifle by the bullet?

94. The maximum load that can safely be supported by a rope in an overhead hoist is 400 N. What is the maximum acceleration that can safely be given to a 25-kilogram object being hoisted vertically upward?

95. The coefficient of static friction between a 9.0-kilogram object and a horizontal surface is 0.45. Would a force of 35 N applied horizontally cause the object to move from rest? If so, what would be its acceleration?

96. An object with a mass of 2.0 kg travels with a constant velocity of 4.8 m/s northward. It is then acted on by a force of 6.5 N in the direction of the motion and a force of 8.5 N to the south, both of which continue even after the mass comes momentarily to rest. (a) How far will the object travel before coming to rest? (b) What will be its position 1.5 s after the object comes momentarily to rest?

97. In the operation of a machine, a 5.0-kilogram steel part is to move on a horizontal steel surface. Force is applied to the part downward at an angle of 30° from the horizontal. (a) What is the magnitude of the applied force required to set the part in motion if the surface is dry? (b) If the surface is lubricated, what is the needed force?

98. ⊙ A crate weighing 9.80×10^3 N is pulled up a 37° incline by a force parallel to the plane. If the coefficient of kinetic friction between the crate and the surface of the plane is 0.750, what is the magnitude of the applied force required to move the crate with a constant velocity?

99. For an Atwood machine with suspended masses of 0.30 kg and 0.40 kg, the acceleration of the masses is measured as 0.95 m/s². What is the effective force of friction for the system?

100. A catapult jet plane with a weight of 2.75×10^6 N is ready for takeoff. If its engines supply 6.35×10^6 N of net thrust, how long a runway will the plane need to reach its minimum take-off speed of 285 km/h?

101. A 0.45-kilogram shuffleboard puck is given an initial speed of 4.5 m/s down the flat playing surface. If the coefficient of sliding friction between the puck and the surface is 0.20, how far will the puck slide before coming to rest?

102. A frictionless ramp 135 m long is to be built for a ski jump. If a skier starting from rest at the top is to have a speed no faster than 24 m/s at the bottom, what should be the maximum angle of inclination? Ignore friction.

103. A school bus pulls into an intersection as a car approaches with a speed of 25 km/h on an icy street. Seeing the bus from 26 m away, the driver of the car locks the brakes, causing the car to slide toward the intersection. If the coefficient of kinetic friction between the car's tires and the icy road is 0.10, does the car hit the bus?

104. While catching a baseball that is traveling horizontally with a speed of 15 m/s, the player's glove and arm move straight backward 25 cm from the time of contact to the time the ball comes to rest. If the ball has a mass of 0.14 kg, what is the magnitude of the average force on the ball during that interval?

Work and Energy

CHAPTER 5

5.1 Work Done by a Constant Force
5.2 Work Done by a Variable Force
5.3 The Work–Energy Theorem: Kinetic Energy
5.4 Potential Energy
5.5 The Conservation of Energy
5.6 Power

Insight

- More Broken Records: The Clap Skate

Work and energy can be somewhat slippery concepts. For the most part, our everyday sense of these words accords pretty well with their definitions in physics, but not always. You might think that this crane, for example, must be performing a good deal of work to support the miner and move him around, but that's not necessarily true from a physicist's point of view. Depending on how and where the bucket is moved, the crane might be doing no work at all. It might even be doing *negative* work—a concept that most people can't easily imagine. A physicist would also say that the miner has slightly less energy down here than he would outside the mine—not because he prefers working on his lawn to mining zinc, but because he's a tiny bit closer to the center of the Earth.

This chapter centers on two concepts that are important in both science and everyday life—*work* and *energy*. We commonly think of work as being associated with doing or accomplishing something. Because work makes us physically (and sometimes mentally) tired, we have invented machines and use them to decrease the amount of effort we expend personally. Energy tends to bring to mind the cost of fuel for transportation and heating, or perhaps the food that supplies the energy our bodies need to carry out life processes and to do work.

Although these notions do not really define work and energy, they point us in the right direction. As you might have guessed work and energy are closely related. In physics, as in everyday life, when something possesses energy, it usually has the ability to do work. For example, water rushing through the sluices of a dam has energy of motion, and this energy allows the water to do the work of driving a turbine or dynamo. Conversely, no work can be performed without energy.

Energy exists in various forms: There is mechanical energy, chemical energy, electrical energy, heat energy, nuclear energy, and so on. A transformation from one form to another may take place, but the total amount of energy is *conserved*, or always remains the same. This is the point that makes the concept of energy so useful. When a physically measurable quantity is conserved, it not only gives us an insight that leads to a better understanding of nature, but also usually provides another approach to practical problems. (You will be introduced to other conserved quantities during the course of our study of physics.)

5.1 Work Done by a Constant Force

Objectives: To (a) define mechanical work and (b) compute the work done in various situations.

The word *work* is commonly used in a variety of ways: We go to work; we work on projects; we work at our desks or on computers; we work problems. In physics, however, work has a very specific meaning. Mechanically, **work** involves force and displacement, and we use this word to describe quantitatively what is accomplished when a force moves an object through a distance. In the simplest case of a *constant* force acting on an object:

Work—involves force and displacement

> The work done by a constant force in moving an object is equal to the product of the magnitudes of the displacement and the component of the force parallel to the displacement.

Work then involves moving an object through a distance. A force may be applied, as in •Fig. 5.1a, but *if there is no motion (no displacement), then no work is done*. For a constant force **F** acting *in the same direction* as the displacement **d** (Fig. 5.1b), the work (W) is simply the product of their magnitudes:

$$W = Fd \tag{5.1}$$

(As you might expect, when work is done in Fig. 5.1b, energy is expended. We shall discuss the relationship between work and energy in Section 5.3.)

In general, work is done on an object only by a force, or force *component*, parallel to the line of motion or displacement of that object (Fig. 5.1c). That is, if the force acts at an angle θ to the object's displacement, then $F_\parallel = F \cos\theta$ is the com-

•**FIGURE 5.1 Work done by a constant force**—the product of the magnitudes of the parallel component of force and the displacement   (a) If there is no displacement, no work is done: $W = 0$. (b) For a constant force in the same direction as the displacement, $W = Fd$. (c) For a constant force at an angle to the displacement, $W = (F \cos\theta)d$.

ponent of the force parallel to the displacement. Thus, a more general equation for work done by a constant force is

$$W = F_\parallel d = (F \cos \theta) d \quad \text{work done by a constant force} \quad (5.2)$$

Notice that θ is the angle *between* the force and the displacement vectors; as a reminder, cos θ can be written between the magnitudes of the force and displacement in Eq. 5.2. If $\theta = 0°$ (force and displacement are in the same direction as in Fig. 5.1b), then $W = F(\cos 0°)d = Fd$, so Eq. 5.2 reduces to Eq. 5.1. The perpendicular component of the force, $F_\perp = F \sin \theta$, does no work since there is no displacement in this direction.

The units of work can be determined from the equation $W = Fd$. With force in newtons and displacement in meters, work has the SI unit of newton-meters (N·m). This unit is given the special name **joule** (J):

$$Fd = W$$

$$1 \text{ N·m} \equiv 1 \text{ J}$$

For example, the work done by a force of 25 N in moving an object through a parallel displacement of 2.0 m is $W = Fd = (25 \text{ N})(2.0 \text{ m}) = 50$ N·m, or 50 J.

From the previous equation, we also see that in the British system work would have the unit pound-foot. However, this name is commonly written in reverse. The British standard unit of work is the *foot-pound* (ft·lb).

Remember that *work is a scalar quantity* and, as such, may have a positive or negative value. In Fig. 5.1b, the work would be positive because the force acts in the same direction as the displacement (cos 0° is positive). The work would also be positive in Fig. 5.1c, because a force component acts in the direction of the displacement (cos θ is positive).

However, if the force, or a force component, acts in the opposite direction of the displacement, the work is negative since the cosine term is negative. For example, for $\theta = 180°$ (force opposite displacement), cos 180° = -1, so the work is negative: $W = Fd = (F \cos 180°)d = -Fd$. An example is a braking force that tends to slow down or decelerate an object. Work is done (and energy expended) to accomplish this action. See Learn by Drawing on p. 140.

The product of two vectors (force and displacement) in this case is a special type of vector multiplication and yields a scalar quantity equal to $(F \cos \theta) d$. Thus, work is a scalar—it does not have direction. It can, however, be positive, zero, or negative, depending on the angle.

The joule (J) (pronounced "jool") was named in honor of James Prescott Joule (1818–1889), a British scientist who investigated work and energy.

Example 5.1 ■ Applied Psychology: Mechanical Work

A student holds her psychology textbook, which has a mass of 1.5 kg, out of a second-story dormitory window until her arm is tired; then she releases it (•Fig. 5.2). (a) How much work is done on the book by the student in simply holding it out the window? (b) How much work will have been done by the force of gravity during the time in which the book falls 3.0 m?

Thinking It Through. Analyze the situations in terms of the definition of work, keeping in mind that force and displacement are the key factors.

Solution. Listing the data, we have

Given: $v_o = 0$ (initially at rest)   Find: (a) W (work in holding)
$m = 1.5$ kg   (b) W (work in falling)
$d = 3.0$ m

(a) Even though the student gets tired (because work is performed within the body to maintain muscles in a state of tension), she does *no* mechanical work *on the book* in merely holding it stationary. She exerts an upward force on the book (equal in magnitude to its weight), but the displacement is zero in this case ($d = 0$). Thus, $W = Fd = F \times 0 = 0$ J.

(b) While the book is falling, the net force acting on it is the force of gravity, which is equal in magnitude to the weight of the book: $F = w = mg$ (neglecting air resistance).

•**FIGURE 5.2 Mechanical work requires motion** See Example 5.1.

5.1 Work Done by a Constant Force 139

The displacement is in the same direction as the force ($\theta = 0°$) and has a magnitude of $d = 3.0$ m, so the work done by gravity is

$$W = Fd \cos 0° = (mg)d = (1.5 \text{ kg})(9.8 \text{ m/s}^2)(3.0 \text{ m}) = +44 \text{ J}$$

Follow-up Exercise. How much work is done on a 0.20-kilogram ball by gravity as the ball rises between heights of 2.0 m and 3.0 m?

LEARN BY DRAWING
Determining the Sign of Work

Example 5.2 ■ Yard Work: Parallel Component of a Force

If the person in Fig. 5.1c pushes on the lawn mower with a constant force of 90.0 N at an angle of 40° to the horizontal, how much work does she do in pushing it a horizontal distance of 7.50 m?

Thinking It Through. Only the component of the force parallel to the displacement does work on the mower. Clearly identify the angle between the force and the displacement.

Solution.

Given: $F = 90.0$ N
$\theta = 40°$
$d = 7.50$ m

Find: W (work done by person)

Here, the horizontal component of the applied force, $F \cos \theta$, is parallel to the displacement, so Eq. 5.2 applies:

$$W = (F \cos 40°)d = (90.0 \text{ N})(0.766)(7.50 \text{ m}) = +517 \text{ J}$$

Work is done by the horizontal component of the force, but the vertical component does no work because there is no vertical displacement.

Follow-up Exercise. When you wheel a wheelbarrow on a level surface, the force you apply has an upward component. Is work done by this component?

We commonly specify which force is doing work *on* which object. For example, the force of gravity does work on a falling object, such as the book in Example 5.1. Also, when you lift an object, *you* do work *on* the object. We sometimes describe this as doing work *against* gravity, because the force of gravity acts in the direction opposite that of the applied lift force and opposes it.

In both Examples 5.1 and 5.2, work was done by a single constant force. If more than one force acts on an object, the work done by each can be calculated separately. The *total* or *net work* is then the work done by all the forces, or the scalar sum of those quantities of work, as shown in the following Example.

Example 5.3 ■ Total or Net Work

A 0.75-kilogram block slides with a uniform velocity down a 20° inclined plane (●Fig. 5.3). (a) How much work is done by the force of friction on the block as it slides the total length of the plane? (b) What is the net work done on the block? (c) Discuss the net work done if the angle of incline is adjusted so that the block accelerates down the plane.

Thinking It Through. (a) The length of the plane can be found by using trigonometry, so this part boils down to finding the force of friction. (b) The net work is the sum of all the work done by the individual forces. (c) If there is an acceleration, Newton's second law applies, and this involves a net force.

Solution. We list what is given, but equally importantly we want to know specifically what is to be found.

Given: $m = 0.75$ kg
$\theta = 20°$
$L = 1.2$ m (from figure)

Find: (a) W_f (work done by friction)
(b) W_{net} (net work)
(c) W (discuss work with block accelerating)

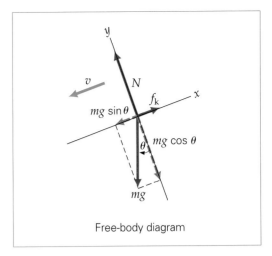

•**FIGURE 5.3 Total or net work**
See Example 5.3.

(a) Note from Fig. 5.3 that only two forces do work because there are only two forces parallel to the motion: f_k, the force of kinetic friction; and $mg \sin \theta$, the component of the block's weight acting down the plane. The normal force and $mg \cos \theta$, the components of the block's weight acting perpendicular to the plane, do no work on the block. (Why?) We first find the work done by the frictional force:

$$W_f = f_k d \cos 180° = -f_k d = -\mu_k N d$$

where $f_k = \mu_k N$. The angle 180° indicates that the force and displacement are in opposite directions. (It is common in such cases to write $W_f = -f_k d$ directly, since friction typically opposes motion.) The magnitude of the normal force N is equal to that of the component of the block's weight perpendicular to the surface (these are equal and opposite forces. Why?):

Note: Recall the discussion of friction in Section 4.6.

$$N = mg \cos \theta$$

The distance d the block slides down the plane can be found by using trigonometry. Note that $\cos \theta = L/d$, so

$$d = \frac{L}{\cos \theta}$$

But what is μ_k? It is not given (nor are the types of surfaces specified). This is an example in which a key piece of data is given to you indirectly in a problem. Note that the block slides at a uniform (constant) velocity down the plane. In Chapter 4, you learned that for uniform motion ($a = 0$) down a plane, $\mu_k = \tan \theta$. Thus,

$$W_f = -\mu_k N d = -(\tan \theta)(mg \cos \theta)\left(\frac{L}{\cos \theta}\right) = -mgL \tan 20°$$
$$= -(0.75 \text{ kg})(9.8 \text{ m/s}^2)(1.2 \text{ m})(0.364) = -3.2 \text{ J}$$

(b) To find the net work, we need to calculate the work done by gravity and then add it to our result in part (a). Since $F_\parallel$ for gravity is just $mg \sin \theta$, we have

$$W_g = F_\parallel d = (mg \sin \theta)\left(\frac{L}{\cos \theta}\right) = mgL \tan 20° = +3.2 \text{ J}$$

where the calculation is the same as in (a) except for the sign. Then,

$$W_{net} = W_g + W_f = +3.2 \text{ J} + (-3.2 \text{ J}) = 0$$

Remember that work is a scalar quantity, so scalar addition is used to find net work.

(c) If the block accelerates down the plane, then from Newton's second law, $F - mg \sin \theta - f_k = ma$. The component of the gravitational force is greater than the opposing frictional force, so there is net work done on the block because now $|W_g| > |W_f|$. You might be wondering what the effect of nonzero net work is. As you will learn shortly, it causes a change in the amount of energy an object has.

Follow-up Exercise. In part (c) of this Example, is it possible for the frictional work to be greater in magnitude than the gravitational work? What would this mean in terms of the block's speed?

Problem-Solving Hint

Note how in part (a) of Example 5.3, the equation for W_f was simplified by using the algebraic expressions for N and d instead of computing these quantities initially. It is a good rule of thumb not to plug numbers into an equation until you have to. Simplifying an equation through cancellation is easier with symbols and saves computation time.

5.2 Work Done by a Variable Force

Objectives: To (a) differentiate between work done by constant and variable forces and (b) compute the work done by a spring force.

The discussion in the preceding section was limited to work done by constant forces. In general, however, forces are variable; that is, they change with time and/or position. For example, someone pushes harder and harder on an object to overcome the force of static friction, until the applied force exceeds $f_{s_{max}}$. However, the force of static friction does no work because there is no motion or displacement.

An example of a variable force doing work in stretching a spring is illustrated in •Fig. 5.4. As a spring is stretched (or compressed) farther and farther, the restoring force of the spring (the force that opposes the stretching) gets greater and an increasing applied force is required. For most springs, the applied force F is di-

In Fig 5.4, the hand applies a variable force F in stretching the spring. At the same time, the spring exerts an equal and opposite force, F_s, on the hand.

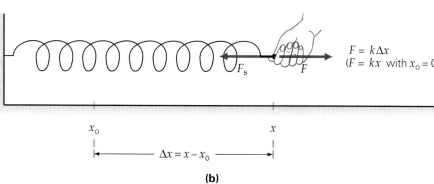

•**FIGURE 5.4 Spring force** (a) An applied force F stretches the spring, and the spring exerts an equal and opposite force F_s on the hand. (b) The magnitude of the force depends on the change in the spring's length. This is often referenced to the position of a mass on the end of the spring.

rectly proportional to the change in length of the spring from its unstretched length. In equation form, this relationship is expressed as

$$F = k\Delta x = k(x - x_o)$$

or, if we chose $x_o = 0$,

$$F = kx$$

where x now represents the distance the spring is stretched (or compressed) from its unstretched length. As can be seen, the force varies with x. We describe this by saying that the *force is a function of position*.

The k in this equation is a constant of proportionality and is commonly called the **spring constant**, or **force constant**. The greater the value of k, the stiffer or stronger is the spring. As you should be able to prove to yourself, the SI units of k are newtons per meter (N/m).

k, the spring constant or force constant

The relationship expressed by the equation $F = kx$ holds only for ideal springs. Real springs approximate this linear relationship between force and displacement within certain limits. For example, if a spring is stretched beyond a certain point, called its *elastic limit*, the spring will be permanently deformed and $F = kx$ will no longer apply.

Note that a spring exerts a force that is equal and opposite to the external applied force F. Thus

$$F_s = -k\Delta x = -k(x - x_o)$$

or, with $x_o = 0$,

$$F_s = -kx \quad \text{ideal spring force} \quad (5.3)$$

Hooke's law

The minus sign indicates that this spring force acts in the direction opposite to the displacement whether the spring is stretched or compressed. This equation is a form of what is known as *Hooke's law*, named after Robert Hooke, a contemporary of Newton.

To compute the work done by variable forces generally requires calculus. But we are fortunate that the spring force can be computed by using the average force. An analogous case involving average velocity, $\bar{v} = (v + v_o)/2$, was used in the study of kinematics in Chapter 2, and it is instructive to explore this analogy graphically.

Plots of F versus x and v versus t are shown in •Fig. 5.5. The graphs have straight-line slopes of k and a, respectively, with $F = kx$ and $v = at$. [The applied force F, rather than the spring force F_s, is plotted to avoid a negative slope ($F_s = -kx$) and thereby allow analogous graphs.]

By the same reasoning given for v in Section 2.3, the average force $\bar{F}$ can be expressed as

$$\bar{F} = \frac{F + F_o}{2}$$

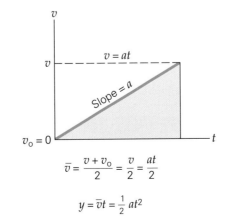

•**FIGURE 5.5 Work done by a uniformly variable force** The work done by a uniformly varying force of the form $F = kx$ is $W = \frac{1}{2}kx^2$. A plot of this special case of a variable force is graphically analogous to a plot of v versus t for a uniformly varying velocity starting at rest ($v_o = 0$): $v = at$. The work (W) and the distance (y) are equal to the areas under the respective straight lines.

or, since $F_o = 0$ (why?),

$$\bar{F} = \frac{F}{2}$$

Thus, the work done in stretching or compressing the spring an amount x from its unstretched length is

$$W = \bar{F}x = \frac{Fx}{2}$$

Since $F = kx$, the work done is

$$W = \tfrac{1}{2}kx^2 \quad \begin{matrix} \text{work done in stretching} \\ \text{(or compressing)} \\ \text{a spring from } x_o = 0 \end{matrix} \quad (5.4)$$

Note: We have tried to show the physical basis of Eq. 5.4.

Note that the work done is just the area under the curve in Fig. 5.5.

Example 5.4 ■ Determining the Spring Constant

A 0.15-kilogram mass is suspended from a vertical spring and descends a distance of 4.6 cm, after which it hangs at rest (•Fig. 5.6). An additional 0.50-kilogram mass is then suspended from the first. What is the total extension of the spring? (Neglect the mass of the spring.)

Thinking It Through. The spring constant k appears in Eq. 5.3. Therefore to find the value of k for a particular instance, the spring force and distance the spring is stretched (or compressed) must be known.

Solution. The data given are as follows:

Given: $m_1 = 0.15$ kg
$\quad\quad\quad x_1 = 4.6$ cm $= 0.046$ m
$\quad\quad\quad m_2 = 0.50$ kg

Find: x (total stretch length)

The total stretch distance is given by $F = kx$, where F is the applied force, which in this case is the weight of the mass suspended on the spring. However, the spring constant, k, is not given. This may be found from the data pertaining to the suspension of m_1 and resulting displacement x_1. (This is a common method of determining spring constants.) As seen in Fig. 5.6a, the magnitudes of the weight force and the restoring spring force are equal since $a = 0$, so we may write (with directional minus signs neglected)

$$F_s = m_1 g = kx_1$$

and

$$k = \frac{m_1 g}{x_1} = \frac{(0.15 \text{ kg})(9.8 \text{ m/s}^2)}{0.046 \text{ m}} = 32 \text{ N/m}$$

Then, knowing k, we find the total extension of the spring from the balanced force situation shown in Fig. 5.6b:

$$F = (m_1 + m_2)g = kx$$

Thus,

$$x = \frac{(m_1 + m_2)g}{k} = \frac{(0.15 \text{ kg} + 0.50 \text{ kg})(9.8 \text{ m/s}^2)}{32 \text{ N/m}} = 0.20 \text{ m (or 20 cm)}$$

Follow-up Exercise. How much work is done by gravity in stretching the spring through both displacements in Example 5.4?

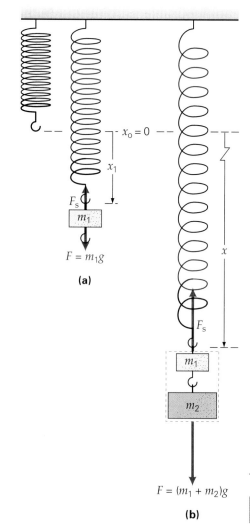

•**FIGURE 5.6** Determining the spring constant and the work done in stretching a spring See Example 5.4.

Problem-Solving Hint

The reference position x_o for the change in length of a spring is arbitrary and is usually chosen for convenience. *The important quantity in computing work is the position difference Δx, or the net change in the length of the spring.* As shown in •Fig. 5.7 for a mass suspended on a spring, x_o can be referenced to the unloaded length of the spring

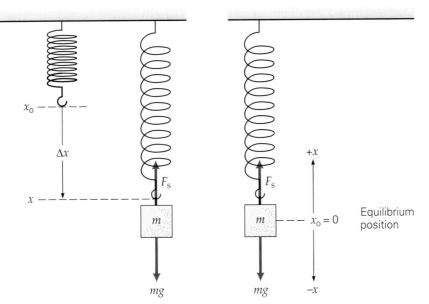

•**FIGURE 5.7 Displacement reference**
The reference position x_o is arbitrary and is usually chosen for convenience. It may be **(a)** at the end of the spring at its unloaded position or **(b)** at the equilibrium position when a mass is suspended on the spring. The latter is particularly convenient in cases in which the mass oscillates up and down on the spring.

or to the loaded position, which may be taken as the zero position for convenience. In Example 5.4, x_o was referenced to the end of the unloaded spring. Also, with the displacement only in one direction, downward was designated as the $+x$ direction to avoid minus signs.

When the net force on the suspended mass is zero, the mass is said to be at its equilibrium position (as in Fig. 5.7a with m_1 suspended). This position, rather than the unloaded length, may be taken as a zero reference ($x = x_o = 0$, Fig. 5.7b). The equilibrium position is a convenient reference point for a case in which the mass oscillates up and down on the spring. (We will describe this motion in Chapter 13.) Notice that in general there are both positive and negative directions. The chosen signs are arbitrary but must be adhered to during the whole problem. Also, since the displacement is in the vertical direction, the x's are often replaced by y's.

5.3 The Work–Energy Theorem: Kinetic Energy

Objectives: To (a) explain the work–energy theorem and (b) apply it in solving problems.

Now that we have an operational definition of work, we are ready to look at how work is related to energy. Energy is one of the most important concepts in science. We describe it as a quantity that objects or systems possess. Basically, work is something that is *done on* objects, whereas energy is something that objects *have*.

One form of energy that is closely associated with work is *kinetic energy*. (Another form of energy, *potential energy*, will be described in Section 5.4.) Consider an object at rest on a frictionless surface. Let a horizontal force act on the object and set it in motion. Work is done *on* the object, but where does the work "go," so to speak? It goes into setting the object into motion or changing its *kinetic* conditions. Because of its motion, we say the object has energy—kinetic energy, which gives it the capability to do work.

For a constant net force doing work on a moving object, as illustrated in •Fig. 5.8, the force does an amount of work, $W = Fx$. But what are the kinematic effects? The force causes the object to accelerate, and from Equation 2.10, $v^2 = v_o^2 + 2ax$,

$$a = \frac{v^2 - v_o^2}{2x}$$

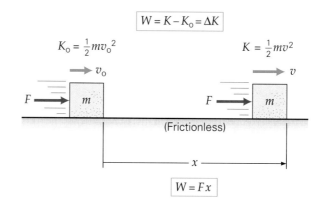

•**FIGURE 5.8 The relationship of work and kinetic energy** The work done on a block in moving along a horizontal frictionless surface is equal to the change in its kinetic energy, $W = \Delta K$.

where v_o may or may not be zero and Δx can be written as x if $x_o = 0$. Writing the force in its Newton's second law form ($F = ma$) and then substituting the expression for a from the previous equation gives

$$F = ma = m\left(\frac{v^2 - v_o^2}{2x}\right)$$

Using this expression in the equation for work, we get

$$W = Fx = m\left(\frac{v^2 - v_o^2}{2x}\right)x$$
$$= \tfrac{1}{2}mv^2 - \tfrac{1}{2}mv_o^2$$

It is convenient to define $\tfrac{1}{2}mv^2$ as the **kinetic energy** (K).

Kinetic energy—the energy of motion

$$K = \tfrac{1}{2}mv^2 \quad \textit{kinetic energy} \quad (5.5)$$

SI unit of energy: joule (J)

Kinetic energy is often called the energy of motion. Note that it is directly proportional to the square of the (instantaneous) speed of a moving object.

Then, in terms of kinetic energy, the previous work expression can be written as follows:

$$W = \tfrac{1}{2}mv^2 - \tfrac{1}{2}mv_o^2 = K - K_o = \Delta K$$

or

Work–energy theorem

$$W = \Delta K \quad (5.6)$$

where it is understood that W is the net work if more than one force acts on the object, as shown in Example 5.3. This equation is called the **work–energy theorem**; it relates the work done on an object to the change in its kinetic energy. That is, *the net work done on a body by an external net force is equal to the change in kinetic energy of the body.* Both work and energy have the units of joules, and both are *scalar* quantities.

Keep in mind that the work–energy theorem is true in general for varying forces and not just for the special case considered in deriving Eq. 5.6. As a concrete example, recall that in Example 5.1 the force of gravity did +44 J of work on a book that fell from rest through a distance of $y = 3.0$ m. At that position and instant, the falling book had 44 J of kinetic energy. Since $v_o = 0$ in this case, $v^2 = 2gy$, and $gy = v^2/2$. Substituting this expression into the equation for the work done on the falling book by gravity, we get

$$W = Fd = mgy = \frac{mv^2}{2} = K$$

where $K_o = 0$. Thus the kinetic energy gained by the book is equal to the net work done on it: 44 J in this instance. (As an exercise, confirm this fact by calculating the speed of the book and evaluating its kinetic energy.)

What the work–energy theorem tells us is that when work is done, there is change in or a transfer of energy. In general then, we might say that *work is a measure of the transfer of kinetic energy.* For example, a force doing work on an object that causes the object to speed up gives rise to an increase in the object's kinetic energy. Conversely, work done by the force of kinetic friction may cause a moving object to slow down and decrease its kinetic energy. So, for a change in or transfer of kinetic energy, there must be a net force and net work, as Eq. 5.6 tells us.

When an object is in motion, it possesses kinetic energy and has the capability to do work. For example, a moving automobile has kinetic energy and can do work in crumpling a fender in a fender-bender—not *useful* work in that case, but still work. Other examples are shown in •Fig. 5.9.

Example 5.5 ■ A Game of Shuffleboard: The Work–Energy Theorem

A shuffleboard player (•Fig. 5.10) pushes a 0.25-kilogram puck, initially at rest, such that a constant horizontal force of 6.0 N acts on it through a distance of 0.50 m. (Neglect friction.) (a) What are the kinetic energy and the speed of the puck when the force is removed? (b) How much work would be required to bring the puck to rest?

Thinking It Through. Apply the work–energy theorem. If you can find the work done, you know the change in kinetic energy, and vice versa.

Solution. Listing the given data as usual, we have

Given: $m = 0.25$ kg Find: (a) K (kinetic energy)
$F = 6.0$ N v (speed)
$d = 0.50$ m   (b) W (work done in stopping puck)
$v_o = 0$

(a) Since the speed is not known, we cannot compute the kinetic energy ($K = \frac{1}{2}mv^2$) directly. However, kinetic energy is related to *work* by the work–energy theorem. The work done on the puck is

$$W = Fd = (6.0 \text{ N})(0.50 \text{ m}) = +3.0 \text{ J}$$

Then, by the work–energy theorem,

$$W = \Delta K = K - K_o = 3.0 \text{ J}$$

But $K_o = \frac{1}{2}mv_o^2 = 0$ because $v_o = 0$, so

$$K = 3.0 \text{ J}$$

(a)

(b)

•**FIGURE 5.9 Kinetic energy and work** (a) A moving object, such as a wrecking ball, possesses kinetic energy and so can do work. (b) A great deal of kinetic energy was converted to work when a large meteorite made this crater in Arizona, perhaps 10 000 y ago. The crater is 1200 m in diameter and 180 m deep.

•**FIGURE 5.10 Work and kinetic energy** See Example 5.5.

5.3 The Work–Energy Theorem: Kinetic Energy   147

The speed can be found from the kinetic energy. Since $K = \frac{1}{2}mv^2$, we have

$$v = \sqrt{\frac{2K}{m}} = \sqrt{\frac{2(3.0\text{ J})}{0.25\text{ kg}}} = 4.9 \text{ m/s}$$

(b) As you might guess, the work required to bring the puck to rest is equal to its kinetic energy (the amount of energy that must be "removed" from the puck to stop its motion). To confirm this equality, we essentially perform the reverse of the previous calculation, with $v_o = 4.9$ m/s and $v = 0$:

$$W = K - K_o = 0 - K_o = -\tfrac{1}{2}mv_o^2 = -\tfrac{1}{2}(0.25\text{ kg})(4.9\text{ m/s})^2 = -3.0 \text{ J}$$

The minus sign indicates that the puck loses energy as it slows down. The work is done *against* the motion of the puck; that is, the opposing force is in a direction opposite that of the motion. (In a real situation, the opposing force would be friction.)

Follow-up Exercise. Suppose the puck in this Example had twice the final speed. Would it then take twice as much work to stop the puck?

Problem-Solving Hint

Notice how work–energy considerations were used to find speed in Example 5.5. This could be done in another way. First, the acceleration could be found from $a = F/m$, and then the kinematic equation $v^2 = v_o^2 + 2ax$ could be used to find v (where $x = d$ in the equation).

The point is that many problems can be solved in different ways, and finding the fastest and most efficient way is often the key to success. As our discussion of energy progresses, you will see how useful and powerful the notions of work and energy are, both as theoretical concepts and as practical tools for solving many kinds of problems.

Conceptual Example 5.6

Kinetic Energy: Mass versus Speed

In a football game, a 140-kilogram guard runs at a speed of 4.0 m/s, and a 70-kilogram free safety moves at 8.0 m/s. Then, it is correct to say that (a) both players have the same kinetic energy, (b) the safety has twice as much kinetic energy as the guard, (c) the guard has twice as much kinetic energy as the safety, (d) the safety has four times as much kinetic energy as the guard.

Reasoning and Answer. The kinetic energy of a body depends on both its mass and its speed. You might think that, with half the mass but twice the speed, the safety would have the same kinetic energy as the guard, but this is not the case. Kinetic energy, $K = \frac{1}{2}mv^2$, is directly proportional to the mass but proportional to the *square* of the speed. Thus, halving the mass decreases the kinetic energy by a factor of two; so if the two athletes had equal speeds, the safety would have half as much kinetic energy as the guard.

However, doubling the speed increases the kinetic energy, not by a factor of 2 but by a factor of 2^2, or 4. Thus, the safety, with half the mass but twice the speed, would have $\tfrac{1}{2} \times 4 = 2$ times as much kinetic energy as the guard, and the answer is (b).

Note that to answer this question it was not necessary to calculate the kinetic energy of each player. We can do so, however, to check our conclusions:

$$K_s = \tfrac{1}{2}m_s v_s^2 = \tfrac{1}{2}(70\text{ kg})(8.0\text{ m/s})^2 = 2.2 \times 10^3 \text{ J}$$
$$K_g = \tfrac{1}{2}m_g v_g^2 = \tfrac{1}{2}(140\text{ kg})(4.0\text{ m/s})^2 = 1.1 \times 10^3 \text{ J}$$

Thus we see explicitly that our answer was correct.

Follow-up Exercise. Suppose that the safety's speed were only 50 percent greater than the guard's, or 6.0 m/s. Which athlete would then have the greater kinetic energy, and how many times greater?

Problem-Solving Hint

Note that the work–energy theorem relates the work done to the *change* in the kinetic energy. Often we have $v_o = 0$ and $K_o = 0$, so $W = \Delta K = K$. But take care! You *cannot* simply use the square of the change in speed $(\Delta v)^2$ to calculate ΔK, as you might at first think. In terms of speed, we have

$$W = \Delta K = K - K_o = \tfrac{1}{2}mv^2 - \tfrac{1}{2}mv_o^2 = \tfrac{1}{2}m(v^2 - v_o^2)$$

But $v^2 - v_o^2$ is *not* the same as $(v - v_o)^2 = (\Delta v)^2$, since $(v - v_o)^2 = v^2 - 2vv_o + v_o^2$. Hence the work or change in kinetic energy is **not** equal to $\tfrac{1}{2}m(v - v_o)^2 = \tfrac{1}{2}m(\Delta v)^2$.

What this means is that you must compute the kinetic energy of an object at one point or time (using the instantaneous speed to get the instantaneous kinetic energy) and also at another point or time. Then the quantities are subtracted to find the change in kinetic energy, or the work. Alternatively, you can find the difference of the *squares* of the speeds $(v^2 - v_o^2)$ first in computing the change.

Conceptual Example 5.7

An Accelerating Car: Speed and Kinetic Energy

A car traveling at 5.0 m/s speeds up to 10 m/s, with an increase in kinetic energy that requires work W_1. Then, the speed is increased from 10 m/s to 15 m/s, requiring work W_2. How do the amounts of work compare? (a) $W_1 > W_2$, (b) $W_1 = W_2$, (c) $W_2 > W_1$.

Reasoning and Answer. As noted above, the work–energy theorem relates the work done to the *change* in the kinetic energy. Since the speeds have the same increment in each case ($\Delta v = 5.0$ m/s), it might appear that (b) would be the answer. However, keep in mind that the work is equal to the *change* in kinetic energy and involves $(v_2^2 - v_1^2)$, not $(\Delta v)^2 = (v_2 - v_1)^2$.

So, the greater the speed of an object, the greater its kinetic energy, and we would expect the *difference* in kinetic energy in changing speeds (or work required) to be greater for higher speeds for the same Δv. Thus, (c) is the answer.

We can confirm this conclusion by calculating the actual changes in the kinetic energy in each of these cases:

$$W_1 = \tfrac{1}{2}m(v^2 - v_o^2) = \tfrac{1}{2}m[(10 \text{ m/s})^2 - (5.0 \text{ m/s})^2] = \tfrac{1}{2}m(75 \text{ m}^2/\text{s}^2)$$

Similarly,

$$W_2 = \tfrac{1}{2}m(v^2 - v_o^2) = \tfrac{1}{2}m[(15 \text{ m/s})^2 - (10 \text{ m/s})^2] = \tfrac{1}{2}m(125 \text{ m}^2/\text{s}^2)$$

And since m is the same for each case, we see that $W_2 > W_1$, which confirms our expectations. The main point is that the Δv values are the same, but it requires more work to increase the kinetic energy of an object at higher speeds.

Follow-up Exercise. Suppose the car speeds up a third time, from 15 m/s to 20 m/s, a change requiring work W_3. How does the work done in this increment compare with W_2?

5.4 Potential Energy

Objectives: To (a) explain how potential energy depends on position and
(b) compute values of gravitational potential energy.

An object in motion has kinetic energy. However, whether an object is in motion or not, it may have another form of energy—potential energy. As the name implies, an object having potential energy has the *potential* to do work. You can probably think of many examples: a compressed spring, a drawn bow, water held back by a dam, a wrecking ball poised to drop. In all such cases, the potential to do work derives from the position or configuration of bodies. The spring has energy because it is compressed, the bow because it is drawn, the water and the ball

(a)

(b)

•**FIGURE 5.11 Potential energy**
Potential energy has many forms.
(a) Work must be done to bend the bow, giving it potential energy. That energy is converted into kinetic energy when the arrow is released. (b) Gravitational potential energy is converted into kinetic energy when an object falls. (Where did the gravitational potential energy of the water and the diver come from?)

because they have been lifted above the surface of the Earth (•Fig. 5.11). Consequently, **potential energy** (U) is often called the energy of position (and/or configuration).

In a sense, potential energy can be thought of as stored work. You have already seen an example of potential energy in Section 5.2 when work was done in compressing a spring from its equilibrium position. Recall that the work done in such a case is $W = \frac{1}{2}kx^2$ (with $x_o = 0$). Note that the amount of work done depends on the amount of compression (x). Because work is done, there is a *change* of position and a *change* in potential energy (ΔU), which is equal to the work done *by the applied force* in compressing (or stretching) the spring:

$$W = \Delta U = U - U_o = \tfrac{1}{2}kx^2 - \tfrac{1}{2}kx_o^2$$

Thus, with $x_o = 0$ and $U_o = 0$, as they are commonly taken to be for convenience, the *potential energy of a spring* is

$$U = \tfrac{1}{2}kx^2 \qquad \text{potential energy of a spring} \qquad (5.7)$$

SI unit of energy: joule (J)

[Since the potential energy has x^2 as one term, the previous problem-solving hint applies when $x_o \neq 0$; that is, $x^2 - x_o^2 \neq (x - x_o)^2$.]

Perhaps the most common type of potential energy is **gravitational potential energy**. In this case, position refers to the height of an object above some reference point, such as the floor or the ground. Suppose that an object of mass m is lifted a distance Δh (•Fig. 5.12). Work is done against the force of gravity, and an applied force at least equal to the object's weight is necessary to lift it: $F = w = mg$. The work done in lifting is then equal to the change in potential energy. Expressed in equation form, since there is no overall change in kinetic energy, we have

work = change in potential energy

•**FIGURE 5.12 Gravitational potential energy**
The work done in lifting an object is equal to the change in gravitational potential energy, $W = F\Delta h = mg(h - h_o)$.

or

$$W = F\,\Delta h = \Delta U = U - U_o = mgh - mgh_o$$

And, with the common choices of $h_o = 0$ and $U_o = 0$, the gravitational potential energy is

$$U = mgh \qquad \textit{gravitational potential energy} \qquad (5.8)$$

Gravitational potential energy

SI unit of energy: joule (J)

Because we commonly take upward to be the $+y$ direction, Eq. 5.8 is often written $U = mgy$. (Eq. 5.8 is the gravitational potential energy on or near the Earth's surface where g is considered to be constant. A more general form of gravitational potential energy shall be given in Section 7.5.)

Example 5.8 ■ A Thrown Ball: Kinetic Energy and Gravitational Potential Energy

A 0.50-kilogram ball is thrown vertically upward with an initial velocity of 10 m/s (•Fig. 5.13). (a) What is the change in the ball's kinetic energy between the starting point and its maximum height? (b) What is the change in the ball's potential energy between its starting point and the maximum height? (Neglect air resistance.)

Thinking It Through. Kinetic energy is lost and gravitational potential energy is gained as the ball travels upward.

Solution. Studying Fig. 5.13 and listing the data, we have

Given: $m = 0.50$ kg Find: (a) ΔK (change in kinetic energy)
$v_o = 10$ m/s (b) ΔU (the change in potential energy
$a = g$ between y_o and y_{max})

(a) To find the *change* in kinetic energy, we first compute the kinetic energy at each point. We know the initial velocity v_o, and at the maximum height, $v = 0$, so $K = 0$. Thus,

$$\Delta K = K - K_o = 0 - K_o = -\tfrac{1}{2}mv_o^2 = -\tfrac{1}{2}(0.50\text{ kg})(10\text{ m/s})^2 = -25\text{ J}$$

That is, the ball loses 25 J of kinetic energy as negative work is done on it by the force of gravity (force and displacement are in opposite directions).

(b) To find the change in potential energy we need to know the ball's height above its starting point when $v = 0$. Using the equation $v^2 = v_o^2 - 2gy$ to find y_{max},

$$y_{max} = \frac{v_o^2}{2g} = \frac{(10\text{ m/s})^2}{2(9.8\text{ m/s}^2)} = 5.1\text{ m}$$

Then, with $y_o = 0$ and $U_o = 0$, $\Delta U = U - U_o = U - 0$, and

$$\Delta U = U = mgy_{max} = (0.50\text{ kg})(9.8\text{ m/s}^2)(5.1\text{ m}) = +25\text{ J}$$

The potential energy increases by 25 J, as might be expected. Notice that this is the change in potential energy with respect to the release point, which was taken as the zero reference point ($y_o = 0$).

Follow-up Exercise. In this Example, what are the total changes in the ball's kinetic and potential energies when it returns to the starting point?

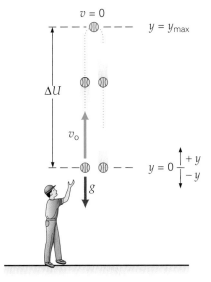

•**FIGURE 5.13 Kinetic and potential energies** See Example 5.8. (The ball is displaced sideways for clarity.)

Zero Reference Point

An important point is illustrated in Example 5.8: namely, the choice of a zero reference point. Potential energy is the energy of *position*, and the potential energy at a particular position (U) is referenced to the potential energy at some other position (U_o). The reference position or point is arbitrary, as is the origin of a set of coordinate axes for analyzing a system. They are usually chosen with convenience in

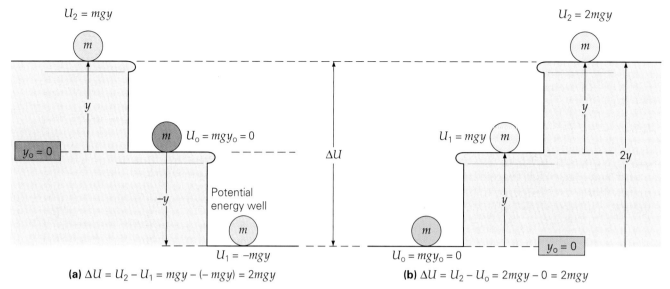

•**FIGURE 5.14 Reference point and change in potential energy** (a) The choice of a reference point (zero height) is arbitrary and may give rise to a negative potential energy. An object is said to be in a potential energy well in this case. (b) The well may be avoided by selecting a new zero reference. Note that the difference, or *change*, in potential energy (ΔU) associated with the two positions is the same regardless of the reference point. There is no physical difference even though there are two coordinate systems and two different zero reference points.

mind, for example, $y_o = 0$ or $h_o = 0$. The value of the potential energy at a particular position depends on the reference point used. However, the *difference or change in potential energy associated with two positions is the same regardless of the reference position.*

In Example 5.8 if ground level had been taken as the zero reference point, then U_o at the release point would not have been zero. However, U at the maximum height would have been greater and $\Delta U = U - U_o$ would have been the same. This concept is illustrated in •Fig. 5.14. Note that the potential energy can be negative. When an object has a negative potential energy, it is said to be in a potential energy *well*, which is analogous to being in an actual well. Work is needed to raise the object to a higher position in the well or to get it out of the well.

Also, for gravitational potential energy, the path by which an object is raised (or lowered) makes no difference (•Fig. 5.15). That is, *the change in gravitational potential energy is independent of path.* As illustrated in the figure, an object raised to a height h has a change in potential energy of $\Delta U = mgh$ whether it is lifted vertically or moved along an inclined plane. This is because the force of gravity always acts downward and only vertical displacement (or vertical component) is involved in doing work against gravity and changing the potential energy.

•**FIGURE 5.15 Path-independence of the change in gravitational potential energy** When it is resting on the table, the crate has the same potential energy relative to the floor regardless of how it got there. Only the vertical component of the force that moves the crate does work against gravity in lifting it vertically or moving it up an inclined plane or ramp.

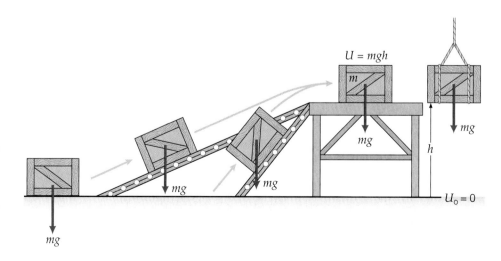

5.5 The Conservation of Energy

Objectives: To (a) distinguish between conservative and nonconservative forces and (b) explain their effects on the conservation of energy.

Conservation laws are the cornerstones of physics, both theoretically and practically. Most scientists would probably name the conservation of energy as the most profound and far-reaching of these important laws. When we say something is *conserved*, we mean that it is constant, or has a constant value. Because so many things continually change in physical processes, conserved quantities are extremely helpful in our attempts to understand and describe the universe. Keep in mind, though, that quantities are generally conserved only under special conditions.

One of the most important conservation laws is that concerning conservation of energy. (You may have seen this coming in Example 5.8.) A familiar statement is that the total energy of the universe is conserved. This is true because the whole universe is taken to be a system. A *system* is defined as a definite quantity of matter enclosed by boundaries, either real or imaginary. In effect, the universe is the largest possible closed, or isolated, system we can imagine. Within a *closed system*, particles can interact with each other but have absolutely no interaction with anything outside. In general then, the amount of energy remains constant when no mechanical work is done on or by the system, and no energy is transmitted to or from the system (including thermal energy and radiation).

Note: A system is a physical situation with real or imaginary boundaries. A classroom might be considered a system, and so might an arbitrary cubic meter of air.

Thus, the **law of conservation of total energy** may be stated:

> The total energy of an isolated system is always conserved.

Conservation of total energy

Within such a system, energy may be converted from one form to another, but the total amount of all forms of energy is constant or unchanged. Total energy can never be created or destroyed.

Conservative and Nonconservative Forces

We can make a general distinction among systems by considering two categories of forces that may act within them: conservative and nonconservative forces. You have already been introduced to a couple of conservative forces: the force due to gravity and the spring force. A classic nonconservative force, friction, was considered in Chapter 4. Here's the distinction:

Note: Friction was discussed in Section 4.6.

> A force is said to be conservative if the work done by or against it in moving an object is independent of the object's path.

What this means is that the work done by a **conservative force** depends only on the initial and final positions of an object.

Conservative force—work independent of path

The concept of conservative and nonconservative forces is sometimes difficult to comprehend at first. Because this concept is so important in the conservation of energy, let's consider some illustrative examples to increase our understanding.

First, what does *independent of path* mean? An example was given in Fig. 5.15, where work was done against the conservative force of gravity. The figure illustrates that the work done in moving a crate onto a table does not depend on the *path* but only on the initial and final positions. The magnitude of the work done is equal to the change in potential energy, and in fact, *the concept of potential energy is associated only with conservative forces*. A change in potential energy can be defined in terms of the work done by a conservative force.

Conversely, work done against a **nonconservative force**, such as friction, *does* depend on path. Given the same frictional conditions, it would take more work (against friction) to move the crate up the longer ramp, so the work would depend on path—the longer the path, the more frictional work done. It should therefore be evident that the *total* work done is not equal to a change in potential energy. Some work was done against a nonconservative force, in addition to the conservative gravitational force. In this case, the energy associated with the work done against friction would be converted to heat energy. Hence, in a sense, a conservative force allows you to conserve or store all of the energy as potential energy, whereas a nonconservative force does not.

To further highlight this idea, consider moving the crate around the table top. Here, work is done only against the nonconservative frictional force (why?). Certainly the work done in moving the crate between two points depends on the path taken. It would be greater, for example, if we moved the crate around the table's edges before coming to the destination point than if we took a straight-line path.

Another approach to explain the distinction between conservative and nonconservative forces is through an equivalent statement of the previous definition:

Another way of describing a conservative force

A force is conservative if the work done by or against it in moving an object through a round trip is zero.

Consider a book resting on a table. It has gravitational potential energy $U = mgh$, relative to some reference point $h_o = 0$. You could drop it on the floor, pick it up, and place it back at its original position; or you could pick it up from the table and carry it around with you all day, then place it back at its original position. Both are round trips, and the potential energy of the book is the same when it is returned to its original position. Thus, the change in its potential energy or the work done by the conservative force of gravity is $\Delta U = W = 0$. However, if you pushed the book around on the tabletop and eventually back to its original position, the work done against the nonconservative force of friction would depend on the path (the longer the path, the more work done). The work done would not be stored but lost as heat and sound.

Notice that for the *conservative* gravitational force, the force and displacement are sometimes in the same direction (in which case positive work is done) and sometimes in opposite directions (in which case negative work is done against the force) during a round trip. Think of the simple case of the book falling to the floor and being placed back on the table. With positive and negative work, the total work done by gravity can be zero.

However, for only a *nonconservative* force of kinetic friction, which always opposes the motion or is in the opposite direction to the displacement, the total work done in a round trip can *never* be zero and is always negative (energy lost). But don't get the idea that nonconservative forces only take energy away from a system. On the contrary, we often supply nonconservative pushes and pulls (forces) that add to the energy of a system, such as when you push a stalled car.

Conservation of Total Mechanical Energy

The idea of a conservative force allows us to extend the conservation of energy to the special case of mechanical energy, which greatly helps us better analyze many physical situations. The sum of the kinetic and potential energies is called the **total mechanical energy**:

Mechanical energy—kinetic plus potential

$$E = K + U \quad (5.9)$$

$$\underset{\text{mechanical energy}}{\text{total}} = \underset{\text{energy}}{\text{kinetic}} + \underset{\text{energy}}{\text{potential}}$$

For a **conservative system** (one in which only conservative forces do work), the total mechanical energy is constant, or conserved; that is,

$$E = E_o$$

$$K + U = K_o + U_o$$

or

$$\tfrac{1}{2}mv^2 + U = \tfrac{1}{2}mv_o^2 + U_o \quad (5.10)$$

•**FIGURE 5.16 Conservative and nonconservative systems** (a) The work done by a conservative force exchanges energy between kinetic and potential forms: that is, $\Delta K + \Delta U = 0$. This means that $K_o + U_o = K + U$, or $E_o = E$, and the total mechanical energy is conserved (none is lost or gained). (b) For a nonconservative force, not all the work goes into the mechanical energy exchange, but some energy (ΔE) is generally lost, such as to friction. The mechanical energy is not conserved in this case. (c) A graphical energy summary for the nonconservative case. (Keep in mind that a nonconservative force doing work may instead add energy to a system.) How would the energy summary for a conservative system look?

This is a mathematical statement of the **law of the conservation of mechanical energy**:

> In a conservative system, the sum of all types of kinetic energy and potential energy is constant and equals the total mechanical energy of the system.

Conservation of mechanical energy

The kinetic and potential energies in a conservative system may change, but their sum is always constant. This is illustrated in •Fig. 5.16a.

Notice for a conservative system, when work is done and energy is transferred, we have

$$\underbrace{K + U}_{\text{final energy}} = \underbrace{K_o + U_o}_{\text{initial energy}}$$

This equation can be rewritten as

$$(K - K_o) + (U - U_o) = 0 \tag{5.11}$$

or as

$$\Delta K + \Delta U = 0 \quad \text{(only when nonconservative forces do no work)}$$

This expression tells us that these quantities are related in seesaw fashion: If there is a decrease in potential energy, then the kinetic energy must increase by an equal amount to keep the sum equal to zero. However, in a nonconservative system, some mechanical energy is generally lost (for example, to the heat of friction), and thus $\Delta K + \Delta U < 0$. Such a situation is illustrated in Fig. 5.16b. In terms of the total mechanical energy, $\Delta E = E - E_o$, where ΔE is the amount of energy lost from the system. But, keep in mind, as pointed out above, a nonconservative force may instead add energy to a system.

Examples 5.9 through 5.11 illustrate the conservation of mechanical energy for some conservative systems.

Example 5.9 ■ Look Out Below! Conservation of Mechanical Energy

A painter on a scaffold drops a 1.50-kilogram can of paint from a height of 6.00 m. (a) What is the kinetic energy of the can when it is at a height of 4.00 m? (b) With what speed will the can hit the ground? (Neglect air resistance.)

Thinking It Through. Conservation of energy: The initial total mechanical energy can be found, and potential energy decreases as kinetic energy (and speed) increases.

Solution. Listing what is given and what we are to find, we have

Given: $m = 1.50$ kg *Find:* (a) K (kinetic energy at h)
$h_o = 6.00$ m (b) v (speed hitting the ground)
$h = 4.00$ m
$v_o = 0$

(a) First, it is convenient to find the can's total mechanical energy, since this quantity is conserved while it is falling (why?). Initially, with $v_o = 0$, the can's total mechanical energy is all potential. Taking the ground as the zero reference point, we have

$$E = K + U = 0 + mgh_o = (1.50 \text{ kg})(9.80 \text{ m/s}^2)(6.00 \text{ m}) = 88.2 \text{ J}$$

The relation $E = K + U$ continues to hold while the can is falling, but now we know what E is. Rearranging the equation, we have $K = E - U$ and can find U at $h = 4.00$ m:

$$K = E - U = E - mgh = 88.2 \text{ J} - (1.50 \text{ kg})(9.80 \text{ m/s}^2)(4.00 \text{ m}) = 29.4 \text{ J}$$

Alternatively, we could have computed the change in (in this case, the loss of) potential energy, ΔU. Whatever potential energy was lost must have been gained as kinetic energy (Eq. 5.11). Then,

$$\Delta K + \Delta U = 0$$

$$(K - K_o) + (U - U_o) = (K - K_o) + (mgh - mgh_o) = 0$$

With $K_o = 0$ (since $v_o = 0$),

$$K = mg(h_o - h) = (1.50 \text{ kg})(9.8 \text{ m/s}^2)(6.00 \text{ m} - 4.00 \text{ m}) = 29.4 \text{ J}$$

(b) Just before the can strikes the ground ($h = 0$, $U = 0$), the total mechanical energy is all kinetic, or

$$E = K = \tfrac{1}{2}mv^2$$

Thus,

$$v = \sqrt{\frac{2E}{m}} = \sqrt{\frac{2(88.2 \text{ J})}{1.50 \text{ kg}}} = 10.8 \text{ m/s}$$

Basically, all of the potential energy of a free-falling object released from some height h is converted into kinetic energy just before the object hits the ground, so

$$|\Delta K| = |\Delta U|$$

$$\tfrac{1}{2}mv^2 = mgh$$

or

$$v = \sqrt{2gh}$$

Note that the mass cancels and is not a consideration. This result is also obtained from the kinematic equation $v^2 = v_o^2 + 2gy$ (Eq. 2.10′), with $v_o = 0$ and $y = h$.

Follow-up Exercise. A fellow painter on the ground wishes to toss a paintbrush vertically upward a distance of 5.0 m to his partner on the scaffold. Use conservation of mechanical energy methods to determine the minimum speed that he must give to the brush.

Conceptual Example 5.10

A Matter of Direction? Speed and Conservation of Energy

Three balls of equal mass m are projected with the same speed in different directions as shown in •Fig. 5.17. If air resistance is neglected, which ball would you expect to strike the ground with the greatest speed? (a) ball 1, (b) ball 2, (c) ball 3, (d) all strike with the same speed.

Reasoning and Answer. All of the balls have the same initial kinetic energy, $K_o = \tfrac{1}{2}mv_o^2$. (Recall that energy is a scalar quantity, and the different directions of projections do not produce any difference in the kinetic energies.) Regardless of their trajectories, all of the balls ultimately descend a distance h relative to their common starting point, so they all

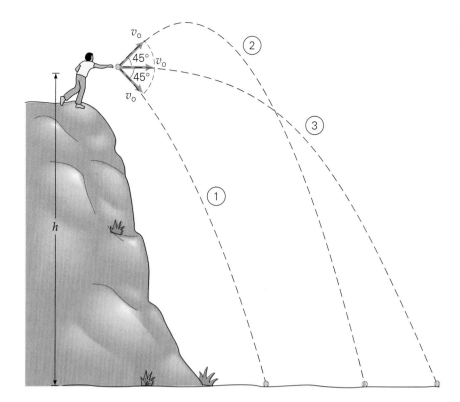

•FIGURE 5.17 **Speed and energy** See Conceptual Example 5.10.

lose the same amount of potential energy. (U is energy of *position* and thus is *independent of path*—see Fig. 5.15.)

By the law of conservation of energy, the amount of potential energy each ball loses is equal to the amount of kinetic energy it gains. Since all balls start with the same amount of kinetic energy and gain the same amount of kinetic energy, all three will have equal kinetic energies just before striking the ground. This means that their speeds must be equal, so the answer is (d).

Note that although balls 1 and 2 are projected at 45° angles, this is not a factor. Since the change in potential energy is independent of path, it is independent of the projection angle. The vertical distance between the starting point and the ground is the same (h) for projectiles at any angle. (Note: Although the strike speeds are equal, the *time* it takes each ball to reach the ground is different.)

Follow-up Exercise. Would the balls strike the ground with different speeds if their masses were different?

Example 5.11 ■ Conservative Forces: Mechanical Energy of a Spring

A 0.30-kilogram mass sliding on a horizontal frictionless surface with a speed of 2.5 m/s, as depicted in •Fig. 5.18, strikes a light spring that has a spring constant of 3.0×10^3 N/m. (a) What is the total mechanical energy of the system? (b) What is the kinetic energy (K_1) of the mass when the spring is compressed a distance $x_1 = 1.0$ cm? (Assume that no energy is lost in the collision.)

Thinking It Through. (a) All of the total energy is initially kinetic. (b) The total energy is the same as in (a), but it is now divided between kinetic energy and spring potential energy.

Solution.

Given: $m = 0.30$ kg
$\quad\quad v_o = 2.5$ m/s
$\quad\quad k = 3.0 \times 10^3$ N/m
$\quad\quad x_1 = 1.0$ cm $= 0.010$ m

Find: (a) E (total mechanical energy)
(b) K_1 (kinetic energy)

•FIGURE 5.18 **Conservative force and the mechanical energy of a spring** See Example 5.11.

(a) Before the mass makes contact with the spring, the total mechanical energy of the system is all in the form of kinetic energy, and

$$E = K_o = \tfrac{1}{2}mv_o^2 = \tfrac{1}{2}(0.30 \text{ kg})(2.5 \text{ m/s})^2 = 0.94 \text{ J}$$

Since the system is conservative (no energy is lost), this is the total energy at any time.

(b) When the spring is compressed a distance x_1, it has potential energy $U_1 = \tfrac{1}{2}kx_1^2$, and

$$E = K_1 + U_1 = K_1 + \tfrac{1}{2}kx_1^2$$

Solving for K_1, we have

$$K_1 = E - \tfrac{1}{2}kx_1^2$$
$$= 0.94 \text{ J} - \tfrac{1}{2}(3.0 \times 10^3 \text{ N/m})(0.010 \text{ m})^2 = 0.94 \text{ J} - 0.15 \text{ J} = 0.79 \text{ J}$$

Follow-up Exercise. How far will the spring in Example 5.11 be compressed when the mass comes to a stop?

See Learn by Drawing on this page for another example of energy exchange.

LEARN BY DRAWING

Energy Exchanges: A Falling Ball

Both the physical situation and the graphs of gravitational potential energy (U_g), kinetic energy (K), and spring potential energy (U_s) are drawn to scale. (Air resistance, the mass of the spring, and any energy loss in the collision are assumed to be negligible.) Why is the spring energy only one-quarter of the total when the spring is half-way compressed?

Total Energy and Nonconservative Forces

In the preceding examples, we ignored the force of friction, which is probably the most common nonconservative force. In general, both conservative and nonconservative forces can do work on objects. However, as you know, when nonconservative forces do work, the mechanical energy is not conserved. Mechanical energy is "lost" through the work done by a nonconservative force such as friction.

You might think that we can no longer use an energy approach to analyze problems involving nonconservative forces, since mechanical energy can be lost or dissipated (•Fig. 5.19). However, in some instances we can use the conservation of total energy to find out how much energy was lost to the work done by a nonconservative force. Suppose an object initially has mechanical energy and that nonconservative forces do an amount of work (W_{nc}) on it, which is equal to the mechanical energy that is lost or not conserved. Starting with the work–energy theorem, we have

$$W = \Delta K = K - K_o$$

In general, the net work (W) may be done by both conservative forces (W_c) and nonconservative forces (W_{nc}), so we may write

$$W_c + W_{nc} = K - K_o \qquad (5.12)$$

But recall that the work done by conservative forces is independent of path and depends only on the initial and final positions, as does the change in potential energy. Thus, $W_c = U_o - U$, and Eq. 5.12 then becomes

$$W_{nc} = K - K_o - (U_o - U)$$
$$= (K + U) - (K_o + U_o)$$
$$W_{nc} = E - E_o = \Delta E \qquad (5.13)$$

Hence, the work done by the nonconservative forces acting on a system is equal to the change in mechanical energy. Notice that for dissipative forces such as friction, $E_o > E$. Thus the change is negative, indicating a decrease in mechanical energy. The following Example illustrates this concept.

•**FIGURE 5.19 Nonconservative force and energy loss** Friction is a nonconservative force—when it is present, mechanical energy is not necessarily conserved. Can you tell from the photo what is happening to the work being done by the motor on the grinding wheel after it is converted into rotational kinetic energy?

Example 5.12 ■ Downhill Racer: Nonconservative Force

A skier with a mass of 80 kg starts from rest and skis down a slope from an elevation of 110 m (•Fig. 5.20). The speed of the skier at the bottom of the slope is 20 m/s. (a) Show that the system is nonconservative. (b) How much work is done by the nonconservative force of friction?

Thinking It Through. (a) If the system is nonconservative, then $E_o \neq E$, and these quantities can be computed. (b) We cannot determine the work from force–distance considerations, but W_{nc} is equal to the difference in total energies (Eq. 5.13).

•**FIGURE 5.20 Work done by a nonconservative force** See Example 5.12.

Solution.

Given: $m = 80$ kg
$v_o = 0$
$v = 20$ m/s
$h = 110$ m

Find: (a) Show that E is not constant
(b) W_{nc} (work done by friction)

(a) If the system is conservative, the total mechanical energy is constant. Taking $U_o = 0$ at the bottom of the hill, we find the initial energy at the top of the hill to be

$$E_o = U = mgh = (80 \text{ kg})(9.8 \text{ m/s}^2)(110 \text{ m}) = 8.6 \times 10^4 \text{ J}$$

And at the bottom of the slope,

$$E = K = \tfrac{1}{2}mv^2 = \tfrac{1}{2}(80 \text{ kg})(20 \text{ m/s})^2 = 1.6 \times 10^4 \text{ J}$$

Therefore, $E_o \neq E$, so this system is not conservative.

(b) The amount of work done by the nonconservative force of friction is equal to the change in the mechanical energy, or to the amount of mechanical energy lost (Eq. 5.13):

$$W_{nc} = E - E_o = (1.6 \times 10^4 \text{ J}) - (8.6 \times 10^4 \text{ J}) = -7.0 \times 10^4 \text{ J}$$

This is over 80% of the initial energy. (Where did this energy actually go?)

Note: Free fall is discussed in Section 2.5.

Follow-up Exercise. In free fall, air resistance is neglected; but for skydivers, air resistance has a very practical effect. Typically, a skydiver descends about 450 m before reaching a terminal velocity (Section 4.6) of 60 m/s. (a) What is the percentage energy loss to nonconservative forces during this descent? (b) Show that after terminal velocity is reached, the rate of energy loss is given by $(60mg)$ J/s, where m is the mass of the skydiver.

Note that in a nonconservative system the *total energy* (*not* total mechanical energy) is conserved (including heat and so on), but not all of it is available for mechanical work. For a conservative system, you get back what you put in, so to speak. That is, if you do work on the system, the transferred energy is available to do work. But keep in mind that conservative systems are idealizations, because all real systems are nonconservative to some degree. However, working with ideal conservative systems gives us an understanding of the conservation of energy.

During this course of study, you will be concerned with other forms of energy, such as thermal, electrical, nuclear, and chemical energies. In general, on the microscopic and submicroscopic levels, these forms of energy can be described in terms of kinetic energy and potential energy. Also, you will learn that mass is a form of energy and that the law of the conservation of energy must take this form into account to be applied to the analysis of nuclear reactions.

5.6 Power

Objectives: To (a) define power and (b) describe mechanical efficiency.

A particular task may require a certain amount of work, but that work might be done over different lengths of time or at different rates. For example, suppose that you have to mow a lawn. This takes a certain amount of work, but you might do the job in a half hour, or you might take an hour or two. There's a practical distinction to be made here. There is usually not only an interest in the amount of work done, but also an interest in how fast it is done—that is, the rate at which it is done. *The rate of doing work* is called **power**.

The average power is the work done divided by the time it takes to do the work, or work per unit of time:

Power: the time rate of doing work

$$\overline{P} = \frac{W}{t - t_o}$$

If t_o is taken to be zero,

$$\overline{P} = \frac{W}{t} \qquad (5.14)$$

If the work is done by a constant force (magnitude F) while an object moves through a displacement (magnitude d), then

$$\overline{P} = \frac{W}{t} = \frac{Fd}{t} = F\overline{v} \qquad (5.15)$$

SI unit of power: J/s or watt (W)

where it is assumed that the force is in the direction of the displacement. Here $\overline{v}$ is the magnitude of the average velocity. If the velocity is constant, then $\overline{P} = P = Fv$. If the force and displacement are not in the same direction, then we can write

$$\overline{P} = \frac{Fd \cos\theta}{t} = F\overline{v}\cos\theta \qquad (5.16)$$

where θ is the angle between the force and displacement.

As you can see from Eq. 5.15, the SI unit of power is joules per second (J/s), but this unit is given another name:

$$1 \text{ J/s} \equiv 1 \text{ watt (W)}$$

The watt (W), unit of power

The SI power unit is called a **watt (W)** in honor of James Watt (1736–1819), a Scottish engineer who developed one of the first practical steam engines. A common unit of electrical power is the *kilowatt* (kW).

The British unit of power is foot-pounds per second (ft·lb/s). However, a larger unit, the **horsepower (hp)** is more commonly used:

$$1 \text{ hp} \equiv 550 \text{ ft·lb/s} = 746 \text{ W}$$

In Watt's time, steam engines were replacing horses for work in mines and mills. To characterize the performance of his new engine, which was more efficient than other existing ones, Watt used the average rate at which a horse could do work as a unit—a horsepower.

Power tells you how fast work is being done *or* how fast energy is transferred. For example, motors have power ratings (commonly given in horsepower). A 1-horsepower motor can do a given amount of work in half the time that a $\frac{1}{2}$-horsepower motor would take, or twice the work in the same amount of time. That is, a 1-horsepower motor is twice as "powerful" as a $\frac{1}{2}$-horsepower motor (see Example 5.14).

Example 5.13 ■ A Crane Hoist: Work and Power

A crane hoist like the one shown in •Fig. 5.21 lifts a load of 1.0 metric ton a vertical distance of 25 m in 9.0 s at a constant velocity. How much useful work is done by the hoist each second?

Thinking It Through. The useful work done each (per) second is power output, so this is what is to be found.

Solution.

Given: $m = 1.0$ metric ton
$\quad\quad\quad = 1.0 \times 10^3$ kg
$\quad\quad h = 25$ m
$\quad\quad t = 9.0$ s

Find: W per second ($=$ power)

Keep in mind that the work per unit time (work per second) is power, so this is what we need to compute. Since the load moves with a constant velocity, $\overline{P} = P$. (Why?) The work is done against gravity, so $F = mg$, and

$$P = \frac{W}{t} = \frac{Fd}{t} = \frac{mgh}{t}$$

$$= \frac{(1.0 \times 10^3 \text{ kg})(9.8 \text{ m/s}^2)(25 \text{ m})}{9.0 \text{ s}} = 2.7 \times 10^4 \text{ W (or 27 kW)}$$

•FIGURE 5.21 Power delivery See Example 5.13.

Thus, since a watt (W) is a joule per second (J/s), the hoist did 2.7×10^4 J of work each second. Note that the velocity has a magnitude of $v = d/t = 25$ m$/9.0$ s $= 2.8$ m/s, and the power is $P = Fv$.

Follow-up Exercise. If the hoist motor of the crane in this Example is rated at 70 hp, what percentage of this power output goes into useful work?

Example 5.14 ■ Cleaning Up: Work and Time

The motors of two vacuum cleaners have net power outputs of 1.00 hp and 0.500 hp, respectively. (a) How much work in joules can each motor do in 3.00 min? (b) How long does it take for each motor to do 97.0 kJ of work?

Thinking It Through. (a) Since power is work/time ($P = W/t$), the work can be computed. Note that power is given in horsepower units. (b) This is another application of Eq. 5.15.

Solution.

Given: $P_1 = 1.00$ hp $= 746$ W Find: (a) W (work)
$P_2 = 0.500$ hp $= 373$ W   (b) t (time)
$t = 3.00$ min $= 180$ s
$W = 97.0$ kJ $= 97.0 \times 10^3$ J

(a) Since $P = W/t$,

$$W_1 = P_1 t = (746 \text{ W})(180 \text{ s}) = 1.34 \times 10^5 \text{ J}$$

$$W_2 = P_2 t = (373 \text{ W})(180 \text{ s}) = 0.67 \times 10^5 \text{ J}$$

Note that in the same time the smaller motor does half the work the larger one does, as you would expect.

(b) The times are given by $t = W/P$, and for the same amount of work,

$$t_1 = \frac{W}{P_1} = \frac{97.0 \times 10^3 \text{ J}}{746 \text{ W}} = 130 \text{ s}$$

$$t_2 = \frac{W}{P_2} = \frac{97.0 \times 10^3 \text{ J}}{373 \text{ W}} = 260 \text{ s}$$

Note that the smaller motor takes twice as long as the larger one to do the same amount of work.

Follow-up Exercise. (a) A 10-horsepower motor breaks down and is temporarily replaced with a 5-horsepower motor. What can you say about the rate of work output? (b) Suppose the situation were reversed, a 5-horsepower motor replaced with a 10-horsepower motor?

A real-life example of power, energy, and physics is given in the Insight on p. 163.

Efficiency

Machines and motors are quite common, and we often talk about their efficiency. Efficiency involves work, energy, and/or power. Simple and complex machines that do work have mechanical parts that move, so some input energy is always lost because of friction or some other cause (perhaps in the form of sound). Thus, not all of the input energy goes into doing useful work.

Mechanical efficiency is essentially a measure of what you get out for what you put in—that is, the *useful* work output for the energy input. **Efficiency (ε)** is given as a fraction (or percentage):

Insight

More Broken Records: The Clap Skate

At the 1998 Winter Olympics in Nagano, Japan, speed skating records were broken and new ones set—largely because of an innovation in ice skates (and applied physics). This new skate technology, invented by Dutch researchers in biomechanics, is commonly called the *clap skate*. The skates are designed to increase the amount of time they are in contact with the ice and therefore also the length of the skater's stride.

The new skates have a spring-loaded hinge on the toe that allows the blade to pull away from the heel (Fig. 1). With a longer stride, the amount of work done by the skater's leg muscle is increased, providing more kinetic energy and speed. Toward the end of the stride, the blade returns to the boot with a "clap" sound when the foot is lifted off the ice—hence the name clap skate. (The skates' inventors called them "slap skates," however, because the skates enables a skater to "slap on" an extra amount of work with each stride.)

Traditional skates require skaters to push from side to side. However, with claps, a new skating technique must be learned, with an emphasis on pushing straight back during the stride to keep the blade on the ice longer. Will the clap skate replace the traditional skate? Probably not completely. On a long track against the clock, the clapping is no more than an annoyance. But on a relatively short curved track, where skaters compete against each other, a surprise move to pass another skater would be well announced—clap, clap, clap—no surprise.

FIGURE 1   Clap, clap, clap   The new clap skate is helping skaters set records. A hinge allows the heel to lift off the blade so that the blade stays on the ice longer, thus providing the skater with more kinetic energy and speed.

$$\varepsilon = \frac{\text{work output}}{\text{energy input}} \; (\times 100\%) = \frac{W_{\text{out}}}{E_{\text{in}}} \; (\times 100\%) \qquad (5.17)$$

Efficiency—What you get out for what you put in

Efficiency is a unitless quantity

For example, if a machine has a 100-joule (energy) input and a 50-joule (work) output, then its efficiency is

$$\varepsilon = \frac{W_{\text{out}}}{E_{\text{in}}} = \frac{50 \text{ J}}{100 \text{ J}} = 0.50 \; (\times 100\%) = 50\%$$

An efficiency of 0.50, or 50%, means that half of the energy input is lost because of friction or some other cause and doesn't serve its intended purpose.

You can also write efficiency in terms of power (P):

$$\varepsilon = \frac{P_{\text{out}}}{P_{\text{in}}} \; (\times 100\%) \qquad (5.18)$$

Note that if both terms of the ratio in Eq. 5.17 are divided by time (t), $W_{\text{out}}/t = P_{\text{out}}$ and $E_{\text{in}}/t = P_{\text{in}}$.

Example 5.15 ■ Home Improvement: Mechanical Efficiency and Work Output

The motor of an electric drill with an efficiency of 80% has a power input of 600 W. How much useful work is done by the drill in a time of 30 s?

Thinking It Through.   This is an application of Eq. 5.18 and the definition of power.

Solution.

Given: $\varepsilon = 80\% = 0.80$ *Find:* W_{out} (work output)
$P_{in} = 600$ W
$t = 30$ s

Given the efficiency and power input, we can readily find the power output (P_{out}) from Eq. 5.18, and this is related to the work output ($P_{out} = W_{out}/t$). First, we rearrange Eq. 5.18:

$$P_{out} = \varepsilon P_{in} = (0.80)(600 \text{ W}) = 4.8 \times 10^2 \text{ W}$$

Then,

$$W_{out} = P_{out}t = (4.8 \times 10^2 \text{ W})(30 \text{ s}) = 1.4 \times 10^4 \text{ J}$$

Follow-up Exercise. (a) Is it possible to have a mechanical efficiency of 100%? (b) What would an efficiency of greater than 100% imply?

TABLE 5.1 Typical Efficiencies of Some Machines

Machine	Efficiency (%)
Compressor	85
Electric motor	70–95
Automobile	<15
Human muscle*	20–25
Steam locomotive	5–10

*Technically not a machine, but used to perform work.

The typical efficiencies of some machines are given in Table 5.1. You may be surprised by the relatively low efficiency of the automobile. Much of the energy input (from gasoline combustion) is lost as exhaust heat and through the cooling system (more than 60%), and friction accounts for a great deal more. Less than 15% of the input energy is converted to useful work that goes into propelling the vehicle. Air conditioning, power steering, radio, and tape and CD players are nice, but they use energy and also contribute to the car's decrease in efficiency.

Chapter Review

Important Terms

work   138
joule (unit)   139
spring constant (force constant)   143
kinetic energy   146
work–energy theorem   146
potential energy   150

gravitational potential energy   150
law of conservation of total energy   153
conservative force   153
nonconservative force   153
total mechanical energy   154
conservative system   154

law of conservation of mechanical energy   155
power   160
watt (unit)   161
horsepower (unit)   161
efficiency   162

Important Concepts

- Work done by a constant force is the product of the displacement and the component of the force parallel to the displacement.
- Work done by a variable force requires advanced mathematics. A simple example for which we can use average values is the work done by a spring force.
- Kinetic energy is the energy of motion. By the work–energy theorem, the net work done on an object (by a net external force) is equal to the change in the kinetic energy of the object.
- Potential energy is the energy of position and/or configuration. Most common is gravitational poten-

tial energy, associated with the Earth's gravitational attraction.
- Conservation of energy: The total energy of the universe or of an isolated system is always conserved. The total mechanical energy (kinetic plus potential) is constant in a conservative system.
- In systems with nonconservative forces, total mechanical energy is changed. If friction is the nonconservative force, then total mechanical energy is lost.
- Power is the time rate of doing work (or expending energy).
- Efficiency relates work output to energy (work) input.

Important Equations

Work:
$$W = (F \cos \theta) d \quad (5.2)$$

Hooke's Law (spring force):
$$F_s = -kx \quad (5.3)$$

Work Done in Stretching or Compressing a Spring (from Equilibrium):
$$W = \tfrac{1}{2} k x^2 \quad (5.4)$$

Kinetic Energy:
$$K = \tfrac{1}{2} m v^2 \quad (5.5)$$

Work–Energy Theorem:
$$W = K - K_o = \Delta K \quad (5.6)$$

Elastic (Spring) Potential Energy:
$$U = \tfrac{1}{2} k x^2 \quad (\text{with } x_o = 0) \quad (5.7)$$

Gravitational Potential Energy:
$$U = mgh \quad (\text{with } h_o = 0) \quad (5.8)$$

Conservation of Mechanical Energy:
$$\tfrac{1}{2} m v^2 + U = \tfrac{1}{2} m v_o^2 + U_o \quad (5.10)$$

Conservation of Energy (with a Nonconservative Force):
$$W_{nc} = E - E_o = \Delta E \quad (5.13)$$

Average Power:
$$\overline{P} = \frac{W}{t} = \frac{Fd}{t} = F\overline{v} \quad (5.15)$$

(constant force in direction of d and v)

$$\overline{P} = \frac{Fd \cos \theta}{t} \quad (5.16)$$

(force acts at an angle θ between d and v)

Efficiency (percent):
$$\varepsilon = \frac{W_{out}}{E_{in}} \ (\times 100\%) \quad (5.17)$$
$$= \frac{P_{out}}{P_{in}} \ (\times 100\%) \quad (5.18)$$

Exercises

5.1 Work Done by a Constant Force

1. The units of work are (a) N·m, (b) kg·m²/s², (c) J, (d) all of the preceding.

2. If you push against a wall, are you doing any work? Explain.

3. (a) As a weightlifter strains preparing to lift a barbell (•Fig. 5.22a), is he doing work? Why or why not? (b) In raising the barbell above his head, is he doing work? Explain. (c) In holding the barbell above his head (Fig. 5.22b), is he doing more work, no work, or the same amount of work as in lifting the barbell? Explain. (d) If the weightlifter drops the barbell, is work done on the barbell? Explain what happens in this situation.

(a) (b)

•FIGURE 5.22 **Man at work?** See Exercise 3.

4. You are carrying a backpack across the campus. What is the work done by your carrying force on the backpack? Explain.

5. A jet plane flies in a vertical circular loop. In what regions of the loop is the work done by the plane's weight positive and/or negative? Is the work constant? If not, are there maximum and minimum instantaneous values? Explain.

6. ■ ◐ A person does 50 J of work in moving a 30-kilogram box over a 10-meter distance on a horizontal surface. What is the minimum force required?

7. ■ A 5.0-kilogram box slides a 10-meter distance on ice. If the coefficient of kinetic friction if 0.20, what is the work done by the friction force?

8. ■ You carry a 15-kilogram bucket at constant speed from the ground floor to the third floor in the Science building. If the third floor is 10 m above the ground, how much work do you do on the bucket?

9. ■ ◐ A boy pushing a lawnmower on a level lawn applies a constant force of 250 N at an angle of 30° downward from the horizontal. How far does the boy push the mower in doing 1.44×10^3 J of work?

10. ■■ A 3.00-kilogram block slides down a frictionless plane inclined 20° to the horizontal. If the length of the plane's

surface is 1.50 m, how much work is done, and by what force?

11. ■■ Suppose the coefficient of kinetic friction between the block and the plane in Exercise 10 is 0.275. What would be the net work done in this case?

12. ■■ The formula for work can be written $W = F_\parallel d$, where $F_\parallel$ is the component of the force parallel to the displacement. Show that work is also given by $Fd_\parallel$, where $d_\parallel$ is the component of the displacement parallel to the force.

13. ■■ A hot-air balloon with a mass of 500 kg ascends at a constant rate of 1.50 m/s for 20.0 s. How much work is done by the upward buoyant force? (Neglect air resistance.)

14. ■■ A father pulls his young daughter on a sled with a constant velocity on a level surface through a distance of 10 m, as illustrated in •Fig. 5.23a. If the total mass of the sled and the girl is 35 kg and the coefficient of kinetic friction between the sled runners and the snow is 0.20, how much work does the father do?

(a)

(b)

•FIGURE 5.23 **Fun and work** See Exercises 14 and 15.

15. ■■ A father pushes horizontally on his daughter's sled to move it up a snowy incline (Fig. 5.23b). If the sled moves with a constant velocity up the hill, how much work does he do in moving it from the bottom to the top of the hill? (Some necessary data are given in Exercise 14.)

16. ■■ A boy pulls a 20-kilogram box with a 50-newton force at 37° above a horizontal surface. If the coefficient of kinetic friction between the box and the horizontal surface is 0.15 and the box is pulled a distance of 25 m, what is (a) the work done by the boy? (b) the work done by the friction force? (c) the net work done on the box?

17. ■■■ A 500-kilogram helicopter ascends from the ground with an acceleration of 2.00 m/s². Over a 5.00-second interval, what is (a) the work done by the lifting force? (b) the work done by the gravitational force? (c) the net work done on the helicopter?

18. ■■■ A 50-kilogram crate slides down a 5.0-meter loading ramp that is inclined at an angle of 25° to the horizontal. A worker pushes on the crate parallel to the ramp surface so that it slides down with a constant velocity. If the coefficient of kinetic friction between the crate and the ramp is 0.33, how much work is done by (a) the worker, (b) the force of friction, and (c) the force of gravity? (d) What is the net work done on the crate?

5.2 Work Done by a Variable Force

19. The work done by a variable force of the form $F = kx$ is equal to (a) kx^2 (b) kx, (c) the area under the F versus x curve, (d) none of these.

20. If a spring is compressed 2.0 cm from its equilibrium position and then compressed an additional 4.0 cm, how much more work is done in the second compression than in the first? (a) the same amount, (b) twice as much, (c) four times as much, (d) eight times as much.

21. Does it take the same amount of work to stretch the second centimeter as the first centimeter of a spring from its equilibrium position? Explain.

22. ■ To measure the spring constant of a certain spring, a student applies a 4.0-newton force; the spring stretches by 5.0 cm. What is the spring constant?

23. ■ If a 10-newton force is used to compress a spring with a spring constant of 4.0×10^2 N/m, what is the resulting spring compression?

24. ■ A spring has a spring constant of 40 N/m. How much work is required to stretch the spring 2.0 cm from its equilibrium position?

25. ■ If it takes 400 J of work to stretch a spring 8.00 cm, what is the spring constant?

26. ■ A student suspends a 0.25-kilogram mass on a spring, which stretches a distance of 5.0 cm. The student then removes the mass and pulls down the spring, stretching it and doing 10 J of work. How far is the spring stretched as a result of the work done by the student?

27. ■■ When a 75-gram mass is suspended from a vertical spring, the spring is stretched from a length of 4.0 cm to a length of 7.0 cm. If the mass is then pulled downward an additional 10 cm, what is the total work done against the spring force?

28. ■■ A particular spring has a force constant of 2.5×10^3 N/m. (a) How much work is done in stretching the relaxed spring

by 6.0 cm? (b) How much more work is done in stretching the spring an additional 2.0 cm?

29. ■■ For the spring in Exercise 28, how much mass would have to be suspended from the vertical spring to stretch it (a) the first 6.0 cm and (b) the additional 2.0 cm?

30. ■■ A particular force is described by the equation $\mathbf{F} = (60 \text{ N/m})x \ \hat{\mathbf{x}}$. How much work is done when this force pushes a box horizontally between (a) $x_o = 0$ and $x = 0.15$ m, and (b) $x_o = 0.15$ m and $x = 0.25$ m?

31. ■■ Compute the work done by the variable force in the F versus x graph in •Fig. 5.24. [Hint: The area of a triangle is $A = \frac{1}{2}$ altitude × base.]

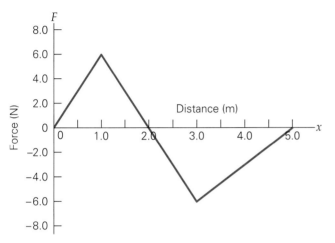

•FIGURE 5.24 How much work is done? See Exercise 31.

5.3 The Work–Energy Theorem: Kinetic Energy

32. If the angle between the force and the displacement is greater than 90°, (a) work is positive, (b) work is negative, (c) the change in energy is zero, (d) energy is either gained or lost.

33. •Figure 5.25 is a close-up of the clap skates discussed in the Insight on p. 163. The skates have a spring-loaded hinged toe at the front of the boot and a plunger system at the heel. As the skater strides, the heel lifts from the blade—unlike traditional skates—lengthening the skater's stride. Explain how these skates can improve performance.

34. ● Which of the following objects has the smallest kinetic energy: (a) an object of mass $4m$ and speed v; (b) an object of mass $3m$ and speed $2v$; (c) an object of mass $2m$ and speed $3v$; (d) an object of mass m and speed $4v$?

35. A 0.50-kilogram ball with a speed of 10 m/s hits a wall and bounces back with only half the original speed. How much kinetic energy is lost in the ball's collision with the wall?

36. ■ A large car and a small car, both traveling with the same speed, skid to a stop with the same coefficient of friction. Use the work–energy theorem to determine how their stopping distances compare.

37. ■ A large car of mass $2m$ travels with speed v, and a small car of mass m travels with speed $2v$. Both skid to a stop with the same coefficient of friction. How do their stopping distances compare? (Use the work–energy theorem, not Newton's laws.)

38. ■ ● An automobile with a mass of 1200 kg travels at a speed of 90 km/h. (a) What is its kinetic energy? (b) What is the net work that would be required to bring it to rest?

39. ■ A constant net force of 75 N acts on an object initially at rest and acts through a parallel distance of 0.60 m. (a) What is the final kinetic energy of the object? (b) If the object has a mass of 0.20 kg, what is its final speed?

40. ■ ● A 3.0-gram bullet traveling at 350 m/s hits a tree and slows down uniformly while penetrating a distance of 12 cm. What was the force exerted on the bullet in bringing it to rest?

41. ■■ ● A 0.60-kilogram object with an initial velocity of 3.0 m/s in the positive x direction is acted on by a force in the direction of motion. The force does +2.5 J of work. What is the final velocity of the object?

42. ■■ The stopping distance of a vehicle is an important safety factor. Assuming a constant braking force, use the work–energy theorem to show that a vehicle's stopping distance is proportional to the square of its initial speed. If an automobile traveling at 45 km/h is brought to a stop in 50 m, what would be the stopping distance for an initial speed of 90 km/h?

43. ■■■ If the work required to speed a car up from 10 km/h to 20 km/h is 5.0×10^3 J. What would be the work required to increase its speed from 20 km/h to 30 km/h?

5.4 Potential Energy

44. The potential energy change of a spring is proportional to (a) k^2, (b) $(x - x_o)^2$, (c) x_o^2, (d) $x^2 - x_o^2$.

•FIGURE 5.25 Clap to records See Exercise 33.

45. A change in gravitational potential energy (a) is always positive, (b) depends on the reference point, (c) depends on the path, (d) depends only on the initial and final heights.

46. Sketch a plot of U versus x for a mass oscillating on a spring between the limits of $-A$ and $+A$.

47. ■ What is the gravitational potential energy, relative to the ground, of a 1.0-kilogram box at the top of a 50-meter building?

48. ■ To store exactly 1.0 J of potential energy in a spring for which $k = 45$ N/m, how much would the spring have to be stretched beyond its equilibrium length?

49. ■ How much more gravitational potential energy does a 1.0-kilogram hammer have when it is on a shelf 1.5 m high than when it is on a shelf 0.90 m high?

50. ■ You are told that the gravitational potential energy of a 2.0-kilogram object decreased by 10 J. What physically happened to the object? Can you determine its height?

51. ■■ A 0.20-kilogram stone is thrown vertically upward with an initial velocity of 7.5 m/s from a starting point 1.2 m above the ground. (a) What is the potential energy of the stone at its maximum height relative to the ground? (b) What is the change in the potential energy of the stone between its launch point and its maximum height?

52. ■■ A 60-kilogram diver dives off a board that is 5.0 m above the surface of water; he touches the bottom of a pool 3.0 m below the surface of the water. What are (a) the potential energies of the diver relative to the water surface when he is on the board and at the bottom of the pool? (b) the change in the diver's potential energy relative to the board, to the water surface, and to the bottom of the pool?

53. ■■ The floor of a basement is 3.0 m below ground level, and the floor of the attic is 4.5 m above ground level. (a) What are the potential energies of 1.5-kilogram objects in the basement and attic, relative to ground level? (b) If one such object in the attic were brought to the basement, what would be the change in potential energy relative to the attic, ground, and basement? [Hint: Make a sketch.]

54. ■■■ A student has six textbooks, each with a thickness of 4.0 cm and a weight of 30 N. What is the minimum work the student would have to do to stack the books on top of one another with each initially on the table surface?

5.5 The Conservation of Energy

55. ● If a nonconservative force acts on an object, (a) the object's kinetic energy is always greater than its potential energy, (b) the mechanical energy is conserved, (c) its kinetic energy is conserved, (d) the total energy is conserved.

56. The speed of a pendulum is greatest (a) before the pendulum reaches the bottom of its swing, (b) when its acceleration is the greatest, (c) when its potential energy is the least, (d) when its kinetic energy is a minimum.

57. A bowling ball suspended from a ceiling is displaced from the vertical position to one side and released from rest just in front of a student's nose (●Figure 5.26). If the student doesn't move, why won't the bowling ball hit his nose?

●FIGURE 5.26 **In the face?** See Exercise 57.

58. Discuss all the different energy conversions involved in a pole vault, as shown in ●Fig. 5.27. (Include the vaulter's running start. Where does the energy for this come from?)

59. An energy transfer question: When jumping straight upward from the ground, you can achieve only the same maximum height with each jump. However, on a trampoline, you can jump higher with each bounce. Why?

●FIGURE 5.27 **Energy conversion(s)** See Exercise 58.

There's a limit on the trampoline too. What determines this limit?

60. ■ A person standing on a bridge at a height of 115 m above a river drops a 0.250-kilogram rock. (a) What is the rock's mechanical energy at the time of release relative to the river surface? (b) What are the rock's kinetic, potential, and mechanical energies after it has fallen 75.0 m? (c) Just before the rock hits the water, what are its speed and total mechanical energy? (d) Repeat (a)–(c) but take the reference point ($h = 0$) at the elevation where the rock is released.

61. ■ A 0.300-kilogram ball is thrown vertically upward with an initial speed of 10.0 m/s. If the initial potential energy is called zero, find the ball's kinetic, potential, and mechanical energies (a) at its initial position, (b) at 2.50 m above the initial position, (c) at its maximum height.

62. ■ What is the maximum height reached by the ball in Exercise 61?

63. ■■ A 0.50-kilogram ball thrown vertically upward has an initial kinetic energy of 80 J. What are (a) its kinetic and potential energies when it has traveled three-quarters of the distance to its maximum height? (b) its speed at this point? (c) its potential energy at the maximum height (assume a reference is chosen to be zero at the launch point)?

64. ■■ A girl swings back and forth on a swing whose ropes are 4.00 m long. The maximum height she reaches is 2.00 m above the ground. At the lowest point of the swing, she is 0.500 m above the ground. What is the girl's maximum speed, and where does she attain it?

65. ■■ When a certain rubber ball is dropped from a height of 1.25 m onto a hard surface, it loses 18.0% of its mechanical energy on each bounce. How high will the ball bounce (a) on the first bounce? (b) on the second bounce? (c) With what speed would the ball have to be thrown downward to make it reach its original height on the first bounce?

66. ■■ A skier coasts down a very smooth slope as shown in •Fig. 5.28. If the speed of the skier at point A is 5.0 m/s, what is his speed at point B?

•FIGURE 5.28 **Down the slope** See Exercises 66 and 71.

67. ■■ A roller coaster rolls on a track as shown in •Fig. 5.29. If the speed of the roller coaster at point A is 5.0 m/s, (a) what is its speed at point B? (b) Will it reach point C? (c) What speed at A is required for the roller coaster to reach point C?

•FIGURE 5.29 **Enough energy?** See Exercise 67.

68. ■■ A simple pendulum has a length of 0.75 m and a bob whose mass is 0.15 kg. The bob is released from an angle of 25° relative to a vertical reference line (•Fig. 5.30). (a) Show that the vertical height of the bob when released is $h = L(1 - \cos 25°)$. (b) What is the kinetic energy of the bob when the string is at an angle of 9.0°? (c) What is the speed of the bob at the bottom of the swing? (Neglect friction and the mass of the string.)

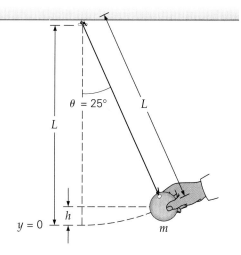

•FIGURE 5.30 **A pendulum swings** See Exercises 68 and 69.

69. ■■ Suppose the simple pendulum in Exercise 68 were released from an angle of 60°. (a) What would be the speed of the bob at the bottom of the swing? (b) To what height would the bob swing on the other side? (c) What angle of release would give half the speed of the 60° release angle at the bottom of the swing?

70. ■■ A 1.5-kilogram box sliding with a speed of 12 m/s on a frictionless surface approaches a horizontal spring (see Fig. 5.10). If the spring has a spring constant of 2000 N/m, (a) how far will the spring be compressed in stopping the box? (b) How far will the spring be

compressed when the box's speed is reduced to half of its initial speed?

71. ■■■ 🌐 In Exercise 66, the skier has a mass of 60 kg. If the force of friction retards the motion by doing 2500 J of work, what is his speed at point B?

5.6 Power

72. Which of the following is not a unit of power? (a) J/s, (b) W·s, (c) ft·lb/s, (d) hp

73. If you check your electricity bill, you will note that you are paying the power company for so many kilowatt-hours (kWh). Are you really paying for power? Explain.

74. (a) Does efficiency describe how fast work is done? Explain. (b) Does a machine with a higher efficiency always do work faster than one with a lower efficiency? Explain.

75. Two students who weigh the same start at the same ground floor location at the same time to go to the same classroom on the third floor by different routes. If they arrive at different times, which student will have expended more power? Explain.

76. ■ What is the power in watts of a motor rated at 1/4 hp?

77. ■ Marcia consumes 8.4×10^6 J (2000 food calories) of energy per day while maintaining a constant weight. What is the average power she produces in a day?

78. ■ A 1500-kilogram race car can go from zero to 90 km/h in 5.0 s. What average power is required to do this?

79. ■ The two 0.50-kilogram weights of a cuckoo clock descend 1.5 m in a 3-day period. At what rate is gravitational potential energy decreased?

80. ■ A 60-kilogram woman runs up a staircase 15 m high (vertically) in 20 s. (a) How much power does she expend? (b) What is the horsepower rating?

81. ■■ An electric motor with a 2.0-horsepower output drives a machine with an efficiency of 45%. What is the energy output of the machine per second?

82. ■■ Water is lifted out of a well 30.0 m deep by a motor rated at 1 hp. Assuming 100% efficiency, how many kilograms of water can be lifted in 1 min?

83. ■■ In Exercise 82, if the motor has an efficiency of 83% and 230 kg of water is to be lifted in 1 min, what would be the required horsepower rating?

84. ■■■ A 3250-kilogram aircraft takes 12.5 min to achieve its cruising altitude of 10.0 km and cruising speed of 850 km/h. If the plane's engines deliver, on the average, 1500 hp of power during this time, what is the efficiency of the engines?

85. ■■■ A sled and driver with a total mass of 120 kg are pulled up a hill with a 15° incline by a horse, as illustrated in •Fig. 5.31. (a) If the overall retarding frictional force is 950 N and the sled moves up the hill with a constant velocity of 5.0 km/h, what is the power output of the horse? (Express in horsepower, of course. Note the magnitude of your answer and explain.) (b) Suppose that in a spurt of energy the horse accelerates the sled uniformly from 5.0 km/h to 20 km/h in 5.0 s. What is the horse's maximum instantaneous power output? Assume the same force of friction.

•FIGURE 5.31 **A one-horse open sleigh** See Exercise 85.

Additional Exercises

86. An electron ($m = 9.11 \times 10^{-31}$ kg) has 8.00×10^{-17} J of kinetic energy. What is the speed of the electron?

87. A large electric motor with an efficiency of 75% has a power output of 1.5 hp. If the motor is run steadily and the cost of electricity is $0.12/kWh, how much does it cost, to the nearest penny, to run the motor for 2.0 h?

88. 🌐 A 28-kilogram child slides down a playground slide from a height of 3.0 m above the bottom of the slide. If her speed at the bottom is 2.5 m/s, how much energy is lost to friction?

89. In planing a piece of wood 35 cm long, a carpenter applies a force of 40 N to a plane at a downward angle of 25° to the horizontal. How much work is done by the carpenter?

90. A spring with a force constant of 50 N/m is stretched from 0 to 20 cm. Is more work required to stretch the spring from 10 cm to 20 cm than from 0 to 10 cm? If so, how much more? Justify your answer mathematically.

91. A 120-kilogram sleigh is pulled by one horse at a constant velocity for a distance of 0.75 km on a level snowy surface. The coefficient of kinetic friction between the sleigh runners and the snow is 0.25. (a) Calculate the work done by the horse. (b) Calculate the work done by friction.

92. A water slide has a height of 4.0 m. The people coming down the slide shoot out horizontally at the bottom,

which is a distance of 1.5 m above the water in the swimming pool. If a person starts down the slide from rest, neglecting frictional losses, how far from a point directly below the bottom of the slide does the person land? Does it make any difference whether the person is a small child or an adult?

93. A hiker plans to swing on a rope across a ravine in the mountains, as illustrated in •Fig. 5.32, and to drop when she is just above the far edge. (a) At what horizontal speed should she be moving when she starts to swing? (b) When would she be in danger of falling into the ravine? Explain.

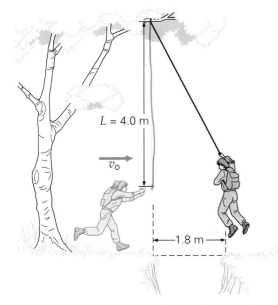

•FIGURE 5.32 **Can she make it?** See Exercise 93.

94. A tractor pulls a wagon with a constant force of 700 N at a constant speed of 20.0 km/h. (a) How much work is done by the tractor in 3.50 min? (b) What is the tractor's power output?

95. A sports car weighs one-third as much as a large luxury car. (a) If the sports car is traveling at a speed of 90 km/h, at what speed would the larger car have to travel to have the same kinetic energy? (b) Suppose that the large car is traveling at a speed that gives it half the kinetic energy of the sports car. What is the speed of the more massive car?

96. A ball with a mass of 0.360 kg is dropped from a height of 1.20 m above the top of a fixed vertical spring, whose force constant is 350 N/m. (a) What is the maximum distance the spring is compressed by the ball? (Neglect energy loss due to the collision.) (b) What is the speed of the ball when the spring has been compressed 5.00 cm?

97. A student on a sled starts from rest at a vertical height of 25 m above the horizontal base of the hill and slides down. If the sled and the student have a speed of 20 m/s at the bottom of the hill, is mechanical energy conserved? Justify your answer.

98. A constant horizontal force of 30 N is required to move a box with a constant speed along a rough horizontal surface. If the force does work at a rate of 50 W, (a) what is the box's speed? (b) How much work is done by the force in 2.5 s?

99. In 10 s, a 70-kilogram student runs up two flights of stairs whose combined vertical height is 8.0 m. Compute the student's power output in doing work against gravity in (a) watts and (b) horsepower.

CHAPTER 6

Momentum and Collisions

6.1 Linear Momentum
6.2 Impulse
6.3 The Conservation of Linear Momentum
6.4 Elastic and Inelastic Collisions
6.5 Center of Mass
6.6 Jet Propulsion and Rockets

Insight
- The Automobile Air Bag

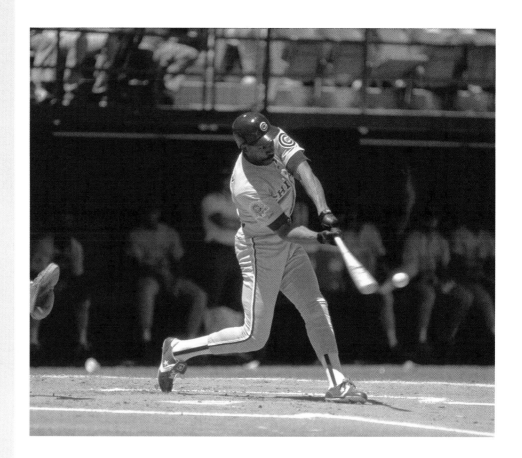

Tomorrow, the sportscasters may say that in this instant the momentum of the entire game changed. As a result of this clutch hit, one team gained momentum while their opponents lost it. But regardless of the effect on the team, it's clear that the momentum of the *ball* must have changed dramatically in the instant before this photograph was taken. The ball was traveling from right to left, probably at a pretty good rate of speed—and thus with lots of momentum. But a collision with several pounds of hardwood—with plenty of momentum of its own—changed its trajectory in a fraction of a second. A fan might say that the batter turned it around; after studying Chapter 4, you might say that the force he applied gave it a large negative acceleration, reversing its velocity vector (and probably increasing its magnitude as well). Yet, if you summed up the momentum of the ball and bat just before the collision and just afterward, you'd discover that although both the ball and the bat had momentum changes, the total momentum never changed!

If you were bowling and the ball bounced off the pins and rolled back toward you, you would probably be very surprised. But why? What leads us to expect that the ball will send the pins flying and continue on its way, rather than rebounding? You might say that the momentum of the ball carries it onward even after the collision (and you would be right)—but what does that really mean? In this chapter, you will study the concept of *momentum* and learn how it is particularly useful in analyzing motion and collisions.

6.1 Linear Momentum

Objective: To compute linear momentum and the components of momentum.

The term *momentum* may bring to mind a football player running down the field, knocking down players who are trying to stop him. Or you might have heard someone say that a team lost its momentum (and so lost the game). Such everyday usages give some insight into the meaning of momentum. They suggest the idea of mass in motion and therefore of inertia. We tend to think of heavy or massive objects in motion as having a great deal of momentum, even if they move very slowly. However, according to the technical definition of momentum, a light object can have just as much momentum as a heavier one, and sometimes more.

Newton referred to what modern physicists term **linear momentum** as "the quantity of motion ... arising from velocity and the quantity of matter conjointly." In other words, the momentum of a body is proportional to both its mass and its velocity. By definition,

The linear momentum of an object is the product of its mass and velocity:

$$\mathbf{p} = m\mathbf{v} \tag{6.1}$$

SI unit of momentum: kilogram-meters per second (kg·m/s)

It is common to refer to linear momentum as simply momentum. From Eq. 6.1, you can see that the SI units for momentum are kilogram-meters per second. Momentum is a vector quantity having the same direction as the velocity. Like velocity, it can be resolved into rectangular components: $\mathbf{p}_x = m\mathbf{v}_x$ and $\mathbf{p}_y = m\mathbf{v}_y$.

Equation 6.1 expresses the momentum of a single object or particle. For a system of more than one particle, the **total linear momentum** of the system is the vector sum of the momenta (plural of momentum) of the individual particles:

$$\mathbf{P} = \mathbf{p}_1 + \mathbf{p}_2 + \mathbf{p}_3 + \cdots = \Sigma_i \mathbf{p}_i \tag{6.2}$$

Note: The momentum vector of a single object is in the direction of the object's velocity.

Total linear momentum—a vector sum

Example 6.1 ■ Momentum: Mass *and* Velocity

A 100-kilogram football player runs with a velocity of 4.0 m/s straight down the field. A 1.0-kilogram artillery shell leaves the barrel of a gun with a muzzle velocity of 500 m/s. Which has the greater momentum (magnitude)—the football player or the shell?

Thinking It Through. Given the mass and velocity of an object, the momentum can be calculated from Eq. 6.1.

Solution. As usual, we first list the given data and what we are to find, using the subscripts "p" and "s" to refer to the player and shell, respectively.

Given: $m_p = 100$ kg
$\qquad v_p = 4.0$ m/s
$\qquad m_s = 1.0$ kg
$\qquad v_s = 500$ m/s

Find: p_p and p_s (magnitudes of the momenta)

The magnitude of the momentum of the player is

$$p_p = m_p v_p = (100 \text{ kg})(4.0 \text{ m/s}) = 4.0 \times 10^2 \text{ kg·m/s}$$

and that of the shell is

$$p_s = m_s v_s = (1.0 \text{ kg})(500 \text{ m/s}) = 5.0 \times 10^2 \text{ kg·m/s}$$

Thus, the much lighter, or less massive, shell has the greater momentum. Remember: The magnitude of momentum depends on *both* the mass and the magnitude of the velocity.

(a)

(b)

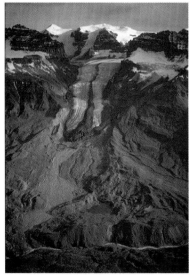

(c)

FIGURE 6.1 Three moving objects: a comparison of momenta and kinetic energies (a) A .22-caliber bullet shattering a ball point pen; (b) a cruise ship, (c) a glacier, Glacier Bay, Alaska. See Example 6.2.

Follow-up Exercise. What would the football player's speed have to be for his momentum to have the same magnitude as the artillery shell's? Would this speed be realistic?

Example 6.2 ■ Momentum versus Kinetic Energy: Some Ballpark Comparisons

(a) Consider the three objects shown in •Fig. 6.1—a .22-caliber bullet, a cruise ship, and a glacier. Assuming each to be moving at its normal speed, which do you think has the greatest linear momentum? The least? Which has the greatest kinetic energy? The least? (List your choices before going on.)

	Greatest	Least
Momentum	_____	_____
Kinetic energy	_____	_____

(b) Using the estimates given below as your data, calculate the magnitudes of the momenta and the kinetic energies of the three objects. How accurate were your original answers—any surprises?

Bullet: A typical .22-caliber bullet would have a weight of about 30 grains and a muzzle velocity of about 1300 ft/s. (A grain, abbreviated gr, is an old British unit used for pharmaceuticals, such as 5-grain aspirin tablets; 1 lb = 7000 gr.)

Ship: A ship like the one shown would have a weight of about 70 000 tons and a speed of about 20 knots. (A knot is another old unit, still commonly used in nautical contexts; 1 knot = 1.15 mi/h.)

Glacier: The glacier might be 1 km wide, 10 km long, and 250 m deep, and it might move at a rate of 1 m per day. (There is much variation among glaciers. Therefore these figures must involve more assumptions and rougher estimates than those for the bullet or ship. For example, we are assuming a uniform, rectangular cross-sectional area for the glacier. The depth is particularly difficult to estimate from a photograph; a minimum value is given by the fact that glaciers must be at least 50–60 m thick before they can "flow." Observed speeds range from a few centimeters to as much as 40 m a day for valley glaciers such as the one in Fig. 6.1c. The value we have chosen is considered a typical one.)

Solution. Converting our data to metric units, we have:

Bullet: $m_b = 30 \text{ gr} \left(\frac{1 \text{ lb}}{7000 \text{ gr}}\right)\left(\frac{1 \text{ kg}}{2.2 \text{ lb}}\right) = 0.0019 \text{ kg}$

$v_b = 1.3 \times 10^3 \text{ ft/s} \left(\frac{0.305 \text{ m/s}}{\text{ft/s}}\right) = 4.0 \times 10^2 \text{ m/s}$

Ship: $m_s = 7.0 \times 10^4 \text{ ton} \left(\frac{2.0 \times 10^3 \text{ lb}}{\text{ton}}\right)\left(\frac{1 \text{ kg}}{2.2 \text{ lb}}\right) = 6.4 \times 10^7 \text{ kg}$

$v_s = 20 \text{ knots} \left(\frac{1.15 \text{ mi/h}}{\text{knot}}\right)\left(\frac{0.447 \text{ m/s}}{\text{mi/h}}\right) = 10 \text{ m/s}$

Glacier: width = 10^3 m, length = 10^4 m, depth = 2.5×10^2 m;

$v_g = 1 \text{ m/day} \left(\frac{1 \text{ day}}{86\,400 \text{ s}}\right) = 1.2 \times 10^{-5} \text{ m/s}$

We know the equations for kinetic energy, $K = \frac{1}{2}mv^2$ (Eq. 5.5), and momentum, $p = mv$ (Eq. 6.1). We have all the speeds and masses except for m_g, the mass of the glacier. To calculate this we need to know the density of ice, since $m = \rho V$ (Eq. 1.1). The density of ice is less than that of water, but the two are not very different, so we will use the density of water, $1.0 \times 10^3 \text{ kg/m}^3$, to simplify the calculations. (The actual density of ice is $0.92 \times 10^3 \text{ kg/m}^3$, but most of our other data in this Example are also approximations or estimates, so this shortcut should produce results that are good enough for our present purposes.)

$$m_g = \rho V = \rho(\ell \times w \times d)$$
$$\approx (1.0 \times 10^3 \text{ kg/m}^3)(10^4 \text{ m})(10^3 \text{ m})(2.5 \times 10^2 \text{ m}) = 2.5 \times 10^{12} \text{ kg}$$

Then, calculating the magnitudes of the momenta of the objects, we have

Bullet: $p_b = m_b v_b = (0.0019 \text{ kg})(4.0 \times 10^2 \text{ m/s}) = 0.76 \text{ kg·m/s}$

Ship: $p_s = m_s v_s = (6.4 \times 10^7 \text{ kg})(10 \text{ m/s}) = 6.4 \times 10^8 \text{ kg·m/s}$

Glacier: $p_g = m_g v_g = (2.5 \times 10^{12} \text{ kg})(1.2 \times 10^{-5} \text{ m/s}) = 3.0 \times 10^7 \text{ kg·m/s}$

Thus, the glacier has about 40 million times as much momentum as the bullet, and the ship nearly a billion times as much. The enormous differences in the masses of these objects outweigh the differences in their speeds.

When we turn to the kinetic energies, however, the dependence on the *square* of the speed (Example 5.6) produces results that might surprise you:

Bullet: $K_b = \frac{1}{2} m_b v_b^2 = \frac{1}{2}(0.0019 \text{ kg})(4.0 \times 10^2 \text{ m/s})^2 = 1.5 \times 10^2 \text{ J}$

Ship: $K_s = \frac{1}{2} m_s v_s^2 = \frac{1}{2}(6.4 \times 10^7 \text{ kg})(10 \text{ m/s})^2 = 3.2 \times 10^9 \text{ J}$

Glacier: $K_g = \frac{1}{2} m_g v_g^2 = \frac{1}{2}(2.5 \times 10^{12} \text{ kg})(1.2 \times 10^{-5} \text{ m/s})^2 = 1.8 \times 10^2 \text{ J}$

The bullet actually has about the same kinetic energy as the glacier—but the ship has more than 10 million times as much kinetic energy as either of them!

Problem-Solving Strategies: "Ballpark" and Order-Of-Magnitude Calculations

Calculations such as we have performed in Example 6.2 are sometimes known as "ballpark" or "back of the envelope" calculations. In performing such calculations we know that, because the data are only approximate (they are not known precisely, or they are being rounded off to simple whole numbers for the purpose of rapid calculation), the answer can't be taken as very accurate (see Section 1.6). However, it is often precise enough for our needs. In Example 6.2, some of the data were unavoidable approximations (the glacier isn't a simple rectangular object), whereas others, such as the density of ice, were deliberately approximated to simplify calculation. Nevertheless, this method gave us a good idea of the relative momenta and kinetic energies of the three objects.

Physicists often find it convenient to use a formalized version of the "ballpark" approach, commonly known as *order-of-magnitude calculation*. In this method, all values are rounded off to the nearest "order of magnitude," or power of 10: 10^2, 10^3, 10^4, and so on. Since it is very easy to multiply and divide powers of 10, this shortcut makes calculation very fast. However, although sometimes just the powers of 10 can be used in a calculation, it is usually a good idea to retain one significant figure of the prefix. This is particularly true when quantities are squared or raised to higher powers, as in the kinetic energy calculations. For instance, $v_s = 4 \times 10^2$ m/s or $\approx 10^2$ m/s. Then $v_s^2 = (4 \times 10^2 \text{ m/s})^2 = 16 \times 10^4 \text{ m}^2/\text{s}^2 \approx 10^5 \text{ m}^2/\text{s}^2$. However, if we had used just 10^2 m/s, we would have obtained 10^4 m^2/s^2.

In order-of-magnitude calculations, we can generally expect the answer to be correct to the nearest order of magnitude. In other words, if we get an answer of 1000, the true value might really be 750 or 1200, but we can be pretty confident that it is closer to 1000 than to 100 or to 10 000.

Example: A light-year—a unit used in astronomy—is the distance light travels in 1 y in a vacuum. The speed of light is $c = 3.00 \times 10^8$ m/s. Approximately how many meters are there in a light-year?

Solution: From the Conversion Factors Table inside the front cover (or simple calculation), we know that there are 3.16×10^7 s in 1 year. Using the relationship distance = speed × time,

$$d = ct = (3 \times 10^8 \text{ m/s})(3 \times 10^7 \text{ s}) \approx 10^{16} \text{ m}$$

Note that the product of the powers of 10 gives 10^{15}, but $3 \times 3 = 9$, so the prefixes give about another order of magnitude. Thus $(10 \times 10^{15}) = 10^{16}$ is the order of magnitude of the number of meters in a light-year. (The actual value is 9.48×10^{15} m.)

Example 6.3 ■ Total Momentum: A Vector Sum

What is the total momentum for each of the systems of particles illustrated in •Fig. 6.2?

Thinking It Through. The total momentum is the vector sum of the individual momenta (Eq. 6.2). This can be computed for each component.

Solution.

Given: Magnitudes and directions of momenta from Fig. 6.2

Find: (a) Total momentum (**P**)
(b) Total momentum (**P**)

(a) The total momentum of a system is the vector sum of the momenta of the individual particles, so

$$\mathbf{P} = \mathbf{p}_1 + \mathbf{p}_2 = (2.0 \text{ kg·m/s})\hat{\mathbf{x}} + (3.0 \text{ kg·m/s})\hat{\mathbf{x}} = (5.0 \text{ kg·m/s})\hat{\mathbf{x}} \quad (+x \text{ direction})$$

(b) Computing the total momenta in the x and y directions gives

$$\mathbf{P}_x = \mathbf{p}_1 + \mathbf{p}_2 = (5.0 \text{ kg·m/s})\hat{\mathbf{x}} + (-8.0 \text{ kg·m/s})\hat{\mathbf{x}}$$
$$= -(3.0 \text{ kg·m/s})\hat{\mathbf{x}} \quad (-x \text{ direction})$$
$$\mathbf{P}_y = \mathbf{p}_3 = (4.0 \text{ kg·m/s})\hat{\mathbf{y}} \quad (+y \text{ direction})$$

Then

$$\mathbf{P} = \mathbf{P}_x + \mathbf{P}_y = (-3.0 \text{ kg·m/s})\hat{\mathbf{x}} + (4.0 \text{ kg·m/s})\hat{\mathbf{y}}$$

Follow-up Exercise. In Example 6.3, if $\mathbf{p}_1$ and $\mathbf{p}_2$ in (a) were added to $\mathbf{p}_2$ and $\mathbf{p}_3$ in (b), what would be the total momentum?

(a) $\mathbf{P} = \mathbf{p}_1 + \mathbf{p}_2$

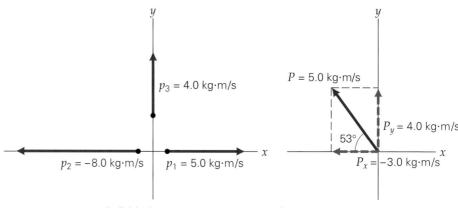

(b) $\mathbf{P} = \mathbf{p}_1 + \mathbf{p}_2 + \mathbf{p}_3$

•**FIGURE 6.2 Total momentum** The total momentum of a system of particles is the vector sum of the individual momenta. See Example 6.3.

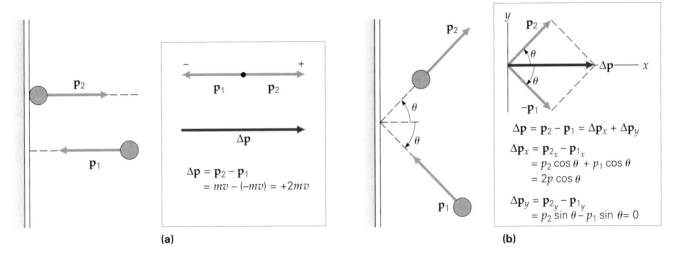

•**FIGURE 6.3 Change in momentum** The change in momentum is given by the *difference* in the momentum vectors. (a) Here the vector sum is zero, but the vector *difference*, or change in momentum, is not. (The particles are displaced for convenience.) (b) The change in momentum is found by computing the change in the components.

In Example 6.3a, the momenta were along the coordinate axes and thus added straightforwardly. If the motion of one (or more) of the particles is not along an axis, its momentum may be broken up, or resolved, into rectangular components and added to the component sums as in Example 6.3b—just as you learned to do with force components in Chapter 4. For example, suppose you were given a momentum vector with a magnitude of 5.0 kg·m/s at an angle of 53° relative to the negative x axis (like the total momentum vector in Fig. 6.2b). Then, its rectangular components are

Note: Review Example 4.4 and Fig. 4.10.

$$\mathbf{p}_x = -p \cos 53° \hat{\mathbf{x}} = -(5.0 \text{ kg·m/s})(0.60)\hat{\mathbf{x}} = (-3.0 \text{ kg·m/s})\hat{\mathbf{x}}$$
$$\mathbf{p}_y = p \sin 53° \hat{\mathbf{x}} = (5.0 \text{ kg·m/s})(0.80)\hat{\mathbf{y}} = (4.0 \text{ kg·m/s})\hat{\mathbf{y}}$$

where the signs give the directions. Each momentum vector would be broken down thus and the rectangular components along the axis in each dimension added separately to find the vector sum.

Since momentum is a vector, a change in momentum can result from a change in magnitude and/or direction. Examples of changes in the momenta of particles because of changes of direction on collision are illustrated in •Fig. 6.3. In the figure, the magnitude of a particle's momentum is taken to be the same before and after collision (as indicated by the arrows of equal length). Figure 6.3a illustrates a direct rebound—a 180° change in direction. Note that the change in momentum ($\Delta \mathbf{p}$) is the vector difference and that directional signs for the vectors are important. Figure 6.3b shows a glancing collision, for which the change in momentum is given by the component differences.

Force and Momentum

As you know from Chapter 4, a change in velocity (an acceleration) requires a force. Similarly, since momentum is directly related to velocity (as well as mass), a change in momentum also requires a force. In fact, Newton originally expressed his second law of motion in terms of momentum rather than acceleration. The force-momentum relationship may be seen by starting with $\mathbf{F} = m\mathbf{a}$ and using $\mathbf{a} = (\mathbf{v} - \mathbf{v}_o)/\Delta t$, where the mass is assumed to be constant:

$$\mathbf{F} = m\mathbf{a} = \frac{m(\mathbf{v} - \mathbf{v}_o)}{\Delta t} = \frac{m\mathbf{v} - m\mathbf{v}_o}{\Delta t} = \frac{\mathbf{p} - \mathbf{p}_o}{\Delta t} = \frac{\Delta \mathbf{p}}{\Delta t}$$

•FIGURE 6.4 **Change in the momentum of a projectile** The total momentum vector of a projectile is tangential to its path (as is its velocity); this vector changes in magnitude and direction because of the action of an external force (gravity). The x component of the momentum is constant. (Why?)

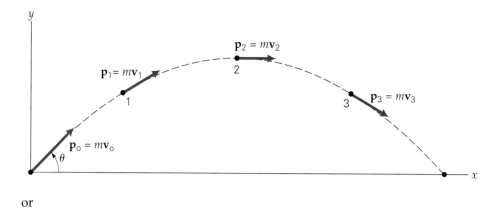

Newton's second law of motion in terms of momentum

or

$$\mathbf{F} = \frac{\Delta \mathbf{p}}{\Delta t} \tag{6.3}$$

where **F** is the *average* force if the acceleration is not constant or the *instantaneous* force if Δt goes to zero.

Expressed in this form, Newton's second law states that *the net external force acting on an object is equal to the time rate of change of the object's momentum.* It is easily seen from the development of Eq. 6.3 that the equations $\mathbf{F} = m\mathbf{a}$ and $\mathbf{F} = \Delta \mathbf{p}/\Delta t$ are equivalent if the mass is constant. In some situations, however, the mass may vary. This will not be a consideration here in our discussion of particle collisions, but a special case will be given later in the chapter. The more general form of Newton's second law, Eq. 6.3, is true even if the mass varies.

Just as the equation $\mathbf{F} = m\mathbf{a}$ indicates that an acceleration is evidence of a net force, the equation $\mathbf{F} = \Delta \mathbf{p}/\Delta t$ indicates that *a change in momentum is evidence of a net force.* For example, as illustrated in •Fig. 6.4, the momentum of a projectile is tangential to the projectile's parabolic path and changes in both magnitude and direction. The change in momentum indicates that there is a net force acting on the projectile, which you know is the force of gravity. Changes in momentum were illustrated in Fig. 6.3. Can you identify the forces in these cases? Think in terms of Newton's third law.

6.2 Impulse

Objectives: To relate (a) impulse and momentum, and (b) kinetic energy and momentum.

When two objects—such as a hammer and a nail, a golf club and a golf ball, or even two cars—collide, they exert large forces on one another for a short period of time (•Fig. 6.5a). The force is not constant in this case. However, Newton's second law in momentum form is still useful for analyzing such situations by using average values. Written in this form, the law states that the *average* force is equal to the time rate of change of momentum, $\bar{\mathbf{F}} = \Delta \mathbf{p}/\Delta t$ (Eq. 6.3). Rewriting the equation to express the change in momentum, we have

$$\bar{\mathbf{F}} \Delta t = \Delta \mathbf{p} = \mathbf{p} - \mathbf{p}_o \tag{6.4}$$

The term $\bar{\mathbf{F}} \Delta t$ is known as the **impulse** of the force:

$$\text{Impulse} = \bar{\mathbf{F}} \Delta t = \Delta \mathbf{p} = m\mathbf{v} - m\mathbf{v}_o \tag{6.5}$$

SI unit of impulse and momentum: newton-seconds (N·s)

Thus, *the impulse exerted on a body is equal to the change in the body's momentum.* This statement is referred to as the **impulse–momentum theorem**. Impulse has units of newton-seconds (N·s), which are also units of momentum by Eq. 6.4 and Eq. 6.1 (1 kg·m/s = 1 N·s).

•FIGURE 6.5 **Collision impulse** (a) Collision impulse causes the football to be deformed. (b) The impulse is the area under the curve of an *F* versus *t* graph. Note that the impulse force on the ball is not constant but rises to a maximum.

In Chapter 5 you learned that by the work–energy theorem ($W = F \Delta x = \Delta K$), the area under an F (net force) versus x curve is equal to the net work or change in kinetic energy. Similarly, the area under an F versus t curve is equal to the impulse or the change in momentum (Fig. 6.5b). An impulse force varies with time and is not a constant force, as Eq. 6.4 might suggest. However, in general, it is convenient to talk about the equivalent *constant* average force $\overline{F}$ acting over a time interval Δt to give the same impulse (same area under the force versus time curve), as shown in •Fig. 6.6.

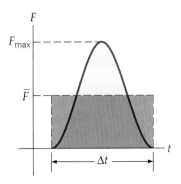

•**FIGURE 6.6 Average impulse force** The area under the average force curve ($\overline{F} \Delta t$, within the dashed red lines) is the same as the area under the F versus t curve, which is usually difficult to evaluate.

Example 6.4 ■ Teeing Off: The Impulse–Momentum Theorem

A golfer drives a 0.10-kilogram ball from an elevated tee, giving the ball an initial horizontal speed of 40 m/s (about 90 mi/h). The club and the ball are in contact for 1.0 ms (millisecond). What is the average force exerted by the club on the ball during this time?

Thinking It Through. The average force is equal to the time rate of change of momentum (Eq. 6.5 rearranged).

Solution.

Given: $m = 0.10$ kg *Find:* $\overline{F}$ (average force)
$v = 40$ m/s
$v_o = 0$
$\Delta t = 1.0$ ms $= 1.0 \times 10^{-3}$ s

Notice that the mass and the initial and final velocities are given, so the change in momentum can be found. Then, the average force can be computed from the impulse-momentum theorem:

$$\overline{F} \Delta t = p - p_o = mv - mv_o$$

Thus,

$$\overline{F} = \frac{mv - mv_o}{\Delta t} = \frac{(0.10 \text{ kg})(40 \text{ m/s}) - 0}{1.0 \times 10^{-3} \text{ s}} = 4000 \text{ N} \quad (\text{or } 900 \text{ lb})$$

The force is in the direction of the acceleration and is the *average* force. The instantaneous force is even greater near the midpoint of time interval of the collision (Δt in Fig. 6.6).

Follow-up Exercise. Suppose the golfer in Example 6.4 drives a ball with the same average force but "follows through" on the swing so as to increase the contact time to 1.5 ms. What effect would this have on the initial horizontal speed of the drive?

(a)

(b)

•**FIGURE 6.7 Adjust the impulse** (a) The change in momentum in catching the ball is a constant mv_o. If the ball is stopped quickly (small Δt), the impulse force is large (big $\overline{F}$) and stings the catcher's bare hands. (b) Increasing contact time (large Δt) by moving the hands with the ball reduces the impulse force and makes catching more enjoyable.

Example 6.4 illustrates the large internal forces that colliding objects can exert on one another during short contact times. However, in some instances the impulse may be manipulated to reduce the force. Suppose there is a fixed change in momentum in a given situation. Then, since $\Delta p = \overline{F} \Delta t$, if Δt could be made longer, the average impulse force $\overline{F}$ would be reduced.

You have probably tried to minimize the impulse force on occasion. For example, when jumping from a height onto a hard surface, you try not to land stiff-legged. The abrupt stop (small Δt) would apply a large impulse force to your leg bones and joints and could cause injury. If you bend your knees as you land, the impulse is constant and vertically upward opposite your velocity ($\overline{F} \Delta t = \Delta p = -mv_o$, with the final velocity zero). Thus increasing the time interval Δt makes the impulse force smaller.

Similarly, in catching a fast-moving, hard ball, you quickly learn not to catch it with your arms rigid but rather to move your hands with the ball. This movement increases the contact time and reduces the impulse force and the "sting" (•Fig. 6.7).

Insight

The Automobile Air Bag

A dark, rainy night—a car goes out of control and hits a big tree head-on! But the driver walks away with only minor injuries because he had his seatbelt buckled and had a car equipped with air bags. Air bags, along with seatbelts, are safety devices designed to prevent injuries to passengers in the front seat in automobile collisions. (Backseat air bags are also available.)

When a car collides with something basically immovable, such as a tree or a bridge abutment, or has a head-on collision with another vehicle, it stops almost instantaneously. If the front-seat passengers have not buckled up (and there are no air bags), they keep moving until acted on by an external force (Newton's first law). For the driver, this force is supplied by the steering wheel and column; for the passenger, by the dashboard and/or windshield.

Even when everyone has buckled up, there can be injuries. Seatbelts absorb energy by stretching and spread the force over a wide area to reduce the pressure. However, if a car is going fast enough and hits something truly immovable, there may be too much energy for the belts to absorb. This is where the air bag comes in. The bag inflates automatically on hard impact (Fig. 1), cushioning the driver (and front-seat passenger if both sides are equipped with air bags). In terms of impulse, the air bag increases the stopping contact time—the fraction of a second it takes your head to sink into the inflated bag is many times longer than the instant in which your head would be stopped by the dashboard. A longer contact time means a reduced impact force and thus much less likelihood of an injury. (Because the bag is large, the total impact force is also spread over a greater area of the body, so the force on any one part of the body is less.)

An interesting point is the inflating mechanism of an air bag. Think of how little time elapses between the front-end impact and the driver hitting the steering column in the collision of a fast-moving automobile. We say such a collision takes place instantaneously, yet during this time the air bag must be inflated! How is this done?

First, the air bag is equipped with sensors that detect the sharp deceleration associated with a head-on collision the instant it begins. If the deceleration exceeds the sensors' threshold settings, an electric current in an igniter in the air bag sets

FIGURE 1 Impulse and safety An automobile air bag increases the contact time, thereby decreasing the impulse force that could cause injury.

off a chemical explosion that generates gas to inflate the bag. The complete process from sensing to full inflation takes only on the order of 25 *thousandths* of a second (0.025 s).

The sensors' signals go first to a control unit, which determines whether a frontal collision rather than a system malfunction is occurring. (Accidental deployment could be dangerous as well as costly.) Typically, the control unit compares signals from two different sensors for collision verification. The unit is equipped with its own power source, since the car's battery and alternator are usually destroyed in a hard front-end collision. Sensing a collision, the control unit completes the circuit to the air bag igniter. It initiates a chemical reaction in a sodium azide (NaN_3) propellant. Gas (mostly nitrogen) is generated at an explosive rate, which inflates the air bag. The bag itself is made of thin nylon that is covered with cornstarch. The cornstarch acts as a lubricant to help the bag unfold smoothly on inflation.

Not only is sodium azide explosive, but it is also poisonous. Thus, you should not open a collapsed air bag, and if a car is junked, undeployed air bags must be disposed of

---

Another example in which the contact time is increased to decrease the impulse force is given in the Insight on this page.

In other instances, the *applied* impulse force may be relatively constant and the contact time (Δt) deliberately increased to produce a greater impulse, and thus a greater change in momentum ($\overline{F} \Delta t = \Delta p$). This is the principle of "following through" in sports, for example, when hitting a ball with a bat or driving a golf ball. In the latter case (•Fig. 6.8a), assuming that the golfer supplies the same average force with each swing, the longer the contact time, the greater will be the impulse or the momentum the ball receives. That is, with $\overline{F} \Delta t = mv$ (since $v_o = 0$), the greater the value of Δt, the greater will be the final velocity of the ball. (This was illustrated in Follow-up Exercise 6.4.) As you learned in Section 3.4, a greater projection velocity gives a projectile a greater range. (What

Insight

properly. Other, nonpoisonous, compounds are being developed for use as air bag inflators.

In general, air bags offer protection only if the occupants are thrown forward, because the bags are designed to deploy only in front-end collisions. They are of little use in side-impact crashes. Side air bags are now available in some automobiles.

Recent Concerns

In some cases, however, the deployment of air bags has been shown to cause injuries and deaths. An air bag is not a soft, fluffy pillow. When activated, it is ejected out of its compartment at speeds up to 320 km/h (200 mi/h) and could hit a person close by with enough force to cause severe injury and even death. Adults are advised to sit at least 13 cm (10 in.) from the air bag compartment to allow a margin of safety from the 5–8 cm (2–3 in.) injury "risk zone." Seats should be adjusted to allow the proper safety distance.*

The most serious concern involves children. Children may get too close to the dashboard if they are not buckled in or not buckled in securely with an adjustable shoulder harness (Fig. 2a). Using a rear-facing child seat in the front passenger seat is also dangerous if the air bag inflates (Fig. 2b). According to government recommendations, 1) *children 12 years old and younger should ride buckled up in the back seat,* and 2) *a rear-facing child seat should <u>never</u> be put in front of an air bag.*

To avoid the possibility of air bag injury, it is now possible to have a cut-off switch installed in an automobile to turn off one or both of the front air bags, pending approval by the appropriate government safety agency. Air bag deactivation may be authorized for the following reasons:

- A rear-facing child seat must be placed in the front seat because the car has no back seat or one that is too small.
- A child 12 years old or younger must ride in the front seat because of a medical condition that requires frequent monitoring.
- An individual who drives (or rides in the front seat) has a medical condition that would make it safer to have the air bag(s) turned off.

*Guidelines from the National Highway Traffic Safety Administration

- Drivers who must sit within several centimeters of the air bag, typically because of extremely short stature (137 cm—or 4 ft, 6 in.—or less).

There may be specific problems in some instances, but air bags do save many lives. All new passenger cars are now required to have dual air bags. But even if your car is equipped with air bags, <u>*always*</u> remember to buckle up. (Maybe we should make that Newton's fourth law of motion!)

(a)

(b)

FIGURE 2 The wrong ways (a) Children who must ride in the front seat for medical monitoring should be buckled in securely so as not to get too close to the air bag in the dashboard. (b) Rear-facing child seats should never be used in the front passenger seat because injury could occur if the airbag is activated.

---

angle of projection should a golfer try to achieve when driving the ball on a level fairway?)

In some instances, a long follow-through that increases the contact time may be used to improve control (Fig. 6.8b).

The word *impulse* implies that the impulse force acts only briefly or quickly (like an "impulsive" person), and this is true in many instances. However, the definition of impulse places no limit on the time interval over which the force may act. Technically, a comet at its closest approach to the Sun is involved in a collision, because in physics collision forces do not have to be contact forces. Basically, a **collision** is an interaction of objects in which there is an exchange of momentum and energy. As you might expect from the work–energy theorem and the impulse–momentum theorem, momentum and kinetic energy are directly related.

Collision—exchange of momentum and energy

•**FIGURE 6.8 Increasing the contact time** (a) A golfer follows through on a drive swing. One reason is to increase the contact time so that the ball receives greater impulse and momentum. (b) The follow-through on a long putt increases the contact time for greater impulse and momentum, but the main reason is for directional control. Notice that the putter is in contact with the ball for a time equivalent to about 4 flash intervals.

A little algebraic manipulation of the equation for kinetic energy (Eq. 5.5) allows us to express kinetic energy in terms of the *magnitude* of momentum:

Kinetic energy and momentum

$$K = \tfrac{1}{2}mv^2 = \frac{(mv)^2}{2m} = \frac{p^2}{2m} \tag{6.6}$$

Thus, kinetic energy and momentum are intimately related. The conservation of these quantities is important in collisions, as we shall see later in the chapter.

6.3 The Conservation of Linear Momentum

Objectives: To (a) explain the conditions for the conservation of linear momentum and (b) apply it to physical situations.

Like total mechanical energy, the momentum of a body or system is a conserved quantity under certain conditions. This fact allows us to analyze a wide range of situations and solve many problems readily. The conservation of momentum is one of the most important principles in physics. In particular, it is used to analyze the collision of objects ranging from subatomic particles to automobiles in traffic accidents.

For the linear momentum of a particle or an object to be conserved (to remain constant with time), one condition must hold. This condition is apparent from the momentum form of Newton's second law (Eq. 6.3). If the net force acting on a particle is zero, that is,

$$\mathbf{F} = \frac{\Delta \mathbf{p}}{\Delta t} = 0$$

then,

$$\Delta \mathbf{p} = 0 = \mathbf{p} - \mathbf{p}_o$$

where $\mathbf{p}_o$ is the initial momentum and $\mathbf{p}$ is the momentum at some later time. Since they are equal, the momentum is conserved:

$$\mathbf{p} = \mathbf{p}_o \tag{6.7}$$

or

$$m\mathbf{v} = m\mathbf{v}_o$$

Note that this conservation is consistent with Newton's first law: An object remains at rest ($\mathbf{p} = 0$), or in motion with a *uniform* velocity (constant $\mathbf{p}$), unless acted on by a net external force.

FIGURE 6.9 An internal force and the conservation of momentum The spring force is an internal force, so the momentum of the system is conserved. See Example 6.5.

The conservation of momentum can be extended to a system of particles if we write Newton's second law in terms of the sums (resultants) of the forces acting on the system and of the momenta of the particles: $\mathbf{F} = \Sigma \mathbf{F}_i$ and $\mathbf{P} = \Sigma \mathbf{p}_i = \Sigma m \mathbf{v}_i$. A more complicated argument similar to the development of Eq. 6.7 can be made for systems of particles. It turns out that if there is no net external force acting *on the system*, then $\mathbf{P} = \mathbf{P}_o$. This generalized condition is referred to as the law of **conservation of linear momentum**:

Note: A lowercase **p** means individual momentum. A capital **P** means total system momentum. Both are vectors.

Conservation of momentum—
no net external force

There are other ways to express this condition. For example, recall from Chapter 5 that a *closed*, or *isolated*, system is one on which no external work is done. Therefore there is no net external force acting on such a system, so the total linear momentum of an isolated system is conserved.

Within a system, internal forces may be acting, for example, when particles collide. These are force pairs of Newton's third law, and there is a good reason such forces are not explicitly referred to in the condition for the conservation of momentum. By Newton's third law, these internal forces are equal and opposite and vectorially cancel each other. Thus, the net internal force of a system is zero.

Note: Third-law force pairs were discussed in Section 4.4.

An important point to understand, however, is that the momenta of *individual* particles or objects within a system may change. But in the absence of a net external force, the *vector sum* of all the momenta remains the same. If the objects are initially at rest (total momentum is zero) and then are set in motion as the result of internal forces, the total momentum must still add to zero. This principle is illustrated in •Fig. 6.9 and analyzed in Example 6.5. Objects in an isolated system may transfer momentum among themselves, but the total momentum after the changes must add up to the initial value, assuming friction is negligible. (See Demonstration 3.)

The conservation of momentum is often a powerful and convenient tool for analyzing situations involving motion. Its application is illustrated in the following Examples.

Example 6.5 ■ Before and After: Conservation of Momentum

Two masses, $m_1 = 1.0$ kg and $m_2 = 2.0$ kg, are held on either side of a light compressed spring by a light string joining them, as shown in Fig. 6.9. The string is burned (negligible external force), and the masses move apart on the frictionless surface, with m_1 having a velocity of 1.8 m/s to the left. What is the velocity of m_2?

Thinking It Through. With no net external force (the weights are canceled by normal forces), the total momentum of the system is conserved. It is initially zero, so afterward the momentum of m_2 is *equal* and *opposite* that of m_1. (Vector addition gives zero total momentum.) (The term *light* indicates that the masses of the spring and string can be ignored.)

Solution. Listing the masses and speed given, we have

Given: $m_1 = 1.0$ kg
$m_2 = 2.0$ kg
$v_1 = 1.8$ m/s (left)

Find: v_2 (velocity—speed and direction)

Here the system consists of the two masses and the spring. Since the spring force is internal to the system, the momentum of the system is conserved. It should be apparent that the initial total momentum of the system ($\mathbf{P}_o$) is zero, and therefore the final momentum must also be zero. Thus, we may write

$$\mathbf{P}_o = \mathbf{P} = 0 \quad \text{and} \quad \mathbf{P} = \mathbf{p}_1 + \mathbf{p}_2 = 0$$

(The spring does not come into the equations because its velocity is zero.) Thus,

$$\mathbf{p}_2 = -\mathbf{p}_1$$

which means that the momenta of m_1 and m_2 are equal and opposite. Using directional signs (with $+x$ to the right in the figure), we get

$$p_2 - p_1 = m_2 v_2 - m_1 v_1 = 0$$

or in terms of magnitudes

$$m_2 v_2 = m_1 v_1$$

and

$$v_2 = \left(\frac{m_1}{m_2}\right) v_1 = \left(\frac{1.0 \text{ kg}}{2.0 \text{ kg}}\right)(1.8 \text{ m/s}) = 0.90 \text{ m/s}$$

Thus, the velocity of m_2 is 0.90 m/s in the positive x direction, or to the right in the figure. This is half v_1, as you might have expected, since m_2 has twice the mass of m_1.

Follow-up Exercise. (a) Suppose that the large block in Fig. 6.9 were attached to the Earth's surface so that the block could not move when the string was burned. Would momentum be conserved in this case? Explain. (b) Two girls, each having a mass of 50 kg, stand at rest on skateboards with negligible friction. The first tosses a 2.5-kilogram ball to the second. If the speed of the ball is 10 m/s, what is the speed of each girl after the ball is caught, and what is the momentum of the ball before it is tossed, while it is in the air, and after it is caught?

Demonstration 3 ■ Internal Forces (and Conservation of Momentum?)

In both cases, the initial total momentum is zero. What would happen after the actions (and reactions) begin? Actually, there is an external force (what is it?), so the momentum is not conserved and the motion is limited.

(a)

(a) Getting ready to push off. Note that the forces are internal to the system.

(b)

(b) Play ball! Here a ball is used to exchange momentum between parts of the system.

Conceptual Example 6.6

How Far? Conservation of Momentum with External Forces

Consider the situation in Fig. 6.9 again. Let's assume that there is friction and that the blocks are made from different materials such that the constant frictional forces acting on the moving blocks have the same magnitude. (Why would they have to be different materials?) Then, under these conditions, on coming to rest, (a) both blocks have traveled the same distance, (b) the lighter block has traveled twice the distance of the heavier one, (c) the lighter block has traveled four times the distance of the heavier one.

Reasoning and Answer. The spring force is internal and initially the frictional forces cancel within the system, so the linear momentum is conserved and we can get the initial speeds as in Example 6.5. However, let's solve this example algebraically rather than by using numbers. (It's usually simpler to work with symbols.) To define the notation, we will say that $m_1 = m$ and $m_2 = 2m$ and will specify the initial velocities, v_{1_o} and v_{2_o}, to distinguish them from the final velocities, $v_1 = v_2 = 0$. (Properly defined notation is important.) Then, using the conservation of momentum as above, we have

$$v_{1_o} = \left(\frac{m_2}{m_1}\right)v_{2_o} = \left(\frac{2m}{m}\right)v_{2_o} = 2v_{2_o}$$

as shown previously with numbers.

With the same constant frictional decelerating force, each block has a constant deceleration, so we can use a kinematic equation from Chapter 2: $v^2 = v_o^2 + 2ax$. With $v = 0$, the magnitude of x is given by

$$v^2 = 0 = v_o^2 - 2ax \quad \text{and} \quad x = \frac{v_o^2}{2a}$$

The distance, then, is proportional to the *square* of the initial speed. So, with twice the initial speed, does the lighter block travel four times the distance of the heavier block, making the answer (c)? We had better look more closely. Note that the acceleration also appears in the expression for the distance. The frictional forces (f) on each block have the same magnitude, but since the masses of the blocks are different, so are the accelerations. After the spring releases, the net force on each block is friction, f. Hence the acceleration of each is $a = f/m$, and the general expression for x becomes

$$x = \frac{mv_o^2}{2f}$$

Now we see that the distance is proportional not only to v^2 but also to m. Forming a ratio of x_1 and x_2, and using known ratio values, we get

$$\frac{x_1}{x_2} = \left(\frac{m_1}{m_2}\right)\left(\frac{v_{1_o}}{v_{2_o}}\right)^2 = (\tfrac{1}{2})(2)^2 = 2$$

Thus, $x_1 = 2x_2$, or the lighter block travels twice the distance of the heavier one, so the answer is (b).

Follow-up Exercise. If the magnitude of the constant frictional force on the lighter block were twice that on the heavier block, would the blocks then travel the same distance? Justify your answer.

Example 6.7 ■ A Glancing Collision: Components of Momentum

A moving shuffleboard puck has a glancing collision with a stationary one of the same mass, as shown in •Fig. 6.10. If friction is negligible, what are the speeds of the pucks after the collision?

Thinking It Through. Even in a glancing collision, the momentum is conserved (with no net external force; the puck weights are canceled by normal forces). If the total momentum is conserved, then each *component* of the momentum is conserved.

•**FIGURE 6.10 A glancing collision** Momentum is conserved in an isolated system. The motion in two dimensions may be analyzed in terms of the components of momentum, which are also conserved. See Example 6.7. Note that v_1 and v_2 form a 90° angle *only* if masses m_1 and m_2 are equal.

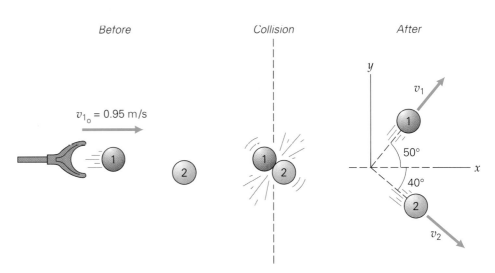

Solution.

Given: $v_{1_o} = 0.95$ m/s   Find: v_1 and v_2 (speeds after collision)
$\quad\quad\theta_1 = 50°$
$\quad\quad\theta_2 = 40°$
$\quad\quad m_1 = m_2$

The collision forces of the two-puck system are internal forces, so the total momentum is conserved (the external force of friction is negligible for the very brief collision time). This means that the rectangular components of momentum are also conserved. That is, $\mathbf{P} = \mathbf{P}_x + \mathbf{P}_y$ is a constant, so $\mathbf{P}_x$ and $\mathbf{P}_y$ must be constant. For the x component of the momentum with signs indicating directions (positive axes shown in figure),

$$\overset{before}{p_{x_o}} = \overset{after}{p_{x_1} + p_{x_2}}$$

or

$$m_1 v_{1_o} = m_1 v_1 \cos 50° + m_2 v_2 \cos 40° \quad (1)$$

For the y component,

$$0 = p_{y_1} - p_{y_2}$$

or

$$0 = m_1 v_1 \sin 50° - m_2 v_2 \sin 40° \quad (2)$$

where the minus sign indicates that the component of puck 2 is in the negative y direction. With two equations and two unknowns, the rest is basic algebra. The masses cancel in Eq. 2 since they are equal ($m_1 = m_2$), and

$$v_2 = \left(\frac{\sin 50°}{\sin 40°}\right) v_1 = \left(\frac{0.766}{0.643}\right) v_1 = 1.19 v_1 \quad (3)$$

Substituting for v_2 in Eq. 1 and canceling the masses (as above) gives

$$v_{1_o} = v_1 \cos 50° + (1.19 v_1)(\cos 40°)$$

and

$$v_1 = \frac{v_{1_o}}{\cos 50° + (1.19)(\cos 40°)} = \frac{0.95 \text{ m/s}}{0.643 + (1.19)(0.766)} = 0.61 \text{ m/s}$$

Using this value in Eq. 3 gives

$$v_2 = 1.19 v_1 = 1.19(0.61 \text{ m/s}) = 0.73 \text{ m/s}$$

The directions of v_1 and v_2 are given by θ_1 and θ_2, respectively.

Follow-up Exercise. (a) Is the kinetic energy conserved in the collision in this Example? Explain. (b) In such a collision, is it possible for both θ_1 and θ_2 to lie above the $+x$ axis? Why or why not?

The conservation of momentum is one of the most important principles in physics. As mentioned previously, it is used to analyze the collisions of objects ranging from subatomic particles to automobiles in traffic accidents. In many instances, external forces may be acting on the objects, which means that the momentum is not conserved.

But, as you will learn in the next section, the conservation of momentum often allows a good approximation *over the short time of a collision* because the internal forces, for which momentum is conserved, are much greater than the external forces. For example, external forces such as gravity and friction also act on colliding objects but are often relatively small compared to the internal forces. Therefore, if the objects interact for only a brief time, the effects of the external forces may be negligible compared to those of the large internal forces.

6.4 Elastic and Inelastic Collisions

Objective: To describe the conditions on kinetic energy and momentum in elastic and inelastic collisions.

Taking a closer look at collisions in terms of the conservation of momentum is simpler if we consider an isolated system, such as a system of particles (or balls) involved in head-on collisions. For simplicity, we will consider only collisions in one dimension. Such collisions can also be analyzed in terms of the conservation of energy. On the basis of what happens to the total kinetic energy, we can define two types of collisions: *elastic* and *inelastic*.

In an **elastic collision** (•Fig. 6.11a), the total kinetic energy is conserved. That is, the *total* kinetic energy of all the objects of the system after the collision is the same as their *total* kinetic energy before the collision. Kinetic energy may be traded between objects of a system, but the total remains constant. That is,

$$\begin{array}{c} \text{total } K \text{ after} = \text{total } K \text{ before} \\ K_f = K_i \end{array} \quad \begin{array}{c} \textit{condition for an} \\ \textit{elastic collision} \end{array} \quad (6.8)$$

Elastic collision—total kinetic energy conserved

During such a collision, some or all of the initial kinetic energy is temporarily converted to potential energy as the objects are deformed. But, after the maximum deformations, the objects elastically spring back to their original shapes, and the system regains all of its original kinetic energy. For example, two steel balls or two billiard balls may have a nearly elastic collision with each ball having the same shape afterward as before; that is, there is no permanent deformation.

Note: In reality, only atoms and subatomic particles have truly elastic collisions, but some larger hard objects have nearly elastic collisions in which the kinetic energy is approximately conserved.

In an **inelastic collision** (Fig. 6.11b), total kinetic energy is *not* conserved. For example, one or more of the colliding objects may not spring back to its original shape, or heat or sound may be generated. In such interactions, work is done

Inelastic collision—total kinetic energy not conserved

(a)   (b)

•**FIGURE 6.11 Collisions**
(a) Elastic collisions (b) An inelastic collision.

by nonconservative forces (Section 5.5), such as friction, and some kinetic energy is lost:

$$K_f < K_i \quad \begin{array}{c}\text{total K after} < \text{total K before} \\ \text{condition for an} \\ \text{inelastic collision}\end{array} \quad (6.9)$$

For example, a hollow aluminum ball that collides with a solid steel ball may be dented. Permanently deforming the ball takes work, and that work is done at the expense of the original kinetic energy of the system.

For isolated systems, momentum is conserved in both elastic and inelastic collisions. For an inelastic collision, only an amount of kinetic energy consistent with the conservation of momentum may be lost. It may seem strange that kinetic energy can be lost and momentum still be conserved, but this fact provides insight into the difference between scalar and vector quantities.

Energy and Momentum in Inelastic Collisions

To see how momentum can remain constant while the kinetic energy changes (decreases) in inelastic collisions, consider the examples illustrated in •Fig. 6.12. In Fig. 6.12a, two balls of equal mass ($m_1 = m_2$) approach each other with equal and opposite velocities ($v_{1_o} = -v_{2_o}$). Hence, the total momentum before the collision is vectorially zero, but the scalar kinetic energy is *not* zero. After the collision the balls are stuck together and stationary, so the total momentum is unchanged—still zero. Momentum is conserved because the forces of collision are internal to the system of the two balls; thus there is no net external force on the system. The total kinetic energy, however, has decreased to zero. In this case, the kinetic energy went into the work done in permanently deforming the balls. Some energy may also have gone into doing work against friction (producing heat) or been lost in some other way (for example, in producing sound).

It should be noted that the balls need not come to rest after collision. In a less inelastic collision, the balls may recoil in opposite directions at reduced but equal speeds. The momentum would be conserved (still equal to zero—why?), but the

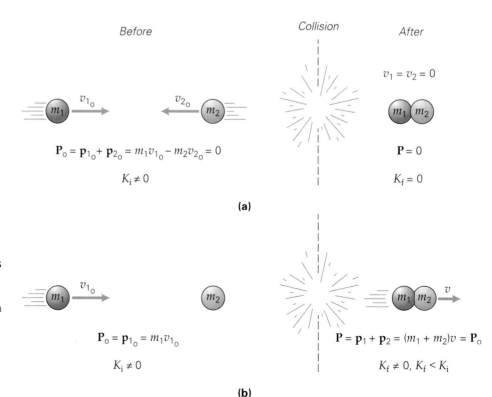

•**FIGURE 6.12 Inelastic collisions** In inelastic collisions, momentum is conserved but kinetic energy is not. Collisions like the ones shown here, in which the objects stick together, are called completely or totally inelastic collisions. The maximum amount of kinetic energy lost is consistent with the conservation of momentum.

kinetic energy would again not be conserved. Under all conditions, the amount of kinetic energy lost must be consistent with momentum conservation.

In Fig. 6.12b, one ball is initially at rest as the other approaches. The balls stick together after collision but are still in motion. Both of these cases are examples of a **completely inelastic collision**, in which the objects stick together and hence have the same velocity after colliding.

Assume that the balls in Fig. 6.12b have different masses. Since the momentum is conserved even in inelastic collisions,

$$\underbrace{m_1 v_o}_{\text{before}} = \underbrace{(m_1 + m_2)v}_{\text{after}}$$

and

$$v = \left(\frac{m_1}{m_1 + m_2}\right) v_o \qquad \begin{array}{l}\text{m_2 initially at rest}\\ \text{(completely inelastic}\\ \text{collision only)}\end{array} \qquad (6.10)$$

Thus, v is less than v_o, since $m_1/(m_1 + m_2)$ must be less than 1. Now let us consider how much kinetic energy has been lost. Initially, $K_i = \tfrac{1}{2} m_1 v_o^2$; finally, after the collision,

$$K_f = \tfrac{1}{2}(m_1 + m_2)v^2$$

Substituting for v from Eq. 6.10 and simplifying the result, we have

$$K_f = \tfrac{1}{2}(m_1 + m_2)\left(\frac{m_1 v_o}{m_1 + m_2}\right)^2 = \frac{\tfrac{1}{2} m_1^2 v_o^2}{m_1 + m_2} = \left(\frac{m_1}{m_1 + m_2}\right)\tfrac{1}{2} m_1 v_o^2 = \left(\frac{m_1}{m_1 + m_2}\right) K_i$$

and

$$\frac{K_f}{K_i} = \frac{m_1}{m_1 + m_2} \qquad \begin{array}{l}\text{m_2 initially at rest}\\ \text{(completely inelastic}\\ \text{collision only)}\end{array} \qquad (6.11)$$

Equation 6.11 gives the fractional amount of the initial kinetic energy that the system has after a completely inelastic collision. For example, if the masses of the balls are equal ($m_1 = m_2$), then $m_1/(m_1 + m_2) = \tfrac{1}{2}$ and $K_f/K_i = \tfrac{1}{2}$, or $K_f = K_i/2$. That is, only half of the initial kinetic energy is lost (consistent with the conservation of momentum).

Note that all the kinetic energy cannot be lost in this case no matter what the masses of the balls are. The momentum after collision cannot be zero, since it was not zero initially. Thus, the balls must be moving and must have at least some kinetic energy.

Example 6.8 ■ Stuck Together: Completely Inelastic Collision

A 1.0-kilogram ball with a speed of 4.5 m/s strikes a 2.0-kilogram stationary ball. If the collision is completely inelastic, (a) what are the speeds of the balls after the collision? (b) What percentage of the initial kinetic energy do they have after the collision? (c) What is the total momentum after the collision?

Thinking It Through. Recall the definition of a *completely inelastic collision*. The balls stick together after collision; kinetic energy is *not* conserved, but total momentum is.

Solution. Using the labeling as in the preceding discussion, we have

Given: $m_1 = 1.0$ kg Find: (a) v (speed after collision)
$m_2 = 2.0$ kg
$v_o = 4.5$ m/s
(b) $\dfrac{K_f}{K_i}(\times 100\%)$
(c) $\mathbf{P}_f$ (total momentum after collision)

(a) The momentum is conserved and

$$\mathbf{P} = \mathbf{P}_o$$

or
$$(m_1 + m_2)v = m_1 v_o$$

The balls stick together and have the same speed after collision. This speed is then

$$v = \left(\frac{m_1}{m_1 + m_2}\right)v_o = \left(\frac{1.0 \text{ kg}}{1.0 \text{ kg} + 2.0 \text{ kg}}\right)(4.5 \text{ m/s}) = 1.5 \text{ m/s}$$

(b) The fractional part of the initial kinetic energy that the balls have after the completely inelastic collision is given by Eq. 6.11. Notice that this fraction, as given by the masses, is the same as that for the speeds (Eq. 6.10) in this special case. By inspection, we can write

$$\frac{K_f}{K_i} = \frac{m_1}{m_1 + m_2} = \frac{1.0 \text{ kg}}{1.0 \text{ kg} + 2.0 \text{ kg}} = \frac{1}{3} = 0.33 \, (\times 100\%) = 33\%$$

Let's show this explicitly:

$$\frac{K_f}{K_i} = \frac{\frac{1}{2}(m_1 + m_2)v^2}{\frac{1}{2}m_1 v_o^2} = \frac{\frac{1}{2}(1.0 \text{ kg} + 2.0 \text{ kg})(1.5 \text{ m/s})^2}{\frac{1}{2}(1.0 \text{ kg})(4.5 \text{ m/s})^2} = 0.33 \, (= 33\%)$$

Keep in mind that Eq. 6.11 applies *only* to *completely* inelastic collisions. For other types of collisions, the initial and final values of the kinetic energy must be computed explicitly.

(c) The momentum is conserved in all collisions (in the absence of external forces), so the total momentum after collision is the same as before. That value is the momentum of the incident ball, with a magnitude of

$$P_f = p_{1_o} = m_1 v_o = (1.0 \text{ kg})(4.5 \text{ m/s}) = 4.5 \text{ kg} \cdot \text{m/s}$$

and in the same direction.

Follow-up Exercise. A small hard-metal ball of mass m collides with a larger, stationary soft-metal ball of mass M. A *minimum* amount of work W is required to make a dent in the larger ball. If the smaller ball initially has kinetic energy $K = W$, will the larger ball be dented in a completely inelastic collision?

Energy and Momentum in Elastic Collisions

For a general *elastic* collision of two objects,

$$\text{Conservation of } K: \quad \overbrace{\tfrac{1}{2}m_1 v_{1_o}^2 + \tfrac{1}{2}m_2 v_{2_o}^2}^{\text{before}} = \overbrace{\tfrac{1}{2}m_1 v_1^2 + \tfrac{1}{2}m_2 v_2^2}^{\text{after}} \quad (6.12a)$$

Linear momentum is conserved even if the collision is inelastic, so we can write

$$\text{Conservation of } \mathbf{P}: \quad m_1 \mathbf{v}_{1_o} + m_2 \mathbf{v}_{2_o} = m_1 \mathbf{v}_1 + m_2 \mathbf{v}_2 \quad (6.12b)$$

Thus for head-on elastic collisions, knowing the masses of the objects and the initial velocities allows you to find the final velocities.

A common collision situation is that in which one of the objects is initially stationary and then is hit head-on (•Fig. 6.13). Assume that the motion of both of the balls after the collision is to the right, or in the positive direction, as shown in the figure. If this is not the case, the math will tell you. (How?)

This is a one-dimensional situation. The condition for an elastic, head-on collision in this situation is

$$\tfrac{1}{2}m_1 v_{1_o}^2 = \tfrac{1}{2}m_1 v_1^2 + \tfrac{1}{2}m_2 v_2^2 \quad (6.13a)$$

and conservation of linear momentum gives

$$m_1 v_{1_o} = m_1 v_1 + m_2 v_2 \quad (6.13b)$$

Rearranging these equations gives

$$m_1(v_{1_o}^2 - v_1^2) = m_2 v_2^2 \quad (6.13a')$$

and

$$m_1(v_{1_o} - v_1) = m_2 v_2 \quad (6.13b')$$

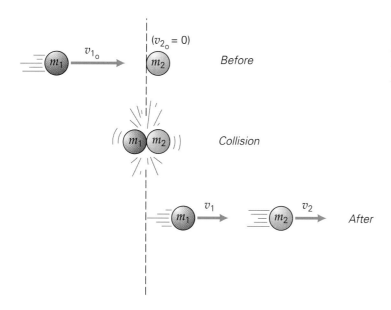

•FIGURE 6.13 **Elastic collision** For an elastic collision between two bodies, one of which is initially at rest, the velocities after the collision depend on the relative masses of the bodies.

Then, using the algebraic relationship $x^2 - y^2 = (x + y)(x - y)$ and dividing the first equation by the second, we get

$$\frac{m_1(v_{1_o} + v_1)(v_{1_o} - v_1)}{m_1(v_{1_o} - v_1)} = \frac{m_2 v_2^2}{m_2 v_2} \quad \frac{(\text{Eq. 6.13a}')}{(\text{Eq. 6.13b}')} \; (divide)$$

or

$$v_{1_o} + v_1 = v_2 \tag{6.14}$$

Equation 6.14 can now be used to eliminate v_1 or v_2 from Eq. 6.13b. Thus, each of these final velocities can be expressed in terms of the initial velocity (v_{1_o}):

$$v_1 = \left(\frac{m_1 - m_2}{m_1 + m_2}\right) v_{1_o} \tag{6.15}$$

$$v_2 = \left(\frac{2m_1}{m_1 + m_2}\right) v_{1_o} \tag{6.16}$$

(final velocities for elastic, head-on collision with m_2 initially stationary)

Note that the final velocities depend on mass *ratios* of the objects. Looking at Eq. 6.15, you can see that if m_1 is greater than m_2, then v_1 is positive, or in the same direction as the initial velocity of the incoming ball, as was assumed in Fig. 6.13. However, if m_2 is greater than m_1, then v_1 is negative, and the incoming ball recoils in the opposite direction after collision. Equation 6.16 shows that v_2 is always in the same direction as the velocity of the incoming ball.

You can also get some general ideas about what happens after such a collision by considering three possible situations for the relative masses of the objects, illustrated in •Fig. 6.14.

Case 1. $m_1 = m_2$ (Fig. 6.14a). From Eqs. 6.15 and 6.16,

$$v_1 = 0 \quad \text{and} \quad v_2 = v_{1_o}$$

Elastic head-on collision, $m_1 = m_2$

That is, if the masses of the colliding objects are equal, the objects simply exchange momentum and energy. The incoming ball is stopped on collision, and the originally stationary ball moves off with the same velocity as that of the incoming ball, thus obviously conserving system kinetic energy and momentum. (A real-world example would be two billiard balls.)

Case 2. $m_1 \gg m_2$ (m_1 very much greater than m_2, Fig. 6.14b). In this case, m_2 can be ignored in the addition and subtraction with m_1 in Eqs. 6.15 and 6.16, and

$$v_1 \approx v_{1_o} \quad \text{and} \quad v_2 \approx 2v_{1_o}$$

Elastic head-on collision, $m_1 \gg m_2$

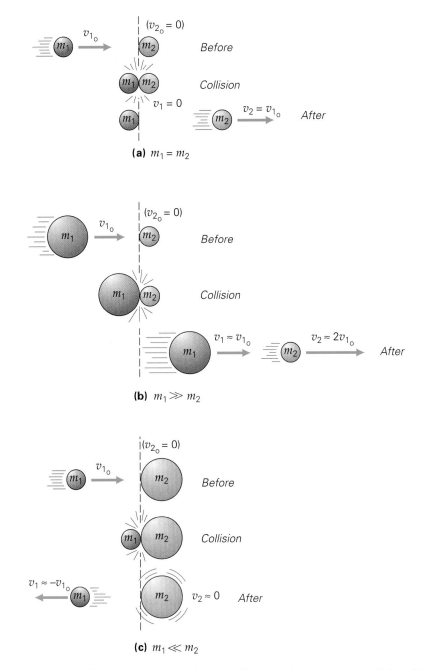

•**FIGURE 6.14 Special cases of head-on elastic collisions** (a) When a moving object collides elastically with a stationary object of equal mass, there is a complete exchange of momentum and energy. (b) When a very massive moving object collides elastically with a much less massive stationary object, the very massive object continues to move essentially as before, and the less massive object is given a velocity almost twice the initial velocity of the large mass. (c) When a moving object of small mass collides elastically with a very massive stationary object, the incoming object recoils in the opposite direction with approximately the same speed and the very massive object remains essentially stationary.

Elastic head-on collision, $m_1 \ll m_2$

This tells you that if a very massive object collides with a stationary light object, the massive object is slowed down only slightly by the collision, and the light object is knocked away with a velocity almost twice that of the initial velocity of the massive object. (Think of a bowling ball hitting a pin.)

Case 3. $m_1 \ll m_2$ (m_1 very much less than m_2, Fig. 6.14c). Here, m_1 can be ignored in the addition and subtraction with m_2 in Eqs. 6.15 and 6.16, and

$$v_1 \approx -v_{1_o} \quad \text{and} \quad v_2 \approx 0$$

(In the second equation, the approximation $m_1/m_2 \approx 0$ is made.) Thus, if a light object collides with a massive stationary one, the massive object remains *almost* stationary, and the light object recoils backward with approximately the same speed it had before collision. An extreme case of this type is similar to a particle striking a solid, immovable wall (see Fig. 6.3). If the wall is anchored to the ground, then the Earth and wall recoil together, but this is totally unnoticeable (why?). Total momentum, however is still conserved. For the case in Fig. 6.14c, the massive ball must move a bit after the collision to conserve momentum. (Think of throwing a Superball at a stationary bowling ball.)

Example 6.9 ■ Elastic Collision: Conservation of Momentum and Kinetic Energy

A 0.30-kilogram object with a speed of 2.0 m/s in the positive x direction has a head-on elastic collision with a stationary 0.70-kilogram object located at $x = 0$. What is the distance separating the objects 2.5 s after the collision?

Thinking It Through. The incoming object is less massive than the stationary one. Even though it is not considerably less massive, we might expect the objects to separate in opposite directions after collision. Equations 6.15 and 6.16 will give the velocities; knowing the time, we can find the distances traveled or the separation distance after collision.

Solution. Using the previous notation, we have

Given: $m_1 = 0.30$ kg
$v_{1_o} = 2.0$ m/s
$m_2 = 0.70$ kg
$v_{2_o} = 0$
$t = 2.5$ s

Find: $\Delta x = x_2 - x_1$ (separation distance 2.5 s after collision)

From Eqs. 6.15 and 6.16, the velocities after collision are

$$v_1 = \left(\frac{m_1 - m_2}{m_1 + m_2}\right)v_{1_o} = \left(\frac{0.30 \text{ kg} - 0.70 \text{ kg}}{0.30 \text{ kg} + 0.70 \text{ kg}}\right)(2.0 \text{ m/s}) = -0.80 \text{ m/s}$$

$$v_2 = \left(\frac{2m_1}{m_1 + m_2}\right)v_{1_o} = \left[\frac{2(0.30 \text{ kg})}{0.30 \text{ kg} + 0.70 \text{ kg}}\right](2.0 \text{ m/s}) = 1.2 \text{ m/s}$$

Here m_1 is less than m_2, but not *so much* less that it could be ignored, as was done in part (c) of Fig. 6.14.

The objects are separating after collision, and their positions relative to the collision point ($x_o = 0$) are

$$x_1 = v_1 t = (-0.80 \text{ m/s})(2.5 \text{ s}) = -2.0 \text{ m}$$

$$x_2 = v_2 t = (1.2 \text{ m/s})(2.5 \text{ s}) = 3.0 \text{ m}$$

So

$$\Delta x = x_2 - x_1 = 3.0 \text{ m} - (-2.0 \text{ m}) = 5.0 \text{ m}$$

The objects are 5.0 m apart at that time.

Follow-up Exercise. Suppose the objects in this Example had equal masses. What would be the separation distance 2.5 s after collision in this case?

Example 6.10 ■ Spare! Conservation of Momentum and Kinetic Energy

A 7.1-kilogram bowling ball with a speed of 6.0 m/s has a head-on elastic collision with a stationary 1.6-kilogram pin. (a) What is the velocity of each object after the collision? (b) What is the total momentum after the collision?

Thinking It Through. If the ball and the pin have an elastic collision, then Eqs. 6.15 and 6.16 apply and will give the velocities. (The ball is quite a bit more massive than the pin, so as in Case 2 above, the ball and the pin will go in same direction after collision—as you would expect.) The total momentum is conserved, so we can calculate it from the initial total momentum. (Why?)

Solution. Listing the data with the same notation as before, we have

Given: $m_1 = 7.1$ kg
$v_{1_o} = 6.0$ m/s
$m_2 = 1.6$ kg
$v_{2_o} = 0$

Find: (a) $\mathbf{v}_1$ and $\mathbf{v}_2$ (velocities after collision)
(b) $\mathbf{P}$ (total momentum after collision)

(a) The velocities are given directly by Eqs. 6.15 and 6.16, and the direction of the incoming ball is taken as positive:

$$v_1 = \left(\frac{m_1 - m_2}{m_1 + m_2}\right)v_{1_o} = \left(\frac{7.1 \text{ kg} - 1.6 \text{ kg}}{7.1 \text{ kg} + 1.6 \text{ kg}}\right)(6.0 \text{ m/s}) = 3.8 \text{ m/s}$$

$$v_2 = \left(\frac{2m_1}{m_1 + m_2}\right)v_{1_o} = \left[\frac{2(7.1 \text{ kg})}{(7.1 \text{ kg} + 1.6 \text{ kg})}\right](6.0 \text{ m/s}) = 9.8 \text{ m/s}$$

So both objects move in the same direction. Here m_1 is greater than m_2, but not *so much* greater that m_2 could be ignored, as was done in part (b) of Fig. 6.14.

(b) Momentum is conserved in elastic (and inelastic) collisions, so the total momentum afterward is the same as that before the collision. Thus the total momentum is that of the incident ball, in the same direction and with a magnitude of

$$P_f = P_{1_o} = m_1 v_{1_o} = (7.1 \text{ kg})(6.0 \text{ m/s}) = 43 \text{ kg·m/s}$$

Follow-up Exercise. In general, for head-on, elastic collisions with one of the objects initially stationary, is it possible for the velocities to be the same after collision? Explain.

Conceptual Example 6.11

Two In, One Out?

A novelty collision device, as shown in •Fig. 6.15, consists of five identical metal balls. When one ball swings in, after multiple collisions, one ball swings out at the other end of the row of balls. When two balls swing in, two swing out; when three swing in, three swing out, and so on—always the same number out as in.

Suppose that two balls, each of mass m, swing in at velocity v and collide with the next ball. Why doesn't one ball swing out at the other end with a velocity $2v$?

Reasoning and Answer. The collisions along the horizontal row of balls are approximately elastic. Two balls in and one out with twice the velocity wouldn't violate the conservation of momentum: $(2m)v = m(2v)$. However, there's another condition if we assume elastic collisions—the conservation of kinetic energy. Let's check to see if this is so for this case:

$$\underset{\text{before}}{K_o} = \underset{\text{after}}{K}$$

$$\tfrac{1}{2}(2m)v^2 \overset{?}{=} \tfrac{1}{2}m(2v)^2$$

$$mv^2 \neq 2mv^2$$

Hence, the kinetic energy would *not* be conserved if this happened, and the equation is telling us that this situation violates established physical principles and does not occur. Note that there's a big violation—more energy out than in.

Follow-up Exercise. Suppose the first ball of mass m were replaced with a ball of mass $2m$. When this ball is pulled back and allowed to swing in, how many balls will swing out? [*Hint*: Think about the analogous situation in Fig. 6.14 and remember that the balls in the row are actually colliding. It may help to think of them as being separated.]

•**FIGURE 6.15 In and out** See Conceptual Example 6.11.

6.5 Center of Mass

Objectives: To (a) explain the concept of the center of mass and compute its location for simple systems, and (b) describe how the center of mass and center of gravity are related.

The conservation of total momentum gives us a method of analyzing a "system of particles." Such a system may be virtually anything—a volume of gas, water in a container, or a baseball. Another important concept allows us to analyze the overall motion of a system of particles. It involves representing the whole system as a single particle. This concept will be introduced here and applied in more detail in the upcoming chapters.

If no net external force acts on a particle, we have seen that its linear momentum is constant. Similarly, if no net external force acts on a *system* of particles, the linear momentum of the system is constant. This similarity implies that a system of particles might be represented by an *equivalent* single particle. Moving rigid objects, such as balls, automobiles, and so forth, are essentially systems of particles and can be effectively represented by equivalent single particles when we analyze motion. Such representation is done through the concept of the **center of mass (CM)**:

> The center of mass is the point at which all of the mass of an object or system may be considered to be concentrated, for the purposes of linear or translational motion only.

Even if a rigid object is rotating, an important result (beyond the scope of this text to derive) is that the center of mass moves as though it were a particle (•Fig. 6.16). The center of mass is sometimes described as the *balance point* of a solid object. For example, if you balance a meterstick on your finger, the center of mass of the stick is located directly above your finger and all of the mass (or weight) seems to be concentrated there.

An expression similar to Newton's second law for a single particle applies to a *system* when the center of mass is used:

$$\mathbf{F}_{\text{net external}} = M\mathbf{A}_{\text{CM}} \qquad (6.17)$$

where $\mathbf{F}$ is the net external force on the system, M is the total mass of the system or the sum of the masses of the particles of the system ($M = m_1 + m_2 + m_3 + \cdots + m_n$, where the system has n particles), and $\mathbf{A}_{\text{CM}}$ is the acceleration of the center of mass of the system. In words, Eq. 6.17 says that the *center of mass* of a system of particles moves as though all the mass of the system were concentrated there and acted on by the resultant of the external forces. Note that the movement of the parts of the system is *not* predicted by Eq. 6.17.

It follows that *if the net external force on a system is zero*, the total linear momentum of the center of mass is conserved (stays constant) because

$$\mathbf{F} = M\mathbf{A}_{\text{CM}} = M(\Delta \mathbf{V}_{\text{CM}}/\Delta t) = \Delta(M\mathbf{V}_{\text{CM}})/\Delta t = \Delta \mathbf{P}/\Delta t \qquad (6.18)$$

Thus the total momentum of the system, $\mathbf{P} = M\mathbf{V}_{\text{CM}}$, is constant. Since M is constant (why?), in this case $\mathbf{V}_{\text{CM}}$ is a constant. Thus the center of mass either moves with a constant velocity or remains at rest.

•**FIGURE 6.16 Center of mass** (a) The center of mass of this sliding wrench moves as though it were a particle. Note the white dot on the wrench that marks the center of mass. (b) Even after exploding, the center of mass of a fireworks projectile would follow a parabolic path in the absence of air resistance.

(a)

(b)

•**FIGURE 6.17 System of particles in one dimension** Where is the system's center of mass? See Example 6.12.

Although you may more readily visualize the center of mass of a solid object, the concept of the center of mass applies to any system of particles or objects, even a quantity of gas. For a system of n particles arranged in one dimension, along the x axis (•Fig. 6.17), the location of the center of mass is given by

$$X_{CM} = \frac{m_1 x_1 + m_2 x_2 + m_3 x_3 + \cdots + m_n x_n}{m_1 + m_2 + m_3 + \cdots + m_n} \quad (6.19)$$

That is, X_{CM} is the x coordinate of the center of mass of a system of particles. In shorthand notation (using signs to indicate vector directions),

$$X_{CM} = \frac{\Sigma_i m_i x_i}{M} \quad (6.20)$$

where Σ_i is the summation of the products $m_i x_i$ for n particles ($i = 1, 2, 3, \ldots, n$). If $\Sigma_i m_i x_i = 0$, then $X_{CM} = 0$, and the center of mass of the one-dimensional system is located at the origin.

Other coordinates of the center of mass for systems of particles are similarly defined. For a two-dimensional distribution of masses, the coordinates of the center of mass are (X_{CM}, Y_{CM}).

Example 6.12 ■ Finding the Center of Mass: A Summation Process

Three masses, 2.0 kg, 3.0 kg, and 6.0 kg, are located at positions (3.0, 0), (6.0, 0), and (−4.0, 0), respectively, in meters from the origin (•Fig. 6.17). Where is the center of mass of this system?

Thinking It Through. The masses and the positions are given, so the definition of X_{CM} (Eq. 6.20) applies. However, keep in mind that the positions are vector displacements from the origin and these are indicated by the appropriate signs (+ or −).

Solution.

Given: $m_1 = 2.0$ kg
$m_2 = 3.0$ kg
$m_3 = 6.0$ kg
$x_1 = 3.0$ m
$x_2 = 6.0$ m
$x_3 = -4.0$ m

Find: X_{CM} (CM coordinate)

Then, simply performing the summation as indicated in Eq. 6.20,

$$X_{CM} = \frac{\Sigma_i m_i x_i}{M}$$

$$= \frac{(2.0 \text{ kg})(3.0 \text{ m}) + (3.0 \text{ kg})(6.0 \text{ m}) + (6.0 \text{ kg})(-4.0 \text{ m})}{2.0 \text{ kg} + 3.0 \text{ kg} + 6.0 \text{ kg}} = 0$$

The center of mass is at the origin.

Follow-up Exercise. Describe a single particle that could replace the system in this Example in order to analyze the translational part of the motion of the system as a whole.

Example 6.13 ■ A Dumbbell: Center of Mass Revisited

A dumbbell (•Fig. 6.18) has a connecting bar of negligible mass. Find the location of the center of mass (a) if m_1 and m_2 are each 5.0 kg, and (b) if m_1 is 5.0 kg and m_2 is 10.0 kg.

Thinking It Through. This Example shows how the location of the center of mass depends on mass distribution. In (b), you might expect the center of mass to be located closer to the more massive end of the dumbbell.

Solution.

Given: $x_1 = 0.20$ m
$x_2 = 0.90$ m
$y_1 = y_2 = 0.10$ m
(a) $m_1 = m_2 = 5.0$ kg
(b) $m_1 = 5.0$ kg
$m_2 = 10.0$ kg

Find: (a) (X_{CM}, Y_{CM}) (CM coordinates)
(b) (X_{CM}, Y_{CM})

Note that each mass is considered to be a particle located at the center of the sphere (its center of mass).

(a) Finding X_{CM} gives

$$X_{CM} = \frac{m_1 x_1 + m_2 x_2}{m_1 + m_2}$$

$$= \frac{(5.0 \text{ kg})(0.20 \text{ m}) + (5.0 \text{ kg})(0.90 \text{ m})}{5.0 \text{ kg} + 5.0 \text{ kg}} = 0.55 \text{ m}$$

Similarly, we find that $Y_{CM} = 0.10$ m. (You might have seen this right away, since each center of mass is at this height.) The center of mass of the dumbbell is then located at $(X_{CM}, Y_{CM}) = (0.55 \text{ m}, 0.10 \text{ m})$, or midway between the end masses.

(b) With $m_2 = 10.0$ kg,

$$X_{CM} = \frac{m_1 x_1 + m_2 x_2}{m_1 + m_2}$$

$$= \frac{(5.0 \text{ kg})(0.20 \text{ m}) + (10.0 \text{ kg})(0.90 \text{ m})}{5.0 \text{ kg} + 10.0 \text{ kg}} = 0.67 \text{ m}$$

which is $\frac{2}{3}$ of the way between the masses. (Note that the distance of the CM from the center of m_1 is $\Delta x = 0.67 \text{ m} - 0.20 \text{ m} = 0.47$ m; with the distance $L = 0.70$ m between the centers of the masses, then $\Delta x/L = 0.47 \text{ m}/0.70 \text{ m} = 0.67$, or $\frac{2}{3}$.) You might expect the balance point of the dumbbell in this case to be closer to m_2. The y coordinate of the center of mass is again $Y_{CM} = 0.10$ m, as you can prove for yourself.

Follow-up Exercise. In part (b) of this Example, take the origin of the coordinate axes to be at the point where m_1 touches the x axis. What are the coordinates of the CM in this case, and how does its location compare to that found in the Example?

In Example 6.13, when the value of one of the masses changed, the x coordinate of the center of mass changed. You might have expected the y coordinate to change also. However, the centers of the end masses were still at the same height, and Y_{CM} remained the same. To increase Y_{CM}, one or both of the end masses would have to be in a higher position.

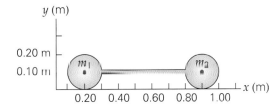

•FIGURE 6.18 Location of the center of mass See Example 6.13.

Now let's see how the center of mass can be applied to a realistic situation.

Example 6.14 ■ Internal Motion: Where's the Center of Mass?

A 75.0-kilogram man stands in the far end of a 50.0-kilogram boat 100 m from the shore, as illustrated in •Fig. 6.19. If he walks to the other end of the 6.00-meter-long boat, how far is the man then from the shore? Neglect friction, and assume that the center of mass of the boat is at its midpoint.

Thinking It Through. The answer is *not* 100 m − 6.00 m = 94.0 m, because the boat moves as the man walks. Why? With no net external force, the acceleration of the center of mass of the man–boat system is zero (Eq. 6.17), and so is the total momentum by Eq. 6.18 ($\mathbf{P} = M\mathbf{V}_{CM} = 0$). Hence, the velocity of the center of mass of the system is zero, or the center of mass is stationary and remains so to conserve system momentum; that is, X_{CM} (initial) = X_{CM} (final).

Solution. Taking the shore as the origin ($x = 0$), we have

Given: $m_m = 75.0$ kg
$x_{m_i} = 100$ m
$m_b = 50.0$ kg
$x_{b_i} = 94.0$ m + 3.00 m = 97.0 m
(where the location of the boat is taken at its center of mass)

Find: x_{m_f} (distance of man from shore)

Note that if we take the man's final position to be a distance x_{m_f} from the shore, then the final position of the boat's center of mass will be $x_{b_f} = x_{m_f} + 3.00$ m, since he will be at the front of the boat.
Then initially,

$$X_{CM_i} = \frac{m_m x_{m_i} + m_b x_{b_i}}{m_m + m_b}$$

$$= \frac{(75.0 \text{ kg})(100 \text{ m}) + (50.0 \text{ kg})(97.0 \text{ m})}{75.0 \text{ kg} + 50.0 \text{ kg}} = 98.8 \text{ m}$$

And finally,

$$X_{CM_f} = \frac{m_m x_{m_f} + m_b x_{b_f}}{m_m + m_b}$$

$$= \frac{(75.0 \text{ kg})x_{m_f} + (50.0 \text{ kg})(x_{m_f} + 3.00 \text{ m})}{75.0 \text{ kg} + 50.0 \text{ kg}} = 98.8 \text{ m}$$

Here $X_{CM_f} = 98.8$ m = X_{CM_i} since the CM does not move. Then, solving for x_{m_f}, we get

$$(125 \text{ kg})(98.8 \text{ m}) = (125 \text{ kg})x_{m_f} + (50.0 \text{ kg})(3.00 \text{ m})$$

and

$$x_{m_f} = 97.6 \text{ m}$$

from the shore.

Follow-up Exercise. Suppose the man then walks back to his original position at the opposite end of the boat. Would he be 100 m from shore?

•**FIGURE 6.19 Walking toward shore** See Example 6.14.

Center of Gravity

As you know, mass and weight are directly related. Closely associated with the center of mass is the concept of the **center of gravity** (**CG**), the point where all of the weight of an object may be considered to be concentrated in representing the object as a particle. If the acceleration due to gravity is constant both in magnitude and direction over the extent of the object, Eq. 6.20 can be rewritten as

$$MgX_{CM} = \Sigma_i m_i g x_i \tag{6.21}$$

Then, the weight, Mg, acts as if it were concentrated at X_{CM}, and the center of mass and the center of gravity coincide. As you may have noticed, the location of the center of gravity was implied in some previous figures in Chapter 4, where the vector arrows for weight ($\mathbf{w} = m\mathbf{g}$) were drawn from a point at or near the center of an object.

In some cases, the center of mass or the center of gravity of an object may be located by symmetry. For example, for a spherical object that is homogeneous (the mass is distributed evenly throughout), the center of mass is at the geometrical center (or center of symmetry). In Example 6.13a, where the end masses of the dumbbell were equal, it was probably apparent that the center of mass was midway between them.

The location of the center of mass of an irregularly shaped object is not so evident and is usually difficult to calculate (even with advanced mathematical methods that are beyond the scope of this book). In some instances, the center of mass may be located experimentally. For example, the center of mass of a flat, irregularly shaped object can be determined experimentally by suspending it freely from different points (•Fig. 6.20). A moment's thought should convince you that the center of mass (or center of gravity) always lies vertically below the point of suspension. Since the center of mass is defined as the point at which all the mass of a body can be considered to be concentrated, this is analogous to a particle of

(a)

(b)

•**FIGURE 6.20 Location of the center of mass by suspension** (a) The center of mass of a flat, irregularly shaped object can be found by suspending the object from two or more points. The CM (and CG) lies on a vertical line under any point of suspension, so the intersection of two such lines marks its location midway through the thickness of the body. The sheet could be balanced horizontally at this point. Why? (b) The process is illustrated with a cutout map of the United States. Note that a plumb line dropped from any other point (third photo) does in fact pass through the CM as located in the first two photos.

•FIGURE 6.21 **The center of mass may be located outside of a body** The center of mass may lie either inside or outside of a body, depending on its mass distribution. (a) For a uniform ring, the center of mass is at its center. (b) For an L-shaped object, if the mass distribution is uniform and the legs are of equal length, the center of mass lies on the diagonal between the legs.

•FIGURE 6.22 **Center of gravity** By arching her body, this high jumper can get over the bar even though her center of gravity passes beneath it.

mass suspended from a string. Suspending the object from two or more points and marking the vertical lines on which the center of mass must lie locates the center of mass as the intersection of the lines.

The center of mass of an object may lie outside the body of the object (•Fig. 6.21). The center of mass of a homogeneous ring is at the ring's center. The mass in any section of the ring is compensated by the mass in an equivalent section directly across the ring, and by symmetry the center of mass is at the center. For an L-shaped object with equal legs, the center of mass lies on a line that makes a 45° angle with both legs. Its location can easily be determined by suspending the L from a point on one of the legs and noting where a vertical line from that point intersects the diagonal line.

Keep in mind that the location of the center of mass or center of gravity of an object depends on the distribution of mass. Therefore, for a flexible object such as the human body, the position of the center of gravity changes as the object changes configuration (mass distribution). For example, when a person raises both arms overhead, his or her center of gravity is raised several centimeters. For a high jumper going over a bar, the center of gravity lies outside the arched body (•Fig. 6.22). In fact, the center of gravity passes *beneath* the bar. This is done purposefully because work must be done to raise the center of gravity, and only the jumper's body has to clear the bar, not the CG.

6.6 Jet Propulsion and Rockets

Objective: To apply the conservation of momentum in the explanation of jet propulsion and the operation of rockets.

The word *jet* is sometimes used to refer to a stream of liquid or gas emitted at a high speed—for example, a jet of water from a fountain or a jet of air from an automobile tire. **Jet propulsion** is the application of such jets to the production of motion. This concept usually brings to mind jet planes and rockets, but squid and octopi propel themselves by squirting jets of water.

You have probably tried the simple application of blowing up a balloon and releasing it. Lacking any guidance or rigid exhaust system, the balloon zig-zags around, driven by the escaping air. In terms of Newton's third law, the air is forced out by the contraction of the stretched balloon—that is, the balloon exerts a force on the air. Thus there must be an equal and opposite reaction force exerted by the air on the balloon. It is this force that propels the balloon on its erratic path.

Jet propulsion is explained by Newton's third law, and in the absence of external forces, the conservation of momentum also applies. You may understand this better by considering the recoil of a rifle, taking the rifle and the bullet as an isolated system (•Fig. 6.23). Initially, the total momentum of this system is zero. When the rifle is fired (by remote control to avoid external forces), the expansion of the gases from the exploding charge accelerates the bullet down the barrel. These gases push backward on the rifle as well, producing a recoil force (the "kick" experienced by a person firing a weapon). Since the initial momentum of the system is zero and the force of the expanding gas is an internal force, the momenta of the bullet and of the rifle must be exactly equal and opposite at any instant. After the bullet leaves the barrel, there is no propelling force, so the bullet and the rifle move with constant velocities (unless acted on by a net external force such as gravity or air resistance).

Similarly, the thrust of a rocket is created by exhausting the gas from burning fuel out the rear of the rocket. The expanding gas exerts a net force on the rocket that propels the rocket in the forward direction (•Fig. 6.24a,b). The rocket exerts a reaction force on the gas, so all of the gas is directed out the exhaust nozzle. If the rocket is at rest when the engines are turned on and there are no external forces (as in deep space, where friction is zero and gravitational forces are negligible), then the instantaneous momentum of the exhaust gas is equal and opposite to that of the rocket. The numerous exhaust gas molecules have small masses and high

•FIGURE 6.23 **Conservation of momentum** (a) Before the rifle is fired, the total momentum of the rifle and bullet (as an isolated system) is zero. (b) During firing, there are equal and opposite internal forces, and the instantaneous total momentum of the rifle-bullet system remains zero (neglecting external forces, such as arise when a rifle is being held). (c) When the bullet leaves the barrel, the total momentum of the system is still zero.

velocities, and the rocket has a much larger mass and a smaller velocity. (See Demonstration 4.)

Unlike a rifle firing a single shot, which has negligible mass, a rocket continuously loses mass when burning fuel (it is more like a machine gun). Thus, the rocket is a system for which the mass is not constant. As the mass of the rocket decreases, it accelerates more easily. Multistage rockets take advantage of this fact. The hull of a burnt-out stage is jettisoned to give a further in-flight reduction in mass (Fig. 6.24c). The payload (cargo) is typically a very small part of the initial mass of rockets for space flights.

Suppose that the purpose of a space flight is to land a payload on the Moon. At some point on the journey, the gravitational attraction of the Moon will become greater than that of Earth, and the spacecraft will accelerate toward the Moon. A soft landing is desirable, so the spacecraft must be slowed down enough to go into orbit around the Moon. This is done by using the rocket engines to apply a *reverse thrust*, or braking thrust. The spacecraft is maneuvered through a 180° angle, or turned around, which is quite easy to do in space. The rocket engines are then fired, expelling the exhaust gas toward the Moon and supplying a braking action.

(a)

•FIGURE 6.24 **Jet propulsion and mass reduction** (a) A rocket burning fuel is continuously losing mass and so becomes easier to accelerate. The resulting force on the rocket (the thrust) depends on the product of the rate of change of its mass with time and the velocity of the exhaust gases: $(\Delta m/\Delta t)\mathbf{v}_{ex}$. Since the mass is decreasing, $(\Delta m/\Delta t)$ is negative and the thrust is opposite $\mathbf{v}_{ex}$. (b) The space shuttle uses a multistage rocket. Both the two booster rockets and the huge external fuel tank are jettisoned in flight. (c) The first and second stages of a *Saturn V* rocket separating after 148 s of burn time.

(b)

(c)

6.6 •Jet Propulsion and Rockets 201

Demonstration 4 ■ Jet Propulsion

A demonstration of Newton's third law and the conservation of momentum. For a rocket or jet engine, thrust is obtained by burning fuel and exhausting the gas out the rear of the engine. Here a fire extinguisher exhausting CO_2 takes the place of the engine.

(a) Ignition and blast off!

(b) Away we go—equal and opposite forces.

You have experienced a reverse thrust effect if you have flown in a commercial jet. In this instance, however, the craft is not turned around. Instead, after touchdown, the jet engines are revved up and a braking action can be felt. Ordinarily, revving up the engines accelerates the plane forward. The reverse thrust is accomplished by activating thrust reversers in the engines that deflect the exhaust gases forward (•Fig. 6.25). The gas experiences an impulse force and a change in momentum in the forward direction (see Fig. 6.3b), and the engine and the aircraft have an equal and opposite momentum change and braking impulse force.

Question: There are no end-of-chapter exercises for this section, so test your knowledge with this one: Astronauts use hand-held maneuvering devices (small rockets) to move around on space walks. Describe how these rockets would be used. Is there any danger on an untethered space walk?

Normal operation

Thrust reverser activated

•**FIGURE 6.25 Reverse thrust** Thrust reversers are activated on jet engines on landing to help slow the plane. The gas experiences an impulse force and a change in momentum in the forward direction, and the plane experiences an equal and opposite momentum change and a braking impulse force.

Chapter Review

Important Terms

linear momentum   173
total linear momentum   173
impulse   178
impulse–momentum theorem   178

conservation of linear
   momentum   183
elastic collision   187
inelastic collision   187

completely inelastic collision   189
center of mass (CM)   195
center of gravity (CG)   199
jet propulsion   200

Important Concepts

- The linear momentum of a particle is a vector and is defined as the product of mass and velocity.
- The total linear momentum of a system is the vector sum of the momenta of the individual particles.
- In the absence of a net external force, the total linear momentum of a system is conserved.
- The impulse–momentum theorem relates impulse to the change in momentum.
- A collision is elastic if the total kinetic energy is conserved. Momentum is conserved in both elastic and inelastic collisions. In a completely inelastic collision, objects stick together after impact.
- The center of mass is the point at which all of the mass of an object or system may be considered to be concentrated. (The center of gravity is the point where all the weight may be considered to be concentrated.)

Important Equations

Linear Momentum:
$$\mathbf{p} = m\mathbf{v} \tag{6.1}$$

Total Linear Momentum of a System:
$$\mathbf{P} = \mathbf{p}_1 + \mathbf{p}_2 + \mathbf{p}_3 + \cdots = \Sigma_i \mathbf{p}_i \tag{6.2}$$

Newton's Second Law in Terms of Momentum:
$$\mathbf{F} = \frac{\Delta \mathbf{p}}{\Delta t} \tag{6.3}$$

Impulse–Momentum Theorem:
$$\text{Impulse} = \bar{\mathbf{F}}\, \Delta t = \Delta \mathbf{p} = m\mathbf{v} - m\mathbf{v}_o \tag{6.5}$$

Conditions for an Elastic Collision:
$$\mathbf{P}_f = \mathbf{P}_i \tag{6.8}$$
$$K_f = K_i$$

Conditions for an Inelastic Collision:
$$\mathbf{P}_f = \mathbf{P}_i \tag{6.9}$$
$$K_f < K_i$$

Final Velocities in Head-On Two-Body Elastic Collisions ($v_{2_o} = 0$):
$$v_1 = \left(\frac{m_1 - m_2}{m_1 + m_2}\right) v_{1_o} \tag{6.15}$$

$$v_2 = \left(\frac{2 m_1}{m_1 + m_2}\right) v_{1_o} \tag{6.16}$$

Coordinate of the Center of Mass (using signs for directions):
$$X_{\text{CM}} = \frac{\Sigma_i m_i x_i}{M} \tag{6.20}$$

Exercises

6.1 Linear Momentum

1. Linear momentum has units of (a) N/m, (b) kg·m/s, (c) N/s, (d) all of these.

2. Linear momentum is (a) always conserved, (b) a scalar quantity, (c) a vector quantity, (d) unrelated to force.

3. Does a fast-running running back always have more linear momentum than a slow-moving, more massive lineman? Explain.

4. ■ If a woman of mass 60 kg is riding in a car traveling at 90 km/h, what is the magnitude of her linear momentum relative to (a) the ground and (b) the car?

5. ■ The linear momentum of a runner in a 100-meter dash is 7.5×10^2 kg·m/s. If the runner's speed is 10 m/s, what is his mass?

6. ■ Find the magnitude of the linear momentum of (a) a 7.1-kilogram bowling ball traveling at 12 m/s and (b) a 1200-kilogram automobile traveling at 90 km/h.

7. ■ Who has more linear momentum, a 70-kilogram running back running at 8.5 m/s or a 120-kilogram lineman moving at 5.0 m/s?

8. ■ How fast would a 1200-kilogram car travel if it had the same linear momentum as a 1500-kilogram truck traveling at 90 km/h?

9. ■ A ball of mass 3.0 kg has a linear momentum of 12 kg·m/s. What is the kinetic energy of the ball?

10. ■■ A 0.150-kilogram baseball traveling with a horizontal speed of 4.50 m/s is hit by a bat and then moves with a speed of 34.7 m/s in the opposite direction. What is the change in the ball's momentum?

11. ■■ A 15.0-gram rubber bullet hits a wall with a speed of 150 m/s. If the bullet bounces straight back with a speed of 120 m/s, what is the change in momentum of the bullet?

12. ■■ If two protons approach each other with speeds of 340 m/s and 450 m/s, respectively, what is the total

momentum of the two-particle system? [Hint: Find the mass of a proton in the Table inside the back cover.]

13. ■■ If a 0.50-kilogram ball is dropped from a height of 10 m, what is the momentum of the ball (a) 0.75 s after being released and (b) just before it hits the ground?

14. ■■ Taking the density of air to be 1.29 kg/m^3, what is the magnitude of the linear momentum of a cubic meter of air moving with a wind speed of (a) 36 km/h and (b) 74 mi/h (the wind speed at which a tropical storm becomes a hurricane).

15. Two runners, of mass 70 kg and 60 kg, respectively, running on a straight track have a total linear momentum of 350 kg·m/s. If the heavier runner is running at 2.0 m/s, what are the possible velocities of the lighter runner?

16. ■■ A 0.20-kilogram billiard ball traveling at a speed of 15 m/s strikes the side rail of the table at an angle of 60° (•Fig. 6.26). If the ball rebounds at the same speed and angle, what is the change in its momentum?

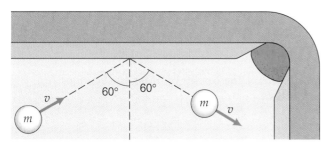

•FIGURE 6.26 Glancing collision See Exercises 16, 17, and 36.

17. ■■ Suppose that the billiard ball in Fig. 6.26 approaches the rail at a speed of 15 m/s and an angle of 60° but rebounds at a speed of 10 m/s and an angle of 50°. What is the change in momentum now? [Hint: Use components.]

18. ■■ A person pushes a 10-kilogram box as it starts from rest and accelerates to a speed of 4.0 m/s. If the force acts on the box for a time of 2.5 s, what is the force exerted by the person? (Ignore friction.)

19. ■■ A loaded tractor-trailer with a total mass of 5000 kg traveling at 3.0 km/h coasts into a loading dock and collides, coming to a stop in 0.64 s. What is the magnitude of the average force exerted on the truck by the dock?

20. ■■ A 2.0-kilogram mud ball drops from rest at a height of 15 m. If the impact between the ball and the ground lasts 0.50 s, what is the average force exerted by the ball on the ground?

21. ■■■ At a basketball game, a 120-pound cheerleader is tossed vertically upward with a speed of 4.50 m/s by another cheerleader. (a) What is the cheerleader's change in momentum from the time she is released to just before being caught if she is caught at the height at which she was released? (b) Would there be any difference in this change if she were caught 0.25 m below the point of release? If so, what is the change then?

6.2 Impulse

22. Impulse is equal to (a) $F \Delta x$, (b) the change in kinetic energy, (c) the change in momentum, (d) $\Delta p / \Delta t$.

23. A karate student tries *not* to follow through in order to break a board as shown in •Fig. 6.27. How can the abrupt stoppage of the fist (by not following through) generate so much force?

•FIGURE 6.27 A karate punch See Exercise 23.

24. Explain the difference for each of the following pairs of actions in terms of impulse: (a) a golfer's long drive and a short chip shot, (b) a boxer's jab and knock-out punches, (c) a baseball player's bunting action and home-run swing.

25. Explain the principle behind (a) the use of Styrofoam as packing material to prevent objects from breaking and (b) the padding of dashboards in cars to prevent injuries.

26. ■ When tossed upward and hit horizontally by a batter, a 0.20-kilogram softball receives an impulse of 3.0 N·s. With what horizontal speed does the ball leave the bat?

27. ■ An automobile with a linear momentum of 3.0×10^4 kg·m/s is brought to a stop in 5.0 s. What is the magnitude of the average braking force?

28. ■ A pool player imparts an impulse of 3.2 N·s to a stationary 0.25-kilogram cue ball with a cue stick. What is the speed of the ball after impact?

29. ■■ For the karate punch in Exercise 23, assume that the fist has a mass of 0.35 kg and that the speeds of the fist just before and after hitting the board are 10 m/s and zero, respectively. What is the average force exerted by the fist on the board if (a) the fist follows through, so the contact time is 3.0 ms, or (b) the fist stops abruptly, so the contact time is only 0.30 ms.

30. ■■ The spout of a water tower used to fill fire-fighting tank trucks is 7.0 m above the ground. If the spout is inadvertently left open and water runs out vertically at a rate of 3.0 kg/s, what is the average force exerted by the water on the ground?

31. ■■ A 0.45-kilogram volleyball is hit with a horizontal velocity of 4.0 m/s over the net. One of the players on the

front line jumps up and hits it back with a horizontal velocity of 7.0 m/s. If the contact time is 0.040 s, what was the average force applied by the player?

32. ■■ A 1.0-kilogram ball is thrown horizontally with a velocity of 15 m/s against a wall. If the ball rebounds with a velocity of 13 m/s and the contact time is 0.020 s, what is the force on the ball by the wall?

33. ■■ In bare hands with his arms rigidly extended, a boy catches a 0.16-kilogram baseball coming directly toward him at a speed of 25 m/s. He emits an audible "ouch!" because the ball stings his hands. He learns quickly to move his hands with the ball as he catches it. If the contact time of the collision is increased from 3.5 ms to 8.5 ms in this way, how do the magnitudes of the average impulse forces compare?

34. ■■ A one-dimensional impulse force acts on a 3.0-kilogram object as diagrammed in •Fig. 6.28. Find (a) the magnitude of the impulse given to the object, (b) the magnitude of the average force, and (c) the final speed if the object had an initial velocity of 6.0 m/s opposite the impulse.

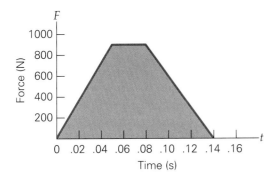

•FIGURE 6.28 **Force versus time graph** See Exercise 34.

35. ■■ A 0.35-kilogram piece of putty is dropped from a height of 2.5 m above a flat surface. When it hits the surface, the putty comes to rest in 0.30 s. What is the average force exerted on the putty by the surface?

36. ■■ If the billiard ball in Fig. 6.26 is in contact with the rail for 0.010 s, what is the magnitude of the average force exerted on the ball? (See Exercise 16.)

37. ■■■ A mother holds her infant son on her lap rather than putting him in a child seat. In a car crash of only 25 mph (40 km/h), the baby, weighing 12 lb (of mass 5.5 kg) will require his mother's arms to exert a force of 240 lb. to avoid injury. How long does the crash last?

6.3 The Conservation of Linear Momentum

38. The linear momentum of an object is conserved if (a) the force acting on the object is conservative, (b) there is a single, unbalanced internal force acting on the object, (c) the mechanical energy is conserved, (d) none of these.

39. A fan boat of the type used in swampy and marshy areas is shown in •Fig. 6.29. Using linear momentum conservation, explain the principle of its propulsion. What would happen if a sail were installed on the boat behind the fan?

•FIGURE 6.29 **Fan propulsion** See Exercise 39.

40. Internal forces do not affect the conservation of momentum of a system because (a) they cancel each other, (b) their effects are canceled by external forces, (c) they can never produce a change in velocity, (d) Newton's second law is not applicable to them.

41. ● Imagine yourself standing in the middle of a frozen lake. The ice is so smooth that it is frictionless. How could you get to shore? (You couldn't walk. Why?)

42. ■ ● A 60-kilogram astronaut floating in space outside the capsule throws his 0.50-kilogram hammer such that it moves with a speed of 10 m/s relative to the capsule. What happens to the astronaut?

43. ■ In a pairs figure-skating competition, a 65-kilogram male and his 45-kilogram partner stand together on skates on the ice. If they push apart and the female has a velocity of 1.5 m/s eastward, what is the velocity of the male? (Neglect friction.)

44. ■ To get off a frozen, frictionless lake, a 70.0-kilogram person throws a 0.150-kilogram object horizontally, directly away from the shore, with a speed of 2.00 m/s. If the person is 5.00 m from the shore, how long does it take for him to reach it?

45. ■■ A bullet of mass 100 g is fired horizontally into a 14.9-kilogram block of wood resting on a horizontal surface, and the bullet becomes embedded in the block. If the muzzle speed of the bullet is 250 m/s, what is the velocity of the block containing the embedded bullet immediately after the impact? (Neglect surface friction.)

46. ■■ A 3.0-kilogram object initially at rest explodes and splits into three fragments. One fragment has a mass of 0.50 kg and flies off along the negative x axis at a speed of 2.8 m/s, and another has a mass of 1.3 kg and flies off along the negative y axis at a speed of 1.5 m/s. What are the speed and direction of the third fragment?

47. ■■ Suppose that the 3.0-kilogram object in Exercise 46 is initially traveling at a speed of 2.5 m/s in the positive

x direction. What will the speed and direction of the third fragment be in this case?

48. ■■ Identical cars hit each other and lock bumpers. In each of the following cases, what are the speeds of the cars immediately after coupling? (a) A car moving with a speed of 90 km/h approaches a stationary one. (b) Two cars approach each other with speeds of 90 km/h and 120 km/h, respectively. (c) Two cars travel in the same direction with speeds of 90 km/h and 120 km/h, respectively.

49. ■■ ● A 1200-kilogram car moving to the right with a speed of 25 m/s collides with a 1500-kilogram truck and locks bumpers with the truck. Calculate the velocity of the combination after the collision if the truck is initially (a) at rest, (b) moving to the right with a speed of 20 m/s, (c) moving to the left with a speed of 20 m/s.

50. ■■ A 10-gram bullet moving horizontally at 400 m/s penetrates a 3.0-kilogram wood block resting on a frictionless horizontal surface. If the bullet slows down to 300 m/s after emerging from the block, what is the speed of the block immediately after the bullet emerges (•Fig. 6.30)?

•FIGURE 6.30 Momentum transfer? See Exercise 50.

51. ■■ A 1600-kilogram (empty) truck rolls with a speed of 2.5 m/s under a loading bin, and a mass of 3500 kg is deposited in the truck. What is the truck's speed immediately after loading?

52. ■■■ A projectile that is fired from a gun has an initial velocity of 90.0 km/h at an angle of 60.0° above the horizontal. When the projectile is at the top of its trajectory, an internal explosion causes it to separate into two fragments of equal mass. One of the fragments falls straight downward as though it has been released from rest. How far does the other fragment land from the gun?

53. ■■■ A *ballistic pendulum* is a device used to measure the velocity of a projectile—for example, the muzzle velocity of a rifle bullet. The projectile is shot horizontally into, and becomes embedded in, the bob of a pendulum as illustrated in •Fig. 6.31. The pendulum swings upward to some height (h), which is measured. The masses of the block and the bullet are known. Using the laws of momentum and energy, show that the initial speed of the projectile is given by $v_o = [(m + M)/m]\sqrt{2gh}$.

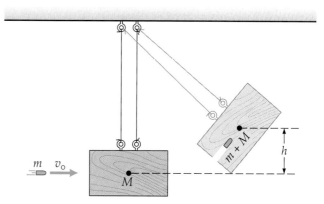

•FIGURE 6.31 A ballistic pendulum See Exercises 53, 70, and 71.

6.4 Elastic and Inelastic Collisions

54. Which of the following is *not* conserved in an inelastic collision: (a) momentum, (b) mass, (c) kinetic energy, (d) total energy?

55. ● In a head-on elastic collision, mass m_1 strikes a stationary mass m_2. The energy transfer is complete if (a) $m_1 = m_2$, (b) $m_1 \gg m_2$, (c) $m_1 \ll m_2$, (d) the masses stick together.

56. A large mass and a small mass have a head-on elastic collision. Which mass receives the greater impulse: (a) the large mass, (b) the small mass, or (c) the impulses are equal?

57. Since $K = p^2/2m$, how can kinetic energy be lost in an inelastic collision yet the total momentum still be conserved? Explain.

58. ■■ ● A 4.0-kilogram ball with a velocity of 4.0 m/s in the $+x$ direction elastically collides head-on with a stationary 2.0-kilogram ball. What are the velocities of the balls after the collision?

59. ■■ A ball with a mass of 0.10 kg is traveling with a velocity of 0.50 m/s in the $+x$ direction and collides head-on with a 5.0-kilogram ball that was at rest. Find the velocities of the balls after the collision; assume that it is elastic.

60. ■■ For the apparatus in Fig. 6.15, show that one ball swinging in with speed v_o will not cause two balls to swing out with speed $v_o/2$.

61. ■■ A proton of mass m moving with a speed of 3.0×10^6 m/s undergoes a head-on elastic collision with an alpha particle of mass $4m$, which is initially at rest. What are the velocities of the two particles after the collision?

62. ■■ ● Two balls with masses of 2.0 kg and 6.0 kg travel toward each other at speeds of 12 m/s and 4.0 m/s, respectively. If the balls have a head-on inelastic collision

and the 2.0-kilogram ball recoils with a speed of 8.0 m/s, how much kinetic energy is lost in the collision?

63. ■■ A 1500-kilogram truck moving with a speed of 25 m/s runs into the rear of a 1200-kilogram stopped car. If the collision is perfectly inelastic, what is the kinetic energy lost in the collision?

64. ■■ Two balls approach each other as shown in •Fig. 6.32, where $m = 2.0$ kg, $v = 3.0$ m/s, $M = 4.0$ kg, $V = 5.0$ m/s. If the balls collide and stick together at the origin, (a) what are the components of the velocity v' of the balls after collision? (b) What is the angle θ?

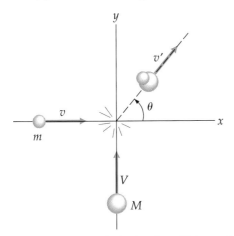

•FIGURE 6.32 **A completely inelastic collision** See Exercise 64.

65. ■■ A 1500-kilogram car traveling east at 90.0 km/h and a 3000-kilogram minivan traveling south at 60.0 km/h collide at a perpendicular intersection. Assuming that the collision is completely inelastic, what is the velocity of the vehicles immediately after collision?

66. ■■ A cue ball traveling at 0.75 m/s hits the stationary 8-ball, which moves off with a speed of 0.25 m/s at an angle of 37° relative to the cue ball's initial direction. Assuming that the collision is inelastic, at what angle will the cue ball be deflected, and what will be its speed?

67. ■■ A fellow student states that the total momentum of a three-particle system ($m_1 = 0.25$ kg, $m_2 = 0.20$ kg, and $m_3 = 0.33$ kg) is initially zero and that after an inelastic triple collision, he calculates the particles to have velocities of 4.0 m/s at 0°, 6.0 m/s at 120°, and 2.5 m/s at 230°, respectively, measured from the $+x$ direction. Do you agree with his calculations? If not, assuming the first two answers to be correct, what should be the momentum of the third particle so the total is zero?

68. ■■ A freight car of mass 25 000 kg rolls down an inclined track through a vertical distance of 2.5 m. At the bottom of the incline, on a level track, the car collides and couples with an identical freight car that was at rest. What percentage of the initial kinetic energy is lost in the collision?

69. ■■■ In an elastic head-on collision with a stationary target particle, a moving particle recoils at one-third of its incident speed. What is (a) the ratio of the particle masses (m_1/m_2)? (b) the speed of the target particle after the collision in terms of the incoming particle's initial speed?

70. ■■■ Show that the fraction of kinetic energy lost in a ballistic pendulum collision (as in Fig. 6.31) is equal to $M/(m + M)$.

71. ■■■ A 10-gram bullet is fired horizontally into, and becomes embedded in, a suspended block of wood with a mass of 0.890 kg (see Fig. 6.31). (a) How does the speed of the block with the embedded bullet immediately after the collision compare with the initial speed (v_o)? (b) If the block with the embedded bullet swings upward and its center of mass is raised 0.40 m, what was the initial speed of the bullet? (c) Was the collision elastic? If not, what percentage of the initial kinetic energy was lost?

72. ■■■ A moving billiard ball collides with an identical stationary one, and the incoming ball is deflected at an angle of 45° from its original direction. Show that if the collision is elastic, both balls will have the same speed afterward and will move at 90° relative to each other.

73. ■■■ (a) For an elastic, two-body head-on collision, show that in general $v_2 - v_1 = -(v_{2_o} - v_{1_o})$. That is, the relative speed of recession after the collision is the same as the relative speed of approach before it. (b) In general, a collision is completely inelastic, completely elastic, or somewhere in between. The degree of elasticity is sometimes expressed as the *coefficient of restitution* (e), defined as the ratio of the relative velocities of recession and approach: $v_2 - v_1 = -e(v_{2_o} - v_{1_o})$. What are the values of e for an elastic collision and a completely inelastic collision?

74. ■■■ The coefficient of restitution (see Exercise 73) for steel colliding with steel is 0.95. If a steel ball is dropped from a height h_o above a steel plate, to what height will the ball rebound?

6.5 Center of Mass

75. •Figure 6.33 shows a performer walking on a tightrope. The long pole he holds curves down at the ends. What effect does this pole have on the center of mass of the performer–pole system? [In Chapter 8, you will learn why

•FIGURE 6.33 **Tightrope walking** See Exercise 75.

this effect and another effect caused by the long pole will make walking on a tightrope easier.]

76. The center of mass of an object (a) always lies at the center of the object, (b) is at the location of the most massive particle in the object, (c) always lies within the object, (d) none of these.

77. The center of mass and center of gravity of an object coincide if (a) the object is flat, (b) there is a uniform mass distribution, (c) they both lie inside the object, (d) the acceleration due to gravity is constant.

78. A spacecraft is initially at rest in free space, and then its rocket engines are fired. Describe the motion of the center of mass of the system.

79. ■ (a) The center of mass of a system consisting of two 0.10-kilogram particles is located at the origin. If one of the particles is at (0, 0.45 m), where is the other? (b) If they are moved so their center of mass is at (0.25 m, 0.15 m), can you tell where the particles are located?

80. ■ The centers of a 4.0-kilogram sphere and a 7.5-kilogram sphere are separated by a distance of 1.5 m. Where is the center of mass of the system?

81. ■ (a) Find the center of mass of the Earth–Moon system. [Hint: Use Appendix I and consider the distance between the two to be measured from their centers.] (b) Where is that center of mass relative to the surface of the Earth?

82. ■■ Find the center of mass of a system composed of three spherical objects with masses of 3.0 kg, 2.0 kg, and 4.0 kg and centers located at (−6.0 m, 0), (1.0 m, 0), and (3.0 m, 0), respectively.

83. ■■ For the system described in Exercise 82, where would a fourth sphere with a mass of 0.50 kg have to be located for the center of mass of the system to be at the origin?

84. ■■ A uniform 3.0-kilogram rod of length 5.0 m supports two point masses of 4.0 kg and 6.0 kg, respectively (•Fig. 6.34). Where is the center of mass of the system?

•FIGURE 6.34 Find the center of mass See Exercise 84.

85. ■■ A uniform sheet of metal measures 25 cm by 25 cm. If a circular piece with a radius of 5.0 cm is cut from the center of the sheet, where is its center of mass now?

86. ■■ Locate the center of mass of the system shown in •Fig. 6.35 (a) if all of the masses are equal; (b) if $m_2 = m_4 = 2m_1 = 2m_3$; (c) if $m_1 = 1.0$ kg, $m_2 = 2.0$ kg, $m_3 = 3.0$ kg, and $m_4 = 4.0$ kg.

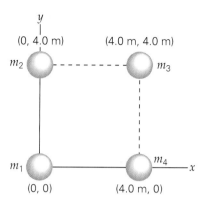

•FIGURE 6.35 Where's the center of mass? See Exercise 86.

87. ■■ A system of two masses has a center of mass given by X_{CM_1}. Another system of three masses has a center of mass given by X_{CM_2}. Show that if all five masses are considered to be one system, the center of mass of that combined system is not $X_{CM} = X_{CM_1} + X_{CM_2}$.

88. ■■ A 100-kilogram astronaut (mass includes space gear) on a space walk is 5.0 m from a 3000-kilogram space capsule at the full length of her safety cord. To return to the capsule, she pulls herself along the cord. Where do the astronaut and capsule meet?

89. ■■ Two skaters with masses of 65 kg and 45 kg, respectively, stand 8.0 m apart; each holds one end of a piece of rope. (a) If they pull themselves along the rope until they meet, how far does each skater travel? (Neglect friction.) (b) If only the 45-kilogram skater pulls along the rope until she meets her friend (who just holds onto the rope), how far does each skater travel?

90. ■■■ Three particles, each with a mass of 0.25 kg, are located at (−4.0 m, 0), (2.0 m, 0), and (0, 3.0 m) and are acted on by forces $\mathbf{F}_1 = (-3.0 \text{ N})\hat{\mathbf{y}}$, $\mathbf{F}_2 = (5.0 \text{ N})\hat{\mathbf{y}}$, and $\mathbf{F}_3 = (4.0 \text{ N})\hat{\mathbf{x}}$, respectively. Find the acceleration (magnitude and direction) of the center of mass. [Hint: Consider the components of that acceleration.]

Additional Exercises

91. ● Two objects have the same momentum. (a) Will they always have the same kinetic energy? (b) What can you say definitely about their motion?

92. A 1.0-kilogram object moving at 10 m/s collides with a stationary 2.0-kilogram object as shown in •Fig. 6.36. If the collision is perfectly inelastic, how far along the inclined plane will the combined system travel? Neglect friction.

93. Near the end of a chess game, each player has three pieces left: a knight (N), the queen (Q), and the king (K), as shown in •Fig. 6.37 (where B stands for black and W for white). If the queen has 3.0 times the mass of a knight

•FIGURE 6.36 **How far up?** See Exercise 92.

and the king has 4.0 times the mass of a knight, find (a) the location (coordinates X_{CM} and Y_{CM}) of the CM of the black pieces and (b) the location of the CM of the white pieces. (c) Find the location of the CM for the system of all of the pieces, regardless of color. (Taking the lower left corner of the board as the origin of the coordinate system, express the coordinates in terms of d, the length of a side of the board's squares; assume that each piece is located at the center of the square it occupies.)

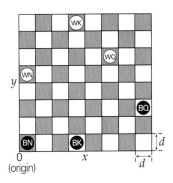

•FIGURE 6.37 **No checkmate yet** See Exercises 93 and 94.

94. In Exercise 93, the locations of the centers of mass of the black and white chess pieces were found in (a) and (b). To find the CM for all of the pieces, could these centers of mass be treated as the locations of point particles where the respective total masses (black and white) were considered concentrated? Check and find out.

95 Two balls of equal mass (0.50 kg) approach the origin along the positive x and y axes at the same speed (3.3 m/s). (a) What is the total momentum of the system? (b) Will the balls necessarily collide at the origin? What is the total momentum of the system after both balls have passed through the origin?

96. Two identical billiard balls approach each other at the same speed (2.0 m/s). At what speeds do they rebound after a head-on elastic collision?

97. A truck with a mass of 2400 kg travels at a constant speed of 90 km/h. (a) What is the magnitude of the truck's linear momentum? (b) What average force would be required to stop the truck in 8.0 s?

98. A 15 000-newton automobile travels at a speed of 45 km/h northward along a street, and a 7500-newton sports car travels at a speed of 60 km/h eastward along an intersecting street. (a) If neither driver brakes and the cars collide at the intersection and lock together, what will the velocity of the cars be immediately after the collision? (b) What percentage of the initial kinetic energy will be lost in the collision?

99. For a movie scene, a 75-kilogram stunt man drops from a tree onto a 50-kilogram sled moving with a velocity of 10 m/s toward the shore on a frozen lake. (a) What is the speed of the sled after the stunt man is on board? (b) If the sled hits the bank and stops but the stunt man keeps on going, with what speed does he leave the sled? (Neglect friction.)

100. During a snowball fight, a 0.25-kilogram snowball traveling at a speed of 14 m/s hits a student in the back of the head. (a) What is the impulse? Is this an elastic collision? (b) If the contact time is 0.10 s, what is the average force on the student's head?

101. A 2.5-kilogram block sliding with a constant velocity of 6.0 m/s on a frictionless horizontal surface approaches a stationary 6.5-kilogram block. (a) If the blocks suffer a completely inelastic collision, what is their velocity after it? (b) How much mechanical energy is lost in the completely inelastic collision?

102. A 90-kilogram astronaut is stranded in space at a point 6.0 m from his spaceship. To get back, he throws a 0.50-kilogram piece of equipment so that it moves at a speed of 4.0 m/s directly away from the spaceship. How long will it take him to reach the ship?

103. In nuclear reactors, subatomic particles called neutrons are slowed down by allowing them to collide with the atoms of a moderator material, such as a carbon atom, which is 12 times more massive than a neutron. (a) In a head-on elastic collision with a carbon atom, what percentage energy is lost by a neutron? (b) If the neutron has an initial speed of 1.5×10^7 m/s, what will its speed be after collision?

104. A 70-kilogram athlete achieves a height of 2.25 m in a high jump. Considering the jumper and the Earth as an isolated system, with what speed does the Earth initially move as the jumper launches himself upward?

105. A uniform, flat piece of metal is shaped like an equilateral triangle with sides that are 30 cm long. What are the coordinates of the center of mass in the xy plane if one apex is at the origin and one side is along the y axis?

CHAPTER 7
Circular Motion and Gravitation

7.1 Angular Measure
7.2 Angular Speed and Velocity
7.3 Uniform Circular Motion and Centripetal Acceleration
7.4 Angular Acceleration
7.5 Newton's Law of Gravitation
7.6 Kepler's Laws and Earth Satellites

Insights

- The Centrifuge: Separating Blood Components
- Space Exploration: Gravity Assists
- Space Colonies And Artificial Gravity

People often say that rides like this "defy gravity." Of course, you know that in reality gravity cannot be defied; it commands respect. There is nothing that will shield you from it and no place in the universe where you can go to be entirely free of it. What, then, keeps these people in their seats, despite the tug of Earth's ever-present gravity? You may be surprised to find that if you can answer this question, you'll also be able to understand what keeps satellites from falling to Earth and keeps the Earth in orbit around the Sun.

Circular motion is everywhere, from atoms to galaxies, from flagella of bacteria to Ferris wheels. Two terms are frequently used to describe such motion. In general, we say that an object *rotates* when the axis of rotation lies within the body and that it *revolves* when the axis is outside the body. Thus, the Earth rotates on its axis and revolves about the Sun.

When a body rotates on its axis, all the particles of the body move in circular paths about the body's axis of rotation. For example, all the particles that make up a compact disc travel in circles about the hub of the CD player. In fact, as a "particle" on Earth, you are continuously in circular motion about the Earth's rotational axis.

Circular motion is motion in two dimensions, and so it can be described by rectangular components as used in Chapter 3. However, it is usually more convenient to describe circular motion in terms of angular quantities that will be introduced in this chapter. Being familiar with the description of circular motion will make the study of rotating rigid bodies much easier, as you will find in Chapter 8.

Gravity plays a large role in determining the motions of the planets, since it supplies the force necessary to maintain their nearly circular orbits. This chapter will consider Newton's law of gravitation, which describes this fundamental force, and will analyze planetary motion in terms of this and related basic laws. The same considerations will help you understand the motions of Earth satellites, which include one natural one (the Moon) and many artificial ones.

7.1 Angular Measure

Objectives: To (a) define units of angular measure and (b) show how angular measure is related to circular arc length.

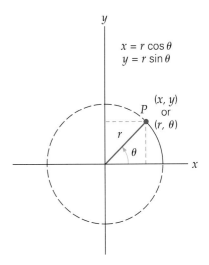

•FIGURE 7.1 **Polar coordinates**
A point may be described by polar coordinates instead of Cartesian coordinates—(r, θ) instead of (x, y). For a circle, θ is the angular distance and r is the radial distance. The two types of coordinates are related by the transformation equations $x = r \cos \theta$ and $y = r \sin \theta$.

Motion is described as a time rate of change of position. As you might guess, *angular* speed and velocity also involve a time rate of change of position, which is expressed by an *angle*. Consider a particle traveling in a circular path, as shown in •Fig. 7.1. At a particular instant, the particle's position (P) may be designated by the Cartesian coordinates x and y. However, the position may also be designated by the polar coordinates r and θ. The distance r extends from the origin, and the angle θ is commonly measured counterclockwise from the x axis. The transformation equations that relate one set of coordinates to the other are

$$x = r \cos \theta \tag{7.1a}$$

$$y = r \sin \theta \tag{7.1b}$$

Polar coordinates, r and θ

as can be seen from the x and y components of point P in Fig. 7.1.

Note that r is the same for any point on a given circle. As a particle travels in a circle, the value of r is constant and only θ changes with time. Thus, circular motion can be described by using one polar coordinate (θ) that changes with time, instead of two Cartesian coordinates (x and y), both of which change with time.

Analogous to linear displacement is **angular displacement**, which is given by

$$\Delta \theta = \theta - \theta_o \tag{7.2}$$

or simply $\Delta \theta = \theta$ when we choose $\theta_o = 0°$. A unit commonly used to express angular displacement (or angles) is the degree (°); there are 360° in one complete circle, or revolution. Each degree is divided into 60 minutes and each minute into 60 seconds. (These divisions have no relationship to time units.)

It is important to be able to relate the angular description of circular motion to the orbital or tangential description—that is, to relate the angular displacement to the arc length s. The *arc length* is the distance traveled along the circular path, and the angle θ is said to *subtend* (define) the arc length. A unit that is very convenient for relating angle to arc length is the **radian (rad)**, which is defined as the angle subtending an arc length (s) that is equal to the radius (r) (•Fig. 7.2a).

To get a general relationship between radians and degrees, let's consider the distance around a complete circle (360°). Basically, a radian is the angle subtending a circular arc equal in length to the radius. Thus for one full circle with $s = 2\pi r$ (the circumference), there is a total of $\theta = 2\pi r/r = 2\pi$ rad, or

$$2\pi \text{ rad} = 360°$$

This relationship can be used to obtain convenient conversions of common angles (Table 7.1). Thus,

$$1 \text{ rad} = 360°/2\pi = 57.3°$$

TABLE 7.1 **Equivalent Degree and Radian Measures**

Degrees	Radians
360°	2π
180°	π
90°	$\pi/2$
60°	$\pi/3$
57.3°	1
45°	$\pi/4$
30°	$\pi/6$

•**FIGURE 7.2 Radian measure**
(a) Angular displacement may be measured either in degrees or in radians (rad). An angle θ is subtended by an arc length s. When $s = r$, the angle subtending s is defined to be 1 rad. More generally, $\theta = s/r$ with θ in radians. (b) A radian is approximately equal to 57.3°, and the circular sector marked off by this angle is similar to an equilateral triangle with sides r.

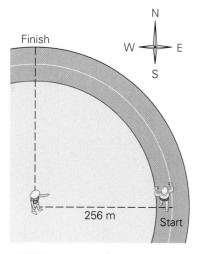

•**FIGURE 7.3 Arc length—found by means of radians** See Example 7.1.

(to 3 significant figures). A radian is thus slightly smaller than one of the angles of an equilateral triangle, as Fig. 7.2b shows. Notice in Table 7.1 that the angles in radians are expressed in terms of π explicitly, for convenience.

Similarly, the number of radians subtended by an arbitrary arc length s is equal to the number of radii that will go into s, or the number of radians $\theta = s/r$. Thus we can write

$$s = r\theta \qquad (7.3)$$

which is an important relationship between the circular arc length s and its radius r. Notice that since $\theta = s/r$, the angle in radians is the ratio of two lengths. This means that a radian measure is a pure number—that is, it is dimensionless and so has no units.

Example 7.1 ■ Finding Arc Length: Using Radian Measure

A spectator standing at the center of a circular running track observes a runner start a practice race 256 m due east of her own position (•Fig. 7.3). The runner runs on the same track to the finish line, which is located due north of the observer's position. What is the distance of the run?

Thinking It Through. Note that the subtending angle of the section of circular track is $\theta = 90°$. The arc length (s) can be found, since the radius r of the circle is known.

Solution. Listing what is given and what is to be found, we have

Given: $r = 256$ m   *Find:* s (arc length)
$\theta = 90° = \pi/2$ rad

Simply using Eq. 7.3 to find the arc length, we get

$$s = r\theta = (256 \text{ m})\left(\frac{\pi}{2}\right) = 402 \text{ m}$$

Note that the rad unit is omitted, but the equation is dimensionally correct. Why?

Follow-up Exercise. What would be the distance of one complete lap around the track in this Example?

Example 7.2 ■ How Far Away? A Useful Approximation

A sailor measures the length of a distant tanker as an angular distance of 1° 9′ (1 degree, 9 minutes) with a divided circle, as illustrated in •Fig. 7.4a. He knows from the shipping charts that the tanker is 150 m in length. Approximately how far away is the tanker?

Thinking It Through. Note in the Learn by Drawing on p. 213 that for small angles, the arc length approximates the vertical leg of the triangle, or $s \approx y$. Hence, if we know the length and the angle, we can find the radial distance, which is approximately equal to the tanker's distance.

Solution. To approximate the distance, we take the ship's length to be nearly equal to the arc length subtended by the measured angle. This is a good approximation for small angles. The data are then

Given: $\theta = 1° 9′$ *Find:* r (radial distance)
$s = 150$ m

Equation 7.3 can be used to find r, but first we must convert θ to radians. Changing the minutes to degrees, we get $9′\left(\dfrac{1°}{60′}\right) = 0.15°$, and $\theta = 1.15°$. Then,

$$\theta = 1.15°\left(\frac{2\pi \text{ rad}}{360°}\right) = 0.0201 \text{ rad}$$

212 CHAPTER 7 Circular Motion and Gravitation

•**FIGURE 7.4 Angular distance**
For small angles, the arc length is approximately a straight line, or the chord length. Knowing the length of the tanker, we can find how far away it is by measuring its angular size. See Example 7.2.

LEARN BY DRAWING

The Small-Angle Approximation

θ *not* small:

$\theta \text{ (in rad)} = \dfrac{s}{r}$

$\sin\theta = \dfrac{y}{r} \qquad \tan\theta = \dfrac{y}{x}$

θ small:

$y \longrightarrow s$
$x \longrightarrow r$

$\theta \text{ (in rad)} = \dfrac{s}{r} \approx \dfrac{y}{r} \approx \dfrac{y}{x}$

$\theta \text{ (in rad)} \approx \sin\theta \approx \tan\theta$

(Note that any of the equivalent relationships in Table 7.1 could be used for the conversion, such as π rad/180° or even 1 rad/57.3°.) Then,

$$r = \dfrac{s}{\theta} = \dfrac{150 \text{ m}}{0.0201} = 7.46 \times 10^3 \text{ m} = 7.46 \text{ km}$$

As noted, the distance r is an approximation, obtained by assuming that for small angles the arc length s and the straight-line chord length L are very nearly the same length (Fig. 7.4b). How good an approximation is this? To see, let's compute the perpendicular distance d to the ship. From the geometry, we have $\tan\theta/2 = L/2d$, so

$$d = \dfrac{L}{2\tan\theta/2} = \dfrac{150 \text{ m}}{2\tan(1.15°/2)} = 7.47 \times 10^3 \text{ m} = 7.47 \text{ km}$$

This is a pretty good approximation—the values are nearly equal.

Follow-up Exercise. As pointed out, the approximation in this Example was for *small* angles. You might wonder what is small. To investigate this question, what would be the percentage error of the approximated distance to the tanker for an angle of 20°?

Problem-Solving Hint

In computing trigonometric functions such as $\tan\theta$ or $\sin\theta$, the angle may be expressed in degrees or radians; for example, $\sin 30° = \sin(\pi/6 \text{ rad}) = \sin(0.524 \text{ rad}) = 0.500$. When finding trig functions with a calculator, note that there is usually a way to change the angle entry between "deg" and "rad" modes. Calculators commonly are set in the degree mode, so if you want to find the value of, say, $\sin(1.22 \text{ rad})$, first change to the "rad" mode and enter 1.22; $\sin(1.22 \text{ rad}) = 0.939$.

Your calculator may have a third, "grad," mode. The grad is a little-used angular unit. A grad is 1/100 of a right (90°) angle; that is, there are 100 grads in a right angle.

7.2 Angular Speed and Velocity

Objectives: To (a) describe and compute angular speed and velocity, and (b) explain their relationship to tangential speed.

The description of circular motion in angular form is analogous to the description of linear motion. In fact, you'll notice that the equations are almost mathematically identical, with different symbols being used to indicate that the quantities have different meanings. The lower case Greek letter omega is used to represent **average angular speed** ($\bar{\omega}$), the angular displacement divided by the total time to travel the distance:

Definition of: Angular speed

$$\bar{\omega} = \frac{\Delta\theta}{\Delta t} = \frac{\theta - \theta_o}{t - t_o} \quad (7.4)$$

We say the units of angular speed are radians per second; technically they are s^{-1} since the radian is unitless, but it is useful to keep the rad. The *instantaneous angular speed* is given for an extremely small time interval, Δt approaches zero.

It is common to take θ_o and t_o to be zero when possible, and we generally write

$$\bar{\omega} = \frac{\theta}{t} \quad \text{or} \quad \theta = \bar{\omega}t \quad (7.5)$$

SI unit of angular speed: radians per second (rad/s or s^{-1})

As in the linear case, if the angular speed is *constant*, then $\bar{\omega} = \omega$.

Another common descriptive unit for angular speed is revolutions per minute (rpm); for example, a CD (compact disk) rotates at a speed of 200–500 rpm (depending on the location of the track). This nonstandard unit of revolutions per minute is readily converted to radians per second, since 1 revolution = 2π rad.

The **average and instantaneous angular velocities** are analogous to their linear counterparts. Angular velocity is associated with angular displacement. Both are vectors and so have direction; however, this directionality may be specified in a special way. In one-dimensional, or linear, motion, a particle can go only in one direction or the other (+ or −), so the displacement and velocity vectors can have only these two directions. In the angular case, a particle moves one way or the other, but the motion is along its *circular path*. Thus, the angular displacement and angular velocity vectors of a particle in circular motion can have only two directions, which correspond to going around the circular path with either increasing or decreasing angular displacement from θ_o—that is, clockwise or counterclockwise. Let's focus on the angular velocity vector, **ω**. (The direction of the angular displacement will be the same as that of the angular velocity. Why?)

The direction of the angular velocity vector is given by a *right-hand rule*, illustrated in •Fig. 7.5a. When the fingers of your right hand are curled in the direction of circular motion, your extended thumb points in the direction of **ω**. Note that circular motion can be in only one of two circular *senses*, clockwise or counterclockwise. Plus and minus signs can be used to distinguish circular rotation directions relative to the angular velocity vector. It is customary to take a counterclockwise rotation as positive (+), since positive angular distance (and displacement) is conventionally measured counterclockwise from the positive *x* axis.

Question: Why not just designate the direction of the angular velocity vector to be either clockwise or counterclockwise?

Answer: Clockwise (cw) and counterclockwise (ccw) are directional senses or indications rather than actual directions. These rotational senses are like right and left. If you faced another person and were asked if some object is on the right or

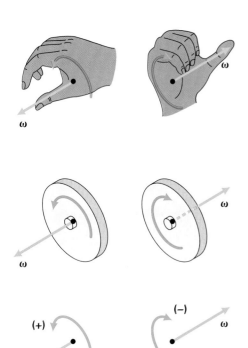

•**FIGURE 7.5 Angular velocity** The direction of the angular velocity vector for an object in rotational motion is given by a right-hand rule: When the fingers of the right hand are curled in the direction of the rotation, the extended thumb points in the direction of the angular velocity vector. Circular senses or directions are commonly indicated by **(a)** plus and **(b)** minus signs.

214 CHAPTER 7 Circular Motion and Gravitation

left, your answers would disagree. Similarly, if you held this book up toward a person facing you and rotated it, would it be rotating cw or ccw?

We can use cw and ccw to indicate rotational "directions" when they are specified relative to a reference, for example, the positive x axis as in the preceding discussion. Referring to Fig. 7.5, imagine yourself being first on one side of one of the rotating disks and then on the other. Which way is the disk rotating from each vantage point, cw or ccw? Then, apply the right-hand rule on both sides. You should find that the angular velocity vector direction is the same for both locations (because it is referenced to the right hand). Relative to this vector—for example, looking at the tip—there is no ambiguity in using + and − to indicate rotational senses or directions.

Relationship between Tangential and Angular Speeds

A particle moving in a circle has an instantaneous velocity tangential to its circular path. For a constant angular velocity and speed, the particle's orbital or **tangential speed** v (the magnitude of the tangential velocity) is also constant. How the angular and tangential speeds are related is revealed by starting with Eq. 7.3 ($s = r\theta$) and Eq. 7.5 ($\theta = \omega t$):

$$s = r\theta = r(\omega t)$$

The arc length, or distance, is also given by

$$s = vt$$

Combining the equations for s gives

$$v = r\omega \quad \text{tangential speed relative to angular speed for circular motion} \quad (7.6)$$

where ω is in radians per second. Equation 7.6 holds in general for instantaneous tangential and angular speeds for solid or rigid body rotation about a fixed axis even when ω might vary with time.

Note that all the particles of an object rotating with constant angular velocity have the same angular speed, but the tangential speeds are different at different distances from the axis of rotation (•Fig. 7.6 and Demonstration 5).

Demonstration 5 ■ Constant Angular Velocity, but Not Tangential Velocity

This demonstration of uniform circular motion shows that tangential velocities are different at different radii. Neon bulbs are placed at different radii on a spinning wheel. The bulbs light up every 1/120 s for the same period of time. As seen in the photo, the bulbs at the greater radii have greater tangential speeds since they have longer lighted path lengths for the same time periods.

(a)

(b)

•FIGURE 7.6 **Tangential and angular speeds** (a) Tangential and angular speeds are related by $v = r\omega$, with ω in radians per second. Note that all of the particles of an object rotating about a fixed axis travel in circles. All the particles have the same angular speed (ω), but particles at different distances from the axis of rotation have different tangential speeds. (b) Sparks from a grinding wheel provide a graphic illustration of instantaneous tangential velocity. (Why do the paths curve slightly?)

•**FIGURE 7.7 Around they go—at the same speeds?** See Example 7.3.

Example 7.3 ■ Merry-Go-Rounds: Do Some Go Faster Than Others?

An amusement park merry-go-round (•Fig. 7.7) at its constant operational speed makes one complete rotation in 45 s. Two children are on horses, one at 3.0 m from the center of the ride and the other farther out, 6.0 m from the center. What are (a) the angular speed and (b) the tangential speed of each child?

Thinking It Through. The angular speed of each child is the same, since both make a complete rotation in the same time. However, the tangential speeds will be different, because the radii are different. That is, the child at the greater radius travels in a larger circle during the rotation time and so must travel faster.

Solution.

Given: $\theta = 2\pi$ rad (one rotation) *Find:* (a) ω_1 and ω_2 (angular speeds)
$t = 45$ s (b) v_1 and v_2 (tangential speeds)
$r_1 = 3.0$ m
$r_2 = 6.0$ m

(a) It should be apparent that $\omega_1 = \omega_2$—that both riders rotate at the same angular speed. All points on the merry-go-round travel through 2π rad in the time it takes to make one rotation. The angular speed can be found from Eq. 7.5 (constant ω), and

$$\omega = \frac{\theta}{t} = \frac{2\pi \text{ rad}}{45 \text{ s}} = 0.14 \text{ rad/s}$$

where the unitless radian is included for clarity. Hence, $\omega = \omega_1 = \omega_2 = 0.14$ rad/s.

(b) The tangential speed is different at different locations on the merry-go-round. All of the "particles" making up the merry-go-round go through one rotation in the same time. Therefore, the farther a point is from the center, the longer its circular path will be, and its tangential speed will be greater, as Eq. 7.6 indicates (see also Fig. 7.6a). Thus,

$$v_1 = r_1\omega = (3.0 \text{ m})(0.14 \text{ rad/s}) = 0.42 \text{ m/s}$$

$$v_2 = r_2\omega = (6.0 \text{ m})(0.14 \text{ rad/s}) = 0.84 \text{ m/s}$$

(Note that the unitless radian has been dropped from the answer.)

Thus a rider on the outer part of the ride has a greater tangential speed than one closer to the center, as you should expect. The outer rider must travel faster than an inner one to make one rotation in the same time.

Follow-up Exercise. On an old 45-rpm record album, the beginning track is 8.0 cm from the center, and the end track is 5.0 cm from the center. (a) What are the angular speeds and the tangential speeds at these distances when the record is spinning at 45 rpm? (b) Why on oval race tracks do inside and outside runners have different starting points, such that some runners start "ahead" of others?

Whenever a tangential or linear quantity is calculated from an angular quantity, the angular unit radians is dropped in the final answer. When an angular quantity is asked for, the unit radians is usually included in the final answer for clarity.

Period and Frequency

Some other quantities commonly used to describe circular motion are period and frequency. The **period** (T) is the time it takes for an object in circular motion to make one complete revolution, or *cycle*. For example, the period of revolution of the Earth about the Sun is 1 year, and the period of the Earth's axial rotation is 24 hours. The standard unit of period is the second (s). Descriptively, the period is sometimes given in seconds per revolution (s/rev) or seconds per cycle (s/cycle).

Closely related to the period is the **frequency** (f), which is the number of revolutions, or cycles, made in a given time, generally a second. For example, if a particle traveling uniformly in a circular orbit makes 5.0 revolutions in 1.0 s, the frequency (of revolution) is $f = 5.0$ rev/1.0 s = 5.0 rev/s, or 5.0 cycles/s (cps). *Revolution* and *cycle* are merely descriptive terms and not part of the frequency unit. The unit of frequency is inverse seconds (1/s, or s^{-1}), which is called the **hertz** (Hz) in the SI.

The hertz (Hz), a unit of frequency, is named for Heinrich Hertz (1857–1894), a German physicist and pioneering investigator of electromagnetic waves, which are also characterized by frequency.

Since the descriptive units for frequency and period are inverses of one another (cycles/second and seconds/cycle), it follows that the two quantities are related by

$$f = \frac{1}{T} \quad \text{frequency and period} \quad (7.7)$$

Note: Frequency (f) and period (T) are inversely related.

SI unit of frequency: hertz (Hz, 1/s or s^{-1})

where the period is in seconds and the frequency is in hertz or inverse seconds.

The frequency can also be related to the angular speed. For uniform circular motion, the orbital speed can be written as $v = 2\pi r/T$: that is, the distance traveled in one revolution divided by the time for one revolution (1 period). Similarly, for the angular case, since a distance of 2π rad is traveled in 1 period (by definition of the period), we have

$$\omega = \frac{2\pi}{T} = 2\pi f \quad \text{angular speed in terms of period and frequency} \quad (7.8)$$

Example 7.4 ■ Frequency and Period: An Inverse Relationship

A compact disc (CD) rotates in a player at a constant speed of 200 rpm. What are the CD's (a) frequency and (b) period of revolution?

Thinking It Through. We can use the relationships among the frequency (f), the period (T), and angular frequency (ω)—Eqs. 7.7 and 7.8.

Solution. The angular speed is not in standard units, and so must be converted. Revolutions per minute (rpm) can be converted to radians per second (rad/s).

Given: $\omega = 200$ rpm $\left(\dfrac{2\pi \text{ rad}}{1 \text{ rev}}\right)\left(\dfrac{1 \text{ min}}{60 \text{ s}}\right)$ **Find:** (a) f (frequency)
$= 20.9$ rad/s   (b) T (period)

(A convenient conversion facator is 1 rev/min $= \pi/30$ rad/s. Can you see how this is obtained?)

(a) Rearranging Eq. 7.8 and solving for f, we get

$$f = \frac{\omega}{2\pi} = \frac{20.9 \text{ rad/s}}{2\pi} = 3.33 \text{ Hz}$$

The units of 2π are rad/cycle or revolution, so the result is in cycles/second or inverse seconds, which is the hertz.

(b) Equation 7.8 could be used to find T, but Eq. 7.7 is a bit simpler:

$$T = \frac{1}{f} = \frac{1}{3.33 \text{ Hz}} = 0.300 \text{ s}$$

Thus, it takes 0.300 s for the CD to make one revolution. (Notice that with Hz $= 1/s$, the equation is dimensionally correct.)

Follow-up Exercise. If the period of a particular CD is 0.500 s, what is its angular speed in revolutions per minute?

7.3 Uniform Circular Motion and Centripetal Acceleration

Objectives: To (a) explain why there is a centripetal acceleration in constant or uniform circular motion, and (b) compute centripetal acceleration.

A simple, but important, type of circular motion is **uniform circular motion**, which occurs when an object moves at a constant speed in a circular path. An example of this movement is a car going around a circular track (•Fig. 7.8). The motion of

Uniform circular motion—constant speed but not constant velocity

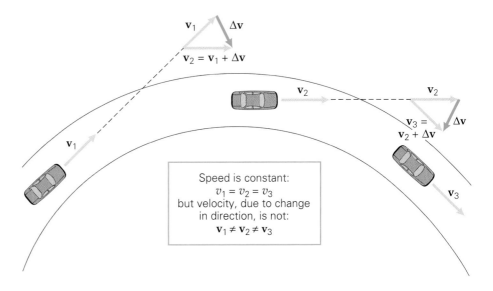

•**FIGURE 7.8 Uniform circular motion** The speed of an object in uniform circular motion is constant, but the object's velocity changes in the direction of motion. Thus, there is an acceleration.

Note: Review the discussion of curvilinear motion in Section 3.1.

Centripetal (center-seeking) acceleration: always directed toward the center of the circle

the Moon around the Earth is approximated by uniform circular motion. Such motion is curvilinear, so you know from the discussion in Chapter 3 that there must be an acceleration. But what is its magnitude and direction?

Centripetal Acceleration

The acceleration of uniform circular motion is not in the same direction as the instantaneous velocity (which is tangent to the circular path at any point). If it were, the object would speed up and the motion wouldn't be uniform. Recall that acceleration is the time rate of change of velocity and that velocity has both magnitude and direction. In uniform circular motion, the direction of the velocity is continuously changing, which is a clue to the direction of the acceleration (see Fig. 7.8).

This directionality is shown more explicitly in •Fig. 7.9a. The velocity vectors at the beginning and end of a time interval give the change in velocity, or $\Delta\mathbf{v}$, via vector addition (or subtraction). The vector triangles for several instantaneous velocities are shown in the figure. All of the instantaneous velocity vectors have the same magnitude or length (constant speed), but they differ in direction. Note that since $\Delta\mathbf{v}$ is not zero, there must be an acceleration ($\mathbf{a} = \Delta\mathbf{v}/\Delta t$).

As illustrated in Fig. 7.9b, as Δt (or $\Delta\theta$) becomes smaller, $\Delta\mathbf{v}$ points more toward the center of the circular path. As Δt approaches zero, the instantaneous change in the velocity, and therefore the acceleration, is exactly toward the center of the circle. As a result, the acceleration in uniform circular motion is called **centripetal acceleration**, which means center-seeking acceleration (from Latin *centri*, "center," and *petere*, "to fall toward" or "to seek").

Without the centripetal acceleration, the motion would not be in a curved path but in a straight line. The centripetal acceleration must be directed radially inward, that is, with no component in the direction of the (tangential) velocity, or else the magnitude of that velocity would change (•Fig. 7.10). Note that for an object in uniform circular motion, the direction of the centripetal acceleration is continuously changing. In terms of x and y components, a_x and a_y are not constant. Can you describe how this differs from projectile motion?

The magnitude of the centripetal acceleration can be deduced from the small shaded triangles in Fig. 7.9. (For very short time intervals, the arc length of Δs is almost a straight line—the chord.) These two triangles are similar because each has a pair of equal sides surrounding the same angle $\Delta\theta$. (Note that the velocity vectors have the same magnitude.) Thus, Δv is to v as Δs is to r, which we can write as

$$\frac{\Delta v}{v} \approx \frac{\Delta s}{r}$$

• **FIGURE 7.9 Analysis of centripetal acceleration**
(a) The velocity vector of an object in uniform circular motion is constantly changing direction. (b) As Δt, the time interval for $\Delta \theta$, is taken smaller and smaller and approaches zero, Δv (the change in the velocity, and therefore an acceleration) is directed toward the center of the circle. The centripetal, or center-seeking, acceleration has a magnitude of $a_c = v^2/r$.

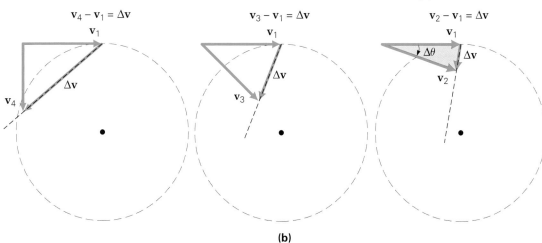

The arc length Δs is the distance traveled in time Δt, thus $\Delta s = v\,\Delta t$:

$$\frac{\Delta v}{v} \approx \frac{\Delta s}{r} = \frac{v\,\Delta t}{r}$$

and

$$\frac{\Delta v}{\Delta t} \approx \frac{v^2}{r}$$

Then, as Δt approaches zero, this approximation becomes exact. The instantaneous centripetal acceleration, $\Delta v/\Delta t = a_c$, thus has a magnitude

$$a_c = \frac{v^2}{r} \qquad \text{magnitude of centripetal acceleration} \qquad (7.9)$$

Using Eq. 7.6 ($v = r\omega$), the centripetal acceleration equation can be written in terms of the angular speed:

$$a_c = \frac{v^2}{r} = \frac{(r\omega)^2}{r} = r\omega^2 \qquad \text{centripetal acceleration in terms of angular speed} \qquad (7.10)$$

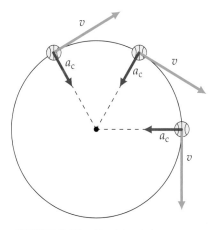

• **FIGURE 7.10 Centripetal acceleration** For an object in uniform circular motion, the centripetal acceleration is directed radially inward. There is no acceleration component in the tangential direction; if there were, the magnitude of the velocity (tangential *speed*) would change.

Example 7.5 ■ A Satellite in Orbit: Centripetal Acceleration

A space station is in a circular orbit about the Earth at an altitude h of 5.0×10^2 km. If the station makes one revolution every 95 min, what are its (a) orbital speed and (b) centripetal acceleration?

Note: Centripetal acceleration depends on tangential speed (v) and radius (r).

7.3 Uniform Circular Motion and Centripetal Acceleration

Thinking It Through. From the given data, we can find the length of the circular path (by calculating the radius); and from the period, we can find the orbital speed. The centripetal acceleration can then be computed by using Eq. 7.9.

Solution.

Given: $h = 5.0 \times 10^2$ km
$ = 5.0 \times 10^5$ m
$t = T = 95$ min $= 5.7 \times 10^3$ s

Find: (a) v (tangential, or orbital, speed)
(b) a_c (centripetal acceleration)

(a) The radius of the circular orbit is not h but $R_E + h$, where R_E is the radius of the Earth, 6.4×10^6 m (see Appendix III):

$$r = R_E + h = (6.4 \times 10^6 \text{ m}) + (5.0 \times 10^5 \text{ m}) = 6.9 \times 10^6 \text{ m}$$

The station travels the circumference of its circular orbit ($2\pi r$) in the given time, so the tangential speed is

$$v = \frac{2\pi r}{T} = \frac{2\pi (6.9 \times 10^6 \text{ m})}{5.7 \times 10^3 \text{ s}} = 7.6 \times 10^3 \text{ m/s}$$

(b) Then the centripetal acceleration is (from Eq. 7.9)

$$a_c = \frac{v^2}{r} = \frac{(7.6 \times 10^3 \text{ m/s})^2}{6.9 \times 10^6 \text{ m}} = 8.4 \text{ m/s}^2$$

directed toward the center of the orbit. Alternatively, we could have computed the angular speed, $\omega = 2\pi/T$, and used Eq. 7.10, $a_c = r\omega^2$.

As you might suspect, this centripetal acceleration is supplied by the Earth's gravitational force—it is the acceleration due to gravity at an altitude of 500 km.

Follow-up Exercise. (a) How does the acceleration found in this Example compare percentagewise with the acceleration due to gravity on Earth's surface? (b) Would an astronaut in the space station be "weightless"?

A practical application of centripetal acceleration is discussed in the Insight on this page and illustrated in the Example that follows.

Insight

The Centrifuge: Separating Blood Components

The centrifuge is a rotating machine used to separate particles of different sizes and densities suspended in a liquid (or a gas). For example, cream is separated from milk by centrifuging, and blood components are separated in centrifuges in medical laboratories (see Fig. 7.11).

Blood components will eventually settle toward the bottom of a tube in layers—a process called sedimentation—under the influence of normal gravity alone. The viscous drag of the plasma on the particles is analogous to (but much greater than) the air resistance that determines the terminal velocity of falling objects (Section 4.6). Red blood cells settle in the bottom layer of the plasma because they reach a greater terminal velocity than do the white blood cells and platelets and so reach the bottom sooner. However, gravitational sedimentation is generally very slow.

Since clinicians cannot afford to wait a long time to see the fractional volume of red cells in the blood, centrifugation is used to speed up the sedimentation process. Tubes are spun horizontally. The resistance of the fluid medium on the particles supplies the centripetal acceleration that keeps them moving in slowly widening circles as they settle toward the bottom of the tube. The bottom of the tube itself must exert a strong force on the contents as a whole and must be strong enough so as not to break.

Laboratory centrifuges commonly operate at speeds sufficient to produce centripetal accelerations thousands of times larger than g (see Example 7.6). Since the principle of the centrifuge involves centripetal acceleration, perhaps "centripuge" would be a more descriptive name.

Example 7.6 ■ A Centrifuge: Centripetal Acceleration

A laboratory centrifuge like that shown in •Fig. 7.11 operates at a rotational speed of 12 000 rpm. (a) What is the magnitude of the centripetal acceleration of a red blood cell at a radial distance of 8.00 cm from the centrifuge's axis of rotation? (b) How does this acceleration compare to g?

Thinking It Through. Here, the angular speed and the radius are given, so the magnitude of the centripetal acceleration can be computed directly from Eq. 7.10. The result can be compared to g by using $g = 9.80$ m/s^2.

Solution. The data are as follows:

Given: $\omega = 1.20 \times 10^4$ rpm $\left(\dfrac{\pi/30 \text{ rad/s}}{\text{rpm}}\right)$ *Find:* (a) a_c
 $= 1.26 \times 10^3$ rad/s (b) a_c compared to g
 $r = 8.00$ cm $= 0.0800$ m

where we convert the angular speed to radians per second.

(a) The centripetal acceleration is found from Eq. 7.10:
$$a_c = r\omega^2 = (0.0800 \text{ m})(1.26 \times 10^3 \text{ rad/s})^2 = 1.27 \times 10^5 \text{ m/s}^2$$

(b) Using the relationship 1 g = 9.80 m/s^2 to express a_c in terms of g, we get
$$a_c = (1.27 \times 10^5 \text{ m/s}^2)\left(\dfrac{1\,g}{9.80 \text{ m/s}^2}\right) = 1.30 \times 10^4\,g\ (= 13\,000\,g!)$$

Follow-up Exercise. What angular speed in revolutions per minute would give an acceleration of 1 g at the radial distance in this Example?

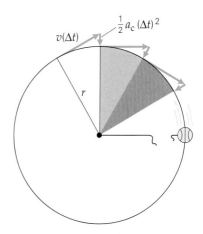

•**FIGURE 7.12 Displacement components, uniform circular motion** In a short time, Δt, an object travels a distance $v\,\Delta t$ tangentially and a distance $\frac{1}{2}a_c(\Delta t)^2$ radially. Thus, the object is essentially "falling" toward the center of the circle (per the short, inward arrows) as it travels tangentially (per the long, outward arrows). As Δt approaches zero, these components describe a circular path. (What happens if the string breaks? See Conceptual Example 7.7.)

The concept of centripetal acceleration may be clearer if you look at the displacement components of uniform circular motion for a short time interval, Δt (•Fig. 7.12). The rectangular components in this case are directed tangentially and radially. An object travels a distance $v\,\Delta t$ tangent to the circle, and at the same time it is displaced a distance $\frac{1}{2}a_c(\Delta t)^2$ radially toward the center of the circle. As Δt approaches zero, a circular path is described. Because of its centripetal acceleration, the Moon in a sense is continuously "falling" toward the Earth. Fortunately, it never gets here because it is traveling tangentially at the same time.

Suppose that an automobile moves into a level, circular curve. To negotiate the curve, the car must have a centripetal acceleration determined by the equation $a_c = v^2/r$. This acceleration is supplied by the force of friction between the tires and the road. Tires act radially (outwardly) on the road, and the reaction (friction) force to this outward force acts inward on the tires.

However, this (static, why?) friction has a maximum limiting value. If the speed of the car is high enough, the friction will not be sufficient to supply the necessary centripetal acceleration, and the car will skid outward from the center of the curve. If the car moves onto a wet or icy spot, the friction between the tires and the road may be reduced, allowing the car to skid at an even lower speed. (Banking a curve also helps vehicles negotiate the curve. See Exercises 42 and 55.)

Conceptual Example 7.7

Breaking Away

A ball attached to a string is swung with uniform motion in a horizontal circle above a person's head (•Fig. 7.13a). If the string breaks, which of the trajectories shown in Fig. 7.13b (viewed from above) would the ball follow?

•**FIGURE 7.11 Centrifuge** Centrifuges are used to separate particles of different sizes and densities suspended in liquids. For example, red and white blood cells can be separated from each other and from the plasma that makes up the liquid portion of the blood.

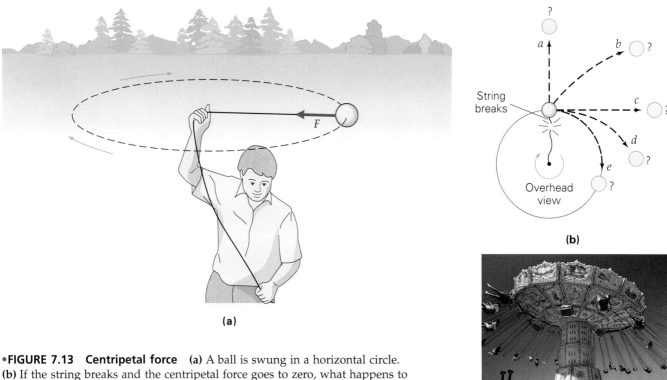

•**FIGURE 7.13 Centripetal force** (a) A ball is swung in a horizontal circle. (b) If the string breaks and the centripetal force goes to zero, what happens to the ball? See Conceptual Example 7.7. (c) What happens to the passengers on this amusement park ride as the angular velocity increases?

Reasoning and Answer. When the string breaks, the centripetal force goes to zero. Newton's first law states that if no force acts on an object in motion, the object will continue to move in a straight line. This rules out b, d, and e.

It should be evident from Figs. 7.6, 7.9, 7.10, and 7.11 that at any instant (including the instant when the string breaks), the isolated ball has a horizontal, tangential velocity. The downward force of gravity acts on it, but this force affects only its vertical motion, which is not visible in Fig. 7.13b. The ball thus flies off tangentially and is essentially a horizontal projectile (with $v_{x_o} = v$, $v_{y_o} = 0$, and $a_y = -g$). Viewed from above, the ball would follow the path labeled c.

Follow-up Exercise. If you swing a ball in a horizontal circle about your head, can the string be exactly horizontal? (See Fig. 7.13a.) Explain your answer. *Hint: Analyze the forces acting on the ball.*

Centripetal Force

To provide an acceleration, there must be a net force. Thus to produce a centripetal (inward) acceleration, we must have a **centripetal force** (net inward force). Expressing the magnitude of this force in terms of Newton's second law ($F = ma$) and inserting the expression for centripetal acceleration from Eq. 7.9, we can write

$$F_c = ma_c = \frac{mv^2}{r} \qquad \begin{array}{l} \textit{magnitude of} \\ \textit{centripetal force} \end{array} \qquad (7.11)$$

The centripetal force, like the centripetal acceleration, is directed toward the center of the circular path.

Keep in mind that in general a force applied at an angle to the direction of motion of an object produces changes in the magnitude *and* direction of the velocity. However, when a force is continuously applied at an angle of 90° to the direction of motion (as is centripetal force), only the direction of the velocity changes. Also notice that because it is always perpendicular to the direction of motion, a centripetal force does no work (why?). Therefore, by the work-energy theorem, a centripetal force does not change the kinetic energy or speed of the object.

As we have seen in Examples 7.6 and 7.7, a centripetal force can be provided by gravitational attraction or by a tension force in a string. In other situations, it could be a static frictional force, as •Fig. 7.14 and Examples 7.8 and 7.9 show.

Example 7.8 ■ Where the Rubber Meets the Road: Friction and Centripetal Force

A car approaches a level, circular curve with a radius of 45.0 m. If the concrete pavement is dry, what is the maximum speed at which the car can negotiate the curve at a constant speed?

Thinking It Through. The car will be in uniform circular motion on the curve, so there must be a centripetal force. This force is supplied by friction, so the maximum frictional force provides the centripetal (net) force when the car is at its maximum tangential speed.

Solution. Writing down what is given and what is to be found, we have

Given: $r = 45.0$ m   *Find:* v (maximum speed)

To go around the curve at a particular speed, the car must have a centripetal acceleration and therefore a centripetal force must act on it. This force is supplied by static friction between the tires and the road. (The tires are not slipping or skidding relative to the road.)

Recall from Chapter 4 that the maximum frictional force is given by $f_{s_{max}} = \mu_s N$, where here N is the magnitude of the normal force on the car and is equal to the weight of the car, mg, on the level road (why?). We may set this equal in magnitude to the expression for centripetal force ($F_c = mv^2/r$) to find the maximum speed. To find $f_{s_{max}}$ we will need the coefficient of friction between rubber and concrete; from Table 4.1, $\mu_s = 1.20$. Then, we can write

$$f_{s_{max}} = F_c$$

$$\mu_s N = \mu_s mg = \frac{mv^2}{r}$$

So

$$v = \sqrt{\mu_s rg} = \sqrt{(1.20)(45.0 \text{ m})(9.80 \text{ m/s}^2)} = 23.0 \text{ m/s}$$

(about 83 km/h, or 52 mi/h).

Follow-up Exercise. Would the centripetal force be the same for all types of vehicles in this Example?

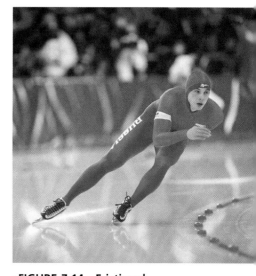

•FIGURE 7.14 **Frictional centripetal force** The centripetal force necessary for the skaters to round the curve is supplied by frictional force. If there were no friction between the skates and the ice, what would happen?

The proper safe speed for driving on a highway curve is an important consideration. The coefficient of friction between tires and road may vary, depending on weather, road conditions, the design of the tires, the amount of tread wear, and so on. When a curved road is designed, safety may be promoted by banking, or inclining, the roadway. This reduces the chances of skidding because the normal force exerted on the car by the road then has a component toward the center of the curve that reduces the need for friction. In fact, for a circular curve with a given banking angle and radius, there is one speed for which no friction is required at all. This condition is used in banking design (see Exercise 55).

Let's look at one more centripetal force example with two objects in uniform circular motion. It will help you understand the motions of satellites in circular orbits in a later section.

Example 7.9 ■ Strung Out: Centripetal Force and Newton's Second Law

Suppose that two masses, $m_1 = 2.5$ kg and $m_2 = 3.5$ kg, respectively, are connected by light strings and are in uniform circular motion on a horizontal frictionless surface as illustrated in •Fig. 7.15, where $r_1 = 1.0$ m and $r_2 = 1.3$ m. The forces acting on the masses

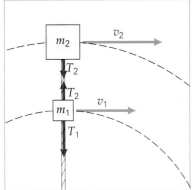

•**FIGURE 7.15 Centripetal force and Newton's second law** See Example 7.9.

are $T_2 = 2.9$ N and $T_1 = 4.5$ N. Find (a) the centripetal accelerations and (b) the magnitudes of the tangential velocities (tangential speeds) of the masses.

Thinking It Through. The centripetal forces on the masses are supplied by the tensions (T_1 and T_2) in the strings. Isolating the masses, the net force on a mass is equal to its centripetal force ($F_c = ma_c$), and a_c for that mass can be found. The tangential speeds can then be found since the radii are known ($a_c = v^2/r$).

Solution.

Given: $r_1 = 1.0$ m and $r_2 = 1.3$ m *Find:* (a) $\mathbf{a}_{c_1}$ and $\mathbf{a}_{c_2}$ (centripetal accelerations)
$m_1 = 2.5$ kg and $m_2 = 3.5$ kg (b) v_1 and v_2 (tangential speeds)
$T_2 = 2.9$ N
$T_1 = 4.5$ N

By isolating m_2 in the figure, you can see that the centripetal force is provided by the tension in the string ($\mathbf{T}_2$ is the only force acting on m_2 toward the center of its circular path). Thus,

$$T_2 = m_2 a_{c_2}$$

and

$$a_{c_2} = \frac{T_2}{m_2} = \frac{2.9 \text{ N}}{3.5 \text{ kg}} = 0.83 \text{ m/s}^2$$

The acceleration is toward the center of the circle. You can find the tangential speed of m_2 from $a_c = v^2/r$:

$$v_2 = \sqrt{a_{c_2} r_2} = \sqrt{(0.83 \text{ m/s}^2)(1.3 \text{ m})} = 1.0 \text{ m/s}$$

The situation is a bit different for m_1. In this case there are two forces acting on m_1: the string tensions $\mathbf{T}_1$ and $-\mathbf{T}_2$. Also, by Newton's second law, in order to have a centripetal acceleration there must be a net force, which is given by the difference in the two tensions. Thus we have

$$F_{\text{net}_1} = +T_1 + (-T_2) = m_1 a_{c_1} = \frac{m_1 v_1^2}{r_1}$$

where the radial direction (toward the center of the circular path) is taken to be positive. Then

$$a_{c_1} = \frac{T_1 - T_2}{m_1} = \frac{4.5 \text{ N} + (-2.9 \text{ N})}{2.5 \text{ kg}} = 0.64 \text{ m/s}^2$$

and

$$v_1 = \sqrt{a_{c_1} r_1} = \sqrt{(0.64 \text{ m/s}^2)(1.0 \text{ m})} = 0.80 \text{ m/s}$$

Follow-up Exercise. Notice in this Example that the centripetal acceleration of m_2 is greater than that of m_1, yet $r_2 > r_1$, and $a_c \propto 1/r$. Is there something wrong here? Explain.

7.4 Angular Acceleration

Objectives: To (a) define angular acceleration and (b) analyze rotational kinematics.

As you might have guessed, another type of acceleration in angular motion is angular acceleration. This is the time rate of change of angular velocity. In the case of circular motion, if there were an angular acceleration, the motion would not be uniform because the speed would be changing. Analogous to the linear case, the magnitude of the **average angular acceleration** ($\bar{\alpha}$) is

$$\bar{\alpha} = \frac{\Delta \omega}{\Delta t}$$

where the bar over the alpha indicates that it is an average. With $t_o = 0$ and if the angular acceleration is constant so that $\bar{\alpha} = \alpha$, we have

$$\alpha = \frac{\omega - \omega_o}{t} \quad \text{(constant angular acceleration)}$$

SI unit of angular acceleration: radians per second squared (rad/s²)

or

$$\omega = \omega_o + \alpha t \quad \text{(constant acceleration only)} \quad (7.12)$$

Angular acceleration—changes in angular velocity.

No boldface symbols are used in Eq. 7.12 because, in general, plus and minus signs will be used to indicate angular directions, as described earlier. As in the case of linear motion, if the angular acceleration increases the angular velocity, both quantities have the same sign, meaning that their vector directions are the same (α is in the same direction as ω as given by the right-hand rule). If the angular acceleration decreases the angular velocity, the two quantities have opposite signs, meaning that their vectors are opposed (α is in the direction opposite to ω as given by the right-hand rule, or is an angular deceleration, so to speak).

Example 7.10 ■ A Rotating CD: Angular Acceleration

A compact disc (CD) accelerates uniformly from rest to its operational speed of 500 rpm in 3.50 s. What is the angular acceleration of the CD (a) during this time? (b) after this time? (c) If the CD comes uniformly to a stop in 4.50 s, what is its angular acceleration?

Thinking It Through. (a) We are given the initial and final angular velocities; the constant (uniform) angular acceleration can thus be calculated (Eq. 7.12) since we know how long the CD accelerates. (b) Keep in mind that the operational speed is constant. (c) Everything is given for Eq. 7.12, but a negative result should be expected. (Why?)

Solution.

Given: $\omega_o = 0$
$\omega = 500 \text{ rpm} \left(\frac{\pi/30 \text{ rad/s}}{1 \text{ rev/min}} \right) = 52.4 \text{ rad/s}$
$t = 3.50 \text{ s}$ (starting up)
$t = 4.50 \text{ s}$ (coming to a stop)

Find: (a) α (during startup)
(b) α (in operation)
(c) α (in coming to a stop)

(a) Using Eq. 7.12, we get

$$\alpha = \frac{\omega - \omega_o}{t} = \frac{52.4 \text{ rad/s} - 0}{3.50 \text{ s}} = 15.0 \text{ rad/s}^2$$

in the direction of the angular velocity.

(b) After the CD reaches its operational speed, the angular velocity remains constant, so $\alpha = 0$.

(c) Again using Eq. 7.12, but this time with $\omega_o = 500$ rpm and $\omega = 0$, we get

$$\alpha = \frac{\omega - \omega_o}{t} = \frac{0 - 52.4 \text{ rad/s}}{4.50 \text{ s}} = -11.6 \text{ rad/s}^2$$

where the minus sign indicates that the angular acceleration is in the direction opposite that of the angular velocity.

Follow-up Exercise. What are the directions of the ω and α vectors in part (a) of this Example if the CD rotates clockwise when viewed from above?

As with arc length and angle ($s = r\theta$) and tangential and angular speeds ($v = r\omega$), there is a relationship between the tangential acceleration and the angular acceleration. The **tangential acceleration** is associated with the tangential speed changes and hence continuously changes direction. The magnitudes of the

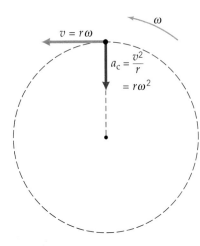

(a) Uniform circular motion
($a_t = r\alpha = 0$)

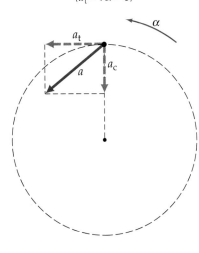

(b) Nonuniform circular motion
($\mathbf{a} = \mathbf{a}_t + \mathbf{a}_c$)

•**FIGURE 7.16 Acceleration and circular motion** (a) In uniform circular motion, there is centripetal acceleration but no angular acceleration ($\alpha = 0$) or tangential acceleration ($a_t = r\alpha = 0$). (b) In nonuniform circular motion, there are angular and tangential accelerations, and the total acceleration is the vector sum of the tangential and centripetal components.

tangential and angular accelerations are related by a factor of r. For circular motion with a constant radius r:

$$a_t = \frac{\Delta v}{\Delta t} = \frac{\Delta(r\omega)}{\Delta t} = \frac{r\Delta\omega}{\Delta t} = r\alpha$$

so

$$a_t = r\alpha \quad \begin{array}{l}\text{magnitude of} \\ \text{tangential acceleration}\end{array} \quad (7.13)$$

The tangential acceleration (a_t) is written with a subscript t to distinguish it from the radial, or *centripetal, acceleration* (a_c), which is necessary for circular motion.

For uniform circular motion, there is no angular acceleration ($\alpha = 0$) or tangential acceleration, as can be seen from Eq. 7.13. There is only centripetal acceleration (•Fig. 7.16a).

However, when there is an angular acceleration α (and therefore a tangential acceleration, $a_t = r\alpha$), there is a change in *both* the angular and the tangential velocities ($v = r\omega$). As a result, the centripetal acceleration, $a_c = v^2/r = r\omega^2$, must increase or decrease if the object is to maintain the same circular orbit (that is, if r is to stay the same). When there are both tangential and centripetal accelerations, the total instantaneous acceleration is their vector sum (Fig. 7.16b). The tangential acceleration vector and the centripetal acceleration vector are perpendicular to each other at any instant, and the total acceleration is $\mathbf{a} = a_t\hat{\mathbf{t}} + a_c\hat{\mathbf{r}}$, where $\hat{\mathbf{t}}$ and $\hat{\mathbf{r}}$ are unit vectors directed tangentially and radially inward, respectively. You should be able to find the magnitude of $\mathbf{a}$ and the angle it makes relative to $\mathbf{a}_t$ by using trigonometry (Fig. 7.16b).

By now, the direct correspondence between the linear and angular kinematic equations should be apparent to you. The other angular equations can be derived, as was done for the linear ones in Chapter 2. That development will not be shown; the set of angular equations with their linear counterparts for constant accelerations are listed in Table 7.2. A quick review of Chapter 2 (with a change of symbols) will show you how the angular equations are derived.

Example 7.11 ■ Even Cooking: Rotational Kinematics

A microwave oven has a rotating plate 30 cm in diameter for even cooking. From rest, the plate accelerates at a uniform rate of 0.87 rad/s² for 0.50 s before reaching its constant operational speed. (a) How many revolutions does the plate make before reaching its operational speed? (b) What are the constant angular speed of the plate and the tangential speed at its rim?

Thinking It Through. This Example involves the use of the angular kinematic equations (Table 7.2). In (a), the angular displacement (θ) will give the number of revolutions. For (b), we want to find ω and then $v = r\omega$.

Solution. Listing what is given and what we are to find, we have

Given: $d = 30$ cm, $r = 15$ cm $= 0.15$ m (radius)   *Find:* (a) θ (in revolutions)
$\omega_o = 0$ (at rest)   (b) ω and v (angular and
$\alpha = 0.87$ rad/s²   tangential speeds)
$t = 0.50$ s

(a) To find the angular distance (ω) in radians, we can use Eq. 4 from Table 7.2:

$$\theta = \omega_o t + \tfrac{1}{2}\alpha t^2 = 0 + \tfrac{1}{2}(0.87 \text{ rad/s}^2)(0.50 \text{ s})^2 = 0.11 \text{ rad}$$

Since 2π rad $= 1$ rev (or 360°),

$$0.11 \text{ rad}\left(\frac{1 \text{ rev}}{2\pi \text{ rad}}\right) = 0.018 \text{ rev}$$

so the plate reaches its operational speed in only a small fraction of a revolution.

(b) From Table 7.2, we see that Eq. 3 gives the angular speed, and

$$\omega = \omega_o + \alpha t = 0 + (0.87 \text{ rad/s}^2)(0.50 \text{ s}) = 0.44 \text{ rad/s}$$

Then, Eq. 7.6 gives the tangential speed at the rim radius:

$$v = r\omega = (0.15 \text{ m})(0.44 \text{ rad/s}) = 0.066 \text{ m/s}$$

Follow-up Exercise. When the oven is turned off, the plate makes half of a revolution before stopping, what is its angular acceleration?

7.5 Newton's Law of Gravitation

Objectives: To (a) describe Newton's law of gravitation and how it relates to the acceleration due to gravity, and (b) apply the general formulation of gravitational potential energy.

Another of Isaac Newton's many accomplishments was the formulation of what is called the **universal law of gravitation**. This law is very powerful and fundamental. Without it, for example, we would not understand the cause of tides or know how to put satellites into particular orbits around the Earth. This law allows us to analyze the motions of planets and comets, stars and galaxies. The word *universal* in the name indicates that we believe it to apply everywhere in the universe. (This term highlights the importance of the law, but for brevity it is common to refer simply to Newton's law of gravitation or the law of gravitation.)

Newton's law of gravitation in mathematical form gives a simple relationship for the gravitational interaction between two particles, or point masses, m_1 and m_2, separated by a distance r (•Fig. 7.17). Basically, every particle in the universe has an attractive interaction with every other particle. The forces of mutual interaction are equal and opposite, forming a force pair as described by Newton's third law (Chapter 4).

The gravitational attraction, or force (F), decreases as the the square of the distance (r^2) between two point masses increases; that is, the magnitude of the gravitational force and the distance separating the two particles are related as follows:

$$F \propto \frac{1}{r^2} \tag{7.14}$$

(This type of relationship is called an *inverse-square law*—F is inversely proportional to r^2.)

Newton's law also correctly postulates that the gravitational force, or attraction of a body, depends on the body's mass—the greater the mass, the greater the attraction. However, because gravity is a mutual interaction between masses, it should be directly proportional to both masses—that is, to their product ($F \propto m_1 m_2$).

Hence, Newton's law of gravitation has the form $F \propto m_1 m_2/r^2$. Expressed as an equation, the magnitude of the mutually attractive gravitational forces between two masses is given by

$$F = \frac{Gm_1 m_2}{r^2} \tag{7.15}$$

Newton's universal law of gravitation

where G is a constant called the **universal gravitational constant**:

$$G = 6.67 \times 10^{-11} \text{ N·m}^2/\text{kg}^2$$

This constant is often referred to as "big G" to distinguish it from "little g," the acceleration due to gravity. Note from Eq. 7.15 that F approaches zero only when r is infinitely large. Thus, the gravitational force has, or acts over, an infinite range.

How did Newton come to his conclusions about the force of gravity? Legend has it that his insight came after he observed an apple fall from a tree to

TABLE 7.2 Equations for Linear and Angular Motion with Constant Acceleration*

Linear	Angular	
$x = \bar{v}t$	$\theta = \bar{\omega}t$	(1)
$\bar{v} = \dfrac{v + v_o}{2}$	$\bar{\omega} = \dfrac{\omega + \omega_o}{2}$	(2)
$v = v_o + at$	$\omega = \omega_o + \alpha t$	(3)
$x = v_o t + \frac{1}{2}at^2$	$\theta = \omega_o t + \frac{1}{2}\alpha t^2$	(4)
$v^2 = v_o^2 + 2ax$	$\omega^2 = \omega_o^2 + 2\alpha\theta$	(5)

*For these equations, $x_o = 0$, $\theta_o = 0$, and $t_o = 0$. The first equation in each column is general, that is, is not limited to situations where the acceleration is constant.

(a) Point masses

(b) Homogeneous spheres

$$F = \frac{Gm_1 m_2}{r^2}$$

•FIGURE 7.17 **Universal law of gravitation** (a) Any two particles, or point masses, are gravitationally attracted to each other with a force that has a magnitude given by Newton's universal law of gravitation. (b) For homogeneous spheres, the masses may be considered to be concentrated at their centers.

Note: The symbol g is reserved for the acceleration due to gravity at the Earth's surface; a_g is more generally the acceleration due to gravity at some greater radial distance.

the ground. Newton had been wondering what supplied the centripetal force to keep the Moon in orbit and might have had this thought: "If gravity attracts an apple toward the Earth, perhaps it also attracts the Moon, and the Moon is falling, or accelerating toward the Earth, under the influence of gravity" (•Fig. 7.18).

Whether or not the legendary apple did the trick, Newton assumed that the Moon and the Earth were attracted to each other and could be treated as point masses, with their total masses concentrated at their centers. The inverse-square relationship had been speculated on by some of his contemporaries. Newton's achievement was demonstrating that the relationship could be deduced from one of Johannes Kepler's laws of planetary motion (Section 7.6).

Newton expressed Eq. 7.15 as a proportion ($F \propto m_1 m_2 / r^2$) because he did not know the value of G. It was not until 1798 (71 years after Newton's death) that the value of the universal gravitational constant was experimentally determined by an English physicist, Henry Cavendish. Cavendish used a sensitive balance to measure the gravitational force between separated spherical masses. If F, r, and the m's are known, big G can be computed from Eq. 7.15.

As mentioned earlier, Newton considered the nearly spherical Earth and Moon to be point masses located at their respective centers. It took him some years, using mathematical methods he developed, to prove that this is the case for spherical, *homogeneous* objects.* The general concept is illustrated in •Fig. 7.19.

Example 7.12 ■ Gravitational Attraction between the Earth and the Moon: A Centripetal Force

Estimate the magnitude of the mutual gravitational force between the Earth and the Moon. (You can assume that the Earth and the Moon are homogeneous spheres.)

Thinking It Through. This is an application of Eq. 7.15 for which the masses and the distance must be looked up. (See the Appendices.)

Solution. No data are given, so they must be available from references.

Given: (from Appendix III)
$M_E = 6.0 \times 10^{24}$ kg
$m_M = 7.4 \times 10^{22}$ kg
$r_{EM} = 3.8 \times 10^8$ m

Find: F (gravitational force)

The average distance from the Earth to the Moon (r_{EM}) is taken to be the distance from the center of one to the center of the other. Using Eq. 7.15, we get

$$F = \frac{Gm_1 m_2}{r^2} = \frac{GM_E m_M}{r_{EM}^2}$$

$$= \frac{(6.67 \times 10^{-11}\,\text{N·m}^2/\text{kg}^2)(6.0 \times 10^{24}\,\text{kg})(7.4 \times 10^{22}\,\text{kg})}{(3.8 \times 10^8\,\text{m})^2}$$

$$= 2.1 \times 10^{20}\,\text{N}$$

This is the magnitude of the centripetal force that keeps the Moon revolving in its orbit around the Earth. It is a very large force, but our Moon is a very massive object, with a correspondingly large inertia to overcome.

Follow-up Exercise. With what acceleration is the Moon accelerating toward Earth?

The acceleration due to gravity at a particular distance from a planet can also be investigated by using Newton's second law of motion and his law of gravitation. The magnitude of the acceleration due to gravity, which we will generally write as a_g, at a distance r from the center of a spherical mass M is found by set-

---

*For a homogeneous sphere, the equivalent point mass is located at the center of mass. However, this is a special case. The center of gravitational force and the center of mass of a configuration of particles or an object do not generally coincide.

•**FIGURE 7.18 Gravitational insight?** **(a)** Newton developed his law of gravitation while studying the orbital motion of the Moon. According to legend, his thinking was spurred when he observed an apple falling from a tree. He supposedly wondered whether the force causing the apple to accelerate toward the ground could extend to the Moon and cause it to fall toward Earth, that is, supply its orbital centripetal acceleration. **(b)** A popular version of the Newton legend—that he was actually hit on the head by a falling apple while sitting under the tree—inspired this cartoon.

ting the force of gravitational attraction due to that spherical mass equal to ma_g, the net force on an object at distance r:

$$ma_g = \frac{GmM}{r^2}$$

Then, the acceleration due to gravity at any distance r from the planet's center is

$$a_g = \frac{GM}{r^2} \qquad (7.16)$$

(a) (b)

•**FIGURE 7.19 Uniform spherical masses** **(a)** Gravity acts between any two particles. The resultant gravitational force exerted on an object outside a homogeneous sphere by two particles at symmetric locations within the sphere is directed toward the center of the sphere. **(b)** Because of the sphere's symmetry and uniform mass distribution, the net effect is as though all the mass of the sphere were concentrated as a particle at its center. For this special case, the gravitational center of force and center of mass coincide, but this is not generally true for other objects. (Only a few of the red force arrows are shown because of space.)

Notice that a_g is proportional to $1/r^2$, so the farther away an object is from the planet, the smaller its acceleration due to gravity and the smaller the attractive force (ma_g) on the object. The force or weight vector is directed toward the center of the planet.

Equation 7.16 can be applied to the Moon or any planet. For example, taking the Earth to be a point mass M_E located at its center and R_E as its radius, we have the acceleration due to gravity (g) at the Earth's surface by setting the distance $r = R_E$:

$$a_{g_E} = g = \frac{GM_E}{R_E^2} \tag{7.17}$$

This equation has several interesting implications. First, it reveals that taking g to be constant everywhere on the surface of the Earth involves assuming that the Earth has a homogeneous mass distribution and that the distance from the center of the Earth to any location on its surface is the same. These two assumptions are not exactly true. Therefore, taking g to be a constant is only an approximation, but one that works pretty well for most situations.

Also, you can see why the acceleration due to gravity is the same for all free-falling objects, that is, independent of the mass of the object. The mass of the object doesn't appear in Eq. 7.17, so all objects in free fall accelerate at the same rate.

Mass of the Earth

Finally, if you're observant, you'll notice that Eq. 7.17 can be used to compute the mass of the Earth. All of the other quantities in the equation are measurable and their values are known, so M_E can readily be calculated. This is what Cavendish did after he determined the value of G experimentally. Then he also found the average density of the Earth. (What do you think this showed? Think of the difference between the Earth's crust and its metallic core. How does the density of a crustal rock compare to Cavendish's average density of the Earth?)

The acceleration due to gravity does vary with altitude. At a distance h above the Earth's surface, we have $r = R_E + h$. The acceleration is then given by

$$a_g = \frac{GM_E}{(R_E + h)^2} \tag{7.18}$$

Problem-Solving Hint

When comparing accelerations due to gravity or gravitational forces, you will often find it convenient to work with ratios. For example, comparing a_g to g (Eqs. 7.16 and 7.17) for the Earth gives

$$\frac{a_g}{g} = \frac{GM_E/r^2}{GM_E/R_E^2} = \frac{R_E^2}{r^2} = \left(\frac{R_E}{r}\right)^2 \quad \text{or} \quad \frac{a_g}{g} = \left(\frac{R_E}{r}\right)^2$$

Note how the constants cancel out. Taking $r = R_E + h$, you can easily compute a_g/g, or the acceleration due to gravity at some altitude h above the Earth compared to g on the Earth's surface (9.80 m/s^2).

Because R_E is very large compared to readily attainable everyday altitudes above the Earth's surface, the acceleration due to gravity does not decrease very rapidly as we ascend. At an altitude of 16 km (10 mi), $a_g/g = 0.99$—a_g is still 99% of the value of g at the Earth's surface. At an altitude of 160 km (100 mi), a_g is 95% of g.

Another aspect of the decrease of g with altitude concerns potential energy. In Chapter 5, you learned that $U = mgh$ for an object at a height h above some zero reference point, since g is essentially constant near the Earth's surface. This potential energy is equal to the work done in raising the object a distance h above the Earth's surface in a *uniform* gravitational field. But what if the change in altitude is so large that g cannot be considered constant while work is done in moving an object, such as a satellite? In this case, the equation $U = mgh$ doesn't apply. In general, it can be shown (using mathematical methods that are beyond the scope of this book) that the **gravitational potential energy** of two point masses separated by a distance r is given by

Gravitational potential energy

$$U = -\frac{Gm_1m_2}{r} \quad (7.19)$$

The minus sign in Eq. 7.19 arises from the choice of the zero reference point (the point where $U = 0$), which is $r = \infty$.

In terms of the Earth and a mass m at an altitude h above the Earth's surface,

$$U = -\frac{Gm_1m_2}{r} = -\frac{GmM_E}{R_E + h} \quad (7.20)$$

where r is the distance separating the Earth's center and the mass. What this means is that on Earth we are in a negative gravitational potential energy well (•Fig. 7.20) that extends to infinity because the force of gravity has an infinite range. As h increases, so does U. That is, U becomes *less negative*, or gets closer to zero, corresponding to a higher position in the potential well. The same is true for the finite well approximation given by $U = mgh$, but over long distances, for which Eq. 7.20 applies, the change in U is not linear; it varies as $1/r$.

Thus, when gravity does negative work (an object moves higher in the well) or gravity does positive work (an object falls lower in the well), there is a *change* in potential energy. As with finite potential energy wells, this change in energy is usually what's important in analyzing situations.

Note: Review Fig. 5.14, which illustrates negative potential energy.

•**FIGURE 7.20 Gravitational potential energy well** On Earth, we are in a negative gravitational potential energy well. As with an actual well or hole in the ground, work must be done against gravity to get higher in the well. The potential energy of an object increases as it moves higher in the well. This means that the value of U becomes less negative. The top of the Earth's gravitational well is at infinity, where the gravitational potential energy is, by choice, zero.

Example 7.13 ■ Different Orbits: Change in Gravitational Potential Energy

Two 50-kilogram satellites are in circular orbits about the Earth at altitudes of 1000 km (about 620 mi) and 36 000 km (about 22 300 mi), respectively. The lower one monitors particles about to enter the atmosphere, and the higher geosynchronous one (synchronous with the Earth's rotation) takes weather pictures from its stationary position with respect to the Earth's surface. What is the difference in the gravitational potential energies of the two satellites in their respective orbits?

Thinking It Through. The potential energies of the satellites are given by Eq. 7.19, where the greater the radial distance (r), the smaller or less negative is U. Thus the satellite with the greater r is higher in the gravitational potential energy well and has more potential energy.

Note: For communications, many satellites are launched into a circular orbit above the equator at an altitude of about 36 000 km. Satellites there are *synchronous* to Earth's rotation. That is, they remain "fixed" over one point on the equator; to an observer on Earth, they are always seen in the same position in the sky.

Solution. Listing the data so that we can better see what's given, we have

Given: $m = 50$ kg
$h_1 = 1000$ km $= 1.0 \times 10^6$ m
$h_2 = 36\,000$ km $= 36 \times 10^6$ m
$M_E = 6.0 \times 10^{24}$ kg (from table inside back cover)
$R_E = 6.4 \times 10^6$ m

Find: ΔU (potential energy difference)

The difference in the gravitational potential energy can be computed directly from Eq. 7.20. Keep in mind that the potential energy is the energy of position, so we compute the potential energies for each position or altitude and subtract. Thus we have

$$\Delta U = U_2 - U_1 = -\frac{GmM_E}{R_E + h_2} - \left(-\frac{GmM_E}{R_E + h_1}\right) = GmM_E\left(\frac{1}{R_E + h_1} - \frac{1}{R_E + h_2}\right)$$

$$= (6.67 \times 10^{-11} \text{ N·m}^2/\text{kg}^2)(50 \text{ kg})(6.0 \times 10^{24} \text{ kg})$$

$$\times \left[\frac{1}{(6.4 \times 10^6 \text{ m}) + (1.0 \times 10^6 \text{ m})} - \frac{1}{(6.4 \times 10^6 \text{ m}) + (36 \times 10^6 \text{ m})}\right]$$

$$= +2.2 \times 10^9 \text{ J}$$

Because ΔU is positive, m_2 is higher in the gravitational potential energy well than m_1. Note that even though both U_1 and U_2 are negative, U_2 is "more positive" or "less negative" and closer to zero.

Follow-up Exercise. (a) Suppose that the altitude of the higher satellite in this Example were doubled, to 72 000 km. Would the difference in the gravitational potential energies of the two satellites then be twice as great? Justify your answer. (b) Note that U_2 has a *smaller* negative value than U_1 (since $h_2 > h_1$). What does this mean?

Using the gravitational potential energy (Eq. 7.19) gives the equation for the total mechanical energy a different form than it had in Chapter 5. For example, the total mechanical energy of a mass m_1 moving near a stationary mass m_2 is

$$E = K + U = \tfrac{1}{2}m_1v^2 - \frac{Gm_1m_2}{r} \tag{7.21}$$

(Note that the potential energy $U = -Gm_1m_2/r$ is negative in this case and *not* written as *mgh*.) This equation and the principle of the conservation of energy can be applied to the Earth moving about the Sun, by neglecting other gravitational forces. The Earth's orbit is not quite circular but slightly elliptical. At *perihelion* (the point of the Earth's closest approach to the Sun), the mutual gravitational potential energy is less (a larger negative number) than it is at *aphelion* (the point farthest from the Sun). Therefore, as can be seen from Eq. 7.21 in the form $\tfrac{1}{2}m_1v^2 = E + Gm_1m_2/r$, where E is constant, the Earth's kinetic energy and orbital speed are greatest at perihelion (the smallest value of r) and least at aphelion (the greatest value of r). Or, in general, the Earth's orbital speed is greater when it is nearer the Sun than when it is farther away.

Mutual gravitational potential energy also applies to a group, or *configuration*, of more than two masses. That is, there is gravitational potential energy due to the masses being in a configuration. This is because work was done in bringing them together. Suppose that there is a single fixed mass m_1, and another mass m_2 is brought close to m_1 from an infinite distance (where $U = 0$). The work done against the attractive force of gravity is negative (why?) and equal to the change in the mutual potential energy of the masses, which are now separated by a distance r_{12}; that is, $U_{12} = -Gm_1m_2/r_{12}$.

If a third mass m_3 is brought close to the other two fixed masses, there are then two forces of gravity acting on m_3, so $U_{13} = -Gm_1m_3/r_{13}$ and $U_{23} = -Gm_2m_3/r_{23}$. The total gravitational potential energy of the configuration is therefore

$$U = U_{12} + U_{13} + U_{23}$$
$$= -\frac{Gm_1m_2}{r_{12}} - \frac{Gm_1m_3}{r_{13}} - \frac{Gm_2m_3}{r_{23}} \quad (7.22)$$

A fourth mass could be brought in, but this development should be sufficient to suggest that the total gravitational potential energy of a configuration of particles is equal to the sum of the individual potential energies for all pairs of particles.

Example 7.14 ■ Total Gravitational Potential Energy: Energy of Configuration

Three masses are in a configuration as shown in •Fig. 7.21. What is their total gravitational potential energy?

Thinking It Through. Equation 7.22 applies, but be sure to keep your masses and their distances distinct.

Solution. From the figure we have,

Given: $m_1 = 1.0$ kg   Find: U (total gravitational potential)
$m_2 = 2.0$ kg
$m_3 = 2.0$ kg
$r_{12} = 3.0$ m
$r_{13} = 4.0$ m
$r_{23} = 5.0$ m (3-4-5-right triangle)

We can use Eq. 7.22 directly since only three masses are used in this example. (Note that Eq. 7.22 can be extended to four, five, ... or any number of masses.)

$$U = U_{12} + U_{13} + U_{23}$$
$$= -\frac{Gm_1m_2}{r_{12}} - \frac{Gm_1m_3}{r_{13}} - \frac{Gm_2m_3}{r_{23}}$$
$$= (6.67 \times 10^{-11} \text{ N·m}^2/\text{kg}^2)$$
$$\times \left[-\frac{(1.0 \text{ kg})(2.0 \text{ kg})}{3.0 \text{ m}} - \frac{(1.0 \text{ kg})(2.0 \text{ kg})}{4.0 \text{ m}} - \frac{(2.0 \text{ kg})(2.0 \text{ kg})}{5.0 \text{ m}} \right]$$
$$= -1.3 \times 10^{-10} \text{ J}$$

Follow-up Exercise. Explain what the *negative* potential energy in this Example means physically.

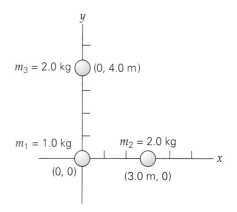

•**FIGURE 7.21** Total gravitational potential energy See Example 7.14.

The effects of gravity are familiar to us, perhaps indirectly. When we lift an object, we may think of it as being heavy, but we are working against gravity. Gravity causes rocks to tumble down and causes mudslides. But we often put gravity to use. For example, fluids from bottles for intravenous injections flow

Insight

Space Exploration: Gravity Assists

As you read this, the *Cassini* spacecraft is on its way to Saturn via Jupiter, after having made two Venus flybys and one Earth flyby (Fig. 1).* What is going on here? Why was the spacecraft launched toward Venus, an inner planet in order to go to Saturn, an outer planet?

Although space probes can be launched from Earth with the current rocket technology, there are limitations—namely, fuel versus payload: The more fuel, the smaller the payload. Using rockets alone, planetary spacecraft are realistically limited to visiting Venus, Mars, and Jupiter. The other planets could not be reached by a spacecraft of reasonable size without taking decades to get there.

So, how will *Cassini* get to Saturn in 2004, almost 7 years after its 1997 launch? By using gravity in a clever scheme called *gravity assist*. Using gravity assists, missions to all of the planets in our solar system are possible. Rocket energy is needed to get a spacecraft to the first planet, and after that the energy is more or less "free." Basically, during a planetary swingby there is an exchange of energy between the planet and the spacecraft, which enables the spacecraft to increase its velocity relative to the Sun. (This is sometimes called a slingshot effect.)

Gravity assists, or swingbys, are a key factor that has made exploration of the solar system possible. The first swingby was in 1973 for the *Mariner 10* mission, which flew by Venus on its way to Mercury. Since then, there have been many planetary missions with gravity assists—for example, *Voyagers* to the outer planets and *Galileo* to Jupiter. Perhaps most spectacular of all was the *Voyager 2* spacecraft, which made three planetary swingbys (of Jupiter, Saturn, and Uranus) on the way to its encounter with Neptune in 1989. After reaching Neptune, there was another swingby in order to gain momentum and trajectory that sent the spacecraft out of the solar system—and it is still going (Newton's first law).

Let's take a brief look at the physics of this ingenious use of gravity. Imagine the *Cassini* spacecraft making a swingby of Jupiter as illustrated in Fig. 2, where the momenta vectors of the motion are shown. Recall from Chapter 6 that a collision is an interaction of objects in which there is an exchange of momentum and energy. Technically, in a swingby, a spacecraft is having a "collision" with the planet.

As shown in Fig. 2, when the spacecraft approaches from "behind" the planet and leaves in "front" (relative to the planet's motional direction), the gravitational interaction gives rise to a change in momentum—that is, $|\mathbf{p}_2| > |\mathbf{p}_1|$ and the direction is different. Looking at the vector triangle, we see there is a $\Delta \mathbf{p}$ in the general "forward" direction of the spacecraft. Since $\Delta \mathbf{p} \propto \mathbf{F}$, there is a force acting on the craft that gives it a "kick" of energy in that direction. So, there is positive work done ($\Delta W > 0$) and a change in ki-

*Cassini was the French-Italian astronomer who studied Saturn, discovering four of its moons and that Saturn's rings are split into two parts by a narrow gap, now called the Cassini Division. The *Cassini* spacecraft will release a "Huygens" probe to Saturn's moon Titan. This moon was discovered by the Dutch scientist Christian Huygens.

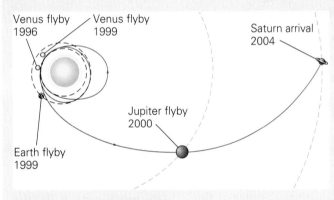

FIGURE 1 Cassini trajectory

because of gravity. An extraterrestrial application of putting gravity to work is given in the Insight on this page.

7.6 Kepler's Laws and Earth Satellites

Objectives: To (a) state and explain Kepler's laws of planetary motion, and
(b) describe the orbits and motions of satellites.

The force of gravity determines the motions of the planets and satellites and holds the solar system (and galaxy) together. A general description of planetary motion had been set forth shortly before Newton's time by the German astronomer and mathematician Johannes Kepler (1571–1630). Kepler was able to formulate three empirical laws from observational data gathered during a 20-year period by the Danish astronomer Tycho Brahe (1546–1601). Brahe's extensive observations of stars and planets were done without the benefit of a telescope, which had not yet been invented. But the observations Brahe made were more accurate than those of most of his contemporaries because he had developed better instrumentation. As a result, he is considered to have been one of the greatest practical astronomers.

Insight

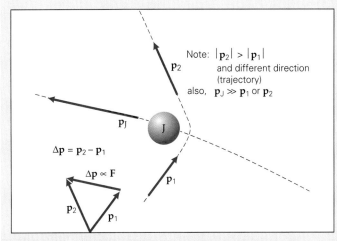

FIGURE 2 Spacecraft on a planetary swingby The gravitational interaction of Jupiter (J) gives the spacecraft a change in momentum (greater magnitude and different direction), as the vector diagram shows. With $\Delta \mathbf{p} \propto \mathbf{F}$, positive work is done on the craft: It has more energy and a greater speed on leaving the region than on entering. By the conservation of momentum, the planet also has a change in momentum, but the motional effects on it are negligible because of its large mass.

netic energy ($\Delta K > 0$, from the work–energy theorem), and the spacecraft leaves with more energy, a greater speed, and a new direction. (If the flyby occurred in the opposite direction, the spacecraft would slow down.)

Momentum and energy are conserved in this elastic collision, and the planet gets an equal and opposite change in momentum, giving a retarding effect. But because the planet's mass is so much larger than the spacecraft, the effect is negligible.

To help you grasp the idea of a gravity assist, consider the analogous roller derby "slingshot maneuver" illustrated in Fig. 3. The skaters interact, and S comes out of the "flyby" with increased speed. Here, the change in momentum on the "slinger," skater J, would probably be noticeable, but that would not be so on Jupiter or any other planet.

FIGURE 3 Skating swingby As an analogy to a planetary swingby, consider a roller derby "slingshot maneuver." Skater J slings skater S, who comes out of the "flyby" with greater speed than she had before. In this case, the change in momentum on skater J, the slinger, would probably be noticeable. (Why?)

Kepler went to Prague to assist Brahe, who was the official mathematician at the court of the Holy Roman Emperor. Brahe died the next year and Kepler succeeded him, inheriting his records of the positions of the planets. Analyzing these data, Kepler announced the first two of his three laws in 1609 (the year Galileo built his first telescope). These laws were applied initially only to Mars. Kepler's third law came 10 y later.

Interestingly enough, Kepler's laws of planetary motion, which took him about 15 y to deduce from observed data, can now be derived theoretically with a page or two of calculations. These three laws apply not only to planets, but to any system composed of a body revolving about a more massive body, to which the inverse-square law of gravitation applies (such as the Moon, artificial Earth satellites, and solar-bound comets).

Kepler's first law (the law of orbits) says that

> Planets move in elliptical orbits with the Sun at one of the focal points.

Kepler's first law: the law of orbits

An ellipse, shown in •Fig. 7.22a, has, in general, an oval shape, resembling a flattened circle. In fact, a circle is a special case of an ellipse in which the focal points,

•**FIGURE 7.22 Kepler's first and second laws of planetary motion** **(a)** In general, an ellipse has an oval shape. The sum of the distances from the focal points F to any point on the ellipse is constant: $r_1 + r_2 = 2a$, where $2a$ is the length of the line joining the two points on the ellipse at the greatest distance from its center, called the major axis. (The line joining the two points closest to the center is $2b$, the minor axis.) Planets revolve about the Sun in elliptical orbits for which the Sun is at one of the focal points and nothing is at the other. **(b)** A line joining the Sun and a planet sweeps out equal areas in equal times. Since $A_1 = A_2$, a planet travels faster along s_1 than along s_2.

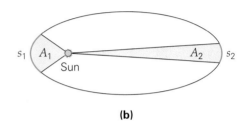

(a)   (b)

or *foci* (plural of focus), are at the same point (the center of the circle). Although the orbits of the planets are elliptical, most do not deviate very much from circles (Mercury and Pluto are notable exceptions; see Appendix III). For example, the difference between the perihelion and aphelion of the Earth (its closest and farthest distances from the Sun, respectively) is about 5 million km. This may sound like a lot, but it is only a little over 3 percent of 150 million km, which is the average distance between the Earth and the Sun.

Kepler's second law (the law of areas) says that

Kepler's second law: the law of areas

A line from the Sun to a planet sweeps out equal areas in equal lengths of time.

This law is illustrated in Fig. 7.22b. Since the time to travel the different orbital distances (s_1 and s_2) is the same, this law tells you that the orbital speed of a planet varies in different parts of its orbit. Because a planet's orbit is elliptical, its orbital speed is greater when it is closer to the Sun than when it is farther away. We used the conservation of energy in Section 7.5 (Eq. 7.21) to deduce this relationship for the Earth.

Kepler's third law (the law of periods) says that

Kepler's third law: the law of periods

The square of the orbital period of a planet is directly proportional to the cube of the average distance of the planet from the Sun; that is, $T^2 \propto r^3$.

Kepler's third law is easily derived for the special case of a planet with a circular orbit, using Newton's law of gravitation. Since the centripetal force is supplied by the force of gravity, the expressions for these forces can be set equal:

$$\underset{\text{centripetal force}}{\frac{m_p v^2}{r}} = \underset{\text{gravitational force}}{\frac{G m_p M_S}{r^2}}$$

and

$$v = \sqrt{\frac{GM_S}{r}}$$

where m_p and M_S are the masses of the planet and the Sun, respectively, and v is the planet's orbital speed. But, $v = 2\pi r/T$ (circumference/period), so

$$\frac{2\pi r}{T} = \sqrt{\frac{GM_S}{r}}$$

Squaring both sides and solving for T^2 gives

$$T^2 = \left(\frac{4\pi^2}{GM_S}\right)r^3$$

or

$$T^2 = Kr^3 \tag{7.23}$$ The law of periods

The constant K for solar system planetary orbits is easily evaluated from orbital data (T and r) for the Earth: $K = 2.97 \times 10^{-19}$ s²/m³. (As an exercise, you might wish to convert K to the more useful units of y²/km³.) Notice that Eq. 7.23 can be used to find the value of K, and if G is known, then the mass of the Sun can be determined. (Note: This value of K does *not* apply to Earth satellites. Why?)

Earth Satellites

We are only about a half a century into the space age. Since the 1950s, numerous unmanned satellites have been put into orbit about the Earth, and now astronauts regularly spend weeks or months in orbiting space laboratories.

Putting a spacecraft into orbit about the Earth (or any planet) is an extremely complex task. However, you can get a basic understanding of the problem from fundamental principles. First, suppose that a projectile could be given the initial speed required to take it just to the top of the Earth's potential energy well. At the exact top of the well, which is an infinite distance away ($r = \infty$), the potential energy is zero. By the conservation of energy and Eq. 7.19,

$$\overset{\text{inital}}{K_o + U_o} = \overset{\text{final}}{K + U}$$

$$\tfrac{1}{2}mv_{esc}^2 - \frac{GmM_E}{R_E} = 0 + 0$$

where v_{esc} is the **escape speed**: the initial speed needed to escape from the surface of a planet or moon. The final energy is zero since the projectile stops at the top of the well (at very large distances, it is barely moving), and $U = 0$ there. Solving for v_{esc} gives

Definition of: Escape speed

$$\tfrac{1}{2}mv_{esc}^2 = \frac{GmM_E}{R_E}$$

and

$$v_{esc} = \sqrt{\frac{2GM_E}{R_E}} \tag{7.24}$$

Since $g = GM_E/R_E^2$ (Eq. 7.17), it is convenient to write

$$v_{esc} = \sqrt{2gR_E} \tag{7.25}$$

Although derived for Earth, this equation may be used generally to find the escape speeds for other planets and our Moon.

Example 7.15 ■ No Return: Escape Speed

What is the escape speed at the Earth's surface?

Thinking It Through. This is a direct application of Eq. 7.25, with the Earth's radius found inside the back cover (or Appendix III).

Solution. Equation 7.25 can be used with known data:

$$v_{esc} = \sqrt{2gR_E} = \sqrt{2(9.80 \text{ m/s}^2)(6.4 \times 10^6 \text{ m})} = 11 \times 10^3 \text{ m/s} = 11 \text{ km/s} \text{ (about 7 mi/s)}$$

Thus, if a projectile or spacecraft could be given an initial upward speed of 11 km/s (about 40 000 km/h, or 25 000 mi/h), it would leave the Earth and not return. Note from the equation that the escape speed is independent of the mass of the spacecraft; this is the initial speed required to send *any* object to the top of the Earth's potential energy well.

Such an initial speed is not attainable instantly at launch but can be acheived in multi-stage steps. For example, in the 1980s probes (*Voyagers*) to the outer planets left Earth never to return and are currently beyond the limits of the solar system.

Follow-up Exercise. How many times greater is the escape speed of the Earth than that of the Moon?

A tangential speed less than the escape speed is required for a satellite to orbit. Consider the centripetal force of a satellite in circular orbit about the Earth. Since the centripetal force on the satellite is supplied by the gravitational attraction between the satelite and the Earth, we may again write

$$F = \frac{mv^2}{r} = \frac{GmM_E}{r^2}$$

Then

$$v = \sqrt{\frac{GM_E}{r}} \tag{7.26}$$

where $r = R_E + h$. For example, suppose that a satellite is in a circular orbit at an altitude of 500 km (about 300 mi); its tangential speed is

$$v = \sqrt{\frac{GM_E}{r}} = \sqrt{\frac{GM_E}{R_E + h}} = \sqrt{\frac{(6.67 \times 10^{-11} \text{ N·m}^2/\text{kg}^2)(6.0 \times 10^{24} \text{ kg})}{(6.4 \times 10^6 \text{ m}) + (5.0 \times 10^5 \text{ m})}}$$

$$= 7.6 \times 10^3 \text{ m/s} = 7.6 \text{ km/s} \text{ (about 4.7 mi/s)}$$

This is about 27 000 km/h, or 17 000 mi/h. As can be seen from Eq. 7.26, the required circular orbital speed *decreases* with altitude.

In practice, a satellite is given a tangential speed by a component of the thrust from a rocket stage (•Fig. 7.23a). The inverse-square relationship of Newton's law of gravitation means that the satellite orbits that are possible about a large mass are ellipses, of which a circular orbit is a special case. This is illustrated in Fig. 7.23b for Earth, using the previously calculated values. If a satellite is not given a sufficient tangential speed, it will fall back to Earth (and possibly be burned up while falling through the atmosphere). If the tangential speed reaches the escape speed, the satellite will leave its orbit and go off into space.

Finally, the total energy of an orbiting satellite in circular orbit is

$$E = K + U = \tfrac{1}{2}mv^2 - \frac{GmM_E}{r} \tag{7.27}$$

Substituting the expression for v from Eq. 7.26 into the kinetic energy term gives

$$E = \frac{GmM_E}{2r} - \frac{GmM_E}{r}$$

Thus,

$$E = -\frac{GmM_E}{2r} \quad \text{total energy of an orbiting satellite} \tag{7.28}$$

Note that the total energy of the satellite is negative: More work is required to put a satellite into a higher orbit, where it has more potential and total energy. The total energy E increases as its *numerical value* becomes smaller—less negative—in go-

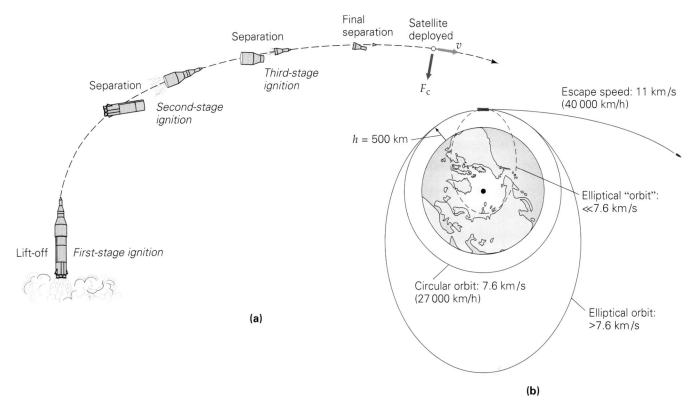

•**FIGURE 7.23 Satellite orbits** (a) A satellite is put into orbit by giving it a tangential speed sufficient for maintaining an orbit at a particular altitude. The higher the orbit, the smaller the tangential speed. (b) At an altitude of 500 km, a tangential speed of 7.6 km/s is required for a circular orbit. With a tangential speed between 7.6 km/s and 11 km/s (the escape speed), the satellite would move out of the circular orbit. Since it would not have the escape speed, it would "fall" around the Earth in an elliptical orbit with the Earth's center at one focal point. A tangential speed less than 7.6 km/s would also give an elliptical path about the center of the Earth; but because the Earth is not a point mass, a certain minimum speed is needed to keep the satellite from striking the Earth's surface.

ing to a higher orbit toward the zero potential at the top of the well. That is, the farther a satellite is from Earth, the greater its total energy.

To help understand why the total energy increases when its value becomes less negative, think of a change in energy from, say, 5.0 J to 10 J. This would be considered an increase in energy. Similarly, a change from -10 J to -5.0 J would be an increase in energy, even though the *absolute* value has decreased:

$$\Delta U = U - U_\text{o} = -5.0\,\text{J} - (-10\,\text{J}) = +5.0\,\text{J}$$

The relationship of speed and energy to orbital radius is summarized in Table 7.3.

Also note from the development of Eq. 7.28 that the kinetic energy of an orbiting satellite is equal to the absolute value of its total energy:

$$K = \frac{GmM_\text{E}}{2r} = |E| \qquad (7.29)$$

TABLE 7.3 Relationship of Radius, Speed, and Energy for Circular Orbital Motion

	Increasing r (larger orbit)	Decreasing r (smaller orbit)
ω	decreases	increases
v	decreases	increases
K	decreases	increases
U	increases (smaller negative value)	decreases (larger negative value)
$E\ (= K + U)$	increases (smaller negative value)	decreases (larger negative value)

Conceptual Example 7.16

Applying the Brakes to Speed Up: Conservation of Energy

A spacecraft is in a circular orbit about the Earth, well outside the atmosphere. Describe what happens to the spacecraft if its retrorockets are fired. (Retrorockets point in the direction of the spacecraft's motion and produce a reverse thrust.)

Reasoning and Answer. The retrorockets apply a braking force, or reverse thrust, to the spacecraft. *Negative* work is done (because the force and displacement are in opposite directions), and the spacecraft loses some energy (the work–energy theorem). That is, the total energy of the spacecraft,

$$E = -\frac{GmM_E}{2r}$$

is reduced. If E decreases (that is, takes on a greater negative value), we can see from the above equation that r must also decrease. Thus, the spacecraft will spiral inward toward the Earth. From Eq. 7.29,

$$K = \tfrac{1}{2}mv^2 = \frac{GmM_E}{2r} \quad \text{or} \quad v^2 = \frac{GM_E}{r}$$

So, if r decreases, v increases, and the spacecraft actually speeds up—even though both U and E decrease, K increases!

Follow-up Exercise. Suppose the retrorockets of an orbiting spacecraft were fired until the craft was within the top layers of the Earth's atmosphere, where air resistance is appreciable. If the rockets were then shut off, what would happen to the spacecraft?

Note: The work–energy theorem is discussed in Section 5.3.

Reverse thrust, provided by the engines of docked cargo ships, is being used to put the Russian space station *Mir* into lower orbits and will ultimately lead to its final destruction. A final thrust will send the station into a decaying orbit and into our atmosphere. Most of the 120-ton *Mir* will burn up in the atmosphere; however, some large pieces are expected to fall into the Pacific Ocean.

Conceptual Example 7.17

Forward Thrust: A Change in Energy

The engines of a spacecraft in circular orbit about the Earth are fired so as to give the craft a *forward* thrust. How will the spacecraft's energy be changed after the firing, assuming another circular orbit is achieved? (a) K increased and U decreased, (b) K decreased and U increased, (c) both K and U increased, (d) the total energy decreased.

Reasoning and Answer. You might think that a forward thrust would cause the spacecraft to speed up and increase its kinetic energy. However, an increase in v would require an increased centripetal force to maintain the spacecraft in orbit. This force is supplied by gravitational attraction, and so could increase only if r were reduced (that is, if the spacecraft moved to a smaller orbit)—something that would not result from the application of a forward thrust. (Draw a diagram to confirm this; in what direction would the thrust act with respect to the circular orbit?)

A forward thrust (in contrast to a reverse thrust in Example 7.16) will send the craft into a larger rather than a smaller orbit, because the forward thrust does positive work on the spacecraft, causing it to rise higher in its potential energy well and increasing its potential energy. With an increase in r, the kinetic energy *decreases* (Eq. 7.29) while the total energy increases by becoming less negative (Eq. 7.28). Hence, the answer is (b). (In actuality, if the engine "burn" is brief, the spacecraft would go into an elliptical orbit with its lowest altitude equal to the radius of the initial circular orbit.)

Follow-up Exercise. Imagine yourself in a spacecraft in a circular orbit, well behind a space station in the same radial orbit. Wishing to dock with the station, what maneuvers would you execute to catch up with it?

The advent of the space age and orbiting satellites has brought us the terms *weightlessness* and *zero gravity*, because astronauts appear to "float" about in orbiting spacecraft (•Fig. 7.24a). However, the terms are misnomers. As mentioned earlier in the chapter, gravity is an infinite-range force, and Earth's gravity acts on a spacecraft and astronauts, supplying the centripetal force necessary to keep them in orbit. Gravity there is not zero, so there must be weight.

A better term to describe the floating effect of astronauts in orbiting spacecraft would be *apparent weightlessness*. They "float" because both the astronauts and the spacecraft are centripetally accelerating (or falling) toward the Earth at the same rate. To help you understand this effect, consider an analogous situation of a person standing on a scale in an elevator (Fig. 7.24b). The "weight"

•**FIGURE 7.24 Apparent weightlessness** (a) An astronaut "floats" in a spacecraft, seemingly in a weightless condition. (b) In a stationary elevator (top), a scale reads true weight. The weight reading is the reaction force N of the scale on the person. If the elevator is descending with an acceleration $a < g$ (middle), the reaction force and apparent weight are less than the true weight. If the elevator were in free fall, $a = g$, the reaction force and indicated weight would be zero since the scale would be falling as fast as the person. (c), (d) "Weightlessness" on Earth.

7.6 Kepler's Laws and Earth Satellites **241**

Insight

Space Colonies and Artificial Gravity

Space colonies, long a staple of science fiction stories and films, are now in the planning stage. However, humans cannot live for long periods of time in a "zero-gravity" environment (see Exercise 71) without detrimental physical effects. It has been suggested that this problem could be avoided if the colony were housed in a huge rotating wheel, which would supply "artificial gravity" (Fig. 1).

As you know, centripetal force is necessary to keep an object in rotational circular motion. On the rotating Earth, that force is supplied by gravity, and we refer to it as weight. Because of our weight, we exert a force on the ground, and the normal force (by Newton's third law) upward on our feet gives us the feeling of "having our feet on solid ground." In a rotating space colony, the situation is somewhat reversed. The rotating colony would supply the centripetal force on the inhabitants, which would be perceived as a weight sensation on the soles of our feet, or artificial gravity (Fig. 2a). Rotation at the proper speed would produce a simulation of normal gravity ($g = 9.80$ m/s^2) within the colony wheel. Note that in the colonists' world, "down" would be outward, toward the periphery of the space station, and "up" would always be inward, toward the axis of rotation.

The inhabitants would experience normal gravitational effects. For example, in Fig. 2a, suppose that the person in

FIGURE 1 Space colony and artificial gravity It has been suggested that a space colony could be housed in a huge, rotating wheel as in this artist's conception. The rotation would supply the "artificial gravity" for the colonists.

measurement that the scale registers is actually the normal force N of the scale on the person. In a non-accelerating elevator ($a = 0$), we have $N = mg = w$, and N is equal to the true weight of the individual. However, suppose the elevator is descending with an acceleration a, and $a < g$. As the vector diagram in the figure shows,

$$mg - N = ma$$

and the *apparent* weight w' is

$$w' = N = m(g - a) < mg$$

where the downward direction is taken as positive in this instance. With a downward acceleration a, we see that N is less than mg, hence the scale indicates that the person weighs less than the true weight. Note that the *apparent acceleration* due to gravity is $g' = g - a$.

Now suppose the elevator were in free fall with $a = g$. As you can see, N and the apparent weight w' would be zero. Essentially, the scale is accelerating or falling at the same rate as the person. The scale may indicate a "weightless" condition, but gravity still acts, as would be noted by the sudden stop at the bottom of the shaft.

(Note that you do not have to become an astronaut or be caught in a falling elevator to experience this condition. Some more familiar examples are shown in Fig. 7.24c,d.)

Space has been called the final frontier. Some day, instead of brief stays in Earth-orbiting spacecraft, we may have permanent space colonies (see the Insight on this page).

Insight

green let go of the ball. Viewed from outside the rotating wheel, you would say the ball has a tangential velocity v as a result of the rotation, and it would go off in a straight line (Newton's first law) until it hits the side of the wheel (as indicated by the dashed line in the figure). However, the inhabitant rotates with the wheel; from that perspective, the ball merely falls to the floor as it would on Earth. Think about it.

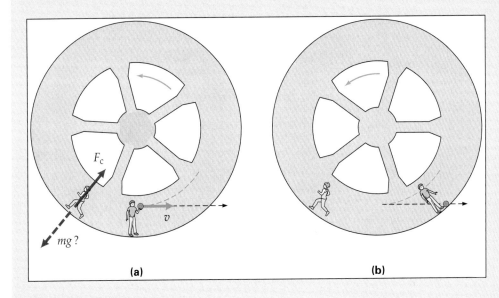

FIGURE 2 Rotating space colony (a) In the frame of reference of someone in a rotating space colony, centripetal force, coming from the normal force N of the floor, would be perceived as weight sensation, or artificial gravity, since we are used to feeling N upward on our feet to balance gravity. Rotation at the proper speed would simulate normal gravity. To an outside observer, a dropped ball would follow a tangential straight-line path (shown as dashed). (b) A colonist would observe the ball to fall downward as in a normal gravitational situation.

Chapter Review

Important Terms

angular displacement 211
radian (rad) 211
average angular speed 214
average angular velocity 214
instantaneous angular velocity 214
tangential speed 215
period 216
frequency 216

hertz (Hz) 216
uniform circular motion 217
centripetal acceleration 218
centripetal force 222
average angular acceleration 224
tangential acceleration 225
universal law of gravitation 227
universal gravitational constant 227

gravitational potential energy 231
Kepler's first law (the law of orbits) 235
Kepler' second law (the law of areas) 236
Kepler's third law (the law of periods) 236
escape speed 237

Important Concepts

- The radian (rad) is a measure of angle; 1 rad is the angle of a circle subtended by an arc length (s) equal to the radius (r).
- Tangential speed (v) and angular speed (ω) for circular motion are directly proportional, with the radius r being the constant of proportionality ($v = r\omega$).
- In uniform circular motion, a centripetal acceleration is required and is always directed toward the center of the circular path.
- A centripetal force (net force directed toward the center of a circle) is a requirement for circular motion.

- Angular acceleration (α) is the time rate of change of angular velocity.
- According to Newton's law of gravitation, every particle attracts every other particle in the universe with a force that is proportional to the masses of both particles and inversely proportional to the square of the distance between them.
- An Earth satellite is in a negative potential energy well. The higher in the well, the greater the potential energy and the less the kinetic energy.

Important Equations

Arc Length (angle in radians):
$$s = r\theta \quad (7.3)$$

Angular Kinematic Equations (for $\theta = 0$ and $t_o = 0$; see Table 7.2 for linear analogues):

$$\theta = \bar{\omega}t \quad \text{(general, not limited to constant acceleration)} \quad (7.5)$$

$$\left.\begin{array}{l} \bar{\omega} = \dfrac{\omega + \omega_o}{2} \\ \omega = \omega_o + \alpha t \\ \theta = \omega_o t + \tfrac{1}{2}\alpha t^2 \\ \omega^2 = \omega_o^2 + 2\alpha\theta \end{array}\right\} \begin{array}{l} \text{constant} \\ \text{acceleration} \\ \text{only} \end{array} \quad \begin{array}{l} (2,\text{ Table } 7.2) \\ (7.12) \\ (4,\text{ Table } 7.2) \\ (5,\text{ Table } 7.2) \end{array}$$

Tangential and Angular Speeds:
$$v = r\omega \quad (7.6)$$

Angular Speed (with uniform circular motion):
$$\omega = \frac{2\pi}{T} = 2\pi f \quad (7.8)$$

Frequency and Period:
$$f = \frac{1}{T} \quad (7.7)$$

Centripetal Acceleration:
$$a_c = \frac{v^2}{r} = r\omega^2 \quad (7.10)$$

Tangential and Angular Accelerations:
$$a_t = r\alpha \quad (7.13)$$

Centripetal Force and Acceleration:
$$F_c = ma_c = \frac{mv^2}{r} \quad (7.11)$$

Newton's Law of Gravitation:
$$F = \frac{Gm_1 m_2}{r^2} \quad (7.15)$$
$$G = 6.67 \times 10^{-11} \text{ N·m}^2/\text{kg}^2$$

Acceleration Due to Gravity at an Altitude h:
$$a_g = \frac{GM_E}{(R_E + h)^2} \quad (7.18)$$

Gravitational Potential Energy of Two Particles:
$$U = -\frac{Gm_1 m_2}{r} \quad (7.19)$$

Kepler's Law of Periods:
$$T^2 = Kr^3 \quad (7.23)$$
(K depends on the mass of the object being orbited; for objects orbiting the Sun, $K = 2.97 \times 10^{-19} \text{ s}^2/\text{m}^3$.)

Escape Speed (from Earth):
$$v_{\text{esc}} = \sqrt{\frac{2GM_E}{r_E}} = \sqrt{2gR_E} \quad (7.24)$$

Energy of Orbiting Satellite Orbiting Earth:
$$E = -\frac{GmM_E}{2r} \quad (7.28)$$
$$K = |E| \quad (7.29)$$

Exercises

7.1 Angular Measure

1. The radian unit is equivalent to a ratio of (a) degree/time, (b) length, (c) length/length, (d) length/time.

2. A proper conversion factor for degrees to radians would be (a) $(\pi/7 \text{ rad})/270°$, (b) $(\pi/2 \text{ rad})/60°$, (c) $(3\pi \text{ rad})/540°$, (d) $(0.50 \text{ rad})/180°$.

3. The Cartesian coordinates of a point on a circle are (1.5 m, 2.0 m). What are the point's polar coordinates (r, θ)?

4. ■ In Cartesian coordinates, the equation of a circle is $x^2 + y^2 = a^2$, where a is the radius of the circle. What is that equation in polar coordinates?

5. ■ Convert the following angles from degrees to radians. Report to 3 significant figures. (a) 15°, (b) 45°, (c) 90°, and (d) 120°.

6. ■ Convert the following angles from radians to degrees. (a) $\pi/6$ rad, (b) $5\pi/12$ rad, (c) $3\pi/4$ rad, and (d) π rad.

7. ■ What is the arc length subtended by an angle of $\pi/4$ rad on a circle with a radius of 6.0 cm?

8. ■ You measure the length of a distant car as an angular distance of 2.0°. If the car is actually 5.0 m long, approximately how far away is the car?

9. ■ A jogger on a circular track that has a radius of 0.250 km runs a distance of 1.00 km. What angular distance does the jogger cover in (a) radians and (b) degrees?

10. ■ Assume that the Earth's orbit around the Sun is circular. In kilometers, what is the approximate orbital distance the Earth travels in 3 months?

11. ■ Using the ratio equation $s/\theta = c/360°$, where c is the circumference of a circle, (a) show that by definition 1 rad is equivalent to 57.3°. (b) Show that 2π rad is equivalent to 360°.

12. ■■ The hour, minute, and second hands on a clock are 0.25 m, 0.30 m, and 0.35 m long, respectively. What are the distances traveled by the tips of the hands in a 30-minute interval?

13. ■■ In Europe, a large circular walking track with a diameter of 0.900 km is marked in angular distances in radians. An American tourist who walks 3.00 mi daily goes to the track. How many radians should he walk?

14. ■■ You ordered a 12-inch-diameter pizza for a party of five. To distribute the pizza evenly, how should you cut it (•Fig. 7.25)?

•FIGURE 7.25 **Tough pizza to cut** See Exercise 14.

15. ■■ (a) In degrees and radians, what is the angular width of a full Moon as viewed from Earth? (b) of a full Earth as viewed from the Moon? [Hint: Use Appendix III.]

16. ■■ To attend the 1996 Summer Olympics, a fan flew from Los Angeles (32° N, 119° W) to Atlanta (33° N, 84° W). Using your knowledge of angular measure, determine the approximate shortest flight distance in kilometers.

17. ■■■ Could a circular pie be cut such that all of the wedge-shaped pieces have an arc length along the outer crust equal to the pie's radius? If not, how many such pieces could you cut, and what would be the angular dimension of the final piece?

18. ■■■ Electrical wire of diameter 0.75 cm is wound on a spool with a radius of 30 cm and a length of 24 cm. (a) Through how many radians must the spool be turned to wrap one even layer of wire? (b) What is the length of this wound wire?

7.2 Angular Speed and Velocity

19. Viewed from above, a turntable is observed to rotate counterclockwise. The angular velocity vector then is (a) tangential to the turntable rim, (b) out of the plane of the turntable, (c) counterclockwise, (d) none of the preceding.

20. The frequency unit of hertz is equivalent to (a) that of the period, (b) cycle, (c) radian/s, (d) s^{-1}.

21. ◯ Do all points on a wheel rotating about a fixed axis have the same angular velocity? the same tangential speed? Explain.

22. When clockwise or counterclockwise is used to describe rotational motion, why is a phrase such as "viewed from above" added?

23. ◯ A wheel is spinning at a constant angular velocity about an axis through its center. Which point(s) on the wheel have the greatest and smallest tangential speeds?

24. ■ A wheel spins at 30 rpm. What is this angular speed expressed in radians per second?

25. ■ A satellite in a circular orbit has a period of 10 h. What is the satellite's frequency in revolutions per day?

26. ■ A race car makes two and a half laps around a circular track in 3.0 min. What is the car's average angular speed?

27. ■ If a ball is rotating with an angular speed of 3.5 rad/s about a central axis, how long does it take for the ball to go through one revolution?

28. ■ What is the period of revolution for (a) a centrifuge rotating at 12 000 rpm and (b) a computer hard disk drive rotating at 10 000?

29. ■■ Determine which has the greater angular speed: particle A, which travels 160° in 2.00 s, or particle B, which travels 4π rad in 8.00 s.

30. ■■ ◯ The tangential speed of a particle on a rotating wheel is 3.0 m/s. If the particle is 0.20 m from the axis of rotation, how long will it take for the particle to go through one revolution?

31. ■■ ◯ A merry-go-round makes 24 revolutions in a 3.0-minute ride. (a) What is its average angular speed? (b) What are the tangential speeds of persons 4.0 m and 5.0 m from the center or axis of rotation?

32. ■■ In 1.5 min, a car moving at constant speed gets halfway around a circular track that has a diameter of 1.0 km. What are the car's (a) angular speed and (b) tangential speed?

33. ■■ Calculate the angular speeds in radians per second of the Earth (a) rotating on its axis and (b) revolving around the Sun.

34. ■■ A CD-ROM in a computer rotates at 500 rpm. How long will it take the disc to make one revolution?

35. ■■■ The driver of a car sets the cruise control and ties the steering wheel so that the car travels at a uniform speed of 15 m/s in a circle with a diameter of 120 m. (a) Through what angular distance does the car move in 4.00 min? (b) What arc distance does it travel in this time?

7.3 Uniform Circular Motion and Centripetal Acceleration

36. In uniform circular motion, there is a (a) constant velocity, (b) constant angular velocity, (c) zero acceleration, (d) net tangential acceleration.

37. If the centripetal force on a particle in uniform circular motion is increased, (a) the tangential speed will remain constant, (b) the tangential speed will decrease, (c) the radius of the circular path will increase, (d) the tangential speed will increase and/or the radius will decrease.

38. Explain why mud flies off a fast-spinning wheel.

39. The spin cycle of a washing machine is used to extract water from recently washed clothes. Explain the physical principle(s) involved.

40. The apparatus illustrated in •Fig. 7.26 is used to demonstrate forces in a rotating system. The floats are in jars of water. When the arm is rotated, which way will the floats move? (Compare with the accelerometer in Fig. 4.28.) Does it make a difference which way the arm is rotated?

•FIGURE 7.26 When set in motion, a rotating system See Exercise 40.

41. When rounding a curve in a fast-moving car, we experience a feeling of being thrown outward (•Fig. 7.27). It is sometimes said that this is because of an outward centrifugal (center-fleeing) force. However, in terms of Newton's laws, this *pseudo*, or *false*, *force* doesn't really exist. Analyze the situation in the figure to show that this is the case. [Hint: Start with Newton's first law.]

•FIGURE 7.27 A center-fleeing force? See Exercise 41.

42. Many race tracks have banked curves, which allow the cars to travel faster than if the curves were flat. Actually, cars could make turns on these banked curves if there were no friction at all. Explain this with the free-body diagram shown in •Fig. 7.28.

•FIGURE 7.28 Banking safety See Exercises 42 and 55.

43. ■ A rotating cylinder about 10 mi in length and 5.0 mi in diameter is designed to be used as a space colony. With what angular speed must it rotate so the residents on it will experience the same acceleration as that due to gravity on Earth?

44. ■ An Indy car with a speed of 120 km/h goes around a level, circular track with a radius of 1.00 km. What is the centripetal acceleration of the car?

45. ■ A wheel of radius 1.5 m rotates at a uniform speed. If a point on the rim of the wheel has a centripetal acceleration of 1.2 m/s^2, what is the point's tangential speed?

46. ■■ The Moon revolves around the Earth in 29.5 days in a nearly circular orbit with a radius of 3.8×10^5 km. Assume that the Moon's motion is uniform. With what acceleration is the Moon falling toward the Earth?

47. ■■ A car with a constant speed of 83.0 km/h enters a circular flat curve with a radius of curvature of 0.400 km. If the friction between the road and the car's tires can supply a centripetal acceleration of 1.25 m/s^2, does the car negotiate the curve safely? Justify your answer.

48. ■■ Imagine that you swing about your head a 0.250-kilogram ball attached to the end of a string. The ball moves at a constant speed in a horizontal circle with a radius of 1.50 m. If it takes 1.20 s for the ball to make one revolution, (a) what is the ball's tangential speed? (b) What centripetal force are you imparting to the ball via the string? (c) Can the string be exactly horizontal?

49. ■■ In Exercise 48, what angle would the string make relative to the horizontal?

50. ■■ A student is to swing a bucket of water in a vertical circle without spilling any (•Fig. 7.29). If the distance from his shoulder to the center of mass of the bucket of water is 1.0 m, what is the minimum speed required to keep the water from coming out of the bucket at the top of the swing?

•FIGURE 7.29 Weightless water? See Exercise 50.

51. ■■ A car of mass 1.5×10^3 kg rounds a circular turn of radius 20 m. If the road is flat and the coefficient of friction is 0.50 between the tires and the road, how fast can the car travel without skidding?

52. ■■ A jet pilot puts an aircraft with a constant speed of 700 km/h into a vertical circular loop with a radius of 2.0 km. (a) At the bottom of the loop, what is the normal force on the seat by the pilot? (b) What is the normal force at the top of the loop?

53. ■■■ A block of mass m slides down an inclined plane into a loop-the-loop of radius r (•Fig. 7.30). (a) Neglecting friction, what is the minimum speed the block must have at the highest point of the loop to stay in the loop? [Hint: What force must act on the block at the top of the loop to keep the block on a circular path?] (b) At what vertical height on the inclined plane (in terms of the radius of the loop) must the block be released if it is to have the required minimum speed at the top of the loop?

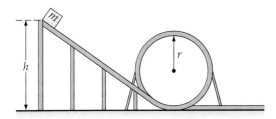

•FIGURE 7.30 **Loop-the-loop** See Exercise 53.

54. ■■■ For a scene in a movie, a stunt driver drives a pickup truck with a length of 4.25 m and a mass of 1.5×10^3 kg around a circular curve with a radius of curvature of 0.333 km (•Fig. 7.31). The truck is to curve off the road, jump across a gully 10 m wide, and land on the other side 2.96 m below the initial side. What is the minimum centripetal acceleration the truck must have in going around the circular curve to clear the gully and land on the other side?

•FIGURE 7.31 **Over the gully** See Exercise 54.

55. ■■■ For the banked curve in Fig. 7.28, the horizontal component of the normal force toward the center of the curve reduces the need for friction to prevent skidding. In fact, for a circular curve banked at a given angle, there is one speed for which no frictional force is required—the centripetal force is supplied completely by the inward component of the normal force. This relationship is used in designing banked curves. (a) Show that the banking angle θ for which no friction is required for a car to negotiate a circular curve safely is given by $\tan\theta = v^2/gr$, where v is the car's speed and r is the radius of curvature. (b) How would the banking angle differ for a more massive truck? (c) Show how friction affects the situation.

56. ■■■ In a theme park, you walk into a giant can-shaped room that can rotate. When the angular speed of the "room" reaches a certain value, the bottom "floor" drops out (•Fig. 7.32). However, you don't fall out! Explain why. If the coefficient of static friction between you and the side wall is 0.30 and the radius of the room is 2.5 m, what is the minimum rotational speed of the room so that you can stay in the bottomless room?

•FIGURE 7.32 **Bottomless room** See Exercise 56.

7.4 Angular Acceleration

57. The angular acceleration in circular motion (a) is equal in magnitude to the tangential acceleration divided by the radius, (b) increases the angular velocity if in the same direction, (c) has units of s^{-2}, (d) all of the preceding.

58. Can you think of an example of a car having both centripetal acceleration and angular acceleration?

59. Is it possible for a car in circular motion to have angular acceleration but not centripetal acceleration?

60. ■ During an acceleration, the angular speed of an engine increases from 700 rpm to 3000 rpm in 3.0 s. What is the average angular acceleration of the engine?

61. ■ A merry-go-round accelerating uniformly from rest achieves its operating speed of 2.5 rpm in five revolutions. What is the magnitude of its angular acceleration?

62. ■■ A $33\frac{1}{3}$-rpm record on a turntable uniformly reaches its operating speed in 2.45 s once the record player is turned on. (a) What is the angular distance traveled during this time? (b) What is the corresponding arc length in feet on the circumference of a 12-inch-diameter record?

63. ■■ A bicycle being repaired is turned upside down, and one wheel is rotated at a rate of 60 rpm. If the wheel slows uniformly to a stop in 15 s, how many revolutions does it make during this time?

64. ■■ A Ferris wheel with a diameter of 35.0 m starts from rest and achieves its maximum operational tangential speed of 2.20 m/s in a time of 15.0 s. (a) What is the magnitude of the wheel's angular acceleration? (b) What is the magnitude of the tangential acceleration after the maximum operational speed is reached?

65. ■■ The blades of a fan running at low speed turn at 250 rpm. When the fan is switched to high speed, the rotation rate increases uniformly to 350 rpm in 5.75 s. (a) What is the magnitude of the angular acceleration of the blades? (b) How many revolutions do the blades go through while the fan is accelerating?

66. ■■ In landing, the rotor of a helicopter slows from an angular speed of 4500 rpm to rest in 5.0 s. How many revolutions does the rotor travel during this time?

67. ■■ A car on a circular track with a radius of 0.30 km accelerates from rest with a constant angular acceleration of magnitude 4.5×10^{-3} rad/s². (a) How long does it take the car to make one lap around the track? (b) What is the total (vector) acceleration of the car when it has completed half of a lap?

68. ■■■ A student investigating circular motion places a dime 10 cm from the center of a $33\frac{1}{3}$-rpm record on a turntable. The record player can accelerate at 1.42 rad/s². The student notes that the dime slides outward 2.25 s after she has switched on the turntable. (a) Why does the dime slide outward? (b) What is the coefficient of static friction between the dime and the record?

7.5 Newton's Law of Gravitation

69. The gravitational force is (a) a linear function of distance, (b) an infinite-range force, (c) applicable only to our solar system, (d) sometimes repulsive.

70. The acceleration due to gravity of an object on the Earth's surface (a) is a universal constant like G, (b) does not depend on the Earth's mass, (c) is directly proportional to the Earth's radius, (d) does not depend on the object's mass.

71. Astronauts in a spacecraft orbiting the Earth or out for a "spacewalk" (•Fig. 7.33) are seen to "float" in midair. This is sometimes referred to as weightlessness or zero gravity (zero g). Are these terms correct? Explain why an astronaut appears to float in or near an orbiting spacecraft.

•FIGURE 7.33 Out for a walk Why does this astronaut seem to "float"? See Exercise 71.

72. If the cup in •Fig. 7.34 were dropped, no water would run out. Explain.

•FIGURE 7.34 Let it go See Exercise 72.

73. Can you determine the mass of the Earth simply by measuring the gravitational acceleration near the Earth's surface? If yes, give the details.

74. (a) A spring scale calibrated in kilograms on the Earth is taken to the Moon to make measurements. Will the scale read correctly? Explain. (b) For a given mass, will the scale reading be more or less at the Earth's equator than at the North Pole? Explain.

75. ■ From the known mass and radius of the Earth, compute the value of the acceleration due to gravity g at the surface of the Earth.

76. ■ Calculate the gravitational force between the Earth and the Moon.

77. ■■ Approximate the gravitational force exerted on the Moon by the Earth and the Sun during (a) a solar eclipse (the Moon is between the Sun and the Earth) and (b) a lunar eclipse (the Earth is between the Sun and the Moon).

78. ■■ Two objects are attracting each other with a 0.90-newton gravitational force. If the distance between them is tripled, what is the new gravitational force between them?

79. ■■ Four identical masses of 2.5 kg each are located at the corners of a square with 1.0 m sides. What is the net force on any one of the masses?

80. ■■ What is the acceleration due to gravity on the top of Mt. Everest? The summit is about 8.80 km above sea level. (Report to 3 significant figures.)

81. ■■ A 75-kilogram man weighs 735 N on Earth's surface. How far above the surface would he have to go to "lose" 10% of his body weight?

82. ■■ For a spacecraft going directly from the Earth to the Moon, beyond what point will lunar gravity begin to dominate; that is, where will the lunar gravitational force be equal in magnitude to the Earth's gravitational force? Are the astronauts on board truly weightless at this point?

83. ■■ Which is greater, the gravitational force exerted on the Earth by the Sun or by the Moon? (Compare these attractions by forming a ratio and giving a factor of how many times greater or smaller one is than the other.)

84. ■■ Assuming that we could get more soil from somewhere, approximately how thick would an added uniform outer layer on the Earth have to be to have $g = 10.0$ m/s^2 exactly? (Take the average density of the Earth to be 5.52 g/cm^3 and $R_E = 6.38 \times 10^3$ km.)

85. ■■■ (a) What is the mutual gravitational potential energy of the configuration shown in •Fig. 7.35 if all the masses are 1.0 kg? (b) What is the gravitational force per unit mass at the center of the configuration?

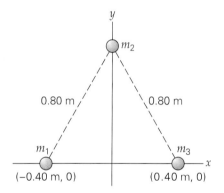

•FIGURE 7.35 **Gravitational potential, gravitational force, and center of mass** See Exercise 85.

7.6 Kepler's Laws and Earth Satellites

86. A new planet is discovered and its period determined. The new planet's distance from the Sun could then be found by using Kepler's (a) first law, (b) second law, (c) third law.

87. 🔵 A planet moves in its orbit about the Sun. (a) Its speed is constant. (b) Its distance from the Sun is constant. (c) It moves faster when it is closer to the Sun. (d) It moves slower when it is closer to the Sun.

88. (a) In one revolution, how much work does the centripetal force do on a satellite in circular orbit about the Earth? (b) A person in a freely falling elevator thinks he can avoid injury by jumping upward just before the elevator strikes the floor. Would this work?

89. (a) In putting satellites into orbits, rockets are launched eastward. Why? (b) In the United States, satellites are launched from Florida. Why not from California, which generally has better weather conditions?

90. ■ 🔵 An instrument package is projected vertically upward to collect data at the top of the Earth's atmosphere (at an altitude of about 800 km). (a) What initial speed is required at the Earth's surface for the package to reach this height? (b) What percentage of the escape speed is this initial speed?

91. ■■ Compute the constant K of Kepler's third law for the orbits of (a) Earth and (b) Venus. (See Appendix III for data.)

92. ■■ Using a development similar to Kepler's law of periods for planets orbiting the Sun, find the required altitude of geosynchronous satellites above the Earth. [Hint: The period of such satellites is the same as that of the Earth.]

93. ■■ Venus has a rotational period of 243 d. What would be the altitude of a geosynchronous satellite for this planet?

94. ■■ The asteroid belt that lies between Mars and Jupiter may be the debris of a planet that broke apart or that was not able to form as a result of Jupiter's strong gravitation. The asteroid belt has a period of 5.0 y. How far from the Sun would this "fifth" planet have been?

Additional Exercises

95. At sunset, the Sun has an angular width of about 0.50°. From the time the lower edge of the Sun just touches the horizon, about how long does it take in minutes to disappear?

96. Ocean tides are primarily produced by the gravitational attraction of the Moon. Explain how this attraction gives rise to two tidal "bulges" on opposite sides of the Earth, resulting in two daily high tides and two daily low tides (•Figure 7.36). [Hint: Consider the inverse-square relation between the distance and the gravitational force acting on

•FIGURE 7.36 **Tides** **(a)** Low tide and **(b)** high tide at the same location off the California coast. See Exercise 96.

the water on opposite sides of the Earth and on the Earth itself.] (You may want to go to the library for a little help on this one.)

97. A 60-kilogram woman prepares to swing out over a pond on a rope (•Fig. 7.37). What tension in the rope is required if she is to maintain a circular arc (a) when she starts her swing by stepping off the platform, (b) when she is 5.0 m above the pond surface, and (c) at the lowest point of her swing? [Hint: Treat the woman and the rope as a simple pendulum.]

•FIGURE 7.37 **A swinging time** See Exercises 97 and 98.

98. Suppose that the woman swinging on the rope in Fig. 7.37 (see Exercise 97) starts her swing at a lower point and has a tangential speed of 8.5 m/s when the rope is at an angle of 30° to the vertical. (a) What is her tangential acceleration at that time? (b) What is the magnitude of the vector sum of the tangential and centripetal accelerations at that time? (c) From what height above the pond surface did she start her swing?

99. An ultracentrifuge (a very fast centrifuge) operates at a speed of 5.00×10^6 rpm. It contains test tubes filled with viruses. (a) What is the centripetal acceleration on a virus at a radial distance of 4.00 cm from the centrifuge's axis of rotation? (b) How does this acceleration compare with g, the acceleration due to gravity?

100. A garden hose wound on a rotating storage cylinder has an outer layer at a distance of 0.45 m from the center of the cylinder. As this layer of wound hose is pulled out, the cylinder makes three revolutions. What is the length of the unwound hose?

101. What is the escape speed of a satellite at an altitude of 750 km in a circular orbit about the Earth ?

102. An astronaut has a weight of 735 N on Earth. What would be the force of gravity on the astronaut in a spacecraft in a circular orbit at an altitude of 450 km?

103. The value of the Kepler constant is $K = 2.97 \times 10^{-19}\,\text{s}^2/\text{m}^3$, and $T^2 = (2.97 \times 10^{-19}\,\text{s}^2/\text{m}^3)r^3$. With nonstandard units, K can be made equal to 1, and the equation can be written as $T^2 = r^3$. Using Earth data, find the units of K that make it equal to 1. [Hint: The average, or mean, distance of the Earth from the Sun is used in astronomy as a unit called an *astronomical unit* (AU).]

104. A pendulum swinging in a circular arc under the influence of gravity, as shown in •Fig. 7.38, has both centripetal and tangential components of acceleration. (a) If the pendulum bob has a speed of 2.7 m/s when the cord makes an angle of $\theta = 15°$ with the vertical, what are the magnitudes of the components at this time? (b) When is the centripetal acceleration a maximum? What is the value of the tangential acceleration at that time?

•FIGURE 7.38 **A swinging pendulum** See Exercise 104.

Rotational Motion and Equilibrium

CHAPTER 8

- **8.1** Rigid Bodies, Translations, and Rotations
- **8.2** Torque, Equilibrium, and Stability
- **8.3** Rotational Dynamics
- **8.4** Rotational Work and Kinetic Energy
- **8.5** Angular Momentum

INSIGHTS

- The Leaning Tower of Pisa
- Stability in Action
- Slide or Roll to a Stop? — Antilock Brakes

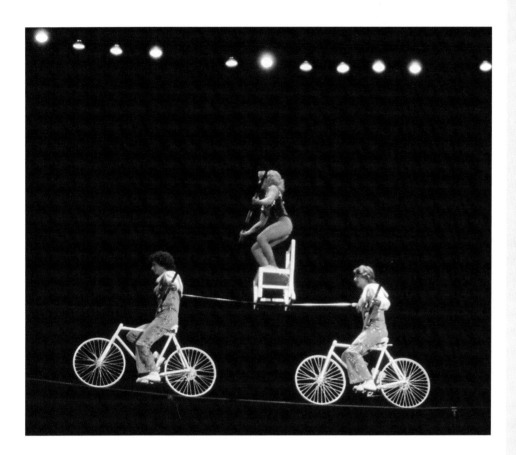

It's always a good idea to keep your equilibrium—but it's more important in some situations than in others! Looking at a photo like this, your first reaction is probably to wonder how these acrobats keep from falling. Presumably, the poles must help—but in what way? If you think about the picture and about what you have already learned, a less obvious but in some ways more interesting question may come to mind: Why do we think they're in danger in the first place? After all, the wire is strong enough to support their weight.

We might say they are in equilibrium. Translational equilibrium ($\Sigma \mathbf{F} = 0$) was discussed in Chapter 4, but here there is another consideration—rotation. Should the performers start to fall (which hopefully they won't), there would initially be a sideways rotation about the wire. To avoid this calamity, another condition must be met—rotational equilibrium, which we will discuss in this chapter.

These tightrope riders are striving to elude rotational motion. But rotational motion is very important in physics because rotating objects are all around us: wheels on vehicles (as on the bicycles in the photo); gears and pulleys in machinery; planets in our solar system; and even many bones in the human body. (Can you think of a few bones that rotate in sockets?)

Fortunately, the equations describing rotational motion can be written as almost direct analogues of those for translational (linear) motion. In Chapter 7, this similarity was pointed out for the kinematic equations. With the addition of equations describing rotational dynamics, you will be able to analyze the general motions of real objects.

8.1 Rigid Bodies, Translations, and Rotations

Objectives: To (a) distinguish between pure translational and pure rotational motions of a rigid body and (b) state the condition(s) for rolling without slipping.

It was convenient initially to consider motions of objects with the understanding that the motion of an object can be represented by a particle located at its center of mass. Rotation, or spinning, was not a consideration then because a particle, a point mass, has no physical dimensions. Rotational motion becomes relevant when we analyze the motion of solid extended objects or rigid bodies, the focus of this chapter.

Definition of:
A rigid body

A **rigid body** is an object or system of particles in which the distances between particles are fixed and remain constant.

A quantity of liquid water is not a rigid body, but the ice that would form if the water were frozen would be. The discussion of rigid body rotation is thus conveniently restricted to solids. Actually, the concept of a rigid body is an idealization. In reality, the particles (atoms and molecules) of a solid vibrate constantly. Also, solids can undergo elastic (and inelastic) deformations (Chapter 6). Even so, most solids can be considered to be rigid bodies for purposes of analyzing rotational motion.

Note: The words "rotation" and "revolution" are commonly used synonymously. In general, this book uses rotation when the axis of rotation goes through the body (the Earth's rotation on its axis, in a period of 24 h) and revolution when the axis is outside the body (the revolution of the Earth about the Sun, in a period of 365 d).

A rigid body may be subject to either or both of two types of motions: translational and rotational. Translational motion is basically the linear motion we studied in previous chapters. If an object has only **translational motion** (•Fig. 8.1a),

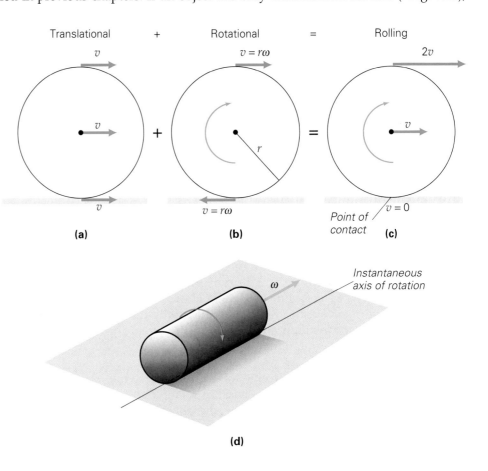

•FIGURE 8.1 Rolling—a combination of translational and rotational motions (a) In pure translational motion, all the particles of an object have the same instantaneous velocity. (b) For pure rotational motion, all the particles of an object have the same instantaneous angular velocity. (c) Rolling is a combination of translational and rotational motions. Summing the velocity vectors for these two motions shows that the point of contact (for a sphere) or the line of contact (for a cylinder) is instantaneously at rest. (d) The line of contact (or, for a sphere, a line through the point of contact) is called the instantaneous axis of rotation. Note that the center of mass of a rolling object on a level surface moves linearly and remains over the point or line of contact.

every particle has the same instantaneous velocity, which means there is no rotation. (Why?)

An object may have only **rotational motion**, motion about a fixed axis (Fig. 8.1b). In this case, all the particles of an object have the same instantaneous angular velocity and travel in circles about the axis of rotation. Although the axis of rotation of an object is commonly taken through its center of mass, this is not always the case. For example, you might pivot a meterstick through one end and have rotational motion about an axis through that end.

General rigid body motion is a combination of translational and rotational motions. When you throw a ball, the translational motion is described by the motion of its center of mass (as in projectile motion). But the ball may also spin, or rotate, and usually does. A common example of rigid body motion involving both translation and rotation is rolling, as illustrated in Fig. 8.1c. The combined motion of any point or particle is given by the vector sum of its instantaneous velocity vectors. (Three points or particles are shown in the figure—top, middle, and bottom). At each instant, a rolling object rotates about an **instantaneous axis of rotation** through its point of contact with the surface (for a sphere) or along its line of contact with the surface (for a cylinder; Fig. 8.1d). The location of this axis changes with time. However, note in Fig. 8.1c that the point or line of contact of the body with the surface is instantaneously at rest (zero velocity), as can be seen from the vector addition of the combined motions at that point.

When an object rolls without slipping, the translational and rotational motions are related simply as discussed in Chapter 7. For example, when a uniform ball (or cylinder) rolls in a straight line on a flat surface, it turns through an angle θ, and a point on the object that was initially in contact with the surface moves through an arc distance s (•Fig. 8.2). From Chapter 7, you know that $s = r\theta$. The center of mass of the ball is directly over the point of contact and moves a linear distance s. Then

$$v_{CM} = \frac{s}{t} = \frac{r\theta}{t} = r\omega$$

In terms of the speed of the center of mass and the angular speed ω, the *condition for rolling without slipping* is

$$v_{CM} = r\omega \qquad \begin{array}{c}\text{condition for rolling}\\ \text{without slipping}\end{array} \qquad (8.1)$$

(This expression is generally written as $v = r\omega$, with the understanding that v is the magnitude of the velocity of the center of mass.) The condition is also expressed by

$$s = r\theta \qquad (8.1a)$$

itself, where s is the distance an object rolls, or the distance the center of mass moves.

Carrying Eq. 8.1 one step further, we can write an expression for the time rate of change of the velocity: $v_{CM}/t = r\omega/t$ (starting from rest). This yields an equation for *accelerated rolling without slipping*:

$$a_{CM} = \frac{v_{CM}}{t} = \frac{r\omega}{t} = r\alpha \qquad (8.1b)$$

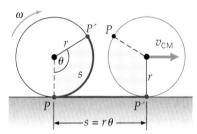

•**FIGURE 8.2 Rolling without slipping** As an object rolls without slipping, the length of the arc between two points of contact on the circumference is equal to the linear distance traveled—think of paint coming off a roller. This distance is $s = r\theta$. The speed of the center of mass is $v_{CM} = r\omega$.

Example 8.1 ■ Slipping or Not Slipping?: That Is the Question

A cylinder with a radius of 12 cm rolls with an instantaneous angular speed of 0.75 rad/s down an inclined plane. (a) If the center of mass of the cylinder travels at a speed of 0.10 m/s at this time, does the cylinder roll without slipping? (b) The cylinder goes onto a level surface and rolls without slipping with the same speed of the center of mass as in part (a). Assuming this speed remains constant for 2.0 s, through what angle does the cylinder turn during this time?

Thinking It Through. Part (a) is a simple test of Eq. 8.1, which must be satisfied for rolling without slipping. In part (b), Eq. 8.1 is valid and can be used to find the angular speed. With the speed and time, we can find the angular distance or the angle through which the cylinder rotated.

Solution. We list the data as usual:

Given: $r = 12$ cm $= 0.12$ m *Find:* (a) if cylinder is slipping
$\omega = 0.75$ rad/s   (b) θ (angle of rotation)
$v_{CM} = 0.10$ m/s
$t = 2.0$ s

(a) If the data satisfy Eq. 8.1, the cylinder will be rolling without slipping:

$$v_{CM} = r\omega = (0.12 \text{ m})(0.75 \text{ rad/s}) = 0.090 \text{ m/s} \neq 0.10 \text{ m/s}$$

Hence, the cylinder is slipping as it goes down the plane. The center of mass is traveling too fast for the nonslipping, rolling condition.

(b) For rolling without slipping on the level surface, the angular speed is given by Eq. 8.1:

$$\omega = \frac{v_{CM}}{r} = \frac{0.10 \text{ m/s}}{0.12 \text{ m}} = 0.83 \text{ rad/s}$$

(Note that the unit of angular speed is 1/s, or s^{-1}, but we add the unitless radian, or rad, for clarity.) Since this speed is constant for the time period, we have

$$\theta = \omega t = (0.83 \text{ rad/s})(2.0 \text{ s}) = 1.7 \text{ rad}$$

The cylinder makes a little over one-quarter of a rotation (right?). Check it yourself.

Follow-up Exercise. How far does the cylinder travel linearly in part (b) of this Example? Find this distance by using two different methods.

8.2 Torque, Equilibrium, and Stability

Objectives: To (a) define torque, (b) apply the conditions for mechanical equilibrium, and (c) describe the relationship between the location of the center of gravity and stability.

Torque

As with translational motion, a force is necessary to produce a change in rotational motion. The rate of change of motion depends not only on the magnitude of the force, but also on the perpendicular distance of its line of action from the axis of rotation, $r_\perp$ (•Fig. 8.3a,b). The line of action of a force is a line extending through the force vector arrow—that is, the line along which the force acts.

Figure 8.3 shows that $r_\perp = r \sin \theta$, where r is the straight-line distance between the axis of rotation and the point at which the force acts, and θ is the angle between the line of r and the force vector **F**. This perpendicular distance $r_\perp$ is called the **moment arm** or **lever arm**.

Torque: the product of a lever arm and a force

The product of the force and the lever arm is called **torque**, τ (from the Latin *torquere*, meaning "to twist"). The magnitude of the torque provided by the force is

$$\tau = r_\perp F = rF \sin \theta \tag{8.2}$$

SI unit of torque: meter-newton (m·N)

The SI units of torque are meters times newtons (m·N). These are the same as the units of work, $W = Fd$ (N·m, or J). However, we will write the units of torque in reverse order as m·N to avoid confusion.

Note that rotational motion is not *always* produced when a force acts on a stationary rigid body. Notice from Eq. 8.2 that $\tau = 0$ when $\theta = 0°$—that is, when the force acts through the axis of rotation (Fig. 8.3c). Also, when $\theta = 90°$, the torque is maximum and the force acts perpendicularly to r, as when you push a door open. Keep in mind that the motion or angular acceleration depends on *where* a perpendicular force is applied (and therefore on the length of the lever arm). As a practical example, think of applying a force to a heavy glass door that swings in and out. Where you apply the force makes a great difference in how easily the door opens, or rotates (through the hinges) on its axis. Have you ever tried to open such a door and inadvertently pushed on the side near the hinges? This force produces a small torque and little or no rotational motion.

We can think of torque in rotational motion as the analogue of force in translational motion. An unbalanced or net force changes translational motion, and an unbalanced or net torque changes rotational motion. Torque, the product of a force and moment arm—both vectors—is itself a vector. Its direction is always perpendicular to the plane of the force and moment arm vectors and is given by a *right-hand rule* similar to that for angular velocity given in Section 7.2. If the fingers of the right hand are curled around the axis of rotation in the direction that the torque would produce a rotational (angular) acceleration, the extended thumb points in the direction of the torque. A sign convention, as in the case of linear motion, can be used to represent torque directions, as we will soon see.

(a) Counterclockwise torque

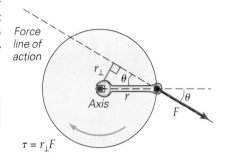

(b) Smaller clockwise torque

Example 8.2 ■ Lifting and Holding: Muscle Torque at Work

In our bodies, torques produced by the contraction of our muscles cause some bones to rotate at joints. For example, when you lift something with your forearm, a torque is applied by the biceps muscle on the lower arm (•Fig. 8.4). With the axis of rotation through the elbow joint and the muscle attached 4.0 cm from the joint, what are the magnitudes of the muscle torques for cases (a) and (b) in Fig. 8.4 if the muscle exerts a force of 600 N?

Thinking It Through. As in many rotational situations, it is important to know the orientations of the r and F vectors so that we can find the angle *between* them to determine the lever arm. Note in the insert in Fig. 8.4a that if the tails of the **r** and **F** vectors were put together, the angle between them would be greater than 90°.

Solution. First we list the data given here and in the figure. This Example demonstrates an important point. Recall that θ is the angle *between* the radial vector **r** and the force **F**.

Given: $r = 4.0$ cm $= 0.040$ m   Find: (a) τ (muscle torque magnitude)
$F = 600$ N   (b) τ (muscle torque magnitude)
(a) $\theta = 30° + 90°$
(b) $\theta = 90°$

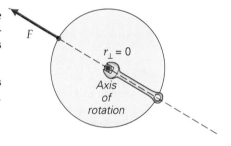

(c) Zero torque

•**FIGURE 8.3 Torque and moment arm** (a) The perpendicular distance $r_\perp$ from the axis of rotation to the line of action of a force is called the moment arm (or lever arm) and is equal to $r \sin \theta$. The torque, or twisting force, that produces rotational motion is given by $\tau = r_\perp F$. (b) The same force in the opposite direction with a smaller moment arm produces a smaller torque in the opposite direction. Note that $r_\perp F = rF_\perp$, or $(r \sin \theta)F = r(F \sin \theta)$. (c) When a force acts through the axis of rotation, $r_\perp = 0$ and $\tau = 0$.

(a) In this case, **r** is directed along the forearm, so the angle between the **r** and **F** vectors is $\theta = 30° + 90°$. Thus, $r \sin \theta = r \sin(30° + 90°) = r \cos 30°$. (See Appendix I for

(a) Starting to lift

(b) Holding

•**FIGURE 8.4 Human torque** See Example 8.1.

•**FIGURE 8.5 Equilibrium and forces** Forces with lines of action through the same point are said to be concurrent. The resultants of the concurrent forces acting on the objects in **(a)** and **(b)** are zero, and the objects are in equilibrium, because the net torque *and* net force are zero. In **(c)**, the object is in *translational* equilibrium, but it will undergo angular acceleration; it is *not* in rotational equilibrium.

this and other useful trigonometric relationships.) Or, just notice from the figure inset that $r_\perp = r \cos 30°$ in this case. Using Eq. 8.2, we have

$$\tau = r_\perp F = rF \sin(30° + 90°) = rF \cos 30°$$
$$= (0.040 \text{ m})(600 \text{ N}) \cos 30° = 21 \text{ m·N}$$

at that instant.

(b) Here, the lever arm and the line of action for the force are perpendicular ($\theta = 90°$), and $r_\perp = r \sin 90° = r$. Then,

$$\tau = r_\perp F = rF = (0.040 \text{ m})(600 \text{ N}) = 24 \text{ m·N}$$

Follow-up Exercise. In (a) of this Example, there must have been a net torque, since the ball was lifted by a rotation of the forearm. In (b), the ball is just being held and there is no rotational acceleration, so there is no net torque on the system. Identify the other torque(s) in each case.

Before considering rotational dynamics with net torques and rotational motions, let's first look at the situation in which the forces and torques acting on a body are balanced or in equilibrium.

Equilibrium

In general, equilibrium means that things are in balance or are stable. This definition applies in the mechanical sense to forces and torques. Unbalanced forces produce translational accelerations, but *balanced* forces produce the condition we call *translational equilibrium*. Similarly, unbalanced torques produce rotational accelerations, and *balanced* torques produce *rotational equilibrium*.

According to Newton's first law of motion, when the sum of the forces acting on a body is zero, the body remains either at rest (static) or in motion with a constant velocity. In either case, the body is said to be in **translational equilibrium**. Stated another way, the *condition for translational equilibrium* is that the net force on a body is zero, $\Sigma \mathbf{F}_i = 0$. In other words, the vector sum of all forces equals zero: $\Sigma \mathbf{F}_i = \mathbf{F}_1 + \mathbf{F}_2 + \mathbf{F}_3 + \cdots = 0$. It should be apparent that this condition is satisfied for the situations illustrated in •Fig. 8.5a,b. Forces with lines of action through the same point are called **concurrent forces**. When these forces vectorially add to zero as in (a) and (b), the body is in translational equilibrium.

But what about the situation pictured in Fig. 8.5c? Here $\Sigma \mathbf{F}_i = 0$, but the opposing forces will cause the object to rotate, and it will clearly not be in a state of static equilibrium. (Such a pair of equal and opposite forces not having the same line of action is called a *couple*.) Thus, the condition $\Sigma \mathbf{F}_i = 0$ is a necessary, but *not sufficient*, condition for static equilibrium.

Since $\Sigma \mathbf{F}_i = 0$ is the condition for translational equilibrium, you might predict (and correctly so) that $\Sigma \tau_i = 0$ is the *condition for rotational equilibrium*. That is, if the sum of the *torques* acting on an object is zero, then the object is in **rotational equilibrium**—it is rotationally at rest or rotates with a constant angular velocity.

Thus we see that there are two equilibrium conditions. Taken together, they define what is called **mechanical equilibrium**:

A body is said to be in mechanical equilibrium when the conditions for both translational and rotational equilibrium are satisfied:

Conditions for mechanical equilibrium

$$\Sigma \mathbf{F}_i = 0 \quad \text{(for translational equilibrium)} \tag{8.3}$$
$$\Sigma \tau_i = 0 \quad \text{(for rotational equilibrium)}$$

A rigid body in mechanical equilibrium may be either at rest or moving with a constant linear and/or angular velocity. An example of the latter is an object rolling without slipping on a level surface with the center of mass having a constant velocity. This is an ideal condition, however, because there is always some friction in reality.

Of greater practical interest is **static equilibrium**, the condition that exists when a rigid body remains at rest. There are many instances in which we do not want things to move, and this absence of motion can occur only if the equilibrium conditions are satisfied. It is particularly comforting to know, for example, that a bridge you are crossing is in static equilibrium and not subject to translational or rotational motions.

Let's consider examples of static translational equilibrium and static rotational equilibrium separately, and then one in which both apply.

Example 8.3 ■ Translational Static Equilibrium: No Translational Motion

A picture hangs motionless on a wall as shown in •Fig. 8.6a. If it has a mass of 5.0 kg, what are the tension forces in the wires?

Thinking It Through. Since the picture remains motionless, it must be in static equilibrium, so applying the conditions for mechanical equilibrium should give equations that yield the tensions. Note that all the forces (tension and weight forces) are concurrent—that is, their lines of action pass through a common point, the nail. Because of this, the condition for rotational equilibrium ($\Sigma \tau_i = 0$) is automatically satisfied—with respect to an axis of rotation along the nail, the moment arms ($r_\perp$) of the forces are zero. Thus, we need consider only translational equilibrium.

Solution.

Given: $\theta = 45°$ Find: T_1 and T_2
        $m = 5.0$ kg

You will find it helpful to isolate the forces acting on the picture in a free-body diagram as was done in Chapter 4 for force problems (Fig. 8.6b). The diagram shows the concurrent forces acting through their common point. Note that we have moved all the force vectors to the common point, which is taken as the origin of the coordinate axes. The weight force mg acts downward.

With the system in static equilibrium, the net force is zero; that is, $\Sigma \mathbf{F}_i = 0$. Thus, the sums of the rectangular components are also zero: $\Sigma \mathbf{F}_{x_i} = 0$ and $\Sigma \mathbf{F}_{y_i} = 0$. Then (using ± for direction), we get

$$\Sigma \mathbf{F}_{x_i}: \quad T_1 \cos 45° - T_2 \cos 45° = 0$$

This just tells you that the x components of the forces are equal and opposite and that $T_1 = T_2 = T$. For the y components, with two upward components of $T \sin 45°$, we have

$$\Sigma \mathbf{F}_{y_i}: \quad T \sin 45° + T \sin 45° - mg = 0$$

(a)

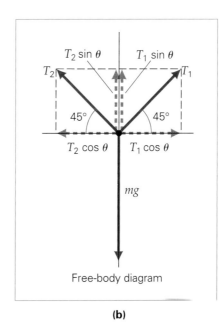

Free-body diagram

(b)

•**FIGURE 8.6 Translational static equilibrium** (a) Since the picture hangs motionless on the wall, the sum of the forces acting on it must be zero. The forces are concurrent, with their lines of action passing through a common point at the nail. (b) This is shown in the free-body diagram, where all the forces have been represented as acting at this point. T_1 and T_2 have been moved to the nail point for convenience. Note, however, that these are the forces acting on the *picture*, not on the nail. See Example 8.3.

or
$$2T \sin 45° - mg = 0$$
Thus,
$$T = \frac{mg}{2 \sin 45°} = \frac{(5.0 \text{ kg})(9.8 \text{ m/s}^2)}{2(0.707)} = 35 \text{ N}$$

Follow-up Exercise. Analyze the situation in Fig. 8.6 that would result if the wires were shortened such that the angles were decreased. Carry this to the limit where the angles approach zero.

As pointed out earlier, torque is a vector and direction is important. Similar to linear motion (Chapter 2), in which we used plus and minus signs to express opposite directions (for example, $+x$ and $-x$), we designate torque directions as being plus and minus depending on the rotational acceleration they tend to produce. The rotational "directions" are taken as clockwise or counterclockwise around the axis of rotation. A torque that tends to produce a counterclockwise rotation will be taken as positive (+), and a torque that tends to produce a clockwise acceleration will be taken as negative (−).

This convention is arbitrary. Remember only that the magnitude of the sum of the counterclockwise torques must equal that of the clockwise torques to have rotational equilibrium. To illustrate, let's apply our convention to the situation shown in •Fig. 8.7.

Example 8.4 ■ Rotational Static Equilibrium: No Rotational Motion

Three masses are suspended from a meterstick as shown in Fig. 8.7a. How much mass must be suspended on the right side for the system to be in static equilibrium? (Neglect the mass of the meterstick.)

Thinking It Through. As the free-body diagram (Fig. 8.7b) shows, the translational equilibrium condition will be satisfied with the upward reaction force N balancing the downward weight forces, so long as the stick is horizontal. But N is not known if m_3 is unknown, so applying the condition for rotational equilibrium should give the required value of m_3. (Note the lever arms are measured from the pivot point, or the center of the meterstick.)

Solution. From the figure, we have

Given: $m_1 = 25$ g   Find: m_3 (unknown mass)
$r_1 = 50$ cm
$m_2 = 75$ g
$r_2 = 30$ cm
$r_3 = 35$ cm

Because the condition for translational equilibrium ($\Sigma \mathbf{F}_i = 0$) is satisfied, $N - Mg = 0$, or $N = Mg$, where M is the total mass. This is true no matter what the total mass may be—that is, regardless of how much mass we add for m_3. However, unless the proper mass for m_3 is placed on the right side, the stick will experience a net torque and rotate.

Notice that the masses on the left side produce torques that tend to rotate the stick counterclockwise, and the mass on the right side produces a torque that tends to rotate it clockwise. We apply the condition for rotational equilibrium by summing the torques about an axis. Let's take this to be through the center of the stick at the 50-centimeter position, or point A in Fig. 8.7b. Then, noting that N passes through the axis of rotation ($r_\perp = 0$) and produces no torque, we have

$$\Sigma \tau_i: \quad \tau_1 + \tau_2 + \tau_3 = +r_1 F_1 + r_2 F_2 + r_3 F_3 \quad \text{(using our sign convention for torque vectors)}$$
$$= r_1(m_1 g) + r_2(m_2 g) - r_3(m_3 g) = 0$$

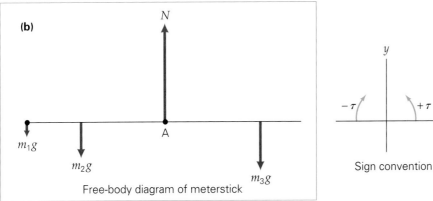

•**FIGURE 8.7 Rotational static equilibrium** For the meterstick to be in rotational equilibrium, the sum of the torques acting about any selected axis must be zero. (The mass of the meterstick is considered negligible.) See Example 8.4.

Noting that the g's cancel out and solving for m_3, we have

$$m_3 = \frac{m_1 r_1 + m_2 r_2}{r_3} = \frac{(25 \text{ g})(50 \text{ cm}) + (75 \text{ g})(30 \text{ cm})}{35 \text{ cm}} = 100 \text{ g}$$

where it was convenient not to convert to standard units. (The mass of the stick was neglected. If the stick is uniform, however, its mass will not affect the equilibrium as long as the pivot point is at the 50-centimeter mark. Why?)

Follow-up Exercise. The axis of rotation could have been taken through any point along the stick. That is, if a system is in static rotational equilibrium, the condition $\Sigma \tau_i = 0$ holds for *any* axis of rotation. Demonstrate this for the system in this Example by taking the axis of rotation through the left end of the stick ($x = 0$).

In general, the conditions for both translational and rotational equilibrium need to be written explicitly to solve a statics problem. Example 8.5 is one such case.

Example 8.5 ■ Static Equilibrium: No Translation, No Rotation

A ladder with a mass of 15 kg rests against a smooth wall (•Fig. 8.8a). A painter who has a mass of 78 kg stands on the ladder as shown in the figure. What frictional force must act on the bottom of the ladder to keep it from slipping?

Thinking It Through. Here we have a variety of forces and torques. However, the ladder will not slip as long as the conditions for static equilibrium are satisfied. Summing both the forces and torques to zero should enable us to solve for the necessary frictional force. Also, as we will see, choosing a convenient axis of rotation for summing the torques can simplify the torque equation.

•**FIGURE 8.8 Static equilibrium**
For the painter's sake, the ladder has to be in static equilibrium; that is, both the sum of the forces and the sum of the torques must be zero. See Example 8.5.

(a)

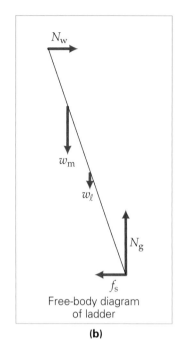

Free-body diagram of ladder

(b)

Solution.

Given: $m_\ell = 15$ kg
$m_m = 78$ kg
Distances given in figure

Find: f_s (force of static friction)

Because the wall is smooth, there is negligible friction between it and the ladder, and only the normal reaction force of the wall (N_w) acts on the ladder at this point (Fig. 8.8b).
  In applying the conditions for static equilibrium, you are free to choose any axis of rotation for the rotational condition. (The conditions must hold for any part of a system for static equilibrium; that is, there can't be motion in any part of the system.) Note that choosing an axis at the end of the ladder where it touches the ground eliminates the torques due to f_s and N_g since the moment arms are zero. Then you can write three equations (using mg for w):

$$\Sigma F_{x_i}: \; N_w - f_s = 0$$
$$\Sigma F_{y_i}: \; N_g - m_m g - m_\ell g = 0$$

and

$$\Sigma \tau_i: \; (m_\ell g)x_1 + (m_m g)x_2 + (-N_w y) = 0$$

The weight of the ladder is considered to be concentrated at its center of gravity. Solving the third equation for N_w and substituting the given values for the masses and distances gives

$$N_w = \frac{(m_\ell g)x_1 + (m_m g)x_2}{y}$$

$$= \frac{(15 \text{ kg})(9.8 \text{ m/s}^2)(1.0 \text{ m}) + (78 \text{ kg})(9.8 \text{ m/s}^2)(1.6 \text{ m})}{5.6 \text{ m}} = 2.4 \times 10^2 \text{ N}$$

From the first equation then,

$$f_s = N_w = 2.4 \times 10^2 \text{ N}$$

Follow-up Exercise. In this Example, would the frictional force between the ladder and the ground (call it f_{s_1}) remain the same if there were friction between the wall and the ladder (call it f_{s_2})? Justify your answer.

> **Problem-Solving Hint**
>
> As the preceding Examples have shown, a good procedure to follow in working problems involving static equilibrium is as follows:
> 1. Sketch a space diagram of the problem.
> 2. Draw a free-body diagram, showing and labeling all external forces and normally resolving the forces into x and y components.
> 3. Apply the equilibrium conditions. Sum the forces: $\Sigma \mathbf{F}_i = 0$, or usually in component form $\Sigma \mathbf{F}_{x_i} = 0$ and $\Sigma \mathbf{F}_{y_i} = 0$. Sum the torques: $\Sigma \boldsymbol{\tau}_i = 0$, remembering to select an appropriate axis of rotation to reduce the number of terms as much as possible. (But be sure to include the unknown force.) Use $\pm$ sign conventions for both $\mathbf{F}$ and $\boldsymbol{\tau}$.
> 4. Solve for the unknown quantities.

Stability and Center of Gravity

The equilibrium of a particle or a rigid body can also be described as being stable or unstable in a gravitational field. For rigid bodies, these categories of equilibria are conveniently analyzed in terms of the center of gravity. Recall from Chapter 6 that the **center of gravity** is the point at which all the weight of an object may be considered to be concentrated in representing it as a particle. When the acceleration due to gravity is constant, the center of gravity and the center of mass coincide.

Note: Review Section 6.5.

If an object is in **stable equilibrium**, any small displacement results in a restoring force or torque, which tends to return the object to its original equilibrium position. As illustrated in •Fig. 8.9a, a ball in a bowl is in stable equilibrium. Analogously, the center of gravity of an extended body in stable equilibrium is

Stable equilibrium—a restoring torque

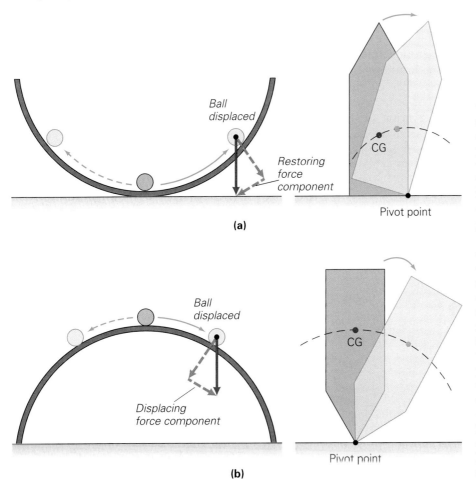

•**FIGURE 8.9 Stable and unstable equilibria** (a) When an object is in stable equilibrium, any small displacement from an equilibrium position results in a force or torque that tends to return the object to that position. A ball in a bowl (left) returns to the bottom after being displaced. Analogously, the center of gravity (CG) of an extended object (right) can be thought of as being in a potential energy bowl—a small displacement raises the CG, increasing the object's potential energy. (b) For an object in unstable equilibrium, any small displacement from its equilibrium position results in a force or torque that tends to take it farther away from that position. The ball on top of an overturned bowl (left) is in unstable equilibrium. For an extended object (right), the CG can be thought of as being on an inverted potential energy bowl—a small displacement lowers the CG, decreasing the object's potential energy.

essentially in a potential energy bowl. Any slight displacement raises its center of gravity, and a restoring force tends to return it to the position of minimum potential energy. This force actually produces a restoring torque that is due to a component of the weight force and that tends to rotate the object about a pivot point back to its original position.

Unstable equilibrium—a torque that toppples

For an object in **unstable equilibrium**, any small displacement from equilibrium results in a force (and a torque) that tends to rotate the object farther away from its equilibrium position. This situation is illustrated in Fig. 8.9b. Note that the center of gravity is at the top of an overturned, or inverted, potential energy bowl; that is, the potential energy is at a maximum in this case.

Small displacements or slight disturbances have profound effects on objects in unstable equilibrium—it doesn't take much to cause such an object to change its position. Yet, even if the angular displacement of an object in stable equilibrium is quite substantial, the object will still be restored to its equilibrium position. An object lying on its long side can be rotated through quite a distance and will still fall back to that original position. This is another way of expressing the *condition of stable equilibrium*:

Condition for stable equilibrium

An object is in stable equilibrium as long as its center of gravity after a small displacement still lies above and inside its original base of support. That is, the line of action of the weight force of the center of gravity intersects the original base of support.

When this is the case, there will always be a restoring torque (•Fig. 8.10a). However, when the center of gravity or center of mass falls outside the base of support, over goes the object—because of a gravitational torque that rotates it away from its equilibrium position (Fig. 8.10b).

Rigid bodies with wide bases and low centers of gravity are therefore most stable and least likely to tip over. This relationship is evident in the design of high-speed race cars, which have wide wheel bases and centers of gravity close to the ground (•Fig. 8.11).

The location of the center of gravity of the human body has an effect on certain physical abilities. For example, women can generally bend over and touch their toes or touch their palms to the floor more easily than can men, who often fall over trying. On the average, men have higher centers of gravity (larger shoulders) than do women (larger pelvises), so it is more likely that a man's center of gravity will be outside his base of support when he bends over. (See Demonstration 6.) See the Insight on p. 264 for another real-life example of equilibrium and stability.

(a) Stable Equilibrium

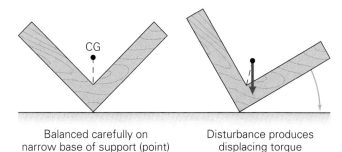

(b) Unstable Equilibrium

•**FIGURE 8.10 Examples of stable and unstable equilibria** (a) When the center of gravity is above and inside an object's base of support, the object is in stable equilibrium (there is a restoring torque). Note how the line of action of the weight of the center of gravity (CG) intersects the original base of support after the displacement. (b) When the center of gravity lies outside the base of support or the line of action of the weight does not intersect the original base of support, the object is unstable (there is a displacing torque).

(a)   (b)   (c)

•**FIGURE 8.11 Stable, stabler, stablest** (a) The acrobat's base of support is very narrow—the small area of head-to-head contact. As long as his center of gravity remains above this area, he is in equilibrium, but a displacement of only a few inches would probably be enough to topple him. (Why he is in a spread-eagle position will become clearer in Section 8.3.) **(b)** This balancing rock in Idaho has a slightly larger base of support than the acrobat in (a); still, a displacement of its center of gravity by a few feet would send it crashing down. **(c)** Race cars are very stable because of their wide wheel bases and low center of gravity.

Demonstration 6 ■ Stability and Center of Gravity

A demonstration to show that women generally have more stability than men—because of their mass distribution and the location of their center of gravity.

(a) An object is placed just beyond the fingertips of a man kneeling with elbows against his knees. Maintaining his balance, he can topple the object with his nose.

(b) With hands now clasped behind his back, the man's center of gravity shifts beyond his base of support (knees) as he bends over to topple the object, and he will topple instead.

(c) A woman can generally do it both ways.

(d) Because her center of gravity is positioned lower in the body, it does not move outside the base of support (past her knees) as she bends with hands clasped behind her back.

8.2 Torque, Equilibrium, and Stability

Insight

The Leaning Tower of Pisa: Stability in Question

The Leaning Tower of Pisa is a famous Italian landmark from which Galileo allegedly performed experiments (see the Chapter 2 Insight on p. 52). Recently, the Tower of Pisa has been in the spotlight—in attempts to keep it from falling. The Tower started leaning before its completion in A.D. 1350 because of soft subsoil. It was closed to the public in 1990, with a lean of about 5.5° off the perpendicular (about 5 m at the top) and an average lean increase of about 1.2 mm a year.

Cement was injected into the base in the 1930s, but the lean accelerated. There has been an international effort to save the Tower. From 1993 to 1995, a concrete ring was built at the base and 620 tons of lead counterweight was placed on it (Fig. 1). Within months, the incline was corrected by 2 to 3 cm. But the construction of a second ring jolted the tower, and overnight it tilted 2.5 mm. With that big scare, an additional 230 tons of lead weight was added.

Current plans call for soil to be removed from beneath the high side of the Tower to balance the natural settling (and to keep the center of gravity above the base of support). However, plastic-coated steel cables will be attached to the Tower to support the structure should there be any sudden shifts (Fig. 2). Should the settling continue, long steel cables connected to ground anchors would be installed to prevent further tilting. It is hoped that these measures will correct the tilt by about 0.5°—enough to save the Tower but still have the leaning attraction.

FIGURE 1 Hold it stable! Tons of lead counterweight being used to help correct the Tower's lean.

FIGURE 2 Keep the center of gravity over the base of support. Efforts will be made to stabilize the Tower by removing soil from beneath the high side. For safety in case of sudden shifts, steel support cables will be attached.

Example 8.6 ■ Stack Them Up: Center of Gravity

Uniform, identical bricks 20 cm long are stacked so that 4.0 cm of each brick extends beyond the brick beneath, as shown in •Fig. 8.12a. How many bricks can be stacked in this way before the stack falls over?

Thinking It Through. As each brick is added, the center of mass (or center of gravity) of the stack moves to the right. The stack will be stable as long as the combined center of mass (CM) is over the base of support—the bottom brick. All of the bricks have the same mass, and the center of mass of each is located at its midpoint. So, the horizontal location of the stack's CM must be computed as bricks are added. The location of the CM was discussed in Chapter 6 (see Eq. 6.19).

Solution.

Given: length = 20 cm *Find:* maximum number of bricks
 displacement, each brick = 4.0 cm with stability

Taking the origin to be at the center of the bottom brick, the horizontal coordinate of the center of mass (or center of gravity) for the first two bricks in the stack is given by Eq. 6.19, where $m_1 = m_2 = m$ and x_2 is the displacement of the second brick:

$$X_{CM_2} = \frac{mx_1 + mx_2}{m + m} = \frac{m(x_1 + x_2)}{2m} = \frac{x_1 + x_2}{2} = \frac{0 + 4.0 \text{ cm}}{2} = 2.0 \text{ cm}$$

The masses of the bricks cancel out (since they are all the same). For three bricks,

$$X_{CM_3} = \frac{m(x_1 + x_2 + x_3)}{3m} = \frac{0 + 4.0 \text{ cm} + 8.0 \text{ cm}}{3} = 4.0 \text{ cm}$$

For four bricks,

$$X_{CM_4} = \frac{m(x_1 + x_2 + x_3 + x_4)}{4m} = \frac{0 + 4.0 \text{ cm} + 8.0 \text{ cm} + 12.0 \text{ cm}}{4} = 6.0 \text{ cm}$$

and so on.

This series of results shows that the center of mass of the stack moves horizontally 2.0 cm for each brick added to the bottom one. For a stack of six bricks, the center of mass is 10 cm from the origin and directly over the edge of the bottom brick (2.0 cm × 5 *added* bricks = 10 cm, which is half the length of the bottom brick), so the stack is in unstable equilibrium. The stack may not topple if the sixth brick is positioned very carefully, but it is doubtful that this could be done in practice. A seventh brick would definitely cause the stack to fall.

Follow-up Exercise. If the bricks in this Example were stacked so that alternately 4.0 cm and 6.0 cm extended beyond the brick beneath, how many bricks could be stacked in this manner before the stack toppled?

For another case of stability, see the Insight on p. 266.

8.3 Rotational Dynamics

Objectives: To (a) describe the moment of inertia of a rigid body and (b) apply the rotational form of Newton's second law to physical situations.

Moment of Inertia

Torque is the rotational analogue of force in linear motion, and a net torque produces rotational motion. To analyze this relationship, consider a constant net force acting on a particle of mass m (•Fig. 8.13). The magnitude of the torque on the particle is

$$\tau = r_\perp F = rF_\perp = rma_\perp = mr^2\alpha \quad \text{torque on a particle} \quad (8.4)$$

where $a_\perp = r\alpha$ is the tangential acceleration (a_t, Eq. 7.13). For the rotation of a rigid body about a fixed axis, this equation can be applied to each particle and the results summed over the entire body (n particles) to find the total torque. Since all the particles of a rotating body have the same angular acceleration, we can simply add the individual torque magnitudes:

$$\Sigma\tau = \tau_1 + \tau_2 + \tau_3 + \cdots + \tau_n$$
$$= m_1 r_1^2 \alpha + m_2 r_2^2 \alpha + m_3 r_3^2 \alpha + \cdots + m_n r_n^2 \alpha$$
$$= (m_1 r_1^2 + m_2 r_2^2 + m_3 r_3^2 + \cdots + m_n r_n^2)\alpha$$
$$\Sigma\tau = \left(\sum_{i=1}^{n} m_i r_i^2\right)\alpha \quad (8.5)$$

But for a rigid body, the masses (m_i's) and the distances from the axis of rotation (r_i's) are constant. Therefore, the quantity in the parentheses in Eq. 8.5 is constant and it is called the **moment of inertia**, I:

$$I = \Sigma m_i r_i^2 \quad \text{moment of inertia} \quad (8.6)$$

SI unit of moment of inertia: kilogram-meters squared (kg m²)

•**FIGURE 8.12 Stack them up!** (a) How many bricks can be stacked like this before the stack falls? See Example 8.6. (b) Try a similar experiment with books.

Torque on a particle

•**FIGURE 8.13 Torque on a particle** The magnitude of the torque on a particle of mass m is $\tau = mr^2\alpha$.

8.2 Rotational Dynamics 265

Insight

Stability in Action

When riding a bicycle and going around a curve or making a circular turn on a level surface, a rider instinctively leans into the curve. Why? We might think that leaning over, rather than remaining upright, is more likely to cause a spill. However, leaning really does increase stability—it's all a matter of torques.

When a bicycle or vehicle goes around a level circular curve, a centripetal force is needed, as we learned in Chapter 7. This force is generally supplied by the force of static friction between the tires and the road. As illustrated in Fig. 1a, the reaction force **R** of the ground on the bicycle provides the required centripetal force F_c to round the curve and the normal force **N**.

Suppose the rider tried to go around the curve with these forces while remaining upright as shown in Fig. 1a. Note that the line of action of **R** does not go through the system's center of gravity (indicated by a dot). Considering an axis of rotation through this point, there would be a counterclockwise torque that would tend to rotate the bicycle in such a way that the wheels would slide inward underneath the rider. However, if the rider leans inward at the proper angle (Fig. 1b), both the line of action of **R** and the weight force go through the center of gravity and there is no rotational instability (as the gentleman on the bicycle well knew).

(Being alert, you might observe there is still a torque on the rider. Indeed, when the rider leans into the curve, the weight force gives rise to a torque about an axis through the point of contact with the ground. This, along with the turning of the handle bars, causes the bicycle to turn. If the bicycle were not moving, there would be a rotation about this axis and the bicycle and rider would fall over.)

The need to lean into a curve is readily apparent in bicycle and motorcycle races on level tracks. Things can be made easier for the riders if tracks or roadways are banked to provide a natural lean (recall Section 7.3 and Exercise 7.55).

FIGURE 1 Leaning into a curve When rounding a curve or making a turn, a bicycle rider must lean into the curve. (This rider could have told you why.)

Thus, the magnitude of the net torque can be conveniently written as

$$\tau = I\alpha \qquad \text{torque on a rigid body} \qquad (8.7)$$

Rotational form of Newton's second law

Note: τ means τ_{net}

This is the *rotational form of Newton's second law* (**F** = m**a** and **τ** = I**α**, in vector form). Keep in mind that *net* forces and torques (**F**$_{net}$ and **τ**$_{net}$) are necessary to produce accelerations, although this is not indicated explicitly in Eq. 8.7; τ means *net* torque.

As you might infer from a comparison of these two forms of Newton's law, the moment of inertia I is a measure of *rotational inertia*, or a body's tendency to resist change in its rotational motion. Although I is constant for a rigid body and is the rotational analogue of mass, you must keep in mind that, unlike the mass of a particle, the moment of inertia is referenced to a particular axis and can have different values for different axes.

The moment of inertia also depends on the mass distribution *relative* to the axis of rotation. It is easier (that is, it takes less torque) to give an object an angular acceleration about some axes than about others. The following Example illustrates this point.

Example 8.7 ■ Rotational Inertia: Mass Distribution and Axis of Rotation

Find the moment of inertia about the axis indicated for each of the one-dimensional dumbbell configurations in •Fig. 8.14. (Consider the mass of the connecting bar to be negligible, and report to three significant figures.)

Thinking It Through. This is a direct application of Eq. 8.6 for cases with different masses and distances. It will show that the moment of inertia of an object depends on the axis of rotation and on the mass distribution relative to the axis of rotation. The sum for I will include only two terms (two masses).

Solution.

Given: Values of m and r from the figure.   *Find:* $I = \Sigma m_i r_i^2$

With $I = m_1 r_1^2 + m_2 r_2^2$,

(a) $I = (30 \text{ kg})(0.50 \text{ m})^2 + (30 \text{ kg})(0.50 \text{ m})^2 = 15 \text{ kg} \cdot \text{m}^2$
(b) $I = (40 \text{ kg})(0.50 \text{ m})^2 + (10 \text{ kg})(0.50 \text{ m})^2 = 12.5 \text{ kg} \cdot \text{m}^2$
(c) $I = (30 \text{ kg})(1.5 \text{ m})^2 + (30 \text{ kg})(1.5 \text{ m})^2 = 135 \text{ kg} \cdot \text{m}^2$
(d) $I = (30 \text{ kg})(0 \text{ m})^2 + (30 \text{ kg})(3.0 \text{ m})^2 = 270 \text{ kg} \cdot \text{m}^2$
(e) $I = (40 \text{ kg})(0 \text{ m})^2 + (10 \text{ kg})(3.0 \text{ m})^2 = 90 \text{ kg} \cdot \text{m}^2$

This Example clearly shows how the moment of inertia depends on the mass *and* its distribution relative to a particular axis of rotation. In general, the moment of inertia is larger the farther the mass is from the axis of rotation. This principle is important in the design of flywheels, which are used in automobiles to keep the engine running smoothly between cylinder firings. The mass of a flywheel is concentrated near the rim, giving a large moment of inertia, which resists changes in motion.

Follow-up Exercise. In parts (d) and (e) of this Example, would the moments of inertia be different if the axis of rotation were through m_2? Explain.

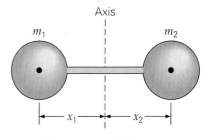

(a) $m_1 = m_2 = 30$ kg
$x_1 = x_2 = 0.50$ m

(b) $m_1 = 40$ kg, $m_2 = 10$ kg
$x_1 = x_2 = 0.50$ m

(c) $m_1 = m_2 = 30$ kg
$x_1 = x_2 = 1.5$ m

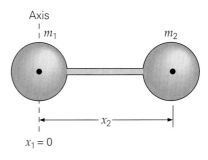

(d) $m_1 = m_2 = 30$ kg
$x_1 = 0, x_2 = 3.0$ m

(e) $m_1 = 40$ kg, $m_2 = 10$ kg
$x_1 = 0, x_2 = 3.0$ m

•**FIGURE 8.14 Moment of inertia** The moment of inertia depends on the distribution of mass relative to a particular axis of rotation and, in general, has a different value for each axis. This difference reflects the fact that objects are easier or more difficult to rotate about certain axes. See Example 8.7.

Balancing Act: Locating the Center of Gravity

A rod with a movable ball, like that shown in •Fig. 8.15, is more easily balanced if the ball is in a higher position. This is because with the ball in a higher position, (a) the system has a higher center of gravity and more stability; (b) when the center of gravity is off the vertical, there is less torque and smaller angular acceleration; (c) the moment of inertia about the axis of rotation is larger; (d) the center of gravity is closer to the axis of rotation.

Reasoning and Answer. With the ball at any position and the rod vertical, the system is in unstable equilibrium. We saw in Section 8.2 that rigid bodies with wide bases and *low* centers of gravity are more stable, so answer (a) isn't correct. Any slight movement will cause the rod to rotate about an axis through its point of contact. With the center of gravity (CG) at a higher position and off the vertical, there would be a greater lever arm (and thus a *greater* torque), so (b) too is incorrect. Also, with the ball in a higher position, the center of gravity is *farther* from the axis of rotation, which eliminates (d). However, moving the CG farther from the axis of rotation has an interesting consequence—a greater moment of inertia or resistance to change in rotational motion.

With the ball in a higher position, as the rod starts to rotate there is a greater torque, but the increased moment of inertia produces an even greater resistance to rotational motion. With the rod rotationally accelerating more slowly, you have more time to adjust your

•**FIGURE 8.15 Greater stability with a higher center of gravity?** See Conceptual Example 8.8.

hand and bring your finger and the axis of rotation under the center of gravity. The torque is then zero and the rod is again in equilibrium, albeit unstable. Thus, the answer is (c).

We can understand this result more clearly if we consider the angular acceleration of the rod as it starts to tilt. The smaller the angular acceleration, the more time is available to adjust your hand under the rod to balance it. The angular acceleration of the rod is given by Eq. 8.7: $\alpha = \tau/I$.

Now let us look at how the torque, τ, and moment of inertia, I, vary with r, the distance of the system's CG from the axis of rotation (through the balancing finger). Neglecting the mass of the rod and considering the CG to be simply that of the balls, $\tau = rF = rmg \sin\theta$ (Eq. 8.2) and $I = mr^2$ (Eq. 8.5). In other words, τ increases with r while I increases with r^2. Combining these relationships, we can see that α is proportional to $1/r$. Thus we have our paradoxical result: The farther the rod's CG from the axis of rotation, the more slowly it tilts and the easier it is to balance.

Follow-up Exercise. When walking on a thin bar or rail, such as a railroad rail, you have probably found that it helps to hold your arms outstretched. Similarly, tightrope walkers often carry long poles, as in the chapter-opening photo. How does this help a performer to maintain balance?

As Example 8.8 shows, the moment of inertia is an important consideration in rotational motion. By changing the axis of rotation and the relative mass distribution, the value of I can be changed and the motion affected. You were probably told to do this if you played softball or baseball as a child. When at bat, children are often instructed to "choke up" on the bat—to move their hands farther up on the handle.

Now you know why. In doing so, the axis of rotaion is moved closer to the more massive end of the bat (or its center of mass). Hence the moment of inertia is decreased (smaller r in the mr^2 term). Then when a swing is taken, the angular acceleration is greater. The bat gets around quicker, and the chance of hitting the ball before it goes past is greater. A batter has only a fraction of second to swing, and with $\theta = \frac{1}{2}\alpha t^2$, a larger α allows the quicker rotation of the bat through an angle of approximately 90°.

Parallel Axis Theorem

The calculations for the moments of inertia of most extended rigid bodies require math that is beyond the scope of this book. The results for some common shapes are given in •Fig. 8.16. The rotational axes are generally taken along axes of symmetry, that is, running through the center of mass so as to give a symmetrical mass distribution. An exception is the rod with an axis of rotation through one end (Fig. 8.16c). This axis is parallel to an axis of rotation through the center of mass of the rod (Fig. 8.16b). The moment of inertia about such a parallel axis is given by a useful theorem called the **parallel axis theorem**:

Note: $I = I_{CM}$ when $d = 0$, which is the minimum I value.

$$I = I_{CM} + Md^2 \qquad (8.8)$$

where I is the moment of inertia about an axis that is parallel to one through the center of mass, I_{CM}, and at a distance d from it, and M is the total mass of the body (•Fig. 8.17). For the axis through the end of the rod (Fig. 8.16c), the moment of inertia is obtained by applying the parallel axis theorem to the thin rod in Fig. 8.16b:

$$I = I_{CM} + Md^2 = \tfrac{1}{12}ML^2 + M\left(\frac{L}{2}\right)^2 = \tfrac{1}{12}ML^2 + \tfrac{1}{4}ML^2 = \tfrac{1}{3}ML^2$$

Applications of Rotational Dynamics

The rotational form of Newton's second law allows us to analyze dynamic rotational situations. Examples 8.9 and 8.10 illustrate how this is done. In such situations, it is very important to make certain that all the data are properly listed to help with the increasing number of variables.

(a) Particle

(b) Thin rod

(c) Thin rod

(d) Thin cylindrical shell, hoop, or ring

(e) Solid cylinder or disk

(f) Annular cylinder

(g) Solid sphere about any diameter

(h) Thin spherical shell

(i) Rectangular plate

(j) Thin rectangular sheet

(k) Thin rectangular sheet

•FIGURE 8.16 Moments of inertia for some uniform objects with common shapes

Example 8.9 ■ Opening the Door: Torque In Action

A student opens a 12-kilogram door by applying a constant force of 40 N at a perpendicular distance of 0.90 m from the hinges (•Fig. 8.18). If the door is 2.0 m in height and 1.0 m wide, what is the magnitude of the angular acceleration of the door? (Assume that the door rotates freely on its hinges.)

Thinking It Through. From the given information we can calculate the applied torque. To find the angular acceleration of the door, we need to know its moment of inertia. This can be calculated since we know mass and dimensions.

Solution. From the problem, we can list the following:

Given: $M = 12$ kg
$F = 40$ N
$r_\perp = r = 0.90$ m
$h = 2.0$ m (door height)
$w = 1.0$ m (door width)

Find: α (magnitude of angular acceleration)

•FIGURE 8.17 Parallel axis theorem The moment of inertia about an axis parallel to another through the center of mass of a body is $I = I_{CM} + Md^2$, where M is the total mass of the body and d is the distance between the two axes.

8.3 Rotational Dynamics **269**

We need to apply the rotational form of Newton's second law (Eq. 8.7): $\tau = r_\perp F = I\alpha$. The problem boils down to finding the moment of inertia of the door.

Looking at Fig. 8.16, we see that part (k) applies to a door (treated as a uniform rectangle) rotating on hinges, so $I = \frac{1}{3}ML^2$, where $L = w$, the width of the door. We must simply do the calculations:

$$\tau = I\alpha$$

or

$$\alpha = \frac{\tau}{I} = \frac{r_\perp F}{\frac{1}{3}ML^2} = \frac{3rF}{Mw^2} = \frac{3(0.90 \text{ m})(40 \text{ N})}{(12 \text{ kg})(1.0 \text{ m})^2} = 9.0 \text{ rad/s}^2$$

Follow-up Exercise. In this Example, if the constant torque were applied through an angular distance of 45° and then removed, how long would it take the door to swing completely open (90°)?

In problems involving pulleys in Chapter 4, the mass (and inertia) of the pulley was always neglected to simplify things. Now we know how to include this and can treat pulleys more realistically.

Example 8.10 ■ Pulleys Have Mass Too: Taking Account of Pulley Inertia

A block of mass m hangs from a string wrapped around a frictionless, disk-shaped pulley of mass M and radius R, as shown in •Fig. 8.19. If the block descends from rest under the influence of gravity, what is the magnitude of the linear acceleration of the block? (Neglect the mass of the string.)

Thinking It Through. Real pulleys have mass and rotational inertia, which affect the motion. The suspended mass via the string applies a torque to the pulley. Here we use the rotational form of Newton's second law to find the angular acceleration of the pulley and then its tangential acceleration, which is the same in magnitude as the linear acceleration of the block. (Why?)

Solution. The linear acceleration of the block depends on the angular acceleration of the pulley, so we look at the pulley system first. The pulley is treated as a disk and thus has a moment of inertia of $I = \frac{1}{2}MR^2$ (Fig. 8.16e). It experiences a torque due to the tension force in the string (T). With $\tau = I\alpha$ (considering only the upper dashed box in Fig. 8.19),

$$\tau = r_\perp F = RT = I\alpha = \tfrac{1}{2}MR^2\alpha$$

so

$$\alpha = \frac{2T}{MR}$$

The linear acceleration of the block and the angular acceleration of the pulley are related by $a = R\alpha$, so

$$a = R\alpha = \frac{2T}{M} \qquad (1)$$

But T is unknown. Looking at the descending mass (the lower dashed box) and summing the forces in the vertical direction (positive in the direction of motion) gives

$$mg - T = ma$$

or

$$T = mg - ma \qquad (2)$$

Using Eq. 2 to eliminate T from Eq. 1 gives

$$a = \frac{2T}{M} = \frac{2(mg - ma)}{M}$$

Solving for a, we get

$$a = \frac{2mg}{(2m + M)} \qquad (3)$$

•**FIGURE 8.18 Torque in action** See Example 8.9.

Note that if $M \to 0$ (as for ideal, massless pulleys in previous chapters), then $I \to 0$, and $a = g$ (from Eq. 3). Here, however, $M \neq 0$, so we have $a < g$. (Why?)

Follow-up Exercise. Pulleys can be analyzed even more realistically. In Example 8.10, friction was neglected, but practically a frictional torque (τ_f) exists and could be included. What would be the form of the angular acceleration in this case? Show your result to be dimensionally correct.

In pulley problems, we also neglect the mass of the string, which still gives a good approximation if the string is relatively light. Taking the mass of the string into account would give a continuously varying mass hanging on the pulley, thus producing a variable torque. Such a problem is beyond the scope of this book.

Suppose you had masses suspended from each side of a pulley. Here, you'd have to compute the net torque. If you didn't know the values of the masses, or which way the pulley would rotate, then you could simply assume a direction. As in the linear case, if the result came out with the opposite sign, this would indicate that you had assumed the wrong direction.

Problem-Solving Hint

For problems such as those of Examples 8.9 and 8.10, dealing with coupled rotational and translational motions, keep in mind that, with no rope slippage, the magnitudes of the accelerations are usually related by $a = r\alpha$, while $v = r\omega$ relates the magnitudes of the velocities at any instant of time. Applying Newton's second law (in rotational or linear form) to different parts of the system gives equations that can be combined by using such relationships. Also, for rolling without slipping, $a = r\alpha$ and $v = r\omega$ relate the angular quantities to the linear motion of the center of mass.

Another application of rotational dynamics is the analysis of motion on an inclined plane. Previously, we worked with block-shaped objects sliding down planes. Now we can generalize to include objects that can roll.

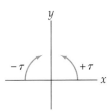

•**FIGURE 8.19 Pulley with inertia** Taking the mass, or rotational inertia, of a pulley into account allows a more realistic description of the motion. See Example 8.10.

Example 8.11 ■ Rolling Down: Torque and Acceleration

A solid, rigid spherical ball of mass M and radius R is released at the top of a hard-surfaced inclined plane. The ball rolls without slipping, with only static friction between it and the plane (•Fig. 8.20a). What is the acceleration of the ball's center of mass?

Thinking It Through. Since the ball rolls without slipping, there is no relative motion between it and the plane at the point of contact (through which the instantaneous axis of rotation passes). At this point the force of static friction (f_s) acts on the ball. Without this force, the ball would slide down the plane. This force is perpendicular to the instantaneous axis of rotation, or directed up the plane. Detailed analyses are given in the solution for *two axes of rotaion*, just to show that the results are the same, as they should be.

Solution. You can analyze the motion by using two different axes of rotation. First, let's consider an axis through the center of mass of the ball (the central dot in Fig. 8.20a) and perpendicular to the page so that we can see the effect of the frictional torque. The acceleration of the ball's center of mass is down the plane, and the ball's angular acceleration is about an axis through that center.

Since the reaction force (the normal force N) and the weight force act through the axis of rotation (see the vector diagram in Fig. 8.20b), they produce no torques, and the angular acceleration is produced entirely by the frictional torque due to f_s.

Applying Newton's second law, $\tau = I\alpha$, and using the moment of inertia about an axis through the ball's center of mass as found in Fig. 8.16g ($I = \tfrac{2}{5}MR^2$), we have

$$\tau = Rf_s = I\alpha = (\tfrac{2}{5}MR^2)\left(\frac{a}{R}\right)$$

(a)

(b)

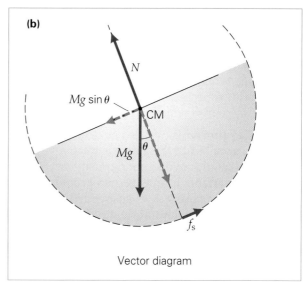

Vector diagram

•**FIGURE 8.20 Rolling down an inclined plane** For a rigid ball rolling down a hard-surfaced inclined plane, the force of static friction acts on the ball in response to the parallel component of the weight force. See Example 8.11.

where we substituted a/R for α. Then, solving for f_s,

$$f_s = \tfrac{2}{5}Ma$$

Looking at the translational motion of the center of mass and summing the forces acting parallel to the plane (see the vector diagram; positive in the direction of motion) gives

$$Mg \sin \theta - f_s = Ma$$

or

$$Mg \sin \theta - \tfrac{2}{5}Ma = Ma$$

Thus,

$$a = \tfrac{5}{7}g \sin \theta$$

Here is another situation in which the acceleration does not depend on M or R. Therefore, uniform spheres of any size or density will roll down the same plane with the same acceleration (assuming no slippage).

Before leaving this Example, let's look at the situation from another perspective. Consider analyzing the rotational motion about the instantaneous axis of rotation, which runs through the point of contact of the sphere and the plane's surface. This point is instantaneously at rest, as noted earlier. As can be seen from Fig. 8.20, both the normal force and the frictional force act through this point and so produce no torque. The torque on the ball from this perspective comes from the component of the weight force parallel to the plane, $Mg \sin \theta$ (see the vector diagram). The torque is then

$$\tau = rF_\perp = R(Mg \sin \theta) = I_i \alpha \qquad (1)$$

where I_i is the moment of inertia *about the instantaneous axis of rotation*. Note that this is *not* equal to the moment of inertia used above, $I_{CM} = \tfrac{2}{5}MR^2$, which is the moment of inertia about an axis *through the center of mass* of the ball. You can find the moment of inertia about the instantaneous axis from the parallel axis theorem:

$$I_i = I_{CM} + Md^2 = \tfrac{2}{5}MR^2 + MR^2 = \tfrac{7}{5}MR^2 \qquad (2)$$

Substituting for I_i from Eq. 2 into Eq. 1 gives

$$R(Mg \sin \theta) = I_i \alpha = (\tfrac{7}{5}MR^2)\left(\frac{a}{R}\right)$$

and

$$a = \tfrac{5}{7}g \sin \theta$$

This is the same as the acceleration found for the other axis of rotation (as it should be).

Follow-up Exercise. That uniform spheres of any size or density will roll down the same plane with the same acceleration is shown in this Example. Will different uniform objects with different shapes (for example, cylinders or hoops) roll down this incline with the same acceleration as a sphere?

Applying a Torque: Which Way Does the Yo-yo Roll?

The string of a yo-yo sitting on a level surface is pulled as shown in •Fig. 8.21. Will the yo-yo roll (a) toward the person or (b) away from the person?

Reasoning and Answer. Apply the physics we have just studied to the situation. Note that the instantaneous axis of rotation is along the line of contact of the yo-yo with the surface. If you had a stick standing vertically in place of the **r** vector and pulled on a string attached to its top in the direction of **F**, which way would the stick rotate? Clockwise (about its instantaneous axis of rotation). The yo-yo reacts similarly—that is, it rolls in the direction of the pull, so the answer is (a). (Get a yo-yo and try it if you're a nonbeliever.)

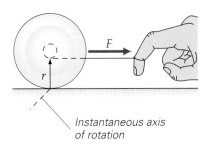

•**FIGURE 8.21 Pulling the yo-yo's string** See Conceptual Example 8.12.

There is more interesting physics in our yo-yo situation. The pull force is not the only force acting on the yo-yo; there are three others. Do they contribute torques? Let's identify these forces. There's the vertical weight of the yo-yo (gravitational attraction) and the normal force from the surface. Also, there is a horizontal force of static friction between the yo-yo and the surface (otherwise the yo-yo would slide rather than roll). But, all these forces act through the line of contact or through the instantaneous axis of rotation, so there are no torques here (why?).

What would happen if we increased the angle of the string or pull force (relative to the horizontal) as illustrated in •Fig. 8.22a? The yo-yo would still roll to the right. But note in Fig. 8.22b that, at some critical angle θ_c, the line of force goes through the axis of rotation and the net force and net torque on the yo-yo are zero, so it is in equilibrium.

If we exceed this critical angle (Fig. 8.22c), the yo-yo will begin to roll counter-clockwise, or to the left. Note that the line of action of the force is on the other side of the axis of rotation than that in Fig. 8.22a and that the lever arm ($r_\perp$) has changed directions, resulting in a reversed net torque direction. (There is a lot of physics in a yo-yo.)

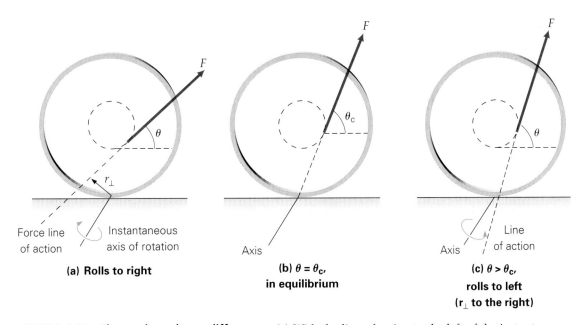

•**FIGURE 8.22 The angle makes a difference** (a) With the line of action to the left of the instantaneous axis, the yo-yo rolls to the right. (b) At a critical angle θ_c, the line of action passes through the axis and the yo-yo is in equilibrium. (c) When the line of action is to the right of the axis, the yo-yo rolls to the left. See Conceptual Example 8.12.

Follow-up Exercise. Suppose you set the yo-yo string at the critical angle, with the string over a round, horizontal bar at the appropriate height, and you suspended a weight on the end of the string to supply the force for the equilibrium condition. What would happen if you then pulled the yo-yo toward you, away from its equilibrium position, and released it?

8.4 Rotational Work and Kinetic Energy

Objectives: To discuss, explain, and use the rotational forms of (a) work, (b) kinetic energy, and (c) power.

This section gives the rotational analogues associated with work and kinetic energy for constant torques. Because their development is very similar to that given for their linear counterparts, detailed discussion is not needed.

Rotational Work. We can go directly from work done by a force to work done by a torque since they are related ($\tau = r_\perp F$). For rotational motion, **rotational work** $W = Fs$ for a single force F acting tangentially along an arc length s:

$$W = Fs = F(r_\perp \theta) = \tau\theta$$

where θ is in radians. Thus, for a single torque acting through an angle of rotation θ,

Rotational work

$$W = \tau\theta \qquad (8.9)$$

In this book, both torque (τ) and angular displacement (θ) vectors are almost always along the fixed axis of rotation, so you will not need to be concerned about parallel components as you were for translational work. The torque and angular displacement may be in opposite directions, in which case the torque does negative work and slows down the rotation of the body. Negative rotational work is analogous to F and d being in opposite directions for translational motion.

Rotational Power. An expression for **rotational power**, the rotational analogue of power (the time rate of doing work), is easily obtained from Eq. 8.9:

Rotational power

$$P = \frac{W}{t} = \tau\left(\frac{\theta}{t}\right) = \tau\omega \qquad (8.10)$$

The Work–Energy Theorem and Kinetic Energy

The relationship between the net rotational work and change in rotational kinetic energy can be derived as follows, starting with the equation for rotational work:

$$W = \tau\theta = I\alpha\theta$$

Since our torques are due only to constant forces, we get a constant α. For a constant angular acceleration, $\omega^2 = \omega_o^2 + 2\alpha\theta$, and therefore

$$W = I\left(\frac{\omega^2 - \omega_o^2}{2}\right) = \tfrac{1}{2}I\omega^2 - \tfrac{1}{2}I\omega_o^2$$

Thus, using Eq. 5.6, we get

Rotational analogue of work–energy theorem

$$W = \tfrac{1}{2}I\omega^2 - \tfrac{1}{2}I\omega_o^2 = K - K_o = \Delta K \qquad (8.11)$$

Here, the expression for **rotational kinetic energy**, K, is

Rotational kinetic energy

$$K = \tfrac{1}{2}I\omega^2 \qquad (8.12)$$

Thus, the net rotational work is equal to the change in the rotational kinetic energy. Consequently, to change the *rotational* kinetic energy of an object, a net torque must be applied.

TABLE 8.1 Translational and Rotational Quantities and Equations

Translational		Rotational	
Force:	$\mathbf{F}$	Torque (magnitude):	$\tau = rF \sin\theta$
Mass (inertia):	m	Moment of inertia:	$I = \Sigma m_i r_i^2$
Newton's second law:	$\mathbf{F} = m\mathbf{a}$	Newton's second law:	$\tau = I\alpha$
Work:	$W = Fd$	Work:	$W = \tau\theta$
Power:	$P = Fv$	Power:	$P = \tau\omega$
Kinetic energy:	$K = \tfrac{1}{2}mv^2$	Kinetic energy:	$K = \tfrac{1}{2}I\omega^2$
Work–energy theorem:	$W = \tfrac{1}{2}mv^2 - \tfrac{1}{2}mv_o^2 = \Delta K$	Work–energy theorem:	$W = \tfrac{1}{2}I\omega^2 - \tfrac{1}{2}I\omega_o^2 = \Delta K$
Momentum:	$\mathbf{p} = m\mathbf{v}$	Momentum:	$\mathbf{L} = I\boldsymbol{\omega}$

It is possible to derive the kinetic energy of a rotating rigid body (on a fixed axis) directly. Summing the instantaneous tangential velocities and kinetic energies of the body's individual particles relative to the fixed axis gives

$$K = \tfrac{1}{2}\Sigma m_i v_i^2 = \tfrac{1}{2}(\Sigma m_i r_i^2)\omega^2 = \tfrac{1}{2}I\omega^2$$

where $v_i = r_i\omega$. Thus, Eq. 8.12 doesn't represent a new form of energy. It is simply another expression for kinetic energy, in a form that is more convenient for rigid-body rotation.

A summary of the analogous equations for translational and rotational motions is given in Table 8.1. (The table also contains momentum, which we shall discuss in Section 8.5.)

When an object has both translational and rotational motions, its total kinetic energy may be divided into parts to reflect this. For example, for a cylinder rolling without slipping on a level surface, the motion is purely rotational relative to the instantaneous axis of rotation, which is instantaneously at rest (see Example 8.12). The kinetic energy is

$$K = \tfrac{1}{2}I_i\omega^2$$

where I_i is the moment of inertia about the instantaneous axis. This moment of inertia is given by the parallel axis theorem (Eq. 8.8): $I_i = I_{CM} + MR^2$, where R is the radius of the cylinder. Then

$$K = \tfrac{1}{2}I_i\omega^2 = \tfrac{1}{2}(I_{CM} + MR^2)\omega^2 = \tfrac{1}{2}I_{CM}\omega^2 + \tfrac{1}{2}MR^2\omega^2$$

But since there is no slipping, $v_{CM} = R\omega$, and

$$\underbrace{K}_{\substack{\text{total} \\ \text{KE}}} = \underbrace{\tfrac{1}{2}I_{CM}\omega^2}_{\substack{\text{rotational} \\ \text{KE}}} + \underbrace{\tfrac{1}{2}Mv_{CM}^2}_{\substack{\text{translational} \\ \text{KE}}} \qquad \text{(rolling, no slipping)} \quad (8.13)$$

Rolling kinetic energy

Note that although a cylinder was used as an example here, this is a general result and applies to any object that is rolling without slipping. *Thus, the total kinetic energy of such an object is the sum of two contributions: the translational kinetic energy of the center of mass and the rotational kinetic energy relative to a horizontal axis through the center of mass.*

Note: A rolling body has both translational kinetic energy and rotational kinetic energy.

Example 8.13 ■ Division of Energy: Rotational and Translational

A uniform, solid 1.0-kilogram cylinder rolls without slipping at a speed of 1.8 m/s on a flat surface. (a) What is the total kinetic energy of the cylinder? (b) What percentage of this total is rotational kinetic energy?

Thinking It Through. The cylinder has both rotational and translational kinetic energies, so Eq. 8.13 applies, and its terms are related by the condition of rolling without slipping.

Solution.

Given: $M = 1.0$ kg
$v_{CM} = 1.8$ m/s
$I_{CM} = \frac{1}{2}MR^2$ (from Fig. 8.16e)

Find: (a) K (total kinetic energy)
(b) $\frac{K_r}{K}$ ($\times 100\%$) (percentage of rotational energy)

(a) The cylinder rolls without slipping, so the condition $v_{CM} = R\omega$ applies. Then, the total kinetic energy is the sum of the rotational kinetic energy K_r and the translational kinetic energy of the center of mass K_{CM} (Eq. 8.13):

$$K = \tfrac{1}{2}(\tfrac{1}{2}MR^2)\left(\frac{v_{CM}}{R}\right)^2 + \tfrac{1}{2}Mv_{CM}^2 = \tfrac{1}{4}Mv_{CM}^2 + \tfrac{1}{2}Mv_{CM}^2$$

$$= \tfrac{3}{4}Mv_{CM}^2 = \tfrac{3}{4}(1.0 \text{ kg})(1.8 \text{ m/s})^2 = 2.4 \text{ J}$$

(b) The rotational kinetic energy K_r of the cylinder is the first term of the preceding equation, so, forming a ratio in symbol form, we get

$$\frac{K_r}{K} = \frac{\tfrac{1}{4}Mv_{CM}^2}{\tfrac{3}{4}Mv_{CM}^2} = \tfrac{1}{3}(\times 100\%) = 33\%$$

Thus, the total kinetic energy of the cylinder is made up of rotational and translational parts, with one-third being rotational.

Note that the radius of the cylinder was not needed and that neither the mass nor any numerical values were needed in part (b). Because we used a ratio, these quantities canceled out. *Don't* think that this exact division of energy is a general result, however. It is easy to show that the percentage is different for objects with different moments of inertia. For example, you should expect a rolling sphere to have a smaller percentage of rotational kinetic energy than a cylinder has because the sphere has a smaller moment of inertia ($I = \tfrac{2}{5}MR^2$).

Follow-up Exercise. Potential energy can be brought into the act by applying the conservation of energy to an object rolling up or down an inclined plane. In this Example, suppose that the cylinder rolled up a 20° inclined plane without slipping. (a) At what vertical height on the plane does the cylinder stop? (b) To find the height in part (a), you probably equated the initial total kinetic energy to the final gravitational potential energy. That is, the total kinetic energy was reduced by the work done by gravity. However, a frictional force also acts (to prevent slipping—see Fig. 8.20). Is there not work done here too?

Example 8.14 ■ Rolling Down versus Sliding Down: Which Is Faster?

A uniform cylindrical hoop is released from rest at a height of 0.25 m near the top of an inclined plane (•Fig. 8.23). If the cylinder rolls down the plane without slipping and there is no energy loss due to friction, what is the linear speed of the cylinder's center of mass at the bottom of the incline?

Thinking It Through. Here, potential energy is converted into kinetic energy—both rotational and translational. The conservation of (mechanical) energy applies.

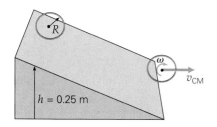

•**FIGURE 8.23 Rolling motion and energy** When an object rolls down an inclined plane, potential energy is converted to translational *and* rotational kinetic energies. This makes the rolling slower than frictionless sliding. See Example 8.14.

Solution.

Given: $h = 0.25$ m
$I_{CM} = MR^2$ (from Fig. 8.16d)

Find: v_{CM} (speed of CM)

Because the total mechanical energy of the cylinder is conserved, you can write

$$E_o = E$$

or, since $v_o = 0$ at the top of the incline, we have

$$\underbrace{Mgh}_{\text{at rest}} = \underbrace{\tfrac{1}{2}I_{CM}\omega^2 + \tfrac{1}{2}Mv_{CM}^2}_{\text{at bottom of incline}}$$

Using the rolling condition $v_{CM} = R\omega$ gives

$$Mgh = \tfrac{1}{2}(MR^2)\left(\frac{v_{CM}}{R}\right)^2 + \tfrac{1}{2}Mv_{CM}^2 = Mv_{CM}^2$$

Solving for v_{CM}, we get

$$v_{CM} = \sqrt{gh} = \sqrt{(9.80 \text{ m/s}^2)(0.25 \text{ m})} = 1.6 \text{ m/s}$$

Again, not much numerical information was needed here. Note that the hoop rolls down from the same height as the solid cylinder rolled up in the Example 8.13 Follow-up Exercise, yet the speed of the hoop is less than that of the cylinder at the bottom of the incline. You should know why—differences in moment of inertia.

Follow-up Exercise. Suppose the inclined plane in this Example were frictionless and the hoop slid down the plane instead of rolling. How would the speed at the bottom compare in this case? Why are the speeds different?

As Example 8.14 shows, v_{CM} is independent of M and R. The masses and radii cancel out, so all objects of a particular shape (with the same formula for the moment of inertia) roll with the same speed, regardless of size or density. But the rolling speed does vary with the moment of inertia, which varies with the shape. Therefore, rigid bodies with different shapes roll with different speeds. For example, if you released a cylindrical hoop, a solid cylinder, and a uniform sphere at the same time from the top of an inclined plane, the sphere would win the race to the bottom, followed by the cylinder, with the hoop coming in last—every time!

You can try this experiment with a couple of food cans or other cylindrical containers—one full of some solid material (a rigid body) and one empty with the ends cut out—and a smooth, solid ball. Remember that the masses and the radii make no difference. You might think that an annular cylinder (a hollow cylinder with inner and outer radii that vary appreciably—Fig. 8.16f) would be a possible front-runner, or front-roller, in such a race, but it wouldn't win. The rolling race down an incline is fixed even when you vary the masses and the radii. (See Exercise 76.)

Another aspect of rolling is discussed in the Insight on p. 278.

8.5 Angular Momentum

Objectives: To (a) define angular momentum and (b) apply the conservation of angular momentum to physical situations.

Another important quantity in rotational motion is angular momentum. Recall from Section 6.1 how the linear momentum of an object is changed by a force. Similarly, angular momentum is associated with torque. As we have seen, torque is the product of a moment arm and a force. In a similar manner, **angular momentum (L)** is the product of a moment arm and a linear momentum. For a particle of mass m, the magnitude of the linear momentum is $p = mv$ and $v = r\omega$. The magnitude of the angular momentum is

$$L = r_\perp p = mr_\perp v = mr_\perp^2 \omega \quad \begin{array}{c}\text{single-particle}\\ \text{angular momentum}\end{array} \quad (8.14)$$

SI unit of angular momentum: kilogram-meters squared per second ($kg \cdot m^2/s$)

where v is the tangential speed of the particle, $r_\perp$ the moment arm, and ω the angular speed. For circular motion, $r_\perp = r$, since $\mathbf{v}$ is perpendicular to $\mathbf{r}$.

For a system of particles comprising a rigid body, the total magnitude of the angular momentum is

$$L = (\Sigma m_i r_i^2)\omega = I\omega \quad \begin{array}{c}\text{rigid-body}\\ \text{angular momentum}\end{array} \quad (8.15)$$

Insight

Slide or Roll to a Stop?—Antilock Brakes

In an emergency situation while driving, you may instinctively jam on the brakes, trying to come to a quick stop—that is, to stop in the shortest distance. But with the wheels locked, the car skids, or slides, to a stop, often out of control. In this case, the force of sliding friction is acting on the wheels.

To prevent this, you may have learned to pump the brakes in order to roll rather than slide to a stop, particularly on wet or icy roads. Many newer automobiles have computerized antilock braking systems (ABS) that do this automatically. When the brakes are applied firmly and the car begins to slide, sensors in the wheels note this and a computer takes over control of the braking system. It momentarily releases the brakes and then varies the brake-fluid pressure with a pumping action (up to thirteen times per second!) so that the wheels will continue to roll without slipping.

In the absence of sliding, both rolling friction and static friction act. In many cases, however, the force of rolling friction is small and only static friction need be taken into account, as will be done here. The ABS works to keep the static friction near the maximum, $f_s \approx f_{s_{max}}$, which you can't do easily by foot.

Does sliding instead of rolling make a big difference in the stopping distance for an automobile? We can calculate the difference by assuming that rolling friction is negligible. Although the external force of static friction does no work to dissipate energy in slowing a car (this is done internally by friction on the brake pads), it does determine whether the wheels roll or slide.

In Example 2.9, a vehicle stopping distance was given by

$$x = \frac{v_0^2}{2a}$$

By Newton's second law, the net force in the horizontal direction is

$$F = f = \mu N = \mu mg = ma$$

and the stopping acceleration is then

$$a = \mu g$$

Thus,

$$x = \frac{v_0^2}{2\mu g} \quad (1)$$

But, as was noted in Chapter 4, the coefficient of sliding (kinetic) friction is generally less than that of static friction; that is, $\mu_k < \mu_s$. The general difference between rolling and sliding stops can be seen by using the same initial velocity (v_0 for both cases). Then, using Eq. 1 to form a ratio,

$$\frac{x_{roll}}{x_{slide}} = \frac{\mu_k}{\mu_s} \quad or \quad x_{roll} = \left(\frac{\mu_k}{\mu_s}\right) x_{slide}$$

From Table 4.1, the value of μ_k for rubber on wet concrete is 0.60, and the value of μ_s for these surfaces is 0.80. Using these values for a comparison of the stopping distances gives

$$x_{roll} = \left(\frac{0.60}{0.80}\right) x_{slide} = (0.75) x_{slide}$$

The car comes to a rolling stop in 75 percent of the distance required for a sliding stop, for example, 15 m instead of 20 m. Although this may vary for different conditions, it could be an important, perhaps life-saving, difference.

---

which, for rotation about a fixed axis, in vector notation, is

Rigid body angular momentum

$$\mathbf{L} = I\boldsymbol{\omega} \quad (8.16)$$

Thus, **L** is in the direction of the angular velocity vector ($\boldsymbol{\omega}$) as given by the right-hand rule.

Note: Review Eq. 6.3, Section 6.1.

For linear motion, the momentum of a system is related to force by $\mathbf{F} = \Delta \mathbf{P}/\Delta t$. Angular momentum is analogously related to net torque:

$$\boldsymbol{\tau} = I\boldsymbol{\alpha} = \frac{I\Delta\boldsymbol{\omega}}{\Delta t} = \frac{\Delta(I\boldsymbol{\omega})}{\Delta t} = \frac{\Delta \mathbf{L}}{\Delta t}$$

That is,

Net torque: the time rate of change of angular momentum

$$\boldsymbol{\tau} = \frac{\Delta \mathbf{L}}{\Delta t} \quad (8.17)$$

Thus, the net torque is equal to *the time rate of change of angular momentum*. In other words, a net torque causes a *change* in angular momentum.

Conservation of Angular Momentum

Equation 8.17 was derived by using $\boldsymbol{\tau} = I\boldsymbol{\alpha}$, which applies to a rigid system of particles or a rigid body having a constant moment of inertia. However, Eq. 8.17 is a general one that also applies to a nonrigid system of particles. In such a system, there may be a change in the mass distribution and a change in the moment of inertia. As a result, there may be an angular acceleration even in the absence of a net torque. How can this be?

If the net torque on a system is zero, then, by Eq. 8.17, $\tau = \Delta L/\Delta t = 0$, and

$$\Delta L = L - L_o = I\omega - I_o\omega_o = 0$$

or

$$I\omega = I_o\omega_o \qquad (8.18)$$

Thus, the condition for the **conservation of angular momentum** is as follows:

In the absence of an external, unbalanced torque, the total (vector) angular momentum of a system is conserved (remains constant).

As with total linear momentum, the internal torques arising from internal forces cancel out.

For a rigid body with a constant moment of inertia (that is, $I = I_o$), the angular speed remains constant ($\omega = \omega_o$) in the absence of a net torque. But it is possible for the moment of inertia to change in some systems, giving rise to a change in the angular speed, as the following Example illustrates.

Conservation of angular momentum

Note: Angular momentum is conserved when the net torque is zero. (L stays fixed.) This is the third conservation law in mechanics.

Example 8.15 ■ Pull It Down: Conservation of Angular Momentum

A small ball at the end of a string that passes through a tube is swung in a circle, as illustrated in •Fig. 8.24. When the string is pulled downward through the tube, the angular speed of the ball increases. (a) Is this caused by a torque due to the pulling force? (b) If the ball is initially swung at a speed of 2.8 m/s in a circle with a radius of 0.30 m, what will be its tangential speed if the string is pulled down far enough to reduce the radius of the circle to 0.15 m?

Thinking It Through. (a) A force is applied to the ball via the string, but consider the axis of rotation. (b) In the absence of a net torque, the angular momentum is conserved (Eq. 8.18), and the tangential speed is related to the angular speed by $v = r\omega$.

Solution.

Given: $r_1 = 0.30$ m
$r_2 = 0.15$ m
$v_1 = 2.8$ m/s

Find: (a) Cause of the increase in angular speed
(b) v_2 (final tangential speed)

•FIGURE 8.24 **Conservation of angular momentum** When the string is pulled downward through the tube, the revolving ball speeds up. See Example 8.15.

(a) The change in the angular velocity, or the angular acceleration, is not caused by a torque due to the pulling force. The force on the ball as transmitted by the string acts through the axis of rotation, and therefore the torque is zero. As the rotating portion of the string is shortened, the moment of inertia of the ball ($I = mr^2$, from Fig. 8.16a) decreases. This causes the ball to speed up, since in the absence of an external torque, the angular momentum of the ball is conserved.

(b) Since the angular momentum is conserved, we have

$$I_o \omega_o = I \omega$$

Then, using $I = mr^2$ and $\omega = v/r$ gives

$$mr_1 v_1 = mr_2 v_2$$

and

$$v_2 = \left(\frac{r_1}{r_2}\right) v_1 = \left(\frac{0.30 \text{ m}}{0.15 \text{ m}}\right) 2.8 \text{ m/s} = 5.6 \text{ m/s}$$

When the radial distance is shortened, the ball speeds up.

Follow-up Exercise. Let's look at the situation in this Example in terms of work and energy. If the inital speed is the same and the vertical pulling force is 7.8 N, what is the final speed of the 0.10-kilogram ball?

Example 8.15 should help you understand Kepler's law of areas (Chapter 7). A planet's angular momentum is essentially conserved, since the Sun exerts no torque on it (why?) and we can neglect the torques from other planets (why?). Thus, when a planet is closer to the Sun in its elliptical orbit and so has a shorter moment arm, its speed is greater by the conservation of angular momentum. Similarly, when an orbiting satellite's altitude varies during the course of an elliptical orbit about a planet, the satellite speeds up or slows down in accordance with the same principle.

Real-Life Angular Momentum

A popular lecture demonstration of the conservation of angular momentum is shown in •Fig. 8.25a,b. A person sitting on a stool that rotates holds weights with his arms outstretched and is started slowly rotating. An external torque to start this rotation must be supplied by someone else, because the person on the stool cannot initiate the motion by himself. (Why not?) Once rotating, if the person brings his arms inward, the angular speed increases and he spins much faster. Extending his arms again slows him down. Can you explain this?

If L is constant, what happens to ω when I is made smaller by reducing r? The angular speed must increase to compensate and keep L constant. Ice skaters do dizzying spins by pulling in their arms to reduce their moment of inertia. Similarly, a diver spins during a high dive by tucking in, greatly decreasing his or her moment of inertia. The enormous wind speeds of tornadoes and hurricanes represent another example of the same effect (Fig. 8.25c).

Angular momentum, **L**, is a vector, and when it is conserved or constant, its magnitude *and* direction must remain unchanged. Thus, when no external torques act, the direction of **L** is fixed in space. This is the principle of passing a football accurately as well as of the gyrocompass (•Fig. 8.26). A football is normally passed with a spiraling rotation. This spin, or gyroscopic action, stabilizes the ball's spin axis in the direction of motion. Similarly, rifle bullets are set spinning by the rifling in the barrel for directional stability.

The **L** vector of a spinning gyroscope in the compass is set in a particular direction (usually north). In the absence of external torques, the compass direction remains fixed, even though its carrier (an airplane or ship, for example) changes directions. You may have played with a toy gyroscope that is set spinning and placed on a pedestal. In a "sleeping" condition, the gyro stands straight up with its angular

(a)

(b)

(c)

•**FIGURE 8.25 Moment of inertia change** (a) Spinning slowly with masses in outstretched arms, the man's moment of inertia is relatively large (masses are farther from the axis of rotation). Note that he is isolated, with no external torques (neglecting friction), so angular momentum, $L = I\omega$, is conserved. (b) Pulling his arms inward decreases the moment of inertia (why?). Consequently ω must increase, and he goes into a dizzy spin. (c) The same principle helps explain the violence of the winds that spiral around the center of a hurricane. As air rushes in toward the center of the storm, where the pressure is low, its rotational velocity must increase for angular momentum to be conserved.

(a)

(b)

•**FIGURE 8.26 Constant direction of angular momentum** When angular momentum is conserved, its direction is constant in space. (a) This can be demonstrated by a passed football. (b) Gyroscopic action also occurs in a gyroscope, a rotating wheel that is universally mounted on gimbals (rings) so that it is free to turn about any axis. When the frame moves, the wheel maintains its direction. This is the principle of the gyrocompass.

momentum vector fixed in space for some time. The gyro's center of gravity is on the axis of rotation, so there is no net torque due to its weight.

However, the gyroscope eventually slows down because of friction, causing **L** to tilt. In doing so, you may have noticed how the spin axis revolves, or *precesses*, about the vertical axis. It revolves tilted over, so to speak (•Fig. 8.27a). Since the gyroscope precesses, the angular momentum vector **L** is no longer constant in direction, indicating that a torque must be acting to give a change (Δ**L**) with time. As can be seen from the figure, the torque arises from the vertical component of the weight force, since the center of gravity no longer lies directly above the point of support or on the vertical axis of rotation. The instantaneous torque is such that the gyroscope's axis moves or precesses about the vertical axis.

In a similar manner, the Earth's rotational axis precesses. The Earth's spin axis is tilted $23\frac{1}{2}°$ with respect to a line perpendicular to the plane of its revolution about the Sun; the axis precesses about this line (Fig. 8.27b). The precession is due to slight gravitational torques exerted on the Earth by the Sun and the Moon. The torques arise because the Earth is not perfectly spherical—it has an equatorial bulge.

The period of the precession of the Earth's axis is about 26 000 years, so it has little day-to-day effect. However, it does have an interesting long-term effect. Polaris will not always be (nor has it always been) the North Star, that is, the star toward which the Earth's axis of rotation points. About 5000 years ago, Alpha Draconis was the North Star, and 5000 years from now it will be Alpha Cephei, which is at an angular distance of about 68° away from Polaris on the circle described by the precession of the Earth's axis.

There are some other long-term torque effects on the Earth and the Moon. Did you know that the Earth's daily spin rate is slowing down and hence the days are getting longer? Also, that the Moon is receding, or getting farther away, from the Earth? This is primarily due to ocean tidal friction, which gives rise to a torque. As a result, the Earth's spin angular momentum, and therefore its rate of rotation, is changing. The slowing rate of rotation causes the average day to be longer; this century will be about 25 seconds longer than the previous.

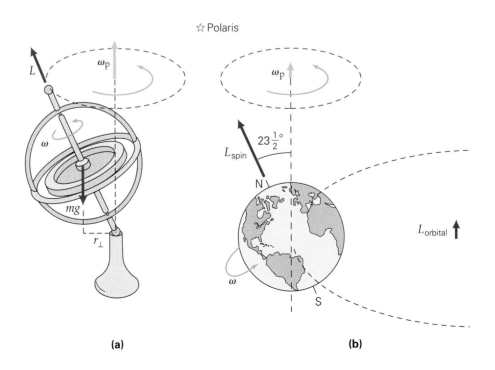

•**FIGURE 8.27 Precession** An external torque causes a change in angular momentum. **(a)** For a spinning gyroscope, this change is directional, and the axis of rotation precesses at angular acceleration ω_p about a vertical line. (The torque due to the weight force would point out of the page as drawn here, as would Δ**L**.) Note that although there is a torque that would topple a nonspinning gyroscope, a spinning gyroscope doesn't fall. **(b)** Similarly, the Earth's axis precesses because of gravitational torques caused by the Sun and the Moon. We don't notice this motion because the period of precession is about 26 000 y.

But this is an average rate. At times, the Earth's rotation speeds up for relatively short periods. This increase is thought to be associated with the rotational inertia of the liquid layer of the Earth's core (see the Insight in Chapter 13).

The tidal torque on the Earth results chiefly from the Moon's gravitational attraction, which is the main cause of ocean tides. This torque is *internal* to the Earth–Moon system, so the total angular momentum of that system is conserved. Since the Earth is losing angular momentum, the Moon must be gaining angular momentum to keep the total angular momentum of the system constant. The Earth loses rotational (spin) angular momentum, and the Moon gains orbital angular momentum. As a result, the Moon drifts slightly farther from Earth and its orbital speed decreases. The Moon moves away from the Earth at about 1 cm per lunar revolution, or lunar month (calculated from the decrease rate of the Earth's rotation). Thus, the Moon moves in a slowly widening spiral.

Finally, a common example in which angular momentum is an important consideration—the helicopter. What would happen if a helicopter had a single rotor? Since the motor supplying the torque is internal, the angular momentum would be conserved. Initially, $\mathbf{L} = 0$, hence to conserve the total system (rotor plus body) angular momentum, the separate angular momenta of the rotor and body would have to be in opposite directions to cancel. Thus on takeoff, the rotor would rotate one way and the helicopter body the other, which is not a desirable situation.

To prevent this situation, helicopters have two rotors. Large helicopters have two overlapping rotors (•Fig. 8.28a). The oppositely rotating rotors cancel each other's angular momenta, so the helicopter body does not have to rotate to provide canceling angular momentum. The rotors are offset at different heights so that they do not collide.

Small helicopters with a single overhead rotor have small "anti-torque" tail rotors (Fig. 8.28b). The tail rotor produces a thrust like a propeller and supplies the torque to counterbalance the torque produced by the overhead rotor. The tail rotor also helps in steering the craft. By increasing or decreasing the tail rotor's thrust, the helicopter turns (rotates) one way or the other.

(a)

(b)

•**FIGURE 8.28 Different rotors**

Chapter Review

Important Terms

rigid body 252
translational motion 252
rotational motion 253
instantaneous axis of rotation 253
moment (lever arm) 254
torque 254
translational equilibrium 256
concurrent forces 256
rotational equilibrium 256
mechanical equilibrium 256
static equilibrium 257
center of gravity 261
stable equilibrium 261
unstable equilibrium 262
moment of inertia 265
parallel axis theorem 268
rotational work 274
rotational power 274
rotational kinetic energy 274
angular momentum 277
conservation of angular momentum 279

Important Concepts

- In pure translational motion, all the particles of an object have the same instantaneous velocity.
- In pure rotational motion (about a fixed axis), all the particles of an object have the same instantaneous angular velocity.
- Torque, the rotational analogue of force, is the product of a force and a moment arm, or lever arm.
- Mechanical equilibrium requires that the summation of the forces be zero (translational equilibrium) and that the summation of the torques be zero (rotational equilibrium).

- An object is in stable equilibrium as long as its center of gravity, upon small displacement, lies above and inside its original base of support.
- Moment of inertia (I) is the rotational analogue of mass in the rotational form of Newton's second law, $\boldsymbol{\tau} = I\boldsymbol{\alpha}$.
- Angular momentum, the product of a moment arm and linear momentum as well as the product of a moment of inertia and angular velocity, is conserved in the absence of an external, unbalanced torque.

Important Equations

Condition for Rolling without Slipping:
$$v_{CM} = r\omega \tag{8.1}$$
(or $s = r\theta$ or $a_{CM} = r\alpha$)

Torque (magnitude):
$$\tau = r_\perp F = rF \sin\theta \tag{8.2}$$

Conditions for Translational and Rotational Mechanical Equilibria, Respectively:
$$\Sigma \mathbf{F}_i = 0 \quad \text{and} \quad \Sigma \boldsymbol{\tau}_i = 0 \tag{8.3}$$

Torque on a Particle (magnitude):
$$\tau = r_\perp F = rF_\perp = rma_\perp = mr^2\alpha \tag{8.4}$$

Moment of Inertia:
$$I = \Sigma m_i r_i^2 \tag{8.6}$$

Rotational Form of Newton's Second Law (magnitude):
$$\tau = I\alpha \tag{8.7}$$

Parallel Axis Theorem:
$$I = I_{CM} + Md^2 \tag{8.8}$$

Rotational Work:
$$W = \tau\theta \tag{8.9}$$

Rotational Power:
$$P = \tau\omega \tag{8.10}$$

Work–Energy Theorem:
$$W = \Delta K = \tfrac{1}{2}I\omega^2 - \tfrac{1}{2}I\omega_o^2 \tag{8.11}$$

Rotational Kinetic Energy:
$$K = \tfrac{1}{2}I\omega^2 \tag{8.12}$$

Kinetic Energy of a Rolling Object:
$$K = \tfrac{1}{2}I_{CM}\omega^2 + \tfrac{1}{2}Mv_{CM}^2 \tag{8.13}$$

Angular Momentum of a Particle in Circular Motion (magnitude):
$$L = r_\perp p = mr_\perp v = mr_\perp^2 \omega \tag{8.14}$$

Angular Momentum of a Rigid Body:
$$\mathbf{L} = I\boldsymbol{\omega} \tag{8.16}$$

Torque in Terms of Angular Momentum:
$$\boldsymbol{\tau} = \frac{\Delta \mathbf{L}}{\Delta t} \tag{8.17}$$

Conservation of Angular Momentum (with $\tau = 0$):
$$I\boldsymbol{\omega} = I_o\boldsymbol{\omega}_o \tag{8.18}$$

Exercises

8.1 Rigid Bodies, Translations, and Rotations

1. In pure rotational motion of a rigid body, (a) all the particles of the body have the same angular velocity, (b) all the particles of the body have the same tangential velocity, (c) the acceleration is always zero, (d) there are always two simultaneous axes of rotation.

2. The condition for rolling without slipping is (a) $a = r\omega^2$, (b) $v = r\omega$, (c) $F = ma$, (d) $a_c = v^2/r$.

3. Suppose someone in your physics class says that it is possible for a rigid body to have translational motion and rotational motion at the same time. Would you agree? Why?

4. For a rolling cylinder, what would happen if v were less than $r\omega$? Is it possible for v to be greater than $r\omega$? Explain.

5. For the tires on your skidding car, (a) $v = r\omega$, (b) $v > r\omega$, (c) $v < r\omega$.

6. ■ A rope goes over a circular pulley with a radius 6.5 cm. If the pulley makes four revolutions without the rope slipping, what length of rope passes over the pulley?

7. ■ A wheel rolls five revolutions on a horizontal surface without slipping. If the center of the wheel moves 3.2 m, what is the radius of the wheel?

8. ■ A wheel rolls uniformly on level ground without slipping. A piece of mud on the wheel flies off when it is at the 9 o'clock position (rear of wheel). Describe the subsequent motion of the mud.

9. ■ A circular disk with a radius of 0.25 m rolls without slipping with an angular speed of 2.0 rad/s on a level surface. (a) What is the linear speed of the center of mass of the disk? (b) What is the instantaneous tangential speed of the top of the disk?

10. ■■ A ball with a radius of 15 cm rolls on a level surface; the translational speed of the center of mass is 0.25 m/s. What is the angular speed about the center of mass if the ball rolls without slipping?

11. ■■ A disk with a radius of 0.15 m rotates through 270° as it travels 0.71 m. Does the disk roll without slipping?

12. ■■ Show that $a = r\alpha$ is also a condition for rolling without slipping.

13. ■■■ A cylinder with a diameter of 20 cm rolls with an angular speed of 0.50 rad/s on a level surface. If the cylinder experiences a uniform tangential acceleration of 0.018 m/s^2 without slipping until its angular speed is 1.25 rad/s, through how many complete revolutions does the cylinder rotate during the acceleration time?

8.2 Torque, Equilibrium, and Stability

14. A net torque is possible when (a) all forces act through the axis of rotation, (b) $\Sigma F_i = 0$, (c) a body is in rotational equilibrium, (d) a body remains in unstable equilibrium.

15. If an object in unstable equilibrium is slightly displaced, (a) its potential energy will decrease, (b) the center of gravity is directly above the axis of rotation, (c) no gravitational work is done, (d) stable equilibrium follows.

16. Explain the balancing acts in •Fig. 8.29. Where are the centers of gravity?

•**FIGURE 8.29 Balancing acts** See Exercise 16. *Left:* A toothpick on the rim of the glass supports a fork and spoon. *Right:* (*top*) A hinged board is mounted on a pole. (*bottom*) The handle of a hammer is placed in a loop that runs around the board; the hinged part of the board remains horizontal.

17. •Figure 8.30 shows a toy clown walking on a tightrope. Explain how this brainless toy can perform such a difficult task. Is the clown in stable equilibrium? What happens if

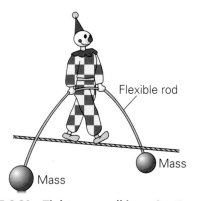

•**FIGURE 8.30 Tightrope walking** See Exercise 17.

the weights are removed? [Hint: Think about the center of gravity.]

18. ■ The drain plug on a car's engine is required to be tightened to a torque of 25 m·N. If you are using a 0.15-meter-long wrench to change the oil of that engine, what is the minimum force you need to apply to loosen the plug?

19. ■ In Exercise 18, due to space limitations (you are crawling under the car), you cannot apply the force perpendicularly to the length of the wrench. If your force makes a 30° angle to that length, what is the force required?

20. ■ In Fig. 8.4a, if the arm makes a 37° angle to the horizontal and a torque of 18 m·N is to be produced, what is the force required by the biceps muscle?

21. ■ How many different positions of stable equilibrium and unstable equilibrium are there for a cube? Consider each surface, edge, and corner to be a different position.

22. ■ The pedals of a bicycle rotate in a circle with a diameter of 40 cm. What is the maximum torque a 55-kilogram rider can apply by putting all of her weight on one pedal?

23. ■ A 35-kilogram child sits on a uniform seesaw of negligible mass; she is 2.0 m from the pivot point (or fulcrum). How far from the pivot point on the other side will her 30-kilogram playmate have to sit for the seesaw to be in equilibrium?

24. ■ A uniform meterstick pivoted at its center, as in Example 8.4, has a 100-gram mass suspended at the 25.0-centimeter position. (a) At what position should a 75.0-gram mass be suspended to put the system in equilibrium? (b) What mass would have to be suspended at the 90.0-centimeter position for the system to be in equilibrium?

25. ■■ Show that the balanced meterstick in Example 8.4 is in static rotational equilibrium about a horizontal axis through the zero end of the stick.

26. ■■ Telephone and electrical lines are allowed to sag between poles so that the tension will not be too great when something hits or sits on the line. Suppose that a line were stretched almost perfectly horizontally between two poles that are 30 m apart. If a 0.25-kilogram bird perches on the wire midway between the poles and the wire sags 1.0 cm, what would the tension in the wire be?

27. ■■ In •Fig. 8.31, what is the force F_m supplied by the deltoid muscle so as to hold up the outstretched arm if the mass of the arm is 3.0 kg? (F_j is the joint force on the bone of the upper arm—the humerus.)

28. ■■ A variation of *Russell traction* (•Fig. 8.32) supports the lower leg when in a cast. Suppose that the patient's leg and cast have a combined mass of 15.0 kg and m_1 is 4.50 kg. (a) What is the reaction force of the leg muscles to the traction? (b) What must m_2 be to keep the leg horizontal?

•FIGURE 8.31 **Arm in static equilibrium** See Exercise 27.

•FIGURE 8.32 **Static traction** See Exercise 28.

29. ■■ A uniform meterstick weighing 5.0 N is pivoted so it can rotate about a horizontal axis through one end (•Fig. 8.33). If a 0.15-kilogram mass is suspended 75 cm from the pivoted end, what is the tension in the string?

•FIGURE 8.33 **What's the tension?** See Exercise 29.

30. ■■ In Example 8.5, if the mass of the painter is 65 kg, what frictional force must act on the bottom of the ladder to keep it from slipping?

31. ■■ An artist wishes to construct a birds-and-bees mobile, as shown in •Fig. 8.34. If the mass of the bee on the lower left is 0.10 kg and each vertical support string has a length of 30 cm, what are the masses of the other birds and bees? (Neglect the masses of the bars and strings.)

•FIGURE 8.34 **Birds and bees** See Exercise 31.

32. ■■ (a) How many uniform, identical textbooks of width 25.0 cm can be stacked on top of each other on a level surface without the stack falling over if each successive book is displaced 3.00 cm in width relative to the book below it? (b) If the books are 5.00 cm thick, what will be the height of the center of mass of the stack above the level surface?

33. ■■ If four metersticks were stacked on a table with 10 cm, 15 cm, 30 cm, and 50 cm, respectively, hanging over the edge, as shown in •Fig. 8.35, would the metersticks remain on the table?

•FIGURE 8.35 **Will they fall off?** See Exercise 33.

34. ■■ A 10.0-kilogram solid uniform cube with 0.500-meter sides rests on a level surface. What is the minimum amount of work necessary to put the cube into an unstable equilibrium position?

35. ■■■ While standing on a long board resting on a scaffold, a 70-kilogram painter paints the side of a house (•Fig. 8.36). If the mass of the board is 15 kg, how close to the end can the painter stand without tipping the board over?

•FIGURE 8.36 **Not too far!** See Exercises 35 and 36.

36. ■■■ Suppose that the board in Fig. 8.36 were suspended from vertical ropes attached to each end instead of resting on scaffolding. If the painter stood 1.5 m from one end of the board, what would the tensions in the ropes be? (See Exercise 35 for additional data.)

37. ■■■ The forces acting on Einstein and the bicycle (Fig. 1 of the Insight on p. 266) are the total weight of Einstein and the bicycle (mg) at the center of gravity of the system, the normal force (N) by the road, and the force of static friction (f_s) on the tires exerted by the road. (a) Explain why the angle of lean θ the bicycle makes with the vertical must be given by $\tan\theta = f_s/N$ if Einstein is to maintain balance. (b) The angle θ in the picture is about 11°. What is the minimum coefficient of static friction μ_s between the road and the tires? (c) If the radius of the circular path is 6.5 m, what is the maximum speed of Einstein's bicycle? [Hint: The net torque about the center of gravity must be zero for rotational equilibrium.]

8.3 Rotational Dynamics

38. The moment of inertia of a rigid body (a) depends on the axis of rotation, (b) cannot be zero, (c) depends on mass distribution, (d) all of the preceding.

39. Which of the following best describes the physical quantity called torque: (a) rotational analogue of force, (b) energy due to rotation, (c) rate of change of linear momentum, (d) force that is tangent to a circle?

40. (a) Does the moment of inertia of a rigid body depend in any way on the center of mass? Explain. (b) Can a moment of inertia have a negative value? If so, explain what this would mean.

41. Why does the moment of inertia of a rigid body have different values for different axes of rotation? What does this mean physically?

42. When a hard-boiled egg laying on a table is given a quick torque (spin), it will rise up and spin on one end like a top. A raw egg will not. Why the difference?

43. Why does jerking a paper towel from a roll cause the paper to tear better than pulling it smoothly? Will the amount of paper on the roll affect the results?

44. (a) Why is it easier to balance a meterstick vertically on your finger than on a pencil? (b) A softball, a volleyball, and a basketball are released at the same time from the top of an inclined plane. Give the results of the race.

45. ■ For the system of masses shown in •Fig. 8.37, find the moment of inertia about (a) the x axis, (b) the y axis, (c) an axis through the origin perpendicular to the page (z axis). Neglect the masses of the connecting rods.

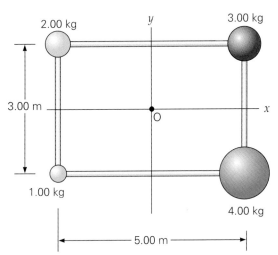

•FIGURE 8.37 Moments of inertia about different axes See Exercise 45.

46. ■ A fixed 0.15-kilogram solid-disk pulley with a radius of 0.075 m is acted on by a net torque of 6.4 m·N. What is the angular acceleration of the pulley?

47. ■ What net torque is needed to give a uniform 20-kilogram solid ball with a radius of 0.20 m and angular acceleration of 2.0 rad/s^2?

48. ■ A light meterstick is loaded with bodies of masses 2.0 kg and 4.0 kg at the 30 cm and 75 cm positions, respectively. What is (a) the moment of inertia about an axis through the 0-centimeter end of the meterstick? (b) the moment of inertia about an axis through the center of mass of the system? (c) Use the parallel axis theorem to find the moment of inertia about the axis through the 0-centimeter end of the stick, and compare the result with that of part (a).

49. ■■ Two masses are joined by a light rod as shown in •Fig. 8.38. Find the moment of inertia of the system about an axis perpendicular to the rod through (a) the center of the rod and (b) the center of mass. (c) Is the moment of inertia about the center of mass the minimum?

•FIGURE 8.38 Moments of inertia about axes through different centers See Exercise 49.

50. ■■ A 2000-kilogram Ferris wheel accelerates from rest to an angular speed of 2.0 rad/s in 12 s. Approximating the Ferris wheel as a circular disk with a radius of 30 m, what is the net torque on the wheel?

51. ■■ A 15-kilogram uniform sphere with a radius of 15 cm rotates about an axis tangent to its surface at 3.0 rad/s. A constant torque of 10 m·N then increases the rotational speed to 7.5 rad/s. Through what angle does the sphere rotate while accelerating?

52. ■■ A 10-kilogram solid disk of radius 0.50 m is rotated about an axis through its center. If the disk accelerates from rest to an angular speed of 3.0 rad/s while rotating 2.0 revolutions, what net torque is required?

53. ■■ If the Earth rotated about a north–south axis tangent to the Earth at the Equator, how much more moment of inertia would it have compared to that about an axis through its center?

54. ■■ A 0.25-kilogram disk-shaped machine part with a radius of 6.0 cm rotates eccentrically (off-centered) about an axis that is normal to its flat surface and located two-thirds of the distance from the center to the circumference. (a) What torque is required to rotate the part about this off-center axis with an angular acceleration of 2.0 rad/s^2? (b) How much greater is this torque than the torque required to rotate the part with the same angular acceleration about a parallel axis through the center?

55. ■■ Two masses are suspended from a pulley as shown in •Fig. 8.39 (the Atwood machine revisited, Chapter 4). The pulley has a mass of 0.20 kg, a radius of 0.15 m, and a constant torque of 0.35 m·N due to the friction between the rotating pulley and its axle. What is the magnitude of the acceleration of the suspended masses if $m_1 = 0.40$ kg and $m_2 = 0.80$ kg? (Neglect the mass of the string.)

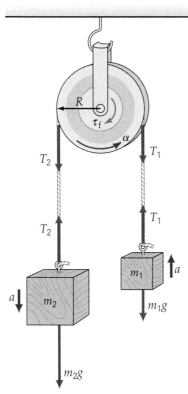

•**FIGURE 8.39 The Atwood machine revisited** See Exercise 55.

56. ■■ For the system in •Fig. 8.40, $m_1 = 8.0$ kg, $m_2 = 3.0$ kg, $\theta = 30°$, and the radius and mass of the pulley are 0.10 m and 0.10 kg, respectively. (a) What is the acceleration of the masses? (Neglect friction and the string's mass.) (b) The pulley has a constant frictional torque of 0.050 m·N when the system is in motion, what is the acceleration? [Hint: Isolate the forces. The tensions in the strings are different. Why?]

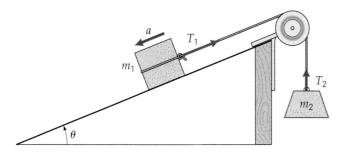

•**FIGURE 8.40 Inclined plane and pulley** See Exercise 56.

57. ■■ What is the magnitude of the tangential force that must be applied to the rim of a 2.0-kilogram wheel that is disk-shaped and has a radius of 0.50 m in order to give it an angular acceleration of 4.8 rad/s²? (Neglect friction.)

58. ■■ A man starts his lawn mower by applying to the starter rope a constant tangential force of 150 N to the 0.30-kilogram disk-shaped flywheel. The diameter of the flywheel is 18 cm. What is the angular speed of the wheel after it has turned through one revolution? (Neglect friction and motor compression.)

59. ■■ A meterstick, pivoted about a horizontal axis through the 0-centimeter end, is held in a horizontal position and let go. (a) What is the tangential acceleration of the 100-centimeter position? Are you surprised by this result? (b) Which position has a tangential acceleration equal to the acceleration due to gravity?

60. ■■ Pennies are placed every 10 cm on a meterstick. One end is put on a table and the other end is held horizontally with a finger as shown in •Fig. 8.41. If the finger is pulled away, what happens to the pennies? [Hint: See Exercise 59.]

•**FIGURE 8.41 Money left behind?** See Exercise 60.

61. ■■■ A uniform 2.0-kilogram cylinder of radius 0.15 m is suspended by two strings wrapped around it (•Fig. 8.42). As the cylinder descends, the strings unwind from it. What is the acceleration of the center of mass of the cylinder? (Neglect the mass of the string.)

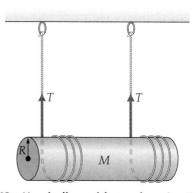

•**FIGURE 8.42 Unwinding with gravity** See Exercise 61.

62. ■■■ A uniform hoop rolls without slipping down a 15° inclined plane. What is the acceleration of the hoop's center of mass?

63. ■■■ In Fig. 8.20, what would be the maximum angle of incline (in terms of the coefficient of static friction) for the ball to roll without slipping down the incline?

8.4 Rotational Work and Kinetic Energy

64. From $W = \tau\theta$, the unit(s) of rotational work is (a) watt, (b) N·m, (c) kg·rad/s², (d) N·rad.

65. A bowling ball rolls without slipping on a flat surface. The ball has (a) rotational kinetic energy, (b) translational kinetic energy, (c) both translational and rotational kinetic energies.

66. ■ A constant retarding torque of 12 m·N stops a rolling wheel of diameter 0.80 m in a distance of 15 m. How much work is done by the torque?

67. ■ A person opens a door by applying a 15-newton force perpendicular to it at a distance 0.90 m from the hinges. The door is pushed wide open (to 120°) in 2.0 s. (a) How much work was done? (b) What was the average power delivered?

68. ■ A constant torque of 10 m·N is applied to a 10-kilogram uniform disk of radius 0.20 m. What is the angular speed of the disk about an axis through its center after it rotates 2.0 revolutions from rest?

69. ■ A 2.5-kilogram pulley of radius 0.15 m is pivoted about an axis through its center. What constant torque is required so the pulley will reach an angular speed of 25 rad/s after rotating 3.0 revolutions, starting from rest?

70. ■ Use the conservation of mechanical energy to find the linear speed of the descending mass ($m = 1.0$ kg) of Fig. 8.19 after it has descended a vertical distance of 2.0 m from rest. (For the pulley, $M = 0.30$ kg and $R = 0.15$ m. Neglect friction and the mass of the string.)

71. ■■ A sphere with a radius of 15 cm rolls on a level surface with a constant angular speed of 10 rad/s. To what height on a 30° inclined plane will the sphere roll before coming to rest? (Neglect frictional losses.)

72. ■■ A thin 1.0-meter-long rod pivoted at one end falls (rotates) from a horizontal position, starting from rest and with no friction. What is the angular speed of the rod when it is vertical? [Hint: Consider the center of mass and use the conservation of mechanical energy.]

73. ■■ A uniform sphere and a uniform cylinder with the same mass and radius roll at the same velocity side by side on a level surface without slipping. If the sphere and the cylinder approach an inclined plane and roll up it without slipping, will they be at the same height on the plane when they come to a stop? If not, what will be the percentage height difference?

74. ■■ ⊙ A hoop starts from rest at a height 1.2 m above the base of an inclined plane and rolls down under the influence of gravity. What is the linear speed of the hoop's center of mass just as the hoop leaves the incline and rolls onto a horizontal surface? (Neglect friction.)

75. ■■ An industrial flywheel with a moment of inertia of 4.25×10^2 kg·m² rotates with a speed of 7500 rpm. (a) How much work is required to bring it to rest? (b) If this work is done uniformly in 1.5 min, how much power is expended?

76. ■■ ⊙ A cylindrical hoop, a cylinder, and a sphere of equal radius and mass are released at the same time from the top of an inclined plane. Using the conservation of mechanical energy, show that the sphere always gets to the bottom of the incline first with a faster speed and that the hoop always arrives last with the slowest speed.

77. ■■ ⊙ For the following objects, which roll without slipping, determine the rotational kinetic energy about the center of mass as a percentage of the total instantaneous kinetic energy: (a) solid sphere, (b) a thin spherical shell, (c) a thin cylindrical shell.

78. ■■ A 0.050-kilogram phonograph record with a radius 0.15 m drops onto a turntable and is soon rotating at $33\frac{1}{3}$ rpm. How much work must be supplied to get the record to rotate at this speed, and what supplies it?

79. ■■■ A steel ball rolls down an incline into a loop-the-loop of radius R (•Fig. 8.43a). (a) What minimum speed must the ball have at the top of the loop in order to stay on the track? (b) At what vertical height (h) on the incline, in terms of the radius of the loop, must the ball be released in order for it to have the required minimum speed at the top of the loop? (Neglect frictional losses.) (c) Figure 8.43b shows the loop-the-loop of a roller coaster. What are the sensations of the riders if the roller coaster has the minimum speed or greater at the top of the loop? [Hint: In case the speed is below the minimum, seat and shoulder straps hold the riders in.]

(a)

(b)

•FIGURE 8.43 **Loop-the-loop and rotational speed** See Exercise 79.

8.5 Angular Momentum

80. The units of angular momentum are (a) N·m, (b) kg·m/s², (c) kg·m²/s, (d) J·m.

81. A child stands on the edge of a rotating (hand-driven) merry-go-round. He then starts to walk toward the center of the merry-go-round. A dangerous situation results. Why?

82. The release of vast amounts of carbon dioxide may increase the Earth's average temperature through the greenhouse effect and cause the polar ice caps to melt. If this occurred

and the ocean level rose substantially, what effect would it have on the Earth's rotation and on the length of the day?

83. An ice skater goes into a fast spin by tucking her arms in (•Fig. 8.44a), and a high-platform diver often draws her legs up against her chest (Fig. 8.44b). Why are the arms and legs put into these positions in each case?

•**FIGURE 8.44 Faster rotation** See Exercise 83.

84. In the classroom demonstration illustrated in •Fig. 8.45, a person on a rotating stool holds a rotating bicycle wheel by handles attached to the wheel. When the wheel is held horizontally, she rotates one way (clockwise as viewed from above). When the wheel is turned over, she rotates in the opposite direction. Explain why this occurs. [Hint: Consider angular momentum vectors.]

85. Cats usually land on their feet when they fall, even if held upside down when dropped (•Fig. 8.46). While a cat is falling, there is no external torque and its center of mass falls as a particle. How can cats turn themselves over while falling?

•**FIGURE 8.46 One down, eight to go** See Exercise 85.

•**FIGURE 8.45 A double rotation** See Exercise 84.

86. ■ What is the angular momentum of a 2.0-gram particle moving counterclockwise (as viewed from above) with an angular speed of 5π rad/s in a horizontal circle of radius 15 cm? (Give magnitude and direction.)

87. ■ A 10-kilogram rotating disk of radius 0.25 m has an angular momentum of 0.45 kg·m^2/s. What is its angular speed?

88. ■■ Compute the ratio of the Earth's orbital angular momentum and its rotational angular momentum. Are these momenta in the same direction?

89. ■■ The period of the Moon's rotation is the same as the period of its revolution: 27.3 days (sidereal). What is the angular momentum for each, rotation and revolution? (Because the periods are equal, we see only one side of the Moon from Earth.)

90. ■■ Circular disks are used in automobile clutches and transmissions. When a rotating disk couples with a stationary one through friction, energy from the rotating disk is transferred to the stationary one. If a disk rotating at 800 rpm couples with a stationary disk with three times the moment of inertia, what is the angular speed of the combination?

91. ■■ ◉ A rotating spherical star in part of its life cycle expands to six times its normal volume. Assuming the mass remains constant and uniformly distributed inside the star, how is the period of rotation affected?

92. ■■ ◉ A skater has a moment of inertia of 100 kg·m² when his arms are outstretched and a moment of inertia of 75 kg·m² when his arms are tucked in close to his chest. If he starts to spin at an angular speed of 2.0 rps (revolutions per second) with his arms outstretched, what will his angular speed be when they are tucked in?

93. ■■ ◉ An ice skater spinning with outstretched arms has an angular speed of 4.0 rad/s. She then tucks in her arms, decreasing her moment of inertia by 7.5%. (a) What is the resulting angular speed? (b) By what factor does the skater's kinetic energy change? (Neglect any frictional effects.) (c) Where does the extra kinetic energy come from?

94. ■■■ A comet approaches the Sun as illustrated in •Fig. 8.47 and is deflected by the Sun's gravitational attraction. This event is considered a collision, and b is called the impact parameter. Find the distance of closest approach (d) in terms of the impact parameter and the velocities (v_o at large distances and v at closest approach). Consider the radius of the Sun to be negligible compared to d. (As the figure shows, the tail of a comet always "points" away from the Sun.)

95. ■■■ A 0.50-kilogram kitten stands on the edge of a lazy Susan (a turntable) of mass 1.5 kg and radius 0.30 m. Assume that the lazy Susan has frictionless bearings and is initially at rest. (a) What will happen if the kitten starts walking around the edge of the lazy Susan? (b) If the kitten walks at a speed of 0.25 m/s relative to the ground, what will be the angular speed of the lazy Susan? (c) When the kitten has walked completely around the edge and is back at its starting point, will that point be above the same point on the ground as it was at the start? If not, where is the kitten relative to the starting ground point? (Speculate on what might happen if everyone on Earth suddenly started to run eastward. What effect might this have on the length of a day?)

Additional Exercises

96. A 10 000-kilogram bridge of length 10 m is supported at both ends. If a 2000-kilogram car is parked on the bridge 3.0 m from the left support, what are the supporting forces at the left and right ends?

97. A spring is twisted 60° about its linear axis (axis of symmetry) by a torque of 100 m·N. How much torsional, or rotational, energy is stored in the spring? [Hint: Let the restoring torque of the spring be $\tau = k\theta$, the rotational analogue of $F = kx$ (Hooke's law—see Chapter 5) with the minus sign omitted.]

98. ◉ Prove that a thin hoop or ring starting from rest will roll more slowly down an inclined plane than an annular cylinder. (Remember that masses and radii do not have to be taken into account.)

99. A 10-kilogram pulley of radius 0.50 m is connected to a 5.0-kilogram mass through a string as shown in •Fig. 8.48. If the mass is let go from rest, (a) what is its linear acceleration? (b) What is the angular acceleration of the pulley?

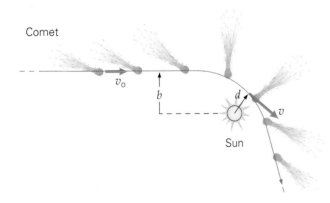

•FIGURE 8.47 A comet "collision" See Exercise 94.

•FIGURE 8.48 Linear and angular accelerations See Exercise 99.

100. A bicycle wheel has a diameter of 0.80 m and a mass of 2.0 kg. (a) If the bicycle travels with a linear speed of 1.5 m/s, what is the angular speed of the wheel? (b) What is the angular momentum of the wheel? (Consider the wheel to be a thin ring.)

101. A piece of machinery with a moment of inertia of 2.6 kg·m^2 rotates with a constant angular speed of 4.0 rad/s when experiencing a frictional torque of 0.56 m·N. What is the net torque acting on the piece?

102. The location of a person's center of gravity relative to his or her height can be found by using the arrangement shown in •Fig. 8.49. The scales are initially adjusted to zero with the board alone. Locate the center of gravity of the person relative to the horizontal dimension. Would you expect the location of the center of gravity in other dimensions to be exactly at the midway points? Explain.

•**FIGURE 8.49 Locating the center of gravity** See Exercise 102.

103. Using bricks identical to those in Example 8.6, you are asked to create a stack of nine bricks (maximum) without its toppling, with each added brick displaced an equal distance horizontally in the same direction. (a) What is the maximum displacement that will allow this? (b) What is the height of the center of mass of the stack if the uniform bricks are 8.0 cm thick?

104. The *radius of gyration* (k) is defined as the distance from an axis of rotation at which all the mass of an object would have to be concentrated to give the same moment of inertia as the actual mass distribution would have—that is, $I = Mk^2$. Take the object to be a rotating particle of mass M. From the axes shown in Fig. 8.16, determine the radius of gyration for (a) a uniform sphere, (b) a uniform cylinder, and (c) a particle.

105. A 25-kilogram child jumps onto the rim of a 100-kilogram rotating disk of radius 2.0 m. If the angular speed of the disk before the child's jump was 2.0 rad/s, what is the angular speed of the disk–child system?

106. A mass is suspended by two cords as shown in •Fig. 8.50. What are the tensions in the cords?

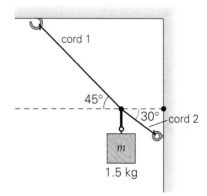

•**FIGURE 8.50 Strung-out equilibrium** See Exercises 106 and 107.

107. If the cord attached to the vertical wall in Fig. 8.50 were horizontal (instead of at a 30° angle), what would the tensions in the cords be?

108. The outer back wheels of an oversize tractor-trailer rig are 3.66 m apart. When the trailer is loaded, its center of gravity is equidistant from its sides and 3.58 m above the road surface. At what road angle will the parked trailer be in unstable equilibrium?

109. A 0.50-kilogram thin rectangular plate has dimensions of 30 cm by 40 cm. For which axis of rotation does the plate have the greater rotational inertia, an axis tangent to a smaller side at its midpoint or an axis tangent to a larger side at its midpoint? How many times more? (Both axes are perpendicular to the plate surface.)

CHAPTER 9

Solids and Fluids

9.1 Solids and Elastic Moduli
9.2 Fluids: Pressure and Pascal's Principle
9.3 Buoyancy and Archimedes' Principle
9.4 Fluid Dynamics and Bernoulli's Equation
*9.5 Surface Tension; Viscosity and Poiseuille's Law

Insights

- Feat of Strength or Knowledge of Materials?
- Blood Pressure
- Throwing a Curveball

Like the person hang-gliding, we exist at the intersection of three realms. We walk on the solid surface of the Earth, and in our daily lives use solid objects of all sorts, from scissors to computers. But we are surrounded by fluids—liquids and gases—on which we are completely dependent. Without the water that we drink, we could survive for a few days at most; without the oxygen in the air that we breathe, we could not live for more than a few minutes. Indeed, we ourselves are not nearly as solid as we like to think. By far the most abundant substance in our bodies is water, and it is in the watery environment of our cells that all the chemical processes on which life depends take place.

On the basis of general physical distinctions, matter is commonly divided into these three phases: solid, liquid, and gas. A solid has a definite shape and volume. A liquid has a fairly definite volume but assumes the shape of its container. A gas takes on the shape and volume of its container. Solids and liquids are sometimes called *condensed matter*. We will utilize a different classification scheme and consider matter in terms of solids and fluids. Liquids and gases are referred to collectively as fluids. A **fluid** is a substance that can flow; liquids and gases qualify, but solids do not.

A simplistic description of solids is that they are made up of particles called atoms that are held rigidly together by interatomic forces. This concept of an ideal rigid body was used in Chapter 8 to describe rotational motion. Real solid bodies are not absolutely rigid and can be elastically deformed by external forces. Elasticity usually brings to mind a rubber band or spring that will resume its original dimensions even after being greatly deformed. In fact, all materials are elastic to some degree, even very hard steel. But, as you will learn, there's a limit to such deformation—an *elastic limit*.

Fluids, however, have little or no elastic response to force—a force merely causes an unconfined fluid to flow. This chapter pays particular attention to the behavior of fluids, shedding light on such questions as why icebergs and ocean liners float, why paper towels absorb spills, and what the 10W–40 on a can of motor oil means. You'll also discover why the person in the photo can neither float like a balloon nor fly like a hummingbird yet can, with the aid of a suitably shaped piece of plastic, soar like an eagle.

Because of their fluidity, liquids and gases have many properties in common, and it is convenient to study them together. There are important differences as well: For example, liquids are not very compressible, whereas gases are easily compressed, as we will shortly see.

9.1 Solids and Elastic Moduli

Objectives: To (a) distinguish between stress and strain and (b) use elastic moduli to compute dimensional changes.

As stated previously, all solid materials are elastic to some degree. That is, a body that is slightly deformed by an applied force will return to its original dimensions or shape when the force is removed. The deformation may not be noticeable for many materials, but it's there.

You may be able to visualize why materials are elastic if you think in terms of the simplistic model of a solid in •Fig. 9.1. The atoms of the solid substance are imagined to be held together by springs. The elasticity of the springs represents the resilient nature of the interatomic forces. The springs resist permanent deformation, as do the forces between atoms.

The elastic properties of solids are commonly discussed in terms of stress and strain. **Stress** is a measure of the force causing a deformation. **Strain** is a relative measure of the deformation a stress causes. Quantitatively, *stress is the applied force per unit cross-sectional area*:

$$\text{stress} = \frac{F}{A} \qquad (9.1)$$

SI unit of stress: newtons per square meter (N/m^2)

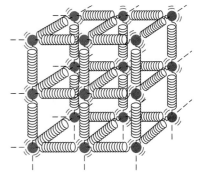

•**FIGURE 9.1 A springy solid** The elastic nature of interatomic forces is indicated by simplistically representing them as springs, which, like the forces, resist deformation.

Here F is the magnitude of the applied force normal (perpendicular) to the cross-sectional area. Equation 9.1 shows that the SI units for stress are newtons per square meter (N/m^2).

As illustrated in •Fig. 9.2, a force applied to the ends of a rod gives rise to either a *tensile stress* (an elongating tension) or a *compressional stress* (a shortening tension), depending on the direction of the force. In both these cases, the *tensile strain* is the ratio of the change in length to the original length:

$$\text{strain} = \frac{\text{change in length}}{\text{original length}} = \frac{\Delta L}{L_o} \qquad (9.2)$$

Strain is a unitless quantity

•**FIGURE 9.2 Tensile and compressional stress and strain** Tensile and compressional stresses are due to forces applied normally to the surface area of the ends of bodies. **(a)** A tension, or tensile stress, tends to increase the length of an object. **(b)** A compressional stress tends to shorten the length.

where $\Delta L = L - L_o$. Note that strain is a unitless quantity (length/length). It is the *fractional change* in length. For example, if the strain is 0.05, the material has changed in length by 5% of its original length.

As might be expected, the resulting strain is proportional to the applied stress; that is, strain $\propto$ stress. For relatively small stresses, this is a direct proportion. The constant of proportionality, which depends on the nature of the material, is called the **elastic modulus**. Thus,

$$\text{stress} = \text{elastic modulus} \times \text{strain}$$

or

Definition of:
Elastic modulus

$$\text{elastic modulus} = \frac{\text{stress}}{\text{strain}} \quad (9.3)$$

SI unit of elastic modulus: newtons per square meter (N/m^2)

That is, the elastic modulus is the stress divided by the strain, or the ratio of stress to strain. The elastic modulus has the same units as stress. Why?

Three general types of elastic moduli (plural of modulus) are associated with stresses that produce changes in length, shape, or volume. These are called *Young's modulus, shear modulus,* and *bulk modulus,* respectively.

Change in Length: Young's Modulus

•Figure 9.3 is a graph of the tensile stress versus the strain for a typical metal rod. The curve is a straight line up to a point called the *proportional limit*. Beyond this point, the strain begins to increase more rapidly to another critical point called the **elastic limit**. If the tension is removed at this point, the material will return

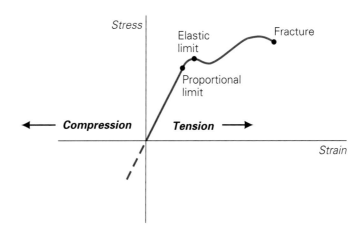

•**FIGURE 9.3 Stress versus strain** A plot of stress versus strain for a typical metal rod is a straight line up to the proportional limit. Then elastic deformation continues until the elastic limit is reached. Beyond that, the rod will be permanently deformed and will eventually fracture or break.

TABLE 9.1 Elastic Moduli for Various Materials (in N/m²)			
Substance	Young's Modulus (Y)	Shear Modulus (S)	Bulk Modulus (B)
Solids			
Aluminum	7.0×10^{10}	2.5×10^{10}	7.0×10^{10}
Bone (limb)	1.5×10^{10}	8.0×10^{10}	
Brass	9.0×10^{10}	3.5×10^{10}	7.5×10^{10}
Copper	11×10^{10}	3.8×10^{10}	12×10^{10}
Glass	5.7×10^{10}	2.4×10^{10}	4.0×10^{10}
Iron	15×10^{10}	6.0×10^{10}	12×10^{10}
Steel	20×10^{10}	8.2×10^{10}	15×10^{10}
Liquids			
Alcohol, ethyl			1.0×10^{9}
Glycerin			4.5×10^{9}
Mercury			26×10^{9}
Water			2.2×10^{9}

to its original length. If the tension is applied beyond the elastic limit and then removed, the material will recover somewhat but will retain some permanent deformation.

The straight-line part of the graph shows a direct proportionality between stress and strain. This relationship, first formalized by Robert Hooke in 1678, is now known as *Hooke's law*. (It is the same general relationship as that given for a spring in Section 5.2—see Fig. 5.5.) The elastic modulus for a tension or a compression is called **Young's modulus** (Y):

$$\underbrace{\frac{F}{A}}_{\text{stress}} = Y \underbrace{\left(\frac{\Delta L}{L_o}\right)}_{\text{strain}} \quad \text{or} \quad Y = \frac{F/A}{\Delta L/L_o} \tag{9.4}$$

SI unit of Young's modulus: newtons per square meter (N/m²)

Thomas Young (1773–1829) was a British physicist who investigated the mechanical properties of materials and optical phenomena.

The units of Young's modulus are the same as of stress, newtons per square meter (N/m²), since the ratio of lengths is unitless. Some typical values of Young's modulus are given in Table 9.1.

To obtain a conceptual or physical understanding of Young's modulus, let's solve Eq. 9.4 for ΔL:

$$\Delta L = \left(\frac{FL_o}{A}\right)\frac{1}{Y} \quad \text{or} \quad \Delta L \propto \frac{1}{Y}$$

Hence we see that the larger the modulus of a material, the smaller its relative change in length (with other parameters being equal).

Example 9.1 ■ Pulling My Leg: Under a Lot of Stress

The femur (upper leg bone) is the longest and strongest bone in the body. Taking a typical femur to be approximately circular with a radius of 2.0 cm, how much force would be required to extend the bone by 0.010%?

Thinking It Through. We can see that Eq. 9.4 should apply, but where does the percentage increase fit in? Recognize that the $\Delta L/L_o$ term is the *fractional* increase in length. For example, if you had a spring with a length of 10 cm (L_o) and you stretched it 1.0 cm (ΔL), then $\Delta L/L_o = 1.0 \text{ cm}/10 \text{ cm} = 0.10$. This ratio can readily be changed to a percentage, and we would say the spring's length was increased be 10%. So, the percentage increase is really just the value of the $\Delta L/L_o$ term (multiplied by 100%).

Solution. Listing the data, we have

Given: $r = 2.0$ cm $= 0.020$ m
$\Delta L/L_o = 0.010\% = 1.0 \times 10^{-4}$
$Y = 1.5 \times 10^{10}$ N/m^2 (from Table 9.1)

Find: tensile force F

Using Eq. 9.4, we have

$$F = Y(\Delta L/L_o)A = Y(\Delta L/L_o)\pi r^2$$
$$= (1.5 \times 10^{10} \text{ N/m}^2)(1.0 \times 10^{-4})\pi(0.020 \text{ m})^2 = 1.9 \times 10^3 \text{ N}$$

How much force is this? Quite a bit—more than 400 lb.; the femur is a pretty strong bone.

Follow-up Exercise. A total mass of 16 kg is suspended from a steel wire with a diameter of 0.10 cm. (a) By what percentage does the length of the wire increase? (b) The tensile or ultimate strength is the maximum stress a material can support before breaking or fracturing. If the tensile strength of the steel wire in (a) is 4.9×10^8 N/m^2, how much mass could be suspended before it would break?

Change in Shape: Shear Modulus

Another way an elastic body can be deformed is by a *shear stress*. In this case, the deformation is due to an applied force that is tangential to the surface area (•Fig. 9.4). A change in shape results without a change in volume. The *shear strain* is given by x/h, where x is the relative displacement of the faces and h is the distance between them.

The shear strain is sometimes defined in terms of the *shear angle* (ϕ). As Fig. 9.4 shows, tan $\phi = x/h$. But the shear angle is usually quite small, so a good approximation is tan $\phi \approx \phi \approx x/h$, where the angle ϕ is in radians. (If $\phi = 10°$, for example, there is only 1.0% difference between ϕ and tan ϕ.) The **shear modulus** (sometimes called the modulus of rigidity) is then

$$S = \frac{F/A}{x/h} \approx \frac{F/A}{\phi} \tag{9.5}$$

SI unit of shear modulus: newtons per square meter (N/m^2)

Note in Table 9.1 that the shear modulus is generally less than Young's modulus; S is approximately $Y/3$ for many materials, which indicates that there is a greater response to a shear stress than to a tensile stress. Also, note the inverse relationship of $\phi \propto 1/S$, similar to that pointed out previously for Young's modulus.

A shear stress may also be of the torsional type, resulting from the twisting action of a torque. For example, a torsional shear stress may shear off the head of a bolt that is being tightened.

Liquids do not have shear moduli (or Young's moduli), hence the gaps in Table 9.1. A shear stress cannot be effectively applied to a liquid or a gas. It is often said that *fluids cannot support a shear*. Why?

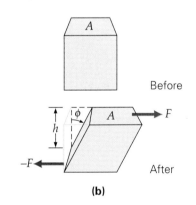

•**FIGURE 9.4 Shear stress and strain** A shear stress is produced when a force is applied tangentially to a surface area. The strain is measured in terms of the relative displacement of the object's faces or the shear angle ϕ.

Change in Volume: Bulk Modulus

Suppose that a force directed inward acts over the entire surface of a body (•Fig. 9.5). Such a *volume stress* is often applied by pressure transmitted by a fluid (Section 9.2). An elastic material will be compressed by a volume stress; that is, it will show a change in volume but not in general shape in response to a pressure change Δp. (Pressure is the force per unit area, as we shall see in Section 9.2.) The change in pressure is equal to the volume stress, or $\Delta p = F/A$. The *volume strain* is the ratio of the volume change (ΔV) to the original volume (V_o). The **bulk modulus** (B) is then

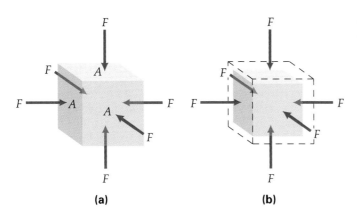

•**FIGURE 9.5 Volume stress and strain** (a) A volume stress is applied when a normal force acts over the entire surface area, as shown here for a cube. This type of stress most commonly occurs for gases. (b) The resulting strain is a change in volume.

$$B = \frac{F/A}{-\Delta V/V_o} = -\frac{\Delta p}{\Delta V/V_o} \qquad (9.6)$$

SI unit of bulk modulus: newtons per square meter (N/m²)

The minus sign is introduced to make B a positive quantity, since $\Delta V = V - V_o$ is negative for an increase in external pressure. And, similar to the previous moduli relationships, $\Delta V \propto 1/B$.

Bulk moduli are listed for solids *and* liquids in Table 9.1. Gases have bulk moduli too, since they can be compressed. For a gas, it is common to talk about the reciprocal of the bulk modulus, which is called the **compressibility** (k):

$$k = \frac{1}{B} \qquad \text{(for gases)} \qquad (9.7)$$

The change in volume ΔV is thus directly proportional to the compressibility k.

Solids and liquids are relatively incompressible and have small values of compressibility. Conversely, gases are easily compressed and have large compressibilities, which vary with pressure and temperature.

Example 9.2 ■ Compressing a Liquid: Volume Stress and Bulk Modulus

By how much should the pressure on a liter of water be changed to compress it by 0.10%?

Thinking It Through. Similar to the fractional change in length $\Delta L/L_o$, the fractional change in volume is given by $-\Delta V/V_o$, which may be expressed as a percentage. The pressure change can then be found from Eq. 9.6. Compression implies a negative ΔV.

Solution. Listing the data, we have

Given: $-\Delta V/V_o = 0.0010$ (or 0.10%)   Find: $\Delta p = F/A$
$V_o = 1.0\text{ L} = 1000\text{ cm}^3$
$B_{H_2O} = 2.2 \times 10^9\text{ N/m}^2$ (from Table 9.1)

Note that $-\Delta V/V_o$ is the *fractional* change in the volume; for a compression of 0.10%, it is 0.0010 (no units, a ratio of volumes). With $V_o = 1000\text{ cm}^3$, the volume reduction is

$$-\Delta V = 0.0010 V_o = 0.0010(1000\text{ cm}^3) = 1.0\text{ cm}^3$$

However, the change in volume is not needed. The fractional change, as listed in the Given data, can be used directly in Eq. 9.6 to find the pressure increase:

$$\Delta p = B\left(\frac{-\Delta V}{V_o}\right) = (2.2 \times 10^9\text{ N/m}^2)(0.0010) = +2.2 \times 10^6\text{ N/m}^2$$

(This increase is about 22 times normal atmospheric pressure!)

Follow-up Exercise. If an extra $1.0 \times 10^6\text{N/m}^2$ of pressure above normal atmospheric pressure is applied to a half liter of water, what is the change in the water's volume?

Insight

Feat of Strength or Knowledge of Materials?

The breaking of wooden boards, concrete blocks, or similar materials with a bare hand or foot is an impressive demonstration often done by karate experts (Fig. 1). The physics of this feat can be analyzed in terms of the properties of the materials involved.

In delivering the blow, the expert imparts a large impulse force (Chapter 6) to the top board or block, which bends under the pressure. (Note that the objects are struck midway between the end supports.) At the same time, the bones of the hand are being compressed as well. Fortunately, human bone can withstand more compressive force than can wood, tile, or concrete, which is why the expert's bones aren't damaged. (The ultimate compressive strength of bone is at least four times greater than that of concrete.*)

When each board or block is hit, the upper surface is compressed and the lower surface is elongated, or subjected to a tension force. Wood, tile, and concrete are weaker under tension than under compression. (The ultimate tensile strength of concrete is only about a twentieth of its ultimate compressive strength.) Thus, the board or block begins to crack at the bottom surface first. The crack propagates from the underside *toward* the hand, widens, and becomes a complete break. Thus the hand never actually "cuts through" the board or block.

The amount of force required to break a board or block in this way depends on several factors. For a wooden board, these are the type of wood, the width and thickness, and the distance between the end supports. Also, the edge of the hand must strike the board parallel to the grain of the wood. Similar considerations apply for a tile or a concrete block. Because tile and concrete are more rigid than wood, more force is generally required to break these substances.

Some karate experts are able to break through a stack of boards or tiles and more than one concrete block. The force required does not increase by a factor equal to the number of objects. High-speed photography shows that the hand makes contact with only one or two boards or tiles at the top of the stack. Then, as each object breaks, it collides with and breaks through the one below it.

Even though the properties of the materials seem to guarantee the success of this demonstration, you should not attempt the feat unless you are an expert and know what you're doing. The board or block might win.

FIGURE 1 Feat of strength or knowledge of materials? Breaking wooden boards, concrete blocks, or roofing tiles (as shown here) with a karate blow depends on both the physical strength of the expert and the strength of the material. Wood, concrete, and tile have different maximum tensile and compressional stresses, but the maximum compressional strength of bone is greater than any of these. (This photo, from a high-speed photography sequence, was made toward the end of the blow.)

*Ultimate strength is the maximum strength a material can stand before it breaks or fractures.

9.2 Fluids: Pressure and Pascal's Principle

Objectives: To (a) explain the pressure–depth relationship and (b) state Pascal's principle and describe how it is used in practical applications.

A force can be applied to a solid at a point of contact, but this won't work with a fluid, as we saw in the preceding discussion on bulk moduli. With fluids, a force must be applied over an area. Such an application of force is expressed in terms of **pressure**, or the *force per unit area*:

Pressure—force per unit area

$$p = \frac{F}{A} \quad (9.8a)$$

SI unit of pressure: newtons per square meter (N/m^2) or pascal (Pa)

The force in this equation is understood to be acting normally (perpendicularly) to the surface area. F may be the perpendicular component of a force that acts at an angle to the surface (•Fig. 9.6).

As Fig. 9.6 shows, in the more general case we should write

$$p = \frac{F_\perp}{A} = \frac{F \cos \theta}{A} \quad (9.8b)$$

Pressure is a scalar quantity (with magnitude only), even though the force producing it or the force produced within a fluid by a pressure does have direction and so is a vector.

Pressure has SI units of newtons per square meter (N/m²). This combined unit is given the special name **pascal (Pa)** in honor of the French scientist and philosopher Blaise Pascal (1623–1662), who studied fluids and pressure:

$$1 \text{ Pa} = 1 \text{ N/m}^2$$

In the British system, a common unit of pressure is pounds per square inch (lb/in², or psi). Other units, some of which will be introduced later, are used in special applications.

Before going on, it may be helpful to review density, which is an important consideration in the study of fluids. Recall from Chapter 1 that the density (ρ) of a substance is defined (in Eq. 1.1) as

$$\text{density} = \frac{\text{mass}}{\text{volume}}$$

$$\rho = \frac{m}{V}$$

SI unit of density: kilograms per cubic meter (kg/m³)
(common cgs unit: grams per cubic centimeter, or g/cm³)

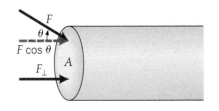

$$p = \frac{F_\perp}{A} = \frac{F \cos \theta}{A}$$

•**FIGURE 9.6 Pressure** Pressure is usually written $p = F/A$, where it is understood that F is the force or component of force normal to the surface. In general, then, $p = (F \cos \theta)/A$.

That is, density is the mass per *unit* volume. The densities of some common substances are given in Table 9.2.

Water has a density of 1.00×10^3 kg/m³ (or 1.00 g/cm³), from the original definition of the kilogram (Chapter 1). Mercury has a density of 13.6×10^3 kg/m³ (or 13.6 g/cm³). Hence, mercury is denser than water—13.6 times more dense. Gasoline is less dense than water. (See Table 9.2.)

We say that density is a measure of the compactness of the matter of a substance—the greater the density, the more matter or mass in a given volume. Notice how density quantifies the amount or mass per volume. It would be incorrect to say that mercury is "heavier" than water. Why? Because you could have a large volume of water that would be heavier than some much smaller volume of mercury.

Pressure and Depth

If you have gone diving, you well know that pressure increases with depth, having felt the increased pressure on your eardrums. An opposite effect is commonly felt when you fly in a plane or ride in a car going up a mountain. With increasing altitude, your ears may "pop" because of reduced external air pressure.

TABLE 9.2 Densities of Some Common Substances (in kg/m³)

Substance	Density (ρ)	Substance	Density (ρ)	Substance	Density (ρ)
Solids		*Liquids*		*Gases**	
Aluminum	2.7×10^3	Alcohol, ethyl	0.79×10^3	Air	1.29
Brass	8.7×10^3	Alcohol, methyl	0.82×10^3	Helium	0.18
Copper	8.9×10^3	Blood, whole	1.05×10^3	Hydrogen	0.090
Glass	2.6×10^3	Blood plasma	1.03×10^3	Oxygen	1.43
Gold	19.3×10^3	Gasoline	0.68×10^3	Water vapor (100°C)	0.63
Ice	0.92×10^3	Kerosene	0.82×10^3		
Iron	7.9×10^3	Mercury	13.6×10^3		
Lead	11.4×10^3	Sea water (4°C)	1.03×10^3		
Silver	10.5×10^3	Water, fresh (4°C)	1.00×10^3		
Steel	7.8×10^3				
Wood, oak	0.81×10^3				

*At 0°C and 1 atm unless otherwise specified.

•**FIGURE 9.7 Pressure and depth** The extra pressure at a depth h in a liquid is due to the weight of the liquid above: $p = \rho g h$, where ρ is the density of the liquid (assumed to be constant). This is shown for an imaginary rectangular column of liquid.

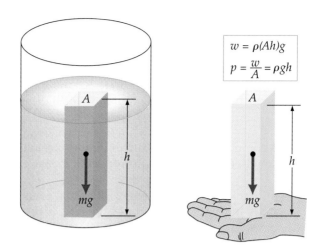

How the pressure in a fluid varies with depth can be demonstrated by considering a container of liquid at rest. Imagine you can isolate a rectangular column of water as shown in •Fig. 9.7. Then, the force on the bottom of the container below the column (or the hand) is equal to the weight of the liquid making up the column: $F = w = mg$. Since density is $\rho = m/V$, the mass in the column is equal to the density times the volume; that is, $m = \rho V$. (The liquid is assumed incompressible, so ρ is constant.)

The volume of the isolated liquid column is equal to the height of the column times the area of its base, $V = hA$. Thus we can write

$$F = w = mg = \rho V g = \rho g h A$$

With $p = F/A$, the pressure at a depth h due to the weight of the column is

$$p = \rho g h \qquad (9.9)$$

This is a general result for incompressible liquids. The pressure is the same everywhere on a horizontal plane at a depth h (with ρ and g constant). Also note that Eq. 9.9 is independent of the base area of the rectangular column. We could have taken the whole cylindrical column of the liquid in the container in Fig. 9.7 and gotten the same result.

The derivation of Eq. 9.9 did not take into account pressure being applied to the open surface of the liquid. This factor adds to the pressure at a depth h to give a *total* pressure of

Pressure–depth relationship

$$p = p_\text{o} + \rho g h \qquad \begin{array}{l}\text{(incompressible fluid} \\ \text{at constant density)}\end{array} \qquad (9.10)$$

where p_o is the pressure applied to the liquid surface (that is, at $h = 0$).

For an open container, p_o is atmospheric pressure, or the weight (force) per area due to the gases in the atmosphere above the liquid's surface. The average atmospheric pressure at sea level is sometimes used as a unit, called an **atmosphere (atm)**:

$$1 \text{ atm} \equiv 101.325 \text{ kPa} = 1.01325 \times 10^5 \text{ N/m}^2 \approx 14.7 \text{ lb/in}^2$$

How atmospheric pressure is measured will be described shortly.

Example 9.3 ■ A Scuba Diver: Pressure and Force

(a) What is the total pressure on the back of a scuba diver in a lake at a depth of 8.00 m?
(b) What is the force on the diver's back due to the water alone, taking the surface of the back to be a rectangle 60.0 cm by 50.0 cm?

Thinking It Through. (a) This is a direct application of Eq. 9.10 in which p_o is taken as the atmospheric pressure p_a. (b) Knowing the area and the pressure due to the water, the force can be found from the definition of pressure, $p = F/A$.

Solution.

Given: $h = 8.00$ m
$A = 60.0$ cm $\times$ 50.0 cm
$= 0.600$ m $\times$ 0.500 m $= 0.300$ m^2
$\rho_{H_2O} = 1.00 \times 10^3$ kg/m^3 (from Table 9.2)
$p_a = 1.01 \times 10^5$ N/m^2

Find: (a) p (total pressure)
(b) F (force due to water)

(a) The total pressure is the sum of the pressure due to the water and the atmospheric pressure (p_a). By Eq. 9.10, this is

$$p = p_a + \rho g h$$
$$= (1.01 \times 10^5 \text{ N/m}^2) + (1.00 \times 10^3 \text{ kg/m}^3)(9.80 \text{ m/s}^2)(8.00 \text{ m})$$
$$= (1.01 \times 10^5 \text{ N/m}^2) + (0.784 \times 10^5 \text{ N/m}^2) = 1.79 \times 10^5 \text{ N/m}^2 \text{ (or Pa)}$$
$$\text{(expressed in atmospheres)} \approx 1.8 \text{ atm}$$

(b) The pressure p_w due to the water alone is the $\rho g h$ portion of the preceding equation, so $p_w = 0.784 \times 10^5$ N/m^2. Then, $p_w = F/A$, and

$$F = p_w A = (0.784 \times 10^5 \text{ N/m}^2)(0.300 \text{ m}^2)$$
$$= 2.35 \times 10^4 \text{ N} \quad \text{(or } 5.29 \times 10^3 \text{ lb—about 2.6 tons!)}$$

Follow-up Exercise. You might question the answer to part (b) of this Example—how could the diver support such a force? To get a better idea of the forces our bodies can support, what would be the force on the diver's back at the water surface from atmospheric pressure alone? How do you suppose our bodies can support such forces or pressures?

•**FIGURE 9.8 Pascal's principle** The pressure applied at point A is fully transmitted to all parts of the fluid and to the walls of the container. There is also pressure due to the weight of the fluid above at different depths (for instance, $\rho g h/2$ at C and $\rho g h$ at D).

Pascal's Principle

When the pressure (for example, air pressure) is increased on the entire open surface of an incompressible liquid at rest, the pressure at any point in the liquid or on the boundary surfaces increases by the same amount. The effect is the same if pressure is applied by means of a piston to any surface of an enclosed fluid (•Fig. 9.8). The transmission of pressure in fluids was studied by Blaise Pascal (after whom the SI pressure unit is named), and the observed effect is called **Pascal's principle**:

> Pressure applied to an enclosed fluid is transmitted undiminished to every point in the fluid and to the walls of the container.

Pascal's principle

For an incompressible liquid, the pressure change is transmitted essentially instantaneously. For a gas, a pressure change will generally be accompanied by a change in volume and/or temperature, but after equilibrium has been reestablished, Pascal's principle remains valid.

Common practical applications of Pascal's principle include the hydraulic braking systems used on automobiles. A force on the brake pedal transmits a force to the wheel brake cylinder. Similarly, hydraulic lifts and jacks are used to raise automobiles and other heavy objects (•Fig. 9.9).

Using Pascal's principle, we can show how such systems not only allow us to transmit force from one place to another, but also to multiply that force. The input pressure p_i, supplied by compressed air for a garage lift, for example, gives an input force F_i on a small piston area A_i (Fig. 9.9). The full magnitude of the pressure is transmitted to the output piston, which has an area A_o. Since $p_i = p_o$,

$$\frac{F_i}{A_i} = \frac{F_o}{A_o}$$

•**FIGURE 9.9 The hydraulic lift** Because the input and output pressures are equal (Pascal's principle), a small input force gives a large output force proportional to the ratio of the piston areas.

and

$$F_o = \left(\frac{A_o}{A_i}\right) F_i \quad \begin{array}{l} \text{hydraulic} \\ \text{force multiplication} \end{array} \quad (9.11)$$

With A_o larger than A_i, then F_o will be larger than F_i. The input force is greatly multiplied.

Example 9.4 ■ The Hydraulic Lift: Pascal's Principle

A garage lift has input and lift (output) pistons with diameters of 10 cm and 30 cm, respectively. The lift is used to hold up a car with a weight of 1.4×10^4 N. (a) What is the force on the input piston? (b) What pressure is applied to the input piston?

Thinking It Through. (a) Pascal's principle, as expressed in the hydraulic Eq. 9.11, has four variables, and three are given (areas via diameters). (b) The pressure is simply $p = F/A$.

Solution.

Given: $d_i = 10$ cm $= 0.10$ m   *Find:* (a) F_i (input force)
$d_o = 30$ cm $= 0.30$ m   (b) p_i (input pressure)
$F_o = 1.4 \times 10^4$ N

(a) Rearranging Eq. 9.11 and using $A = \pi r^2 = \pi d^2/4$ for the circular piston ($r = d/2$) gives

$$F_i = \left(\frac{A_i}{A_o}\right) F_o = \left(\frac{\pi d_i^2/4}{\pi d_o^2/4}\right) F_o = \left(\frac{d_i}{d_o}\right)^2 F_o$$

$$F_i = \left(\frac{0.10 \text{ m}}{0.30 \text{ m}}\right)^2 F_o = \frac{F_o}{9} = \frac{1.4 \times 10^4 \text{ N}}{9} = 1.6 \times 10^3 \text{ N}$$

The input force is one-ninth of the output force, or there was a force multiplication of 9 (that is, $F_o = 9F_i$).

(Note that we didn't really need to write the complete expressions for the areas. We know that the area of a circle is proportional to the square of its diameter. If the ratio of the piston diameters is 3 to 1, the ratio of their areas must therefore be 9 to 1, and we could have used this ratio directly in Eq. 9.11.)

(b) Then we apply Eq. 9.8a:

$$p_i = \frac{F_i}{A_i} = \frac{F_i}{\pi r_i^2} = \frac{F_i}{\pi(d_i/2)^2} = \frac{1.6 \times 10^3 \text{ N}}{\pi(0.10 \text{ m})^2/4}$$

$$= 2.0 \times 10^5 \text{ N/m}^2 \; (= 200 \text{ kPa})$$

This pressure is about 30 lb/in², a common pressure used in automobile tires and about twice atmospheric pressure (about 100 kPa, or 15 lb/in²).

Follow-up Exercise. Pascal's principle is used in shock absorbers on automobiles and on the landing gear of airplanes. (The polished steel piston rods can be seen above the wheels on aircraft.) In these devices, a large force (the shock produced on hitting a bump in the road or an airport runway at high speed) must be reduced to a safe level. Basically, fluid is forced from a larger diameter cylinder into a smaller diameter cylinder (the reverse of the situation in Fig. 9.9). Suppose that the input piston of a shock absorber on a jet plane has a diameter of 8.0 cm. What would be the diameter of the output cylinder that would reduce the force by a factor of 10?

As Example 9.4 shows, we can relate the force directly to the diameters of the pistons: $F_i = (d_i/d_o)^2 F_o$, or $F_o = (d_o/d_i)^2 F_i$. By making $d_o \gg d_i$ we can get huge factors of force multiplication, as is typical for hydraulic presses, jacks, and earth-moving equipment. (The shiny input piston rods are often visible on front loaders and backhoes.) Inversely, we can get a force reduction by making $d_i > d_o$, as in Follow-up Exercise 9.4.

However, don't think that you are getting something for nothing with large force multiplications. Energy is still a factor, and it can never be multiplied by a machine. (Why not?) Looking at the work involved and assuming the work output is equal to the work input, $W_o = W_i$ (an ideal condition—why?), we have from Eq. 5.1

$$F_o x_o = F_i x_i$$

or

$$F_o = \left(\frac{x_i}{x_o}\right) F_i$$

where x_o and x_i are the output and input distances moved by the respective pistons.

Thus, the output force can be much greater than the input force only if the input distance is much greater than the output distance. For example, if $F_o = 10F_i$, then $x_i = 10x_o$, and the input piston must travel ten times the distance of the output piston. We say that force is multiplied at the expense of distance. Specifically, with the preceding numbers, suppose the brake pads of a simple hydraulic braking system of a car move 5.0 mm when the brakes are applied. Then, the input distance of the pedal is $x_i = 10x_o = 10(5.0 \text{ mm}) = 50 \text{ mm}$—a greater input distance, but the brake pads are applied with 10 times the force applied to the pedal, $F_o = 10F_i$.

Pressure Measurement

Pressure can be measured by a variety of mechanical devices that are often spring-loaded. Another type of instrument, called a manometer, uses a liquid—usually mercury. An *open-tube manometer* is illustrated in •Fig. 9.10a. One end of the U-shaped tube is open to the atmosphere, and the other is connected to the container of gas whose pressure is to be measured. The liquid in the U-tube acts as a reservoir through which pressure is transmitted according to Pascal's principle.

The pressure of the gas (p) is balanced by the weight of the column of liquid (of height h, the difference in the heights of the columns) and the atmospheric pressure (p_a) on the open liquid surface:

$$p = p_a + \rho g h \quad (9.12)$$

The pressure p is called the **absolute pressure**.

You may have experience measuring pressure by means of pressure gauges; a tire gauge used to measure air pressure in automobile tires is a common example (Fig. 9.10b). Such gauges, quite appropriately, measure **gauge pressure**: A pressure gauge registers only the pressure *above* atmospheric pressure. Hence, to get the absolute pressure (p), you have to add the atmospheric pressure (p_a) to the gauge pressure (p_g):

$$p = p_a + p_g$$

•**FIGURE 9.10 Pressure measurement** (a) For an open-tube manometer, the pressure of the gas in the container is balanced by the pressure of the liquid column and atmospheric pressure acting on the open surface of the liquid. The absolute pressure of the gas equals the sum of the atmospheric pressure (p_a) and $\rho g h$, the gauge pressure. (b) A tire gauge measures gauge pressure, the difference between the pressure in the tire and atmospheric pressure: $p_{gauge} = p - p_a$. Thus, if a tire gauge reads 200 kPa (30 lb/in²), the actual pressure within the tire is 1 atm higher, or 300 kPa. (c) A barometer is a closed-tube manometer that is exposed to the atmosphere and thus reads only atmospheric pressure.

For example, suppose your tire gauge reads a pressure of 200 kPa ($\approx$ 30 lb/in²). The absolute pressure within the tire is then $p = p_a + p_g = 101$ kPa + 200 kPa = 301 kPa, where normal atmospheric pressure is about 101 kPa (14.7 lb/in²), as will be shown shortly.

It is the gauge pressure of a tire that keeps it rigid or operational. In terms of the more familiar pounds per square inch (psi, or lb/in²), a tire with a gauge pressure of 30 psi has an absolute pressure of about 45 psi (30 + 15, with atmospheric pressure $\approx$ 15 psi). Hence, the pressure on the inside of the tire is 45 psi, and that on the outside is 15 psi. The Δp of 30 psi keeps the tire inflated. If you open the valve or get a puncture, the internal and external pressures equalize and you have a flat!

Atmospheric pressure itself can be measured with a *barometer*. The principle of a mercury barometer is illustrated in Fig. 9.10c. The device was invented by Evangelista Torricelli (1608–1647), Galileo's successor as professor of mathematics at the university in Florence. A simple barometer consists of a tube filled with mercury that is inverted into a reservoir. Some mercury runs out, but as much as can be supported by the air pressure on the surface of the reservoir pool remains in the tube. This device can be considered to be a *closed-tube manometer*, and the absolute pressure is just the atmospheric pressure, since the gauge pressure (the pressure *above* atmospheric pressure) is zero.

The atmospheric pressure is then equal to the pressure due to the weight of the column of mercury, or

$$p_a = \rho g h \quad (9.13)$$

A *standard atmosphere* is defined as the pressure supporting a column of mercury exactly 76 cm in height at sea level and at 0°C.

Example 9.5 ■ Standard Atmospheric Pressure: Converting to Pascals

If a standard atmosphere supports a column height of exactly 76 cm of mercury (chemical symbol Hg), what is the standard atmospheric pressure in pascals? (The density of mercury is 13.5951×10^3 kg/m^3 at 0°C, and $g = 9.80665$ m/s^2.)

Thinking It Through. This is a direct application of Eq. 9.13 to find standard atmospheric pressure in metric units (pascals or newtons per square meter). Note that the height of the column is given in centimeters.

Solution.

Given: $h = 76$ cm $= 0.76$ m (exact)   Find: p_a (atmospheric pressure)
$\rho_{Hg} = 13.5951 \times 10^3$ kg/m^3
$g = 9.80665$ m/s^2

Using Eq. 9.13, we have

$$p_a = \rho_{Hg} g h = (13.5951 \times 10^3 \text{ kg/m}^3)(9.80665 \text{ m/s}^2)(0.760000 \text{ m})$$
$$= 101325 \text{ N/m}^2 = 1.01325 \times 10^5 \text{ Pa} \quad (\text{or } 101.325 \text{ kPa})$$

Follow-up Exercise. What would be the height of a barometer column for one standard atmosphere if water were used instead of mercury?

Changes in atmospheric pressure can be observed as changes in the height of the mercury column. Atmospheric pressure is commonly reported in terms of the height of the barometer column, and weather forecasters say the barometer is rising or falling. That is, 1 standard atmosphere (atm) of pressure has equivalent values of

$$1 \text{ atm} = 76 \text{ cm Hg} = 760 \text{ mm Hg}$$
$$= 29.92 \text{ in. Hg} \quad (\text{about 30 in. Hg})$$

In honor of Torricelli, a pressure supporting 1 mm of mercury is given the name *torr*:

$$1 \text{ mm Hg} \equiv 1 \text{ torr}$$

and

$$1 \text{ atm} = 760 \text{ torr}$$

Because mercury is very toxic, it is sealed inside a barometer. A safer and less expensive device widely used to measure atmospheric pressure is the *aneroid* (without fluid) *barometer*. In an aneroid barometer, a sensitive metal diaphragm on an evacuated container (something like a drum head) responds to pressure changes, which are indicated on a dial. This is the kind of barometer you frequently find in homes in decorative wall mountings.

Since air is compressible, the atmospheric density and pressure are greatest at the Earth's surface and decrease with altitude. We live at the bottom of the atmosphere but don't notice its pressure very much in our daily activities. Remember that our bodies are composed largely of fluids, which exert a matching outward pressure. Indeed, the external pressure of the atmosphere is so important to our normal functioning that we take it with us wherever we can. The pressurized suits worn by astronauts in space or on the Moon are needed not only to supply oxygen, but also to provide an external pressure similar to that on the Earth's surface.

A very important gauge pressure reading is discussed in the Insight on p. 308. Read it before going on to Example 9.6.

Note: Another unit sometimes used in weather reports is the millibar (mb). By definition, 1 atm = 1.01325×10^5 N/m^2 = 1.01325 bar = 1013.25 mb. Normal atmospheric pressures are around 1000 mb.

Insight

Blood Pressure

Basically, a pump is a machine that transfers mechanical energy to a fluid, thereby increasing the pressure and causing the fluid to flow. There are a wide variety of pumps, but one of interest to everyone is the heart. The heart is a muscular pump that drives blood through the body's circulatory network of arteries, capillaries, and veins. With each pumping cycle, the human heart's interior chambers enlarge and fill with freshly oxygenated blood from the lungs (Fig. 1).

On contraction of chambers called ventricles, blood is forced out through the arteries. Smaller and smaller arteries branch off from the main ones, until the very small capillaries are reached. There food and oxygen being carried by the blood are exchanged with the surrounding tissues, and wastes are picked up. The blood then flows into the veins to complete the circuit back to the heart.

The arterial blood pressure rises and falls in response to the cardiac cycle. That is, when the ventricles contract, forcing blood into the arterial system, the pressure in the arteries increases sharply. The maximum pressure achieved during the ventricular contraction is called the *systolic pressure*. When the ventricles relax, the arterial pressure drops, and the lowest pressure before the next contraction is called *diastolic pressure*. (These pressures are named after two parts of the pumping cycle, the *systole* and the *diastole*.)

The walls of the arteries have considerable elasticity and expand and contract with each pumping cycle. This alternate expansion and contraction can be felt as a *pulse* in an artery near the body surface. For example, the radial artery near the surface of the wrist is commonly used to measure a person's pulse. The pulse rate is equal to the ventricle contraction rate, and hence the pulse rate indicates the heart rate.

FIGURE 1 The heart as a pump The human heart is analogous to a mechanical force pump. Its pumping action, consisting of **(a)** intake and **(b)** output, gives rise to variations in blood pressure.

•**FIGURE 9.11 What height is needed?** See Example 9.6.

Example 9.6 ■ An IV: Gravity Assist

An IV ("*intra*venous" injection) is a quite different type of gravity assist from that discussed for space probes in Chapter 7. Consider a hospital patient who receives an IV under gravity flow as seen in •Fig. 9.11. If the blood gauge pressure in the vein is 20.0 mm Hg, above what height should the bottle be placed for the IV to function properly?

Thinking It Through. The fluid gauge pressure at the bottom of the IV tube must be greater than the pressure in the vein and can be computed from Eq. 9.9. (The liquid is assumed to be incompressible.)

Solution.

Given: $p_v = 20.0$ mm Hg (vein gauge pressure) *Find:* h (height for
$\rho = 1.05 \times 10^3$ kg/m^3 (whole blood $p_v > 20$ mm Hg)
  from Table 9.2)

First, we need to change the common medical unit of mm Hg (or torr) to the SI unit of pascals (Pa, or N/m^2):

$$p_v = (20.0 \text{ mm Hg})[133 \text{ Pa}/(\text{mm Hg})] = 2.66 \times 10^3 \text{ Pa}$$

Then, for the condition of $p > p_v$,

Insight

Taking a person's blood pressure involves measuring the pressure of the blood on the arterial walls. This is done with a *sphygmomanometer* (the Greek word *sphygmo* means "pulse"). An inflatable cuff is used to shut off the blood flow temporarily. The cuff pressure is slowly released, and the artery is monitored with a stethoscope (Fig. 2). A point is reached where blood is just forced through the constricted artery. This flow is turbulent and gives rise to a specific sound with each heartbeat.

When the sound is first heard, the systolic pressure is noted, which is normally about 120 mm Hg. When the turbulent beats disappear because of smooth blood flow, the diastolic reading is taken. The pressure at this point is normally 80 mm Hg. Blood pressure is commonly reported by giving the systolic and diastolic pressures separated by a slash, for example, 120/80 (read as "120 over 80"). Normal blood pressure ranges between 100 and 140 for the systolic pressure and between 70 and 90 for the diastolic pressure. (Blood pressure is a gauge pressure. Why?)

Away from the heart, blood vessels branch to smaller and smaller diameters. As you might expect, pressure in the blood vessels decreases as their diameter decreases. In small veins, such as those in the arm, the blood pressure is on the order of 10 to 20 mm Hg, and there is no systolic–diastolic variation.

High blood pressure is a common health problem. The elastic walls of the arteries expand under the hydraulic force of the blood pumped from the heart. Their elasticity may diminish with age, however. Fatty deposits (of cholesterols) can narrow and roughen the arterial passageways, impeding the blood flow and giving rise to a form of arteriosclerosis, or hardening of the arteries. Because of these defects, the driving pressure must increase to maintain a normal blood flow. The heart must work harder, which places a greater demand on its muscles. A relatively slight decrease in the effective cross-sectional area of a blood vessel has a rather large effect on the flow rate.

FIGURE 2 Measuring blood pressure The pressure is indicated on the gauge in millimeters Hg or torr.

or
$$p = \rho g h > p_v$$

$$h > \frac{p_v}{\rho g} = \frac{(2.66 \times 10^3 \text{ Pa})}{(1.05 \times 10^3 \text{ kg/m}^3)(9.80 \text{ m/s}^2)} = 0.259 \text{ m} \ (\approx 26 \text{ cm})$$

The IV bottle needs to be at least 26 cm above the injection site.

Follow-up Exercise. The normal (gauge) blood pressure range is commonly reported as 120/80 (in millimeters Hg). Why is the blood pressure of 20 mm Hg in this Example so low?

9.3 Buoyancy and Archimedes' Principle

Objectives: To (a) relate the buoyant force and Archimedes' principle and (b) tell whether an object will float in a fluid based on relative densities.

When placed in a fluid, an object will either sink or float. This is most commonly observed with liquids—for example, objects floating or sinking in water. But the same effect occurs in gases. A falling object sinks in the atmosphere, and other bodies float (•Fig. 9.12).

Things float because they are buoyant, or are buoyed up. For example, if you immerse a cork in water and release it, the cork will be buoyed up to the surface and float there. From your knowledge of forces, you know that such motion requires an upward net force on the object. That is, there must be an upward force acting on the object that is greater than its downward weight force. The forces are

•**FIGURE 9.12 Fluid buoyancy** The air is a fluid in which objects such as these dirigibles float. The helium inside the blimps is lighter or less dense than the surrounding air. The blimps are supported by the resulting buoyant forces.

$\Delta p = \rho_f g(h_2 - h_1)$

(a)

(b)

•FIGURE 9.13 Buoyancy and Archimedes' principle (a) A buoyant force arises from the pressure difference between different depths. The pressure on the bottom of the submerged block (p_2) is greater than that on the top (p_1), so there is a (buoyant) force directed upward. (It is shifted for clarity.) (b) Archimedes' principle: The buoyant force on the object is equal to the weight of the volume of fluid it displaces. (The scale is set to read zero when the container is empty.)

equal when the object floats, moves at constant velocity, or remains stationary. The upward force resulting from an object being wholly or partially immersed in a fluid is called the **buoyant force**.

How the buoyant force comes about can be seen by considering a buoyant object being held under the surface of a fluid (•Fig. 9.13a). The pressures on the upper and lower surfaces of the block are $p_1 = \rho_f g h_1$ and $p_2 = \rho_f g h_2$, respectively, where ρ_f is the fluid density. Thus, there is a pressure difference $\Delta p = p_2 - p_1 = \rho_f g (h_2 - h_1)$ between the top and bottom of the block, which gives an upward force (the buoyant force), F_b. This force is balanced by the applied force and the weight of the block.

It is not difficult to derive an expression for the magnitude of the buoyant force. We know that pressure is force per unit area. Thus, if both the top and bottom areas of the block are A, the magnitude of the net buoyant force in terms of the pressure difference is

$$F_b = p_2 A - p_1 A = (\Delta p)A = \rho_f g(h_2 - h_1)A$$

Since $(h_2 - h_1)A$ is the volume of the block, and hence the volume of fluid displaced by the block, V_f, we can write this expression as

$$F_b = \rho_f g V_f$$

But $\rho_f V_f$ is simply the mass of the fluid displaced by the block, m_f. Thus we can write the expression for the buoyant force $F_b = m_f g$: The magnitude of the buoyant force is equal to the weight of the fluid displaced by the block (Fig. 9.13b). This general result is known as **Archimedes' principle**:

A body immersed wholly or partially in a fluid is buoyed up by a force equal in magnitude to the weight of the *volume of fluid* it displaces:

$$F_b = m_f g = \rho_f g V_f \qquad (9.14)$$

Archimedes (287–212 B.C.) was given the task of determining whether a crown made for a certain king was pure gold or contained some other, cheaper metal. Legend has it that the solution to the problem came to him when he was bathing, perhaps from seeing the water level rise when he got into the tub and experiencing the buoyant force on his limbs. In any case, it is said that he was so excited that he ran through the streets of the city shouting "Eureka!" (Greek for "I have found it"). Although Archimedes' solution to the problem involved density and volume (see Exercises 53 and 100), it presumably got him thinking about buoyancy.

Example 9.7 ■ Lighter Than Air: Buoyant Force

What is the buoyant force in air on a spherical helium balloon with a radius of 30 cm if $\rho_{air} = 1.29$ kg/m³? (Neglect the weight of the balloon material.)

Thinking It Through. Helium is termed "lighter" (less dense) than air, so there is a buoyant force for the balloon in air. This force is equal to the weight of the volume of air that balloon displaces. To find this weight, we first find the balloon's volume and then use the density of air to find the air's mass and weight.

Solution.

Given: $r = 30$ cm $= 0.30$ m *Find:* F_b (buoyant force)
$\rho_{air} = 1.29$ kg/m³

The volume of the air in the balloon is

$$V = \tfrac{4}{3}\pi r^3 = \tfrac{4}{3}\pi(0.30 \text{ m})^3 = 0.11 \text{ m}^3$$

Then, by Eq. 9.14, the weight of the air displaced by the balloon's volume, or the magnitude of the upward buoyant force, is

$$F_b = m_{air}g = (\rho_{air}V)g = (1.29 \text{ kg/m}^3)(0.11 \text{ m}^3)(9.8 \text{ m/s}^2) = 1.4 \text{ N}$$

Note that the buoyant force depends on the *density of the fluid* and *the volume of the body*. Shape makes no difference.

Follow-up Exercise. Would the buoyant force on the balloon be greater or lesser if the balloon were submerged in water, and by what factor? (Assume the same volume.)

Conceptual Example 9.8

Weight and Buoyant Force: Archimedes' Principle

An overflow container filled with water, like that shown on the left in Fig. 9.13b, sits on a scale that reads 40 N. (The water level is just below the exit tube in the side of the container.) A 5.0-newton object that is less dense than water (such as a block of wood) is placed in the container. The water it displaces runs out the exit tube into another container that is not on the scale. Will the scale reading then be (a) exactly 45 N, (b) between 40 N and 45 N, (c) exactly 40 N, (d) less than 40 N?

Reasoning and Answer. By Archimedes' principle, the block is buoyed upward with a force equal in magnitude to the weight of the water displaced. Since the block floats, the upward buoyant force must balance the weight of the block and has a magnitude of 5.0 N. Thus a volume of water weighing 5.0 N is displaced from the container as a 5.0-newton weight is added to the container. The scale still reads 40 N, so the answer is (c).

Note that the upward buoyant force and the block's weight force act *on the block*. The reaction force (pressure) of the block *on the water* is transmitted to the bottom of the container (Pascal's principle) and is registered on the scale.

Follow-up Exercise. Would the scale reading still be 40 N if the object had a density greater than that of water?

Buoyancy and Density

We commonly say that helium and hot-air balloons float because they are lighter than air. To be correct, technically we should say they are *less dense than air*. An object's density will tell you whether it will sink or float in a fluid as long as you also know the density of the fluid. Consider a solid uniform object totally immersed in a fluid. The weight of the object is

$$w_o = m_o g = \rho_o V_o g$$

The weight of the volume of fluid displaced, or the magnitude of the buoyant force, is

$$F_b = w_f = m_f g = \rho_f V_f g$$

If the object is *completely submerged*, $V_o = V_f$. Dividing the second equation by the first gives

$$\frac{F_b}{w_o} = \frac{\rho_f}{\rho_o} \quad \text{or} \quad F_b = \left(\frac{\rho_f}{\rho_o}\right) w_o \tag{9.15}$$

Thus, if ρ_o is less than ρ_f, then F_b will be greater than w_o and the object will be buoyed to the surface and float. If ρ_o is greater than ρ_f, then F_b will be less than w_o and the object will sink. If ρ_o equals ρ_f, then F_b will be equal to w_o and the object will remain in equilibrium at any submerged depth (as long as the density of the fluid is constant). If the object is not uniform, such that its density varies over its volume, then the density of the object is its average density.

These three conditions expressed in words are as follows:

An object will float in a fluid if the average density of the object is less than the density of the fluid.

An object will sink in a fluid if the average density of the object is greater than the density of the fluid.

An object will be in equilibrium at any submerged depth in a fluid if the average density of the object and the density of the fluid are equal.

Float or sink? Depends on densities of object and fluid

Demonstration 7 ■ Buoyancy and Density

This demonstration of buoyancy shows that the overall density of a can of Diet Coke is less than that of water, while the density of a can of Classic Coke is greater. Consider the following questions: Does one can have a greater volume of metal? higher gas pressure inside? more fluid volume? Do calories make a difference? Investigate the possibilities to determine thereason(s) for the different densities.

(a) Unopened cans of Coke are dropped into a container of water.

(b) The can of Classic Coke sinks, and the can of Diet Coke floats.

•**FIGURE 9.14 Equal densities and buoyancy** This soft drink contains colored gelatin beads that remain suspended for months with no change.

See Demonstration 7 for examples of the first two conditions and •Fig. 9.14 for the last one.

The densities of some solids and fluids are given in Table 9.2 (p. 301). A quick look will tell you if an object will float in a fluid, regardless of the shape or volume of the object. The conditions stated above also apply to a fluid in a fluid, provided the two are immiscible (do not mix).

In general, the densities of objects or fluids will be assumed to be uniform and constant in this book. (The density of the atmosphere varies with altitude but is relatively constant near the surface of the Earth.) However, in practical applications, as we have seen above, it is the *average* density of an object that often matters with regard to floating and sinking. An ocean liner is overall less dense than water, even though it is made of steel. Most of its volume is occupied by air, so the liner's average density is less than that of water. Similarly, the human body has air-filled spaces, so we float in water.

In some instances, the overall density of an object is purposefully varied. For example, a submarine submerges by flooding its tanks with sea water (called "taking on ballast"), which increases its average density while reducing the buoyant force (less water is displaced). When the sub is to surface, the water is pumped out of the tanks, so the average density of the sub becomes less than that of the surrounding sea water.

Example 9.9 ■ Float or Sink? Comparison of Densities

A uniform solid cube of material 10 cm on each side has a mass of 700 g. (a) Will the cube float in water? (b) If so, how much of its volume will be submerged?

Thinking It Through. (a) The question is whether the density of the cube material is greater or less than that of water, so we compute the cube's density. (b) If the cube floats, then the buoyant force and the cube's weight are equal. Both of these forces are related to volume, so we can write them in terms of volume and equate them.

Solution.

Given: $m = 700 \text{ g} = 0.700 \text{ kg}$
$L = 10 \text{ cm}$
$\rho_w = 1.00 \times 10^3 \text{ kg/m}^3 = 1.00 \text{ g/cm}^3$
(Table 9.2)

Find: (a) Whether the cube will float in water
(b) The percentage of the volume submerged if it does float

It is sometimes convenient to work in cgs units in comparing quantities, particularly when working with ratios. For densities in g/cm³, drop the "$\times 10^3$" from the values given in Table 9.2 for solids and liquids and add "$\times 10^{-3}$" for gases.

(a) The density of the cube material is

$$\rho_c = \frac{m}{V_c} = \frac{m}{L^3} = \frac{700 \text{ g}}{(10 \text{ cm})^3} = 0.70 \text{ g/cm}^3 < \rho_w = 1.00 \text{ g/cm}^3$$

Since ρ_c is less than ρ_w, the cube will float.

(b) The weight of the cube is $w_c = \rho_c g V_c$. When the cube is floating (in equilibrium), its weight is balanced by the buoyant force. That is, $F_b = \rho_w g V_w$, where V_w is the volume of water the submerged part of the cube displaces. Equating the expressions for weight and buoyant force gives

$$\rho_w g V_w = \rho_c g V_c$$

or

$$\frac{V_w}{V_c} = \frac{\rho_c}{\rho_w} = \frac{0.70 \text{ g/cm}^3}{1.00 \text{ g/cm}^3} = 0.70$$

Thus, $V_w = 0.70 V_c$, and 70% of the cube is submerged.

Follow-up Exercise. Most of an iceberg floating in the ocean (•Fig. 9.15) is submerged. What is seen is the proverbial "tip of the iceberg." What percentage of an iceberg's volume is seen above the surface? [Note: Icebergs are frozen *fresh* water floating in cold (salty) water.]

•**FIGURE 9.15 The tip of the iceberg** The vast majority of an iceberg's bulk is underneath the water. This large "berg" was photographed off the Antarctic Peninsula. (Does the submerged ice have to be directly below the exposed tip of the iceberg?)

Conceptual Example 9.10

A Floating Candle: Archimedes' Principle and Buoyant Force

A dripless candle is weighted slightly on the bottom so that it floats upright in a container of water as illustrated in •Fig. 9.16. As the candle burns, (a) the portion of the candle above the water burns down and is extinguished, (b) the candle rises relative to the base of the container, (c) the candle sinks lower relative to the base of the container.

Reasoning and Answer. To answer this, we need to look at the situation in terms of Archimedes' principle and buoyant force. Initially, the buoyant force balances the candle's original weight, so it floats at a certain level. As the candle burns, it gradually loses mass (weight). This loss gives rise to a net upward force, and the candle rises until the buoyant force and weight are again equal (with less candle submerged and a smaller buoyant force). Hence, the answer is (b).

The candle continuously rises and stays lit. To understand the process better, you may find it helpful to think of the candle as being burned in increments or segments and then extrapolate this to continuous burning.

Follow-up Exercise. A person sitting in a small boat floating on a large swimming pool notes the water level on the side of the pool. He then drops several concrete blocks over the side of the boat into the water. Does the level of the pool change? If so, how?

•**FIGURE 9.16 Rise or sink?** See Conceptual Example 9.10.

A quantity called specific gravity is related to density. It is commonly used for liquids but also applies to solids. Basically, it is a comparison of the weight of a volume of a substance with the weight of an equal volume of water. The **specific gravity** of a substance is equal to the ratio of the density of the substance (ρ_s) to the density of water (ρ_w at 4°C, the temperature for maximum density):

Specific gravity: a comparison to water

$$\text{sp. gr.} = \frac{\rho_s}{\rho_w}$$

Because it is a ratio of densities, specific gravity has no units. In cgs units $\rho_w = 1.00 \text{ g/cm}^3$, so

$$\text{sp. gr.} = \frac{\rho_s}{1.00} = \rho_s$$

That is, the specific gravity of a substance is equal to the numerical value of its density *in cgs units*. For example, if a liquid has a density of 1.5 g/cm³, its specific gravity is 1.5, which tells you that it is 1.5 times denser than water. To get density values in grams per cubic centimeter, divide the value in Table 9.2 by 10^3. The specific gravities of automobile coolant and battery electrolyte (both water solutions) are commonly measured so as to determine their relative concentrations of antifreeze and sulfuric acid, respectively.

9.4 Fluid Dynamics and Bernoulli's Equation

Objectives: To (a) identify the simplifications used in describing ideal fluid flow and (b) use the continuity equation and Bernoulli's equation to explain common effects.

In general, fluid motion is difficult to analyze. For example, think of trying to describe the motion of a particle (a molecule) of water in a rushing stream. The overall motion of the stream may be apparent, but a mathematical description of the motion of any one particle of it may be virtually impossible because of eddy currents (small whirlpool motions), the gushing of water over rocks, frictional drag on the stream bottom, and so on. A basic description of fluid flow is conveniently obtained by ignoring such complications and considering an ideal fluid. Actual fluid flow can then be approximated in reference to this simpler theoretical model.

In this simplified approach to fluid dynamics, it is customary to consider four characteristics of an **ideal fluid**. In such a fluid, flow is (1) *steady*, (2) *irrotational*, (3) *nonviscous*, and (4) *incompressible*.

> Condition 1: *Steady flow* means that all the particles of a fluid have the same velocity as they pass a given point.

Steady flow might be called smooth or regular flow. The path of steady flow can be depicted in the form of **streamlines** (•Fig. 9.17). Every particle that passes a particular point moves along a streamline. That is, every particle moves along the same path (streamline) as particles that passed by earlier. Streamlines never cross. If they did, a particle would have alternate paths and abrupt changes in its velocity, in which case the flow would not be steady.

Steady flow requires low velocities. For example, steady flow is approximated by the flow relative to a canoe that is gliding slowly through still water. When the flow velocity is high, eddies tend to appear, especially near boundaries, and the flow becomes turbulent.

Streamlines also indicate the relative magnitude of the fluid velocity. The velocity is greater where the streamlines are closer together. Notice this effect in Fig. 9.17a. The reason for this will be explained shortly.

> Condition 2: *Irrotational flow* means that a fluid element (a small volume of the fluid) has no net angular velocity, which eliminates the possibility of whirlpools and eddy currents (the flow is nonturbulent).

Consider the small paddle wheel in Fig. 9.17a. With a zero net torque, it does not rotate. Thus, the flow is irrotational.

> Condition 3: *Nonviscous flow* means that viscosity is neglected.

•**FIGURE 9.17 Streamline flow** (a) Streamlines never cross and are closer together in regions of greater fluid velocity. The stationary paddle wheel indicates that the flow is irrotational, or without whirlpools and eddy currents. (b) The smoke begins to rise in nearly streamline flow but quickly becomes rotational and turbulent.

Viscosity refers to a fluid's internal friction, or resistance to flow. A truly nonviscous fluid would flow freely with no energy lost within it. Also, there would be no frictional drag between the fluid and the walls containing it. In reality, when a liquid flows through a pipe, the speed is lesser near the walls because of frictional drag and is greater toward the center of the pipe. (Viscosity is discussed in more detail in Section 9.5.)

Condition 4: *Incompressible flow* means that the fluid's density is constant.

Liquids can usually be considered incompressible. Gases, conversely, are quite compressible. Sometimes, however, gases approximate incompressible flow—for example, air flowing relative to the wings of an airplane traveling at low speeds.

Theoretical or ideal fluid flow is not characteristic of most real situations. But the analysis of ideal flow provides results that approximate, or generally describe, a variety of applications. This analysis is not usually derived from Newton's laws but instead from two basic principles: conservation of mass and conservation of energy.

Equation of Continuity

If there are no losses of fluid within a uniform tube, the mass of fluid flowing into the tube in a given time must be equal to the mass flowing out of the tube in the same time (by the conservation of mass). For example, in •Fig. 9.18, the mass (Δm_1) entering the tube during a short time (Δt) is

$$\Delta m_1 = \rho_1 \Delta V_1 = \rho_1 (A_1 v_1 \Delta t)$$

Equation of continuity

where A_1 is the cross-sectional area of the tube at the entrance and, in a time Δt, a fluid particle moves a distance equal to $v_1 \Delta t$. Similarly, the mass leaving the tube in the same time interval is

Flow rate equation

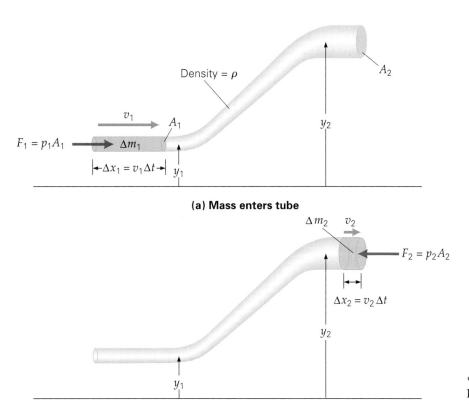

•**FIGURE 9.18 Flow continuity** Ideal fluid flow can be described in terms of the conservation of mass by the equation of continuity.

$$\Delta m_2 = \rho_2 \, \Delta V_2 = \rho_2(A_2 v_2 \, \Delta t)$$

Since the mass is conserved, $\Delta m_1 = \Delta m_2$, and

Equation of continuity

$$\rho_1 A_1 v_1 = \rho_2 A_2 v_2 \quad \text{or} \quad \rho A v = \text{constant} \qquad (9.16)$$

This general result is called the **equation of continuity.**

For an incompressible fluid, the density ρ is constant, so

Flow rate equation

$$A_1 v_1 = A_2 v_2 \quad \text{or} \quad A v = \text{constant} \quad \textit{(for an incompressible fluid)} \qquad (9.17)$$

This is sometimes called the **flow rate equation**, since Av has the standard units of cubic meters per second (m³/s, or volume/time). (In the British system, gallons per minute is often used.) Note that the flow rate equation shows that the fluid velocity is greater where the cross-sectional area of the tube is smaller. That is,

$$v_2 = \left(\frac{A_1}{A_2}\right) v_1$$

and v_2 is greater than v_1 if A_2 is less than A_1. This effect is evident in the common experience that the speed of water is greater from a hose fitted with a nozzle than from the same hose without a nozzle (•Fig. 9.19).

The flow rate equation can be applied to the flow of blood in your body. Blood flows from the heart into the aorta. It then makes a circuit through the circulatory system, passing through arteries, arterioles (small arteries), capillaries, venules (small veins) and back to the heart through veins. The speed is lowest in the capillaries—a contradiction? No. The *total* area of the capillaries is much larger than that of the arteries or veins, so the flow rate equation holds.

•FIGURE 9.19 Flow rate By the flow rate equation, the fluid speed is greater when the cross-sectional area of a tube is smaller. Think of a hose that is equipped with a nozzle such that the cross-sectional area is made smaller.

Example 9.11 ■ Blood Flow: Cholesterol and Plaque

High cholesterol in the blood can cause fatty deposits called plaques to form on the walls of blood vessels. Suppose a plaque reduces the effective radius of an artery by 25%. How does this partial blockage affect the speed of blood through the artery?

Thinking It Through. We will use the flow rate equation (Eq. 9.17), but note that no values of area or speed are given. This indicates that we should use ratios.

Solution. If we take the unclogged artery to have a radius r_1, the plaque then reduces the effective radius to r_2.

Given: $r_2 = 0.75 r_1$ (for a 25% reduction)   *Find:* v_2

Writing the flow rate equation in terms of the radii, we have

$$A_1 v_1 = A_2 v_2$$
$$(\pi r_1^2) v_1 = (\pi r_2^2) v_2$$

Rearranging and canceling, we get

$$v_2 = (r_1/r_2)^2 \, v_1$$

From the given, $r_1/r_2 = 1/0.75$, so

$$v_2 = (1/0.75)^2 \, v_1 = 1.8 v_1$$

Hence, the speed through the clogged artery increases by 80%.

Follow-up Exercise. By how much would the effective radius of an artery have to be reduced to have a 50% increase in the blood speed through it?

Bernoulli's Equation

The conservation of energy or the general work-energy theorem leads to another relationship that has great generality for fluid flow. This relationship was first derived in 1738 by the Swiss mathematician Daniel Bernoulli (1700–1782) and is named for him.

Let's look again at the ideal fluid flowing in the tube in Fig. 9.18. Work is done by the external forces at the ends of the tube; F_1 does positive work (in the same direction as the fluid's motion) and F_2 does negative work (opposite to the fluid motion). The net work done on the system by these forces is then

$$W = F_1 \Delta x_1 - F_2 \Delta x_2 = (p_1 A_1)(v_1 \Delta t) - (p_2 A_2)(v_2 \Delta t)$$

The flow rate equation (Eq. 9.17) requires that $A_1 v_1 = A_2 v_2$, so we may write the work equation as

$$W = A_1 v_1 \Delta t (p_1 - p_2)$$

Recall from the preceding derivation of the equation of continuity that $\Delta m_1 = \rho_1 \Delta V_1 = \rho_1 (A_1 v_1 \Delta t)$, and $\Delta m_1 = \Delta m_2$, so we may write in general

$$W = \frac{\Delta m}{\rho}(p_1 - p_2)$$

The net work done on the system by the external forces (nonconservative work) must be equal to the change in total mechanical energy. That is, $W = \Delta E = \Delta K + \Delta U$. Looking at the change in kinetic energy of an element of mass Δm, we have

$$\Delta K = \tfrac{1}{2} \Delta m (v_2^2 - v_1^2)$$

The corresponding change in gravitational potential energy is

$$\Delta U = \Delta m g (y_2 - y_1)$$

Thus,

$$W = \Delta K + \Delta U$$

$$\frac{\Delta m}{\rho}(p_1 - p_2) = \tfrac{1}{2}\Delta m(v_2^2 - v_1^2) + \Delta m g(y_2 - y_1)$$

Canceling each Δm and rearranging gives the common form of **Bernoulli's equation**:

or
$$p_1 + \tfrac{1}{2}\rho v_1^2 + \rho g y_1 = p_2 + \tfrac{1}{2}\rho v_2^2 + \rho g y_2$$
$$p + \tfrac{1}{2}\rho v^2 + \rho g y = \text{constant} \tag{9.18}$$

Note that in working with a fluid, the terms in Bernoulli's equation are work or energy *per unit volume* (J/m³). That is, $W = F\Delta x = p(A\Delta x) = p\Delta V$, and $p = W/\Delta V$ (work/volume). Similarly, with $\rho = m/V$, we have $\tfrac{1}{2}\rho v^2 = \tfrac{1}{2}mv^2/V$ (energy/volume) and $\rho g y = mgy/V$ (energy/volume).

Note: Compare the derivation of Eq. 5.10 in Section 5.5.

Bernoulli's equation, or principle, can be applied to many situations. If there is horizontal flow ($y_1 = y_2$), then $p + \tfrac{1}{2}\rho v^2 = $ constant, which indicates that the pressure decreases if the fluid speed increases (and vice versa). This effect is illustrated in •Fig. 9.20, where the difference in flow heights through the pipe is considered negligible (so the $\rho g y$ term drops out). (See Demonstration 8.)

Bernoulli effects

Chimneys and smokestacks are tall in order to take advantage of the more consistent and higher wind speeds at greater heights. The faster the wind blows over the top of a chimney, the lower the pressure, and the greater the pressure difference between the bottom and top of the chimney. Thus, the chimney draws

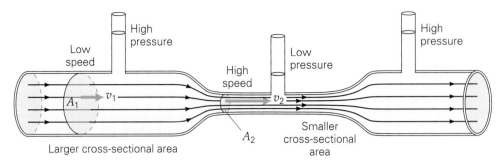

•**FIGURE 9.20 Flow rate and pressure** Taking the horizontal difference in flow heights to be negligible in a constricted pipe, Bernoulli's equation becomes $p + \frac{1}{2}\rho v^2 =$ constant. In a region of smaller cross-sectional area, the flow speed is greater (flow rate equation); from Bernoulli's equation, the pressure in that region is lower than in other regions.

Demonstration 8 ■ Bernoulli-trapped Ball

How does blowing downward through a funnel keep a Ping Pong ball from falling? In this demonstration, seeing is believing in Bernoulli. The air stream above and around the ball (along the wall of the funnel) is moving faster than the air below the ball, and the faster the air moves, the less pressure it exerts. Thus the greater pressure of the slower air below the ball holds it up. Without an air stream, the pressures are equalized (both atmospheric), and the ball falls.

(a) The demonstrator prepares to place the ball in an inverted funnel and to blow downward through the funnel.

(b) With air blowing through, the ball remains "trapped" in the funnel.

(c) When the demonstrator is out of breath, the ball falls.

---

exhaust out better. Bernoulli's equation and the continuity equation ($Av =$ constant) also tell you that if the cross-sectional area of a pipe is reduced, so the velocity of the fluid passing through it is increased, the pressure is reduced.

The Bernoulli effect (as it is sometimes called) is also partially responsible for the lift of an airplane. Ideal air flow over an airfoil or wing is shown in •Fig. 9.21 (turbulence is neglected). The wing is curved on the top side and angled relative to the incident streamlines. As a result, the streamlines above the wing are closer together than those below, which means a higher air speed and lower pressure above the

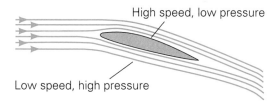

High speed, low pressure

Low speed, high pressure

•**FIGURE 9.21 Airplane lift—Bernoulli in action** Because of the shape and orientation of an airfoil or airplane wing, the air streamlines are closer together, and the air speed is greater, above the wing than below it. By Bernoulli's principle, a pressure difference results that supplies an upward force, or lift.

wing. With a higher pressure on the bottom of the wing, there is a net upward force, or *lift*. Note in the figure that the streamlines leaving the wing curve downward. This curving reflects the fact that since the wing is acquiring upward momentum, the air molecules must acquire an equal downward component of momentum.

Suppose a fluid is at rest ($v_2 = v_1 = 0$). Bernoulli's equation then becomes

$$p_2 - p_1 = \rho g(y_1 - y_2)$$

This is the pressure–depth relationship derived earlier (Eq. 9.10).

Example 9.12 ■ Flow Rate from a Tank: Bernoulli's Equation

Water escapes through a small hole in a tank as depicted in •Fig. 9.22. What is the initial flow rate of the water out of the tank?

Thinking It Through. There is an appreciable difference in flow heights (y) across the tank. Approximations can be made to delete other variables in Bernoulli's equation; for example, the atmospheric pressures on the tank water surface and the hole are nearly equal.

Solution. By Bernoulli's equation,

$$p_1 + \tfrac{1}{2}\rho v_1^2 + \rho g y_1 = p_2 + \tfrac{1}{2}\rho v_2^2 + \rho g y_2$$

where $y_2 - y_1$ is the height of the liquid surface above the hole. The atmospheric pressures acting on the open surface and at the hole, p_1 and p_2, respectively, are essentially equal and cancel from the equation, as does the density, giving

$$v_1^2 - v_2^2 = 2g(y_2 - y_1)$$

The equation of continuity says that $A_1 v_1 = A_2 v_2$, where A_2 is the cross-sectional area of the tank and A_1 is that of the hole. Since A_2 is much greater than A_1, then v_1 is much greater than v_2. So a good approximation is

$$v_1^2 = 2g(y_2 - y_1) \quad \text{or} \quad v_1 = \sqrt{2g(y_2 - y_1)}$$

The flow rate (volume/time) is then

$$\text{flow rate} = A_1 v_1 = A_1 \sqrt{2g(y_2 - y_1)}$$

Given the area of the hole and the height of the liquid above it, you could find the instantaneous speed of the water coming from the hole and the instantaneous flow rate.

Follow-up Exercise. What would be the percentage change in the initial flow rate from the tank in this Example if the diameter of the small circular hole were increased by 30.0%?

•**FIGURE 9.22 Fluid flow from a tank** The flow rate is given by Bernoulli's equation. See Example 9.12.

Another example of the Bernoulli effect is given in the Insight on p. 320.

Conceptual Example 9.13

A Stream of Water: Smaller and Smaller

You have probably observed that a steady stream of water flowing out of a faucet gets smaller the farther it gets from the faucet. Why does this happen?

Reasoning and Answer. This effect can be explained by Bernoulli's principle. As the water falls, it accelerates and its speed increases. Then, by Bernoulli's principle, the internal liquid pressure inside the stream decreases (see Fig. 9.20). A pressure difference

Insight

Throwing a Curveball

In baseball, a pitcher can cause a baseball to curve as it moves toward home plate by giving it an appropriate spin. Why a curveball curves can be understood in terms of Bernoulli's equation and the viscous properties of air.

If air were an ideal fluid, the spin of a pitched ball would have no effect in changing the direction of its motion. (In Fig. 1a, the streamlines are those that would be seen by an observer moving with the ball.) However, because air is viscous, friction between it and the ball causes a thin (boundary) layer of air to be dragged around by the spinning ball. This is illustrated in Fig. 1b for a relatively slow airstream in which the flow is generally smooth. The speed of the ball relative to the air is thus greater on one side than on the other because the velocities of the ball and of the air are in the same direction on one side (top in Fig. 1b) and in opposite directions on the other. By Bernoulli's equation, the low-velocity side (v') has a greater pressure than the high-velocity side (v''). Thus, the ball experiences a net force toward the low-pressure side and is deflected (curves).

However, baseballs are usually pitched very fast, and another effect applies. With a high air flow speed, the rotation of the ball causes the boundary layer to separate at different points on either side of the ball. This separation gives rise to a turbulent wake that is deflected in the spin direction (Fig. 2). A pressure difference results, and the ball is deflected or curves the opposite way. This is called the Magnus effect (after Gustav Magnus, who observed the effect in 1852). In terms of Newton's third law, the turbulent wake is deflected in one direction and the reaction force deflects the ball in the opposite direction.

Note that the ball in Fig. 2 is rotating counterclockwise and is deflected to the left (toward the top, in this overhead view). To have the ball curve to the right requires a clockwise rotation. In general, the curve is in the direction that the front part of the ball is rotating. With the spin axis in the horizontal plane, the deflecting force may be up or down, which can cause the ball not to drop as much under the influence of gravity, or to "sink" even faster. The spin axis can be oriented in any direction to produce a variety of effects with the deflecting force and gravity.

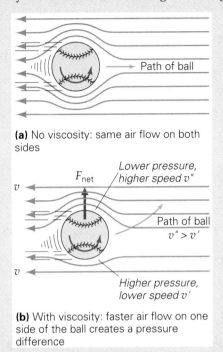

FIGURE 1 Bernoulli effect (a) If air were an ideal fluid (without viscosity) and spin were applied, air flow would be the same on both sides of the ball. (b) Because air has viscosity, some air is dragged around with the ball, so the speed of the ball with respect to the air is greater on one side (top view shown). Pressure on that side is lower, resulting in a net force in that direction.

FIGURE 2 Magnus effect The turbulent wake of a fast-moving ball gives rise to a curving of the ball. In terms of Newton's third law, the wake is deflected in one direction and the ball is deflected in the opposite direction by the reaction force.

between that inside stream and the atmospheric pressure on the outside has been created. As a result, there is an increasing inward force as the stream falls, so it becomes smaller. Eventually, the stream may get so thin that it breaks up into individual droplets.

Follow-up Exercise. The equation of continuity can also be used to explain this stream effect. Give this explanation.

•FIGURE 9.23 Surface tension (a) The net force on a molecule in the interior of a liquid is zero, because it is surrounded by other molecules. However, a molecule at the surface experiences a nonzero net force due to the attractive forces of the neighboring molecules just below the surface. (b) For an object such as a needle to form a surface depression, work must be done, since more interior molecules must be brought to the surface to increase the area. As a result, the surface area acts like a stretched elastic membrane, and the weight force of the object is supported by the upward components of the surface tension. Insects such as this water strider can walk on water because of the upward components of the surface tension, much as you might walk on a large trampoline. Note the depressions in the surface of the liquid where the legs touch it.

*9.5 Surface Tension; Viscosity and Poiseuille's Law

Objectives: To (a) describe the source of surface tension and its effects and
(b) discuss fluid viscosity.

Surface Tension

The molecules of a liquid exert small attractive forces on each other. Even though molecules are electrically neutral overall, there is often some slight asymmetry of charge that gives rise to attractive forces between molecules (called *van der Waals forces*). Within a liquid, where any molecule is completely surrounded by other molecules, the net force is zero (•Fig. 9.23a). However, for molecules at the surface of the liquid, there is no attractive force acting from above the surface. (The effect of air molecules is small and considered negligible.) As a result, the molecules of the surface layer experience net forces due to the attraction of neighboring molecules just below the surface. This inward pull on the surface molecules causes the surface of the liquid to contract and to resist being stretched or broken, a property called **surface tension**.

If a sewing needle is carefully placed on the surface of a bowl of water, the surface acts like an elastic membrane under tension. There is a slight depression in the surface, and molecular forces along the depression act at an angle to the surface (Fig. 9.23b). The vertical components of these forces balance the weight (mg) of the needle, and it "floats" on the surface. Similarly, surface tension supports the weight of a water strider.

The net effect of surface tension is to make the surface area of a liquid as small as possible. That is, a given volume of liquid tends to assume the shape that has the least surface area. As a result, drops of water and soap bubbles have spherical shapes because a sphere has the smallest surface area for a given volume (•Fig. 9.24). In forming a drop or bubble, surface tension pulls the molecules together to minimize the surface area.

Surface tension: resistance to stretching of surface

Viscosity

All real fluids have an internal resistance to flow, which is described as **viscosity**. Viscosity can be considered to be friction between the molecules of a fluid. In liquids, it is caused by short-range cohesive forces and, in gases, by collisions

•**FIGURE 9.24 Surface tension at work** Because of surface tension, **(a)** water droplets and **(b)** soap bubbles tend to assume the shape that minimizes their surface area—that of a sphere.

(a)

(b)

between molecules (see the discussion of air resistance in Section 4.6). The viscous drag for both liquids and gases depends on their velocity and may be directly proportional to it in some cases. However, the relationship varies with the conditions; for example, the drag is approximately proportional to v^2 or v^3 for turbulent flow.

Internal friction causes fluid layers to move relative to each other in response to a shear stress. This layered motion, called *laminar flow*, is characteristic of steady flow for viscous liquids at low velocities (•Fig. 9.25a). At higher velocities, the flow becomes rotational, or *turbulent*, and difficult to analyze.

Since there are shear stresses and shear strains (deformation) in laminar flow, the viscous property of a fluid can be described by a coefficient, as elastic moduli were used in Section 9.1. Viscosity is characterized by a *coefficient of viscosity*, η (the Greek letter eta), commonly referred to as simply the **viscosity**.

Viscosity: resistance to flow

The coefficient of viscosity is, in effect, the ratio of the shear stress to the rate of change of the shear strain (since motion is involved). When the equation for the ratio is written, unit analysis shows that the SI unit of viscosity is pascal-seconds (Pa·s). This combined unit is called the *poiseuille* (Pl), in honor of the French scientist Jean Poiseuille (1799–1869), who studied the flow of liquids, particularly blood. The cgs unit of viscosity is the *poise* (P). A smaller multiple, the *centipoise* (cP), is widely used because of its convenient size; 1 Pl = 10^3 cP.

The viscosities of some fluids are listed in Table 9.3. The larger the viscosity of a liquid, which is easier to visualize than that of a gas, the greater the shear stress required to get the liquid layers to slide along each other. Note, for example, the large viscosity of glycerin compared to that of water.

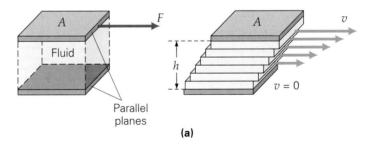
(a)

•**FIGURE 9.25 Laminar flow** **(a)** A shear stress causes fluid layers to move over each other in laminar flow. The shear force and the flow rate depend on the viscosity of the fluid. **(b)** For laminar flow through a pipe, the fluid speed is less near the walls of the pipe than near the center because of frictional drag between the walls and the fluid.

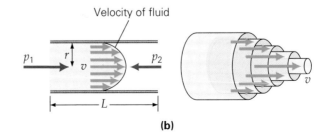
(b)

TABLE 9.3 Viscosities of Various Fluids*

Fluid	Viscosity (η) Poiseuille (Pl)	Centipoise (cP)
Liquids		
Alcohol, ethyl	1.2×10^{-3}	1.2
Blood, whole (37°C)	1.7×10^{-3}	1.7
Blood plasma (37°C)	2.5×10^{-3}	2.5
Glycerin	1.5	1.5×10^{3}
Mercury	1.55×10^{-3}	1.55
Oil, light machine	1.1	1.1×10^{3}
Water	1.00×10^{-3}	1.00
Gases		
Air	1.9×10^{-5}	1.9×10^{-2}
Oxygen	2.2×10^{-5}	2.2×10^{-2}

*At 20°C unless otherwise indicated. 1 poiseuille (Pl) = 10^3 centipoise (cP).

As you might expect, viscosity, and thus fluid flow, varies with temperature, which is evident from the old saying "slow as molasses in January." A familiar application is the viscosity grading of motor oil used in automobiles. In winter, a low-viscosity, or relatively thin, oil should be used (such as SAE grade 10W or 20W) because it will flow more readily, particularly when the engine is cold at start-up. In summer, a higher-viscosity, or thicker, oil is used (SAE 30, 40, or even 50).

Note: SAE stands for Society of Automotive Engineers, an organization that designates the grades of motor oils based on flow rate or viscosity.

Seasonal changes in the grade of motor oil are not necessary if you use the multigrade year-round oils. These oils contain additives called viscosity improvers, which are polymers whose molecules are long, coiled chains. A temperature increase causes the molecules to uncoil and intertwine with each other. Thus, the normal decrease in viscosity is counteracted. The action is reversed on cooling, and the oil maintains a relatively small viscosity range over a large temperature range. Such motor oils are graded as, for example, SAE 10W-40 (or 10W-40 for short).

Poiseuille's Law Viscosity makes analyzing fluid flow difficult. For example, when a fluid flows through a pipe, there is frictional drag between the liquid and the walls, and the fluid velocity is greater toward the center of the pipe (Fig. 9.25b).

In practice, this effect makes a difference in the *average flow rate Q* of a fluid, which is given by $Q = A\bar{v} = \Delta V / \Delta t$ (see Eq. 9.17) and describes the volume (ΔV) of fluid flowing past a given point in a time Δt. The SI unit of flow rate is cubic meters per second (m³/s). The flow rate depends on the properties of the fluid and the dimensions of the pipe as well as on the pressure difference (Δp) between the ends of the pipe.

Jean Poiseuille studied flow in pipes and tubes, assuming constant viscosity and steady or laminar flow. He derived the following relationship, known as **Poiseuille's law**, for the flow rate:

$$Q = \frac{\pi r^4 \Delta p}{8\eta L} \quad (9.19)$$

Poiseuille's law—viscous flow

where r is the radius of the pipe and L is its length.

As should be expected, the flow rate is inversely proportional to the viscosity (η) and the length of the pipe. Also as expected, the flow rate is directly proportional to the pressure difference, Δp, between the ends of the pipe. Somewhat surprisingly, however, the flow rate is proportional to r^4, which makes it more highly dependent on the radius of the tube than we might have thought.

An application of fluid flow was considered for a medical "IV" in Example 9.6. However, Poiseuille's law, which incorporates flow rate, provides more reality in this application, as the following Example shows.

Example 9.14 ■ Poiseuille's Law: A Blood Transfusion

A hospital patient needs a blood transfusion, which will be administered through a vein in the arm via a gravity IV. The physician wishes to have 500 cc of whole blood delivered over a period of 10 min by using an 18-gauge needle with a length of 50 mm and an inner diameter of 1.0 mm. At what height above the arm should the bag of blood be hung? (Assume a venous blood pressure of 15 mm Hg.)

Thinking It Through. This is an application of Poiseuille's law (Eq. 9.19) to find the pressure needed at the inlet of the needle that will provide the required flow rate (Q). Note that $\Delta p = p_i - p_o$ (inlet pressure minus outlet pressure). Knowing the inlet pressure, we can find the required height of the bag as in Example 9.6. *Caution*: There are a lot of nonstandard units, and some quantities are assumed to be known from tables.

Solution. First we write the given (and known) quantities, converting to standard SI units as we go:

Given: $\Delta V = 500$ cc $= 500$ cm^3 (1 m^3/10^6 cm^3) $= 5.00 \times 10^{-4}$ m^3 *Find*: h (height of bag)
$\Delta t = 10$ min $= 600$ s $= 6.00 \times 10^2$ s
$L = 50$ mm $= 5.0 \times 10^{-2}$ m
$d = 1.0$ mm, or $r = 0.50$ mm $= 5.0 \times 10^{-4}$ m
$p_o = 15$ mm Hg $= 15$ torr (133 Pa/torr) $= 2.0 \times 10^3$ Pa
$\eta = 1.7 \times 10^{-3}$ Pl (whole blood, from Table 9.3)

The flow rate Q is

$$Q = \Delta V/\Delta t = 5.00 \times 10^{-4} \text{ m}^3/(6.00 \times 10^2 \text{ s}) = 8.33 \times 10^{-7} \text{ m}^3/\text{s}$$

We apply this value to Eq. 9.19 and solve for Δp:

$$\Delta p = \frac{8\eta L Q}{\pi r^4} = \frac{8(1.7 \times 10^{-3} \text{ Pl})(5.0 \times 10^{-2} \text{ m})(8.33 \times 10^{-7} \text{ m}^3/\text{s})}{\pi (5.0 \times 10^{-4} \text{ m})^4} = 2.9 \times 10^3 \text{ Pa}$$

With $\Delta p = p_i - p_o$, we have

$$p_i = \Delta p + p_o = (2.9 \times 10^3 \text{ Pa}) + (2.0 \times 10^3 \text{ Pa}) = 4.9 \times 10^3 \text{ Pa}$$

Then, to find the height of the bag that will deliver this pressure, we use $p_i = \rho g h$ (where $\rho_{\text{blood}} = 1.05 \times 10^3$ kg/m^3, from Table 9.2), so

$$h = \frac{p_i}{\rho g} = \frac{4.9 \times 10^3 \text{ Pa}}{(1.05 \times 10^3 \text{ kg/m}^3)(9.80 \text{ m/s}^2)} = 0.48 \text{ m}$$

Hence, for the prescribed flow rate, the "pint" of blood should be hung about 48 cm above the needle in the arm.

Follow-up Exercise. Suppose the physician wants to follow up the blood transfusion with 500 cc of saline solution at the same rate of flow. At what height should the saline bag be placed? (The "isotonic" saline solution administered by IV is a 0.85% aqueous salt solution, which has the same salt concentration as do body cells. To a good approximation, saline has the same density as water.)

•FIGURE 9.26   IV technology The mechanism of intravenous injection is still a gravity assist, but IV flow rates are now commonly controlled and monitored by machines.

Gravity-flow IVs are still used, but with modern technology, the flow rates of IVs are now often controlled and monitored by machines (•Fig. 9.26).

Chapter Review

Important Terms

fluid *294*
stress *295*
strain *295*
elastic modulus *296*
elastic limit *296*
Young's modulus *297*
shear modulus *298*
bulk modulus *298*
compressibility *299*

pressure *300*
pascal (Pa) *301*
atmosphere (atm) *302*
Pascal's principle *303*
absolute pressure *305*
gauge pressure *305*
buoyant force *310*
Archimedes' principle *310*
specific gravity *313*

ideal fluid *314*
streamlines *314*
equation of continuity *316*
flow rate equation *316*
Bernoulli's equation *317*
*surface tension *321*
*viscosity *322*
*Poiseuille's law *323*

Important Concepts

- Elastic modulus is the ratio of stress to strain.
- Young's modulus is the ratio of tensile stress to strain. (The stress may also be compressional.)
- Pressure applied to an enclosed fluid is transmitted undiminished to every point in the fluid and to the walls of the container (Pascal's principle).
- A body immersed wholly or partially in a fluid is buoyed up by a force equal in magnitude to the weight of the volume of fluid it displaces (Archimedes' principle).
- An object will float (sink) in a fluid if the average density of the object is less (greater) than the density of the fluid.
- Bernoulli's equation is basically a mathematical statement of the conservation of energy for a fluid.

Important Equations

Stress:

$$\text{stress} = \frac{F}{A} \quad (9.1)$$

Strain (tensile):

$$\text{strain} = \frac{\Delta L}{L_o} = \frac{L - L_o}{L_o} \quad (9.2)$$

Young's Modulus:

$$Y = \frac{F/A}{\Delta L/L_o} \quad (9.4)$$

Shear Modulus:

$$S = \frac{F/A}{x/h} \approx \frac{F/A}{\phi} \quad (9.5)$$

Bulk Modulus:

$$B = \frac{F/A}{-\Delta V/V_o} = -\frac{\Delta p}{\Delta V/V_o} \quad (9.6)$$

Compressibility:

$$k = \frac{1}{B} \quad (9.7)$$

Pressure:

$$p = \frac{F}{A} \quad (9.8a)$$

Pressure–Depth Equation (for an incompressible fluid at constant density):

$$p = p_o + \rho g h \quad (9.10)$$

Archimedes' Principle:

$$F_b = m_f g = \rho_f g V_f \quad (9.14)$$

Equation of Continuity:

$$\rho_1 A_1 v_1 = \rho_2 A_2 v_2 \quad \text{or} \quad \rho A v = \text{constant} \quad (9.16)$$

Flow Rate Equation (for an incompressible fluid):

$$A_1 v_1 = A_2 v_2 \quad \text{or} \quad A v = \text{constant} \quad (9.17)$$

Bernoulli's Equation:

$$p_1 + \tfrac{1}{2}\rho v_1^2 + \rho g y_1 = p_2 + \tfrac{1}{2}\rho v_2^2 + \rho g y_2$$

or

$$p + \tfrac{1}{2}\rho v^2 + \rho g y = \text{constant} \quad (9.18)$$

Poiseuille's Law:

$$Q = \frac{\pi r^4 \Delta p}{8 \eta L} \quad (9.19)$$

Exercises

9.1 Solids and Elastic Moduli

(Use as many significant figures as you need to show small changes.)

1. The pressure on an elastic body is described by (a) a modulus, (b) work, (c) stress, (d) strain.

2. Shear moduli are not zero for (a) solids, (b) liquids, (c) gases, (d) all of these.

3. One material has a greater Young's modulus than another. What does this tell you?

4. Why are scissors sometimes called shears? Is this a descriptive name in the physical sense?

5. Write the general form of Hooke's law, and find the units of the "spring constant" for elastic deformation.

6. ■ A 50-kilogram mass is suspended from a cable with a diameter of 2.0 cm. What is the stress on the cable?

7. ■ A 5.0-meter-long rod is stretched 0.10 m by a force. What is the strain in the rod?

8. ■ A 250-newton force is applied at a 37° angle to the surface of the end of a square bar. That surface is 4.0 cm on a side. What are (a) the compressional stress and (b) the shear stress on the bar?

9. ■■ A metal wire 1.0 mm in diameter and 2.0 m long hangs vertically with a 6.0-kilogram mass suspended from it. If the wire stretches 1.4 mm under the tension, what is the value of Young's modulus for the metal?

10. ■■ A 5.0-kilogram mass is supported by an aluminum wire of length 2.0 m and diameter 2.0 mm. How much will the wire stretch?

11. ■■ A copper wire has a length of 5.0 m and a diameter of 3.0 mm. Under what compressional load will its length decrease by 0.30 mm?

12. ■■ Each steel rail in a railroad is 8.0 m long and has a cross-sectional area of 0.0025 m². No gap has been left between the rails, and on a hot day, each rail tries to thermally expand 3.0×10^{-3} m. What is the force on the ends of a rail? (Is this a good reason to leave expansion gaps between the rails?)

13. ■■ A rectangular steel column (20.0 cm × 15.0 cm) supports a load of mass 12.0 metric tons. If the column was 2.00 m in length before being stressed, what is the decrease in length?

14. ■■ A bimetallic rod as illustrated in •Fig. 9.27 is composed of brass and copper. If the rod is subjected to a compressive force of 5.00×10^4 N, which way will the rod bend? (Justify your answer mathematically.)

•**FIGURE 9.27 Bimetallic rod and mechanical stress** See Exercise 14.

15. ■■ A 500-newton shear force is applied to one face of a cube of aluminum measuring 10 cm on each side. What is the displacement of that face relative to the opposite face?

16. ■■ Two metal posts of aluminum and copper, respectively, of the same size are subjected to equal shear stresses. Which post will show the larger deformation angle and by what factor?

17. ■■ A rectangular block of gelatin of length, width, and height 10 cm, 8.0 cm, and 4.0 cm, respectively, is subjected to a 0.40-newton shear force on its upper surface. If the top surface is displaced 0.30 mm relative to the bottom surface, what is the shear modulus of the gelatin?

18. ■■ Two metal plates are held together by two steel rivets, each of diameter 0.20 cm and length of 1.0 cm. How much force must be applied parallel to the plates to shear off both rivets?

19. ■■ (a) Which of the liquids in Table 9.1 has the greatest compressibility? (b) For equal volumes of ethyl alcohol and water, which would require more pressure to be compressed by 0.10%, and how many times more?

20. ■■■ A brass cube 6.0 cm on each side is placed in a pressure chamber and subjected to a pressure of 1.2×10^7 N/m² on all its surfaces. By how much will each side be compressed under this pressure?

21. ■■■ A 45-kilogram traffic light is suspended from two steel cables of equal length and radii 0.50 cm. If each cable makes a 15° angle with the horizontal, what is the fractional increase in their length due to the weight of the light?

9.2 Fluids: Pressure and Pascal's Principle

22. •Figure 9.28 shows a famous magician's "trick." She lies on a bed of thousands of nails and a sledge hammer is used to pound on an anvil on her chest. The nails will not pierce her skin. Explain why.

•FIGURE 9.28 **The nail mattress** See Exercise 22.

23. For the pressure–depth relationship for a fluid ($p = p_o + \rho g h$), it is assumed that (a) the pressure decreases with depth, (b) a pressure difference depends on the reference point, (c) the fluid density is constant, (d) this applies only to liquids.

24. Two dams form artificial lakes of equal depth. However, one lake backs up 15 km behind the dam, and the other backs up 50 km behind. What effect does the difference in length have on the pressures on the dams?

25. A water dispenser for pets has an inverted plastic bottle, as shown in •Fig. 9.29. (The water is dyed blue for contrast.) When a certain amount of water is drunk from the bowl, more water flows automatically from the bottle into the bowl. The bowl never overflows. Explain the operation of the dispenser. Does the height of the water in the bottle depend on the surface area of the water in the bowl?

•FIGURE 9.29 **Pet barometer** See Exercise 25.

26. (a) Liquid storage cans, such as gasoline cans, generally have capped vent holes. What is the purpose of the vent, and what happens if you forget to remove the cap before you pour the liquid? (b) Explain how a medicine dropper works. (c) Explain how we breathe (inhalation and exhalation).

27. Automobile tires are inflated to about 30 lb/in^2, whereas thin bicycle tires are inflated to 90 to 115 lb/in^2—at least three times higher pressure! Why?

28. Blood pressure is usually measured at the arm. However, suppose the pressure reading were taken on the calf of the leg of a standing person. Would there be a difference in principle? Explain.

29. ■ In his original barometer, Pascal used water instead of mercury. How high would the water column have been? Is this practical?

30. ■ If you dive to 15 m below the surface of a lake, (a) what is the pressure due to the water alone? (b) What is the total or absolute pressure at that depth?

31. ■ In an open U-tube, the pressure of a 15-centimeter water column on one side is balanced by a column of gasoline on the other side. What is the height of the gasoline column?

32. ■ A 75-kilogram athlete does a single-hand handstand. If the area of the hand in contact with the floor is 125 cm^2, what pressure is exerted on the floor?

33. ■ The gauge pressure in both tires of a bicycle is 690 kPa. If the bicycle and the rider have a combined mass of 90.0 kg, what is the area of contact of *each* tire with the ground if each tire supports half the total weight?

34. ■■ What is the pressure exerted on a 15° ski slope by a 90-kilogram skier going down it? The area of contact of the skis with the snow is 0.40 m^2.

35. ■■ In a lecture demonstration, an empty can of dimensions 0.24 m × 0.16 m × 0.10 m is used to demonstrate the force exerted by air pressure. A small quantity of water is poured into the can, and the water is brought to a boil. Then, the can is capped and dipped in cold water. To your surprise, the can is crushed (•Fig. 9.30). Explain. Assume the inside of the can is in a perfect vacuum. What is the force exerted on the can by air pressure?

•FIGURE 9.30 **Air pressure** See Exercise 35.

36. ■■ What is the fractional decrease in pressure when a barometer is raised 35 m to the top of a building? (Assume that the density of air is constant over that distance.)

37. ■■ A student decides to compute the standard barometric reading on top of Mt. Everest (29 028 ft) by assuming the density of air has the same constant density as at sea level. Try this yourself. What does the result tell you?

38. •Figure 9.31 shows a method Pascal used to demonstrate the importance of fluid pressure on fluid depth. An oak barrel with lid of area 0.20 m² is filled with water. A long, thin tube of cross-sectional area 5.0×10^{-5} m² is inserted into a hole at the center of the lid, and water is poured into the tube. When the water reaches 12 m high, the barrel burst. What was (a) the weight of the water in the tube? (b) the pressure of the water on the lid of the barrel? (c) the upward net force on the lid due to the water pressure?

•FIGURE 9.31 Pascal and the bursting barrel See Exercise 38.

39. ■■ In a sample of seawater taken from an oil spill region, it is found that an oil layer 4.0 cm thick floats on 55 cm of water. If the density of the oil is 0.75×10^3 kg/m³, what is the absolute pressure on the bottom of the container?

40. ■■ The pressure exerted by a person's lungs can be measured by having the person blow as hard as possible into one side of a manometer. If a person blowing into one side of an open-tube manometer produces an 80-centimeter difference in the heights of the columns of water in the manometer arms, what is the lung gauge pressure?

41. ■■ In 1960, the U.S. Navy's bathyscaphe *Trieste* (a submersible) descended to a depth of 10 912 m (about 35 000 ft) into the Marianas Trench in the Pacific Ocean. (a) What was the pressure at that depth? (Assume that sea water is incompressible.) (b) What was the force on a circular observation window with a diameter of 15 cm?

42. ■■ The output piston of a hydraulic press has a cross-sectional area of 0.20 m². (a) How much pressure on the input piston is required for the press to generate a force of 1.5×10^6 N? (b) What force is applied to the input piston if that piston has a diameter of 5.0 cm?

43. ■■ A garage hydraulic lift has two pistons, one of cross-sectional area 4.00 cm² and one of cross-sectional area 250 cm². If this lift is designed to lift a 3000-kilogram car, what minimum force must be applied to the small piston?

44. ■■ The force required in Exercise 43 is usually provided by compressed air. If the small piston is circular, what must be the minimum air pressure applied to the small piston?

45. ■■■ A hydraulic balance used to detect small change in mass is shown in •Fig. 9.32. If a mass *m* of 0.25 g is placed on the balance platform, by how much will the height of the water in the smaller cylinder of diameter 1.0 cm have changed when the balance comes to equilibrium?

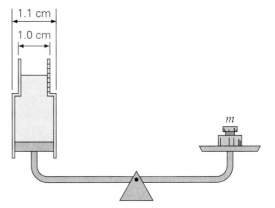

•FIGURE 9.32 A hydraulic balance See Exercise 45.

9.3 Buoyancy and Archimedes' Principle

46. A wood block floats in a swimming pool. The buoyant force on the block by water depends on (a) the volume of water in the pool, (b) the volume of the wood block, (c) the volume of the wood block under water, (d) all of the above.

47. If a submerged object displaces an amount of liquid of greater weight than its own and is then released, the object will (a) rise to the surface and float, (b) sink, (c) remain in equilibrium at its submerged position.

48. ■ (a) What is the most important factor in constructing a life jacket that will keep a person afloat? (b) Why is it so easy to float in the Great Salt Lake in Utah?

49. An ice cube floats in a glass of water. As the ice melts, how does the level of the water in the glass change? Would it make any difference if the ice cube were hollow? Explain.

50. Ocean-going ships in port are loaded to the *Plimsoll mark*, a line indicating the maximum safe loading depth. However, in New Orleans, which is located at the mouth of the Mississippi River and is noted for its brackish

water (partly salt, partly fresh), ships are loaded until the Plimsoll mark is somewhat below the water line. Why?

51. ■ A submarine has a mass of 10 000 metric tons. What mass of water must be displaced for the sub to be in equilibrium just below the ocean surface?

52. ■ A cube 8.5 cm on each side has a mass of 0.65 kg. Will the cube float in water?

53. ■ Suppose that Archimedes found that the king's crown had a mass of 0.750 kg and a volume of 3.980×10^{-5} m^3. (a) How did Archimedes determine the crown's volume? (b) Was the crown pure gold?

54. ■ A rectangular boat as illustrated in •Fig. 9.33 is overloaded such that the water level is just 1.0 cm below the top of the boat. What is the combined mass of the people and the boat?

•FIGURE 9.33 **An overloaded boat** (Not drawn to scale.) See Exercise 54.

55. ■ An aluminum cube 0.15 m on each side is completely submerged in water. What is the buoyant force on it? How does the answer change if the cube is made of steel?

56. ■■ An object has a weight of 8.0 N when in air. However, it apparently weighs only 4.0 N when it is completely submerged in water. What is the density of the object?

57. ■■ When a crown of mass 0.80 kg is submerged in water, the crown's apparent weight is measured to be 7.3 N. Is the crown pure gold?

58. ■■ An irregularly shaped piece of metal has a weight of 0.882 N. It is suspended from a scale, which reads 0.735 N when the piece is submerged in water. What are the volume and density of the piece of metal?

59. ■■ A flat-bottomed rectangular boat is 4.0 m long and 1.5 m wide. If the load is 2000 kg (including the mass of the boat), how much of the boat will be submerged when it floats in a lake?

60. ■■ A block of iron quickly sinks in water, but ships constructed of iron float. A solid cube of iron 1.0 m on each side is made into sheets. From these sheets, to make a hollow cube that will not sink, what should the minimum length of the sides be?

61. ■■ Plans are being made to bring back the zeppelin—a lighter-than-air airship, like the Goodyear blimp, that carries passengers and cargo but is filled with helium, not flammable hydrogen (as was used in the ill-fated *Hindenburg*). One design calls for the ship to be 110 m long and to have a mass comprised of 30.0 metric tons of structure plus the mass of the helium needed to fill it. Assuming the ship's "envelope" to be cylindrical, what would its diameter have to be so as to lift the total weight?

62. ■■■ A girl floats in a lake with 97% of her body beneath the water. What are (a) her mass density and (b) her weight density?

63. ■■■ A bar of 24-karat (pure) gold is supposed to be completely solid. To check this, a scientist takes the bar into the lab. Using a scale, she finds that it has a weight of 9.8 N and an apparent weight of only 9.1 N when completely immersed in water. Is the bar solid?

9.4 Fluid Dynamics and Bernoulli's Equation

64. The speed of blood flow is greater in arteries than in capillaries. However, the flow rate equation (Av = constant) seems to predict that the speed should be greater in the smaller capillaries. Can you resolve this apparent inconsistency?

65. If the speed at some point in a fluid changes with time, the fluid flow is *not* (a) steady, (b) irrotational, (c) incompressible, (d) nonviscous.

66. According to Bernoulli's equation, if the pressure on a liquid is increased, (a) the flow speed always increases, (b) the height of the liquid always increases, (c) both the flow speed and the height of the liquid may increase, (d) none of these.

67. When you suddenly turn on a water faucet in a shower, the shower curtain moves inward. Why?

68. If an Indy car had a flat bottom, it would be very unstable due to the lift it gets when it moves fast (like an airplane wing). To increase friction and stability, the bottom has a concave section called the *Venturi tunnel* (•Fig. 9.34). In

•FIGURE 9.34 **Venturi tunnel** See Exercise 68.

terms of Bernoulli's equation, explain how this concavity supplies extra down force to the car in addition to the front and rear wings (Exercise 4.71).

69. Two common demonstrations of Bernoulli effects: (a) If you hold a narrow strip of paper in front of your mouth and blow over the top surface, the strip will rise (•Fig. 9.35a). (Try it.) Why? (b) A plastic egg is supported vertically by a stream of air from a tube (Fig. 9.35b). The egg will not move away from the midsteam position. Why not?

(a)

(b)

•FIGURE 9.35 Bernoulli effects See Exercise 69.

70. ■ An ideal fluid is moving at 3.0 m/s in a section of a pipe of radius 0.20 m. If the radius in another section is 0.35 m, what is the flow speed there?

71. ■ If the radius of a pipe narrows to half of its original size, what is the ratio of the flow speed in the narrow section to that in the wide section, assuming the same flow rate?

72. ■■ Show that the static pressure–depth equation can be derived from Bernoulli's equation.

73. ■■ The speed of blood in a major artery of diameter 1.0 cm is 4.5 cm/s. (a) What is the flow rate in the artery? (b) If the capillary system has a total cross-sectional area of 2500 cm^2, the average speed of blood through the capillaries is what percentage of that through the major artery? (c) Why must blood flow at low speed through the capillaries?

74. ■■ A room measures 3.0 m by 4.5 m by 6.0 m. If the heating and air conditioning ducts to and from the room are circular, with diameter 0.30 m, and all the air in the room is to be exchanged every 12 min, (a) what is the average flow rate? (b) What is the necessary flow speed in the duct? (Assume that the density of the air is constant.)

75. ■■ The spout heights for the container in •Fig. 9.36 are 10 cm, 20 cm, 30 cm, and 40 cm. The water level is maintained at a 45 cm height by an outside supply. (a) What is the speed of the water out of each hole? (b) Which water stream has the greatest range relative to the base of the container? Justify your answer.

•FIGURE 9.36 Streams as projectiles See Exercise 75.

76. ■■■ In an industrial cooling process, water is circulated through a system. If water is pumped with a speed of 0.45 m/s under a pressure of 400 torr from the first floor through a 6.0-centimeter-diameter pipe, what will be the pressure on the next floor 4.0 m above in a pipe with a diameter of 2.0 cm?

77. ■■■ Water flows at a rate of 25 L/min through a horizontal 7.0-centimeter-diameter pipe under a pressure of 6.0 Pa. At one point, calcium deposits reduce the cross-sectional area of the pipe to 30 cm^2. What is the pressure at this point? (Consider the water to be an ideal fluid.)

78. ■■■ A Venturi meter can be used to measure the flow speed of a liquid. A simple one is shown in •Fig. 9.37.

•FIGURE 9.37 A flow speed meter See Exercise 78.

330 CHAPTER 9 Solids and Fluids

Show that the flow speed of an ideal fluid is given by

$$v_1 = \sqrt{\frac{2g\,\Delta h}{(A_1{}^2/A_2{}^2) - 1}}$$

*9.5 Surface Tension; Viscosity and Poiseuille's Law

79. Water droplets and soap bubbles tend to assume the shape of a sphere. This effect is due to (a) viscosity, (b) surface tension, (c) laminar flow, (d) none of the above.

80. Some insects can walk on water because (a) the density of water is greater than that of the insect, (b) water is viscous, (c) water has surface tension, (d) none of the above.

81. The viscosity of a fluid is due to (a) forces causing friction between the molecules, (b) surface tension, (c) density, (d) none of the above.

82. A motor oil is labeled 10W-40. The numbers 10 and 40 specify the (a) density, (b) surface tension, (c) viscosity, (d) weight of the oil.

83. ■■ A horizontal pipe with an inside diameter of 3.0 cm and a length of 6.0 m carries water, which has a flow rate of 40 L/min. What is the required pressure difference between the ends of the pipe?

84. ■■ A hospital patient receives a transfusion of 500 cc of blood through a needle with a length of 5.0 cm and an inner diameter of 1.0 mm. If the blood bag is suspended 0.85 m above the needle, how long does the transfusion take? (Neglect the viscosity of the blood flowing in the plastic tube between the bag and the needle.)

Additional Exercises

85. A copper wire 100.00 cm long is stretched to 100.02 cm when it supports a certain load. If an aluminum wire of the same diameter is used to support the same load, what should its initial length be if its stretched length is to be 100.02 cm also?

86. A steel cube 0.25 m on each side is suspended from a scale and immersed in water. What will the scale read?

87. A wood cube 0.30 m on each side has a density of 700 kg/m³ and floats level in water. (a) What is the distance from the top of the wood to the water surface? (b) What mass has to be placed on top of the wood so that its top is just at the water level?

88. A 60-kilogram woman balances herself on the heel of one of her high-heeled shoes. If the heel is a square of sides 1.5 cm, what is the pressure exerted on the floor? (Express your answer in pascals, pounds per square inch, and atmospheres.)

89. What total inward force does the atmosphere exert on the body of a person with a surface area of 1.30 m²? (Express your answer in both newtons and pounds.)

90. A scuba diver dives to a depth of 12 m in a lake. If the circular glass plate on the diver's face mask has a diameter of 18 cm, what is the force on it due to the water only?

91. Oil is poured into the open side of an open-tube manometer containing mercury. What is the density of the oil if a column of mercury 5.0 cm high supports a column of oil 80 cm high?

92. If the atmosphere had a constant density equal to its density at the Earth's surface, how high would it extend? What does this tell you?

93. A cylinder has a diameter of 15 cm (•Fig. 9.38). The water level in the cylinder is maintained at a constant height of 0.45 m. If the diameter of the spout pipe is 0.50 cm, how high is h, the vertical stream of water? (Assume the water to be an ideal fluid.)

•FIGURE 9.38 **How high a fountain?** See Exercise 93.

94. A vertical steel beam with a rectangular cross-sectional area of 24 cm² is used to support a sagging floor in a building. If the beam supports a load of 10 000 N, by what percentage is the beam compressed?

95. A container 100 cm deep is filled with water. If a small hole is punched in its side 25 cm from the top, at what initial speed will the water flow from the hole?

96. Show that specific gravity is equivalent to the ratio of densities, given that its strict definition is the ratio of the weight of a given volume of a substance to the weight of an equal volume of water.

97. The flow rate of water through a garden hose is 66 cm³/s, and the hose and nozzle have cross-sectional areas of 6.0 cm² and 1.0 cm², respectively. (a) If the nozzle is held 10 cm above the spigot, what are the flow speeds through the spigot and the nozzle? (b) What is the pressure

difference between these points? (Consider the water to be an ideal fluid.)

98. Ancient stonemasons sometimes split huge blocks of rock by inserting wooden pegs into holes drilled in the rock and then pouring water on the pegs. Can you explain the physics that underlies this technique? [Hint: Think about sponges and paper towels.]

99. A crane lifts from the bottom of a lake a rectangular iron bar whose dimensions are 0.25 m × 0.20 m × 10 m. What is the minimum upward force the crane must supply when the bar is (a) in the water and (b) out of the water?

100. •Figure 9.39 shows a simple laboratory experiment. Calculate the (a) volume and (b) density of the suspended sphere. (Assume that the density of the sphere is uniform and that the liquid in the beaker is water.) (c) Would you be able to make the same determinations if the liquid in the beaker were mercury (see Table 9.2)? Explain.

•**FIGURE 9.39 Dunking a sphere** See Exercise 100.

Temperature

CHAPTER 10

10.1 Temperature and Heat
10.2 The Celsius and Fahrenheit Temperature Scales
10.3 Gas Laws and Absolute Temperature
10.4 Thermal Expansion
10.5 The Kinetic Theory of Gases
*10.6 Kinetic Theory, Diatomic Gases, and the Equipartition Theorem

Insights
- Human Body Temperature
- Physiological Diffusion

Like sailboats, hot-air balloons are low-tech devices in a high-tech world. You can equip a balloon with the latest satellite-linked, computerized navigational system and attempt to fly across the Pacific, but the basic principles that keep you aloft were known and understood centuries ago.

But why must the air in a hot-air balloon be hot? More fundamentally, what do we mean by "hot"? How does hot air differ from cool air? Temperature and heat are frequent subjects of conversation, but if you had to state what the words really mean, you might find yourself at a loss. We use thermometers of all sorts to record temperatures, which provide an objective equivalent for our sensory experience of hot and cold. And we know that a temperature change generally results from the application or removal of heat. Temperature, therefore, is related to heat. But what is heat? In this chapter you'll find that the answers to such simple questions lead to an understanding of some far-reaching physical principles.

An early theory of heat considered it to be a fluidlike substance called caloric (from the Latin word *calor,* meaning "heat") that could be made to flow into and out of a body. Even though this theory has been abandoned, we still speak of heat as flowing from one object to another. Heat is now characterized as energy, and temperature and thermal properties are explained by considering the atomic and molecular behavior of substances. This and the next two chapters examine the nature of temperature and heat in terms of microscopic (molecular) theory and macroscopic observations. Here, you'll explore the nature of heat and the ways in which we measure temperature. You'll also encounter the gas laws, which explain not only the behavior of balloons, but also more important phenomena, such as how our lungs supply us with the oxygen we need to live.

10.1 Temperature and Heat

Objective: To distinguish between temperature and heat.

A good way to begin studying thermal physics is with definitions of temperature and heat. **Temperature** is a relative measure, or indication, of hotness or coldness. A hot stove is said to have a high temperature, and an ice cube to have a low temperature. An object that has a higher temperature than another object does is said to be hotter. Note that hot and cold are relative terms, like tall and short. We can perceive temperature by touch. However, this temperature sense is somewhat unreliable, and its range is too limited to be useful for scientific purposes.

Heat is related to temperature and describes the process of energy transfer from one object to another. That is, **heat** is *the net energy transferred from one object to another because of a temperature difference.* Generally, heat transfer is the result of a temperature difference (though not always, as we shall find in Section 11.4). Thus, heat is energy in transit, so to speak. Once transferred, the energy becomes part of the total energy of the molecules of the object or system, its **internal energy**.

On a microscopic level, temperature is associated with molecular motion. We will show that in kinetic theory (Section 10.5), which treats gas molecules as point particles, temperature is a measure of the average random *translational* kinetic energy of the molecules. However, diatomic molecules and other real substances, besides having such translational "temperature" energy, also may have kinetic energy due to linear vibration and rotation as well as potential energy due to the attractive forces between molecules. These energies would not contribute to the gas's temperature measurement but are definitely part of the internal energy, which is the sum total of all such energies (•Fig. 10.1).

Note that a higher temperature does not necessarily mean that one system has a greater internal energy than another. For example, in a classroom on a cold day, the air temperature is relatively high compared to that of the outdoor air. But all

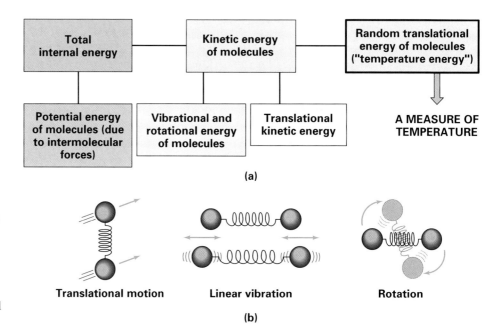

•FIGURE 10.1 Molecular motions (a) Temperature is associated with random translational motion; the internal energy of a system is the total energy. (b) A molecule may move as a whole in translational motion and/or have vibrational and rotational motions.

that cold air outside the classroom has far more internal energy as a whole than does the warm air inside, simply because there is so much *more* of it. If this were not the case, heat pumps would not be practical (Chapter 12). The internal energy of a system also depends on its mass, or the number of molecules in the system.

When heat is transferred between two objects, whether or not they are touching, they are said to be in *thermal contact*. When there is no longer a net heat transfer between objects in thermal contact, they have come to the same temperature and are said to be in *thermal equilibrium*.

10.2 The Celsius and Fahrenheit Temperature Scales

Objectives: To (a) explain how a temperature scale is constructed and (b) convert temperatures from one scale to another.

A measure of temperature is obtained by using a **thermometer**, a device constructed to make evident some property of a substance that changes with temperature. Fortunately, many physical properties of materials change sufficiently with temperature to be used as the bases for thermometers. By far the most obvious and commonly used property is **thermal expansion**, a change in the dimensions or volume of a substance that occurs when the temperature changes.

Almost all substances expand with increasing temperature, but they do so to different extents. Most substances also contract with decreasing temperature. (Thermal expansion refers to both expansion and contraction; contraction is considered to be a negative expansion.) Because some metals expand more than others, a bimetallic strip (strips of two different metals bonded together) can be used to measure temperature changes. As heat is applied, the composite strip will bend away from the side made of the metal that expands more (•Fig. 10.2). Coils formed from such strips are used in dial thermometers and in common household thermostats (•Fig. 10.3).

A common thermometer is the liquid-in-glass type, which is based on the thermal expansion of a liquid. A liquid in a glass bulb expands into a glass stem, rising in a capillary bore (a thin tube). Mercury and alcohol (usually dyed red to make it more visible) are the liquids used in most liquid-in-glass thermometers. These substances were chosen because of their relatively large thermal expansion and because they remain liquids over normal temperature ranges.

Thermometers are calibrated so that a numerical value can be assigned to a given temperature. For the definition of any standard scale or unit, two fixed reference points are needed. Since all substances change dimensions with temperature, an absolute reference for expansion is unavailable. However, the necessary fixed points can be correlated to physical phenomena that always occur at the same temperatures.

(a)

(b)

•**FIGURE 10.3 Bimetallic coil** Bimetallic coils are used in **(a)** dial thermometers and **(b)** household thermostats. In a thermostat, the expansion and contraction of the coil are used to regulate the heating or cooling system, turning it off and on as the room temperature changes.

(a) Initial condition   (b) Heated condition

•**FIGURE 10.2 Thermal expansion** **(a)** A bimetallic strip is made of two strips of different metals bonded together. **(b)** When such a strip is heated, it bends because of unequal expansions of the two metals. Here brass expands more than iron, so the deflection is toward the iron. The deflection of the end of a strip could be used to measure temperature.

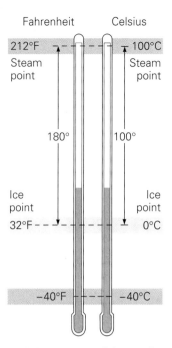

•FIGURE 10.4 **Celsius and Fahrenheit temperature scales** Between the ice and steam fixed points, there are 100 degrees on the Celsius scale and 180 degrees on the Fahrenheit scale. Thus, a Celsius degree is 1.8 times larger than a Fahrenheit degree.

Note: For distinction, a particular temperature measurement, such as $T = 20°C$, is written with °C (pronounced 20 degrees Celsius), whereas a temperature interval, such as $\Delta T = 80°C - 60°C = 20$ C°, is written with C° (pronounced 20 Celsius degrees).

The ice point and the steam point of water are two convenient fixed points. Also commonly known as the freezing and boiling points, these two points are the temperatures at which pure water freezes and boils under a pressure of 1 atm (standard pressure).

The two familiar temperature scales are the **Fahrenheit temperature scale** and the **Celsius temperature scale**. As shown in •Fig. 10.4, the ice and steam points have values of 32°F and 212°F, respectively, on the Fahrenheit scale and 0°C and 100°C on the Celsius scale. On the Fahrenheit scale, there are 180 equal intervals, or degrees, (F°), between the two reference points; on the Celsius scale, there are 100 (C°). Therefore, a Celsius degree is larger than a Fahrenheit degree. Since $180/100 = 1.8$, it is almost twice as large.

A relationship for converting between the two scales can be obtained from a graph of Fahrenheit temperature (T_F) versus Celsius temperature (T_C), such as the one in •Fig. 10.5. The equation of the straight line (in slope–intercept form, $y = mx + b$) is $T_F = (180/100)T_C + 32$, and

$$T_F = \tfrac{9}{5}T_C + 32 \quad \text{or} \quad T_F = 1.8T_C + 32 \qquad \begin{array}{l}\textit{Celsius-to-Fahrenheit}\\ \textit{conversion}\end{array} \quad (10.1)$$

where $\tfrac{9}{5}$ or 1.8 is the slope of the line and 32 is the intercept on the vertical axis. Thus, to change from a Celsius temperature (T_C) to its equivalent Fahrenheit temperature (T_F), you simply multiply the Celsius reading by $\tfrac{9}{5}$ and add 32.

The equation can be solved for T_C to convert from Fahrenheit to Celsius:

$$T_C = \tfrac{5}{9}(T_F - 32) \qquad \begin{array}{l}\textit{Fahrenheit-to-Celsius}\\ \textit{conversion}\end{array} \quad (10.2)$$

Example 10.1 ■ Converting Temperature Scale Readings: Fahrenheit and Celsius

What are (a) 20°C on the Fahrenheit scale and (b) normal body temperature, 98.6°F, on the Celsius scale?

Thinking It Through. This is a direct application of Eqs. 10.1 and 10.2.

Solution. We wish to make the following conversions:

Given: (a) $T_C = 20°C$ *Find:* (a) T_F
         (b) $T_F = 98.6°F$ (b) T_C

•FIGURE 10.5 **Fahrenheit versus Celsius** (a) A plot of Fahrenheit temperature versus Celsius temperature gives a straight line of the general form $y = mx + b$, where $T_F = \tfrac{9}{5}T_C + 32$. (b) The temperature is given here in degrees Celsius. What is the equivalent temperature in degrees Fahrenheit?

(a) Equation 10.1 is for changing Celsius readings to Fahrenheit:

$$T_F = \tfrac{9}{5} T_C + 32 = \tfrac{9}{5}(20) + 32 = 68°F$$

This temperature, often taken to be room temperature, is a good one to remember.

(b) Equation 10.2 changes Fahrenheit to Celsius:

$$T_C = \tfrac{5}{9}(T_F - 32) = \tfrac{5}{9}(98.6 - 32) = \tfrac{5}{9}(66.6)$$
$$= 37.0°C$$

On the Celsius scale, normal body temperature has a whole-number value. Keep in mind that a Celsius degree is 1.8 times (almost twice) as large as a Fahrenheit degree, so a temperature elevation of several degrees on the Celsius scale makes a big difference. For example, a temperature of 40.0°C represents an elevation of 3.0 C° over normal body temperature. This is an increase of $3.0 \times 1.8 = 5.4°$ on the Fahrenheit scale, or a temperature of $98.6 + 5.4 = 104.0°F$. For more on "normal" body temperature, see the Insight on this page.

Follow-up Exercise. Convert the following temperatures: (a) $-40°F$ to Celsius and (b) $-40°C$ to Fahrenheit.

Insight

Human Body Temperature

We commonly take "normal" human body temperature to be 98.6°F (or 37.0°C). The source of this value is a study of human temperature readings done in 1868—more than 130 years ago! A more recent study, done in 1992, notes that the 1868 study used thermometers that were not as accurate as modern electronic thermometers. The new study has some interesting results.

The "normal" human body temperature from oral measurements varies with individuals over a range of about 96°F to 101°F, with an average temperature of 98.2°F. After strenuous exercise, the oral temperature can rise as high as 103°F. When the body is exposed to cold, oral temperatures can fall below 96°F. A rapid drop in temperature of 2 to 3 F° produces uncontrollable shivering. There is not only a contraction of the skeletal muscles, but also of the tiny muscles attached to the hair follicles. The result is "goose bumps."

Your body temperature is typically lower in the morning, after you have slept and your digestive processes are at a low point. "Normal" body temperature generally rises during the day to a peak and then recedes. The study also indicated that women have a slightly higher average temperature than men do (98.4°F versus 98.1°F).

What about the extremes? A fever temperature is typically between 102°F and 104°F. A body temperature above 106°F is extremely dangerous. The enzymes for certain chemical reactions in the body begin to be inactivated, and a total breakdown of body chemistry can result.

On the cold side, decreasing body temperature results in memory lapse and slurred speech, muscular rigidity, erratic heartbeats, loss of consciousness, and, below 78°F, death due to heart failure. However, mild hypothermia (lower-than-normal body temperature) can be beneficial. A decrease in body temperature slows down the body's chemical reactions, and cells use less oxygen than they normally do. This fact is applied in some surgeries (Fig. 1). A patient's body temperature may be lowered significantly to avoid damage to the heart, which must be stopped during such procedures, and to the brain.

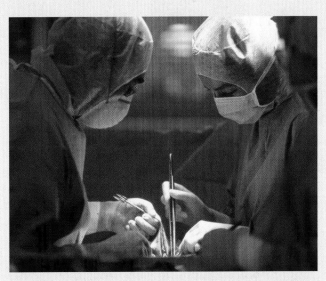

FIGURE 1 Lower than normal During some surgeries, the patient's body temperature is lowered to slow down the body's chemical reactions and to reduce the need for blood to supply oxygen to the tissues.

Problem-Solving Hint

Because Eqs. 10.1 and 10.2 are so similar, it is easy to miswrite them. Since they are equivalent, you need to know only one of the equations—say, Celsius to Fahrenheit (Eq. 10.1, $T_F = \frac{9}{5}T_C + 32$). Solving it for T_C algebraically gives Eq. 10.2. A good way to make sure you have written the conversion equation correctly is to test it with a known temperature, such as the boiling point of water. For example, $T_C = 100°$, so

$$T_F = \tfrac{9}{5}T_C + 32 = \tfrac{9}{5}(100) + 32 = 212°F$$

Thus we know the equation is correct.

Liquid-in-glass thermometers are adequate for many temperature measurements, but problems arise when very accurate determinations are needed. A material may not expand uniformly over a wide temperature range. When calibrated to the ice and steam points, an alcohol thermometer and a mercury thermometer have the same readings at these points. But because alcohol and mercury have different expansion properties, the thermometers will not have exactly the same reading at an intermediate temperature, such as room temperature. For very sensitive temperature measurements and to define intermediate temperatures precisely, some other type of thermometer must be used. One such thermometer is discussed next.

10.3 Gas Laws and Absolute Temperature

Objectives: To (a) describe the ideal gas law and (b) explain how it is used to determine absolute zero.

Whereas different liquid-in-glass thermometers show slightly different readings for temperatures other than fixed points because of the liquids' different expansion properties, a thermometer that uses a gas gives the same readings regardless of the gas used. All gases at very low densities exhibit the same expansion behavior.

The variables that describe the behavior of a given quantity (mass) of gas are pressure, volume, and temperature (p, V, and T). When temperature is held constant, the pressure and volume of a quantity of gas are related as follows:

$$pV = \text{constant} \quad \text{or} \quad p_1 V_1 = p_2 V_2 \quad \text{(at constant temperature)} \quad (10.3)$$

That is, the product of the pressure and volume is a constant. This relationship is known as *Boyle's law*, after Robert Boyle (1627–1691), the English chemist who discovered it. (See Demonstration 9.)

When the pressure is held constant, the volume of a quantity of gas is related to the *absolute* temperature (to be defined shortly):

$$\frac{V}{T} = \text{constant} \quad \text{or} \quad \frac{V_1}{T_1} = \frac{V_2}{T_2} \quad \text{(at constant pressure)} \quad (10.4)$$

•**FIGURE 10.6 Charles' law in action** Demonstrations of the relationship between volume and temperature of a quantity of gas. A weighted balloon, initially at room temperature, is placed in a beaker of water. **(a)** When ice is placed in the beaker and the temperature falls, the balloon's volume is reduced. **(b)** When the water is heated and the temperature rises, the balloon's volume increases.

That is, the ratio of the volume and the temperature is a constant. This relationship is known as *Charles' law*, named for the French scientist Jacques Charles (1747–1823), who made early hot-air balloon flights and was therefore quite interested in the relationship between the volume and temperature of a gas. A popular demonstration of Charles' law is shown in •Fig. 10.6.

Low-density gases obey these laws, which may be combined into a single relationship. Since $pV = $ constant and $V/T = $ constant for a given quantity of gas, then pV/T must also equal a constant. This relationship is the **ideal gas law**:

$$\frac{pV}{T} = \text{constant} \quad \text{or} \quad \frac{p_1 V_1}{T_1} = \frac{p_2 V_2}{T_2} \quad \begin{array}{l}\textit{ideal gas law}\\\textit{(ratio form)}\end{array} \quad (10.5)$$

Demonstration 9 ■ Boyle's Shaving Cream

A demonstration that shows Boyle's law, $p \propto 1/V$, the inverse relationship of pressure (p) and volume (V). Note: A similar effect occurs for a Boyle's marshmallow.

(a) A swirl of shaving cream is placed in a transparent vacuum chamber.

(b) A vacuum pump begins to evacuate the chamber.

(c) As the pressure is reduced, the swirl of cream grows in volume from the expansion of the air bubbles trapped in the cream.

(d) In a dramatic volume reduction, the shaving cream cannot stand up to the sudden increase in pressure when the chamber is vented to the atmosphere.

That is, the ratio pV/T at one time (t_1) is the same as at another time (t_2), or at any other time, as long as the quantity of gas does not change.

This relationship can be written in a more general form that applies not just to a given quantity of a single gas but to any quantity of any dilute gas:

$$\frac{pV}{T} = Nk_B \quad \text{or} \quad pV = Nk_BT \quad \text{ideal gas law} \quad (10.6)$$

Here N is the number of molecules in the sample of gas and k_B is a number known as *Boltzmann's constant*: $k_B = 1.38 \times 10^{-23}$ J/K. The K stands for temperature on the Kelvin scale, discussed below. (Can you show that the units are correct?) Note that the mass of the sample does not appear explicitly in Eq. 10.6. However, a given number of molecules of a gas is proportional to the total mass of the gas. The ideal gas law, sometimes called the *perfect gas law*, applies to all gases at low densities and describes fairly accurately the behavior of most gases at normal densities.

Macroscopic Form of the Ideal Gas Law

Equation 10.6 is a "microscopic" (*micro* means small) form of the ideal gas law in that it refers to the tiny molecules of a gas. (The equation contains N, the number of molecules.) However, the law can be written in a "macroscopic" (*macro* means large) form, which involves quantities that can be measured with everyday laboratory equipment. In this form, we have

$$pV = nRT \quad \text{ideal gas law} \quad (10.7)$$

Ideal gas law (with T on the Kelvin scale)

with the constants nR rather than Nk_B used for convenience. Here, n is the number of moles (mol) of the gas, a quantity defined below, and R is called the *universal gas constant*:

$$R = 8.31 \text{ J/(mol·K)}$$

Note: N is the total number of molecules; N_A is Avogadro's number; $n = N/N_A$ is the number of moles.

In chemistry a **mole** (abbreviated mol) of a substance is defined as the quantity that contains **Avogadro's number** (N_A) of molecules:

$$N_A = 6.02 \times 10^{23} \text{ molecules/mol}$$

Thus, the n and N in the two forms of the ideal gas law are related by $N = nN_A$. From Eq. 10.7 it is easy to show that 1 mol of *any* gas occupies 22.4 L at 0°C and 1 atm. These conditions are known as *standard temperature and pressure* (*STP*).

Equation 10.7 is a practical form of the gas law, because we generally work with measured quantities or moles (n) of gases rather than the number of molecules (N). To use Eq. 10.7, we need to know the number of moles of a gas. The mass of 1 mol of any substance is simply its formula mass expressed in grams. The formula mass is determined from the chemical formula and the atomic masses of the atoms (these are listed in Appendix IV and are commonly rounded to the nearest one-half). For example, water, H_2O, with two hydrogen atoms and one oxygen atom, has a formula mass of $2.0 + 16.0 = 18.0$, because the atomic mass of each hydrogen atom is 1.0 g and that of an oxygen atom is 16.0 g. Thus 1 mol of water has a formula mass of 18.0 g. Similarly, the oxygen we breathe, O_2, has a formula mass of $2 \times 16.0 = 32.0$. Thus a mole of oxygen, 32 g, would occupy 22.4 L at STP.

It is interesting to note that Avogadro's number allows you to compute the mass of a particular type of molecule. For example, suppose you want to know the mass of a water molecule (H_2O). As we have just seen, the formula mass of 1 mol of water is 18.0 g, or 18.0 g/mol. The molecular mass (m) is then given by

$$m = \frac{\text{formula mass (in kilograms)}}{N_A}$$

and, converting grams to kilograms, we have

$$m_{H_2O} = \frac{(18.0 \text{ g/mol})(10^{-3} \text{ kg/g})}{6.02 \times 10^{23} \text{ molecules/mol}} = 2.99 \times 10^{-26} \text{ kg/molecule}$$

The *formula mass* (in kilograms) is equivalent to the mass of 1 mol (in kilograms).

Absolute Zero and the Kelvin Temperature Scale

The important point for the purpose of this section is that the pressure and volume are directly proportional to temperature: $pV \propto T$. This relationship allows a gas to be used to measure temperature in a *constant-volume gas thermometer*. Holding the volume of a gas constant, which can be done easily in a rigid container (•Fig. 10.7), gives $p \propto T$. Thus, with a constant-volume gas thermometer, tem-

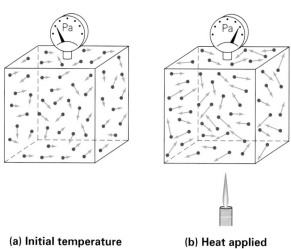

•**FIGURE 10.7 Constant-volume gas thermometer** Such a thermometer indicates temperature as a function of pressure, since for a low-density gas, $p \propto T$. **(a)** At some initial temperature, the pressure reading has a certain value. **(b)** When the gas thermometer is heated, the pressure (and temperature) reading is higher, because on the average, the gas molecules are moving faster.

(a) Initial temperature   **(b) Heat applied**

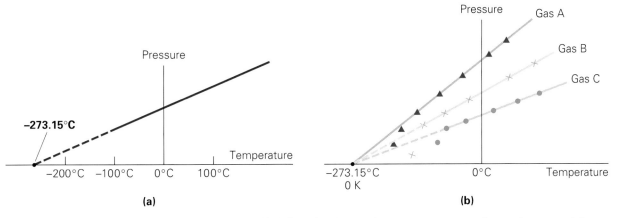

•**FIGURE 10.8 Pressure versus temperature** (a) A low-density gas kept at a constant volume gives a straight line on a p versus T graph: $p = (Nk_B/V)T$. When the line is extended to the zero pressure value, a temperature of $-273.15°C$ is obtained, which is taken to be absolute zero. (b) Extrapolation of lines for all low-density gases indicates the same absolute zero temperature. The actual behavior of gases deviates from this straight-line relationship at low temperatures because they start to liquefy.

perature is read in terms of pressure. A plot of pressure versus temperature gives a straight line in this case (•Fig. 10.8a).

As can be seen in Fig. 10.8b, measurements of real gases (plotted data points) deviate from the values predicted by the ideal gas law at low temperatures. This is because the gases liquefy at low temperatures. However, the relationship is linear over a large temperature range, and it looks as though the pressure might reach zero with decreasing temperature if the gas continued to be gaseous (ideal or perfect).

The absolute minimum temperature for an ideal gas is therefore inferred by extrapolating, or extending the straight line to the axis, as in Fig. 10.8b. This temperature is found to be $-273.15°C$ and is designated as **absolute zero**. Absolute zero is believed to be the lower limit of temperature, but it has never been attained. In fact, there is a physical law that says it never can be (Chapter 12).

There is no known upper limit to temperature. For example, the temperatures at the centers of some stars are estimated to be greater than 100 million degrees.

Absolute zero is the foundation of the **Kelvin temperature scale**, named after the British scientist Lord Kelvin.* On this scale, $-273.15°C$ is taken as the zero point—that is, as 0 K (•Fig. 10.9). The size of a single unit of Kelvin temperature is the same as the Celsius degree, so temperatures on these scales are related by

$$T_K = T_C + 273.15 \quad \text{Celsius-to-Kelvin conversion} \quad (10.8)$$

Here T_K is the temperature in **kelvins** (*not* degrees Kelvin; for example, 300 kelvins). The kelvin is abbreviated as K (*not* °K). In most instances, a rounded value can be used in Eq. 10.8: $T_K = T_C + 273$.

The absolute Kelvin scale is the official SI temperature scale; however, the Celsius scale is used in most parts of the world for everyday temperature readings. The absolute temperature in kelvins is used primarily in scientific applications.

Caution: Keep in mind that Kelvin temperatures *must* be used with the ideal gas law. It is a common mistake to use Celsius or Fahrenheit temperatures in that equation. Suppose you used a Celsius temperature of $T = 0°C$ in the gas law. You would have $pV = 0$, which makes no sense.

*Lord Kelvin, born William Thomson (1824–1907), developed devices to improve telegraphy and the compass and was involved in the laying of the first transatlantic cable. When he received his title, it is said that he considered choosing Lord Cable or Lord Compass as his title but decided on Lord Kelvin, after a river that runs near the University of Glasgow in Scotland, where he was a professor of physics for 50 years.

•**FIGURE 10.9 The Kelvin temperature scale** The lowest temperature on the Kelvin scale (corresponding to $-273.15°C$) is absolute zero. A unit interval on the Kelvin scale, called a kelvin and abbreviated K, is equivalent to a temperature change of 1 C°; thus $T_K = T_C + 273.15$. (The constant is usually rounded to 273 for convenience. For example, a temperature of 0°C is equal to 273 kelvins.)

10.3 Gas Laws and Absolute Temperature 341

Also note that there can be no negative temperatures on the Kelvin scale if absolute zero is the lowest possible temperature. That is, the Kelvin scale doesn't have an arbitrary zero temperature somewhere within the scale—0 K is absolute zero, period.

Example 10.2 ■ Deepest Freeze: Absolute Zero on the Fahrenheit Scale

What is absolute zero on the Fahrenheit scale?

Thinking It Through. We must convert 0 K to the Fahrenheit scale. However, the first conversion is to the Celsius scale. (Why?)

Solution.

Given: $T_K = 0$ K   Find: T_F

Temperatures on the Kelvin scale are related directly to Celsius temperatures by $T_K = T_C + 273.15$, so first we convert 0 K to a Celsius value:

$$T_C = T_K - 273.15 = 0 - 273.15 = -273.15°C$$

Then converting to Fahrenheit (Eq. 10.1) gives

$$T_F = \tfrac{9}{5}T_C + 32 = \tfrac{9}{5}(-273.15) + 32 = -459.67°F$$

Thus, absolute zero is about $-460°F$.

Follow-up Exercise. There is an absolute temperature scale associated with the Fahrenheit temperature called the Rankine scale. A Rankine degree is the same size as a Fahrenheit degree, and absolute zero is taken as 0°R (zero degrees Rankine). Write the conversion equations between (a) the Rankine and the Fahrenheit scales, (b) the Rankine and the Celsius scales, and (c) the Rankine and the Kelvin scales.

Initially, gas thermometers were calibrated by using the ice and steam points. The Kelvin scale uses absolute zero and a second fixed point adopted in 1954 by the International Committee on Weights and Measures. This second fixed point is the **triple point of water**, which represents a unique set of conditions where water coexists simultaneously in equilibrium as a solid, a liquid, and a gas. The conditions for the triple point are a pressure of 4.58 mm Hg, or about 610 Pa (760 mm Hg is normal atmospheric pressure), and a temperature taken to be 273.16 K. The kelvin is then defined as 1/273.16 of the temperature at the triple point of water.*

Now let's use the ideal gas law, which requires absolute temperatures.

Example 10.3 ■ The Ideal Gas Law: Using Absolute Temperatures

A quantity of low-density gas in a rigid container is initially at room temperature (20°C) and a particular pressure (p_1). If the gas is heated to a temperature of 60°C, by what factor does the pressure change?

Thinking It Through. A "factor" of change implies a ratio (p_2/p_1), so Eq. 10.5 should apply. Note that the container is rigid, which implies that $V_1 = V_2$.

Solution.

Given: $T_1 = 20°C$ Find: p_2/p_1 (pressure ratio or factor)
$T_2 = 60°C$
$V_1 = V_2$

Since we want the factor of pressure change, we write p_2/p_1 as a ratio. For example, if $p_2/p_1 = 2$, then $p_2 = 2p_1$, or the pressure would change (increase) by a factor of 2. The

---

*The 273.16 value given here for the triple point temperature and the -273.15 value in Eq. 10.8 indicate different things. The number in the equation comes from the fact that $-273.15°C$ is taken as 0 K. The 273.16 K (or 0.01°C) is a different reading on a different temperature scale.

ratio also indicates that we should use the ideal gas law in ratio form. The gas law requires *absolute* temperatures, so we first change the Celsius temperatures to kelvins:

$$T_1 = 20°C + 273 = 293 \text{ K}$$

$$T_2 = 60°C + 273 = 333 \text{ K}$$

where a rounded value of 273 was used in Eq. 10.8 for convenience. Then, using the ideal gas law (Eq. 10.5) in the form $p_2V_2/T_2 = p_1V_1/T_1$, and since $V_1 = V_2$, we have

$$p_2 = \left(\frac{T_2}{T_1}\right)p_1 = \left(\frac{333 \text{ K}}{293 \text{ K}}\right)p_1 = 1.14 p_1$$

Thus, p_2 is 1.14 times p_1; the pressure increases by a factor of 1.14, or by 14 percent. (What would the factor be if the Celsius temperatures were *incorrectly* used? It would be much larger: 60°C/20°C = 3.)

Note: *Always* use Kelvin (absolute) temperatures with the ideal gas law.

Follow-up Exercise. If the gas in this Example is heated when at room temperature so the pressure increases by a factor of 1.26, what is the final Celsius temperature?

Because of its absolute nature, the Kelvin temperature scale has special significance. As we shall see in Section 10.5, the absolute temperature is directly proportional to the internal energy of an ideal gas and so can be used as an indication of that energy. There are no negative values on the absolute scale. Negative absolute temperatures would imply negative internal energy for the gas—a meaningless concept.

Suppose you were asked to double the temperatures of, say, −10°C and 0°C. What would you do? The following Example should help.

Conceptual Example 10.4

Some Like It Hot: Doubling the Temperature

The evening weather report gives the day's high temperature as 10°C and predicts the next day's high to be 20°C. John says that this means it will be twice as warm tomorrow; Mary says it does not. With whom do you agree?

Reasoning and Answer. Keep in mind that temperature gives a relative *indication* of hotness or coldness. Certainly, 20°C would be warmer than 10°C, but just because the numerical value of the temperature is twice as great does not necessarily mean it is twice as warm—only that the air temperature is 10 degrees higher and therefore relatively warmer.

The *absolute* temperature, with an absolute zero, is not doubled: $T_{K_1} = T_{C_1} + 273$ = 10°C + 273 = 283 K and T_{K_2} = 20°C + 273 = 293 K. So the absolute temperature is increased only by a factor of $\Delta T_K/T_{K_1}$ = 10 K/283 K = 0.035 (or 3.5%—not 100%), and we should agree with Mary.

Follow-up Exercise. The weather report gives the day's high temperature as 0°C. If the next day's temperature were double that, what would the temperature be in degrees Celsius? Would this be environmentally possible?

10.4 Thermal Expansion

Objective: To calculate the thermal expansions of solids and liquids.

Changes in the dimensions and volumes of materials are common thermal effects. As you learned earlier, thermal expansion provides a means of temperature measurement. The thermal expansion of gases is generally described by the ideal gas law and is very obvious. Less dramatic, but by no means less important, is the thermal expansion of solids and liquids.

Note: Solids are discussed in Section 9.1.

Thermal expansion results from a change in the average distance separating the atoms of a substance. The atoms are held together by bonding forces, which can be simplistically represented as springs in a simple model of a solid (see Fig. 9.1). The atoms vibrate back and forth; with increased temperature (that is,

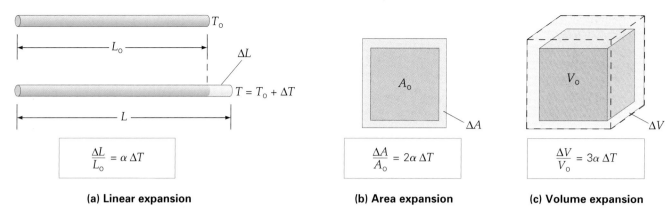

•FIGURE 10.10 Thermal expansion (a) Linear expansion is proportional to the temperature change; that is, the change in length ΔL is proportional to ΔT, and $\Delta L/L_o = \alpha \Delta T$, where α is the thermal coefficient of linear expansion. (b) For isotropic expansion, the thermal coefficient of area expansion is approximately 2α. (c) The thermal coefficient of volume expansion for solids is about 3α.

more internal energy), they become increasingly active and vibrate over greater distances. With wider vibrations in all dimensions, the solid expands as a whole.

The change in one dimension of a solid (length, width, or thickness) is called *linear* expansion. For small temperature changes, linear expansion is approximately proportional to ΔT, or $T - T_o$ (•Fig. 10.10a). The fractional change in length is $(L - L_o)/L_o$, or $\Delta L/L_o$, where L_o is the original length (at the initial temperature). This ratio is related to the change in temperature by

$$\frac{\Delta L}{L_o} = \alpha \Delta T \quad \text{or} \quad \Delta L = \alpha L_o \Delta T \qquad (10.9)$$

where α is the **thermal coefficient of linear expansion**. Note that the unit of α is inverse temperature: inverse Celsius degrees ($1/C°$, or $C°^{-1}$). Values of α for some materials are given in Table 10.1.

A solid may have different coefficients of linear expansion for different directions, but for simplicity this book will assume that the same coefficient applies to all directions (in other words, that solids show *isotropic* expansion). Also, the coefficient of expansion may vary slightly for different temperature ranges. Since this variation is negligible for most common applications, α will be considered to be constant and independent of temperature.

Equation 10.9 can be rewritten to give the final length (L) after a temperature change:

TABLE 10.1 Values of Thermal Expansion Coefficients (in $C°^{-1}$) for Some Materials at 20°C

Material	Coefficient of linear expansion (α)	Material	Coefficient of volume expansion (β)
Aluminum	24×10^{-6}	Alcohol, ethyl	1.1×10^{-4}
Brass	19×10^{-6}	Gasoline	9.5×10^{-4}
Brick or concrete	12×10^{-6}	Glycerin	4.9×10^{-4}
Copper	17×10^{-6}	Mercury	1.8×10^{-4}
Glass, window	9.0×10^{-6}	Water	2.1×10^{-4}
Glass, Pyrex	3.3×10^{-6}		
Gold	14×10^{-6}	Air (and most other gases at 1 atm)	3.5×10^{-3}
Ice	52×10^{-6}		
Iron and steel	12×10^{-6}		

$$\Delta L = \alpha L_o \Delta T$$
$$L - L_o = \alpha L_o \Delta T$$
$$L = L_o + \alpha L_o \Delta T$$

or

$$L = L_o(1 + \alpha \Delta T) \quad (10.10)$$

We can then use Eq. 10.10 to compute the thermal *area* expansion of flat objects. Since area (A) is length squared (L^2) for a square, we get

$$A = L^2 = L_o^2(1 + \alpha \Delta T)^2 = A_o(1 + 2\alpha \Delta T + \alpha^2 \Delta T^2)$$

where A_o is the original area. Because the values of α for solids are much less than 1 ($\sim 10^{-5}$, as shown in Table 10.1), the second-order term (containing α^2) can be dropped with negligible error. As a first-order approximation, then,

$$A = A_o(1 + 2\alpha \Delta T) \quad \text{or} \quad \frac{\Delta A}{A_o} = 2\alpha \Delta T \quad (10.11)$$

Thus, the **thermal coefficient of area expansion** (Fig. 10.10b) is twice as large as the coefficient of linear expansion (that is, it is equal to 2α). See Learn by Drawing.

Similarly, a first-order expression for thermal *volume* expansion is

$$V = V_o(1 + 3\alpha \Delta T) \quad \text{or} \quad \frac{\Delta V}{V_o} = 3\alpha \Delta T \quad (10.12)$$

The **thermal coefficient of volume expansion** (Fig. 10.10c) is equal to 3α.

The equations for thermal expansions are approximations. Even though an equation is a description of a physical relationship, always keep in mind that it may be only an approximation of physical reality and/or may apply only in certain situations. Because calculations of thermal expansion involve such small numbers, you may assume that all quantities are exact.

The thermal expansion of materials is an important consideration in construction. Seams are put in concrete highways and sidewalks to allow room for expansion and prevent cracking. Expansion gaps in large bridge structures and between railroad rails are necessary to prevent damage. Similarly, expansion loops are found in oil pipelines (•Fig. 10.11). The thermal expansion of steel beams and girders can produce tremendous pressures, as the following Example shows.

LEARN BY DRAWING
Thermal Area Expansion

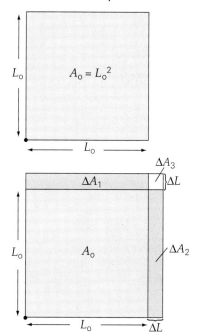

$\Delta A = \Delta A_1 + \Delta A_2 + \Delta A_3$
$\Delta A_1 = \Delta A_2 = L_o \, \Delta L$
$ = L_o \, (\alpha L_o \, \Delta T\,) = \alpha A_o \, \Delta T$
Since ΔA_3 is very small
compared to ΔA_1 and ΔA_2,
$\Delta A \approx 2\alpha A_o \, \Delta T$

Example 10.5 ■ Temperature Rising: Thermal Expansion and Stress

A steel beam is 5.0 m long at a temperature of 20°C (68°F). On a hot day, the temperature rises to 40°C (104°F). (a) What is the change in the beam's length due to thermal expansion? (b) Suppose that the ends of the beam are initially in contact with rigid vertical supports. How much force will the expanded beam exert on the supports if the beam has a cross-sectional area of 60 cm²?

Thinking It Through. (a) This is a direct application of Eq. 10.9. (b) As the constricted beam expands, it applies a stress, and hence a force, to the supports. For a linear expansion, the Young's modulus relationship (Section 9.1) should come into play.

Solution.

Given: $L_o = 5.0$ m
$T_o = 20°C$
$T = 40°C$
$\alpha = 12 \times 10^{-6} \, C°^{-1}$ (from Table 10.1)
$A = 60 \text{ cm}^2 \left(\dfrac{1 \text{ m}}{100 \text{ cm}}\right)^2 = 6.0 \times 10^{-3} \text{ m}^2$

Find: (a) ΔL (change in length)
(b) F (force)

10.4 Thermal Expansion 345

(a) Using Eq. 10.9 to find the change in length with $\Delta T = T - T_o = 40°C - 20°C = 20\,C°$, we get

$$\Delta L = \alpha L_o \Delta T = (12 \times 10^{-6}\,C^{°-1})(5.0\,m)(20\,C°) = 1.2 \times 10^{-3}\,m = 1.2\,mm$$

This may not seem like much of an expansion, but it can give rise to a great deal of force if the beam is constrained and kept from expanding, as part (b) will show.

(b) By Newton's third law, if the beam is kept from expanding, the force the beam exerts on its constraint supports is equal to the force exerted by the supports to prevent the beam from expanding by a length ΔL. This is the same as the force that would be required to compress the beam by that length. Using the Young's modulus form of Hooke's law (Chapter 9) with $Y = 20 \times 10^{10}\,N/m^2$ (Table 9.1), we calculate the stress on the beam as

$$\frac{F}{A} = \frac{Y \Delta L}{L_o} = \frac{(20 \times 10^{10}\,N/m^2)(1.2 \times 10^{-3}\,m)}{5.0\,m} = 4.8 \times 10^7\,N/m^2$$

The force is then

$$F = (4.8 \times 10^7\,N/m^2)A = (4.8 \times 10^7\,N/m^2)(6.0 \times 10^{-3}\,m^2)$$
$$= 2.9 \times 10^5\,N \quad \text{(about 65 000 lb, or 32.5 tons!)}$$

Follow-up Exercise. Expansion gaps between identical steel beams laid end-to-end are specified to be 0.060% of the length of a beam at the installation temperature. With this specification, what is the temperature range for noncontact expansion?

Conceptual Example 10.6

Larger or Smaller? Area Expansion

A circular piece is cut from a flat metal sheet (•Fig. 10.12a). If the sheet is then heated in an oven, the size of the hole will (a) become larger, (b) become smaller, (c) be unchanged.

Reasoning and Answer. It is a common misconception to think that the area of the hole will shrink because of expansion of the metal around it. Think of the piece of metal removed from the hole rather than of the hole itself. This piece would expand with increasing temperature. The metal in the heated sheet reacts as if the piece removed were still part of it. (Think of putting the piece of metal back into the hole and heating, as in Fig. 10.12b. Or consider drawing a circle on an uncut metal sheet and heating it.) So, the answer is (a).

Follow-up Exercise. A circular ring of iron has a tight-fitting metal bar inside it, across its diameter. If heated in an oven to a high temperature, would the circular ring be distorted or bent out of shape, or would it remain circular?

•**FIGURE 10.11 Expansion gaps** (a) Expansion gaps are built into bridge roadways to prevent contact stresses produced by thermal expansion. (b) These loops in oil pipelines serve a similar purpose. As hot oil passes through them, the pipes expand; loops take up the extra length.

Fluids (liquids and gases), like solids, normally expand with increasing temperature. Because fluids have no definite shape, only volume expansion (and not linear or area expansion) is meaningful. The expression is

$$\frac{\Delta V}{V_o} = \beta \Delta T \quad \text{fluid volume expansion} \quad (10.13)$$

where β is the coefficient of volume expansion for fluids. Note in Table 10.1 that the values of β for fluids are typically larger than the values of 3α for solids.

Water exhibits an anomalous volume expansion near its freezing point. The volume of a given amount of water decreases as it is cooled from room temperature, until its temperature reaches 4°C (•Fig. 10.13a). Below 4°C, the volume increases, and therefore the density decreases (Fig. 10.13b). This means that water has a maximum density ($\rho = m/V$) at 4°C (actually, 3.98°C).

When water freezes, the molecules form a hexagonal (six-sided) lattice pattern. (This is why snowflakes have hexagonal shapes.) It is the open structure of this lattice that gives water its almost unique property of expanding on freezing and being less dense as a solid than as a liquid. (This is why ice floats in water and frozen water pipes burst.) The variation of the density of water over the temperature range of 4°C to 0°C indicates that the open lattice structure is beginning to form at about 4°C rather than arising instantaneously at the freezing point.

This property has an important environmental effect: Bodies of water, such as lakes and ponds, freeze at the top first. As a lake cools toward 4°C, water near the surface loses energy to the atmosphere, becomes denser, and sinks. The warmer, less dense water near the bottom rises. However, once the colder water on top reaches temperatures below 4°C, it becomes less dense and remains at the surface, where it freezes. If water did not have this property, lakes and ponds would freeze from the bottom up, which would destroy much of their animal and plant life (and would make ice skating a lot less popular). There would also be no oceanic ice caps at the polar regions. Rather, there would be a thick layer of ice at the bottom of the ocean, covered by a layer of water.

Conceptual Example 10.7

Quick Chill: Temperature and Density

Ice is put into a container of water at room temperature. For faster cooling, the ice (a) should be allowed to float naturally in the water or (b) be pushed to the bottom of the container with a stick and held there.

Reasoning and Answer. When the ice melts, the water in the vicinity is cooled and hence becomes denser (Fig. 10.13b). If the ice is permitted to float at the top, the denser water sinks and the warmer, less dense water rises. This mixing causes the water to cool quickly. If the ice were at the bottom of the container, however, the colder, denser water would remain there, and cooling would be slowed, so the answer is (a).

Follow-up Exercise. Suppose that the density-versus-temperature curve for water (Fig. 10.13b) were inverted, so that it dipped downward. What would this imply for the situation in this Example and the freezing of lakes?

(a) Metal plate with hole

(b) Metal plate without hole

•FIGURE 10.12 **A larger or smaller hole?** See Conceptual Example 10.6.

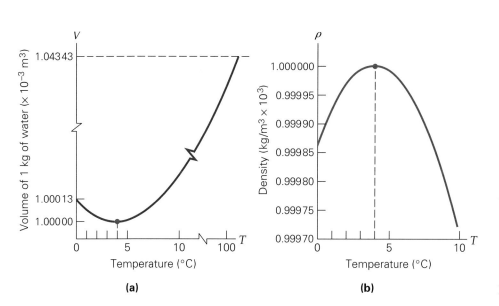

(a)

(b)

•FIGURE 10.13 **Thermal expansion of water** Water exhibits nonlinear expansion behavior near its freezing point. **(a)** Above 4°C (actually, 3.98°C), water expands with increasing temperature, but from 4°C down to 0°C, it expands with decreasing temperature. **(b)** As a result, water has its maximum density near 4°C.

10.5 The Kinetic Theory of Gases

Objectives: To (a) relate kinetic theory and temperature and (b) explain the process of diffusion.

If the molecules of a gas sample are viewed as colliding particles, the laws of mechanics can be applied to each molecule of that gas. We should then be able to describe its characteristics, such as pressure, internal energy, and so on, in terms of molecular motion. Because of the large number of particles involved, however, a statistical approach is employed for such a description.

One of the major accomplishments of theoretical physics was to do exactly this: derive the ideal gas law from mechanical principles. This derivation led to a new interpretation of temperature in terms of the translational kinetic energy of the gas molecules. As a theoretical starting point, the molecules of an ideal gas are viewed as point masses in random motion with relatively large distances separating them.

Our discussion in this book involves only two types of gases: (1) *mon*atomic (single-atom gases), such as He, and (2) *di*atomic (two-atom gas molecules), such as O_2. In either case, the vibrational and rotational motions can be ignored in terms of temperature and pressure, since these quantities depend only on *linear* motion.

Note: Elastic collisions are discussed in Section 6.4.

According to the **kinetic theory of gases**, the molecules undergo perfectly elastic collisions with each other and with the walls of the container. The forces between molecules are considered to act only over a short range, so the molecules interact with each other only *during* collisions. And since the molecules of an ideal gas interact only during collisions, such a gas would remain gaseous at low temperatures; it would not liquefy as real gases do. This gives the theoretical ideal gas its idealness. Colliding molecules lose and gain kinetic energy, with corresponding changes in their potential energy. We ignore this potential energy in summing the total energy of the system because molecules spend a negligible amount of time in collisions.

From Newton's laws of motion, the force on the walls of the container can be calculated from the change in momentum of the gas molecules when they collide with the walls (•Fig. 10.14). If this force is expressed in terms of pressure (force/area), the following equation is obtained (see Appendix II for derivation):

$$pV = \tfrac{1}{3} N m v_{\text{rms}}^2 \qquad (10.14)$$

where V is the volume of the container or gas, N the number of gas molecules in the closed container, and m the mass of a gas molecule. In this expression, v_{rms} is the average speed, but a special kind of average. It is obtained by averaging the squares of the speeds and then taking the square root of the average—that is, $\sqrt{\overline{v^2}} = v_{\text{rms}}$. As a result, v_{rms} is called the *root-mean-square* (*rms*) speed.

Solving Eq. 10.6 for pV and equating that expression with Eq. 10.14 shows how temperature came to be interpreted as a measure of translational kinetic energy:

Average molecular kinetic energy and temperature

$$pV = Nk_B T = \tfrac{1}{3} N m v_{\text{rms}}^2 \quad \text{or} \quad \tfrac{1}{2} m v_{\text{rms}}^2 = \tfrac{3}{2} k_B T \quad \text{(for all ideal gases)} \qquad (10.15)$$

Thus, the temperature of a gas (and that of the walls of the container or a thermometer bulb in thermal equilibrium with the gas) is directly proportional to its average random kinetic energy (per molecule), since $\overline{K} = \tfrac{1}{2} m v_{\text{rms}}^2 = \tfrac{3}{2} k_B T$. (Don't forget that T is the absolute temperature in kelvins.)

Example 10.8 ■ Molecular Speed: Relation to Absolute Temperature

What is the average (rms) speed of a helium atom (He) in a helium balloon at room temperature? (Take the mass of the helium atom to be 6.65×10^{-27} kg.)

Thinking It Through. All the data we need to solve for the average speed in Eq. 10.15 are known.

Solution.

Given: $m = 6.65 \times 10^{-27}$ kg
$T = 20°C$ (room temperature)
$k_B = 1.38 \times 10^{-23}$ J/K (known)

Find: v_{rms} (rms speed)

We will use Eq. 10.15, so we list k_B among the given quantities.

We must change the Celsius temperature to kelvins, but note that k_B has units of J/K. So,

$$T_K = T_C + 273 = 20°C + 273 = 293 \text{ K}$$

where the 273.15 in Eq. 10.8 was rounded to 273 for convenience.

Rearranging Eq. 10.15, we have

$$v_{rms} = \sqrt{\frac{3k_B T}{m}} = \left[\frac{3(1.38 \times 10^{-23} \text{ J/K})(293 \text{ K})}{6.65 \times 10^{-27} \text{ kg}}\right]^{1/2} = 1.35 \times 10^3 \text{ m/s} = 1.35 \text{ km/s}$$

This is over 3000 mi/h—pretty fast!

Follow-up Exercise. In this Example, if the temperature of the gas were increased by 10 C°, what would be the corresponding percentage increases in the average (rms) speed and in the average kinetic energy?

Interestingly, Eq. 10.15 predicts that at absolute zero ($T = 0$ K), all translational molecular motion of a gas would cease. According to classical theory, this would correspond to absolute zero energy. However, modern quantum theory says that there would still be some zero-point motion, and a corresponding minimum *zero-point energy*. Basically, absolute zero is the temperature at which all the energy that *can* be removed from an object has been removed.

Internal Energy of Monatomic Gases

Because the "particles" in an ideal monatomic gas do not vibrate or rotate, as explained previously, the total kinetic energy of all the molecules is equal to the total internal energy of the gas. That is, the gas's internal energy is all "temperature" energy (Section 10.1). With N molecules in a system, we can use Eq. 10.15, the energy per molecule, to write an expression for the total internal energy U:

$$U = N(\tfrac{1}{2}mv_{rms}^2) = \tfrac{3}{2}Nk_B T = \tfrac{3}{2}nRT \quad \text{(for ideal monatomic gases only)} \quad (10.16)$$

Thus, we see that the internal energy of an ideal monatomic gas is directly proportional to its absolute temperature. (In Section 10.6 we will see this is true regardless of the molecular structure of the gas. However, the expression for U will be a bit different for gases that are not monatomic.) This means that if the absolute temperature of a gas is doubled (by heat transfer), for example, from 200 K to 400 K, then its internal energy is also doubled. This does not apply to the Celsius and Fahrenheit temperatures, since their zero points are not referenced to the zero-point energy. If a gas is at a temperature of 0°C (or 0°F) and its internal energy is doubled, the Celsius (or Fahrenheit) temperature is not doubled—doubling zero gives zero, which is not a new temperature.

Diffusion

We depend on our sense of smell to detect odors, such as the smell of smoke from something burning. That you can smell something from a distance implies that molecules get from one place to another in the air, from the source to your nose. This process of random molecular mixing in which particular molecules move from a region where they are present in higher concentration to one where they are in lower concentration is called **diffusion**. Diffusion also occurs readily in liquids; think about what happens to a drop of ink in a glass of water (•Fig. 10.15). It even occurs to some degree in solids.

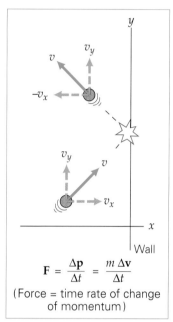

$$F = \frac{\Delta \mathbf{p}}{\Delta t} = \frac{m \Delta \mathbf{v}}{\Delta t}$$

(Force = time rate of change of momentum)

•FIGURE 10.14 Kinetic theory of gases The pressure a gas exerts on the walls of a container is due to the force resulting from the change in momentum of the gas molecules that collide with the wall. The force exerted by an individual molecule is equal to the time rate of change of momentum: $\mathbf{F} = \Delta \mathbf{p}/\Delta t = m\,\Delta \mathbf{v}/\Delta t$, where $\mathbf{p} = m\mathbf{v}$. The sum of the instantaneous normal components of the collision forces gives rise to the average pressure on the wall.

Note: Eq. 10.16 applies only to ideal gases or very dilute real gases.

•**FIGURE 10.15 Diffusion in liquids** Random molecular motion would eventually distribute the dye throughout the water. Here there is some distribution due to mixing, and the ink colors the water after a few minutes. The distribution would take more time by diffusion only.

The rate of diffusion for a particular gas depends on the rms speed of its molecules. Even though gas molecules have large average speeds (Example 10.8), the molecules do not fly from one side of a room to the other. There are frequent collisions, and as a result the molecules "drift" rather slowly. For example, suppose someone opened a bottle of ammonia on the other side of a closed room. It would take some time for the ammonia to diffuse across the room until you could smell it. (Much of the movement that people commonly attribute to diffusion is actually due to air currents.)

Gases can also diffuse through porous materials or permeable membranes. (This process is sometimes referred to as *effusion*.) Energetic molecules enter the material through the pores (openings), and, colliding with the pore walls, they slowly meander through the material. Such gaseous diffusion can be used to separate different gases from a mixture physically.

The kinetic theory of gases says that the average translational kinetic energy (per molecule) is proportional to the absolute temperature of a gas, $\frac{1}{2}mv_{rms}^2 = \frac{3}{2}k_B T$. So, on the average, the molecules of different gases (having different masses) move at different speeds at a given temperature. As you might expect, because they move faster, lighter gas molecules diffuse through the tiny openings of a porous material faster than heavier gas molecules do.

For example, at a particular temperature, molecules of oxygen (O_2) move faster on the average than the more massive molecules of carbon dioxide (CO_2) do. Because of this difference in molecular speed, oxygen can diffuse through a barrier faster than carbon dioxide can. Suppose that a mixture of equal volumes of oxygen and carbon dioxide is contained on one side of a porous barrier (•Fig. 10.16). After a while, some O_2 molecules and CO_2 molecules will have diffused through

•**FIGURE 10.16 Separation by gaseous diffusion** The molecules of both gases diffuse (or effuse) through the porous barrier. But, because oxygen molecules have the greater average speed, more of them pass through. Thus, over time there is a greater concentration of oxygen molecules on the other side of the barrier.

Insight

Physiological Diffusion

Diffusion plays a central role in many life processes. For example, consider a cell membrane in the lung. Such a membrane is permeable to a number of substances, any of which will diffuse through the membrane from a region where its concentration is high to a region where its concentration is low. Most importantly, the lung membrane is permeable to oxygen (O_2), and the transfer of O_2 across the membrane occurs because of a concentration gradient.

The blood carried to the lungs is low in O_2, having given up the oxygen to tissues for metabolism during its circulation through the body. Conversely, the air in the lungs is high in O_2 because there is a continuous exchange of fresh air in the breathing process. As a result of this concentration difference, or gradient, O_2 diffuses from the lung volume into the blood that flows through the lung tissue, and the blood leaving the lungs is high in O_2.

Exchanges between the blood and the tissues occur across capillary walls, and diffusion again is a major factor. The chemical composition of arterial blood is regulated to maintain the proper concentrations of particular solutes (substances dissolved in the blood solution), so diffusion takes place in the appropriate directions across capillary walls. For example, as cells take up O_2 and glucose (blood sugar), the blood continuously brings in fresh supplies of the substances to maintain the concentration gradient needed for diffusion to the cells. The continuous production of carbon dioxide (CO_2) and metabolic wastes in the cells produces concentration gradients in the opposite direction for these substances. They therefore diffuse out of the cells into the blood, to be carried away from the tissue by the circulatory system.

During periods of physical exertion, there is an increase in cellular activity. More O_2 is used up and more CO_2 is produced, thereby increasing the concentration gradients and the diffusion rates. With an increased demand for O_2, how do the lungs respond to provide this to the blood? As you might expect, the rate of diffusion depends on the surface area and thickness of the lung membrane. Deeper breathing during exercise causes the alveoli (small air sacs in the lungs) to increase in volume. Such stretching increases the alveolar surface area and decreases the thickness of the membrane wall.

Also, the heart works harder during exercise, and the blood pressure is raised. The increased pressure forces open capillaries that are normally closed during rest or normal activity. As a result, the total exchange area between the blood and cells is increased. Each of these changes helps expedite the exchange of gases during exercise.

Now you can understand why people with emphysema (a disease involving the breakdown of alveoli walls) or pneumonia (characterized by fluid accumulations within or around the lungs) have difficulty providing enough oxygen to their tissues.

---

the barrier. The diffused gases would still be mixed. After another diffusion stage, however, the oxygen concentration would become greater on the far side of the barrier. Almost pure oxygen can be obtained by repeating the separation process many times. Separation by gaseous diffusion is a key process in obtaining enriched uranium, which was used in the atomic bomb and in nuclear reactors that generate electricity (Chapter 30).

Fluid diffusion is very important to organisms. In plant photosynthesis, carbon dioxide from the air diffuses into leaves, and oxygen and water vapor diffuse out. The diffusion of liquid water across a permeable membrane down a concentration gradient is called **osmosis**; this process is vital in living cells. Osmotic diffusion is also important to kidney functioning: Tubules in the kidneys concentrate waste matter from the blood in much the same way oxygen is removed from mixtures. (See the Insight on this page for other examples of diffusion.)

*10.6 Kinetic Theory, Diatomic Gases, and the Equipartition Theorem

Objectives: To understand (a) the difference between monatomic and diatomic gases, (b) the meaning of the equipartition theorem, and (c) the expression for the internal energy of a diatomic gas.

In the real world, most of the gases we deal with are *not* monatomic gases. Recall from chemistry that monatomic gases are elements known as *noble* or *inert* gases because they do not readily combine with other atoms. These elements are found on the far right side of the periodic chart: helium, neon, argon, krypton, xenon, and radon.

However, the mixture of gases we breathe (collectively known as "air") consists mainly of diatomic molecules of nitrogen (N_2, 78% by volume) and oxygen (O_2, 21% by volume). Both of these gases have two identical atoms chemically bonded together to form a single molecule. How do we deal with these realistic, more complicated molecules in terms of the kinetic theory of gases? There are even more complicated gas molecules consisting of more than two atoms, such as carbon dioxide (CO_2). However, because of their complexity, we will limit our discussion to diatomic molecules.

The Equipartition Theorem

As we saw in Section 10.5, it is only the translational kinetic energy of a gas that determines the gas's temperature. Thus, for any type of gas, regardless of how many atoms make up its molecules, it is *always true* that the average *translational* kinetic energy per molecule is still proportional to the temperature of the gas (Eq. 10.15): $\frac{1}{2}mv_{rms}^2 = \frac{3}{2}k_B T$ (for all gases).

Recall that for monatomic gases, the total internal energy, U, consists solely of translational kinetic energy. For diatomic molecules, this is not true, because a diatomic molecule is free to rotate and vibrate in addition to moving linearly. Therefore, these extra forms of energy must be taken into account. The expression given in Eq. 10.16 ($U = \frac{3}{2}Nk_B T$) for monatomic gases, which assumes that the total energy is due only to translational kinetic energy, therefore does *not* hold for diatomic gases.

Scientists wondered exactly how the expression for the internal energy of a diatomic gas might differ from that for a monatomic gas. In looking at the kinetic theory derivation that led up to Eq. 10.16, they realized that the factor of 3 in this equation was due to the fact that the gas molecules had three independent linear ways (dimensions) of moving. Thus for each molecule, there were three independent ways of possessing kinetic energy: with x, y, or z linear motion. Each independent way a molecule has for possessing energy is called a **degree of freedom**.

According to this scheme, a monatomic gas has only three degrees of freedom, since its molecules can move only linearly and possess kinetic energy in three dimensions. It was reasoned that quite possibly a diatomic gas could vibrate (see Fig. 10.1), thus having vibrational kinetic and potential energies (two additional degrees of freedom). A diatomic molecule might also rotate (•Fig. 10.17).

However, the rotational about the molecular axis (x in the figure) possesses very little energy compared to the other axes. As illustrated in Fig. 10.17, if we assume the atoms to be small spheres on the end of a stick, the moment of inertia (I) about the stick (the molecular axis) is very small and taken to be negligible. Since rotational kinetic energy depends directly on the moment of inertia, rotational motion about this axis provides negligible rotational kinetic energy. Thus, there are only two degrees of freedom associated with diatomic molecule rotation.

On the basis of the new understanding of monatomic gases and their three degrees of freedom, the **equipartition theorem** was proposed. (As the name implies, the total energy of a gas or molecule is "partitioned," or divided, equally for each degree of freedom.) That is,

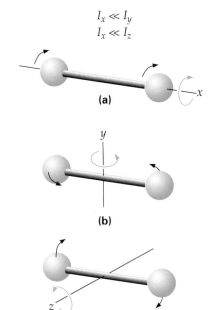

•**FIGURE 10.17 Model of a diatomic gas molecule** If we consider a diatomic molecule as two spheres on the end of a stick, the rotation about **(a)** the molecular axes is very small or negligible compared to **(b)** and **(c)**, the other axes. That is, the moment of inertia about the x axis (I_x) is relatively small and so is the rotational kinetic energy ($K = \frac{1}{2}I\omega^2$).

Definition of:
Equipartition theorem

> On average, the total internal energy (U) of an ideal gas is divided equally among each degree of freedom its molecules possess. Furthermore, each degree of freedom contributes $\frac{1}{2}Nk_B T$ (or $\frac{1}{2}nRT$) to the total internal energy.

The equipartition theorem fits the special case of a monatomic gas, since this theorem predicts that $U = \frac{3}{2}Nk_B T$, which we know to be true. With three degrees of freedom, we have $U = 3(\frac{1}{2}Nk_B T)$, in agreement with the monatomic result given earlier (Eq. 10.16).

The Internal Energy of a Diatomic Gas

Exactly how does the equipartition theorem enable us to calculate the internal energy of a diatomic gas such as oxygen? To do this, we must realize that U must now include all the available degrees of freedom. In addition to the translational degrees of freedom, what other motions are the molecules undergoing? The analysis is complicated and beyond the scope of this text, so we give only the general results. *Typically for normal (room) temperatures, quantum theory predicts (and experiment verifies) that only the rotational motions are important in the degrees of freedom. Diatomic molecules do NOT vibrate at normal temperatures.*

So, the total internal energy of a diatomic gas is comprised of the internal energies due to the three linear degrees of freedom and two rotational degrees of freedom, for a total of five. Hence, we can write

$$U = K_{trans} + K_{rot} = 3(\tfrac{1}{2}nRT) + 2(\tfrac{1}{2}nRT)$$

$$= \tfrac{5}{2}nRT = \tfrac{5}{2}Nk_B T \qquad \text{(for diatomic gases at room temperature)} \qquad (10.17)$$

Thus a monatomic gas sample at normal room temperature has 40% less internal energy than a diatomic gas sample at the same temperature, or equivalently possesses only 60% of the internal energy of the diatomic gas sample.

Example 10.9 ■ Monatomic versus Diatomic: Are Two Atoms Better Than One?

More than 99% of the air we breathe consists of diatomic gases, mainly nitrogen (N_2, 78%) and oxygen (O_2, 21%). There are traces of other gases, one of which is radon (Ra), a monatomic gas arising from radioactive decay of uranium in the ground. [Radon itself is radioactive, which is irrelevant here, but this fact does make radon a possible health danger when concentrated inside a house (as we shall see in Chapter 29).] (a) Calculate the total internal energy of 1.00-mole samples each of oxygen and radon at room temperature (20°C). (b) For each sample, calculate the amount of internal energy associated with molecular *translational* kinetic energy.

Thinking It Through. (a) We have to consider the number of degrees of freedom in a monatomic gas and a diatomic gas in computing the internal energy (U). (b) Only three linear degrees of freedom contribute to the translational kinetic energy portion (U_{trans}) of the internal energy.

Solution. We list the data and convert to kelvins right away because we know that the internal energy is expressed in terms of absolute temperature:

Given: $n = 1.00$ mol
$T = 20°C + 273 = 293$ K

Find: (a) U for O_2 and Ra samples at room temperature
(b) U_{trans} for O_2 and Ra samples at room temperature

(a) Let's compute the total internal energy of the (monatomic) radon sample first, using Eq. 10.16:

$$U_{Ra} = \tfrac{3}{2}nRT = \tfrac{3}{2}(1.00\text{ mol})[8.31\text{ J/(mol·K)}](293\text{ K}) = 3.65 \times 10^3\text{ J}$$

Since this sample is at room temperature, the (diatomic) oxygen will also include internal energy stored as two extra degrees of freedom, due to rotation. Thus we have

$$U_{O_2} = \tfrac{5}{2}nRT = \tfrac{5}{2}(1.00\text{ mol})[8.31\text{ J/(mol·K)}](293\text{ K}) = 6.09 \times 10^3\text{ J}$$

As we have seen, even though there is the same number of molecules in each sample and the temperature is the same, the oxygen sample has almost 70% more energy.

(b) For (monatomic) radon, all the internal energy is in the form of translational kinetic energy; thus the answer is the same as in (a):

$$U_{trans} = U_{Ra} = 3.65 \times 10^3\text{ J}$$

For (diatomic) oxygen, only $\frac{3}{2}nRT$ of the total internal energy ($\frac{5}{2}nRT$) is from translational kinetic energy, so the answer is the same as for radon; that is, $U_{trans} = 3.65 \times 10^3$ J for both gas samples.

Follow-up Exercise. In this Example, how much energy is associated with the rotational motion of the oxygen molecules? Which sample has the highest rms speed? (Note: The mass of one radon atom is about 7 times the mass of an oxygen molecule.) Explain your reasoning.

Chapter Review

Important Terms

temperature 334
heat 334
internal energy 334
thermometer 335
thermal expansion 335
Fahrenheit temperature scale 336
Celsius temperature scale 336
ideal (perfect) gas law 338
mole 340

Avogadro's number 340
absolute zero 341
Kelvin temperature scale 341
kelvins 341
triple point of water 342
thermal coefficient of linear expansion 344
thermal coefficient of area expansion 345

thermal coefficient of volume expansion 345
kinetic theory of gases 348
diffusion 349
osmosis 351
*degree of freedom 352
*equipartition theorem 352

Important Concepts

- Heat is the net energy transferred from one object to another because of temperature differences. Once transferred, the energy becomes part of the internal energy of the object (or system).
- The ideal gas law relates pressure, volume, and absolute temperature.
- Absolute zero (0 K) corresponds to $-273.15°C$.

- Thermal coefficients of expansion relate the fractional change in dimension(s) to a change in temperature.
- According to the kinetic theory of gases, the absolute temperature of a gas is directly proportional to the average random kinetic energy per molecule.

Important Equations

Celsius–Fahrenheit Conversion:

$$T_F = \tfrac{9}{5}T_C + 32 \quad (10.1)$$

$$T_C = \tfrac{5}{9}(T_F - 32) \quad (10.2)$$

Ideal (or perfect) Gas Law (always absolute temperature):

$$pV = Nk_B T \quad \text{or} \quad \frac{p_1 V_1}{T_1} = \frac{p_2 V_2}{T_2} \quad (10.5, 10.6)$$

or

$$pV = nRT \quad (10.7)$$

$$k_B = 1.38 \times 10^{-23} \text{ J/K}$$

$$R = 8.31 \text{ J/(mol·K)}$$

Avogadro's number: $N_A = 6.02 \times 10^{23}$ molecules/mol

Kelvin–Celsius Conversion:

$$T_K = T_C + 273.15 \quad (10.8)$$

Thermal Expansion of Solids:

linear: $\dfrac{\Delta L}{L_o} = \alpha \Delta T \quad$ or $\quad L = L_o(1 + \alpha \Delta T) \quad (10.9, 10.10)$

area: $\dfrac{\Delta A}{A_o} = 2\alpha \Delta T \quad$ or $\quad A = A_o(1 + 2\alpha \Delta T) \quad (10.11)$

volume: $\dfrac{\Delta V}{V_o} = 3\alpha \Delta T \quad$ or $\quad V = V_o(1 + 3\alpha \Delta T) \quad (10.12)$

Thermal Volume Expansion of Fluids:

$$\frac{\Delta V}{V_o} = \beta \Delta T \quad (10.13)$$

Results of Kinetic Theory of Gases:

$$pV = \tfrac{1}{3} N m v_{rms}^2 \quad (10.14)$$

$$\tfrac{1}{2} m v_{rms}^2 = \tfrac{3}{2} k_B T \quad \text{(all gases)} \quad (10.15)$$

$$U = \tfrac{3}{2} N k_B T = \tfrac{3}{2} nRT \quad \text{(ideal monatomic gases only)} \quad (10.16)$$

$$U = \tfrac{5}{2} N k_B T = \tfrac{5}{2} nRT \quad \text{(diatomic gases)} \quad (10.17)$$

Exercises*

10.2 The Celsius and Fahrenheit Temperature Scales

1. Which one of the following is the closest to 15°C? (a) 8.3°F, (b) 27°F, (c) 40°F, (d) 50°F?

2. Which of the following temperature scales has the smaller degree interval, Celsius or Fahrenheit?

3. Heat flows spontaneously from a body at a higher temperature to one at a lower temperature that is in thermal contact with it. Does it always flow from a body with more internal energy to one with less internal energy? Explain.

4. What is the hottest (highest temperature) item in a home? [Hint: Think about this, and maybe a light will come on.]

5. ■ Convert the following to Celsius readings: (a) 1000°F, (b) 0°F, (c) −20°F, (d) −40°F.

6. ■ Convert the following to Fahrenheit readings: (a) 150°C, (b) 32°C, (c) −25°C, (d) −273°C.

7. ■ The boiling point of alcohol is 172°F. What is this temperature on the Celsius scale?

8. ■ A person running a fever has a body temperature of 39.4°C. What is this temperature on the Fahrenheit scale?

9. ■ The highest and lowest recorded air temperatures in the United States are 134°F (Death Valley, California, 1913) and −80°F (Prospect Creek, Alaska, 1971), respectively. What are these temperatures on the Celsius scale?

10. ■ Which is the lower temperature, (a) 245°C or 245°F? (b) 200°C or 375°F?

11. ■ The highest and lowest recorded air temperatures in the world are 58°C (Libya, 1922) and −89°C (Antarctica, 1983). What are these temperatures on the Fahrenheit scale?

12. ■■ Suppose you didn't know that −40°C = −40°F (see Follow-up Exercise 10.1) and are asked if the Celsius and Fahrenheit temperatures are ever equal. How could you find out algebraically? Do it.

13. ■■■ (a) The largest temperature drop recorded in the United States in one day occurred in Browning, Montana, in 1916, when the temperature went from 7°C to −49°C. What is the corresponding change on the Fahrenheit scale? (b) The largest recorded temperature rise occurred in 2 min from −4°F to 45°F (Spearfish, South Dakota, 1943). What is the corresponding change on the Celsius scale?

*Assume all temperatures to be exact, and neglect significant figures for small dimension changes.

14. ■■■ (a) On the plot of Fahrenheit temperature versus Celsius temperature in Fig. 10.5a, the y-intercept is 32°F. What would be the x-intercept value? (b) What would the slope and y-intercept be if the graph were plotted in the opposite way (Celsius versus Fahrenheit)?

10.3 Gas Laws and Absolute Temperature

15. The temperature used in the ideal gas law is (a) Celsius, (b) Fahrenheit, (c) Kelvin, (d) any of the preceding.

16. When the temperature of a quantity of gas is increased, (a) the pressure must increase, (b) the volume must increase, (c) both the pressure and volume must increase, (d) none of the preceding.

17. The temperature of a quantity of gas is decreased. How is the density affected (a) if the pressure is held constant? (b) if the volume is held constant?

18. A type of constant-volume gas thermometer is shown in •Fig. 10.18. Describe how it operates.

•FIGURE 10.18 A type of constant-volume gas thermometer See Exercise 18.

19. Describe how a constant-pressure gas thermometer might be constructed.

20. In terms of the ideal gas law, what would a temperature of absolute zero imply? a negative absolute temperature?

21. Excited about a New Year's Eve party in Times Square, you pumped up 10 balloons in your warm apartment and took them to the cold square. However, you were very disappointed. What happened? Why?

22. Convert the following temperatures to absolute temperatures in kelvins: (a) 0°C, (b) 100°C, (c) 20°C, and (d) −35°C.

23. Convert the following temperatures to degrees Celsius: (a) 0 K, (b) 250 K, (c) 273 K, and (d) 325 K.

24. What are the absolute temperatures in kelvins for (a) −40°F and (b) −40°C?

25. Which is the lower temperature, 300°F or 300 K?

26. The surface temperature of the Sun is about 6000 K. (a) What is this temperature on the Fahrenheit and Celsius scales? (b) The surface temperature is sometimes reported to be 6000°C. Assuming that 6000 K is correct, what is the percentage error of this Celsius value?

27. How many moles are there in (a) 40 g of water, (b) 245 g of H_2SO_4 (sulfuric acid), (c) 138 g of NO_2 (nitrogen dioxide), (d) 56 L of SO_2 (sulfur dioxide) at STP?

28. A constant-volume gas thermometer exerts an absolute pressure of 1000 Pa at room temperature (20°C). If the pressure increases to 1500 Pa, what is the new Celsius temperature?

29. The air in a balloon of volume $0.10\ m^3$ exerts a pressure of 1.4×10^5 Pa. If the volume of the balloon increases to $0.12\ m^3$ at a constant temperature, what is the pressure?

30. Show that 1.00 mol of ideal gas under STP (standard temperature of exactly 0°C and pressure of exactly 1 atm) occupies a volume of $0.0224\ m^3 = 22.4$ L.

31. What is the volume occupied by 4.00 g of hydrogen under a pressure of 2.00 atm and a temperature of 300 K?

32. How many molecules are there in (a) 1.00 L and (b) $1.00\ cm^3$ of a gas at STP?

33. ■■ The Kelvin temperature of an ideal gas is doubled and its volume is halved. How is the pressure affected?

34. ■■ On a warm day (92°F), the air in a balloon occupies a volume of $0.20\ m^3$ and exerts a pressure of $20.0\ lb/in^2$. If the balloon is cooled to 32°F in a refrigerator, the pressure decreases to $14.7\ lb/in^2$. What is the volume of the balloon? (Assume that the air behaves as an ideal gas.)

35. ■■ A steel-belted radial automobile tire is inflated to a gauge pressure of $30.0\ lb/in^2$ when the temperature is 61°F. Later in the day, the temperature rises to 100°F. Assuming that the volume of the tire remains constant, what is the pressure at the elevated temperature? [Hint: Remember that the ideal gas law uses absolute pressure.]

36. ■■ If $2.4\ m^3$ of a gas initially at STP is compressed to $1.6\ m^3$ and its temperature raised to 30°C, what is the final pressure?

37. ■■ The pressure on a low-density gas in a cylinder is kept constant as the temperature of the gas is increased from 10°C to 40°C. (a) Does the gas expand or compress? (b) What is the percentage change in the volume of the gas?

38. ■■■ A diver releases an air bubble of volume $2.0\ cm^3$ from a depth of 15 m below the surface of a lake; there the temperature is 7.0°C. What is the volume of the bubble when it reaches the surface of the lake, where the temperature is 20°C?

10.4 Thermal Expansion

39. The units of the thermal coefficient of linear expansion are (a) m/C°, (b) $m^2/C°$, (c) m·C°, (d) 1/C°.

40. The thermal coefficient of volume expansion for a solid is (a) α, (b) 2α, (c) 3α, (d) α^3.

41. A cube of ice sits on a bimetallic strip at room temperature (•Fig. 10.19). What will happen if (a) the upper strip is aluminum and the lower strip brass or (b) the upper strip is iron and the lower strip copper? (c) If the cube is made of a hot metal rather than ice and the two strips are brass and copper, should the brass or copper be on top to keep the cube from falling off?

•FIGURE 10.19 Which way will the cube go? See Exercise 41.

42. A demonstration of thermal expansion is shown in •Fig. 10.20. Initially, the ball goes through the ring made of the same metal. (a) When the ball is heated, it does not go through the ring. (b) If both the ball and the ring are heated, the ball again goes through the ring. Explain what is being demonstrated.

43. What happens to a volume of water if it is cooled from 4°C to 2°C?

44. A solid metal disk rotates freely, so the conservation of angular momentum applies (Chapter 8). If the disk is heated while it is rotating, will there be any effect on the rate of rotation (the angular speed)?

45. A circular ring of iron has a tight-fitting iron bar across its diameter as illustrated in •Fig. 10.21. If the ring-and-bar arrangement is heated in an oven to a high temperature,

(a)   (b)   (c)

•FIGURE 10.20 **Ball-and-ring expansion** See Exercises 42 and 53.

will the circular ring be distorted? What if the bar is made of aluminum?

•FIGURE 10.21 **Stress out of shape?** See Exercise 45.

46. ■ A steel beam 10 m long is installed in a structure at 20°C. What are its changes in length in millimeters at the temperature extremes of −30°C to 45°C?

47. ■ An aluminum tape measure is accurate at 20°C. If the tape measure is placed in a deep freeze at −5.0°C, what would be the tape measure's percentage error because of thermal contraction? Would it read high or low?

48. ■ Concrete highway slabs are poured in lengths of 10.0 m. How wide in millimeters should the expansion gaps between the slabs be at a temperature of 20°C to ensure that there will be no contact between adjacent slabs over a temperature range of −25°C to 45°C?

49. ■ A man's gold wedding ring has an inner diameter of 2.4 cm at 20°C. If the ring is dropped into boiling water, what will be the change in the inner diameter of the ring?

50. ■■ What temperature change would cause a 0.10% increase in the volume of a quantity of water that was initially at 20°C?

51. ■■ A piece of copper tubing used in plumbing has a length of 60.0 cm and an inner diameter of 1.50 cm at 20°C. When hot water at 85°C flows through the tube, what are (a) the tube's new length and (b) the change in its cross-sectional area? Can the latter affect the flow speed?

52. ■■ A circular piece of diameter 8.00 cm is cut from an aluminum sheet at 20°C. If the sheet is then placed in an oven and heated to 150°C, (a) will the hole be larger or smaller? (b) What will the new area of the hole be?

53. ■■ In Fig. 10.20, the steel ring of diameter 2.5 cm is 0.10 mm smaller in diameter than the steel ball at 20°C. If you have a torch, which should you heat so that the ball will go through the ring? What is the required temperature?

54. ■■ A circular steel plate of radius 0.10 m is cooled from 350°C to 20°C. By what percentage does the plate's area decrease?

55. ■■ One morning when the temperature is 10°C, an employee at a rent-a-car company fills the 25-gallon steel gas tank of an automobile to the top and then parks the car a short distance away. That afternoon, when the temperature is 30°C, gasoline drips from the tank onto the pavement. How much gas will be lost? (Neglect the expansion of the tank.) What is the lesson as to the best time to buy gas (if you drive right away and don't lose it!)?

56. ■■ Show that the density of a solid substance varies with temperature as $\rho = \rho_o(1 - 3\alpha \Delta T)$ or $\rho = \rho_o(1 - \beta \Delta T)$.

57. ■■ The density of mercury is 13.6×10^3 kg/m^3 at 0°C. Calculate its density at 100°C.

58. ■ A copper block has an internal spherical cavity of diameter 10 cm (•Fig. 10.22). The block is heated in an oven from 20°C to 500 K. (a) Does the cavity get larger or smaller? (b) What is the change in the cavity's volume?

•FIGURE 10.22 **A hole in a block** See Exercise 58.

59. ■■■ A brass rod has a circular cross section of radius 0.500 cm. It fits into a circular hole in a copper sheet with a clearance of 0.010 mm completely around it when both it and the sheet are at 20°C. (a) At what temperature will the clearance be zero? (b) Would such a tight fit be possible if the sheet were brass and the rod were copper?

60. ■■■ A Pyrex beaker that has a capacity of 1000 cm^3 at 20°C contains 990 cm^3 of mercury at that temperature. Is there some temperature at which the mercury will completely fill the beaker? Justify your answer. (Assume no mass loss by vaporization.)

10.5 The Kinetic Theory of Gases

61. If the kinetic energy of an average ideal gas molecule in a sample at 20°C doubles, its final temperature must be (a) 10°C, (b) 40°C, (c) 313°C, (d) none of the preceding.

62. If the temperature of a quantity of ideal gas is raised from 20°C to 40°C, its internal energy is (a) doubled, (b) tripled, (c) unchanged, (d) none of the preceding.

63. Equal volumes of helium gas (He) and neon gas (Ne) at the same temperature (and pressure) are on opposite sides of a porous membrane (•Fig. 10.23). Describe what happens after a period of time, and why.

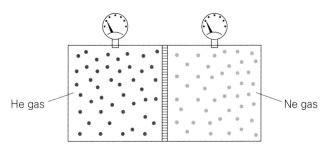

•FIGURE 10.23 **What happens as time passes?** See Exercise 63.

64. Natural gas is odorless; to alert people to gas leaks, the gas company adds an additive with smell. When there is a gas leak, the additive reaches your nose before the gas does. What can you conclude about the masses of the additive molecules and gas molecules?

65. ■ What is the average kinetic energy per molecule of an ideal gas at (a) 20°C and (b) 100°C?

66. ■ If the temperature of an ideal gas is raised from 20°C to 50°C, by how much will the internal energy of 2.0 mol of gas change?

67. ■ (a) What is the average kinetic energy per molecule of an ideal gas at a temperature of 27°C? (b) What is the average (rms) speed of the molecules if the gas is helium? (A helium molecule consists of a single atom of mass 6.65×10^{-27} kg.)

68. ■ What is the average speed of the molecules in low-density oxygen gas at 0°C? (The mass of an oxygen molecule, O_2, is 5.31×10^{-26} kg.)

69. ■■ At a given temperature, which would be greater, the rms speed of oxygen (O_2) or of ozone (O_3), and how many times greater?

70. ■■ If the temperature of an ideal gas is raised from 25°C to 100°C, how many times faster is the new average (rms) speed of the gas molecules?

71. ■■ A quantity of an ideal gas is at 0°C. An equal quantity of another ideal gas is twice as hot. What is its temperature?

*10.6 Kinetic Theory, Diatomic Gases, and the Equipartition Theorem

72. The temperature of a diatomic molecule such as O_2 is a measure of its (a) translational kinetic energy, (b) rotational kinetic energy, (c) vibrational kinetic energy, (d) all of the above.

73. On the average, the total internal energy of a gas is divided equally among (a) each atom, (b) each degree of freedom, (c) linear motion, rotational motion, and vibrational motion, (d) none of the above.

74. ■ If 1.0 mol of an ideal monatomic gas has a total internal energy of 5.0×10^3 J at a certain temperature, what is the total internal energy of 1.0 mol of a diatomic gas at the same temperature?

75. ■ What is the total internal energy of 1.0 mol of O_2 gas at 20°C?

76. ■■ For the average molecule in a sample of N_2 gas at 0°C, what is its (a) translational kinetic energy? (b) rotational kinetic energy? (c) total internal energy?

Additional Exercises

77. In Exercise 55, if the expansion of the steel gas tank is *not* ignored, what is the volume of gas spilled?

78. To conserve energy, thermostats in an office building are set at 78°F in the summer and 65°F in the winter. What would the settings be if the thermostats had a Celsius scale?

79. A copper wire 0.500 m long is at 20°C. If the temperature is increased to 100°C, what is the change in the wire's length?

80. If 2.0 mol of oxygen gas is confined in a 10-liter bottle under a pressure of 6.0 atm, what is the average translational kinetic energy of an oxygen molecule?

81. A mercury thermometer has a uniform capillary bore whose cross-sectional area is 0.012 mm². The volume of the mercury in the thermometer bulb at 10°C is 0.130 cm³. If the temperature is increased to 50°C, how much will the height of the mercury column in the capillary bore change? (Neglect the expansion of the mercury in the bulb and of the glass of the thermometer.)

82. An aluminum cube is 10.0 cm on each side at 20°C. It is heated in an oven to 300°C. (a) What is the cube's new volume? (b) Using this volume, compute the length of each edge of the cube.

83. What temperature increase would produce a stress of 8.0×10^7 N/m² on a rigidly held steel beam?

84. Using the ideal gas law, express the coefficient of volume expansion in terms of temperature and pressure. [Hint: Use different temperatures and pressures.]

85. An ideal gas occupies a container of volume 0.75 L at STP. Find (a) the number of moles and (b) the number of molecules of the gas. (c) If the gas is carbon monoxide (CO), what is its mass?

86. An aluminum tape measure is used to measure a 1.0-meter-long steel beam. At 0°C, the measurement is accurate. What will the measurement be if the temperature increases to 40°C? (Ignore significant figures.)

87. (a) If the temperature drops by 10 C°, what is the corresponding temperature change on the Fahrenheit scale? (b) If the temperature rises by 10 F°, what is the corresponding change on the Celsius scale?

88. A new metal alloy is made into a 50-centimeter rod. (a) Determine the alloy's coefficient of linear expansion if, after being heated from 20°C to 250°C, the rod has elongated by 0.44 mm. (b) Is this a potentially valuable alloy? Explain.

89. Calculate the number of gas molecules in a container of volume 0.20 m³ filled with gas under a partial vacuum of pressure 20 Pa at 20°C.

90. Is there a temperature that has the same numerical value on the Kelvin and Fahrenheit scales? Justify your answer.

91. Show that the coefficient of area expansion for solids is approximately equal to 2α.

92. In the troposphere (the lowest part of the atmosphere), the temperature decreases rather uniformly with altitude at a rate of about 6.5 C°/km; this decrease is called the *lapse rate*. What are the temperatures (a) near the top of the troposphere (which has an average thickness of 11 km) and (b) outside a commercial aircraft flying at a cruising altitude of 34 000 ft? (Assume that the ground temperature is normal room temperature.)

93. A mercury thermometer has a bulb volume of 0.200 cm³ and a capillary bore diameter of 0.065 mm. How far up the bore will the column of mercury move if the overall temperature of the thermometer is increased by 30 C°? (Neglect the expansion of the glass of the thermometer.)

94. On a hot (40°C) day, copper electric wires are used to connect a power box to a home 150 m away. On a cold (−10°C) winter night, what should the minimum extra length of the wire be in centimeters to avoid snapping?

95. A quantity of an ideal gas initially at atmospheric pressure is maintained at a constant temperature while it is compressed to half of its volume. What is the final pressure of the gas?

96. Use algebraic equations (not numbers) to demonstrate the discrepancy shown in Example 10.5.

CHAPTER 11

Heat

11.1 Units of Heat
11.2 Specific Heat
11.3 Phase Changes and Latent Heat
11.4 Heat Transfer

INSIGHTS

- Phase Changes and Ice Skating
- The Greenhouse Effect
- Heat Transfer by Radiation: The Microwave Oven

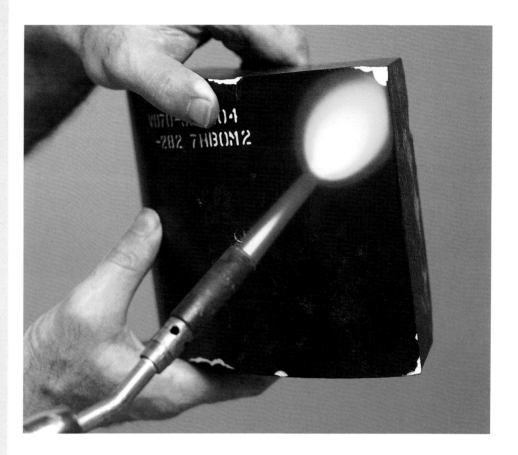

Most of us know to be very careful about handling anything that has recently been in contact with a flame or other source of heat. Yet, while the metal handle of a pot on the stove may be almost as hot as the bottom, a wooden pot handle is probably safe to touch. In other words, not only does heat flow through objects, but the rate of flow can vary widely (as the photograph shows in dramatic fashion). Direct physical contact isn't always necessary for heat to be transmitted: Think of touching the dashboard or the metal seat-belt buckle of a car left in direct sunlight for a few hours.

Think for a moment how crucial heat is in our existence. Our bodies must precisely balance heat loss and heat gain to stay within the narrow temperature range necessary for life. On a larger scale, the average temperature of the Earth, so critical to our environment and the survival of the organisms that inhabit it, is maintained through a similar balance. We are warmed by energy that comes to us across a 150-million-kilometer void. Each day, vast quantities of energy reach our planet's atmosphere and surface from the Sun, only to be radiated away into the cold of space.

These thermal balances are delicate, and any disturbance can have serious consequences. In an individual, sickness can disrupt the balance, producing a chill or fever. Similarly, we worry that an atmospheric buildup of "greenhouse" gases, a product of our industrial society, could give our whole planet a "fever" that would affect all life.

In this chapter you'll learn what heat is and how it is measured, and will study the various processes by which heat passes from one body to another. This knowledge will allow you to explain many everyday things as well as provide a basis for understanding the conversion of thermal energy into useful mechanical work.

11.1 Units of Heat

Objectives: To (a) distinguish the various units of heat and (b) define the mechanical equivalent of heat.

Like work, heat involves a transfer of energy. It is commonly thought that heat is a form of energy, but this is not true in the strictest sense. Rather, **heat** is the name we use to describe a type of net energy *transfer*. When we refer to "heat" or "heat energy" in this book, we mean the addition (or removal) of internal energy to a body or system.

Because heat is energy *in transit*, it is measurable as energy losses or gains and is described by standard energy units like any other quantity of energy. Recall that the SI standard unit of energy is the joule (J), or newton-meter (N·m). Thus, it is correct to say, for example, that 20 J of heat energy is transferred from one body to another.

As usual, we will use SI units, but we define for completeness other commonly used units of heat. A chief one is the **kilocalorie (kcal)** (•Fig. 11.1a):

> One kilocalorie (kcal) is defined as the amount of heat needed to raise the temperature of 1 kg of water 1 C° (from 14.5°C to 15.5°C).

Definition of:
The kilocalorie

The kilocalorie is technically known as the 15° kilocalorie. The temperature range is specified because the energy needed to raise 1 kg of water by 1 C° varies slightly with temperature. It is a minimum between 14.5°C and 15.5°C at a pressure of 1 atm. Over the temperature range in which water is a liquid, the variation is small and can be ignored for most purposes, as we will do.

For smaller quantities, the **calorie (cal)** is sometimes used (1 kcal = 1000 cal). *One calorie is the amount of heat needed to raise the temperature of 1 g of water by 1 C° (from 14.5°C to 15.5°C)* (Fig. 11.1b).

A familiar use of the larger kilocalorie is for specifying the energy values of foods. In this context the word is usually shortened to *Calorie* (Cal). That is, people on diets really count kilocalories; a piece of cake may contain 400 Cal (kcal), or 400 000 cal. The capital C distinguishes the larger kilogram-Calorie, or kilocalorie, from the smaller gram-calorie, or calorie. They are sometimes referred to as "big Calorie" and "little calorie." (In some countries, the joule is used for food values—see •Fig. 11.2.)

•FIGURE 11.1 Units of heat (a) A kilocalorie raises the temperature of 1 kg of water by 1 C°. (b) A calorie raises the temperature of 1 g of water by 1 C°. (c) A Btu raises the temperature of 1 lb of water by 1 F°. (Not drawn to scale.)

A unit of heat that is commonly used in industry is the **British thermal unit (Btu)**. *One Btu is the amount of heat needed to raise the temperature of 1 lb of water by 1 F° (from 63°F to 64°F)* (Fig. 11.1c), and 1 Btu = 252 cal = 0.252 kcal. If you buy an air conditioner or an electric heater, you will find it rated in Btus; for example, window air conditioners range from 4000 to 25 000 Btus. This number is actually Btus *per hour* and specifies how much heat energy the unit will remove (or deliver, in the case of a heater) during that time.

The Mechanical Equivalent of Heat

The idea that heat is a transfer of energy is the result of work by many scientists on the relationship of heat and energy. Some early observations were made by Benjamin Thompson, Count Rumford, while he was supervising the boring of cannon barrels in Germany. Rumford noticed that water put into the bore of the cannon to prevent overheating during drilling boiled away and had to be frequently replenished. He did several experiments, including ones in which he tried to detect "caloric fluid" by changes in the weights of heated substances. He eventually concluded that mechanical work was responsible for the heating of the water.

This conclusion was later proven quantitatively by James Joule, the English scientist after whom the SI unit for work and energy is named. Using the apparatus shown in •Fig. 11.3, Joule demonstrated that when a given amount of mechanical work was done, the water was heated, as indicated by an increase in temperature. He found that for every 4186 J of work done, the temperature of the water rose 1 C° per kg, or that 4186 J was equivalent to 1 kcal:

$$1 \text{ kcal} = 4186 \text{ J} = 4.186 \text{ kJ}$$

This relationship is called the **mechanical equivalent of heat**. It provides a conversion factor between heat and mechanical (standard) units, such as

$$1 \text{ cal} = 4.186 \text{ J}$$

•**FIGURE 11.2 It's a joule** In Australia, diet drinks are labeled as being "low joule." In Germany, the labeling is a bit more specific: "Less than 4 kilojoules (1 kcal) in 0.3 Liter." How does this compare to our diet drinks?

Benjamin Thompson (1753–1814) was born in New England but went to England at the time of the American Revolution. He later worked in Bavaria, where he was made a count. He took the name Count Rumford after his birthplace (now known as Concord, New Hampshire).

•**FIGURE 11.3 Joule's apparatus for measuring the mechanical equivalent of heat** As the weights descend, the paddle wheels churn the water and mechanical energy or work is converted into heat energy. For every 4186 J of work done, the temperature of the water rises 1 C° per kilogram, and 4186 J is equivalent to 1 kilocalorie.

Example 11.1 ■ Stirred, Not Shaken: Mechanical Equivalent of Heat

How much mechanical work in joules would have to be done in Joule's apparatus to raise the temperature of a liter of water from room temperature (20°C) to 25°C?

Thinking It Through. We learned that temperature rises 1 C° per kilogram for every 4186 J of work. The relationship of a liter of water to mass is the key. Temperatures are assumed to be exact.

Solution. Listing the given data and what we want to find, we have

Given: $T_i = 20°C$ *Find:* W (work in joules)
        $T_f = 25°C$
        $V = 1.0 \text{ L}$

where T_i and T_f are the initial and final temperatures, respectively.

From Joule's mechanical equivalent of heat, we know how much heat energy (symbolized Q) it takes to raise the temperature of a certain amount of water 1 C°. Here, the amount of water is given in terms of its volume. However, we know from Chapter 1 that 1.0 L of water has a mass of 1.0 kg.

Since 4.186 kJ raises the temperature of 1 kg of water 1 C°, to raise the temperature of 1 kg of water by $\Delta T = T_f - T_i = 25°C - 20°C = 5 \text{ C°}$ would require five times this amount of heat energy, if we assume no heat loss. This energy is supplied by mechanical work, and

$$W = Q = 4.186 \text{ kJ/C°} \times 5 \text{ C°} = 20.93 \text{ kJ} = 2.093 \times 10^4 \text{ J}$$

Follow-up Exercise. To get an idea of the mechanical equivalent of 5.00 kcal of heat, to what height could a 100-kilogram object (of equivalent weight 220 lb) be lifted by using this amount of energy?

11.2 Specific Heat

Objectives: To (a) describe specific heat and (b) explain how the specific heats of materials are obtained from calorimetry.

Recall from Chapter 10 that when heat is added to a substance, the energy may go to increase the random molecular motion, which results in a temperature change, and also to increase the potential energy associated with the molecular bonds. Different substances have different molecular configurations and bonding patterns. Thus, if equal amounts of heat are added to equal masses of different substances, the resulting temperature changes will *not* generally be the same. For example, suppose that you have 1 kg of water and 1 kg of aluminum and add 1 kcal of heat to each. From the definition of the kilocalorie, you know that the temperature of the water will rise 1 C°. But the temperature of the aluminum will increase by 4.5 C°. The reason for this difference is that more of the added energy goes into the nontranslational ("nontemperature") part of the internal energy of the water than of the aluminum.

Note: Review Section 10.1 and Fig. 10.1.

The amount of heat (Q) required to change the temperature of a substance is proportional to the mass (m) of the substance and to the change in the temperature (ΔT). That is, $Q \propto m \, \Delta T$, or, in equation form,

$$Q = mc\,\Delta T \quad \text{or} \quad c = \frac{Q}{m\,\Delta T} \tag{11.1}$$

Definition of: Specific heat

Here $\Delta T = T_f - T_i$ is the temperature change, or the difference between final temperature (T_f) and the initial temperature (T_i), and c is the *specific heat capacity*, or simply the **specific heat**. It is characteristic of, or *specific* for, a given substance and is independent of m. Thus, the specific heat gives us an indication of a material's internal molecular configuration and bonding.*

Writing Eq. 11.1 as $c = Q/(m\,\Delta T)$ shows us that the units of specific heat are J/(kg·K) [or kcal/(kg·C°)]. The standard SI unit of specific heat is the one with the temperature in kelvins. However, the use of Celsius temperature in this unit [J/(kg·C°)] is commonly accepted because the size of a Celsius degree is the same as the size of a kelvin.

Note that the *specific heat is the amount of heat energy required to raise the temperature of 1 kg of a substance by 1 C°*. The specific heats of some common substances are given in Table 11.1. Specific heat depends somewhat on temperature (and pressure); but over normal temperature intervals, such variations can be neglected and c can be considered constant.

The greater the specific heat of a substance, the more energy must be transferred to it or taken from it to change the temperature of a given mass of it. That is, a substance with a higher specific heat has a greater capacity for heat, or accepts or yields more heat for a given temperature change (and mass).

Water has a relatively large specific heat of 4186 J/(kg·C°), or 1.00 kcal/(kg·C°). (The kilocalorie value is exactly 1.00 because the definition of the kilocalorie states that 1 kcal raises the temperature of 1 kg of water by 1 C°.)

You have been the victim of the high specific heat of water if you have ever burned your mouth on a baked potato or the cheese of a pizza. These foods have a high water content and a large heat capacity, so they don't cool off as quickly as some other foods do.

Note from Eq. 11.1 that when there is a temperature increase, ΔT is positive ($T_f > T_i$) and Q is positive. This corresponds to energy being *added to* a system. Conversely, ΔT and Q are negative when energy is *removed from* a system.

$+Q$ = heat added
$-Q$ = heat removed

---

*This may also be expressed in terms of *molar heat capacity*, in which the mass is expressed in term of moles (Section 10.3) rather than kilograms. This is common in chemical analyses, particularly when working with gases.

TABLE 11.1 Specific Heats of Various Substances (at 20°C and 1 atm)

Substance	Specific heat (c)	
	J/(kg·C°)	kcal/(kg·C°) [or cal/(g·C°)]
Air (50°C)	1050	0.25
Alcohol, ethyl	2430	0.58
Aluminum	920	0.22
Copper	390	0.093
Glass	840	0.20
Ice (−5°C)	2100	0.50
Iron or steel	460	0.11
Lead	130	0.031
Mercury	140	0.033
Soil (average)	1050	0.25
Steam (110°C)	2010	0.48
Water (15°C)	4186	1.00
Wood (average)	1680	0.40

Example 11.2 ■ Warming Up: Specific Heat

How much heat is required to raise the temperature of 0.20 kg of water from 15°C to 45°C?

Thinking It Through. This is a direct application of Eq. 11.1. All quantities are known, including the specific heat of water (Table 11.1).

Solution.

Given: $m = 0.20$ kg *Find:* Q (heat)
$\Delta T = T_f - T_i = 45°C - 15°C = 30$ C°
$c = 4186$ J/(kg·C°)
(from Table 11.1)

Using Eq. 11.1, we get

$$Q = mc\,\Delta T = (0.20 \text{ kg})[4186 \text{ J/(kg·C°)}](30 \text{ C°}) = +2.5 \times 10^4 \text{ J}$$

where the + sign indicates that heat was added to the system.

Follow-up Exercise. How much heat (in joules) would have to be removed from 0.20 kg of water at 50°C in order to lower its temperature to 20°C?

Example 11.3 ■ Cooling Down: Specific Heat Again

A half-liter of water at 30°C is cooled, with the removal of 63 kJ of heat. What is the final temperature of the water?

Thinking It Through. Recall that 1.0 L of water has a mass of 1.0 kg; hence the mass of the water is half that, or 0.50 kg. We want T_f; note that it is part of ΔT in Eq. 11.1.

Solution.

Given: $m = 0.50$ kg *Find:* T_f (final temperature)
$T_i = 30$ C°
$Q = -63$ kJ
(negative because heat energy is removed)
$c = 4.186$ kJ/(kg·C°)
(from Table 11.1)

With relatively large numbers, it can be convenient to work in kilojoules rather than the standard joules. Writing out the ΔT term of Eq. 11.1 gives

$$Q = mc\,\Delta T = mc(T_f - T_i)$$

Solving for T_f gives

$$T_f = \frac{Q}{mc} + T_i = \frac{-63\text{ kJ}}{(0.50\text{ kg})[4.186\text{ kJ}/(\text{kg}\cdot\text{C}°)]} + 30°\text{C} = -30°\text{C} + 30°\text{C} = 0°\text{C}$$

The (liquid) water is at its freezing point; the removal of more heat would cause the water to freeze. However, Eq. 11.1 *does not apply* when the temperature change (ΔT) is an interval that includes a change of phase, as you will learn in Section 11.3.

Follow-up Exercise. Suppose in this Example that 30 kJ of heat were initially removed, then 60 kJ added. What would be the final temperature of the water?

Example 11.4 ■ Heat Capacity: Different Materials

Equal masses of aluminum (Al) and copper (Cu) are at the same temperature. Which will require the greater heat to raise its temperature by a given amount, and how many times greater is this than the heat that would have to be added to the other metal?

Thinking It Through. The first part of the question can be answered from the definition and values of specific heat. The second part is more detailed, but "how many times greater" implies a factor or ratio.

Solution.

Given: $m_{Al} = m_{Cu}$
$\Delta T_{Al} = \Delta T_{Cu}$
$c_{Al} = 920\text{ J}/(\text{kg}\cdot\text{C}°)$
$c_{Cu} = 390\text{ J}/(\text{kg}\cdot\text{C}°)$
(from Table 11.1)

Find: Which material requires the greater value of Q and how many times greater it is

Since aluminum has a larger specific heat, it has a higher heat capacity, and more heat is required to raise its temperature by a given amount.

If we find $x = Q_{Al}/Q_{Cu}$, we can write $Q_{Al} = xQ_{Cu}$, or Q_{Al} is x times greater than Q_{Cu}. To find the ratio we use Eq. 11.1 for each metal and divide one by the other:

$$\frac{Q_{Al}}{Q_{Cu}} = \left(\frac{m_{Al}}{m_{Cu}}\right)\left(\frac{c_{Al}}{c_{Cu}}\right)\left(\frac{\Delta T_{Al}}{\Delta T_{Cu}}\right) = \frac{c_{Al}}{c_{Cu}}$$

where quantities that are given as being equal (mass and ΔT) have been grouped in parentheses so that it can be easily seen that they cancel out. Then,

$$Q_{Al} = \left(\frac{c_{Al}}{c_{Cu}}\right)Q_{Cu} = \left[\frac{920\text{ J}/(\text{kg}\cdot\text{C}°)}{390\text{ J}/(\text{kg}\cdot\text{C}°)}\right]Q_{Cu}$$

$$Q_{Al} = 2.36 Q_{Cu}$$

That is, 2.36 times more heat is required for the aluminum than for the copper.

Follow-up Exercise. To illustrate that the specific heat of water is relatively large, let equal amounts of heat be added to equal masses of water and mercury initially at the same temperature. Which liquid has the greater temperature change, and how many times greater?

Problem-Solving Hint

When the quantities given in a problem are equal, such as the masses and temperature intervals in Example 11.4, it is a good indication that a ratio can be used to cancel out the equal quantities. Also keep in mind that phrases such as "how many times more" imply a factor derived from a ratio.

•FIGURE 11.4 **Calorimetry apparatus** The calorimeter cup (center, with black insulating ring) goes into the larger container. The cover with the thermometer and stirrer is seen at the right. Metal shot or pieces of metal are heated in the small cup with handle that is inserted into the hole at the top of the steam generator on the tripod.

Calorimetry

Calorimetry is the quantitative measure of heat exchange, which allows us to determine the specific heats of substances. Such measurements are made by using a *calorimeter* (cal-oh-RIM-i-ter), an insulated container that allows little heat loss (ideally none). A simple laboratory calorimeter is shown in •Fig. 11.4.

The specific heat of a substance is determined by measuring the quantities in Eq. 11.2 ($Q = mc\,\Delta T$) other than c. A substance of known mass and temperature is put into a quantity of water in a calorimeter. The water is at a different temperature, usually a lower one. The principle of the conservation of energy is then applied to determine c. This procedure is sometimes called the *method of mixtures*. An example follows. Keep in mind that such heat exchange problems are just a matter of "thermal accounting." If something loses heat, something else must gain heat—that is, $\Sigma Q_i = 0$—or the conservation of energy would be violated.

Example 11.5 ■ Calorimetry: The Method of Mixtures

Students in a physics lab are to determine the specific heat of copper experimentally. They heat 0.150 kg of copper shot to 100°C and then carefully pour the hot shot into a calorimeter cup (Fig. 11.4) containing 0.200 kg of water at 20°C. The final temperature of the mixture in the cup is measured to be 25°C. If the aluminum cup has a mass of 0.045 kg, what is the specific heat of copper? (Assume that there is no heat loss to the surroundings.)

Thinking It Through. The conservation of energy is involved: The sum of heats lost = the sum of heats gained. In calorimetry problems, it is important to identify and label all of the quantities so as to keep them straight. Identification of the heat gains and losses is thus very important. You will probably use this method in the laboratory.

Solution. We will use subscripts Cu, H$_2$O, and Al to refer to the copper, water, and aluminum calorimeter cup, respectively, and the subscripts h, i, and f to refer to the temperatures of the hot metal shot, the water (and cup) initially at room temperature, and the final temperature of the system, respectively. With this notation, we have,

Given: $m_{Cu} = 0.150$ kg
$m_{H_2O} = 0.200$ kg
$c_{H_2O} = 4186$ J/(kg·C°)
  (from Table 11.1)
$m_{Al} = 0.045$ kg
$c_{Al} = 920$ J/(kg·C°)
$T_h = 100$°C, $T_i = 20$°C,
  and $T_f = 25$°C

Find: c_{Cu} (specific heat)

If no heat is lost from the system, its energy is conserved: $\Sigma Q_i = 0$, and

$$\Sigma Q_i = Q_{H_2O} + Q_{Al} + Q_{Cu} = 0$$

or

$$Q_{H_2O} + Q_{Al} = -Q_{Cu}$$

(heats gained by water and cup) = (heat lost by metal)

Substituting Eq. 11.1 ($Q = mc\,\Delta T$) for these heats, we have

$$m_{H_2O}c_{H_2O}\Delta T_{H_2O} + m_{Al}c_{Al}\Delta T_{Al} = -m_{Cu}c_{Cu}\Delta T_{Cu}$$

or

$$m_{H_2O}c_{H_2O}(T_f - T_i) + m_{Al}c_{Al}(T_f - T_i) = -m_{Cu}c_{Cu}(T_f - T_h)$$

Here the water and aluminum cup initially at T_i are heated to T_f, so $\Delta T_{H_2O} = \Delta T_{Al} = (T_f - T_i)$; and the copper initailly at T_h is cooled to T_f, so $\Delta T_{Cu} = (T_f - T_h)$. Solving for c_{Cu} gives

$$c_{Cu} = \frac{(m_{H_2O}c_{H_2O} + m_{Al}c_{Al})(T_f - T_i)}{-m_{Cu}(T_f - T_h)}$$

$$= \frac{\{(0.200 \text{ kg})[4186 \text{ J/(kg·C°)}] + (0.045 \text{ kg})[920 \text{ J/(kg·C°)}]\}(25°C - 20°C)}{-0.150 \text{ kg}(25°C - 100°C)}$$

$$= 390 \text{ J/(kg·C°)}$$

Follow-up Exercise. In actual calorimetry experiments, some heat is lost from the system. What effect would this have on the experimentally determined specific heat?

(a) Solid

11.3 Phase Changes and Latent Heat

Objectives: To (a) compare and contrast the three common phases of matter and (b) relate latent heat to phase changes.

Matter normally exists in one of three *phases*: solid, liquid, or gas (•Fig. 11.5). The phase that a substance is in depends on the substance's internal energy (as indicated by its temperature) and the pressure on it. You probably think more readily of adding or removing heat to change the phase of a substance because most of your experience with phase changes has been at normal atmospheric pressure, which is relatively constant.

In the **solid phase**, molecules are held together by attractive forces, or bonds. Simplistically, these bonds can be represented as springs (Fig. 11.5a), as was done in Section 9.1. Adding heat causes increased motion about the molecular equilibrium positions. If enough heat is added to provide sufficient energy to break the intermolecular bonds, the solid undergoes a phase change and becomes a liquid. The temperature at which this occurs is called the **melting point**. The temperature at which a liquid becomes a solid is called the **freezing point**. In general, these temperatures are the same for a given substance, but they can differ slightly.

(b) Liquid

Ice, table salt, and most metals are *crystalline* solids. That is, they have orderly molecular or atomic arrangements. Other substances, however, such as glass, are noncrystalline, or *amorphous*. Instead of melting at a particular temperature, these substances melt over a temperature range. We will be concerned primarily with substances that have definite melting points.

In the **liquid phase**, molecules of a substance are relatively free to move, and a liquid assumes the shape of its container (Fig. 11.5b). In certain liquids, there may be some locally ordered structure over several molecular lengths, giving rise to so-called liquid crystals, such as are used in LCDs (liquid crystal displays) of calculators and clocks (Chapter 24). Adding heat increases the motion of the molecules of a liquid. When the molecules have enough energy to become separated by large distances (compared to their diameters), the liquid changes to the **gaseous phase**, or **vapor phase** (Fig. 11.5c). (The distinction between a gas and a vapor will be made shortly.) This change may occur slowly by the process of evaporation or rapidly at a particular temperature called the **boiling point**. The temperature at which a gas condenses and becomes a liquid is the **condensation point**.

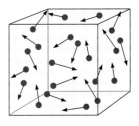

(c) Gas

•**FIGURE 11.5 Three phases of matter** (a) The molecules of a solid are held together by bonds; consequently, a solid has definite shape and volume. (b) The molecules of a liquid can move more freely, so a liquid has a definite volume and assumes the shape of its container. (c) The molecules of a gas interact weakly and are separated by relatively large distances; thus, a gas has no definite shape or volume.

Some solids, such as dry ice (solid carbon dioxide), moth balls, and certain air fresheners, change directly from the solid to the gaseous phase. This process is called **sublimation**. Like the rate of evaporation, the rate of sublimation increases with temperature. A phase change from a gas to a solid is called *deposition*. Frost, for example, is solidified water vapor deposited directly on grass, car windows, and other objects. Frost is not frozen dew, as is sometimes mistakenly assumed.

Note: Solid, liquid, and gas are sometimes referred to as states of matter rather than phases of matter, but the *state* of a system has a different meaning in physics, as you will learn in Chapter 12.

Latent Heat

In general, when heat energy is transferred to a substance, the substance's temperature increases. However, when heat that is added (or removed) causes only a phase change, the temperature of the substance does *not* change. For example, if heat is added to a quantity of ice [$c = 2100$ J/(kg·C°) or 0.50 kcal/(kg·C°)] at $-10°C$, the

Note: Keep in mind that ice can be colder than 0°C.

TABLE 11.2 Temperatures of Phase Changes and Latent Heats for Various Substances (at 1 atm)

Substance	Melting point	L_f J/kg	L_f kcal/kg	Boiling point	L_v J/kg	L_v kcal/kg
Alcohol, ethyl	−114°C	1.0×10^5	25	78°C	8.5×10^5	204
Gold	1063°C	0.645×10^5	15.4	2660°C	15.8×10^5	377
Helium*	—	—	—	−269°C	0.21×10^5	5
Lead	328°C	0.25×10^5	5.9	1744°C	8.67×10^5	207
Mercury	−39°C	0.12×10^5	2.8	357°C	2.7×10^5	65
Nitrogen	−210°C	0.26×10^5	6.1	−196°C	2.0×10^5	48
Oxygen	−219°C	0.14×10^5	3.3	−183°C	2.1×10^5	51
Tungsten	3410°C	1.8×10^5	44	5900°C	48.2×10^5	1150
Water	0°C	3.33×10^5	80	100°C	22.6×10^5	540

*Not a solid at 1 atm pressure; melting point −272°C at 26 atm.

temperature of the ice increases until it reaches its melting point of 0°C. At this point, the addition of more heat does not increase the temperature but causes the ice to melt, or change phase. (The heat must be added slowly so that the ice and melted water remain in thermal equilibrium.) Once the ice is melted, adding more heat will again cause the temperature of the water to rise. A similar situation occurs during the liquid–gas phase change at the boiling point. Adding more heat to boiling water only causes more vaporization, not a temperature increase.

From the earlier description of the molecular nature of the three phases of matter, you can see that during a phase change, the heat energy goes into the work of breaking bonds and separating molecules rather than into increasing the temperature. The heat energy involved in a phase change is called the **latent heat** (L)*, and

Latent heat and phase changes

$$Q = mL \quad \text{or} \quad L = \frac{Q}{m} \tag{11.2}$$

where m is the mass of the substance. As this equation shows, the latent heat has units of joules per kilogram (J/kg) in the SI, or kilocalories per kilogram (kcal/kg).

The latent heat for a solid–liquid phase change is called the **latent heat of fusion** (L_f), and that for a liquid–gas phase change is called the **latent heat of vaporization** (L_v). These are often referred to as simply the heat of fusion and the heat of vaporization. The latent heats of some substances, along with their melting and boiling points, are given in Table 11.2. (The latent heat for the less common solid–gas phase change is called the *latent heat of sublimation* and is symbolized by L_s.) As you might expect from the conservation of energy, the latent heat (in joules per kilogram) is the amount of energy per kilogram given up when the phase change is in the opposite direction, from liquid to solid or gas to liquid.

It is helpful to focus on the fusion and vaporization of water. A plot of temperature versus heat energy for a quantity of water is shown in •Fig. 11.6. Note that when heat is added (or removed) at the temperature of a phase change, 0°C or 100°C, the temperature remains constant. Once the phase change is complete, adding more heat causes the temperature to increase. The slopes of the phase lines in Fig. 11.6 are not all the same, which indicates that the specific heats of the various phases are not all the same. (Why?)

For water, the latent heats of fusion and vaporization are

$L_f = 3.33 \times 10^5$ J/kg (or about 80 kcal/kg)
$L_v = 22.6 \times 10^5$ J/kg (or about 540 kcal/kg)

latent heats for water

*Latent is Latin for "hidden"; Q seems to be disappearing with no change in temperature.

•FIGURE 11.6 **Temperature versus heat for water** As heat is added to the various phases of water, the temperature increases. During a phase change, however, the heat energy does the work of separating the molecules and the temperature remains constant. Note the different slopes of the phase lines, which indicate different values of specific heat. (Not drawn to scale.)

That is, 3.33×10^5 J (or 80 kcal) of heat is needed to melt 1 kg of ice at 0°C, and 22.6×10^5 J (or 540 kcal) of heat is needed to convert 1 kg of water to steam at 100°C (•Fig. 11.7). Note that the latent heat of vaporization is almost 7 times the latent heat of fusion. This indicates that more energy is needed to separate the molecules in going from water to steam than to break up the lattice structure in going from ice to water.

The term "latent" may be understood by considering a situation involving human skin. Since 22.6×10^5 J of heat energy is required to convert 1 kg of boiling water into steam, the conservation of energy tells you that when 1 kg of steam

•FIGURE 11.7 **Phase changes and latent heats** (a) At 0°C, 3.33×10^5 J must be added to 1 kg of ice or removed from 1 kg of water to change its phase. (b) At 100°C, 22.6×10^5 J must be added to 1 kg of water or removed from 1 kg of steam to change its phase.

11.3 Phase Changes and Latent Heat

condenses into water, 22.6×10^5 J of energy must be given up. As a result, burns from steam are usually more serious than those from boiling water (for equal masses). The condensing of the steam on the skin provides an additional 22.6×10^5 J/kg of heat that is seemingly hidden until contact.

Example 11.6 ■ Temperature Changes and Phase Changes: Specific Heat and Latent Heat

Heat is added to 0.500 kg of water at room temperature (20°C). How much heat in joules is required to change the water to steam at 110°C?

Thinking It Through. Three steps are involved: heating water (specific heat), a phase change (latent heat), and heating steam (specific heat).

Solution.

Given: $m = 0.500$ kg
$T_i = 20°C$, $T_f = 110°C$
$L_v = 22.6 \times 10^5$ J/kg (from Table 11.2)
(specific heats from Table 11.1)

Find: Q (total heat in joules)

The temperature intervals and the heats required in the process are shown in •Fig. 11.8. First, the water is heated to the boiling point (100°C), where a phase change occurs, and then the steam is heated to 110°C. The Q's for the various steps are

(heating water) $Q_1 = mc_w \Delta T_1 = (0.500 \text{ kg})[4186 \text{ J/(kg·C°)}](100°C - 20°C) = 1.67 \times 10^5$ J

(phase change) $Q_{L_v} = mL_v = (0.500 \text{ kg})(22.6 \times 10^5 \text{ J/kg}) = 11.3 \times 10^5$ J

(heating steam) $Q_2 = mc_s \Delta T_2 = (0.500 \text{ kg})[2010 \text{ J/(kg·C°)}](110°C - 100°C) = 0.101 \times 10^5$ J

Then, rounding, we get

$$Q_{total} = \Sigma_i Q_i = (1.7 + 11.3 + 0.1) \times 10^5 \text{ J} = +13.1 \times 10^5 \text{ J}$$

Follow-up Exercise. How many joules of heat would have to be removed to condense 1.00 kg of steam at 110°C?

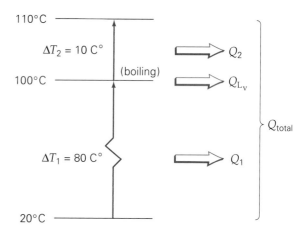

•**FIGURE 11.8 Steps in changing phases** In going from water at 20°C to steam at 110°C, three heat-adding steps are involved. See Example 11.6.

Problem-Solving Hint

Note that you must compute the latent heat at each phase change. It is a common error to use the specific heat equation with a temperature interval that includes a phase change. Also, you cannot assume a complete phase change until you have checked it numerically.

Example 11.7 ■ Water and Ice: Thermal Equilibrium

A 0.30-kilogram piece of ice at 0°C is placed in a liter of water at room temperature (20°C) in an insulated container. Assuming that no heat is lost to the container, what is the final temperature of the water?

Thinking It Through. The final water temperature will be less than room temperature because heat is removed from the water to melt the ice (latent heat). Some ice will melt, but will all of it? To find out, we compare how much latent heat would be required to melt the total mass to how much energy the water has to give up in going to 0°C. Once this is determined, then we will know how to proceed—a mixture of ice and water at 0°C or melting and warming of the ice water to a final temperature that we must determine.

Solution. We list the data as usual:

Given: $m_{ice} = 0.30$ kg
$T_{ice} = 0°C$
$V_{H_2O} = 1.0$ L, so $m_{H_2O} = 1.0$ kg
$T_{H_2O} = 20°C$
(latent heats from Table 11.2)

Find: T_f (final temperature)

We know that 1.0 L of water has a mass (m_{H_2O}) of 1.0 kg and that the water supplies the heat to melt the ice. If all of the ice melts, we could then view the system as being two masses of water at different temperatures, which come to equilibrium at some intermediate temperature.

Since there is no heat loss, it is tempting to write an expression equating the amount of heat lost by the water to the amounts of heat used to melt the ice and to warm up the ice water. But does *all* the ice melt? To melt 0.30 kg of ice requires

Note: This is a matter of "thermal accounting"—if something loses, something else gains.

$$Q_{ice} = m_{ice}L_f = (0.30 \text{ kg})(3.33 \times 10^5 \text{ J/kg}) = 1.0 \times 10^5 \text{ J}$$

Then, looking at the water, you must ask how much heat it can supply to the ice. The maximum amount would be that given up in lowering the temperature of the water to 0°C, which would be a temperature decrease of $\Delta T_{H_2O} = -20$ C°. This *maximum* amount of heat would be

$$Q_{H_2O} = m_{H_2O}c_{H_2O}\Delta T_{H_2O} = (1.0 \text{ kg})[4186 \text{ J/(kg·C°)}](-20 \text{ C°}) = -8.4 \times 10^4 \text{ J}$$

Thus, all of the ice does *not* melt, since the water does not have enough energy to do so. The final temperature of the water is therefore 0°C.

The heat given up by the water ($Q_{H_2O} = 8.4 \times 10^4$ J) will melt a mass of ice equal to

$$m_{ice} = \frac{Q_{H_2O}}{L_f} = \frac{8.4 \times 10^4 \text{ J}}{3.33 \times 10^5 \text{ J/kg}} = 0.25 \text{ kg}$$

Thus, the final mixture would be 0.30 kg − 0.25 kg = 0.05 kg of ice in thermal equilibrium with 1.25 kg (or 1.25 L) of water at $T_f = 0°C$.

Follow-up Exercise. In this Example, what would the initial temperature of the water have to be to melt all of the ice but heat up no further?

Note that the freezing and boiling points of 0°C and 100°C, respectively, for water apply at 1 atm of pressure. The phase-change temperatures vary with pressure. For example, the boiling point of water decreases with decreasing pressure. In fact, a container of water in a vacuum chamber will boil at room temperature. The cooling effect of the boiling (the removal of latent heat) will eventually cause the remaining water to freeze if the vacuum is maintained.

At high altitudes, where there is lower atmospheric pressure, the boiling point of water is lowered. For example, at Pike's Peak, Colorado, at an elevation of about 4300 m, the atmospheric pressure is about 600 torr, and water boils at about 94°C rather than at 100°C. The lower temperature lengthens the cooking time of food. A pressure cooker can be used to reduce the cooking time (at Pike's Peak or at sea level)—by increasing the pressure, a pressure cooker raises the boiling point.

Insight

Phase Changes and Ice Skating

It was once thought that the lowering of the freezing point of water by pressure provided the mechanism that makes ice skating possible. Supposedly, the pressure of the narrow skate blade on the ice would lower the melting point below the ambient temperature. The skater would thus glide on a thin film of water, which quickly refroze when the pressure was removed. However, for water, a pressure of about 120 atm is needed to lower the freezing point from 0°C to −1°C.

A typical skate blade has an area of about 27 cm × 0.40 cm = 11 cm² = 11×10^{-4} m². For a 70-kilogram skater with all the weight on one blade, the pressure would be $p = F/A = mg/A = (70 \text{ kg})(9.8 \text{ m/s})(11 \times 10^{-4} \text{ m}^2) = 6.2 \times 10^5$ N/m², or approximately 6 atm. This is hardly enough, and the outdoor ice skating temperature is usually well below −1°C.

Frictional heating between the skate blade and the ice does contribute to melting. For high-friction surfaces, such as skis on snow, this is the main mechanism. However, for ice skating another factor contributes to the low coefficient of friction of ice. This factor is called *surface melting*, proposed by the English scientist Michael Faraday (1791–1867; Chapter 20). He suggested that a thin layer of liquid normally exists on the surface of a solid even at temperatures well below the solid's melting point. Modern techniques have shown this to be the case for most solids. For ice, the thickness of the water film is about 40 nm near 0°C and about 0.50 nm near −35°C. (Recall that 1 nm = 10^{-9} m.)

Conversely, the freezing point of water decreases with increasing pressure. This inverse relationship is characteristic of only a very few substances, including water (Section 10.4), that expand when they freeze. Pressure and ice skating are discussed in the Insight on this page.

Evaporation

The **evaporation** of water from an open container becomes evident only after a relatively long period of time. This phenomenon can be explained in terms of the kinetic theory of gases (Section 10.5). The molecules in a liquid are in motion at different speeds. A faster-moving molecule near the surface may momentarily leave the liquid. If its speed is not too large, the molecule will return to the liquid because of the attractive forces exerted by the other molecules. Occasionally, however, a molecule has a large enough speed to leave the liquid entirely and become part of the air. The higher the temperature of the liquid, the more likely this is to occur.

The escaping molecules take their energy with them. Since those molecules with greater-than-average energy are the ones most likely to escape, the energy and temperature of the remaining liquid will be reduced. Thus, *evaporation is a cooling process* for the object from which the molecules escape. You have probably noticed this when drying off after a bath or shower. About 2500 kJ is needed to evaporate 1 kg of water from the skin. This energy requirement can be estimated by considering a different process. The amount of heat per kilogram (Q/m) needed to raise the temperature of water from 35°C (approximately skin temperature) to 100°C ($\Delta T = 100°C - 35°C = 65$ C°) is $c \Delta T$ (since $Q = mc \Delta T$). Adding the latent heat of vaporization, which is also the heat per unit mass ($Q = mL_v$ and $L_v = Q/m$), gives

$$\frac{Q}{m} = c \Delta T + L_v = [4186 \text{ J}/(\text{kg} \cdot \text{C°})](65°C) + (22.6 \times 10^5 \text{ J/kg}) \approx 2500 \text{ kJ/kg}$$

Evaporation will never lead to heating to the boiling point (100°C), but this estimate gives the maximum possible cooling effect.

Although evaporation is a relatively slow process, it is often important in preventing our bodies from overheating. Usually, radiation and conduction (discussed in Section 11.4) are sufficient to maintain a rate of heat loss that keeps us comfortable, given the temperature difference between our bodies and our surroundings. However, when the air gets hot and the temperature difference narrows (or disappears), we start to perspire. The evaporation of perspiration helps cool our bodies. On a hot summer day, a person may stand in front of a fan and remark how cool the blowing air feels. But the fan is merely blowing hot air from one place to another. The air feels cool because its flow promotes evaporation, which removes heat energy. Of course, evaporation depends on the humidity (the amount of moisture already in the air). We feel less comfortable on hot humid days because evaporation is reduced. (Why?)

11.4 Heat Transfer

Objectives: To (a) describe the three methods of heat transfer and (b) give practical and/or environmental examples of each.

Since heat is defined as energy in transit, how the transfer takes place is important. Heat moves from place to place by one of three mechanisms: conduction, convection, or radiation.

Conduction

You can keep a pot of coffee hot on a hot stove because heat is conducted through the coffee pot from the burner. The process of **conduction** is visualized as resulting from molecular interactions. Molecules in one part of a body at a higher temperature vibrate faster. They collide with, and transfer some of their energy to, the less energetic molecules located toward the cooler part of the body. In this way, energy is conductively transferred from a higher-temperature region to a lower-temperature region—transfer as a result of a temperature difference.

Solids can be divided into two general categories: metals and nonmetals. *Metals* are good conductors of heat, or **thermal conductors.** Modern theory views metals as having a large number of electrons that are free to move around (not permanently bound to a particular molecule or atom). These free electrons are believed to be primarily responsible for the heat conduction of metals. *Nonmetals*, such as wood or cloth, have relatively few free electrons and are poor heat conductors. A poor heat conductor is called a **thermal insulator.**

In general, the ability of a substance to conduct heat depends on the substance's phase. Gases are poor thermal conductors: Their molecules are relatively far apart, and collisions are therefore infrequent. Liquids are better thermal conductors than gases are because their molecules are closer together and can interact more readily.

Heat conduction can be described quantitatively as the time rate of heat flow ($\Delta Q/\Delta t$) in a material for a given temperature difference (ΔT), as illustrated in •Fig. 11.9. It has been established through experiments that the rate of heat flow through a substance depends on the temperature difference between its boundaries. Heat conduction also depends on the size and shape of the object. In our analysis of heat flow, we will use a uniform slab of the substance.

A moment's thought should convince you that the heat flow through a slab of material is directly proportional to its surface area (A) and inversely proportional to its thickness (d). That is,

$$\frac{\Delta Q}{\Delta t} \propto \frac{A\,\Delta T}{d}$$

The term $\Delta T/d$ is called the *thermal gradient* (the change in temperature per unit of length). Using a constant of proportionality allows us to write the relation as an equation:

$$\frac{\Delta Q}{\Delta t} = \frac{kA\,\Delta T}{d} \qquad (conduction\ only) \qquad (11.3)$$

The constant k, called the **thermal conductivity,** characterizes the heat-conducting ability of a material. The greater the value of k for a material, the more rapidly it will conduct heat. The units of k are joules per meter-second-Celsius degree [J/(m·s·C°),

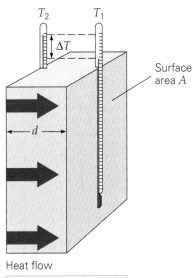

•**FIGURE 11.9 Thermal conduction** Heat conduction is characterized by the time rate of heat flow ($\Delta Q/\Delta t$) in a material with a temperature difference ΔT. For a slab of material, $\Delta Q/\Delta t$ is directly proportional to the cross-sectional area (A) and the thermal conductivity of the material (k); it is inversely proportional to the slab thickness (d).

TABLE 11.3 Thermal Conductivities of Some Substances

Substance	Thermal conductivity (k) J/(m·s·C°)	kcal/(m·s·C°)
Metals		
Aluminum	240	5.73×10^{-2}
Copper	390	9.32×10^{-2}
Iron and steel	46	1.1×10^{-2}
Silver	420	10×10^{-2}
Liquids		
Transformer oil	0.18	4.3×10^{-5}
Water	0.57	14×10^{-5}
Gases		
Air	0.024	0.57×10^{-5}
Hydrogen	0.17	4.1×10^{-5}
Oxygen	0.024	0.57×10^{-5}
Other materials		
Brick	0.71	17×10^{-5}
Concrete	1.3	31×10^{-5}
Cotton	0.075	1.8×10^{-5}
Fiberboard	0.059	1.4×10^{-5}
Floor tile	0.67	16×10^{-5}
Glass (typical)	0.84	20×10^{-5}
Glass wool	0.042	1.0×10^{-5}
Human tissue (average)	0.20	4.8×10^{-5}
Ice	2.2	53×10^{-5}
Styrofoam	0.042	1.0×10^{-5}
Wood, oak	0.15	3.6×10^{-5}
Wood, pine	0.12	2.9×10^{-5}
Vacuum	0	0

or kcal/(m·s·C°)]. The thermal conductivities of various substances are listed in Table 11.3. These values vary slightly over different temperature ranges but can be considered to be constant over normal temperature ranges and differences.

Compare the relatively large thermal conductivities of the good thermal conductors, the metals, with the relatively small thermal conductivities of some good thermal insulators, such as Styrofoam and wood. Some cooking pots have copper coatings on the bottoms (•Fig. 11.10). Being a good conductor of heat, the copper promotes the distribution of heat over the bottom of a pot for even cooking. Conversely, plastic foams are good insulators mainly because they contain pockets of air. Recall that gases are poor conductors, and note the low thermal conductivity of air in Table 11.3.

•FIGURE 11.10 **Copper-bottomed pots** A layer of copper is found on the bottoms of some pots and saucepans. The high heat conductivity of this metal ensures the rapid and even spread of heat from the burner, reducing the likelihood of local "hot spots." (Contrast the chapter-opening photograph.)

Example 11.8 ■ Insulation: Controlling Thermal Conductivity

A room with a pine ceiling that measures 3.0 m × 5.0 m × 2.0 cm thick has a layer of glass wool insulation above it that is 6.0 cm thick (•Fig. 11.11a). On a cold day, the temperature inside the room is 20°C and the temperature in the attic above the room is 8.0°C. Assuming that the temperatures remain constant with a steady heat flow, how much energy does the layer of insulation save in 1.0 h? Assume losses are due to conduction only.

•**FIGURE 11.11 Insulation and thermal conductivity** (a, b) Attics should be insulated to prevent the loss of heat through conduction. See Example 11.8. (c) This thermogram of a house allows us to visualize heat loss. (Blue represents the areas that lose the least heat; white, pink, and red, the most.) What conclusions can you draw from it? (Compare Fig. 11.15.)

Thinking It Through. Here we have two materials, so we must consider Eq. 11.3 for two different thermal conductivities (k). We want to find $\Delta Q/\Delta t$ for the combination so that we can get ΔQ for $\Delta t = 1.0$ h. The situation is a bit complicated because the heat flows through two materials. But we know that at a steady rate, the heat flows must be the same through both. (Why?) To find the energy saved in 1.0 h, we need to compute how much heat is conducted in this time with and without the layer of insulation. Then we will derive a general equation for multimaterial insulations.

Solution. Computing some of the quantities and making conversions as we list the data, we have

Given: $A = 3.0$ m $\times$ 5.0 m $= 15$ m^2 *Find:* Energy saved
$d_1 = 6.0$ cm $= 0.060$ m in 1.0 h
$d_2 = 2.0$ cm $= 0.020$ m
$\Delta T = T_2 - T_1 = 20°\text{C} - 8.0°\text{C} = 12$ C°
$\Delta t = 1.0$ h $= 3.6 \times 10^3$ s
$k_1 = 0.042$ J/(m·s·C°) (glass wool) $\Big\}$ (from Table 11.3)
$k_2 = 0.12$ J/(m·s·C°) (wood)

(In working such problems with lots of given quantities, it is especially important to label all the data correctly.)

First, let's consider how much heat would be conducted in 1.0 h through the wooden ceiling with no insulation present. Since we know Δt, we can rearrange Eq. 11.3 to find ΔQ_c (heat conducted through the wooden ceiling):

$$\Delta Q_c = \left(\frac{k_2 A \Delta T}{d_2}\right)\Delta t$$

$$= \left\{\frac{[0.12 \text{ J/(m·s·C°)}](15 \text{ m}^2)(12 \text{ C°})}{0.020 \text{ m}}\right\}(3.6 \times 10^3 \text{ s}) = 3.9 \times 10^6 \text{ J}$$

Now we need to find the heat conducted through the ceiling *and* the insulation layer together. Let T be the temperature at the interface of the materials and T_2 and T_1 be the warmer and cooler temperatures, respectively (Fig. 11.11b). Then

$$\frac{\Delta Q_1}{\Delta t} = \frac{k_1 A(T - T_1)}{d_1} \quad \text{and} \quad \frac{\Delta Q_2}{\Delta t} = \frac{k_2 A(T_2 - T)}{d_2}$$

We don't know T, but when the conduction is steady, the flow rates are the same for both; that is, $\Delta Q_1/\Delta t = \Delta Q_2/\Delta t$, or

$$\frac{k_1 A(T - T_1)}{d_1} = \frac{k_2 A(T_2 - T)}{d_2}$$

The A's cancel, and solving for T gives

$$T = \frac{k_1 d_2 T_1 + k_2 d_1 T_2}{k_1 d_2 + k_2 d_1}$$

Substituting into either of the flow rate equations and rearranging, we get a general expression for $\Delta Q/\Delta t$ for the combined layers:

$$\frac{\Delta Q}{\Delta t} = \frac{A(T_2 - T_1)}{(d_1/k_1) + (d_2/k_2)} \tag{11.4*}$$

$$= \frac{(15 \text{ m}^2)(12 \text{ C}°)}{[0.060 \text{ m}/0.042 \text{ J}/(\text{m·s·C°})] + [0.020 \text{ m}/0.12 \text{ J}/(\text{m·s·C°})]} = 1.1 \times 10^2 \text{ J/s}$$

In $1.0 \text{ h} = 3.6 \times 10^3 \text{ s}$,

$$\Delta Q = (1.1 \times 10^2 \text{ J/s}) \Delta t = (1.1 \times 10^2 \text{ J/s})(3.6 \times 10^3 \text{ s}) = 4.0 \times 10^5 \text{ J}$$

Thus, the heat loss is decreased by

$$\Delta Q_c - \Delta Q = (39 \times 10^5 \text{ J}) - (4.0 \times 10^5 \text{ J}) = 35 \times 10^5 \text{ J}$$

Ideally, this represents a savings of

$$\frac{35 \times 10^5 \text{ J}}{39 \times 10^5 \text{ J}} (\times 100\%) = 90\%$$

Follow-up Exercise. Suppose that in this Example there is an upstairs room (in the attic) with a pine floor, similar to the lower ceiling, covered with floor tile 0.50 cm thick. What would be the flow rate of heat between floors in this case if the upstairs room temperature is 15°C?

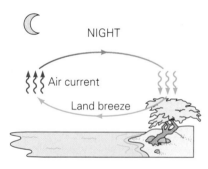

•**FIGURE 11.12 Convection cycles** During the day, natural convection cycles give rise to sea breezes near large bodies of water. At night, the pattern of circulation is reversed and land breezes blow. The temperature differences between land and water are the result of their different specific heats. Water has a much greater specific heat, so the land warms up more quickly during the day. At night, the land cools more quickly, while the water remains warmer because of its greater heat capacity.

Convection

In general, compared to solids, liquids and gases are not good thermal conductors. However, the mobility of molecules in fluids permits heat transfer by another process—convection. **Convection** is heat transfer as a result of mass transfer, which can be natural or forced.

Natural convection cycles occur in liquids and gases. For example, when cold water is in contact with a hot object, such as the bottom of a pot on a stove, the object transfers heat to the water adjacent to the pot by conduction. But the water carries the heat away with it by natural convection, and a cycle is set up in which upper, cold water replaces the rising warm water. Such cycles are important in atmospheric processes, as illustrated in •Fig. 11.12. During the day, the ground heats up more quickly than do large bodies of water, as you may have noticed if you have been to the beach. This occurs both because the water has a greater specific heat than the land and because convection currents disperse the absorbed heat throughout the great volume of water. The air in contact with the warm ground is heated by conduction. That air expands, becoming less dense than in surrounding cooler air. As a result, the warm air rises (air currents) and, to fill the space, other air moves horizontally (winds)—creating a sea breeze near a large body of water. Cooler air descends, and a thermal convection cycle is set up, which transfers heat away from the land. At night, the ground loses its heat more quickly than the water, and the water surface is warmer than the land. As a result, the cycle is reversed.

*This equation can be extended to any number of layers or slabs of materials by adding additional d/k terms to the denominator: $\Delta Q/\Delta t = A(T_2 - T_1)/\Sigma(d_i/k_i)$.

You can see the evidence of convection currents in the air above a hot road surface in the summer and in transparent liquids, such as heated water in a glass container. This is because regions of different temperatures have different densities, which cause a bending, or refraction, of the light that passes through the regions (Chapter 22).

In *forced convection*, the medium of heat transfer is moved mechanically. Note that we can have heat transfer without a temperature difference. In fact, we can have energy transferred from a low-temperature region to a high-temperature region, as in the case of the forced convection of a refrigerator coolant removing energy from the inside of the refrigerator. (The circulating coolant carries heat energy from the inside of the refrigerator, and this heat is given up to the environment, as we will see in Section 12.4.)

Other common examples of forced convection systems are forced-air heating systems in homes (•Fig. 11.13), the human circulatory system, and the cooling system of an automobile engine. The human body does not use all of the energy obtained from food; a great deal is lost. (There's usually a temperature difference between your body and your surroundings.) So that body temperature will stay normal, the internally generated heat energy is transferred close to the skin surface by blood circulation. From the skin, that energy is conducted to the air or lost by radiation (the other heat-transfer mechanism, to be discussed shortly).

Water or some other coolant is circulated (pumped) through most automobile cooling systems (some engines are air-cooled). The fluid medium carries heat to the radiator (a heat exchanger), where forced air flow produced by the fan carries it away. The *radiator* of an automobile is actually misnamed—most of the heat is transferred from it by forced convection rather than by radiation.

•**FIGURE 11.13 Forced convection** Houses are commonly heated by forced convection. Registers or gratings in the floors or walls allow heated air to enter and cooler air to return to the heat source. In older homes, natural convection was (and still is) used.

Conceptual Example 11.9

Foam Insulation: Better Than Air?

Polymer foam insulation is sometimes blown into the space between the inner and outer walls of a house. Since air is a good thermal insulator, why is the foam insulation needed? (a) To prevent loss of heat by conduction, (b) to prevent loss of heat by convection, (c) for fireproofing.

Reasoning and Answer. Polymer foams will generally burn, so (c) isn't likely to be the answer. Air is a poor thermal conductor, even poorer than plastic foam (Styrofoam, Table 11.3), so the answer can't be (a). However, as a gas, it is subject to convection within the wall space. In the winter, the air near the warm inner wall is heated and rises, thus setting up a convection cycle in the space and transferring heat to the cold outer wall. In the summer, with air conditioning, the heat-loss cycle is reversed. Foam blocks such convection cycles. Hence, the answer is (b).

Follow-up Exercise. Thermal underwear and thermal blankets are loosely knitted with lots of small holes. Wouldn't they be more effective if the material were denser?

Radiation

Conduction and convection require some material as a transport medium. The third mechanism of heat transfer needs no medium; it is called **radiation**, which refers to energy transfer by electromagnetic waves (Chapter 20). Heat is transferred to the Earth from the Sun through empty space by radiation. Visible light and other forms of electromagnetic radiation are commonly referred to as radiant energy.

You have experienced heat transfer by radiation if you've ever stood near an open fire (•Fig. 11.14). You can feel the heat on your exposed hands and face. This heat transfer is not due to convection or conduction, since heated air rises and air is a poor conductor. Visible radiation is emitted from the burning material, but most of the heating effect comes from the invisible **infrared radiation** emitted by

•**FIGURE 11.14 Heating by conduction, convection, and radiation** The hands at left are warmed by the convection of rising hot air (and some radiation). The gloved hand at upper right is warmed by conduction. The hands at lower right are warmed by radiation.

Note: Resonance is discussed in Section 13.5.

the glowing embers or coals. You feel this radiation because it is absorbed by water molecules in your skin. (Body tissue is about 85% water.) The water molecule has an internal vibration whose frequency coincides with that of infrared radiation, which is therefore absorbed readily. (This effect is called *resonance absorption*. The electromagnetic wave drives the molecular vibration, and energy is transferred to the molecule, somewhat like pushing a swing.)

Infrared radiation is sometimes referred to as "heat rays." You may have noticed the red infrared lamps used to keep food warm in cafeterias. Heat transfer by infrared radiation is also important in maintaining our planet's warmth by a mechanism known as the *greenhouse effect*. This important environmental topic and another common example of heating by radiation absorption are discussed in the Insights on pp. 379 and 380.

Although infrared radiation is invisible to the human eye, it can be detected by other means. The frequency of the infrared radiation is proportional to the temperature of its source. This relationship is the basis for infrared thermometers, which, using infrared detectors, can measure temperature remotely. Also, some cameras can use special infrared film. A picture taken with this film is an image consisting of contrasting light and dark areas corresponding to regions of higher and lower temperatures. Special instruments that apply such *thermography* are used in industry and medicine; the images they produce are called *thermograms* (•Fig. 11.15).

A new application of thermograms is for security. The system consists of an infrared camera and a computer that identifies an individual by means of the unique heat pattern emitted by the facial blood vessels. The camera takes a picture of the radiation from a person's face, which is compared with an earlier image stored in the computer memory. It is reported that the system can even distinguish between identical twins, whose facial features are slightly different. Also, changes in body temperature from weather conditions or a fever do not affect the identification, as the relative patterns of radiation remain the same.

The rate at which an object radiates energy has been found to be proportional to the fourth power of the absolute temperature (T^4). This relationship is expressed in an equation known as **Stefan's law**:

$$P = \frac{\Delta Q}{\Delta t} = \sigma A e T^4 \qquad (11.5)$$

where P is the power radiated in watts (W) or joules per second (J/s). The symbol σ (the Greek letter sigma) is the *Stefan-Boltzmann constant*: $\sigma = 5.67 \times 10^{-8}$ W/(m²·K⁴). The radiated power is also proportional to the surface area (A) of the object. The **emissivity** (e) is a unitless number between 0 and 1 that is characteristic of the

•**FIGURE 11.15 Thermography** A thermogram of a man holding his arm above his head. The skin temperature varies about 1 C° for each color, from white (hottest) to black (coldest). Note the high temperature of the armpit, due to the proximity of blood vessels to the skin, and the cooler nose and hair.

Insight

The Greenhouse Effect

You may have heard or read about the greenhouse effect in connection with global warming. The *greenhouse effect* helps regulate the Earth's long-term average temperature, which has been fairly constant. A portion of the solar radiation we receive reaches and warms the Earth's surface. The Earth in turn reradiates energy in the form of infrared radiation. The balance between absorption and radiation is a major factor in stabilizing the Earth's temperature.

This balance is affected by the concentration of *greenhouse gases*—primarily water vapor and carbon dioxide (CO_2)—in the atmosphere. As the infrared radiation passes through the atmosphere, some of the radiation is absorbed by the greenhouse gases. These gases are selective absorbers: They absorb radiation at certain wavelengths but not at others (Fig. 1a).

If terrestrial infrared radiation is absorbed, the atmosphere warms, warming the Earth (heat transfer by radiation). This rise in surface temperature causes a shift in the wavelength of the emitted radiation. The wavelength is eventually shifted to a "window" in the absorption spectrum where little or no absorption takes place, and the terrestrial radiation passes through the atmosphere into space. Thus, the Earth loses energy, and its surface temperature decreases.

But with a temperature decrease, the terrestrial radiation shifts to a longer wavelength and is again absorbed by the greenhouse gases. We have a turning on and off, so to speak, similar to the action of a thermostat. Hence, the selective absorption of atmospheric gases plays an important role in maintaining the Earth's average temperature

Why is this phenomenon called the greenhouse effect? The reason is that the atmosphere functions like the glass in a greenhouse. That is, the absorption and transmission properties of glass are similar to those of the atmospheric greenhouse gases—in general, visible radiation is transmitted, but infrared radiation is selectively absorbed (Fig. 1b). We have all observed the warming effect of sunlight passing through glass, for example, in a closed car on a sunny, cold day. Similarly, a greenhouse heats up by absorbing sunlight and trapping the reradiated infrared radiation. Thus, it is quite warm inside on a sunny day, even in winter. (The glass enclosure also keeps warm air from escaping upward, as it does in the atmosphere. In practice, this elimination of heat loss by convection is the chief factor in maintaining an elevated temperature. The temperature of a greenhouse in the summer is controlled by painting the glass panels white so that some of the sunlight is reflected and by opening panels to let some hot air escape.)

The problem is that on Earth, human activities may accentuate greenhouse warming. With the combustion of fuels for heating and industrial processes, vast amounts of CO_2 and other greenhouse gases are vented into the atmosphere. There is concern that the result of this trend will be—or already is—*global warming*: an increase in the Earth's average temperature that could dramatically affect the environment. For example, the climate in many parts of the globe would be altered, with effects on agricultural production and world food supplies that are very difficult to predict. A general rise in temperature could cause partial melting of the polar ice caps. Sea levels would rise, flooding low-lying regions and endangering coastal ports and population centers.

In an actual greenhouse, the visible portion of the sunlight is transmitted through the glass, but the short-wavelength ultraviolet radiation is absorbed. The ultraviolet spectrum from 280 nm to 400 nm is divided into two regions, called UV-A (320–400 nm) and UV-B (280–320 nm). Ordinary window glass does not appreciably transmit UV-B radiation, which is the primary cause of suntans and sunburns (Section 20.4). Hence, although a considerable warming effect occurs through glass from visible radiation, you do not receive a tan or severe sunburn through ordinary window glass because of UV-B absorption. Slight reddening may occur for sensitive skin from the UV-A radiation that is transmitted.

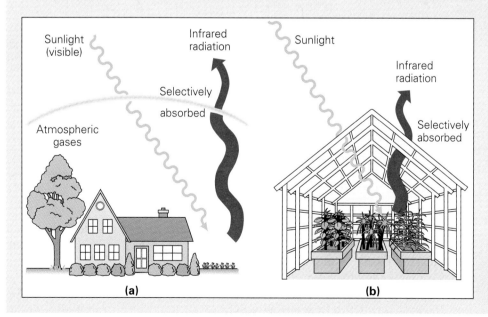

FIGURE 1 The greenhouse effect (a) The greenhouse gases of the atmosphere, particularly water vapor and carbon dioxide, are selective absorbers with absorption properties similar to those of glass used in greenhouses. Visible light is transmitted and heats the surface, while some of the infrared radiation that is re-emitted is absorbed. (b) A greenhouse operates in a similar way.

Insight

Heat Transfer by Radiation: The Microwave Oven

The microwave oven quickly became a common kitchen appliance. It is both time-saving and energy-saving, since the oven doesn't have to be warmed up as does a conventional oven. The principle of operation of the microwave oven is heat transfer by radiation.

Microwaves are a form of electromagnetic radiation; they have a frequency range just below that of infrared radiation (Chapter 21). Like infrared radiation, microwaves are absorbed chiefly by water molecules (in a molecular resonance). Fats and sugars also absorb mcrowaves.

In a microwave oven, the microwaves are generated electronically and distributed by reflection from a metal stirrer or fan and off the metal walls (Fig. 1). Because the walls reflect the radiant energy, they do not get hot.

Microwaves pass through plastic wrap, glass, or dishes made of other "microwave-safe" materials and are absorbed by water molecules in the food, producing rapid heating. (Metal utensils or objects cannot be used in microwave ovens because the microwaves can dislodge electrons from metals, causing sparking and possibly damage.) The microwaves do not penetrate the food completely but are absorbed within 2 to 3 cm of the surface. Heat is then conducted to the interior of the food, just as it is in conventional oven heating. This is why it is advisable to let large items or portions sit for a time after the microwave oven has shut off, so that they will be warmed or cooked throughout.

Since microwaves could be absorbed by water molecules in the skin, causing burns, microwave ovens have several important safety features. The door is tight-fitting so that microwaves cannot leak out. The glass in the door is fitted with a metal shield that has small holes through which you can view the food without opening the door. Microwaves are reflected by this shield and prevented from coming through the glass (essentially, the waves are larger than the holes). Also, there is a mechanism that automatically shuts off the oven when the door is opened. In fact, the oven cannot be turned on when the door is open.

You may be hearing more about microwaves soon. Developmental work is being done on microwave clothes dryers, and small prototype versions are now available.

FIGURE 1 Heat transfer by radiation In a microwave oven, microwaves (a type of electromagnetic radiation) are absorbed by molecules, chiefly those of water (and to a lesser extent, fats and sugars), raising their temperature.

---

material. Dark surfaces have emissivities close to 1, and shiny surfaces have emissivities close to 0. The emissivity of human skin is about 0.70.

Dark surfaces are not only better emitters of radiation, they are also good absorbers. In general, *a good absorber is also a good emitter*. That is, a body that is a good absorber of certain wavelengths of radiation is a good emitter of those same wavelengths. An ideal, or perfect, absorber (and emitter) is referred to as a **blackbody** ($e = 1.0$). Shiny surfaces are poor absorbers, since most of the incident radiation is reflected. This fact can be demonstrated easily as shown in •Fig. 11.16. (Can you see why it is better to wear light-colored clothes in the summer and dark-colored clothes in the winter?)

When an object is in thermal equilibrium with its surroundings, its temperature is constant; thus, it must be emitting and absorbing radiation at the same rate. If the temperatures of the object and its surroundings are different, there must be a net flow of radiant energy. If an object is at a temperature T and its surroundings are at a temperature T_s, the net rate of energy loss or gain per unit time (power) is given by

$$P_{net} = \sigma A e (T_s^4 - T^4) \tag{11.6}$$

Note that if T_s is less than T, then P (or $\Delta Q / \Delta t$) will be negative, indicating a net energy loss. *Keep in mind that the temperatures used in calculating radiated power are the absolute temperatures in kelvins.*

You may have noticed in Chapter 10 that we defined *heat* as the net energy transfer because of a temperature difference. The word *net* here is important. It is possible to have energy transfer between an object and its surroundings, or be-

•**FIGURE 11.16 Good absorber** Black objects are generally good absorbers. The bulb of the thermometer on the right has been painted black. Note the difference in the temperature readings.

tween objects, at the same temperature. Note that if $T_s = T$ (no temperature difference) in Eq. 11.6, there is a continuous exchange of radiant energy (Eq. 11.5 still holds), but there is no *net* change of internal energy. Suppose you had two identical objects relatively near each other with the same temperatures and emissivities. They would constantly be exchanging energy via radiation, even though there is no temperature difference. But on the average, there would be no *net* energy transfer between them.

Example 11.10 ■ Body Heat: Radiant Heat Transfer

Suppose that your skin has an emissivity of 0.70 and that its exposed area is 0.27 m². How much net energy will be radiated per second from this area if the air temperature is 20°C? Assume your skin temperature to be the same as normal body temperature, 37°C. (The average skin temperature is actually less. Why?)

Thinking It Through. Everything is given for us to find P_{net} from Eq. 11.6. We must remember to work in kelvins.

Solution.

Given: $T_s = 20°C + 273 = 293$ K   Find: P_{net} (net power)
$T = 37°C + 273 = 310$ K
$e = 0.70$
$A = 0.27$ m²
$\sigma = 5.67 \times 10^{-8}$ W/(m²·K⁴) (known)

Using Eq. 11.6 directly, we get

$$P_{net} = \sigma A e (T_s^4 - T^4)$$
$$= [5.67 \times 10^{-8} \text{ W/(m}^2\cdot\text{K}^4)](0.27 \text{ m}^2)(0.70)[(293 \text{ K})^4 - (310 \text{ K})^4]$$
$$= -20 \text{ W (or } -20 \text{ J/s)}$$

Since $P_{net} = \Delta Q/\Delta t$, 20 J of energy is radiated, or *lost* (as indicated by the minus sign), each second.

Follow-up Exercise. A piece of cake contains 100 Calories. If all of this energy were converted to body heat, how long would it supply the radiative heat loss computed in this Example?

Problem-Solving Hint

Note that in Example 11.10 the fourth powers of the temperatures were found first and then their difference. It is *not* correct to find the temperature difference and then raise it to the fourth power: $T_s^4 - T^4 \neq (T_s - T)^4$.

Let's consider a thoughtful example of heat transfer.

Conceptual Example 11.11

Solar Panels: Reducing the Heat Transfer

Solar panels are used to collect solar energy to heat water, which may then be used directly to heat a home at night. The panel boxes have black interiors (why?) through which the piping runs to carry the water, and the top side is covered with glass. However, ordinary glass absorbs most of the Sun's ultraviolet radiation. This absorption reduces the heating effect. Wouldn't it be better to leave the glass off the panel boxes?

Reasoning and Answer. Some energy is absorbed by the glass, but the glass serves a useful purpose and saves a lot more energy than it absorbs. As the black interior and piping of the panel box heat up, there could be heat loss by radiation (infrared) and convection. The glass prevents this via the greenhouse effect. It absorbs much of the infrared radiation and keeps the convection inside the solar panel box. (See the Insight on the greenhouse effect, p. 379.)

Follow-up Exercise. Is there any practical reason for window drapes (other than privacy)?

Let's look at a few more real-life examples of heat transfer. In the spring, a late frost could kill the buds on fruit trees in an orchard. To reduce heat transfer, some growers spray water on the trees to form ice before a hard frost occurs. Using ice to save buds? Ice is a relatively poor (and inexpensive) conductor of heat, so it has an insulating effect. It will maintain the bud temperature at 0°C, not going below, and therefore protects the buds.

Another method to protect orchards from freezing is the use of smudge pots, containers in which material is burned to create a dense cloud of combustion products. At night, when the Sun-warmed ground cools off by radiation, the cloud absorbs this heat and reradiates it back to the ground. Thus the ground takes longer to cool, hopefully without reaching freezing temperatures before the Sun comes up. (Recall that frost is the direct condensation of water vapor in the air to ice—not frozen dew.)

Lastly, check out •Fig. 11.17. Why would anyone wear a dark robe in the desert? We learned that dark objects absorb radiation (Fig. 11.16). Wouldn't a white robe be better? A dark robe definitely absorbs more radiant energy and warms the air inside near the body. But note that the robe is open at the bottom. The warm air rises (by buoyancy) and exits at the neck area, and outside cooler air enters the robe at the bottom—a natural-convection air circulation!

•**FIGURE 11.17 A dark robe in the desert?** Dark objects absorb more radiation than do lighter ones, and they are hotter. What's going on here?

Chapter Review

Important Terms

heat   361
kilocalorie (kcal)   361
calorie (cal)   361
British thermal unit (Btu)   362
mechanical equivalent of heat   362
specific heat   363
solid phase   367
melting point   367
freezing point   367
liquid phase   367

gaseous phase (vapor phase)   367
boiling point   367
condensation point   367
sublimation   367
latent heat   368
latent heat of fusion   368
latent heat of vaporization   368
evaporation   372
conduction   373

thermal conductor   373
thermal insulator   373
thermal conductivity   373
convection   376
radiation   377
infrared radiation   377
Stefan's law   378
emissivity   378
blackbody   380

Important Concepts

- As a form of energy in transit, heat has the SI unit of joule (J). Common nonstandard units are the kilocalorie (kcal) and the British thermal unit (Btu).
- The mechanical equivalent of heat relates joules to kilocalories.
- The amount of heat required to raise the temperature of 1 kg of a substance by 1 C° is called the specific heat. The greater the specific heat of a substance, the more heat energy necessary to change the temperature of a given mass.
- Latent heat is the heat energy, in joules per kilogram, associated with a phase change.
- Heat transfer is effected by three methods: conduction, convection, and radiation.

Important Equations

Mechanical Equivalent of Heat:

$$1 \text{ kcal} = 4186 \text{ J} = 4.186 \text{ kJ}$$
$$1 \text{ cal} = 4.186 \text{ J}$$

Specific Heat (specific heat capacity):

$$Q = mc\,\Delta T \quad \text{or} \quad c = \frac{Q}{m\,\Delta T} \quad (11.1)$$

Latent Heat:

$$Q = mL \quad \text{or} \quad L = \frac{Q}{m} \quad (11.2)$$

(for water)

$$L_f = 3.33 \times 10^5 \text{ J/kg (or 80 kcal/kg)}$$
$$L_v = 22.6 \times 10^5 \text{ J/kg (or 540 kcal/kg)}$$

Thermal Conduction:

$$\frac{\Delta Q}{\Delta t} = \frac{kA\,\Delta T}{d} \quad (11.3)$$

Stefan's Law:

$$P = \frac{\Delta Q}{\Delta t} = \sigma A e T^4 \quad (11.5)$$

Net Radiant Power (loss or gain):

$$P_{net} = \sigma A e (T_s^4 - T^4) \quad (11.6)$$

Exercises*

11.1 Units of Heat

1. The SI unit of heat energy is the (a) calorie, (b) kilocalorie, (c) Btu, (d) joule.

2. Which of the following is the largest unit of heat energy: (a) calorie, (b) Btu, (c) joule, (d) kilojoule?

3. ■ A heater supplies 240 Btu of energy. What is this in joules?

4. ■ A person goes on a 1500-Cal-per-day diet to lose weight. What is the equivalent daily allowance expressed in joules?

5. ■ A window air conditioner has a rating of 20 000 Btu (per hour). What would this rating be in watts?

6. ■ What is the mechanical equivalent of heat expressed in British thermal units?

7. ■■ A student eats a Thanksgiving dinner that totaled 2800 Cal. He wants to use up all that energy by lifting a 20-kilogram mass a distance of 1.0 m. (a) How many times must he lift the mass? (b) If he can lift the mass once every 5.0 s, how long does this exercise take (neglecting lowering)?

11.2 Specific Heat

8. The amount of heat necessary to change the temperature of 1 kg of a substance by 1 C° is called the substance's (a) specific heat, (b) latent heat, (c) heat of combustion, (d) mechanical equivalent of heat.

*Neglect heat losses in the exercises unless instructed otherwise, and consider all temperatures to be exact.

9. When you swim in the ocean or a lake at night, the water may feel pleasantly warm even when the air is quite cool. Why?

10. Is it possible to have a negative specific heat? Explain.

11. ● Is it possible to have negative heat? Explain.

12. Near a large body of water, what would you expect the direction of a breeze to be during the night?

13. ● Does heat flow depend on the temperatures of two locations or only on the temperature difference?

14. An astronaut in deep space says, "I'm going to perform an experiment to see how many Btus it takes to increase the temperature of 3.0 lb of water from 60°F to 212°F." What is wrong with this statement, and what does it tell you?

15. ■ ● The temperature of a lead block and a copper block, both 1.0 kg and both at 110 K, is to be raised to 190 K. (a) Which one requires more heat? (b) How much more?

16. ■ ● A 5.0-gram pellet of aluminum at 20°C gains 200 J of heat. What is its final temperature?

17. ■ How many joules heat must be added to 5.0 kg of water at 20°C to bring it to the boiling point?

18. ■■ A 0.250-kilogram cup at 20°C is filled with 0.250 kg of boiling coffee. The cup and the coffee come to thermal equilibrium at 80°C. If no heat is lost, what is the specific heat of the cup material? [Hint: Consider the coffee to be essentially boiling water.]

19. ■ An aluminum serving spoon at 100°C is placed in a Styrofoam cup containing 0.200 kg of water at 20°C. If the final equilibrium temperature is 30°C and no heat is lost to the cup itself, what is the mass of the spoon?

20. ■■ 🔘 How many kilograms of aluminum will experience the same temperature rise as 3.00 kg of copper when the same amount of heat is added to each?

21. ■■ A waterfall is 75 m high. If all of the gravitational potential energy of the water were converted into heat energy, by how much would the temperature of the water increase in going from the top to the bottom of the falls? [Hint: Consider a kilogram of water going over the falls.]

22. ■■ A cube of aluminum 10 cm on each side is cooled from 100°C to 20°C. If the energy removed from the aluminum cube were added to a copper cube of the same size at 20°C, what would be the final temperature of the copper cube?

23. ■■ A camper heats 30 L of water to boiling to get hot water for bathing. What volume of water from a 15°C stream must he then add to get the bath water to 45°C? (Neglect any losses.)

24. ■■ To determine the specific heat of a new metal alloy, 0.150 kg of the substance is heated to 400°C and then placed in a 0.200-kilogram aluminum calorimeter cup containing 0.400 kg of water at 10.0°C. If the final temperature of the mixture is 30.5°C, what is the specific heat of the alloy? (Ignore the calorimeter stirrer and thermometer.)

25. ■■ In a calorimetry experiment, 0.50 kg of a metal at 100°C is added to 0.50 kg of water at 20°C in an aluminum calorimeter cup. The cup has a mass of 0.250 kg. (a) If the final temperature of the mixture is 25°C, what is the specific heat of the metal? (b) What would be the effect on the calculated specific heat value if some water splashed out of the cup when the metal was added?

26. ■■■ An electric immersion heater has a power rating of 1500 W. If the heater is placed in a liter of water at 20°C, how many minutes will it take to bring the water to a boil? (Assume that there is no heat loss except to the water itself.)

27. ■■■ A 0.100-kilogram piece of aluminum at 90°C is immersed in 1.00 kg of water at 20°C. Assuming no heat is lost to the surroundings or container, what is the temperature of the metal and water when they reach thermal equilibrium?

28. ■■■ At what average rate would heat have to be removed from 1.5 L of (a) water and (b) mercury to reduce the liquid's temperature from 20°C to its freezing point in 3.0 min?

11.3 Phase Changes and Latent Heat

29. The SI units of latent heat are (a) 1/C°, (b) J/(kg·C°), (c) J/C°, (d) J/kg.

30. Latent heat is always (a) part of the specific heat, (b) related to the specific heat, (c) the same as the mechanical equivalent of heat, (d) involved in a phase change.

31. Why do different substances have different freezing and boiling points? Would you expect the latent heats to be different for different substances? Explain.

32. You are monitoring the temperature of some cold ice cubes (−5.0°C) in a closed, insulated cup as the ice and cup are heated. Initially, the temperature rises, but it stops at 0°C. After a while it begins rising again. Is there anything wrong with the thermometer? Explain.

33. In general, you can get a more severe burn from steam at 100°C than from an equal mass of hot water at 100°C. Why?

34. ■ How much heat is required to boil away 0.500 kg of water that is initially at 100°C?

35. ■ How much more heat is needed to convert 1.0 kg of water at 100°C to steam at 100°C than to raise the temperature of 1.0 kg of water from 0°C to 100°C?

36. ■ How much heat must be added to 0.75 kg of lead at 20°C to cause it to melt completely?

37. ■ How much energy in joules is required to boil away 0.50 L of liquid nitrogen at −210°C? (Take the density of liquid nitrogen to be 0.80×10^3 kg/m^3.)

38. ■ How much heat in joules is required to boil away 0.50 kg of water initially at 50°C?

39. ■ A mercury diffusion vacuum pump contains 0.015 kg of mercury vapor at a temperature of 630 K. Suppose that you wanted to condense the vapor. How much heat would have to be removed?

40. ■■ (a) A 0.500-kilogram piece of ice at −20°C is converted to steam at 115°C. How much heat must be supplied to do this? (b) To convert the steam back to ice at −5°C, how much heat would have to be removed?

41. ■■ How much ice (at 0°C) must be added to 1.0 kg of water at 100°C so as to end up with all liquid at 20°C?

42. ■■ A 0.60-kilogram piece of ice at −10°C is placed in 0.30 kg of water at 50°C. How much water is there when the system reaches thermal equilibrium?

43. ■■ If 0.050 kg of ice at 0°C is added to 0.300 kg of water at 25°C in a 0.100-kilogram aluminum calorimeter cup, what is the final temperature of the water?

44. ■■ Steam at 100°C is bubbled into 0.250 kg of water at 20°C in a calorimeter cup. How much steam will have been added when the water in the cup reaches 60°C? (Ignore the effect of the cup.)

45. ■■ Ice is added to 0.75 L of tea at 20°C to make very cold iced tea. If enough 0°C ice is added so that the mixture is all liquid, how much liquid is in the pitcher when this occurs?

46. ■■ A volume of 0.50 L of water at 16°C is put into an aluminum ice cube tray of mass 0.250 kg at the same temperature. How much energy must be removed from this system by the refrigerator to turn the water into ice at −8.0°C?

47. ■■■ Suppose 3.0 cm of rain at 10°C falls on a rectangular region of dimensions 2.0 km × 3.0 km. How much energy was released when water vapor condensed to form this amount of rain? Is it a large amount of energy?

48. ■■■ A kilogram of a substance gives a *T*-versus-*Q* graph as shown in •Fig. 11.18. What are the (a) melting and boiling points? In SI units, what are (b) the specific heats of the various phases and (c) the latent heats at the various phase changes?

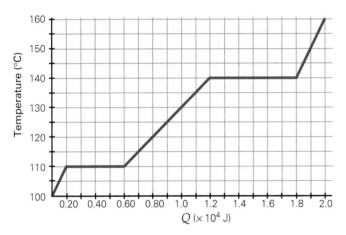

•**FIGURE 11.18   Temperature versus heat**   See Exercise 48.

49. ■■■ In an experiment, a 0.150-kilogram piece of ceramic superconducting material at 20°C is placed in liquid nitrogen (N_2) at its boiling point to cool. The N_2 is in a perfectly insulated flask that allows the gaseous N_2 to escape immediately. How many liters of liquid nitrogen will be boiled away? (Take the specific heat of the ceramic material to be the same as that of glass, and take the density of liquid nitrogen to be 0.80×10^3 kg/m³.)

11.4 Heat Transfer

50. The warming of the atmosphere involves (a) conduction, (b) convection, (c) radiation, (d) all of these.

51. Water is a very poor heat conductor. Why can it be thoroughly heated relatively quickly?

52. A plastic ice tray and a metal ice tray are removed from the same freezer (same initial temperature). However, once your hands touch both, the metal one feels cooler. Why?

53. (a) You blow on a spoonful of hot soup to cool it. Yet, your breath is warm. How does blowing cause the soup to cool? (b) On hot, humid days, ice can build up on the cooling coils of a window air conditioner. Does this help with the room cooling? Explain.

54. Why is the warning shown on the highway road sign in •Fig. 11.19 necessary?

•**FIGURE 11.19   A cold warning**   See Exercise 54.

55. What is the purpose of the fins on a motorcycle radiator or an automobile radiator (•Fig. 11.20)?

•**FIGURE 11.20   Radiator with fins**   See Example 55.

56. Newton's law of cooling (Sir Isaac was a busy man) states that, in general, the rate of heat loss from an object is proportional to the temperature difference between it and the surroundings: $\Delta Q/\Delta t = K\Delta T$, where K is a constant that includes losses by conduction, convection, and radiation. Apply this law to the following question: Would a cup of hot coffee stay hotter longer if you put cream in it right away or waited until you were ready to drink it, at a later time?

57. A Thermos bottle (•Fig. 11.21) keeps cold beverages cold and hot ones hot. It consists of a double-walled, partially evacuated container with silvered walls (mirrored

•FIGURE 11.21 **Thermal insulation** The Thermos bottle uses all three methods of heat transfer. See Exercise 57.

interior). The bottle is constructed to minimize all three mechanisms of heat transfer. Explain how.

58. If you put your hand in a hot oven, the heat is uncomfortable but tolerable. Yet, if you touch a pan in the oven, you burn your hand. Why the difference?

59. ■ How many times faster would heat be conducted from your bare feet by a tile floor than by an oak floor? (Assume that both floors have the same temperature and thickness.)

60. ■ If an object has an emissivity of 0.75 and an area of 0.20 m², how much energy does it radiate outward per second at 20°C?

61. ■■ The glass pane in a window has dimensions of 2.00 m × 1.50 m and is 4.00 mm thick. How much heat in joules will flow through the glass in 1.00 h if there is a temperature difference of 2 C° between the inner and outer surfaces? (Consider conduction only.)

62. ■■ Assuming the human body has a 4.0-centimeter-thick layer of skin tissue and a surface area of 1.5 m², estimate the rate at which heat is conducted from inside the body to the surface if the skin temperature is 33°C. (Assume normal body temperature of 37°C for the interior temperature, and neglect insulating clothes.)

63. ■■ A copper teakettle with a circular bottom 30 cm in diameter has a uniform thickness of 2.5 mm. It sits on a burner whose temperature is 150°C. (a) If the teakettle is full of boiling water, what is the rate of heat conduction through its bottom? (b) Assuming that the heat from the burner is the only heat input, how much water is boiled away in 5.0 min? Is your answer reasonable? If not, explain why.

64. ■■ An aluminum bar and a copper bar of identical cross-sectional area have the same temperature difference between their ends and conduct heat at the same rate. Which bar is longer, and how many times longer?

65. ■■ A Styrofoam box has a surface area of 1.0 m² and a thickness of 2.5 cm. If 5.0 kg of ice at 0°C is stored inside and the outside temperature is 35°C, how long does it take for all the ice to melt? (Consider conduction only.)

66. ■■ The thermal insulation used in building is commonly rated in terms of its *R-value*, defined as d/k, where d is the thickness of the insulation in inches and k is the thermal conductivity. For example, 3.0 in. of foam plastic would have an R-value of $3.0/0.30 = 10$, where, in British units, $k = 0.30$ Btu·in./(ft²·h·F°). This value is expressed as R-10. (a) What does the R-value tell you; that is, how does thermal insulation vary with R-value? (b) What thicknesses of (1) fiberboard and (2) brick would give an R-value of R-10? [Hint: Use known ratios for unit conversion.]

67. ■■ Pine wood 14 in. thick has an R-value of 19. What thickness of (a) glass wool and (b) fiberboard would have the same R-value? (See Exercise 66.)

68. ■■ A large picture window measures 2.0 m by 3.0 m. At what rate will heat be conducted through the window when the room temperature is 20°C and the outside temperature is 0°C (a) if the window has a single pane of glass 4.0 mm thick and (b) if the window has a double pane of glass (a thermopane), and each pane is 2.0 mm thick with an intervening air space of 1.0 mm? (Assume that there is a constant temperature difference.)

69. ■■ The emissivity of an object is 0.60. How many times greater would the outward radiation be from a similar perfectly blackbody at the same temperature?

70. ■■ A house wall is composed of a solid concrete block with outside brick veneer and faced on the inside with fiberboard as illustrated in •Fig. 11.22. If the outside

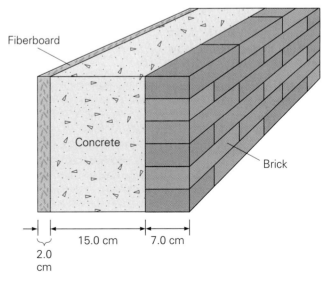

•FIGURE 11.22 **Thermal conductivity and heat loss** See Exercise 70.

temperature on a cold day is −10°C and it is 20°C on the inside, how much energy in joules is conducted through a wall with dimensions of 3.5 m × 5.0 m in 1.0 h?

71. ■■ Suppose you wished to cut the heat loss through the wall in Exercise 70 by 50% by installing insulation. What thickness of Styrofoam would have to be placed between the fiberboard and concrete block to accomplish this?

72. ■■ If the temperature of an object is increased from 20°C to 40°C, how are the object's (a) emissivity and (b) output radiation rate affected? (c) What are the effects if instead the object's temperature in kelvins is doubled?

73. ■■ A steel cylinder of radius 5.0 cm and length 4.0 cm is placed in end-to-end thermal contact with a copper cylinder of the same dimensions. If the free ends of the two cylinders are maintained at constant temperatures of 95°C (steel) and 15°C (copper), how much heat will flow through the cylinders in 20 min?

74. ■■■ For the metal cylinders in Exercise 73, what is the temperature at the interface of the cylinders?

75. ■■■ Solar heating takes advantage of solar collectors such as the type shown in •Fig. 11.23. About 1000 W/m² of incoming solar radiation reaches the Earth's surface during daylight hours on a clear day. (The rest is reflected, scattered, absorbed, and so on.) How much heat energy would be collected, on the average, by the cylindrical collector shown in the figure during 10 h of collection?

•**FIGURE 11.23 Solar collector and solar heating** See Exercise 75.

Additional Exercises

76. A modern engine of alloy construction consists of 25 kg of aluminum and 80 kg of iron. How much heat does it absorb as its temperature increases from 20°C to 120°C as it warms up to operating temperature?

77. In a calorimetry experiment, 0.35 kg of aluminum shot at 100°C is carefully poured into 50 mL of water that has been chilled to 10°C. Neglecting any losses to the container (or otherwise), what is the final equilibrium temperature of the mixture?

78. A lamp filament radiates 100 W of power when the temperature of the surroundings is 20°C and 99.5 W of power when the temperature of the surroundings is 30°C. If the temperature of the filament is the same in each case, what is that temperature on the Celsius scale?

79. A student doing an experiment pours 0.150 kg of heated copper shot into a 0.375-kilogram aluminum calorimeter cup containing 0.200 kg of water at 25°C. The mixture (and the cup) comes to thermal equilibrium at 28°C. What was the initial temperature of the shot?

80. A student mixes 1.0 L of water at 40°C with 1.0 L of ethyl alcohol at 20°C. Assuming that there is no heat loss to the container or the surroundings, what is the final temperature of the mixture? [Hint: See Table 11.1.]

81. After a barrel jump, a 65-kilogram ice skater traveling at 25 km/h glides to a stop. If 40% of the frictional heat generated by the skate blades goes into melting ice at 0°C, how much ice is momentarily melted? Where does the other 60% of the energy go?

82. A 0.030-kilogram lead bullet hits a steel plate, both initially at 20°C. The bullet melts and splatters on impact. (This action has been photographed.) Assuming that the bullet receives 80% of its kinetic energy as heat energy, at what minimum speed must it be traveling to melt on impact?

83. ● Two metals, 0.50 kg of aluminum and 0.50 kg of iron, initially at 20°C are heated to 100°C. Which metal gains more heat, and how much more?

84. Equal amounts of heat are added to different quantities of copper and lead. The temperature of the copper increases by 5.0 C°, and the temperature of the lead by 10 C°. Which piece of metal has the greater mass, and how much greater?

85. How much heat is required to convert 0.75 kg of ice at −10°C to steam at 120°C?

CHAPTER 12

Thermodynamics

12.1 Thermodynamic Systems, States, and Processes
12.2 The First Law of Thermodynamics
12.3 The Second Law of Thermodynamics and Entropy
12.4 Heat Engines and Thermal Pumps
12.5 The Carnot Cycle and Ideal Heat Engines

Insights

- Life, Order, and the Second Law
- The Otto Cycle

At certain locations on Earth, hot springs from deep in the interior rise to the surface. In Yellowstone National Park, they produce boiling pools and geysers such as Old Faithful, shown here. On the coast of Iceland, they warm the ocean, creating a tropical lagoon surrounded by glaciers. Such intriguing settings are popular vacation spots.

But the uses of such hot springs extend beyond recreation. Wherever there is a temperature difference, the potential exists for obtaining useful work. For instance, geothermal power plants draw on geysers as a renewable resource and produce no pollution. In this chapter you'll learn under what conditions, and with what efficiency, heat can be exploited to perform work, in machines as different as automobile engines and home freezers. You'll also find that the laws governing such energy conversions include some of the most general and far-reaching in all of physics.

As the word implies, **thermodynamics** deals with the transfer or the actions (dynamics) of heat (the Greek word for "heat" is *therme*). The development of thermodynamics started about 200 years ago and grew out of efforts to develop heat engines. The steam engine was one of the first of such devices, which convert heat energy to mechanical work. Steam engines in factories and locomotives powered the industrial revolution that changed the world. Although this course of study is primarily concerned with heat and work, thermodynamics is a broad and comprehensive science that includes a great deal more than heat engine theory. In this chapter, you will learn about the two general laws on which thermodynamics is based as well as the concept of entropy.

12.1 Thermodynamic Systems, States, and Processes

Objectives: To (a) define thermodynamic systems and states of systems, and
(b) explain how thermal processes affect such systems.

Definition of:
Thermodynamics system

Thermodynamics is a field that makes use of many special terms, or everyday terms with special meanings, and you will find it useful to become familiar with some of them at the onset. The term **system**, as used in thermodynamics, refers to a definite quantity of matter enclosed by boundaries or surfaces, either real or imaginary. For example, a quantity of gas in the piston-cylinder of an engine has real boundaries, and imaginary boundaries enclose a cubic meter of air in a room. The enclosing boundaries of a system need have no definite shape, nor do they have to enclose a fixed volume. For example, a cylinder of gas could be compressed by advancing the piston.

Occasionally it is necessary to consider systems for which there is a transfer of matter across boundaries. However, for the most part, we will consider systems of constant mass. An important thermodynamic consideration is the interchange of energy between a system and its surroundings. This exchange may occur through a transfer of heat and/or the performance of mechanical work. For example, if a balloon is heated, it expands and work is done against the surrounding atmospheric pressure (and in stretching the balloon).

If heat transfer into or out of the system is impossible, the system is said to be a **thermally isolated system**. Work may be done on a thermally isolated system. For example, a thermally isolated balloon (perhaps surrounded by insulation) can be compressed by an external force or pressure, so work is done on the system; as we know, work results in a transfer of energy.

When heat does enter or leave a system, it is absorbed from or given up to the surroundings or to heat reservoirs. A **heat reservoir** is a system assumed to have an unlimited heat capacity. Thus any amount of heat can be withdrawn from or added to a heat reservoir without appreciably changing its temperature. This is like taking a cup of water from or adding a cup of water to the ocean—you don't change the water level appreciably. Similarly, pouring a warm bottle of water into a cold lake does not noticeably raise the lake's temperature.

State of a System

Just as there are kinematic equations to describe aspects of the motion of an object, there are **equations of state** to describe the conditions of thermodynamic systems. Such an equation expresses a mathematical relationship of the thermodynamic variables of a system. The ideal gas law, $pV = Nk_BT$, is a simple equation of state. This expression establishes a relationship among the pressure (p), volume (V), absolute temperature (T), and mass (N, the number of molecules) of a gas. These quantities are called *state variables*. Different gaseous states have different sets of values for these variables. A set of these variables that satisfies the ideal gas law specifies a state of a system of ideal gas completely as long as the system is in thermal equilibrium and has a uniform temperature.

The state of a given mass of gas in a closed system can be specified by the variables p, V, and T. It is convenient to plot the states according to the thermodynamic coordinates (p, V, T), much as we plot graphs with Cartesian coordinates (x, y, z). A general two-dimensional illustration of such a plot is shown in •Fig. 12.1. (This is not a p-V graph for an ideal gas. Several of these will be shown shortly.)

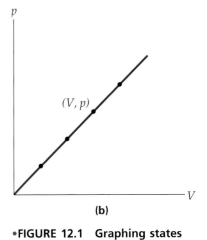

•FIGURE 12.1 Graphing states
(a) On a common Cartesian graph, the coordinates (x, y) represent an individual point. (b) Similarly, on a p-V graph or diagram, the coordinates (V, p) represent a particular state of a system. It is common to say p-V diagram, rather than V-p, to match the order of the coordinates in the ideal gas law (p, V, T). The graph is not that described by the ideal gas law equation of state.

Note: Because p, V, and T are connected by the ideal gas law, specifying any two of these variables automatically tells you the third.

Note: A change of state (of a system) does not necessarily mean a change of phase, such as from liquid to solid. To avoid confusion, always refer to solid, liquid, and gas as *phases* rather than states of matter.

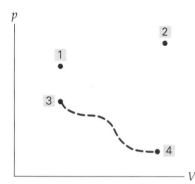

•**FIGURE 12.2 Paths of reversible and irreversible processes** If a gas quickly goes from state 1 to state 2, the process is irreversible since we do not know the "path." If, however, the gas is taken through many closely spaced states (as in going from state 3 to state 4), the process is reversible, at least in principle. Reversible means "exactly retraceable."

Just as the coordinates (x, y) specify individual points on a Cartesian graph, the coordinates (V, p) specify individual *states* on the p-V graph or diagram.

Using the ideal gas law, $pV = Nk_BT$ (Eq. 10.6), we can determine the temperature of a gas if we know its pressure and volume (and the number of molecules or moles in the sample). In other words, on a p-V diagram for a gas, each "coordinate" gives the pressure and volume directly, as well as indirectly the unique temperature for that condition or state of the gas. Thus for most situations involving a gas, only a p-V plot is necessary. In some cases, however, it can be instructive to refer to other coordinate plots such as p-T or T-V.

Processes

A **process** is a change in the state, or the thermodynamic coordinates, of a system. That is, when a system undergoes a process, the set of variables p, V, and T that describe its state completely will, in general, change. For example, we go from state (p_1, V_1, T_1) to state (p_2, V_2, T_2) by a thermodynamic process. Processes are said to be either reversible or irreversible.

Suppose that a system of gas in equilibrium (known p, V, T values) is allowed to expand quickly. The state of the system will change rapidly and unpredictably but will eventually return to equilibrium with another set of thermodynamic coordinates, in another state. On a graph such as a p-V diagram (•Fig. 12.2), we could show the initial state and the final state, but not what happened in between. Since the intermediate states changed so fast, there would be no data describing them. This is called an **irreversible process**—a process for which the intermediate steps are nonequilibrium states. "Irreversible" does not mean that the system can't be taken back to the initial state; it means only that the process path can't be retraced because of the nonequilibrium conditions.

If, however, the gas expands very, very slowly, passing from one known equilibrium state to a neighboring one and eventually arriving at the final state (see Fig. 12.2), then the process path between the initial and final states would be known. Such a process is called a **reversible process**—one whose path is known. In fact, a perfectly reversible process cannot be achieved. All real thermodynamic processes are irreversible to some degree because they follow complicated paths with many intermediate states. However, the concept of an ideal reversible process is useful.

These thermodynamic terms and ideas will now be applied in the laws of thermodynamics.

12.2 The First Law of Thermodynamics

Objectives: To (a) explain the relationship among internal energy, heat, and work as expressed by the first law, and (b) analyze various thermal processes.

The conservation of energy is valid for any system. Thus, the **first law of thermodynamics** is simply a statement of the conservation of energy as applied to thermodynamic systems. Heat, internal energy, and work are the quantities involved in a thermodynamic system. Suppose that some heat (Q) is added to a system. Where does it go? That it serves to increase the system's internal energy (ΔU; see Chapter 10) is certainly a possibility, as when the total molecular kinetic energy of an ideal gas increases with a temperature rise. Another possibility is that it could result in work (W) being done by the system. For example, when a heated gas expands, it does work on its surroundings (think of expanding air in a balloon).

Thus, it is possible for the added heat to go into internal energy, work, or both. Writing this as an equation gives a mathematical expression for the first law:

First law of thermodynamics: conservation of energy

$$Q = \Delta U + W \qquad (12.1)$$

Our sign convention will be as follows: A positive value of heat ($+Q$) means that heat is *added to* the system, and a positive value of work ($+W$) means that work is *done by* the system (for example, the work done by an expanding gas). Negative quantities mean that heat *is removed from the system* (−Q) and work *is done on the system* (−W), for example, when a gas is compressed.

Before going on with some examples of the first law, let's make certain we understand a fundamental difference among U, Q, and W. Any given system in a particular state will have a certain amount of internal energy U. Yet, a system does not possess certain amounts of heat or work. These are what *change* the state of the system and generally change the internal energy. When heat is added to or removed from a system or work is done on or by a system, *thermodynamic processes* occur that can change the system from one state to another, each having a particular internal energy U. That is, the internal energy depends only on the state of the system and not on how it got there.

Hence, to find the change in internal energy, ΔU, in going from one state to another, we need to know only the internal energy of the two states, and $\Delta U = U_2 - U_1$. In other words, the change in the internal energy, ΔU, is independent of the process path; it depends only on the initial and final states. This is analogous to the change in gravitational potential energy (*mgh*) being independent of path (Section 5.4).

The first law can be applied to several processes for a closed system of an ideal gas. In some of these processes, one thermodynamic variable is kept constant. Such processes have names beginning with iso- (from the Greek *isos*, meaning "equal").

Note: Keep in mind that, unless otherwise stated, we are working with constant-mass systems and with ideal gases.

Isobaric Process. A constant-pressure process is called an **isobaric process** (*iso* for equal, *bar* for pressure). An isobaric process for an ideal gas is illustrated in •Fig. 12.3. On a p-V diagram, the path of an isobaric process is along a horizontal line called an *isobar*. When heat is added to the gas in the cylinder, the ratio V/T must remain constant ($V/T = Nk_B/p$ = a constant). The heated gas expands; there is an increase in its volume. The temperature must therefore also increase proportionally, which means the internal energy of the gas increases. (Recall from Chapter 10 that kinetic theory says that the internal energy of an ideal gas is directly proportional to its absolute temperature.)

Note: *Isobaric* means constant *pressure*.

Note: Review Sections 10.1 and 10.5.

Work is done by the gas as it expands, in moving the cylindrical piston in Fig. 12.3. From the definition of work in Section 5.1,

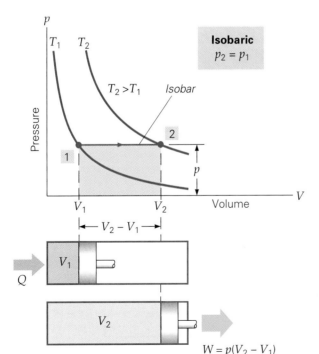

•**FIGURE 12.3 Isobaric (constant-pressure) process** The heat added to the gas in the frictionless piston goes into increasing the internal energy of the gas and doing work (the expanding gas moves the piston): $Q = \Delta U + W$. The work is equal to the area under the process path (from state 1 to state 2 here) on the p-V diagram.

12.2 The First Law of Thermodynamics

$$W = F\Delta x$$

In terms of pressure ($p = F/A$), the force can be written $F = pA$, where A is the area of the piston. Then, the previous equation becomes

$$W = pA\Delta x$$

The change in volume is a right cylinder with end area A and height Δx. Hence $A\Delta x$ is simply the change in the volume of the gas, $A\Delta x = \Delta V = V_2 - V_1$. Thus,

$$W = p\Delta V = p(V_2 - V_1) \quad \textit{(for an isobaric process)} \quad (12.2)$$

In Fig. 12.3, you can see that $p\Delta V$ is equal to area of the rectangle under the isobar on the p-V diagram. For a nonisobaric process (one in which the pressure changes), the work is also equal to the area under the line showing the process path, but the shape is not rectangular. Thus, the work depends on the process path as well as on the initial and final states: There will be different areas under different paths.

The internal energy of a quantity of an ideal gas depends only on its (absolute) temperature, however. Therefore a change in this internal energy is *independent* of the process path and depends only on the initial and final states, or on the difference between the temperatures of these states ($\Delta U = U_2 - U_1 \propto T_2 - T_1$).

Since V_2 is greater than V_1 for an expanding gas, work is done by the system ($+W$). In terms of the first law, then,

$$Q = \Delta U + W = \Delta U + p\Delta V \quad \textit{(for an isobaric process)} \quad (12.3)$$

That is, the heat added to the system goes into increasing the internal energy *and* into work done by the system. If the process were reversed and the gas were being compressed by an external force doing work *on* the system, all the quantities would be negative. Heat would flow out of the system ($-Q$), and the internal energy, and therefore the temperature, of the gas would decrease ($-\Delta U$).

Note: *Isometric* means constant *volume*.

Isometric Process. An **isometric process** (short for iso*volu*metric, or equal-volume, process) is a constant-volume process, sometimes called an *isochoric process*. As illustrated in •Fig. 12.4, the process path on a p-V diagram is along a vertical line, commonly called an *isomet*. No work is done ($W = p\Delta V = 0$, since $\Delta V = 0$), so if heat is added, it goes into increasing the internal energy, and therefore the temperature, of the gas. By the first law,

$$Q = \Delta U + W = \Delta U + 0$$

and thus

$$Q = \Delta U \quad \textit{(for an isometric process)} \quad (12.4)$$

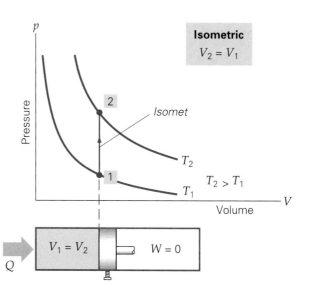

•**FIGURE 12.4 Isometric (constant-volume) process** All of the heat added to the gas goes into increasing the internal energy when the volume is held constant: $Q = \Delta U$. This causes an increase in temperature.

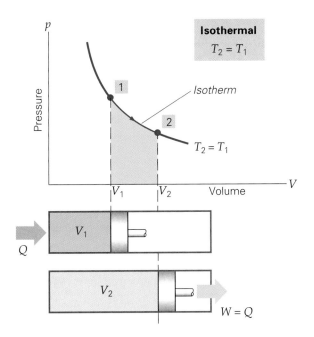

•FIGURE 12.5 Isothermal (constant-temperature) process All of the heat added to the gas goes into doing work (the expanding gas moves the piston): since $\Delta T = 0$, then $\Delta U = 0$ and $Q = W$. As always, the work is equal to the area under the process path on the p-V diagram.

Isothermal Process. An **isothermal process** is a constant-temperature process (•Fig. 12.5). In this case, the process path is along an *isotherm*, or a line of constant temperature. Since $p = (Nk_BT)/V = (\text{constant})/V$ for an ideal-gas isothermal process, an isotherm is a hyperbola on a p-V diagram. (The general form of the equation for a hyperbola is $y = a/x$.)

In going from state 1 to state 2 in Fig. 12.5, heat is added to the system, and both the pressure and volume change in order to keep the temperature constant (pressure decreases and volume increases). The work done by the expanding system ($+W$) is again equal to the area under the process path.

For an isothermal process, the internal energy of the ideal gas remains constant ($\Delta U = 0$), because the temperature is constant. By the first law,

$$Q = \Delta U + W = 0 + W$$

and

$$Q = W \quad \text{(for an isothermal process)} \quad (12.5)$$

Note: *Isothermal* means constant temperature.

Thus, for an ideal gas, in an isothermal process heat energy is converted to mechanical work (or vice versa for the reverse path). See Learn by Drawing on p. 394.

Adiabatic Process. There's one more type of process in which a thermodynamic condition remains constant (•Fig. 12.6). In an **adiabatic process**, no heat is transferred into or out of the system; that is, $Q = 0$ (the Greek word *adiabatos* means "impassable.") The condition $Q = 0$ is satisfied for a thermally isolated system. This is an ideal situation, since there is always some heat transfer in actual system processes. Under actual conditions, we can only approximate adiabatic processes. For example, processes that are nearly adiabatic can take place in a system that is not thermally isolated if they occur rapidly enough that there isn't time for much energy to be transferred into or out of the system.

Adiabatic process: no heat transfer

Note: In practice, adiabatic means *quickly*, before significant heat can flow in or out ($Q = 0$).

As such a system follows the process path, a curve called an *adiabat*, all three thermodynamic coordinates change. For example, suppose a quantity of ideal gas were compressed in a thermally isolated cylindrical piston. If the piston were suddenly released, the gas would expand (that is, p and V would change). Work would be done at the expense of the internal energy of the gas, so the temperature of the gas would drop. By the first law,

$$Q = 0 = \Delta U + W$$

and

$$W = -\Delta U \quad \text{(for an adiabatic expansion)} \quad (12.6)$$

12.2 The First Law of Thermodynamics

LEARN BY DRAWING

Leaning on Isotherms

When you are analyzing the various thermodynamic processes dicussed in this chapter, it is sometimes hard to keep track of the heat flow (Q), work (W), and internal energy change (ΔU), together with their correct signs. One trick that can help with this bookkeeping is to superimpose a series of isotherms on the p-V plot you are working with (as in Figs. 12.2 and 12.4–12.6). This is generally useful even if the situation you are studying does not involve an isothermal process.

Before starting, recall the two important properties of an isothermal process. By definition, an isothermal process is one in which the temperature remains constant:

1. In an isothermal process, ΔU is zero. (Why?)
2. Since T is constant, the ideal gas law (Eq. 10.3) tells us that in an isothermal process, pV must also be constant: $pV = k$. You may recall from algebra that this is the equation of a hyperbola; on a p-V diagram, an isotherm is a hyperbola. The farther from the axes the hyperbola is, the higher the temperature it represents (Fig. 1).

To take advantage of these properties, follow these steps:

- Sketch a set of isotherms for a series of different temperatures on the p-V plot (Fig. 1).
- Then sketch the actual process or processes you are analyzing—for example, those shown in Fig. 2 and Fig. 3. (Take a moment to recall what the names mean: isomet = constant volume, isobar = constant pressure, and adiabat = no heat flow.)
- Finally, use the first law of thermodynamics, $Q = \Delta U + W$, to calculate the final results, including signs. W is the area under the p-V curve for the process represented, and its sign is determined by expansion (+) or compression (−). The sign of ΔT will be clear from the isotherms; they serve as a temperature scale, with progressively higher temperatures represented by isotherms at greater distances from the axes. A rise in T implies a positive value of ΔU, since the internal energy of the gas increases with its temperature. You will thus have all you need to determine the sign of Q and whether heat flows into ($Q > 0$) or out of the system ($Q < 0$).

The following two examples indicate the power of this approach. In Fig. 2, we can decide whether heat flows into or out of the gas during an isobaric expansion. Expansion implies positive work. But what is the direction of the heat flow (or is it zero)? Sketching the isobar, we see that it crosses from lower-temperature isotherms to higher-temperature ones. Hence there is a temperature increase, and ΔU must be positive. Since the gas has both gained internal energy and done positive work, both terms on the right-hand side of the first-law equation are positive, so it follows that Q must be positive. Positive heat flow means "into the gas," so we have our answer.

An isometric process involving a pressure drop is analyzed in Fig. 3. Again, starting with lightly drawn isotherms, we plot a vertical line (constant volume) with an arrow pointing downward to show pressure reduction. Since there is no volume change, the gas does no work. However, we can see that its temperature drops (why?), so its internal energy must decrease. This means that ΔU is negative, hence Q must be negative. Thus heat flows "out of the gas."

As an exercise, try analyzing an adiabatic process via this graphical approach to determine where the work comes from if there is no heat flow into or out of the gas.

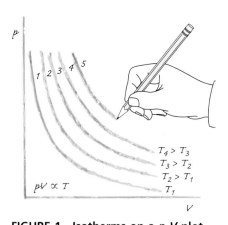

FIGURE 1 Isotherms on a *p-V* plot

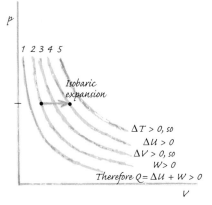

FIGURE 2 An isobaric expansion

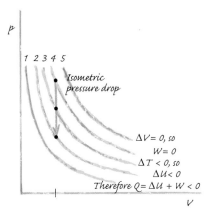

FIGURE 3 An isometric decline in pressure

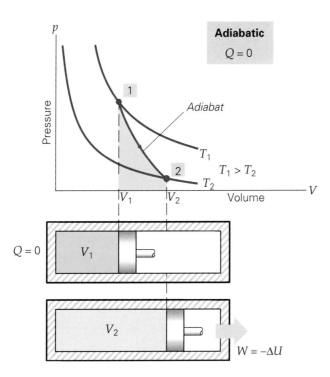

•**FIGURE 12.6 Adiabatic (no heat transfer) process** In an adiabatic process (shown here as a cylinder with heavy insulation), no heat is added to or removed from the system: $Q = 0$. Work is done at the expense of the internal energy: $W = -\Delta U$. The pressure, volume, and temperature all change in the process.

Thus, in an adiabatic expansion, work (the area under the process path) is done by the system and is therefore positive. In turn, ΔU must be negative, corresponding to a decrease in the system's internal energy. Since the internal energy, U, of an ideal gas depends only on its temperature, this decrease in internal energy is evidenced by a decrease in temperature.

Conceptual Example 12.1

Exhaling: Blowing Hot and Cold?

The air in your lungs is warm. This can be shown by putting your bare upper arm near your mouth and blowing air on your arm with your mouth opened widely. If you were to repeat this with your lips puckered, the air would feel (a) warmer, (b) cooler, (c) the same.

Reasoning and Answer. We'll go from answer to reasoning on this one, since it is easy to determine the correct answer experimentally. Try it, and you might be surprised to find that the answer is (b).

The interesting thing is why this occurs. When you blow on your arm with mouth open, you get a gush of warm air (roughly at body temperature). However, when you blow with your lips puckered, the stream of air is compressed. Then, as it emerges, the air expands, doing work against the atmosphere. This work is done at the expense of the internal energy of the air, so the temperature of the air is decreased and it feels cooler. This takes place quickly and is an essentially adiabatic process, at least for a few seconds. Eventually, the cooled air absorbs heat from the surrounding air and its temperature rises.

Follow-up Exercise. After repairing a flat tire in his cool garage, a bicyclist pumps the tire up to full pressure. He then feels the pump and tire, and both feel warmer than before. (Neglect friction. The pump sides and plunger are assumed to be slippery.) Explain why this happens in terms of thermodynamic processes.

Example 12.2 ■ Isobaric Expansion: Change in Internal Energy

An ideal gas occupies a volume of 22.4 L at STP (standard temperature and pressure; see Chapter 10). While absorbing 2.53 kJ of heat from the surroundings, the gas expands isobarically to 32.4 L. What is the change in the internal energy of the gas?

Thinking It Through. Heat is absorbed, and hence Q is positive. Also, for an isobaric process, we can compute the work, W, as the area of a rectangle. Here W will be positive (why?). Lastly, the first law will allow us to compute ΔU.

Solution. Listing the data, along with appropriate conversions, we have

Given: $p_1 = p_2 = 1$ atm $\qquad\qquad$ Find: ΔU (change in internal energy)
$\qquad\qquad\quad = 1.01 \times 10^5$ N/m² (Pa)
$\qquad\quad V_1 = 22.4$ L $= 22.4 \times 10^{-3}$ m³
$\qquad\quad V_2 = 32.4$ L $= 32.4 \times 10^{-3}$ m³
$\qquad\quad T_1 = 0°C = 273$ K
$\qquad\quad Q = 2.53$ kJ $= 2.53 \times 10^3$ J

The change in internal energy is given by the first law: $\Delta U = Q - W$. We know Q, but we must first compute W, the work done by the expanding gas, to find ΔU. Using Eq. 12.2,

$$W = p\Delta V = p(V_2 - V_1)$$
$$= (1.01 \times 10^5 \text{ N/m}^2)[(32.4 \times 10^{-3} \text{ m}^3) - (22.4 \times 10^{-3} \text{ m}^3)] = 1.01 \times 10^3 \text{ J}$$

Then, by the first law,

$$\Delta U = Q - W = (2.53 \times 10^3 \text{ J}) - (1.01 \times 10^3 \text{ J}) = 1.52 \times 10^3 \text{ J}$$

Follow-up Exercise. In this Example, what is the equilibrium temperature, in degrees Celsius, of the gas after the expansion?

12.3 The Second Law of Thermodynamics and Entropy

Objectives: To (a) state and explain the second law of thermodynamics in several forms and (b) explain the concept of entropy.

Suppose that a piece of hot metal is placed in an insulated container of cool water. Heat will be transferred from the metal to the water, and the two will come to thermal equilibrium at some intermediate temperature. For a thermally isolated system, the system's total energy remains constant. Could there be a reverse process in which heat would be transferred from the cooler water to the hot metal? This process would not happen naturally. But if it did, the total energy of the system would remain constant, and the energy transfer would *not* violate energy conservation.

Clearly, there must be another principle that specifies the *direction* in which a process can take place. This principle is embodied in the **second law of thermodynamics**, which says that certain processes do not take place, or have never been observed to take place, even though they are consistent with the first law.

There are many equivalent statements of the second law, which are worded differently according to their application. One applicable to the above situation is

Heat will not flow spontaneously from a colder body to a warmer body.

An alternative statement of the second law involves *thermal cycles*. A thermal cycle typically consists of several separate thermal processes, connected such that they end up back at the starting conditions. If a gas is used as the material, this means that the gas comes back to the same *p-V-T* state in which it started. The second law, stated in terms of a thermal cycle (operating as a heat engine, which we shall discuss in Section 12.4), is as follows:

In a thermal cycle, heat energy cannot be completely transformed into mechanical work.

In general, the second law applies to all forms of energy. It is considered to be true because no one has ever found an exception to it. If it were not valid, a perpetual motion machine could be built. Such a machine could transform heat completely into work and motion (mechanical energy), with no energy loss. The mechanical energy could then be transformed back into heat and be used to reheat the reservoir (again with no loss). Since the processes could be repeated indefinitely, the machine would run perpetually. All of the energy is accounted for, so this

Second law of thermodynamics

Note: Real machines can be made virtually frictionless, but their efficiency is still less than 100%. The second law limit does *not* refer to frictional losses.

situation does not violate the first law. However, it is obvious that real machines are always less than 100% efficient, that the work output is always less than the energy input. Another statement of the second law is therefore:

It is impossible to construct an operational perpetual motion machine.

Many attempts have been made to construct perpetual motion machines, but there have been no known successes.*

It would be convenient to have some way of expressing the direction of a process in terms of the thermodynamic properties of a system. One such property is temperature. In analyzing a conductive heat transfer process, you need to know the temperatures of the system and its surroundings. Knowing the temperature difference allows you to state the direction in which the heat transfer will spontaneously take place, into or out of the system.

Entropy

A property that indicates the *natural direction* of a process was first described by Rudolf Clausius (1822–1888), a German physicist. This property is called **entropy**, a name coined by Clausius. Physically, entropy is a particularly rich, multifaceted concept. Its interpretations are diverse and intriguing. We might draw an analogy to the fable of the Blind Men and the Elephant, in which three blind men describe what each thinks an elephant to be after touching different parts. Thus, a discussion of entropy is likely to resemble a debate:

"Entropy is a measure of a system's ability to do useful work. As a system loses the ability to do work, its entropy increases."

"No, entropy determines the direction of time. It's 'time's arrow' that points out the forward flow of events, thereby distinguishing past events from future ones."

"You're both wrong. Entropy is a measure of disorder. A system naturally moves toward a state of greater disorder or disarray. The more order, the less the system's entropy is."

As you see, entropy has several equally valid physical interpretations. Here we are concerned with the mathematical definition of entropy in terms of thermodynamic properties. The change in a system's entropy (ΔS) when an amount of heat (Q) is added or removed by a reversible process at a constant temperature is

Note: ΔS is positive if a system absorbs heat ($+Q$) and negative if heat leaves ($-Q$). The sign of ΔS is determined by the sign of Q.

$$\Delta S = \frac{Q}{T} \quad \begin{array}{l}\text{change in entropy} \\ \text{(constant temperature only)}\end{array} \quad (12.7)$$

SI unit of entropy: joule per kelvin (J/K)

where T is the Kelvin temperature and the units of entropy are joules per kelvin (J/K).

If the temperature changes during the process, the change in entropy can still be calculated by using advanced mathematics. Our discussions will be limited to isothermal processes or ones with relatively small temperature changes. For the latter, reasonable approximations of entropy changes may be obtained by using average temperatures, as we will see in Example 12.4. First, however, let's look at a relatively simple calculation of entropy change.

Always use kelvins! What would happen if you used the Celsius temperature for a phase change from ice to water at 0°C?

Example 12.3 ■ Change in Entropy: An Isothermal Process

What is the change in entropy of ethyl alcohol when 0.25 kg of it vaporizes at its boiling point of 78°C (latent heat of vaporization $L_v = 1.0 \times 10^5$ J/kg)?

Thinking It Through. A phase change occurs at constant temperature; hence we can apply Eq. 12.7 after converting to kelvins. From Eq. 11.2, we can compute the amount of heat added.

*Although perpetual motion *machines* do not exist, perpetual motion is known to exist—for example, planets in orbit around the Sun.

Solution. From the problem, we have

Given: $m = 0.25$ kg **Find:** ΔS (change in entropy)
$T = 78°C + 273 = 351$ K
$L_v = 1.0 \times 10^5$ J/kg

Since this is a phase change and we are given the latent heat,

$$Q = mL_v = (0.25 \text{ kg})(1.0 \times 10^5 \text{ J/kg}) = 2.5 \times 10^4 \text{ J}$$

Then

$$\Delta S = \frac{Q}{T} = \frac{+2.5 \times 10^4 \text{ J}}{351 \text{ K}} = +71 \text{ J/K}$$

Note that Q is positive because heat is added to the system. The change in entropy, then, is also positive, and the entropy of the alcohol increases.

Follow-up Exercise. What is the change in entropy of water when 1.00 kg of it freezes to form ice at 0°C?

Example 12.4 ■ Warm Spoon into Cool Water: Entropy Change?

A metal spoon at 24°C is placed in 1.00 kg of water at 18°C. The thermally isolated system comes to equilibrium at a temperature of 20°C. Find the approximate change in the entropy of the system.

Thinking It Through. The system is isolated, so we know that $Q_w + Q_m = 0$, where the subscripts w and m stand for water and metal, respectively. We can determine Q_w from the given data, so we have both Q values. Strictly speaking, we cannot use Eq. 12.7 (why?), but since the temperature changes are small, we can get a good approximation for ΔS by using each object's *average* temperature, $\overline{T}$.

Solution. Using subscripts i and f to stand for initial and final, respectively, we have

Given: $T_{m_i} = 24°C$ **Find:** ΔS (change in entropy of the system)
$T_{w_i} = 18°C$
$m_w = 1.00$ kg
$T_f = 20°C$
$c_w = 4186$ J/(kg·C°)
(from Table 11.1)

As in Example 12.3, we need to find the amount of heat (Q) to solve for ΔS. However, in this system there are two Q values: Q_w, the heat gained by the water, and Q_m, the heat lost by the metal.

With $\Delta T_w = T_f - T_{w_i} = 20°C - 18°C = 2.0$ C°, the heat gained by the water is

$$Q_w = m_w c_w \Delta T = (1.00 \text{ kg})[4186 \text{ J/(kg·C°)}](2.0 \text{ C°}) = 8.37 \times 10^3 \text{ J}$$

from Eq. 11.1. This is also the amount of heat *lost* by the metal, since by the conservation of energy, heat gained must equal heat lost. Therefore, $Q_m = -8.37 \times 10^3$ J, where the minus sign indicates a loss of heat.

In this case, we have small temperature changes, so we can approximate the entropy changes by using average temperatures in kelvins:

$$\overline{T}_w = \frac{T_f + T_{w_i}}{2} = \frac{20°C + 18°C}{2} = 19°C = 292 \text{ K}$$

$$\overline{T}_m = \frac{T_{m_i} + T_f}{2} = \frac{24°C + 20°C}{2} = 22°C = 295 \text{ K}$$

We can then use these average temperatures to compute the approximate entropy changes for both the water and the metal:

$$\Delta S_w = \frac{Q_w}{\overline{T}_w} = \frac{+8.37 \times 10^3 \text{ J}}{292 \text{ K}} = +28.7 \text{ J/K}$$

$$\Delta S_m = \frac{Q_m}{\overline{T}_m} = \frac{-8.37 \times 10^3 \text{ J}}{295 \text{ K}} = -28.4 \text{ J/K}$$

The change in the entropy of the *system* is the sum of these results:

$$\Delta S = \Delta S_w + \Delta S_m = +28.7 \text{ J/K} - 28.4 \text{ J/K} = +0.3 \text{ J/K}$$

The entropy of the metal decreased because heat flowed out of it. The entropy of the water increased by a greater amount, so the total entropy of the system overall increased (a positive change).

Follow-up Exercise. What would be the initial temperatures in this Example if the overall entropy change were zero ($\Delta S = 0$)?

Note that the total entropy change of the system in Example 12.4 is positive. In general, the direction of any process is toward an increase in total system entropy. That is, *the entropy of an isolated system never decreases*. Another way to state this observation about entropy is to say that *the entropy of an isolated system increases for every natural process* ($\Delta S > 0$). In coming to an intermediate temperature, the water and spoon in Example 12.4 are undergoing a natural process, so-called because this is what is always observed to occur.

If, in coming to thermal equilibrium with the water, the spoon somehow spontaneously became hotter while the water became cooler, this would certainly be an *unnatural* process ($\Delta S < 0$). Similarly, water at room temperature in an isolated ice cube tray will not naturally turn into ice.

If a system is not isolated, it may undergo a decrease in entropy. For example, if an ice cube tray filled with water, is put into a freezer compartment, the water will freeze, undergoing a decrease in entropy. But there will be a *larger increase in entropy somewhere else* in the environment or the universe. Work has to be done or energy expended to bring about the transfer of heat involved in the change from water to ice. Thus, a more general statement of the second law of thermodynamics in terms of entropy is:

The total entropy of the universe increases in every natural process.

Note: The entropy of a closed system increases for every natural process.

Note: Total energy can neither be created nor destroyed; total entropy can be created but not destroyed.

The entropy of a system is a function of its state. Each state of a system has a particular value of entropy, and a change in entropy depends only on the initial and final states for a process, or $\Delta S = S_f - S_i$. (This is analogous to the change in internal energy for an ideal gas, $\Delta U = U_f - U_i$, which depends only on the initial and final states and *not* on the details of the process.) An entropy change can be represented on a T-S diagram, as shown in •Fig. 12.7. Just as the area under a curve on a p-V diagram for a gas is equal to the work ($W = p\Delta V$), the area under a T-S curve is equal to the heat flow, which is easily seen for the isothermal process in the figure ($Q = T\Delta S$). Like work for a process with variable pressure, heat transferred in a process with variable temperature is also equal to the area under the T-S curve (which, once again, requires advanced mathematics to calculate).

There is a theoretical process for which the entropy is constant. Recall that for an adiabatic process $Q = 0$. Since in this case $\Delta S = Q/T = 0$, there is no change in entropy. Hence, an adiabatic process is a constant-entropy process, or an **isentropic process**. With the inclusion of an ideal adiabatic process, we may say that during any process the entropy of the universe can only increase or remain constant ($\Delta S \geq 0$). How would an isentropic process appear on a T-S diagram?

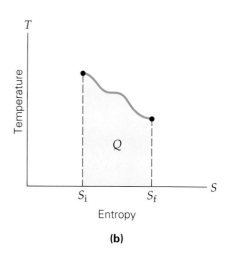

•**FIGURE 12.7 T-S (temperature–entropy) diagrams** When the temperature (T) is plotted versus entropy (S), the area under the process path is equal to the heat (Q) transferred in the process. This is similar to work being equal to the area under a process path on a p-V diagram. **(a)** This T-S diagram is for an isothermal process. (How can you tell?) Note that the area under the path is that of a rectangle. **(b)** This T-S diagram is for a general process in which the temperature varies.

Conceptual Example 12.5

More Order in the System: Entropy Decreasing?

Which of the following involves a *decrease* in the system's entropy: (a) A hot fudge sundae is left uneaten too long, so the ice cream melts and the fudge solidifies; (b) a green plant combines water and carbon dioxide molecules in photosynthesis to make a larger, more complex molecule of sugar; (c) you drop your term paper on the way up the library stairs and find that, when you have gathered up all the pages, they are no longer in sequence; (d) perfume in an opened bottle evaporates and fills the room with scent; (e) a shortage of library personnel makes it increasingly difficult to find the books you want in their proper locations on the shelves?

Reasoning and Answer. It is not difficult to see that both cases (c) and (e) involve a decrease in order and thus an increase in entropy. The same is true, in a slightly less obvious way, of case (d). When the perfume molecules are no longer confined within the restricted space of the bottle but instead are distributed at random throughout the larger volume of air in the room, there is a loss of order—the positions of individual perfume molecules cannot be specified as accurately. Similarly in case (a), the system (fudge plus ice cream) is less highly ordered when the speeds of all the molecules are randomized than when they are divided into two distinct populations, one hot and one cold. (We showed in another way in Example 12.4 how the flow of heat from a warmer body to a colder one entails an increase in entropy.) Hence the answer is (b). In this instance a more complex, highly ordered structure was created from simpler components—a process that will not occur naturally without an input of energy and a larger increase in entropy somewhere else in the universe (see the Insight on p. 401).

Follow-up Exercise. In a garden compost pile, newly clipped grass blades are gradually broken down into a natural fertilizer. Does this process represent an increase or decrease in the total entropy of the grass blades? of the universe?

12.4 Heat Engines and Thermal Pumps

Objectives: To (a) explain the concept of a heat engine and compute thermal efficiency, and (b) explain the concept of a thermal pump and compute coefficient of performance.

Heat engine: converts heat → work

A **heat engine** is any device that converts heat energy to work. Since the second law says that perpetual motion machines are impossible, some of the heat supplied to a heat engine will necessarily be lost. In studying thermodynamics, there is no need to be concerned with the usual mechanical components of an engine, such as pistons, cylinders, and gears. For our purposes, a heat engine is simply a device that takes heat from a high-temperature source (a hot reservoir), converts some of it to useful work, and transfers the rest to its surroundings

Insight

Life, Order, and the Second Law

The second law of thermodynamics is one of the foundations of physics and operates universally—the total entropy of the universe increases in every natural process. Certainly life is a natural process. Yet clearly, there is an increase in order during the development of the human embryo into a mature adult. Moreover, in virtually all life processes larger, more complex molecules and cellular structures are continuously being synthesized from simpler components. Does this mean that living things are exempt from the second law?

To answer this question, we must realize that cells are not closed systems. They are involved in continuous exchanges of matter and energy with their surroundings. Living cells are constantly occupied in the exchange and conversion of energy from one form to another, which is governed by the first law of thermodynamics. For example, cells convert chemical potential energy into kinetic energy (movement) and electrical energy (the basis of nerve impulses). Many plant cells convert sunlight energy into chemical energy through photosynthesis (Fig. 1a). We have taken a hint from nature in attempting to harness the energy in sunlight (Fig. 1b).

In all biological processes, living systems maintain or even increase their organization by causing a larger decrease in the order of their surroundings. The second law is not violated by a *local* decrease in entropy as long as the *total* entropy of the universe increases in the process.

As an example of how cells create order while remaining within the confines of the second law, consider how they use chemical raw materials such as the sugar glucose ($C_6H_{12}O_6$). Cells use carbon atoms from glucose (and other organic compounds) to build up more complex molecules—a decrease in entropy. But they get the energy for these reactions by "burning" a great many glucose molecules as well—oxidizing them to produce many smaller, simpler molecules of water (H_2O) and carbon dioxide (CO_2). This process involves an increase in entropy. The product molecules, being less highly structured than glucose, have greater entropy. Moreover, heat is released by the oxidation process, raising the entropy of the cell's surroundings. Thus a cell "pays" for its increase in order by increasing the disorder of the rest of the universe.

But, you may ask, where did the complex glucose molecule, with its chemical potential energy, come from in the first place? As we mentioned above, photosynthetic cells use the energy of sunlight to make glucose and related compounds. Thus we can look at the flow of energy through organisms in a more general way. As we will see in the Chapter 20 Insight on the greenhouse effect, life on Earth is dependent on receiving radiant energy in the form of sunlight and reradiating infrared energy back into space. In effect, the Sun is a high-temperature source and space is a low-temperature

(a)

(b)

FIGURE 1 Solar collectors (a) Green plants capture trillions of joules of solar energy every day and have been doing so for billions of years. The chemical energy they store in the form of sugars and other complex molecules is used by nearly all organisms on Earth to maintain and increase the highly organized state that we call life. (b) Humans have developed devices (such as these reflective solar energy concentrators) to capture some of the energy of sunlight, although the technology is not yet in widespread use.

sink. The flow of energy from one to another allows living things to decrease entropy locally, creating this island of order that we call the biosphere. Thus biological organisms do not violate the second law, as is sometimes thought; they merely take advantage of a loophole in it.

•**FIGURE 12.8 Work from heat** At this geothermal energy facility in the San Francisco area, natural heat is exploited to do useful work (generating electricity) without the need to "burn" fuel. Water heated by thermal processes beneath the Earth's surface serves as the hot reservoir. (What do you think is the cold reservoir?)

(a cold, or low-temperature, reservoir). For example, most of the turbines that generate the electricity we use (Chapter 20) are heat engines, using input heat from various sources: the burning of chemical fuels (oil, gas, or coal); nuclear reactions (Chapter 30); or the thermal energy already present beneath the Earth's surface (•Figure 12.8). They might be cooled by cold river water, thus losing heat to this low-temperature reservoir. A generalized heat engine is usually represented as shown in •Fig. 12.9a.

Simply adding heat to a gas in a cylindrical piston can produce work in a single process. But since a continuous output is usually wanted, practical heat engines usually operate in a **thermal cycle,** or a series of processes, which brings the engine or system back to its original condition. Cyclic heat engines include steam engines and internal combustion engines, such as automobile engines. An idealized, rectangular thermodynamic cycle is shown in Fig. 12.9b. It consists of two isobars and two isomets. When these processes occur in the sequence indicated, the system goes through a cycle (1-2-3-4-1), returning to its original condition. Recall that for an isobaric process, the work done is equal to the area under the isobar, $W = p\Delta V$, on a p-V diagram (see Fig. 12.3). Here, the work in the 1-2 process is positive and the work in the 3-4 process is negative (why?). Hence, the *net work* (W_{net}) is the area of the rectangle formed by the isobars and isomets.

For a general heat engine cycle, the net work done per cycle is the area enclosed by the process paths that compose the cycle as shown in Fig. 12.9c. When the paths are not straight lines, advanced mathematics is needed to calculate the area. The idea is the same regardless of the details: For a heat engine, the object is to get more positive work out ($+W$, expansion) than must be put back ($-W$, compression). This means a *net positive work* output per cycle.

Conceptual Example 12.6

The Drinking Bird: A Heat Engine

An unusual example of a cyclic heat engine is a novelty item called a drinking bird, shown in •Fig. 12.10. The glass portion of the bird contains a volatile liquid (usually ether). When the absorbent material on the head and beak is initially wet and the bird is released, it begins a slow, periodic motion: dunking its head in the water, bobbing back out for a while, and then repeating the dunking/bobbing cycle. Using the principles of thermodynamics, expain how this "heat engine" works.

•**FIGURE 12.9 Heat engine** (a) Energy flow for a generalized cyclic heat engine. Note that the width of the arrow representing Q_{in} is equal to the combined widths of the arrows representing W_{net} and Q_{out}, reflecting the conservation of energy. (b) This cyclic process consists of two isobars and two isomets. The net work output per cycle is the area of the rectangle formed by the process paths. (c) For a general heat engine cycle, the net work (W_{net}) per cycle is still represented by the area inside the cycle's separate paths.

Reasoning and Answer. When the head and beak are initially wet, evaporation occurs. Thus heat is removed from the head, which lowers its temperature and internal vapor pressure. The higher vapor pressure is now in the body, forcing the liquid into the head. This raises the bird's center of gravity above the pivot point, and a torque causes a forward rotation. The bird then rewets its beak in the glass of water.

The liquid from the body runs into the head, warming it. With the tube not completely filled with liquid and partially open, the pressures in the head and body equalize, so there is no longer a pressure difference. The liquid then drains back into the body. With the lowering of its center of gravity below the pivot point, the bird swings back to the vertical position, and the cycle begins again. The net effect is that heat is transferred from a high-temperature reservoir (the body) to a low-temperature reservoir (the head), with work being done (evidenced by the rotational motion). Gravity and the atmosphere also play roles in the operation of this specialized heat engine.

Follow-up Exercise. Describe how changes in atmospheric conditions such as humidity and pressure would affect the operation of the drinking bird.

Thermal Efficiency

Thermal efficiency is used to rate heat engines. The **thermal efficiency** (ϵ_{th}) of a heat engine is defined as

$$\epsilon_{th} = \frac{\text{net work out}}{\text{heat in}} = \frac{W_{net}}{Q_{in}} \qquad (12.8)$$

The efficiency tells you what you get out (net) for what you put in, so to speak. The higher compression and pressure present in an automotive diesel engine, for example, give it a higher thermal efficiency than a gasoline engine—about 25% as opposed to 15%.* For more about engines, see the Insight on p. 404.

For one cycle of a cyclic heat engine, the work output per cycle W_{net} is determined by the first law as applied to the whole cycle. We now take the symbol Q to mean the *magnitude* of each heat flow and show Q_{out} explicitly as $-Q_{cold}$ (flow out) and Q_{in} as $+Q_{hot}$ (flow in):

$$Q = \Delta U + W \quad \text{or} \quad Q_{hot} - Q_{cold} = 0 + W_{net} \quad \text{(full cycle)}$$

Here we have set ΔU to zero, since the system returns to its original state in completing a cycle ($\Delta T = 0$). The thermal efficiency (per cycle) of a cyclic heat engine is thus given by

•**FIGURE 12.10 A heat engine?** See Conceptual Example 12.6.

$$\epsilon_{th} = \frac{W_{net}}{Q_{hot}} = \frac{Q_{hot} - Q_{cold}}{Q_{hot}} = 1 - \frac{Q_{cold}}{Q_{hot}} \qquad (12.9)$$

Thermal efficiency of a heat engine

In general, each cycle of an engine has the same efficiency.

Like mechanical efficiency, thermal efficiency is a dimensionless fraction and is commonly expressed as a percentage. Equation 12.9 indicates that a heat engine could have 100% efficiency if Q_{cold} were zero. This would mean that no heat energy would be lost and all the heat input would be converted to useful work, which is impossible according to the second law of thermodynamics. In 1851, Lord Kelvin stated this observation, yet another form of the second law:

Note: Here all Q's are positive; heat flow in or out will be shown by a $+$ or $-$ sign.

No heat engine operating in a cycle can convert its heat input completely to work.

The second law, in terms of heat engines

Practically, we see from Eq. 12.9 that to maximize the work output per cycle of a heat engine, we must work to minimize the ratio of Q_{cold}/Q_{hot}, which increases the efficiency.

---

*The thermal efficiency of a heat engine in an automobile depends on many parameters, including the operation of features such as air conditioning. These values are only approximate.

Insight

The Otto Cycle

The internal combustion engine is a cyclic heat engine employed in automobiles and many other devices. Two different cycles of operation are in general use: a two-stroke cycle and a four-stroke cycle. (The number of strokes tells how many up and down motions a piston makes in a cycle; for example, a two-stroke cycle has one upstroke and one downstroke per cycle.) Engines with a two-stroke cycle are commonly used in motorcycles, outboard motors, chain saws, and some lawn mowers.

Most automobile engines utilize a four-stroke cycle. The steps in this cycle are shown in Fig. 1, along with a p-V diagram of the theoretical thermodynamic pro-cesses each cycle approximates. The theoretical cycle is called the Otto cycle, for the German engineer Nickolas Otto (1832–1891), who built one of the first successful gasoline engines.

During the intake stroke (1–2), an isobaric expansion, the air–fuel mixture is admitted at atmospheric pressure through the open valve as the piston drops. This mixture is compressed adiabatically (very quickly) on the compression stroke (2–3), which is followed by ignition (isometric pressure rise) (3–4) and an adiabatic expansion during the power stroke (4–5). Next is an isometric cooling of the system when the piston is at its lowest position (5–2). The final, exhaust stroke is along the isobaric leg of the Otto cycle (2–1). Notice that it takes two cycles of the piston to produce one power stroke.

A two-stroke engine does not have an isobaric leg on its theoretical cycle. The air–fuel mixture is added to the cylinder at the same time as the previously combusted gases are expelled, during the isometric cooling process (5–2). This results in some loss of fuel (in the exhaust gases), which is the major disadvantage of the two-stroke engine. An advantage is that there is one power stroke for every crankshaft revolution in a two-stroke engine, compared to only *one* power stroke for *two* crankshaft revolutions in a four-stroke engine, so a two-stroke engine requires less work and can have higher efficiency.

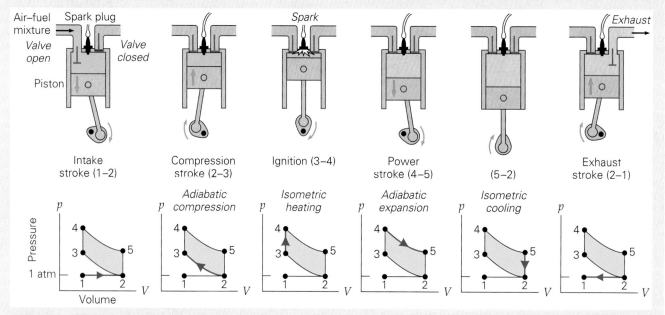

FIGURE 1 The four stroke cycle of a heat engine The process steps for the four-stroke Otto cycle. The piston moves up and down twice each cycle, making a total of four strokes per cycle.

Example 12.7 ■ Thermal Efficiency: What You Get Out of What You Put In

The small, gasoline-powered engine of a leaf blower removes 800 J of heat energy from a high-temperature reservoir and exhausts 700 J to a low-temperature reservoir. What is the engine's thermal efficiency?

Thinking It Through. We can apply the definition of thermal efficiency of a heat engine (Eq. 12.8). We first need W_{net} to determine ϵ_{th}.

Solution.

Given: $Q_{hot} = 800$ J *Find:* ϵ_{th} (thermal efficiency)
$Q_{cold} = 700$ J

Let's calculate W_{net} first from the conservation of energy (first law) applied over one full cycle (so $\Delta U = 0$). Recall that our notation for heat now indicates magnitude only, so we show the minus sign explicity for heat flow out:

$$W_{net} = Q_{hot} - Q_{cold} = 800 \text{ J} - 700 \text{ J} = 100 \text{ J}$$

Therefore, from Eq. 12.8 the thermal efficiency is

$$\epsilon_{th} = \frac{W_{net}}{Q_{hot}} = \frac{100 \text{ J}}{800 \text{ J}} = 0.125 \, (\times 100\%) = 12.5\%$$

Follow-up Exercise. What would be the net output work of the engine in this Example if the efficiency were raised to 15% and the input heat were raised to 1000 J/cycle?

Thermal Pumps: Refrigerators, Air Conditioners, and Heat Pumps

The function performed by a thermal pump is basically the reverse of that of a heat engine. That is, the name **thermal pump** is a generic term for *any* device, including refrigerators, air conditioners, and heat pumps, that transfers heat energy from a low-temperature reservoir to a high-temperature reservoir (•Fig. 12.11a). To do this, there must be work input, since the second law says that heat will not spontaneously flow from a cold body to a hot body.

Note: As used here, the term *thermal pump* is general and does not refer specifically to the devices that are commonly used for heating and cooling of buildings. These will be discussed shortly and are called "heat pumps."

•**FIGURE 12.11 Thermal pumps** (a) An energy-flow diagram for a generalized cyclic thermal pump. The width of the arrow representing Q_{hot}, the heat transferred into the high-temperature reservoir, is equal to the combined widths of the arrows representing W_{in} and Q_{cold}, reflecting the conservation of energy. (b) An air conditioner is an example of a thermal pump. With work input, it transfers heat (Q_{cold}) from a low-temperature reservoir (inside the house) to a high-temperature reservoir (outside). Thermodynamically, it operates like a refrigerator.

Refrigerator operation

A familiar example of a thermal pump is a refrigerator. With work input (from electrical energy), heat is transferred from inside the refrigerator (low-temperature reservoir) to the surroundings (high-temperature reservoir). Have you ever wondered how a refrigerator (or an air conditioner, another thermal pump; Fig. 12.11b) operates? The knowledge you have gained about thermodynamic processes will allow you to understand the basic operation (•Fig. 12.12).

The refrigerant, or heat-transferring medium, is a substance with a relatively low boiling point. Ammonia (boiling point $-33.3°C$ at 1 atm), sulfur dioxide ($-10.1°C$), and Freon ($-29.8°C$) can be used as refrigerants. Until recently Freon was used in most domestic refrigerators. The other compounds were once used, but they were quite dangerous if the system happened to leak. Freon has been replaced by still other refrigerants due to its destruction of the Earth's ozone layer (see the Insight in Chaper 19).

It is convenient to think of the system diagrammed in Fig. 12.12 as having high and low (temperature and pressure) sides. On the low side, heat is transferred to the evaporator coils from inside the refrigerator. The heat causes the refrigerant to boil and is carried away as latent heat of vaporization. The gaseous vapor is drawn into the compressor chamber on the downstroke of the piston and is compressed on the upstroke (work input by compressor motor). The compression increases the temperature of the gas, which is discharged from the compressor as a superheated vapor (on the high side). This vapor condenses in the cooler condenser unit, where circulating air (or water in larger units) carries away the latent heat of condensation and the heat of compression. The condensed liquid collects in a receiver.

An expansion valve maintains a balance between the high and low sides, thereby controlling the rate at which the refrigerant passes back to the low side. On being admitted to the low side, the liquid immediately boils and vaporizes because of the low pressure. This is a cooling process, because the heat of vaporization is supplied by the internal energy of the liquid. The cooled refrigerant is drawn into the evaporator, and the cycle begins again.

•**FIGURE 12.12 Refrigerator operation** Heat (Q_{cold}) is carried away from the interior by the refrigerant as latent heat. This heat energy and that of the work input (W_{in}) are discharged from the condenser to the surroundings (Q_{hot}). A refrigerator can be thought to remove heat (Q_{cold}) from an already cold region *or* as a heat pump adding heat (Q_{hot}) to an already warm area (the kitchen).

In essence, a refrigerator (or air conditioner) pumps heat up a temperature gradient, or "hill." (Think of pumping water up an actual hill against the force of gravity.) The cooling efficiency of this operation depends on the amount of heat extracted from the cold temperature reservoir (the freezer compartment or the inside of a house), Q_{cold}, compared to the work W_{in} needed to do so. Since a practical refrigerator operates in a cycle to provide a continuous removal of heat, $\Delta U = 0$ for the cycle. Then, by the conservation of energy (or first law), $Q_{cold} + W_{in} = Q_{hot}$, where Q_{hot} is the heat ejected to the high temperature reservoir or the outside.

The measure of a refrigerator or air conditioner's performance is defined differently from that of a heat engine, because of the difference in their functions. For these appliances, the efficiency is expressed as a **coefficient of performance (COP)**. Since the purpose is to extract the most heat (Q_{cold}) per unit work input (W_{in}), the coefficient of performance for a refrigerator (or air conditioner), COP_{ref}, is expressed as their ratio:

$$COP_{ref} = \frac{Q_{cold}}{W_{in}} = \frac{Q_{cold}}{Q_{hot} - Q_{cold}} \quad \begin{array}{l}\text{(for refrigerator}\\ \text{or air conditioner)}\end{array} \quad (12.10)$$

Cooling with a refrigerator or air conditioner

where the first-law conservation relationship given above was used to express the input work in terms of the two heats.

Thus, the greater the COP, the better the performance—more heat is extracted for each unit of work done. For normal refrigerator operation, the work input is less than the heat removed, so the COP is greater than 1. The COP's of typical refrigerator cycles range from 3 to 5, depending on operating conditions. This means that the heat removed from the cold reservoir (the refrigerator freezer or the inside of a house) is 3 to 5 times the work needed to remove it.

Refrigerators and air conditioners are generically referred to as thermal pumps because they basically "pump" heat uphill. However, the term **heat pump** is now more specifically applied to the common commercial devices used to cool homes and offices in the summer and to heat them in the winter. The summer operation is that of an air conditioner, as described above. In this mode it cools the inside (of a house) and heats the outdoors. Operating in the heating mode, a heat pump heats the inside and cools the outdoors, usually by taking heat energy from the air. This may seem hard to believe, as it may be quite cold outside. However, keep in mind that the air has internal energy regardless of its temperature.

For a heat pump in its heating mode, the heat input is the major interest, so the COP is defined differently from that of a refrigerator or air conditioner. As you might guess, it is the ratio of Q_{hot} to W_{in} (heat you get out for work you put in):

$$COP_{hp} = \frac{Q_{hot}}{W_{in}} = \frac{Q_{hot}}{Q_{hot} - Q_{cold}} \quad \begin{array}{l}\text{(for heat pump}\\ \text{in heating mode)}\end{array} \quad (12.11)$$

Heating with a heat pump

where again $Q_{cold} + W_{in} = Q_{hot}$. Typical COPs for heat pumps range between 2 and 4, depending on operating conditions.

Compared to electrical heating, heat pumps are efficient. For each unit of electric energy consumed, a heat pump typically pumps in from one and one-half to three times as much heat as direct electric heating systems provide. However, if the outside temperature is very low, a heat pump may not be efficient enough to heat a building adequately. Then it must be supplemented by a conventional heating system. Some heat pumps use water from underground reservoirs, wells, or buried loops of pipe as a heat source and sink. These are more efficient than the ones that use the outside air because water has a larger specific heat than air, and the average temperature difference between the water and the inside air is usually smaller.

Example 12.8 ■ Air Conditioner/Heat Pump: Thermal Switch Hitting

An air conditioner operating in summer extracts 100 J of heat from the interior of the house for every 40 J of electric energy required to operate it. Determine (a) its COP and (b) its COP if it is reversed in the winter, running as a heat pump to move the same amount of heat for the same amount of electric energy.

Thinking It Through. We know the input work and input heat in part (a), so we apply the definition of COP for a refrigerator (Eq. 12.10). Operating in reverse, it is the output heat that is important, so we must find this from the conservation of energy before we can compute the COP for operation as a heat pump.

Solution.

Given: $Q_{cold} = 100$ J *Find:* (a) COP_{ref}
$W_{in} = 40$ J (b) COP_{hp}

(a) Using Eq. 12.10, we determine the COP for this engine operating as an air conditioner:

$$COP_{ref} = \frac{Q_{cold}}{W_{in}} = \frac{100 \text{ J}}{40 \text{ J}} = 2.5$$

(b) When the engine operates as a heat pump, the relevant heat is the output heat, which we can calculate from the conservation of energy:

$$Q_{hot} = Q_{cold} + W_{in} = 100 \text{ J} + 40 \text{ J} = 140 \text{ J}$$

Thus the COP for this engine operating as a heat pump in winter is, from Eq. 12.11,

$$COP_{hp} = \frac{Q_{hot}}{W_{in}} = \frac{140 \text{ J}}{40 \text{ J}} = 3.5$$

Follow-up Exercise. (a) Suppose you redesigned the engine in this Example to perform the same operation but with 25% less work input. What would be the new values of the two COPs? (b) Which COP would have the larger percentage increase?

12.5 The Carnot Cycle and Ideal Heat Engines

Objectives: To (a) explain how the Carnot cycle applies to heat engines, (b) compute the ideal Carnot efficiency, and (c) state the third law of thermodynamics.

Lord Kelvin's statement of the second law of thermodynamics says that any *cyclic* heat engine, regardless of its design, must always exhaust some heat energy (Section 12.4). But how much heat must be lost in the process? In other words, what is the *maximum* possible efficiency of a heat engine? In designing heat engines, engineers strive to make them as efficient as possible, but there must be some theoretical limit, and, according to the second law, it must be less than 100%.

Sadi Carnot (1796–1832), a French engineer, studied this limit. The first thing he sought was the thermodynamic cycle an *ideal* heat engine would use, that is, the most efficient cycle. Carnot found that the ideal heat engine absorbs heat from a *constant* high-temperature reservoir (T_{hot}) and exhausts it to a *constant* low-temperature reservoir (T_{cold}). These are ideally reversible isothermal processes and may be represented as two isotherms on a *p-V* diagram. But what are the processes that complete the cycle and make it the most efficient cycle operating between these temperature extremes? Carnot showed that these are reversible adiabatic processes. As we saw in Section 12.2, they are called adiabats when represented on a graph (●Fig. 12.13a).

Thus, the ideal **Carnot cycle** consists of two isotherms and two adiabats and is conveniently represented on a *T-S* diagram, where it forms a rectangle (Fig. 12.13b). The area under the upper isotherm (1–2) is the heat added to the system from the high-temperature reservoir: $Q_{hot} = T_{hot} \Delta S$. Similarly, the area

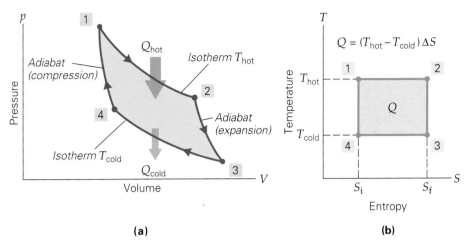

•FIGURE 12.13 The Carnot cycle (a) The Carnot cycle consists of two isotherms and two adiabats. Heat is absorbed during the isothermal expansion and exhausted during the isothermal compression. (b) On a T-S diagram, the Carnot cycle forms a rectangle, the area of which is equal to Q.

under the lower isotherm (3–4) is the heat exhausted: $Q_{cold} = T_{cold}\Delta S$. Here, Q_{hot} and Q_{cold} are magnitudes of heat transfers at *constant* temperatures (T_{hot} and T_{cold}). There is no heat transfer ($Q = 0$) in the adiabatic legs of the cycle. (Why?) This is not typical of practical heat-engine cycles. See Exercises 68 and 69.

The difference between these heat transfers is the work output, which is equal to the area of the rectangle enclosed by the process paths (the shaded area on the diagram):

$$W_{net} = Q_{hot} - Q_{cold} = (T_{hot} - T_{cold})\Delta S$$

Since ΔS is the same for the areas under both isotherms, these expressions can be used to relate the temperatures and heats. That is, since

$$Q_{hot} = T_{hot}\Delta S \quad \text{and} \quad Q_{cold} = T_{cold}\Delta S$$

then

$$\frac{Q_{hot}}{T_{hot}} = \frac{Q_{cold}}{T_{cold}} \quad \text{or} \quad \frac{Q_{hot}}{Q_{cold}} = \frac{T_{hot}}{T_{cold}} \tag{12.12}$$

This equation can be used to express the efficiency of an ideal heat engine in terms of temperature. From Eq. 12.9, this ideal **Carnot efficiency** (ϵ_C) is

$$\epsilon_C = 1 - \frac{Q_{cold}}{Q_{hot}} = 1 - \frac{T_{cold}}{T_{hot}}$$

or

$$\epsilon_C = 1 - \frac{T_{cold}}{T_{hot}} = \frac{T_{hot} - T_{cold}}{T_{hot}} \quad \begin{array}{c}\textit{Carnot efficiency}\\ \textit{(for an ideal}\\ \textit{heat engine)}\end{array} \tag{12.13}$$

Note: Keep in mind that the temperature in the Carnot efficiency is the absolute, or Kelvin, temperature.

where the fractional efficiency is often expressed as a percentage. Note that T_{cold} and T_{hot} are Kelvin temperatures.

The Carnot efficiency expresses the theoretical upper limit on the thermodynamic efficiency of a cyclic heat engine; this limit can never be achieved. A true Carnot engine cannot be built because the necessary reversible processes can only be approximated. Although reversible adiabatic processes can be approached, reversible isothermal processes are virtually impossible to approximate during the heat transfer processes in a real engine.

Note: The Carnot efficiency, which can never be attained, gives an ideal upper limit.

However, the Carnot efficiency does show that the greater the difference in the temperatures of the heat reservoirs, the greater the efficiency. For example, if T_{hot} is twice T_{cold}, or $T_{cold}/T_{hot} = \frac{1}{2} = 0.50$, the Carnot efficiency will be

$$\epsilon_C = 1 - \frac{T_{cold}}{T_{hot}} = 1 - 0.50 = 0.50 \,(\times 100\%) = 50\%$$

But if T_{hot} is four times T_{cold}, or $T_{cold}/T_{hot} = \frac{1}{4} = 0.25$, then

$$\epsilon_C = 1 - \frac{T_{cold}}{T_{hot}} = 1 - 0.25 = 0.75\ (\times 100\%) = 75\%$$

Since a heat engine can never attain 100% thermal efficiency, it is useful to compare its actual efficiency, ϵ_{th}, to its theoretical maximum efficiency, that of a Carnot cycle, ϵ_C. The ratio of these two efficiencies is sometimes called the engine's **relative efficiency**, ϵ_{rel}, defined as

$$\epsilon_{rel} = \frac{\epsilon_{th}}{\epsilon_C} \quad \textit{(relative efficiency of a heat engine)} \quad (12.14)$$

We can express this quantity as a ratio or a percentage by multiplying by 100%. To see the importance of this concept in more detail, study the next Example carefully.

Example 12.9 ■ Carnot Efficiency: The Yardstick of Any Real Engine

An engineer is designing a cyclic heat engine to operate between temperatures of 150°C and 27°C. (a) What is the maximum theoretical efficiency that can be achieved? (b) Suppose the engine, when built, does 100 J of work per cycle for every 500 J of input heat per cycle. What is its relative efficiency?

Thinking It Through. The maximum efficiency for certain high and low temperatures is given by Eq. 12.13. We must convert temperatures to the Kelvin scale. In part (b), we need to calculate the actual efficiency and compare it to our answer in (a).

Solution.

Given: $T_{hot} = 150°C + 273 = 423$ K *Find:* (a) ϵ_C (Carnot efficiency)
$T_{cold} = 27°C + 273 = 300$ K (b) ϵ_{rel} (relative efficiency)
$W_{net} = 100$ J
$Q_{hot} = 500$ J

(a) Using Eq. 12.13, we get

$$\epsilon_C = 1 - \frac{T_{cold}}{T_{hot}} = 1 - \frac{300\ K}{423\ K} = 0.291\ (\times 100\%) = 29.1\%$$

(b) The actual efficiency is, from Eq. 12.8,

$$\epsilon_{th} = \frac{W_{net}}{Q_{hot}} = \frac{100\ J}{500\ J} = 0.200,\ \text{or}\ 20.0\%$$

Thus

$$\epsilon_{rel} = \frac{\epsilon_{th}}{\epsilon_C} = \frac{0.200}{0.291} = 0.687,\ \text{or}\ 68.7\%$$

In other words, the engine is operating at almost 70% of its theoretical maximum. Not bad!

Follow-up Exercise. If the operating high temperature of the engine in this Example were increased to 200°C, what would be the change in the theoretical efficiency?

There are also Carnot COPs for refrigerators and heat pumps. (See Exercise 88.)

The Third Law of Thermodynamics

Another interesting conclusion can be drawn from the expression for the Carnot efficiency (Eq. 12.13). To have ϵ_C equal to 100%, T_{cold} would have to be absolute zero (see Section 10.3). Since a heat engine with 100% efficiency is impossible by the second law, the **third law of thermodynamics** is as follows:

It is impossible to reach a temperature of absolute zero.

Absolute zero has never been observed experimentally. If it were, this would in effect violate the second law. In cryogenic (low-temperature) experiments, scientists have come close to absolute zero—to about 20 nK (2.0×10^{-8} K)—but have never reached it.

Chapter Review

Important Terms

thermodynamics 388
system 389
thermally isolated system 389
heat reservoir 389
equations of state 389
process 390
irreversible process 390
reversible process 390
first law of thermodynamics 390

isobaric process 391
isometric process 392
isothermal process 393
adiabatic process 393
second law of thermodynamics 396
entropy 397
isentropic process 399
heat engine 400
thermal cycle 402

thermal efficiency 403
thermal pump 405
coefficient of performance (COP) 407
heat pump 407
Carnot cycle 408
Carnot (ideal) efficiency 409
relative efficiency 410
third law of thermodynamics 410

Important Concepts

- The state of a system is described by an equation of state. A process is a change in the state of a system. There are ideal reversible processes (changes through equilibrium states) and irreversible processes (changes through nonequilibrium states).
- The first law of thermodynamics is a statement of the conservation of energy for a thermodynamic system.
- The second law of thermodynamics specifies whether a process can take place naturally or the direction of a process.

- The total entropy of the universe increases in every natural process.
- A heat engine is a device that converts heat energy into work. Thermal efficiency is the ratio of work output to heat input.
- A thermal pump is a device that transfers heat energy from a low-temperature reservoir to a high-temperature reservoir. The coefficient of performance (COP) is the ratio of useful heat transferred to the work input required to do this.

Important Equations

First Law of Thermodynamics:
$$Q = \Delta U + W \qquad (12.1)$$

sign convention $\begin{cases} +Q, \text{ heat } added \text{ to system} \\ -Q, \text{ heat } removed \text{ from system} \\ +W, \text{ work done } by \text{ system} \\ -W, \text{ work done } on \text{ system} \end{cases}$

Work Done by an Expanding Gas (constant pressure):
$$W = p\,\Delta V = p(V_2 - V_1) \qquad (12.2)$$

Change in Entropy (constant temperature):
$$\Delta S = \frac{Q}{T} \qquad (12.7)$$

Thermal Efficiency of a Heat Engine:
$$\epsilon_{th} = \frac{W_{net}}{Q_{hot}} = \frac{Q_{hot} - Q_{cold}}{Q_{hot}} = 1 - \frac{Q_{cold}}{Q_{hot}} \qquad (12.9)$$

Coefficients of Performance:
$$COP_{ref} = \frac{Q_{cold}}{W_{in}} = \frac{Q_{cold}}{Q_{hot} - Q_{cold}} \qquad (12.10)$$
(for refrigerator or air conditioner)

$$COP_{hp} = \frac{Q_{hot}}{W_{in}} = \frac{Q_{hot}}{Q_{hot} - Q_{cold}} \qquad (12.11)$$
(for heat pump in heating mode)

Carnot Efficiency of an Ideal Heat Engine:
$$\epsilon_C = 1 - \frac{T_{cold}}{T_{hot}} = \frac{T_{hot} - T_{cold}}{T_{hot}} \qquad (12.13)$$

Exercises*

12.1 Thermodynamic Systems, States, and Processes

1. There may be an exchange of heat with the surroundings for (a) a thermally isolated system, (b) a completely isolated system, (c) a heat reservoir, (d) none of these.

2. Only initial and final states are known for irreversible processes on (a) *p-V* diagrams, (b) *p-T* diagrams, (c) *V-T* diagrams, (d) all of these.

3. On a *p-V* diagram, a reversible process is one (a) whose path is known, (b) whose path is unknown, (c) for which the intermediate steps are nonequilibrium states, (d) none of these.

4. Explain why the *p-V* diagram in Fig. 12.1b is not that of an ideal gas.

12.2 The First Law of Thermodynamics

5. On a *p-V* diagram, sketch a general cyclic process that consists of an isothermal expansion, an isobaric compression, and an isometric process—in that order.

6. On a *p-T* diagram, sketch the general paths for the following reversible processes for an ideal gas: (a) isothermal, (b) isobaric, and (c) isometric.

7. There is no exchange of heat in an (a) isothermal process, (b) adiabatic process, (c) isobaric process, (d) isometric process.

8. According to the first law of thermodynamics, if work is done on a system, then (a) the internal energy of the system must change, (b) heat must be transferred from the system, (c) the internal energy of the system changes and/or heat is transferred from the system, (d) heat is transferred to the system.

9. If heat is added to a system of ideal gas during an isothermal expansion process, (a) work is done on the system, (b) the internal energy decreases, (c) the effect is the same as for an isometric process, (d) none of these.

10. (a) Why does the body of a hand pump become hot when a tire is pumped up? (Neglect friction. The plunger is usually lubricated to prevent air leakage.) (b) Why does the valve of a tire become cold when air is released from the tire?

11. In •Fig. 12.14, the plunger is pushed in quickly in a syringe and the small pieces of paper in the syringe catch fire. Explain by using the first law of thermodynamics.

*Take temperatures to be exact—that is, with no uncertainty.

(Similarly, in a diesel engine, there are no spark plugs. How can the air–fuel mixture be ignited?)

•**FIGURE 12.14 Fire syringe** See Exercise 11.

12. ■ A rigid container contains 1 mol of ideal gas that slowly receives 2.0×10^4 J of heat. What is the change in the internal energy of the gas in joules?

13. ■ A quantity of ideal gas goes through a cyclic process and does 400 J of net work. (a) Is a net amount of heat added to or removed from the system, and how much heat is involved? (b) Does the temperature of the gas increase or decrease over the cycle?

14. ■ The internal energy of an ideal gas system increases by 20 J, while 50 J of heat is removed from the system. Does the system do work, or is work done on the system? How much work is involved?

15. ■ While doing 500 J of work, a system of ideal gas is expanded adiabatically to 1.5 times its volume. (a) Does the temperature of the gas increase or decrease? (b) How much heat is transferred? (c) What is the change in the internal energy of the gas?

16. ■ An ideal gas system expands from 1.0 m³ to 3.0 m³ at STP (standard temperature and pressure). If the system absorbs 5.0×10^4 J of heat in the process, what is the change in internal energy of the system?

17. ■■ A system of gas at low density has an initial pressure of 1.65×10^4 Pa and occupies a volume of 0.20 m³. The slow addition of 1000 J of heat to the system causes it to expand isobarically to a volume of 0.40 m³. (a) How much work is done by the system in the process? (b) Does the internal energy of the system change? If so, by how much?

18. ■■ An Olympic weight lifter lifts 145 kg a vertical distance of 2.1 m. In doing so, his internal energy decreases by 6.0×10^4 J. Consider the weight lifter to be a thermodynamic system. How much heat flows and in what direction?

19. ■■ A system of ideal gas is taken through the reversible processes shown in •Fig. 12.15. In terms of state variables, (a) what is the overall change in the internal energy of the gas? (b) How much work is done by or on the gas? (c) What is the amount of heat transfer in the overall process?

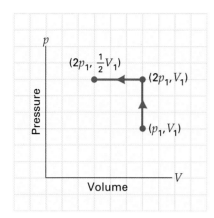

•FIGURE 12.15 A *p-V* diagram for an ideal gas See Exercise 19.

20. ■■ The temperature of 2.0 mol of ideal gas is increased from 150°C to 250°C by two different process paths. In one, 2500 J of heat is added to the system; in the other, 3000 J of heat is added. (a) What is the change in the internal energy of the gas in each case? [Hint: See Eq. 10.16.] (b) In which case is more work done, and how much more?

21. ■■ A gram of water (1.00 cm³) at 100°C is converted to a volume of 1671 cm³ of steam at atmospheric pressure. What is the change in the internal energy of the system?

22. ■■ ⊙ A fixed quantity of gas undergoes the reversible changes illustrated in the *p-V* diagram in •Fig. 12.16. How much work is done in each process?

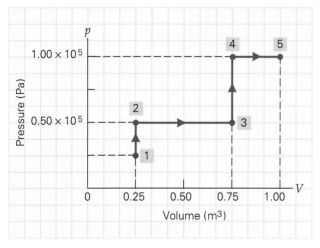

•FIGURE 12.16 A *p-V* diagram and work See Exercises 22 and 23.

23. ■■ ⊙ Suppose that after the final process shown in Fig. 12.16 (see Exercise 22), the pressure of the gas is decreased isometrically from 1.0×10^5 Pa to 0.70×10^5 Pa, and then the gas is compressed isobarically from 1.0 m³ to 0.80 m³. What is the total work done in all of these processes, including 1 through 5?

24. ■■■ A quantity of gas is compressed as shown on the *p-V* diagram in •Fig. 12.17. How much work is done on the system?

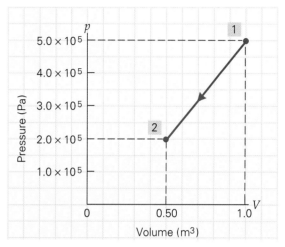

•FIGURE 12.17 A variable *p-V* process and work See Exercise 24.

25. ■■■ ⊙ One mole of ideal gas is taken through the cyclic process shown in •Fig. 12.18. (a) Compute the work involved (*W*) for each of the four processes. (b) Find ΔU, *W*, and *Q* for the complete cycle. (c) What is T_3?

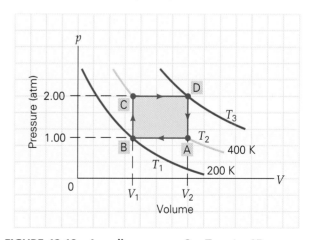

•FIGURE 12.18 A cyclic process See Exercise 25.

12.3 The Second Law of Thermodynamics and Entropy

26. The second law of thermodynamics (a) describes the state of a system, (b) applies only when the first law is satisfied, (c) precludes perpetual motion machines, (d) does not apply to an isolated system.

27. In any natural process, the change in the entropy of the universe is (a) negative, (b) positive, (c) zero, (d) the same as the change in the internal energy.

28. Do the entropies of the systems in the following processes increase or decrease? (a) *ice* melts, (b) *water vapor* condenses, (c) *water* is heated on a stove, (d) *food* is cooled in a refrigerator.

29. In Example 12.4, what would be the implication if the total change in entropy had been negative?

30. When a quantity of hot water is mixed with a quantity of cold water, the combined system comes to thermal equilibrium at some intermediate temperature. How does the entropy of the system change?

31. ■ What change in entropy is associated with the reversible phase change of 1.0 kg of ice to water at 0°C?

32. ■ What is the change in entropy when 0.50 kg of steam condenses to water at 100°C?

33. ■ What is the change in entropy when 0.50 kg of mercury vapor ($L_v = 2.7 \times 10^5$ J/kg) condenses to a liquid at its boiling point of 357°C?

34. ■ Which process has the greater change in entropy: 0.75 kg of ice changing to water at 0°C, or 0.25 kg of water changing to steam at 100°C?

35. ■ One mole of ideal gas goes through an isothermal compression at room temperature. If 7.5×10^3 J of work is done in compressing the gas, what is the change in entropy of the system?

36. ■■ A quantity of an ideal gas initially at STP undergoes a reversible isothermal expansion and does 3.0×10^3 J of work on its surroundings in the process. What is the change in the entropy of the gas?

37. ■■ During a liquid-to-solid phase change of a quantity of a substance, the change in entropy is -4.19×10^3 J/K. If 1.67×10^6 J of heat is removed in the process, what is the freezing point of the substance in degrees Celsius?

38. ■■ An isolated system consists of two reservoirs with temperatures 373 K and 273 K, respectively. If 1000 J of heat were to flow from the cold reservoir to the hot reservoir spontaneously, (a) what would be the change in entropy of the hot reservoir? (b) of the cold reservoir? (c) What would be the total change in entropy of the isolated system? (d) Could the process take place naturally?

39. ■■ Two heat reservoirs at temperatures 200°C and 60°C, respectively, are brought into thermal contact, and 1.50×10^3 J of heat spontaneously flows from one to the other. What is the change in the entropy of the two-reservoir system?

40. ■■ A system of ideal gas at 20°C undergoes an isometric process with an internal energy decrease of 4.50×10^3 J to a final temperature of 16°C. What is the approximate change in entropy in joules per kelvin?

41. ■■ (a) How much heat is transferred in the processes shown on the T-S diagram in •Fig. 12.19? (b) What type of process takes place as the system goes from 2 to 3?

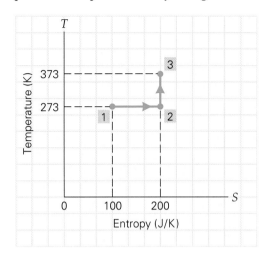

•FIGURE 12.19 **Entropy and heat** See Exercises 41 and 42.

42. ■■ Suppose that the system described by the T-S diagram in Fig. 12.19 is returned to its original state, 1, by a reversible process depicted by a straight line from 3 to 1. What is the change in entropy of the total cycle? How much heat is transferred in the cyclic process?

43. ■■■ A 50.0-gram ice cube at 0°C is placed in 500 mL of water at 20°C. Estimate the change in entropy when all the ice has melted (a) for the ice, (b) for the water, and (c) for the ice–water system.

12.4 Heat Engines and Thermal Pumps*

44. For a cyclic heat engine, (a) $\epsilon_{th} > 1$, (b) $Q_{hot} = W_{out}$, (c) $\Delta U = W_{out}$, (d) $Q_{hot} > Q_{cold}$.

45. A thermal pump (a) is rated by thermal efficiency, (b) requires work input, (c) is not consistent with the second law, (d) violates the first law.

46. Is leaving a refrigerator door open a practical way to air condition a room? Explain.

47. Lord Kelvin's statement of the second law ("No heat engine operating in a cycle can convert its heat input completely to work") as applied to heat engines refers to their operating *in a cycle*. Why is the italicized phrase included?

48. The energy output of a thermal pump is greater than the energy used to operate the pump. Is this device a violation of the first law of thermodynamics?

*Consider efficiencies to be exact.

49. In normal atmospheric convection cycles, colder air from a higher altitude is transferred to a lower, warmer level. Does this process violate the second law? Explain.

50. ■ A gasoline engine is 28% efficient. If the engine absorbs 2000 J of heat in each cycle, (a) what is the work output in each cycle? (b) How much heat is exhausted in each cycle?

51. ■ If an engine does 200 J of work and exhausts 600 J of heat per cycle, what is its efficiency?

52. ■ A heat engine with an efficiency of 40% does 800 J of work each cycle. How much heat per cycle is lost to the surroundings (the low-temperature reservoir)?

53. ■ An internal combustion engine with a thermal efficiency of 15.0% does 2.60×10^4 J of work each cycle. How much heat is lost by the engine in each cycle?

54. ■ The heat output of a particular engine is 7.5×10^3 J per cycle, and the work output is 4.0×10^3 J per cycle. What is the thermal efficiency of the engine?

55. ■■ A steam engine does 4500 J of useful work each cycle but loses 500 J to friction and exhausts 6300 J of heat energy. What is the engine's efficiency?

56. ■■ An engineer redesigns a heat engine and improves its thermal efficiency from 20% to 25%. (a) Does the ratio of the heat output to heat input increase or decrease, and why? (b) What is the change in Q_{cold}/Q_{hot}?

57. ■■ A gasoline engine burns gas as it releases 3.3×10^8 J of heat per hour. (a) What is the energy input during a period of 2.0 h? (b) If the engine delivers 25 kW of power during this time, what is its efficiency?

58. ■■ A refrigerator takes heat from its cold interior at a rate of 7.5 kW when the required work is done at a rate of 2.5 kW. At what rate is heat exhausted to the kitchen?

59. ■■ A refrigerator with a COP of 2.2 removes 4.2×10^5 J of heat from its storage area each cycle. (a) How much heat is exhausted each cycle? (b) What is the total work input in joules for 10 cycles?

60. ■■ A heat pump removes 2.0×10^3 J of heat from the outdoors and delivers 3.5×10^3 J of heat to the inside of a house each cycle. (a) How much work is required per cycle? (b) What is the COP of the pump?

61. ■■ An air conditioner has a COP of 2.75. What is its power rating if it is to remove 1.00×10^7 J of heat in 20 min?

62. ■■ A heat engine has a thermal efficiency of 30.0%. Its heat input each cycle is supplied by the condensation of 8.00 kg of steam at 100°C. (a) What is the work output per cycle? (b) How much heat is lost to the surroundings each cycle?

63. ■■ A heat pump that uses an underground water reservoir as a heat source extracts 2.1×10^5 J each cycle while requiring 3.0×10^4 J of work. (a) How much heat is delivered to the inside of the house? (b) What is the temperature drop of the water when returned to the reservoir if 5.0 kg is used each cycle?

64. ■■■ A coal-fired power plant produces 900 MW of electricity and operates at an overall thermal efficiency of 35%. (a) What is the thermal input (power) to the plant? (b) What is the rate of heat discharge from the plant? (c) Why is the water heated by the discharge cooled in a cooling tower before being ejected into a nearby river?

65. ■■■ A four-stroke engine runs on the Otto cycle and delivers 150 hp. If the efficiency of the engine is 20%, what is (a) the rate of energy input? (b) the rate of heat energy exhausted to the environment?

12.5 The Carnot Cycle and Ideal Heat Engines

66. The Carnot cycle consists of (a) two isobaric and two isothermal processes, (b) two isometric and two adiabatic processes, (c) two adiabatic and two isothermal processes, (d) four arbitrary processes that return the system to its initial state.

67. Which of the following temperature reservoir relationships would have the highest efficiency for a Carnot engine? (a) $T_{cold} = 0.15 T_{hot}$, (b) $T_{cold} = 0.25 T_{hot}$, (c) $T_{cold} = 0.50 T_{hot}$, (d) $T_{cold} = 0.90 T_{hot}$

68. An idealized cycle for a two-stroke internal combustion engine is shown in •Fig. 12.20. It consists of an adiabatic expansion and compression (legs 1 and 3, respectively) and an isometric decompression and compression (legs 2 and 4, respectively). The text stated that in a Carnot cycle, heat transfers occur at constant temperatures. For the cycle in Fig. 12.20, (a) is heat taken in at a constant temperature? (b) Is heat rejected at a constant temperature? (c) Draw a sketch that indicates the areas representing work (W, with

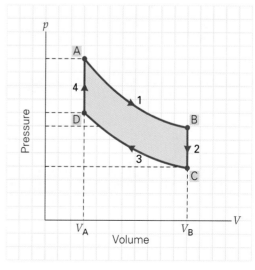

•FIGURE 12.20 **Non-Carnot cycle** Two adiabats and two isomets. See Exercises 68 and 69. (Not drawn to scale.)

signs) during expansion and compression, and draw another sketch for the net work.

69. In Fig. 12.20, legs 1 and 3 are adiabatic processes. Suppose these were isothermal processes instead. (a) Does this change make the cycle a Carnot cycle? Explain. (b) In which legs are there heat transfers? [Indicate directions by signs.] Do these transfers occur at constant temperatures?

70. Automobile engines can be either air-cooled or water-cooled. Which type of engine would you expect to be more efficient, and why?

71. ■ Diesel engines are more efficient than gasoline engines. Which type of engine runs hotter? Why?

72. ■ A steam engine operates between 100°C and 20°C. What is its Carnot efficiency?

73. ■ A Carnot engine with an efficiency of 35% takes in heat from a high-temperature reservoir at 147°C. What is the Celsius temperature of the low-temperature reservoir?

74. ■ What is the temperature of the hot reservoir of a 30% efficient Carnot engine with a 20°C cold reservoir?

75. ■ It has been proposed that the temperature difference in the ocean could be utilized to run a heat engine to generate electricity. In tropical regions, the water temperature is about 25°C at the surface and about 5°C at depth. (a) What would be the maximum theoretical efficiency of such an engine? (b) Would a heat engine with such a low efficiency be practical? Explain.

76. ■ An engineer wants to run a heat engine with an efficiency of 40% between a high-temperature reservoir at 350°C and a low-temperature reservoir. Below what temperature must the low-temperature reservoir be for practical operation of the engine?

77. ■■ An ideal heat engine takes 2.7 × 10⁴ J of heat per cycle from a high-temperature reservoir at 320°C and exhausts some of it to a low-temperature reservoir at 120°C. How much net work is done by the engine per cycle?

78. ■■ A Carnot engine of efficiency 40% operates with a low-temperature reservoir at 50°C and exhausts 1200 J of heat each cycle. What are (a) the heat input per cycle and (b) the Celsius temperature of the high-temperature reservoir?

79. ■■ An ideal heat engine takes in heat from a reservoir at 327°C and has an efficiency of 30%. If the exhaust temperature does not vary and the efficiency is increased to 40%, what would be the increase in the temperature of the hot reservoir?

80. ■■ An inventor claims to have developed a heat engine that, on each cycle, takes in 5.0 × 10⁵ J of heat from a high-temperature reservoir at 400°C and exhausts 2.0 × 10⁵ J to the surroundings at 125°C. Would you invest your money in the production of this engine? Explain.

81. ■■ Equation 12.13 shows that the greater the temperature difference between the reservoirs of a heat engine, the greater the Carnot efficiency. Suppose you had the choice of raising the high-temperature reservoir a certain number of kelvins or lowering the low-temperature reservoir the same number of kelvins. Which would you choose to increase the efficiency? Prove your answer numerically.

82. ■■ A heat engine operates at a relative efficiency of 45% (ϵ_{rel}). If the temperature of the high-temperature and low-temperature reservoirs are 400°C and 100°C, respectively, what are the Carnot efficiency ϵ_C and the thermal efficiency ϵ_{th} of the engine?

83. ■■ A heat engine has half the thermal efficiency (ϵ_{th}) of a Carnot engine operating between temperatures of 100°C and 375°C. (a) What is the relative efficiency of the heat engine? (b) If the heat engine absorbs heat at a rate of 50 kW, at what rate is heat exhausted?

84. ■■ The working substance of a cyclic heat engine is 0.75 kg of an ideal gas. The cycle consists of two isobaric processes and two isometric processes as shown in •Fig. 12.21. What would be the efficiency of a Carnot engine operating with the same high-temperature and low-temperature reservoirs?

•FIGURE 12.21 Thermal efficiency   See Exercise 84.

85. ■■ In each cycle, a Carnot engine takes 800 J of heat from a high-temperature reservoir and discharges 600 J to a low-temperature reservoir. (a) What is the Carnot efficiency of the engine? (b) What is the temperature ratio of the high-temperature reservoir to the low-temperature reservoir?

86. ■■ A Carnot engine operating between reservoirs at 27°C and 227°C does 1500 J of work each cycle. (a) What is the efficiency of the engine? (b) What is the change in entropy of the engine each cycle?

87. ■■ Because of material limitations, the maximum temperature of the superheated steam used in a turbine for electrical generation is about 540°C. (a) If the steam condenser operates at 20°C, what is the ideal efficiency? (b) The actual efficiency is about 35–40%. What does this tell you?

88. ■■■ There is a Carnot coefficient of performance (COP_C) for an ideal or Carnot heat pump. (a) Show that this quantity is given by

$$COP_C = \frac{T_{hot}}{T_{hot} - T_{cold}}$$

(b) What does this quantity tell you about adjusting the temperatures for maximum performance of a heat pump? (Could you guess the equation for the COP_C for a refrigerator?)

89. ■■■ A salesperson tells you that a new refrigerator with a high COP removes 2.6×10^3 J each cycle from the inside of the refrigerator at a temperature of 5.0°C and rejects 2.8×10^3 J into the 30°C kitchen. (a) What is the refrigerator's COP? (b) Is this scenario possible? Justify your answer. (See Exercise 88.)

90. ■■■ A Carnot engine has an efficiency of 40%. Suppose it could be run in reverse as a heat pump. What would be the COP_C of the pump? (See Exercise 88.)

91. ■■■ An ideal heat pump is equivalent to a Carnot engine running in reverse. Show that the input work required for an ideal heat pump is given by

$$W_{in} = Q_{cold}\left[\left(\frac{T_{hot}}{T_{cold}}\right) - 1\right]$$

Additional Exercises

92. ⚫ A thermally isolated quantity of gas with an initial volume of 10 L is compressed isobarically at a pressure of 300 kPa. If 900 J of work is done on the system, what is the final volume of the gas?

93. For the two-step reversible process illustrated in •Fig. 12.22, the same amount of heat is added to and removed from the system. Prove that the total entropy (of the universe) increases in the process.

•FIGURE 12.22 **Entropy increase?** See Exercise 93.

94. Use the general definition of efficiency of a heat engine, $\epsilon_{th} = W_{net}/Q_{in}$, to derive the Carnot efficiency. [Hint: Consider the cycle in terms of entropy, and see Fig. 12.13b.]

95. ⚫ A gas is enclosed in a cylindrical piston with a radius of 12.0 cm. Heat is slowly added to the system while the pressure is maintained at 1.00 atm, and the piston moves 6.00 cm. (a) What type of process is this? (b) If the quantity of heat transferred to the system during the expansion is 420 J, what is the change in the internal energy of the gas?

96. A heat engine operates between a reservoir at 20°C and one at 250°C. What is the maximum theoretical efficiency of the engine? Could this efficiency ever be achieved?

97. Two automobiles with equal masses of 1500 kg are traveling at 90 km/h when they have a head-on, totally inelastic collision at 20°C. Estimate the change in entropy of this process. Assume all the kinetic energy is converted to heat.

98. The energy or useful work lost as a result of a change in the entropy of the universe is given by $W = T_{cold}\Delta S_u$, where T_{cold} is the temperature of the cold reservoir for the process and S_u is the entropy of the universe. (a) Show that for heat (Q) transferred from a high-temperature reservoir to a low-temperature reservoir, $W = Q[1 - (T_{cold}/T_{hot})]$. (b) What is the quantity within the brackets?

99. In an isothermal expansion at 27°C, an ideal gas does 30 J of work. What is the change in entropy of the gas?

100. An ideal gas in a cylinder is compressed adiabatically while 1500 J of work is done in moving a piston in the cylinder. What is the change in the internal energy of the gas? Does the temperature of the gas increase or decrease?

101. In a thermodynamic process, 2500 J of heat is transferred to a system and 1000 J of work is done on the system. What is the change in the internal energy of the system?

102. A heat reservoir supplies heat to melt 0.75 kg of ice at 0°C. Once all of the ice has melted, what is the entropy change?

103. In 1834 Joule performed an important experiment on the internal energy of a gas. Two containers connected by a valve are submerged in a water bath (•Fig. 12.23). Container A is filled with a gas at high pressure, and container B is empty (evacuated). If the valve is opened quickly, the gas in container A undergoes "free expansion" into the evacuated container (in other words, the gas does no work in pushing against "a vacuum"). The temperature of the water bath is the same after the gas has again come to equilibrium. Does the internal energy of the gas change?

•FIGURE 12.23 **Free expansion** See Exercise 103.

CHAPTER 13

Vibrations and Waves

13.1 Simple Harmonic Motion
13.2 Equations of Motion
13.3 Wave Motion
13.4 Wave Properties
13.5 Standing Waves and Resonance

Insights

- Earthquakes, Seismic Waves, and Seismology
- Desirable and Undesirable Resonances

This is what a lot of people probably think of first when they hear the word *wave*. We're all familiar with ocean waves or their smaller relatives, the ripples that form on the surface of a lake or pond when something disturbs the surface. Yet in many ways, the waves that are most important to us, as well as most interesting to physicists, are either invisible or don't look like waves. Sound, for example, is a wave. As a wave traveling through a medium, it shares many properties with water waves. Perhaps most surprisingly, light is a wave. In fact, all electromagnetic radiations are waves—radio waves, microwaves, X-rays, and so on. Whenever you peer through a microscope or put on a pair of glasses or look up at a rainbow, you are experiencing wave energy in the form of light. In Chapter 28 you'll learn how even moving particles have wavelike properties. But first, we need to look at the basic description of waves.

What is a wave, and what properties do waves of *all* kinds have in common? You will find that to understand wave motion, it is necessary to study some other kinds of motion that at first might not seem closely related to waves. It turns out, however, that the motion of a weight bouncing up and down on a spring or the swinging of a pendulum in a grandfather clock provides important keys to an understanding of wave motion, whether it be that of an electron, a Slinky®, or the perfect wave that surfers dream about.

A vibration or oscillation involves back-and-forth motion, such as that of a swinging pendulum. Another example is a ball in a round-bottomed bowl. If the ball is displaced from its equilibrium position, it will roll back and forth, or oscillate, and finally come to rest at the equilibrium position, where its potential energy is lowest.

Recall from Chapter 8 that the ball must be oscillating about a point of stable equilibrium, for which there is a restoring force or torque. This is true in general for objects that undergo vibrating or oscillating motions. In a material medium, the restoring force is provided by intermolecular forces. If a molecule is disturbed, restoring forces exerted by its neighbors tend to return it to its original position, and it begins to oscillate. In so doing, it affects the adjacent molecules, which are in turn set into oscillation.

In this situation, we can see that something happens at point A at time t_1, and this causes a similar happening at a point B at a later time t_2, and so on. This is referred to as *propagation*. But what is propagated by the molecules in a material? It is energy. A single disturbance, which happens when you give the end of a stretched rope a quick shake, gives rise to what is referred to as a *wave pulse*. A continuous, repetitive disturbance gives rise to a continuous propagation of energy that we call wave motion.

13.1 Simple Harmonic Motion

Objectives: To (a) describe simple harmonic motion and (b) relate energy and speed in such motion.

The motion of an oscillating object depends on the restoring force that makes it go back and forth. It is convenient to begin to study such motion by considering the simplest type of force acting along the x axis: a force that is directly proportional to the displacement from equilibrium. One such force is the (ideal) spring force, described by **Hooke's law**,

$$F_s = -kx \qquad (13.1)$$

where k is the spring constant. Recall from Chapter 5 that the minus sign indicates that the force always points in the opposite direction to the displacement, $F_s = k(-x)$. That is, it always tends to restore the spring to its equilibrium position.

Suppose that an object on a horizontal frictionless surface is connected to a spring as shown in •Fig. 13.1. When the object is displaced to one side of its equilibrium position and released, it will move back and forth—that is, it will vibrate or oscillate. Here, an oscillation or a vibration is clearly a *periodic motion*, a motion that repeats itself again and again along the same path. For linear oscillations, like those of an object attached to a spring, the path may be back and forth or up and down. For the angular oscillation of a pendulum, the path is back and forth along a circular arc.

Motion under the influence of the type of force described by Hooke's law is called **simple harmonic motion (SHM)**, because the force is the simplest restoring force and because the motion can be described by harmonic functions (sines and cosines), as you will see later in the chapter. The directed distance of the object in SHM from its equilibrium position is its **displacement**. Note in Fig. 13.1 that the displacement can be either positive or negative ($+x$ or $-x$), which indicates direction. The maximum displacements are $+A$ and $-A$ (Fig. 13.1b,d).

The magnitude of the maximum displacement, or the maximum distance of an object from its equilibrium position, is called the **amplitude** (A). It is a scalar quantity that expresses the distance for both extreme displacements.

Besides the amplitude, two other important quantities for describing an oscillation are its period and frequency. The **period** (T) is the time needed to complete one cycle of motion. A cycle is a *complete* round trip, or motion through a complete

Note: Hooke's law (for an ideal spring) is discussed in Section 5.2.

Note: Here $+x$ signifies spring extension and $-x$ means compression.

(a) Equilibrium

(b) $t = 0$ Just before release

(c) $t = \frac{1}{4}T$

(d) $t = \frac{1}{2}T$

(e) $t = \frac{3}{4}T$

(f) $t = T$

•**FIGURE 13.1 Simple harmonic motion (SHM)** When displaced from **(a)** its equilibrium position $x = 0$ and **(b)** released, the object on a spring undergoes SHM (assuming no frictional losses). The time it takes to complete one cycle is the period of oscillation (T). (Here F_s is the spring force and F_h is the force the hand applies.) **(c)** At $t = T/4$, the object is back at its equilibrium position; **(d)** at $t = T/2$, it is at $x = -A$. **(e)** During the next half cycle, the motion is to the right; **(f)** at $t = T$, the object is back at its initial ($t = 0$) starting position.

TABLE 13.1 Terms Used to Describe SHM

displacement—directed distance of an object ($\pm x$) from its equilibrium position.
amplitude (A)—magnitude of the maximum displacement or the maximum distance of an object from its equilibrium position.
period (T)—the time for one *complete* cycle of motion.
frequency (f)—the number of cycles per second (in hertz or inverse seconds, where $f = 1/T$).

oscillation. For example, if an object starts at $x = A$ (Fig. 13.1b), when it returns to $x = A$ (as in Fig. 13.1f), it will have completed one cycle in a time we call one period. If an object were initially at $x = 0$ when disturbed, then its second return to this point would mark a cycle. In either case, the object would travel a distance of $4A$ during one cycle. Can you show this? (Why a *second* return?)

The **frequency** (f) is the number of cycles per second. The frequency and the period are related by

$$f = \frac{1}{T} \quad \begin{array}{l}\text{frequency}\\ \text{and period}\end{array} \quad (13.2)$$

SI unit of frequency: hertz (Hz), or cycle per second (cycle/s)

The inverse relationship is reflected in the units: The period is the number of seconds per cycle, and the frequency is the number of cycles per second. For example, if $T = \frac{1}{2}$ s/cycle, then $f = 2$ cycles/s.

The standard unit of frequency is the **hertz (Hz)**, which is one cycle per second.* From Eq. 13.2, frequency has the unit inverse seconds (1/s, or s^{-1}), since the period is a measure of time. Although cycle is not really a unit, you might find it convenient at times to express frequency in cycles per second to help with unit analysis. This is similar to the way the radian (rad) is used in the description of circular motion in Sections 7.1 and 7.2.

The preceding terms used to describe SHM are summarized in Table 13.1.

Energy and Speed of a Spring–Mass System in SHM

Recall from Chapter 5 that the potential energy stored in a spring stretched or compressed a distance $\pm x$ from equilibrium is given by

$$U = \tfrac{1}{2}kx^2 \quad (13.3)$$

The *change* in potential energy of an object oscillating on a spring is related to the work done by the spring force, as illustrated and explained in •Fig. 13.2. (Compare Fig. 5.5, Section 5.2.) An object with mass m oscillating on a spring also has kinetic energy. The kinetic and potential energies together give the total mechanical energy of the system:

$$E = K + U = \tfrac{1}{2}mv^2 + \tfrac{1}{2}kx^2 \quad (13.4)$$

When the object is at one of its maximum displacements, $+A$ or $-A$, it is instantaneously at rest ($v = 0$). Thus, all the energy is in the form of potential energy at this location; that is,

$$E = \tfrac{1}{2}m(0)^2 + \tfrac{1}{2}k(\pm A)^2 = \tfrac{1}{2}kA^2$$

or

$$E = \tfrac{1}{2}kA^2 \quad \begin{array}{l}\text{total energy in}\\ \text{SHM of a spring}\end{array} \quad (13.5)$$

---

*The unit is named for Heinrich Hertz (1857–1894), a German physicist and early investigator of electromagnetic waves.

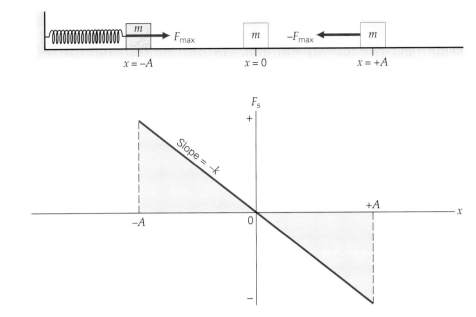

•FIGURE 13.2 **Work and potential energy** The work done *by the spring force* on the oscillating object is equal to the negative change in potential energy $(-\Delta U)$. The magnitude of this change is equal to the area under the spring force (F_s) versus displacement (x) curve (where $F_s = -kx$). The work can be positive or negative. For example, in going from $x = 0$ to $x = A$, the work is negative (F_s and x in opposite directions) and U increases. (Hence the minus sign in $W_F = -\Delta U$.) But, going back from $x = A$ to $x = 0$, the work done by the spring is positive (why?), and U decreases by the same amount. Thus, the overall ΔU is zero. (How about going from $x = 0$ to $x = -A$ and back—or halfway back?)

This result is a general result for SHM:

> The total energy of an object in simple harmonic motion is directly proportional to the square of the amplitude of the motion.

Equation 13.5 allows us to express the velocity of an object oscillating on a spring as a function of position:

$$E = K + U \quad \text{or} \quad \tfrac{1}{2}kA^2 = \tfrac{1}{2}mv^2 + \tfrac{1}{2}kx^2$$

Rearranging, we get

$$v^2 = \frac{k}{m}(A^2 - x^2)$$

so

$$v = \pm\sqrt{\frac{k}{m}(A^2 - x^2)} \qquad (13.6)$$

Velocity of an object in SHM

Note: This discussion will be limited to light springs, the mass of which can be considered negligible.

where the $\pm$ indicates velocity direction. Note that at $x = \pm A$, the velocity is zero since the object is instantaneously at rest when at its maximum displacement from equilibrium.

Note that when the oscillating object passes through the equilibrium position ($x = 0$), the potential energy is zero. At that instant, all the energy is kinetic and the object is traveling at its maximum speed, v_{max} (Fig. 13.1c,e). The energy expression for this case is

$$E = \tfrac{1}{2}kA^2 = \tfrac{1}{2}mv_{max}^2$$

and thus

$$v_{max} = \sqrt{\frac{k}{m}}\,A \qquad \begin{array}{l}\textit{maximum}\\ \textit{speed of mass}\\ \textit{on a spring}\end{array} \qquad (13.7)$$

Consider the use of energy methods in the following Example. You can visualize the continuous tradeoff between kinetic and potential energy in Learn by Drawing on p. 422.

LEARN BY DRAWING

Oscillating in a Parabolic Potential Well

A way to visualize conservation of energy for simple harmonic motion is shown in Fig. 1. The potential energy of a spring–mass system can be sketched on a plot of energy (E) versus position (x). Since $U = \frac{1}{2}kx^2 \propto x^2$, the graph is a *parabola*.

In the absence of nonconservative forces, the total energy of the system, E, is constant. But E is the sum of the kinetic and potential energies. During the oscillations, there is a continuous tradeoff between the two types of energies, but their sum remains constant. Mathematically this is written as $E = K + U$. In Fig. 2, U (shown as a blue arrow) is represented by the vertical distance from the x axis. Since E is constant, independent of x, it will plot as a horizontal line (shown in green). The kinetic energy is the piece of the total energy that is *not* potential energy, that is, $K = E - U$; it can be graphically interpreted (purple arrow) as the vertical distance between the potential energy parabola and the horizontal green total energy line. As the object oscillates on the x axis, the energy tradeoffs can be visualized as the lengths of the two arrows change.

A general location, x_1, is shown in Fig. 2. Neither the kinetic energy nor the potential energy is at its maximum value of E there. These maximum values occur instead at $x = 0$ and $x = \pm A$, respectively. The motion cannot exceed $x = \pm A$ because that would imply a negative kinetic energy, which is physically impossible (why?). The amplitude positions are sometimes called the *endpoints* of the motion because they are the locations where the speed is instantaneously zero and the object reverses direction.

Try using the graphical approach to answer the following questions (and make up some of your own): What do you have to do to E to increase the amplitude of oscillation, and how might you do this? What happens to the amplitude of a real-life system in SHM in the presence of a force such as friction when E decays with time?

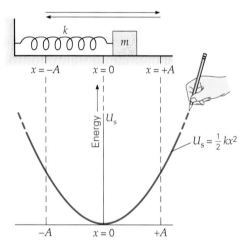

FIGURE 1 The potential energy "well" of a spring–mass system The potential energy of a spring stretched or compressed ($+x$) from its equilibrium position ($x = 0$) is a parabola, since $U_s \propto x^2$. At $x = \pm A$, all of the system's energy is potential.

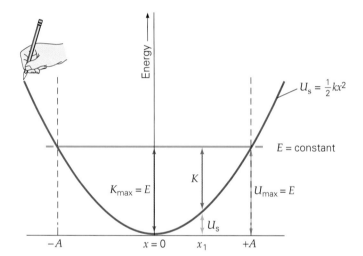

FIGURE 2 Energy transfers as the spring–mass system oscillates The vertical distance from the x axis to the parabola is the system's potential energy. The remainder—the vertical distance between the parabola and the horizontal line representing the system's constant total energy E—is the system's kinetic energy.

Example 13.1 ■ A Block and a Spring: Simple Harmonic Motion

A block with a mass of 0.25 kg sitting on a frictionless surface is connected to a light spring that has a spring constant of 180 N/m (see Fig. 13.1). If the block is displaced 15 cm from its equilibrium position and released, what are (a) the total energy of the system and (b) the speed of the block when it is 10 cm from its equilibrium position?

Thinking It Through. The total energy depends on the spring constant (k) and amplitude (A), which are given. At $x = 10$ cm, we expect the speed to be less than the maximum speed (why?).

Solution. First we list the given data, as usual, and what it is we are to find. The initial displacement *is* the amplitude (why?).

Given: $m = 0.25$ kg
$k = 180$ N/m
$A = 15$ cm $= 0.15$ m
$x = 10$ cm $= 0.10$ m

Find: (a) E (total energy)
(b) v (speed)

(a) The total energy is given directly by Eq. 13.5:

$$E = \tfrac{1}{2}kA^2 = \tfrac{1}{2}(180 \text{ N/m})(0.15 \text{ m})^2 = 2.0 \text{ J}$$

(b) The instantaneous speed of the block at a distance of 10 cm from the equilibrium position is given by Eq. 13.6 without directional signs:

$$v = \sqrt{\frac{k}{m}(A^2 - x^2)} = \sqrt{\frac{180 \text{ N/m}}{0.25 \text{ kg}}[(0.15 \text{ m})^2 - (0.10 \text{ m})^2]} = \sqrt{9.0 \text{ m}^2/\text{s}^2} = 3.0 \text{ m/s}$$

What would the speed be at $x = -10$ cm?

Follow-up Exercise. In part (b) of this Example, the block at $x = 10$ cm is at two-thirds or 67% of its maximum displacement. Is its speed at that position therefore 67% of its maximum speed? Prove your answer mathematically.

The spring constant is commonly determined by placing an object of known mass on the end of the spring and letting it settle vertically into a new equilibrium position. The next Example shows some typical results.

Example 13.2 ■ The Spring Constant: Experimental Determination

When a 0.50-kilogram mass is suspended from a spring, the spring stretches a distance of 10 cm to a new equilibrium position (•Fig. 13.3a). (a) What is the spring constant of the spring? (b) The mass is then pulled down another 5.0 cm and released. What is the highest position of the oscillating mass?

Thinking It Through. At the equilibrium position, the net force on the mass is zero because $a = 0$. In (b) we will use $-y$ to designate "downward," as is customary in vertical problems.

Solution.

Given: $m = 0.50$ kg
$y_o = 10$ cm $= 0.10$ m
$y = -5.0$ cm $= -0.050$ m

Find: (a) k (spring constant)
(b) A (amplitude)

(a) When the suspended mass and the stretched spring are in equilibrium (Fig. 13.3a), the net force on the mass is zero. Thus the weight of the mass and the spring force are equal and opposite. Then, equating their magnitudes,

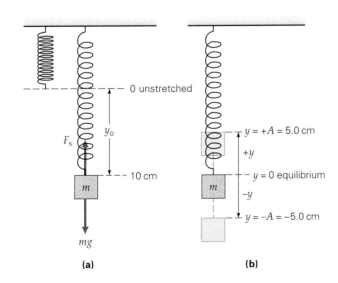

•**FIGURE 13.3 Determination of the spring constant** (a) When an object suspended on a spring is in equilibrium, the two forces on it cancel: $F_s = w$, or $ky_o = mg$. Thus the spring constant k can be computed: $k = mg/y_o$. (b) The zero reference point of SHM of an object suspended on a spring is conveniently taken as the new equilibrium position, as the motion is symmetric about this point. See Example 13.2.

or
$$F_s = w$$

Thus,
$$ky_o = mg$$

$$k = \frac{mg}{y_o} = \frac{(0.50 \text{ kg})(9.8 \text{ m/s}^2)}{0.10 \text{ m}} = 49 \text{ N/m}$$

(b) Once set into motion, the mass oscillates up and down through the equilibrium position. Since the motion is symmetric about this point, it is designated as the zero reference point of the oscillation (Fig. 13.3b). The initial displacement is $-A$, so the highest position of the mass will be 5.0 cm above the equilibrium position $(+A)$.

Follow-up Exercise. How much more potential energy does the spring in this Example have at the bottom of its oscillation that at the top?

13.2 Equations of Motion

Objectives: To (a) understand the equation of motion for SHM and (b) explain what is meant by phase and phase differences.

We refer to the **equation of motion** of an object as the equation that gives the object's position as a function of time. For example, the equation of motion with a constant linear acceleration is $x = v_o t + \frac{1}{2}at^2$, where v_o is the initial velocity (Chapter 2). However, the acceleration is not constant for simple harmonic motion, so the kinematic equations of Chapter 2 do not apply to this case.

The equation of motion for an object in simple harmonic motion can be derived from a relationship between simple harmonic and uniform circular motions. SHM can be simulated by a component of uniform circular motion, as illustrated in •Fig. 13.4. As the illuminated object moves in uniform circular motion (with constant angular speed ω) in a horizontal plane, its shadow moves back and forth horizontally, following the same path as the object on the spring, which is in simple harmonic motion. Since the shadow and the object have the same position at any time, it follows that the equation of *horizontal* motion for the object in circular motion is the same as the equation of motion for a horizontally oscillating object on the spring.

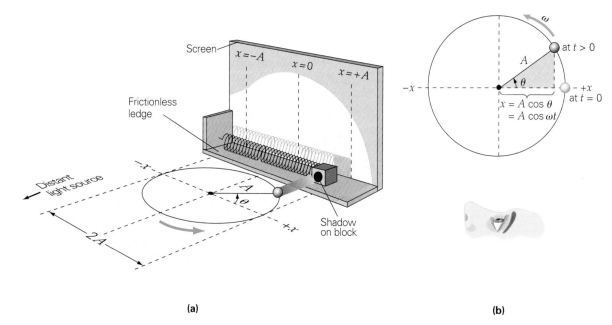

•**FIGURE 13.4 Reference circle for horizontal motion** (a) The shadow of the object in uniform circular motion has the same horizontal motion as the object on the spring in simple harmonic motion. (b) The motion can thus be described by $x = A \cos \theta = A \cos \omega t$ (assuming $x = +A$ at $t = 0$).

From the reference circle in Fig. 13.4b, the x coordinate (position) of the object is given by
$$x = A \cos \theta$$
But the object moves with a constant angular velocity with a magnitude ω. In terms of angular distance θ, assuming $\theta = 0$ at $t = 0$, we have $\theta = \omega t$, so

$$x = A \cos \omega t \qquad \text{(SHM for } x_o = +A\text{)} \qquad (13.8)$$

The angular speed ω (in rad/s) of the *reference circle object* is called the *angular frequency* of the oscillating object, since $\omega = 2\pi f$, where f is the frequency of revolution or rotation (Section 7.2). Figure 13.4 shows that the frequency of the "orbiting" object is the same as the frequency of oscillation of the object on the spring. Thus, we can write Eq. 13.8 alternatively (using $f = 1/T$) as

Note: See Eq. 7.8, Section 7.2.

$$x = A \cos(2\pi f t) = A \cos\left(\frac{2\pi t}{T}\right) \qquad \text{(SHM for } x_o = +A\text{)} \qquad (13.9)$$

Equations 13.8 and 13.9 give three equivalent forms of the equation of motion for an object in simple harmonic motion. Any one of them can be used for convenience, depending on the known parameters. For example, suppose you are given the time t in terms of the period T—say, $t_o = 0$, $t_1 = T/2$, and $t_2 = T$—and are asked to find the position of an object in SHM at these times. In this case, it is convenient to use Eq. 13.9, and

Different forms of the equation of motion

$$t_o = 0 \quad x_o = A \cos[2\pi(0)/T] = A \cos 0 = A$$
$$t_1 = \frac{T}{2} \quad x_1 = A \cos[2\pi(T/2)/T] = A \cos \pi = -A \qquad (13.10)$$
$$t_2 = T \quad x_2 = A \cos[2\pi T/T] = A \cos 2\pi = A$$

Hence, the results tell us that the object was initially at $x = A$, as we knew. One-half period later it was at $x = -A$, or the opposite extreme of its oscillation; and at a time of one period (T), it was back where it started, which is to be expected since the motion is periodic.

The period of an object oscillating on a spring can be expressed in terms of the mass m (or inertia) of the object and the spring constant (or stiffness) k. Our algebraic equation of motion describes the motion once you know the numerical values.

To show the power of the reference circle, let us use it to compute the period of the spring–object system. Note that the time for the object in the reference circle to make one complete "orbit" is exactly the time it takes for the oscillating object to make one complete cycle (see Fig. 13.4). Thus all we need is the time for one reference circle orbit and we have the period of oscillation. Because the reference circle object is in uniform circular motion at a constant speed equal to the maximum speed of oscillation, the object travels a distance of one circumference in one period. Because $t = d/v$, where $t = T$, d is the circumference, and v is $v_{\max}$ given by Equation 13.7,

$$T = \frac{d}{v_{\max}} = \frac{2\pi A}{\sqrt{k/m}\,A}$$

or

$$T = 2\pi \sqrt{\frac{m}{k}} \qquad \begin{array}{l}\textit{period of object}\\ \textit{oscillating on a spring}\end{array} \qquad (13.11)$$

Because the amplitudes canceled out in Eq. 13.11, *the period (and frequency) are independent of the amplitude of the motion.* This statement is a general feature of simple harmonic oscillators—that is, for oscillators driven by a linear restoring force, such as a spring obeying Hooke's law. For example, a pendulum-driven clock that is not properly rewound and is running down would still keep correct time because the period would remain unchanged as its amplitude decreased.

Note: Period and frequency are independent of amplitude for SHM.

We see from Eq. 13.11 that the greater the mass, the longer the period; and the greater the spring constant (the stiffer the spring), the shorter the period. It is the *ratio* of mass to stiffness that determines the period. Thus you can offset a mass increase by using a stiffer spring.

Since $f = 1/T$,

$$f = \frac{1}{2\pi}\sqrt{\frac{k}{m}} \qquad \text{frequency of object oscillating on a spring} \qquad (13.12)$$

Thus, the greater the spring constant (the stiffer the spring), the more frequently the system vibrates, as you might expect.

Also, note that since $\omega = 2\pi f$, we may write

$$\omega = \sqrt{\frac{k}{m}} \qquad \text{anguluar frequency of object oscillating on a spring} \qquad (13.13)$$

As another example, a simple pendulum (a small, heavy object on a string) will undergo simple harmonic motion for small angles of oscillation. Advanced mathematics is needed to show that the period of a simple pendulum oscillating through a small angle ($\theta \leq 10°$) is given, to a good approximation, by

$$T = 2\pi\sqrt{\frac{L}{g}} \qquad \text{period of a simple pendulum} \qquad (13.14)$$

where L is the length of the pendulum and g is the acceleration due to gravity.

An important difference between the spring–mass period and the pendulum period is that the latter is independent of the mass of the bob. Can you explain why? Think about what supplies the restoring force for the pendulum's oscillations. It is gravity. Hence, we would expect the acceleration (along with the velocity and period) to be independent of mass. That is, the gravitational force automatically provides the same acceleration to different bob masses of pendulums with the same length. Thus the acceleration doesn't depend on the mass. We have seen that similar effects occur in free fall (Chapter 2) and for blocks sliding and cylinders rolling down inclines (Chapters 4 and 8, respectively). The following Example demonstrates the usage of the equation of motion for SHM.

Example 13.3 ■ Equation of Motion for a Practical Oscillation: Shaken, Not Stirred

To stir a mixture of liquids consistently and gently, a biologist designs a simple mechanical system consisting of a horizontal spring attached to a beaker on the slippery Teflon® surface. She pours two liquids with a total mass of 100 g into the beaker, which has a mass of 50.0 g. She then pulls on the beaker, stretching the spring. On releasing the beaker from rest ($t = 0$), she sets the system in motion with an amplitude of 5.00 cm. To avoid spilling and to promote gentle mixing, the frequency is designed to be exactly 0.300 Hz. (a) What is the required spring constant? (b) What is the displacement of the beaker from equilibrium at time $t = 2.00$ s?

Thinking It Through. (a) Since we know the frequency and total mass (beaker plus liquids), we can use Eq. 13.12 to solve for the spring constant. (b) The system has an equation of motion given by Eq. 13.8. We must be careful to express the angle in radians.

Solution.

Given: $m = 50.0$ g $+ 100$ g $= 150$ g $= 0.150$ kg (total mass) Find: (a) k (spring constant)
$A = 5.00$ cm $= 5.00 \times 10^{-2}$ m   (b) x (displacement)
$f = 0.300$ Hz
$t = 2.00$ s

(a) We can rearrange Eq. 13.12 to solve for the spring constant:

$$k = 4\pi^2 m f^2 = 4\pi^2 (0.150 \text{ kg})(0.300 \text{ Hz})^2 = 0.533 \text{ N/m}$$

(b) First we calculate the angular frequency, ω, from the relationship between angular frequency and frequency (Eq. 7.8):

$$\omega = 2\pi f = 2\pi(0.300 \text{ Hz}) = 1.88 \text{ rad/s}$$

To find the location of the beaker (its displacement from equilibrium), we use the equation of motion, Eq. 13.8:

$$x = A \cos \omega t = (0.0500 \text{ m}) \cos[(1.88 \text{ rad/s})(2.00 \text{ s})] = (0.0500 \text{ m}) \cos(3.76 \text{ rad})$$

$$= -0.0407 \text{ m} = -4.07 \text{ cm}$$

The minus sign indicates that at this time the spring is compressed 4.07 cm from equilibrium.

Follow-up Exercise. The cone of a loudspeaker moves in SHM in and out a distance of 0.20 mm while vibrating at a frequency of 500 Hz. (a) What is its maximum speed? (b) Write the equation of motion in the form of Eq. 13.8 for the cone, and use it to find the cone's displacement at $t = 1.1 \times 10^{-3}$ s.

Problem-Solving Hint

Note that in the part (b) calculation of Example 13.3, where we have cos (3.76), the angle is in radians, *not* degrees. Don't forget to set your calculator to radians (rather than degrees) when finding the value of a trigonometric function in equations for simple harmonic or circular motion.

Example 13.4 ■ Fun with a Pendulum: Frequency and Period

A helpful older brother takes his sister to play on the swings in the park. He pushes her from behind on each return. Assuming that the swing behaves as a simple pendulum with a length of 2.50 m, (a) what would be the frequency of the oscillations, and (b) what would be the interval between the brother's pushes?

Thinking It Through. (a) The period is given by Eq. 13.14, and the frequency and period are inversely related: $f = 1/T$. (b) Since the brother pushes from one side on each return, he must push once every completed cycle, so the time between his pushes is equal to the swing's period.

Solution.

Given: $L = 2.50$ m   *Find:* (a) f (frequency)
                                  (b) T (period)

(a) We can invert Eq. 13.14 to solve directly for the frequency:

$$f = \frac{1}{T} = \frac{1}{2\pi}\sqrt{\frac{g}{L}} = \frac{1}{2\pi}\sqrt{\frac{9.80 \text{ m/s}^2}{2.50 \text{ m}}} = 0.315 \text{ Hz}$$

(b) The period is then found from the frequency:

$$T = 1/f = 1/(0.315 \text{ Hz}) = 3.17 \text{ s}$$

The brother must push every 3.17 s to maintain a steady swing (and to keep his sister from complaining).

Follow-up Exercise. In this Example, the older brother, a physics buff, carefully measures the period of the swing to be 3.18 s, not 3.17 s. If the length is accurate at 2.50 m, what is the acceleration due to gravity at their location? Using this accurate value of g, do you think the park is at sea level?

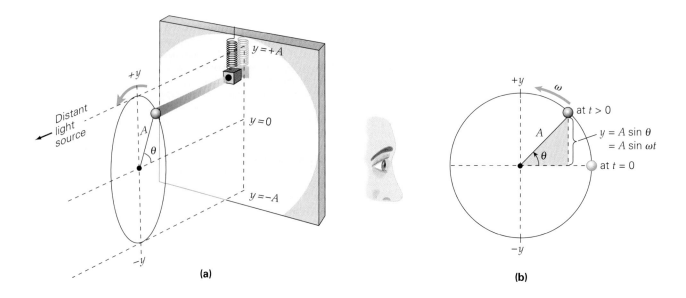

•**FIGURE 13.5 Reference circle for vertical motion** (a) The shadow of the object in uniform circular motion has the same vertical motion as the object oscillating on the spring in simple harmonic motion. (b) The motion can thus be described by $y = A \sin \theta = A \sin \omega t$ (assuming $y = 0$ at $t = 0$).

A vertical version of the reference circle can be used to describe vertical simple harmonic motion, as illustrated in •Fig. 13.5 for a suspended object oscillating on a spring. Suppose that, having been given an upward push, the object initially started its upward motion from its equilibrium position $y = 0$ (recall that $+y$ means upward). In this case, the vertical displacement is given by $y = A \sin \theta$. A development similar to that for the horizontal case leads to this equation of motion:

$$y = A \sin \omega t \quad \begin{array}{l} \textit{initially upward} \\ \textit{motion with } y_o = 0 \end{array}$$

Notice that at $t = 0$, the object is indeed at its equilibrium point ($y = 0$).

However, suppose the object starts initially from its maximum positive displacement position. As shown in •Fig. 13.6, the resulting curve must instead be a cosine function

$$y = A \cos \omega t \quad \begin{array}{l} \textit{vertical motion} \\ \textit{with } y_o = A, v_o = 0 \end{array}$$

This is because only a cosine equation correctly describes the initial conditions— $y = +A$ at $t = 0$.

Thus, the equation of motion for an oscillating object may be either a sine or a cosine function. Both of these functions are referred to as being *sinusoidal*. That is, simple harmonic motion is described by a sinusoidal function of time.

•**FIGURE 13.6 Sinusoidal equation of motion** As time passes, the oscillating object traces out a sinusoidal curve on the moving paper. In this case, $y = A \cos \omega t$, because the object's initial displacement is $y_o = +A$.

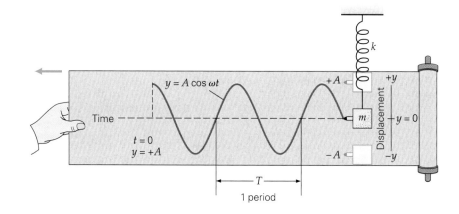

428 CHAPTER 13 Vibrations and Waves

Initial Conditions and Phase

You may be wondering how to decide whether to use a sine or cosine function to describe a particular case of simple harmonic motion. In general, the form of the function is determined by the initial displacement and velocity of the object: the *initial conditions* of the system. These initial conditions are the values of the displacement and velocity at $t = 0$; when, taken together, they tell how the system is initially set into motion.

Note: Initial conditions include both x_o and v_o, the displacement and velocity at $t = 0$.

Let's look at four special cases. If an object in vertical SHM has an initial displacement of $y = 0$ at $t = 0$ and moves initially upward, the equation of motion is $y = A \sin \omega t$ (•Fig. 13.7a). Note that $y = A \cos \omega t$ does not satisfy the initial condition, that is, $y_o = A \cos \omega t = A \cos \omega(0) = A$, since $\cos 0 = 1$.

Suppose that the object is initially released ($t = 0$) from its positive amplitude position ($+A$), as in the case of the object on a spring shown in Fig. 13.6. Here, the equation of motion is $y = A \cos \omega t$ (Fig. 13.7b). This expression satisfies the initial condition: $y_o = A \cos \omega(0) = A$.

The other two cases are $y = 0$ at $t = 0$, with motion initially downward (for an object on a spring) or in the negative direction (for horizontal SHM); and $y = -A$ at $t = 0$, meaning that the object is initially at its negative amplitude

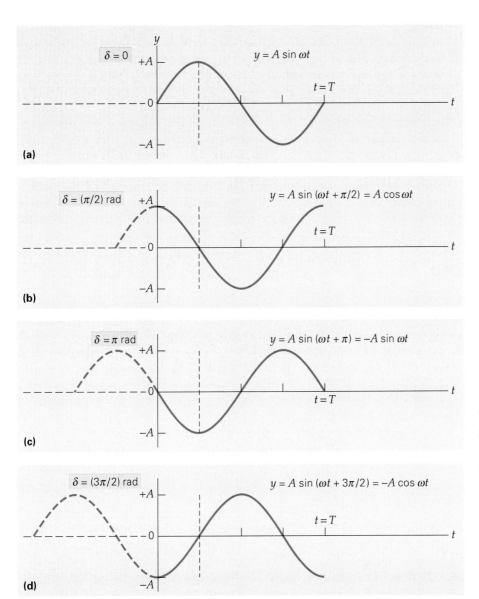

•FIGURE 13.7 Phase differences For the general SHM equation $y = A \sin(\omega t + \delta)$, oscillations are described by either sine or cosine terms for the phase constants shown. The initial displacement and velocity direction determine δ. Each curve is 90° (or $\pi/2$ rad) out of phase with the preceding one. This is equivalent to shifting each successive curve a quarter of a cycle relative to the previous one. Everthing else, including the period, is identical for each curve (why?).

position. These motions are described by $y = -A \sin \omega t$ and $y = -A \cos \omega t$, respectively, as illustrated in Fig. 13.7c and 13.7d.

Note in Fig. 13.7 that if the curves are extended in the negative direction to the horizontal axis (dashed purple lines in figure), they all have the same shape but have been "shifted," so to speak. We describe this change as a shift in *phase*. For the general case, we may write the equation of simple harmonic motion as

General equation of motion for simple harmonic motion

$$y = A \sin(\omega t + \delta) \quad (13.15)$$

where $(\omega t + \delta)$ is the *phase angle* and δ is the **phase constant**. The phase constant essentially shifts the curve to match the appropriate sinusoidal function to the motion and is determined by the initial conditions.

The phase constants for the cases in Fig. 13.7 are given in the figure. For example, for $\delta = 90°$ (or $\pi/2$ rad), the equation of motion is $y = A \cos \omega t$, and an object oscillating up and down on a spring would be initially released at the $+A$ position. You probably recognize that the curve for the $\delta = 90°$ is a cosine. This can be shown algebraically by using the following trigonometric formula: $\sin(a + b) = \sin a \cos b + \cos a \sin b$. Then, with $\delta = 90°$,

$$y = A \sin(\omega t + 90°) = A[\sin \omega t \cos 90° + \cos \omega t \sin 90°] = A \cos \omega t$$

since $\cos 90° = 0$ and $\sin 90° = 1$.

The curves for $\delta = 0$ and $\delta = 90°$ (or $\pi/2$ rad) are said to be 90° *out of phase*, or shifted by a quarter cycle, with respect to one another. Notice in Fig. 13.7 that the curve for $\delta = 90°$ has essentially been shifted to the left by a quarter cycle (90°) from the curve for $\delta = 0$. The cases for $\delta = 180°$ (or π rad) and $\delta = 270°$ (or $3\pi/2$ rad) are each shifted an additional 90° out of phase.

Two objects oscillating at the same frequency *in phase* (having the same δ) will oscillate together. Two that oscillate completely out of phase (180° difference in δ) will always be going in opposite directions or be at opposite maximum displacements.

For each of the four special cases in Fig. 13.7, the equation of motion is either a sine or cosine curve. If the initial displacement is not zero or $\pm A$, then δ is not a multiple of 90° and things get a bit more complicated. Equation 13.15 still applies, but the formula above gives an equation of motion with both sine and cosine terms. See Demonstration 10 for a pictorial view of a sinusoidal oscillation and phase.

Velocity and Acceleration in SHM

Expressions for the velocity and acceleration of an object in SHM can be easily derived from energy and force considerations. We have already derived an expression (Eq. 13.6) for the velocity of an object oscillating on a spring in terms of the displacement. We can now express the velocity as a function of time. For the special case of vertical simple harmonic motion, with the displacement $y = A \sin \omega t$, we can rewrite Eq. 13.6 as

$$v = \sqrt{\frac{k}{m}(A^2 - y^2)} = \sqrt{\frac{k}{m}(A^2 - A^2 \sin^2 \omega t)} = \sqrt{\frac{k}{m}} A\sqrt{1 - \sin^2 \omega t}$$

Since $\omega = \sqrt{k/m}$ (Eq. 13.13) and $\cos \theta = \sqrt{1 - \sin^2 \theta}$ (see Appendix I), we can write this expression as

Note: Maximum speed $v = \omega A$.

$$v = \omega A \cos \omega t \quad \begin{array}{l}\text{vertical velocity} \\ \text{if } v_o \text{ upward, } y_o = 0\end{array} \quad (13.16)$$

where the directional signs are given by the cosine function. Using Newton's second law to find the acceleration with the spring force $F_s = -ky$, we have

Demonstration 10 ■ Simple Harmonic Motion (SHM) and Sinusoidal Oscillation

A demonstration to show that SHM can be represented by a sinusoidal function. A "graph" of the function is generated with an analogue of a strip chart recorder.

(a) A salt-filled funnel oscillates, suspended from two strings.

(b) The salt falls on a black-painted poster board that will be pulled in a direction perpendicular to the plane of the funnel's oscillation.

(c) Away we go.

(d) The salt trail traces out a plot of displacement versus time, or $y = A \sin(\omega t + \delta)$. Note that in this case the phase constant is about $\delta = 90°$ and $y = A \cos \omega t$. (Why?)

$$a = \frac{F_s}{m} = \frac{-ky}{m} = -\frac{k}{m} A \sin \omega t$$

and since $\omega = \sqrt{k/m}$,

$$a = -\omega^2 A \sin \omega t = -\omega^2 y \quad \begin{array}{l} \textit{vertical acceleration} \\ \textit{if } v_o \textit{ upward, } y_o = 0 \end{array} \quad (13.17)$$

Note that the functions for the velocity and acceleration are out of phase with that for the displacement. Since the velocity is 90° out of phase with the displacement, the speed is greatest when $\cos \omega t = \pm 1$ at $y = 0$—when the oscillating object is passing through its equilibrium position. The acceleration is 180° out of phase with the displacement (as indicated by the minus sign on the right-hand side of Eq. 13.17). Therefore, the magnitude of the acceleration is a maximum when $\sin \omega t = \pm 1$ at $y = \pm A$—when the displacement is a maximum, or when the object is at an amplitude position. At any position except the equilibrium position, the directional sign of the acceleration is opposite that of the displacement, as it should be for an acceleration resulting from a restoring force. At the equilibrium position, both the displacement and acceleration are zero. (Can you see why?)

Also note that the acceleration for SHM is not constant with time. Hence, the kinematic equations for acceleration (Chapter 2) *cannot* be used, since they are for constant accelerations.

Note: Maximum acceleration magnitude $a = \omega^2 A$.

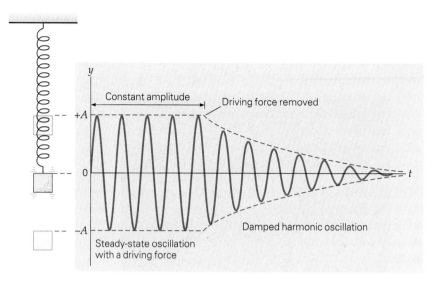

•FIGURE 13.8 **Damped harmonic motion** When a driving force adds energy to a system equal to its energy losses, the oscillation is steady with a constant amplitude. When the driving force is removed, the oscillations decay (that is, are damped), and the amplitude decreases nonlinearly with time.

Damped Harmonic Motion

Simple harmonic motion with a constant amplitude implies that there are no losses of energy, but in practical applications there are always some frictional losses. Therefore, to maintain a constant-amplitude motion, energy must be added to the system by some external driving force, such as someone pushing a swing. Without a driving force, the amplitude and the energy of an oscillator decrease with time, giving rise to **damped harmonic motion** (•Fig. 13.8). The time required for the oscillations to cease, or to damp out, depends on the magnitude and type of the damping force (for example, air resistance).

In many applications involving continuous periodic motion, damping is unwanted and necessitates energy input. However, in some instances, damping is desirable. For example, the dial in a spring-operated bathroom scale oscillates briefly before stopping at a weight reading. If not properly damped, these oscillations would continue for some time and you would have to wait before you could read your weight. Damping is also required for shock absorbers on automobiles and needle indicators on instruments measuring electrical quantities. In California many new buildings incorporate damping mechanisms (giant shock absorbers) to dampen their oscillatory motion after they are set in motion by earthquake waves.

13.3 Wave Motion

Objectives: To (a) describe wave motion in terms of various parameters and
(b) identify different types of waves.

The world is full of waves of various types—some examples are water waves, sound waves, waves generated by earthquakes, and light waves. All waves result from a disturbance, the source of the wave. In this chapter we will be concerned with mechanical waves, or those that are propagated in some medium. (Light waves, which do not require a propagating medium, will be considered in more detail in later chapters.)

When a medium is disturbed, energy is imparted to it. Suppose that energy is added to a material mechanically, such as by impact or (in the case of a gas) by compression. This addition sets some of the particles vibrating. Because the particles are linked by intermolecular forces, the oscillation of each particle affects that of its neighbors. The added energy propagates, or spreads, by means of interactions among the particles of the medium. An analogy for this process is shown in •Fig. 13.9, where the "particles" are dominoes. As each domino falls, it topples

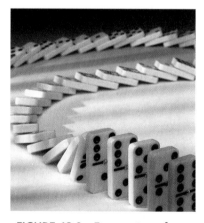

•FIGURE 13.9 **Energy transfer** The propagation of a disturbance, or a transfer of energy through space, is seen in a row of falling dominoes.

the one next to it. Thus energy is transferred from domino to domino, and the disturbance propagates through the medium.

In this case, there is no restoring force between the dominoes, so they do not oscillate as do particles in a continuous material medium. Therefore the disturbance moves in space, but it does not repeat itself in time at any one location.

Similarly, if the end of a stretched rope is given a quick shake, the disturbance transfers energy from the hand to the rope, as illustrated in •Fig. 13.10. The forces acting between the rope "particles" cause them to move in response to the motion of the hand, and a *wave pulse* travels down the rope. Each "particle" goes up and then back down as the pulse passes by. This motion of individual particles and that of the wave pulse propagation as a whole can be observed by tying pieces of ribbon onto the rope (at x_1 and x_2 in Fig. 13.10). As the disturbance passes point x_1, the ribbon rises and falls, as do the rope "particles." Later, the same occurs for the ribbon at x_2, which indicates that the disturbance energy is propagating or traveling along the rope.

In a continuous material medium, particles interact with their neighbors and restoring forces cause them to oscillate when they are disturbed. Thus any disturbance not only propagates through space but may be repeated over and over in time at each position. Such a regular, rhythmic disturbance in both time and space is called a **wave**, and the transfer of energy is said to take place by means of **wave motion**.

A continuous wave motion, or *periodic wave*, requires a disturbance from an oscillating source (•Fig. 13.11). In this case, the particles move up and down continuously. If the driving source is such that a constant amplitude is maintained (the source oscillates in simple harmonic motion), the resulting particle motion is also simple harmonic.

Such periodic wave motion will have sinusoidal forms (sine or cosine) in both time and space. Being sinusoidal in space means that if you took a photograph of the wave at any instant (freezing it in time), you would see a sinusoidal waveform (such as one of the curves in Fig. 13.11). However, if you looked at a single point in space as a wave passed by, you would see a particle of the medium oscillating up and down sinusoidally with time, like the mass on a spring discussed in Section 13.2. (For example, imagine looking through a thin slit at a fixed location on the moving paper in Fig. 13.6. The wave trace would be seen rising and falling like a particle.)

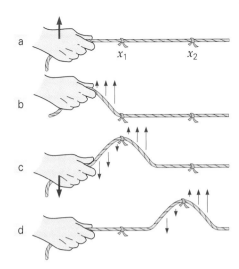

•**FIGURE 13.10 Wave pulse** The hand disturbs the stretched rope in a quick up and down motion, and a wave pulse propagates along the rope. (The red arrows represent the velocities of the hand and of pieces of the rope at different times and locations.) The rope "particles" move up and down as the pulse passes. The energy in the pulse is thus *both* kinetic and potential (elastic and gravitational).

Note: A wave is a combination of oscillations in space and time.

Wave Characteristics

Specific characteristics of sinusoidal waves are used to describe them. As for a particle in simple harmonic motion, the *amplitude* (A) of a wave is the magnitude of the maximum displacement or the maximum distance from the particle equilibrium position (Fig. 13.11). This quantity corresponds to the height of a wave crest or the depth of a trough. Recall from Section 13.2 that, in SHM, the total energy of the oscillator is proportional to the square of the amplitude. Similarly, the energy *transported* by a wave is proportional to the square of its amplitude ($E \propto A^2$). Note the difference—a wave is one way of *transmitting* energy through space, whereas an oscillator's energy is localized in space.

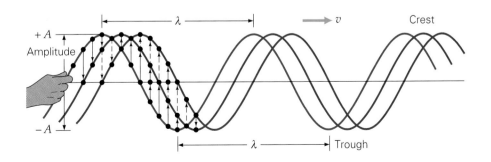

•**FIGURE 13.11 Periodic wave** A continuous harmonic disturbance can set up a sinusoidal wave in stretched rope, and the wave travels down the rope with wave speed v. Note that the rope "particles" oscillate vertically in simple harmonic motion. The distance between two successive points that are in phase (for example, at two crests) on the waveform is the wavelength λ of the wave. Can you tell how much time has elapsed, as a fraction of the period T, between the first (red) and last (blue) waves?

For a periodic wave, the distance between two successive crests (or troughs) is called the **wavelength** (λ) (Fig. 13.11). Actually, it is the distance between any two successive parts of the wave that are in phase (that is, at identical points on the waveform). The crest and trough positions are usually used for convenience. Note that a wavelength corresponds spatially to one cycle. Keep in mind that the wave, not the medium or material, is traveling.

The *frequency* (f) of a wave is the number of cycles per second—that is, the number of complete waveforms, or wavelengths, that pass by a given point during each second. The wave frequency is the same as the frequency of the SHM source that created it.

A periodic wave is said to possess a *period* (T). The period $T = 1/f$ is the time for one complete waveform (a wavelength) to pass by a given point. Since a wave moves, it also has a **wave speed** (or velocity, if direction is specified). Any particular point on the wave (for example, a crest) travels a distance of one wavelength λ in a time of one period T. Then, since $v = d/t$ and $f = 1/T$, we have

$$v = \frac{\lambda}{T} = \lambda f \qquad \text{wave speed} \qquad (13.18)$$

Note that the dimensions are correct (length/time). In general, the wave speed depends on the nature of the medium in addition to the source frequency f.

Example 13.5 ■ Dock of the Bay: Finding Wave Speed

A person on a pier observes a set of incoming waves that have a sinusoidal form with a distance of 1.6 m between the crests. If a wave laps against the pier every 4.0 s, what are (a) the frequency and (b) the speed of the waves?

Thinking It Through. We know the period and wavelength, so we can use the definition of frequency and Eq. 13.18 for wave speed.

Solution. The distance between crests is the wavelength, so we have

Given: $\lambda = 1.6$ m   Find: (a) f (frequency)
$\quad\quad\quad T = 4.0$ s $\quad\quad\quad$ (b) v (wave speed)

(a) The lapping indicates the arrival of a wave crest, so 4.0 s is the wave period—the time it takes to travel one wavelength (the crest-to-crest distance). Then

$$f = \frac{1}{T} = \frac{1}{4.0 \text{ s}} = 0.25 \text{ s}^{-1} = 0.25 \text{ Hz}$$

(b) The frequency or the period can be used in Eq. 13.18 to find the wave speed:

$$v = \lambda f = (1.6 \text{ m})(0.25 \text{ s}^{-1}) = 0.40 \text{ m/s}$$

Alternatively,

$$v = \frac{\lambda}{T} = \frac{1.6 \text{ m}}{4.0 \text{ s}} = 0.40 \text{ m/s}$$

Follow-up Exercise. On another day, the person measures the speed of sinusoidal water waves at 0.25 m/s. (a) How far does a wave crest travel in 2.0 s? (b) If the distance between successive crests is 2.5 m, what is the frequency of these waves?

Types of Waves

Waves may be divided into two types based on the direction of the particles' oscillations relative to the wave velocity. In a **transverse wave**, the particle motion is perpendicular to the direction of the wave velocity. The wave produced in a stretched string (Fig. 13.11) is an example of a transverse wave, as is the wave shown in •Fig. 13.12a. A transverse wave is sometimes called a *shear wave* because the disturbance supplies a force that tends to shear the medium—to separate layers of that medium at a right angle to the direction of the wave velocity. Shear waves can propagate only in solids, since a liquid or a gas cannot support a shear. That is, a liquid or a gas does not have sufficient restoring forces between its particles to propagate a transverse wave.

In a **longitudinal wave**, the particle oscillation is parallel to the direction of the wave velocity. A longitudinal wave can be produced in a stretched spring by moving the coils back and forth along the spring axis (Fig. 13.12b). Alternating pulses of compressions and relaxations move along the spring. A longitudinal wave is sometimes called a *compressional wave*.

Sound waves in air are another example of longitudinal waves. A periodic disturbance produces compressions in the air. Between the compressions are *rarefactions*, regions where the density of the air is reduced, or rarefied. A loudspeaker oscillating back and forth, for example, can create these compressions and rarefactions, which travel out into the air as sound waves. Sound will be discussed in detail in Chapter 14.

Longitudinal waves can propagate in solids, liquids, and gases, since all phases of matter can be compressed to some extent. The propagations of transverse and longitudinal waves in different media give information about the Earth's interior structure, as discussed in the Insight on p. 436.

The sinusoidal profile of water waves might make you think that these are transverse waves. Actually, they reflect a combination of longitudinal and transverse motions (•Fig. 13.13). The particle motion may be nearly circular at the surface but becomes more elliptical with depth, eventually becoming longitudinal. A hundred meters or so below the surface of a large body of water, the wave disturbances have little effect. For example, a submarine at these depths is undisturbed by large waves on the ocean's surface. As a wave approaches shallower water near shore, the water particles have difficulty completing their elliptical paths. When the water becomes too shallow, the particles can no longer move through the bottom parts of their paths and the wave breaks. Its crest falls forward to form breaking surf as the waves' kinetic energy is transformed into potential energy—a water "hill" that eventually topples over.

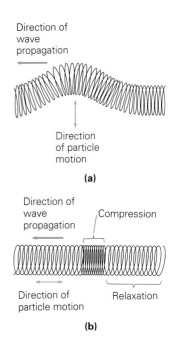

•**FIGURE 13.12 Transverse and longitudinal waves** (a) In a transverse wave, the particle motion is perpendicular to the direction of the wave velocity, as shown here in a spring for a wave moving to the left. (b) In a longitudinal wave, the particle motion is parallel to (or *along*) the direction of the wave velocity. Here a wave pulse also moves to the left. Can you explain the motion of the wave *source* for both types of waves?

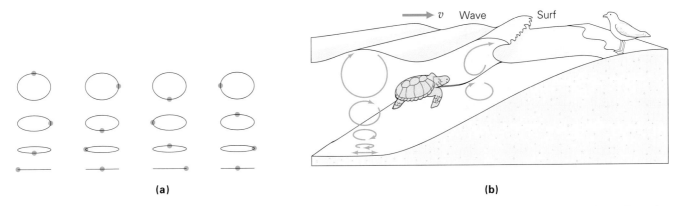

•**FIGURE 13.13 Water waves** Water waves are a combination of longitudinal and transverse motions. (a) At the surface, the water particles move in circles, but their motions become more longitudinal with depth. (b) When a wave approaches the shore, the lower particles are forced into steeper paths until finally the wave breaks or falls over to form surf.

Insight

Earthquakes, Seismic Waves, and Seismology

The Earth's interior structure is still something of a mystery. The deepest mine shafts and drillings extend only a few kilometers into the Earth. Using waves to probe the Earth's structure is one way to investigate it further. Waves generated by earthquakes have proved to be especially useful for this purpose. Seismology is the study of these waves, called *seismic waves*.

Earthquakes are caused by the sudden release of built-up stress along cracks and faults, such as the famous San Andreas Fault in California (Fig. 1). According to the geological theory of plate tectonics, the outer layer of the Earth consists of rigid plates, huge slabs of rock that move very slowly relative to one another. Stresses are continuously built up, particularly along boundaries between plates.

When slippage finally occurs, the energy from this stress-relieving event propagates outward as (seismic) waves from a site below the surface called the *focus*. These are of two general types: surface waves and body waves. *Surface waves*, which move along the Earth's surface, account for most earthquake damage (Fig. 2). *Body waves* travel through the Earth. There are both longitudinal and transverse body waves. The compressional (longitudinal) waves are called *P waves*, and shear (transverse) waves are called *S waves* (Fig. 3). The P and S stand for primary and secondary and indicate the waves' relative speeds (actually, their arrival times at monitoring stations). In general, primary waves travel through materials faster than do secondary waves and are detected first. An earthquake's rating on the Richter scale is related to the energy released in the form of seismic waves.

Seismic stations around the world monitor these waves with sensitive detecting instruments called seismographs (Fig. 4). From the data gathered, we can map the paths of the waves through the Earth and thereby learn about the interior structure. The Earth's interior seems to be divided into three general regions: the crust, the mantle, and the core, which itself has a solid inner region and a liquid outer region.*

The locations of their boundaries are determined in part by the shadow zones. Shadow zones are regions where no

FIGURE 2 Bad vibrations Earthquake damage caused by the major shock that struck Kobe, Japan, in January 1995.

*The crust is in most places about 24–30 km (15–20 mi) thick; the mantle is 2900 km (1800 mi) thick; and the core has a radius of 3450 km (2150 mi). The solid inner core has a radius of about 1200 km (750 mi).

FIGURE 1 The San Andreas Fault A small section of the fault, which runs through the San Francisco Bay area as well as the more rural regions of California shown here.

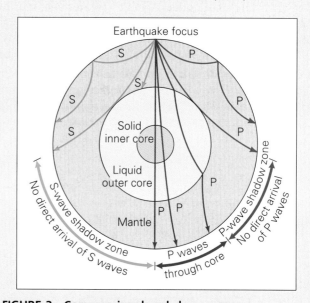

FIGURE 3 Compressional and shear waves Earthquakes produce waves that travel through the Earth. Because transverse S waves are not detected on the opposite side of the Earth, scientists believe that at least part of the Earth's core is a viscous liquid under high pressures and temperatures. The waves bend continuously, or refract, because their speed varies with depth.

Insight

waves of a particular type are detected. These zones appear because although longitudinal waves can travel through solids *or* liquids, transverse waves can travel only through solids. When an earthquake occurs at a particular location, P waves are detected on the other side of the Earth but S waves are not (see Fig. 3). The absence of S waves in a shadow zone leads to the conclusion that the Earth must have a region near its center that is in the liquid phase. This region is a highly viscous metallic liquid—but definitely a liquid, since it does not support a shear (transverse waves are not propagated).

When the transmitted P waves enter and leave the liquid region, they are refracted (bent). This refraction gives rise to a P wave shadow zone, which indicates that only the outer part of the core is liquid. As you will learn in Chapter 19, the combination of a liquid outer core and rotation may be responsible for the Earth's magnetic field.

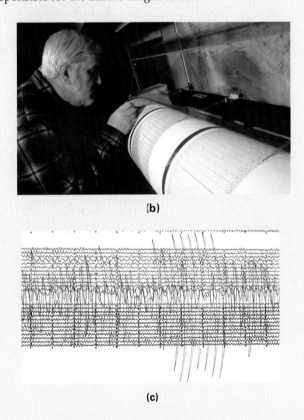

FIGURE 4 The seismograph (a) A simple seismograph. The device records the amplitude of the ground vibrations. The energy of the seismic waves is proportional to the square of their amplitude. (b) A U.S. Geological Survey scientist monitors seismograph readings. (c) Seismograph trace from the destructive Kobe, Japan, earthquake of 1995.

13.4 Wave Properties

Objective: **To explain various wave properties and resulting phenomena.**

Among the properties exhibited by all waves are interference, superposition, reflection, refraction, dispersion, and diffraction. Particles do not share these properties.

Interference and Superposition

When two or more waves meet or pass through the same region of a medium, they pass through each other and proceed without being altered. While they are in the same region, the waves are said to be interfering.

What happens during interference; that is, what does the combined waveform look like? The relatively simple answer is given by the **principle of superposition**:

> At any time, the combined waveform of two or more interfering waves is given by the sum of the displacements of the individual waves at each point in the medium.

The principle of **interference** is illustrated in •Fig. 13.14. The displacement of the combined waveform at any point is given by $y = y_1 + y_2$, where y_1 and y_2 are the

•**FIGURE 13.14 Principle of superposition** When two waves meet, they interfere. The beige tint marks the area where the two waves, moving in opposite directions, overlap and combine. The displacement at any point on the combined wave is equal to the sum of the displacements on the individual waves: $y = y_1 + y_2$.

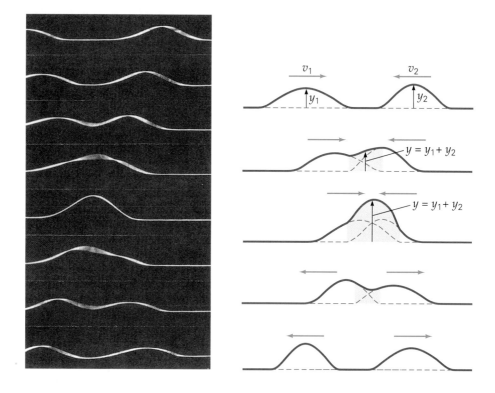

displacements of the individual pulses at that point (directions are indicated by plus and minus signs). Interference, then, is the physical addition of waves. In adding waves, we must take into account the possibility that they are producing disturbances in opposite directions. In other words, we must treat the disturbances in terms of vector addition.

In Fig. 13.14, the vertical displacements of the two pulses are in the same direction, and the amplitude of the combined waveform is greater than that of either pulse. This situation is called **constructive interference**. Conversely, if one pulse has a negative displacement, the two pulses tend to cancel each other when they overlap, and the amplitude of the combined waveform is smaller than that of either pulse. This situation is called **destructive interference**.

Special cases of total constructive and destructive interference are shown in •Fig. 13.15. These are shown for traveling wave pulses of the same width and amplitude. At the instant these interfering waves are exactly in phase (crest coincid-

•**FIGURE 13.15 Interference (a)** When two wave pulses of the same amplitude meet and are in phase, they interfere constructively. When the pulses are exactly superimposed (3), total constructive interference occurs. **(b)** When the interfering pulses are completely out of phase (180°) and are exactly superimposed (3), total destructive interference occurs.

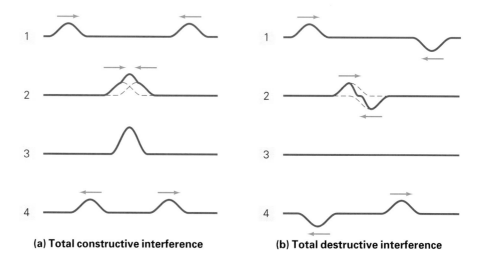

(a) Total constructive interference   (b) Total destructive interference

438 CHAPTER 13 Vibrations and Waves

ing with crest), the amplitude of the combined waveform is twice that of either individual wave. This case is referred to as **total constructive interference**. When these interfering pulses are completely out of phase (180° difference, or crest coinciding with trough), the waveforms momentarily disappear; that is, the amplitude of the combined wave is zero. This case is called **total destructive interference**.

The word *destructive* unfortunately tends to imply that the energy as well as the form of the waves is destroyed. This is not the case. At the point of total destructive interference, when the net wave shape and hence potential energy are zero, the wave energy is stored in the medium completely in the form of kinetic energy.

Reflection, Refraction, Dispersion, and Diffraction

Besides meeting other waves, waves can (and do) meet objects or a boundary with another medium. In such cases, several things may occur. One of these is reflection. **Reflection** occurs when a wave strikes an object or comes to a boundary of another medium and is at least partly diverted back into the original medium. An echo is the reflection of sound waves, and mirrors reflect light waves.

Two cases of reflection are illustrated in •Fig. 13.16. If the end of the string is fixed, the reflected pulse is inverted, or undergoes a 180° phase shift (Fig. 13.16a). This is because the pulse causes the string to exert an upward force on the wall, and the wall exerts an equal and opposite downward force on the string (by Newton's third law). This downward force creates the downward or inverted reflected pulse. If the end of the string is free to move, then the reflected pulse is not inverted (zero or no phase shift). This is illustrated in Fig. 13.16b, where the string is attached to a light ring that can move freely on a smooth pole. The ring is accelerated upward by the front portion of the incoming pulse and then comes back down, thus creating a noninverted reflected pulse.

More generally, when a wave strikes a boundary, the wave is not completely reflected. Instead, some of the wave's energy is reflected and some is transmitted or absorbed. When a wave crosses a boundary into another medium, its speed generally changes because the new material has different characteristics. Entering the medium obliquely (at an angle), the transmitted wave moves in a direction different from that of the incident wave. This phenomenon is called **refraction** (•Fig. 13.17a).

Since refraction depends on changes in wave speed, you might be wondering which physical parameters determine the wave speed. Generally, there are two types of situations. The simplest kind of wave is one whose speed does *not* depend on the wavelength (or frequency) of the wave. All such waves travel at the same speed determined solely by the properties of the medium. These waves are called *nondispersive waves*, because they do not disperse, or spread apart from one another. An example of a nondispersive transverse wave is a wave on a string, where, as we

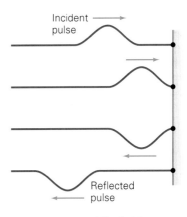

(a) Phase shift of 180°

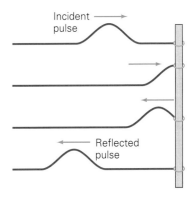

(b) No phase shift

•**FIGURE 13.16 Reflection** (a) When a wave (pulse) on a string is reflected from a fixed boundary, the reflected wave is inverted, or undergoes a 180° phase shift. (b) If the string is free to move at the boundary, there is no phase shift of the reflected wave.

•**FIGURE 13.17 Refraction and dispersion** (a) Refraction of water waves is shown from overhead. As the crests approach the beach, their left edge slows as it enters shallow water first. Thus the whole crest rotates, approaching the beach more or less head on. (b) For dispersive waves, wave speed depends on frequency or wavelength, as shown for surface waves traveling in deep water.

•**FIGURE 13.18 Diffraction**
Diffraction effects are greatest when the opening (or object) is about the same size as or smaller than the wavelength of the waves. **(a)** With an opening much larger than the wavelength of these plane water waves, diffraction is noticeable only near the edges. **(b)** With an opening about the same size as the wavelength of the waves, diffraction produces nearly semi-circular waves.

(a)

(b)

shall see, the speed is determined only by the tension and mass density of the string (Section 13.5). Sound is a nondispersive longitudinal wave; the speed of sound (in air) is determined only by the compressibility and density of the air. Indeed, if sound speed did depend on the frequency, at the back of the symphony hall you might hear the violins well before the clarinets, even though the two sound waves were in perfect synchronization when they left the orchestra pit.

When the wave speed *does* depend on wavelength (or frequency), the waves are said to exhibit **dispersion**—waves of different frequencies spread apart from one another. Although waves of light are nondispersive in a vacuum, when they enter water they become dispersive. Thus in water the different wavelengths of light (seen as colors) separate. This is the basis for rainbow formation, as we shall see in Chapter 22. Water waves are dispersive but only in very deep water (Fig. 13.17b). Under such conditions, longer wavelengths travel faster than shorter ones. Thus if waves are created in a storm region in the deep ocean, the long-wavelength, low-frequency waves reach the shore first. Dispersion will be most important for us in our study of light, but you should remember that waves other than light can also be dispersive under the right conditions.

Diffraction refers to the bending of waves around an edge of an object but is unrelated to refraction. For example, if you stand along an outside wall of a building near the corner, you can hear people talking around the corner. Assuming there are no reflections or air motion (wind), this would not be possible if the sound waves traveled in a straight line. As the sound waves pass the corner, instead of being sharply cut off, they "wrap around" the edge; thus you can hear the sound.

In general, the effects of diffraction are evident only when the size of the diffracting object or opening is about the same as or smaller than the wavelength of the waves. The dependence of diffraction on wavelength and size of the object or opening is illustrated in •Fig. 13.18. For many waves, diffraction is negligible under normal circumstances. For instance, visible light has wavelengths on the order of 10^{-6} m. Such wavelengths are much too small to exhibit diffraction when passing through common-sized openings, such as an eyeglass lens.

Reflection, refraction, dispersion, and diffraction will be considered in more detail for light waves in Chapter 22.

13.5 Standing Waves and Resonance

Objectives: To (a) describe the formation and characteristics of standing waves and (b) explain the phenomenon of resonance.

If you shake one end of a stretched rope, waves travel down it to the fixed end and are reflected back. The waves going down and back interfere. In most cases, the combined waveforms have a changing, jumbled appearance. But if the rope is shaken at just the right frequency, a steady waveform, or series of uniform loops, appears to stand in place along the rope. Appropriately, this phenomenon is called

•**FIGURE 13.19 Standing waves (a)** Standing waves are formed by interfering waves traveling in opposite directions. **(b)** Conditions of destructive and constructive interference recur as each wave travels a distance of $\lambda/4$ in a time of $t = T/4$. The velocities of the rope particles are indicated by the arrows. This gives rise to standing waves with stationary nodes and maximum amplitude antinodes.

(a)

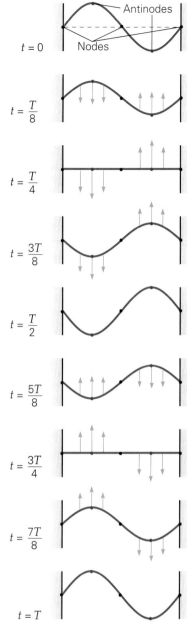

(b)

a **standing wave** (•Fig. 13.19a; see Demonstration 11 on p. 442). It arises because of interference with the reflected waves, which have the same wavelength, amplitude, and speed. Since the two identical waves travel in opposite directions, the net energy flow down the rope is zero. The energy is "standing" in the loops.

Some points on the rope remain stationary at all times and are called **nodes**. At these points, the displacements of the interfering waves are *always* equal and opposite. Thus, by the principle of superposition, the interfering waves must cancel each other completely at these points, and the rope does not undergo displacement there. At all other points, the rope oscillates back and forth at the same frequency. The points of maximum amplitude, where constructive interference is greatest, are called **antinodes**. As you can see in Fig. 13.19b, adjacent antinodes are separated by a half wavelength ($\lambda/2$), or one loop; adjacent nodes are also separated by a half wavelength.

Standing waves can be generated in a rope by more than one driving frequency; the higher the frequency, the more oscillating half-wavelength loops in the rope. The only requirement is that the half wavelengths "fit" the rope length. The frequencies at which large-amplitude standing waves are produced are called **natural frequencies**, or **resonant frequencies**. The resulting standing wave patterns are called *normal*, or *resonant*, *modes of vibration*. In general, all systems that oscillate have one or more natural frequencies, which depend on such factors as mass, elasticity or restoring force, and geometry (boundary conditions). The natural frequencies of a system are sometimes called its *characteristic* frequencies.

A stretched string or rope can be analyzed to determine its natural frequencies. The boundary conditions are that the ends are fixed; thus, there must be a node at each end. The number of closed segments or loops of a standing wave that will fit between the nodes at the ends (along the length of the string) is equal to an integral number of *half*-wavelengths (•Fig. 13.20). Note that $L = \lambda_1/2$, $L = 2(\lambda_2/2)$, $L = 3(\lambda_3/2)$, $L = 4(\lambda_4/2)$, and so on. In general,

$$L = n\left(\frac{\lambda_n}{2}\right) \quad \text{or} \quad \lambda_n = \frac{2L}{n} \quad (\text{for } n = 1, 2, 3, \dots)$$

The natural frequencies of oscillation, where v is the speed of waves on a string, are

$$f_n = \frac{v}{\lambda_n} = n\left(\frac{v}{2L}\right) = nf_1 \quad \text{for } n = 1, 2, 3, \dots \qquad \begin{array}{c}\textit{natural frequencies for}\\ \textit{a stretched string}\end{array} \quad (13.19)$$

The lowest natural frequency ($f_1 = v/2L$ for $n = 1$,) is called the **fundamental frequency**. All of the other natural frequencies are integral multiples of the fundamental frequency: $f_n = nf_1$ (for $n = 1, 2, 3, \dots$). The set of frequencies $f_1, f_2 = 2f_1, f_3 = 3f_1, \dots$, is called a **harmonic series**: f_1 (the fundamental frequency) is the *first harmonic*, f_2 the *second harmonic*, and so on.

Strings fixed at each end are found in stringed musical instruments such as violins, pianos, and guitars. When such a string is excited, the resulting vibration

Demonstration 11 ■ Flame Standing Wave Pattern

Demonstration of a compressional standing wave: a sound wave in a closed tube. The distance between successive high (or low) flames is equal to a half wavelength of the standing sound (compressional) wave in the pipe. What would happen to the waveform if the next lower resonant frequency were tuned in?

(a) Regularly spaced holes in a piece of pipe allow a line of uniformly sized flames when flammable gas is supplied to the pipe. One end of the pipe is closed by a metal plate, and the other end is fitted with a rubber diaphragm and loudspeaker capable of sending sound waves down the tube to reflect off the closed end.

(b) When an audio oscillator driving the speaker is tuned to a resonant frequency of the pipe, standing sound (compressional) waves are created. The highest flames are spaced a half wavelength apart. [See *The Physics Teacher*, 17, 307 (1979).] Note: Some fast, contemporary music with good bass instead of an oscillator output produces a rapidly varying, dancing flame pattern.

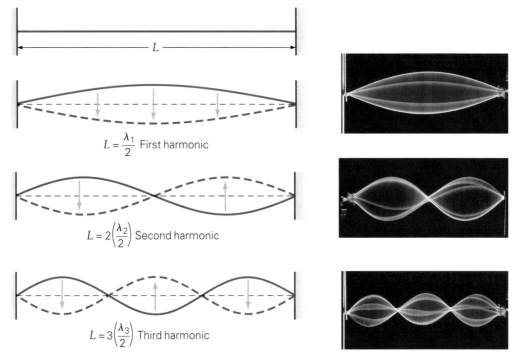

•**FIGURE 13.20 Natural frequencies** A stretched string can have standing waves only at certain frequencies. These correspond to the numbers of half-wavelength loops that will fit along the length of string between the nodes at the fixed ends.

generally includes several harmonics in addition to the fundamental frequency. The number of harmonics depends on how and where the string is excited—that is, plucked, struck, or bowed. It is the combination of harmonic frequencies that gives a particular instrument its characteristic sound quality. (More on this in Chapter 14.) As Eq. 13.19 shows, the fundamental frequency of a stretched string, as well as the other harmonics, depends on the length of the string. Think of how different notes are obtained on a particular string of a violin or a guitar (•Fig. 13.21).

Natural frequencies also depend on other parameters, such as mass and force, which affect the wave speed in the string. For a stretched string, the wave speed (v) can be shown to be

$$v = \sqrt{\frac{F_T}{\mu}} \qquad (13.20)$$

where F_T is the tension in the string and μ is the linear mass density (mass per unit length, $\mu = m/L$). (We use F_T, rather than T of previous chapters, so as not to confuse the tension with period T.) Thus Eq. 13.19 can be written as

$$f_n = n\frac{v}{2L} = \frac{n}{2L}\sqrt{\frac{F_T}{\mu}} = nf_1 \quad \text{(for } n = 1, 2, 3, \ldots\text{)} \qquad (13.21)$$

Note that the greater the linear mass density of a string, the lower its natural frequencies. As you may know, the low-note strings on a violin or guitar are thicker, or more massive, than the high-note strings. By tightening a string, we increase all frequencies of that string. Changing string tension is how violinists, for example, tune their instruments before a performance.

•FIGURE 13.21 **Fundamental frequencies** Performers on string instruments such as the violin or guitar use their fingers to stop or fret the strings. By pressing a string against the fingerboard, the player reduces the amount of its length that is free to vibrate. This reduction changes the resonant frequency of the string and thus the pitch of the tone it produces.

Example 13.6 ■ A Piano String: Fundamental Frequency and Harmonics

A piano string with a length of 1.15 m and a mass of 20.0 g is under a tension of 6.30×10^3 N. (a) What is the fundamental frequency of the string when it is struck? (b) What are the frequencies of the next two harmonics?

Thinking It Through. We have the tension and can calculate the linear mass density from the data. This will give us the fundamental frequency, and from that we can get the harmonics.

Solution.

Given: $L = 1.15$ m
$\quad\quad\quad m = 20.0$ g $= 0.0200$ kg
$\quad\quad\quad F_T = 6.30 \times 10^3$ N

Find: (a) f_1 (fundamental frequency)
(b) f_2 and f_3 (frequencies of next two harmonics)

(a) The linear mass density of the string is

$$\mu = \frac{m}{L} = \frac{0.0200 \text{ kg}}{1.15 \text{ m}} = 0.0174 \text{ kg/m}$$

Then, using Eq. 13.21, we get

$$f_1 = \frac{1}{2L}\sqrt{\frac{F_T}{\mu}} = \frac{1}{2(1.15 \text{ m})}\sqrt{\frac{6.30 \times 10^3 \text{ N}}{0.0174 \text{ kg/m}}} = 262 \text{ Hz}$$

This is approximately the frequency of middle C (C_4) on a piano.

(b) Since $f_2 = 2f_1$ and $f_3 = 3f_1$,

$$f_2 = 2f_1 = 2(262 \text{ Hz}) = 524 \text{ Hz}$$

and

$$f_3 = 3f_1 = 3(262 \text{ Hz}) = 786 \text{ Hz}$$

The second harmonic corresponds approximately to C_5 on a piano, since by definition the frequency doubles with each octave (every eighth white key).

Follow-up Exercise. A musical note is referenced to the fundamental vibrational frequency, or first harmonic. In musical terms, the second harmonic is the first overtone, the third harmonic is the second overtone, and so on. If an instrument has a third overtone with a frequency of 880 Hz, what is the frequency of the first overtone?

Note: Don't be confused by the language: *First* overtone means the first frequency above the fundamental frequency—that is, the *second* harmonic.

Conceptual Example 13.7

Tuning Up: Raising the Frequency of a Guitar String

You wish to raise the fundamental frequency of a guitar string from the A note (220 Hz) below middle C to the A note (440 Hz) above middle C. Would you (a) loosen the string to halve its tension, (b) tighten the string to double the tension, (c) use another string of the same material with half the diameter at the same tension, (d) use another string of the same material with twice the diameter at the same tension?

Reasoning and Answer. The fundamental frequency of a stretched string is given by Eq. 13.21:

$$f = \frac{1}{2L}\sqrt{\frac{F_T}{\mu}} \quad \text{(for } n = 1\text{)}$$

The frequency of the string is thus proportional to the *square root* of the tension force F_T, so loosening the string—that is, decreasing F_T—would not increase the frequency. Nor would doubling the tension double the frequency (because $\sqrt{2F_T} \neq 2\sqrt{F_T}$). Thus, neither (a) nor (b) is the correct answer.

The question then is, how does the frequency vary with the diameter of the string? Clearly, if we stay with the same string material (same density ρ), the greater the diameter of the string, the greater its mass per unit length (μ). Hence, a thinner string will vibrate at a higher frequency, and the answer can only be (c).

We must also confirm that halving the diameter doubles the frequency. Remember that doubling the tension did *not* double the frequency! So let's be careful.

Here μ is the mass per unit length ($\mu = m/L$), and the mass can be written in terms of the volume density (ρ) and the geometry of the string, $m = \rho V$, where V is the volume of the string. The volume can be written in terms of its circular cross-sectional area (assumed constant) and the length of the string, which we treat as a long cylinder:

$$V = AL = \frac{\pi d^2 L}{4}$$

where d is the diameter of the string ($A = \pi r^2 = \pi d^2/4$).

Writing the frequency in terms of these parameters, we have

$$f = \frac{1}{2L}\sqrt{\frac{F_T}{\mu}} = \frac{1}{2L}\sqrt{\frac{F_T L}{m}} = \frac{1}{2L}\sqrt{\frac{4F_T}{\rho \pi d^2}} = \left(\frac{1}{L}\sqrt{\frac{F_T}{\rho \pi}}\right)\frac{1}{d}$$

So, with constant F_T, ρ, and L (the active length of the string between the bridge and neck of a stringed instrument), the frequency is inversely proportional to the diameter of the string:

$$f \propto \frac{1}{d}$$

We need to show explicitly that using a string with half the diameter ($d_2 = d_1/2$) would double the frequency. Forming a ratio shows this quickly, since

$$\frac{f_2}{f_1} = \frac{d_1}{d_2} \quad \text{or} \quad f_2 = \left(\frac{d_1}{d_2}\right)f_1$$

For $f_2 = 2f_1$ (that is, 440 Hz = 2 × 220 Hz), we must have $d_1/d_2 = 2$ or $d_2 = d_1/2$, as expected.

Follow-up Exercise. The fundamental frequency of a violin string is A below middle C (220 Hz). How could you tune this string to middle C (264 Hz) *without* changing strings as was done in this Example?

When an oscillating system is driven at one of its natural, or resonant, frequencies, maximum energy transfer to the system occurs. The system is physically suited to any of its natural frequencies. These are the frequencies at which it "wants" to vibrate, so to speak. The condition of driving a system at a natural frequency is referred to as **resonance**.

A common example of a system in mechanical resonance is someone being pushed on a swing. (For some typical numerical values, see Example 13.4.) Basically, a swing is a simple pendulum and has only one resonant frequency for a given length $[f = (1/2\pi)\sqrt{g/L}]$. If you push the swing with this frequency and in phase with its motion, the amplitude and energy increase (•Fig. 13.22). If you push at a slightly different frequency, the energy transfer is no longer a maximum. (What do you think happens if you push with the resonant frequency but 180° out of phase with the swing's motion?)

Unlike a simple pendulum, a stretched string has many natural frequencies. Almost any driving frequency will cause a disturbance in the string. However, if the frequency of the driving force is not equal to one of the natural frequencies, the resulting wave will be relatively small and jumbled. However, when the frequency of the driving force matches one of the natural frequencies, the maximum amount of energy is transferred to the string. A steady standing wave pattern results with the amplitude at the antinodes becoming relatively large.

Mechanical resonance is not the only type of resonance. When you tune a radio, you are changing the resonant frequency of an electrical circuit (Chapter 21) so that it will be driven by, or will pick up, the frequency signal of the station you want. Other examples of resonance are described in the Insight on this page.

•**FIGURE 13.22 Resonance in the playground** The swing behaves like a pendulum in SHM. To transfer energy efficiently, the woman must time her pushes to its natural frequency.

Insight

Desirable and Undesirable Resonances

When we hold a seashell to our ear, we say we hear the ocean, or a sound like that of the ocean. What causes this is a resonance effect—a desirable one because of the pleasant sound. Soft sounds enter the shell from the environment, and some of these are at the resonant frequencies of the shell. Standing sound waves are set up that can be heard when the shell is held close to the ear. (We shall learn more about sound in Ch. 14.)

The mixture of sound frequencies is not harmonic because of the complicated shape of the shell's interior. As a result, the sound has a variety of frequencies—"white noise," so to speak (analogous to "white light," which is a mixture of all visible frequencies). The resulting sound is similar to that of the ocean.

When a large number of soldiers march over a small bridge, they are generally ordered to break step. The reason is that the marching frequency may correspond to one of the natural frequencies of the bridge and set it into resonant vibration, which could cause it to collapse. This actually occurred on a suspension bridge in England in 1831. The bridge was weak and in need of repair, but the resonance vibrations induced by the marching soldiers crossing the bridge helped it fail sooner—and caused some injuries.

Another incidence of bridge vibration wasn't due to marching soldiers but to the driving force of the wind. On the morning of November 7, 1940, winds with speeds of 40 to 45 mi/h started the main span of the Tacoma Narrows Bridge (in Washington state) vibrating. The bridge, 2800 ft (855 m) long and 39 ft (12m) wide, had been first opened to traffic only 4 months earlier.

During the first month of use, small transverse modes of vibration had been observed. But on November 7, special wind effects drove the bridge in near resonance, and the main span vibrated at a frequency of 36 vib/min and an amplitude of 1.5 ft. At 10 A.M., the main span began to vibrate in a torsional (twisting) mode in two segments at a frequency of 14 vib/min. The wind continued to drive the bridge in res-

onance, and the vibrational amplitude increased. Shortly after 11 A.M., the main span collapsed (Fig. 1).*

"Galloping Gertie" (the nickname given to the bridge) was rebuilt on the same tower foundations. However, the new design made the structure stiffer to increase its resonant frequency so that high winds could not produce unwanted resonance.

*It is doubtful that the gusting of the wind set the bridge into vibration. The wind velocity was moderately steady, and gust fluctuations are normally random. One explanation for the driving source of the oscillations involves the formation of vortices as the wind blew past the bridge. Vortices are like the eddies that form in water at the end of oars when a boat is rowed. The wind blowing over and under the bridge formed vortices that rotated in opposite directions. The formation and "shedding" of the vortices (like eddies coming off oars) would have imparted energy to the bridge, and if the frequency of this action were near a natural frequency, a standing wave would have been set up.

FIGURE 1 Galloping Gertie The collapse of the Tacoma Narrows Bridge on November 7, 1940, is captured in this frame from a movie camera.

Chapter Review

Important Terms

Hooke's law 419
simple harmonic motion (SHM) 419
displacement 419
amplitude 419
period 419
frequency 420
hertz 420
equation of motion 424
phase constant 430
damped harmonic motion 432
wave 433
wave motion 433
wavelength 434
wave speed 434
transverse wave 434
longitudinal wave 435
principle of superposition 437
interference 437
constructive interference 438
destructive interference 438
total constructive interference 439
total destructive interference 439
reflection 439
refraction 439
dispersion 440
diffraction 440
standing wave 441
node 441
antinode 441
natural (resonant) frequencies 441
fundamental frequency 441
harmonic series 441
resonance 444

Important Concepts

- Simple harmonic motion (SHM) requires a restoring force directly proportional to the displacement, such as the spring force.
- In general, the total energy of an object in SHM is directly proportional to the square of the amplitude.
- The form of an equation of motion for an object in SHM depends on the object's initial displacement and velocity, or initial conditions expressed in terms of a phase constant.
- A wave is a disturbance in time and space; energy is transferred or propagated by wave motion.
- At any time, the combined waveform of two or more interfering waves is given by the sum of the displacements of the individual waves at each point in the medium.
- At natural frequencies, standing waves can result from the interference of two waves of identical wavelength, amplitude, and speed traveling in opposite directions on a string.

Important Equations

Hooke's Law:
$$F_s = -kx \qquad (13.1)$$

Frequency and Period for SHM:
$$f = \frac{1}{T} \qquad (13.2)$$

Total Energy of a Spring and Mass in SHM:
$$E = \tfrac{1}{2}kA^2 = \tfrac{1}{2}mv^2 + \tfrac{1}{2}kx^2 \qquad (13.4\text{--}5)$$

Velocity of Oscillating Mass on a Spring:
$$v = \pm\sqrt{\frac{k}{m}(A^2 - x^2)} \qquad (13.6)$$

Period of Mass Oscillating on a Spring:
$$T = 2\pi\sqrt{\frac{m}{k}} \qquad (13.11)$$

Angular Frequency of a Mass Oscillating on a Spring:
$$\omega = 2\pi f = \sqrt{\frac{k}{m}} \qquad (13.13)$$

Period of a Simple Pendulum (small-angle approximation):
$$T = 2\pi\sqrt{\frac{L}{g}} \qquad (13.14)$$

Displacement of a Mass in SHM:
$$y = A\sin(\omega t + \delta) \qquad (13.15)$$
$$(\text{with } \delta = 0) \quad y = A\sin\omega t = A\sin 2\pi f t = A\sin\frac{2\pi t}{T}$$

Velocity of a Mass in SHM ($\delta = 0$):
$$v = \omega A \cos\omega t \qquad (13.16)$$

Acceleration of a Mass in SHM ($\delta = 0$):
$$a = -\omega^2 A \sin\omega t = -\omega^2 y \qquad (13.17)$$

Wave Speed:
$$v = \frac{\lambda}{T} = \lambda f \qquad (13.18)$$

Natural Frequencies in a Stretched String:
$$f_n = n\left(\frac{v}{2L}\right) = \frac{n}{2L}\sqrt{\frac{F_T}{\mu}} \quad (\text{for } n = 1, 2, 3, \ldots) \qquad (13.21)$$

Exercises

13.1 Simple Harmonic Motion

1. A particle in SHM (a) has variable amplitude, (b) has a restoring force in the form of Hooke's law, (c) has a frequency directly proportional to its period, (d) has its position represented graphically by $x(t) = at + b$.

2. The maximum kinetic energy of a spring–mass system in SHM is equal to (a) A, (b) A^2, (c) kA, (d) $kA^2/2$.

3. If the amplitude of an object in SHM is doubled, how are (a) the energy and (b) the maximum speed affected?

4. How does the speed of an object in SHM change as it approaches its equilibrium position? Explain.

5. If the period of a system in SHM is doubled, its frequency is (a) doubled, (b) halved, (c) four times as large, (d) one-quarter as large.

6. When a particle in SHM is at the equilibrium position, the potential energy of the system is (a) zero, (b) maximum, (c) negative, (d) none of the above.

7. ■ A particle oscillates in SHM with an amplitude A. What is the total *distance* it travels in one period?

8. ■ A 1.0-kilogram object oscillating on a spring completes a cycle every 0.50 s. What is the frequency of this oscillation?

9. ■ A particle in simple harmonic motion has a frequency of 40 Hz. What is the period of this oscillation?

10. ■ The frequency of a simple harmonic oscillator is doubled from 0.25 Hz to 0.50 Hz. What is the change in its period?

11. ■ What is the spring constant of a spring that stretches 6.0 cm when a 0.25-kilogram mass is suspended from it?

12. ■ An object of mass 0.50 kg is attached to a spring, with spring constant 10 N/m. If the object is stretched 0.050 m from the equilibrium position and released, what is its maximum speed?

13. ■■ Atoms in a solid are in continuous vibrational motion due to thermal energy. At room temperature, the amplitude of these atomic vibrations is typically about 10^{-9} cm, and their frequency is on the order of 10^{12} Hz. (a) What is the approximate period of oscillation of a typical atom? (b) What is its maximum speed?

14. ■■ A 0.500-kilogram object resting on a horizontal frictionless surface is attached to a spring with a spring constant of 150 N/m. If the object is pulled 0.150 m from its equilibrium position and released, what are the magnitude of force on the object and its acceleration at (a) $t = 0$, (b) $x = 0.050$ m, and (c) $x = 0$?

15. ■■ (a) At what position is the magnitude of the force on an object in a spring–object system maximum: $x = 0$, $x = -A$, or $x = +A$? (b) How about its speed? (c) the magnitude of its acceleration?

16. ■■ In a lab experiment, you are given a spring with a spring constant of 12 N/m. What mass would you suspend on the spring to have an oscillation period of 0.91 s when in SHM?

17. ■■ If the oscillating spring in Exercise 16 had an amplitude of 15 cm, (a) what would be its maximum speed, and (b) where would this occur? (c) What would be the speed at a half-amplitude position?

18. ■■ If the mass of an object attached to a spring is doubled, what is the ratio of the new period to the old one?

19. ■■ If the spring constant in a spring–mass system is doubled, what is the ratio of the new period to the old one?

20. ■■ A 0.25-kilogram mass on a vertical spring oscillates at a frequency of 1.0 Hz. If a 0.50-kilogram mass were instead suspended from the same spring, what would be the frequency of oscillation?

21. ■■ The oscillations of two oscillating spring–object systems are graphed in •Fig. 13.23. If the mass of the object in (a) is 4 times that in (b), which system has more energy, and how many times more?

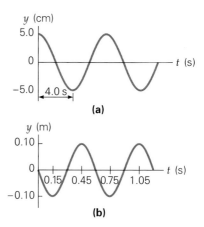

•**FIGURE 13.23 Wave energy and equation of motion** See Exercises 21, 47, and 48.

22. ■■ A 0.25-kilogram object suspended on a light spring is released from a position 15 cm above the stretched equilibrium position. The spring has a spring constant of 80 N/m. (a) What is the total (spring) energy of the system? (Neglect gravitational potential energy.) (b) Does this energy depend on the mass of the object? Explain.

23. ■ What is the speed of the object in Exercise 22 when the object is (a) 5.0 cm above its equilibrium position and (b) 5.0 cm below its equilibrium position? (c) What is the object's maximum speed, and where does this occur?

24. ■■■ A 75-kilogram circus performer jumps from a height 5.0 m onto a trampoline and stretches it 0.30 m. Assume that the trampoline obeys Hooke's law. (a) How far will it stretch if the performer jumps from a height 8.0 m? (b) How far will it stretch if the performer stands still on it while taking a bow?

25. ■■■ A 0.250-kilogram ball is dropped from a height of 10.0 cm onto a spring as illustrated in •Fig. 13.24. If the spring has a spring constant of 60.0 N/m, (a) what distance will the spring be compressed? (Neglect energy loss during collision.) (b) On recoiling upward, how high will the ball go?

•FIGURE 13.24 **How far down?** See Exercise 25.

13.2 Equations of Motion

26. The equation of motion for a particle in SHM (a) is always a cosine function, (b) reflects damping action, (c) is independent of the initial conditions, (d) gives the position of the particle as a function of time.

27. ◯ If the length of a pendulum is doubled, what is the ratio of the new period to the old one?

28. The apparatus in Fig. 13.6 demonstrates that the motion of an object on a spring can be described by a sinusoidal function of time. How could this be demonstrated for a pendulum?

29. Could simple harmonic motion be described by a tangent function? Explain.

30. ◯ Would the period of a pendulum in an upward accelerating elevator be increased or decreased from that of the period in a non-accelerating elevator? Explain.

31. ■ What mass on a spring with a spring constant of 10 N/m will oscillate with a period of 2.0 s?

32. ■ A 0.50-kilogram mass oscillates in simple harmonic motion on a spring with a spring constant of 200 N/m. What are (a) the period and (b) the frequency of the oscillation?

33. ■ ◯ The simple pendulum in a grandfather clock is 1.0 m long. What are (a) the period and (b) the frequency of this pendulum?

34. ■ A breeze sets into oscillation a lamp suspended from the ceiling. If the period is 1.0 s, what is the distance from the ceiling to the lamp at the lowest point? Assume the lamp acts as a simple pendulum.

35. ■ ◯ Write the general equation of motion for an object that is on a horizontal frictionless surface and is connected to a spring at equilibrium (a) if the object is initially given a quick push outward and (b) if the object is pulled away from the spring and released.

36. ■ ◯ The equation of motion for a vertical SHM oscillator is given by $y = (0.10 \text{ m}) \sin 100t$. What are the (a) amplitude, (b) frequency, and (c) period of this motion?

37. ■ ◯ The displacement of an object in SHM is given by $y = (5.0 \text{ cm}) \sin 20\pi t$. What are the (a) amplitude, (b) frequency, and (c) period of oscillation?

38. ■ ◯ If a SHM oscillator's displacement obeys the equation $y = (0.25 \text{ m}) \cos 314t$, where y is in meters and t in seconds, what is the position of the oscillator at (a) $t = 0$, (b) $t = 5.0$ s, and (c) $t = 15$ s?

39. ■■ Show that the total energy of a spring–object system in simple harmonic motion is given by $\frac{1}{2}m\omega^2 A^2$.

40. ■■ Show that for a pendulum to oscillate at the same frequency as an object on a spring, the pendulum's length must be $L = mg/k$.

41. ■■ ◯ Students use a simple pendulum with a length of 36.90 cm to measure the acceleration of gravity at the location of their school. If the period of the pendulum is 1.220 s, what is the experimental value of g at the school?

42. ■■ ◯ The equation of motion of a particle in vertical simple harmonic motion is given by $y = (10 \text{ cm}) \sin 0.50t$. What are the particle's (a) displacement, (b) velocity, and (c) acceleration at $t = 1.0$ s?

43. ■■ A 0.15-kilogram block oscillates vertically in simple harmonic motion on a spring with a spring constant 6.0 N/m. The block was initially raised 8.0 cm from equilibrium. (a) Write the equation of motion. (b) Determine the phase constant. (c) What is the displacement of the block at $t = 0.50$ s?

44. ■■ Two objects of equal mass oscillate on light springs, the second with a spring constant twice that of the first. Which system will have the greater period, and how many times greater?

45. ■■ If a grandfather clock were taken to the Moon, where the acceleration due to gravity is only one-sixth (assume exact) of that on the Earth, by what factor will the period of oscillation change? Does it increase or decrease? (See Exercise 33.)

46. ■■ A 0.075-kilogram mass oscillates on a light spring, and another 0.075-kilogram mass oscillates as the bob of a simple pendulum in SHM. If the periods of the oscillations are equal and the pendulum length is 0.30 m, what would be the spring constant of the spring?

47. ■■ The motion of a particle is described by the curve in Fig. 13.23a. Write the equation of motion in terms of a cosine function.

48. ■■ The motion of a 0.35-kilogram mass oscillating on a light spring is described by the curve in Fig. 13.23b. (a) Write the equation for its displacement as a function of time. (b) What is the spring constant of the spring?

49. ■■■ The forces acting on a simple pendulum are shown in •Fig. 13.25. (a) Show that for the small-angle approximation ($\sin\theta \approx \theta$), the force producing the motion has the same form as Hooke's law. (b) Show by analogy with an object on a spring that the period of a simple pendulum is given by $T = 2\pi\sqrt{L/g}$. [Hint: Think of the effective spring constant.]

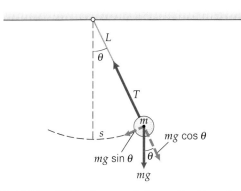

•FIGURE 13.25 **SHM of a pendulum** See Exercise 49.

50. ■■■ A grandfather clock uses a pendulum that is 75 cm long. It is accidentally broken, and when repaired the length is shorter by 2.0 mm. Considering the clock's pendulum to be a simple pendulum, (a) will the repaired clock gain or lose time? (b) By how much will the repaired clock differ from the correct time (taken to be the time determined by the original pendulum in 24 h)? (c) If the pendulum string were metal, would temperature make a difference in the timekeeping of the clock? Explain.

13.3 Wave Motion

51. Wave motion in a material involves (a) the propagation of a disturbance, (b) interparticle interactions, (c) the transfer of energy, (d) all of the preceding.

52. For a periodic wave propagating at a speed v, the following relationship holds: (a) $\lambda = v/f$, (b) $v = \lambda/f$, (c) $v = \lambda f^2$, (d) $f = \lambda/v$.

53. What type(s) of wave(s) will propagate through (a) solids, (b) liquids, and (c) gases?

54. •Figure 13.26 shows snapshots of two waves. Identify whether each is transverse or longitudinal.

•FIGURE 13.26 **Transverse or longitudinal?** See Exercise 54.

55. ■ A longitudinal sound wave has a speed of 340 m/s in air. If this wave produces a tone with a frequency of 1000 Hz, what is the wavelength?

56. ■ A transverse wave has a wavelength of 0.50 m and a frequency of 20 Hz. What is the wave speed?

57. ■ A student reading his physics book on a lake dock notices that the distance between two incoming wave crests is about 2.4 m, and he then measures the time of arrival between wave crests to be 1.6 s. What is the approximate speed of the waves?

58. ■ Light waves travel in a vacuum at a speed of 300 000 km/s. The frequency of visible light is about 5×10^{14} Hz. What is the approximate wavelength of the light?

59. ■ A certain laser emits light of wavelength 6.33×10^{-7} m. What is the frequency of this light in a vacuum?

60. ■■ The range of sound frequencies audible to the human ear extends from about 20 Hz to 20 kHz. If the speed of sound in air is 345 m/s, what are the limits of this audible range expressed in wavelengths?

61. ■■ The AM frequencies on a radio dial range from 550 kHz to 1600 kHz, and the FM frequencies range from 88.0 MHz to 108 MHz. All of these radio waves travel at a speed of 3.00×10^8 m/s (speed of light). What are the wavelength ranges of (a) the AM band and (b) the FM band in terms of wavelengths?

62. ■■ A sonar generator on a submarine on the ocean surface produces periodic ultrasonic waves at a frequency of 2.50 MHz. The wavelength of the waves in sea water is 4.80×10^{-4} m. When the generator is directed downward, an echo reflected from the ocean floor is received 10.0 s later. How deep is the ocean at that point? (Assume wavelength is constant at all depths.)

63. ■■ In watching a transverse wave go by, a woman notes that 13 crests go by in a time of 3.0 s. If she measures the distance between two successive crests to be 0.75 m and if the first and the last points that pass her are crests, what is the speed of the wave?

64. ■■ A wave traveling in the $+x$ direction at a certain time is shown in •Fig. 13.27a. The particle displacement at a particular location in the medium through which the wave travels is shown in Fig. 13.27b. (a) What is the amplitude of the traveling wave? (b) What is the wave speed?

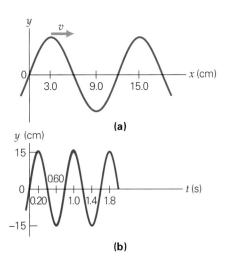

•FIGURE 13.27 **How high and how fast?** See Exercise 64.

65. ■■ Assume that P and S (primary and secondary) waves from an earthquake with a focus near the Earth's surface travel through the Earth at nearly constant average speeds of 8.0 km/s and 6.0 km/s, respectively. Assume that there is no deflection or refraction of the waves. (a) How long is the delay between the arrivals of successive waves at a seismic monitoring station located 90° in latitude from the epicenter (the spot on the surface directly above the focus) of the quake? (b) Do the waves cross the boundary of the mantle? (c) How long does it take for the waves to arrive at a monitoring station on the opposite side of the Earth?

66. ■■■ The speed of longitudinal waves traveling in a long solid rod is given by $v = \sqrt{Y/\rho}$, where Y is Young's modulus and ρ is the density of the solid. If a disturbance has a frequency of 40 Hz, what is the wavelength of the waves it produces in (a) an aluminum rod and (b) a copper rod? [Hint: See Tables 9.1 and 9.2.]

67. ■■■ As noted in Exercise 66, the speed of longitudinal waves in a solid rod is given by $v = \sqrt{Y/\rho}$. Fred strikes a steel train rail with a hammer at a frequency of 0.50 Hz, and Wilma puts her ear to the rail 1.0 km away. (a) How long after the first strike does Wilma hear the sound? (b) What is the time interval between successive sound pulses she hears? [Hint: See Tables 9.1 and 9.2]

13.4 Wave Properties

68. When waves meet each other and interfere, the resultant waveform is determined by (a) reflection, (b) refraction, (c) diffraction, (d) superposition.

69. Refraction (a) involves constructive interference, (b) refers to a change in direction at media interfaces, (c) is synonymous with diffraction, (d) occurs only for mechanical waves, not for light.

70. What is destroyed when destructive interference occurs?

71. Dolphins determine the location of prey by emitting ultrasonic sound (this is called *echolocation*). Which wave phenomenon is involved?

72. If sound waves were dispersive (that is, the speed of sound depended on its frequency), what would be the consequences if you were in a concert hall?

13.5 Standing Waves and Resonance

73. For two traveling waves to form standing waves, the waves must have the same (a) frequency, (b) amplitude, (c) speed, (d) all of the preceding.

74. When a stretched violin string oscillates in its third harmonic mode, then the standing wave in the string will exhibit (a) 3 wavelengths, (b) 1/3 wavelength, (c) 3/2 wavelengths, (d) 2 wavelengths.

75. A child's swing (a pendulum) has only one natural frequency f_o, yet it can be kept going by pushes at a frequency of $f_o/2$. How is this possible?

76. A guitar string fixed at both ends is vibrating in its fourth harmonic. How many nodes are there on the string?

77. ■ The fundamental frequency of a stretched string is 100 Hz. What are the frequencies of (a) the second harmonic and (b) the third harmonic?

78. ■ If the frequency of the third harmonic of a vibrating string is 450 Hz, what is the frequency of the first harmonic?

79. ■ 🔘 A standing wave is formed in a stretched string that is 3.0 m long. What are the wavelengths of (a) the first harmonic and (b) the third harmonic?

80. ■■ 🔘 Will a standing wave be formed in a 4.0-meter length of stretched string that transmits waves at a speed of 12 m/s if it is driven at a frequency of (a) 15 Hz or (b) 20 Hz?

81. ■■ Two waves of equal amplitude and wavelength of 0.80 m travel with a speed of 250 m/s in opposite directions in a string. If the string is 2.0 m long, which harmonic mode is the standing wave set up in the string?

82. ■■ A piece of rubber tubing with a linear mass density of 0.125 kg/m is stretched by a force of 9.00 N. (a) What will be the transverse wave speed in the tubing? (b) If the stretched tubing has a length of 10.0 m, what are its natural frequencies?

83. ■■ Find the first four harmonics for a string that is 2.0 m long, has a linear mass density of 2.5×10^{-2} kg/m, and is under a tension of 40 N.

84. ■■ Two stretched strings have the same tension and linear mass density. Are any of the first six harmonics of the strings equal if the string lengths are (a) 1.0 m and 3.0 m or (b) 1.5 m and 2.0 m, respectively?

85. ■■ A violin string is tuned to 440 Hz (fundamental frequency, or first harmonic). When playing the instrument, the violinist puts a finger down on the string one-eighth of the string length from the neck end. What is the frequency of the string when played like this?

86. ■■■ A thin, flexible metal rod is 1.0 m long. It is clamped at one end to a table, and the other end can vibrate freely. What are the natural frequencies of the rod if the wave speed in the material is 3.5×10^3 m/s?

87. ■■■ In a common laboratory experiment on standing waves, the waves are produced in a stretched string by an electrical vibrator that oscillates at 60 Hz (•Fig. 13.28). The string runs over a pulley, and a hanger is suspended from the end. The tension in the string is varied by adding weights to the hanger. If the active length of string (the part that vibrates) is 1.5 m and this length of string has a mass of 0.10 g, what masses must be suspended to produce the first four harmonics in that length?

•FIGURE 13.28 **Standing waves in strings** Twin vibrating strings with standing waves. This demonstration model allows you to vary string tension, string length, and string type (linear mass density). Also, the vibration frequency can be adjusted. See Exercise 87.

Additional Exercises

88. A 0.10-kilogram block suspended on a spring is pulled to 8.0 cm below its equilibrium position and released. When the block passes through the equilibrium position, it has a speed of 0.40 m/s. What is the speed of the block when it is 4.0 cm from the equilibrium position?

89. ● A standing wave has nodes at $x = 0$ cm, $x = 6.0$ cm, $x = 12$ cm, and $x = 18$ cm. (a) What is the wavelength of the waves that are interfering to produce this standing wave? (b) At what positions are the antinodes?

90. A simple pendulum makes 5.0 complete cycles in 10 s. (a) What are the frequency and period of this pendulum? (b) What is its length?

91. ● The motion of an object is $x = (0.10 \text{ m}) \cos 50t$. What are the magnitudes of the object's maximum velocity and maximum acceleration?

92. A steel piano wire is 60 cm long and has a mass of 3.0 g. If the tension in the wire is 550 N, what are the fundamental frequency and wavelength?

93. ● A stretched string exhibits four equal loops in a standing wave when driven at a frequency of 420 Hz. What driving frequency will set up a wave with two equal loops?

94. On a violin, a correctly tuned A string has a frequency of 440 Hz. If an A string produces sound at 450 Hz when under a tension of 500 N, what should the tension be to produce the correct frequency?

95. If the tension in a string fixed at both ends is doubled, what will happen to the (a) wavelength and (b) frequency of the first harmonic?

96. What is the length of a string whose third harmonic standing wave has a frequency of 27 Hz and along which waves travel at 45 m/s?

97. ● An object in vertical simple harmonic motion is described by $y = (0.20 \text{ cm}) \sin 1.8\pi t$. What is the speed of the object at $t = 10$ s?

98. During an earthquake, the corner of a tall building oscillates with an amplitude of 20 cm at 0.50 Hz. What are the magnitudes of (a) the maximum displacement, (b) the maximum velocity, and (c) the maximum acceleration of the corner of that building?

99. An object resting on a horizontal frictionless surface is connected to a fixed spring. The object is displaced 16 cm from its equilibrium position and released. At $t = 0.50$ s, it is 8.0 cm from its equilibrium position (and has not passed through it yet). What is the object's period of oscillation?

CHAPTER 14

Sound

14.1 Sound Waves
14.2 The Speed of Sound
14.3 Sound Intensity and Sound Intensity Level
14.4 Sound Phenomena
14.5 The Doppler Effect
14.6 Musical Instruments and Sound Characteristics

Insights
- Speech and Hearing
- Doppler Radar

Good vibrations, clearly! We owe a lot to sound waves. Not only do they provide us with one of our main sources of enjoyment in the form of music, they also bring us a wealth of vital information about our environment, from the chime of a doorbell to the warning shrill of a police siren to the song of a mockingbird. Indeed, they are the basis for our major form of communication: speech. They can also constitute a highly irritating distraction (noise). But sound waves become music, speech, or noise only when our ears perceive them. Physically, sound is simply waves that propagate in solids, liquids, and gases. Without a medium, there can be no sound; in a vacuum such as outer space, there is utter silence.

This distinction between the sensory and physical meanings of sound gives you a way to answer the old philosophical question, If a tree falls in the forest where there is no one to hear it, does it make a sound? The answer depends on how sound is defined—it is no if we are thinking in terms of sensory hearing, but yes if we are considering only the physical waves. Since sound waves are all around us most of the time, we are exposed to many interesting sound phenomena. You'll explore some of the most important of these in this chapter.

- FIGURE 14.1 **Vibrations make waves** (a) A vibrating tuning fork disturbs the air, producing alternating high-pressure regions (condensations) and low-pressure regions (rarefactions), which form sound waves. (b) After being picked up by a microphone, the pressure variations are converted to electrical signals. When these signals are displayed on an oscilloscope, the sinusoidal waveform is evident.

14.1 Sound Waves

Objectives: To (a) define sound and (b) explain the sound frequency spectrum.

For sound waves to exist, there must be a disturbance or vibrations in some medium. This disturbance may be the clapping of hands or the skidding of tires as a car comes to a sudden stop. Underwater you can hear the click of rocks against one another. If you put your ear to a thin wall, you can hear sounds from the other side through it. **Sound waves** in gases and liquids (fluids) are primarily longitudinal waves. However, sound disturbances moving through solids can have both longitudinal and transverse components. The intermolecular interactions in solids are much stronger than in fluids and allow transverse components to propagate.

The characteristics of sound waves can be visualized by considering those produced by a tuning fork (•Fig. 14.1). A tuning fork is essentially a metal bar bent into a U shape. The prongs, or tines, vibrate when struck. The fork vibrates at its fundamental frequency (with an antinode at the end of each tine), so a single tone is heard. (A *tone* is sound with a definite frequency.) The vibrations disturb the air, producing alternating high-pressure regions called *condensations* and low-pressure regions called *rarefactions*. As the fork vibrates, these disturbances propagate outward, and a series of them can be described by a sinusoidal wave.

As the disturbances traveling through the air reach the ear, the eardrum (a thin membrane) is set into vibration by the pressure variations. On the other side of the eardrum, tiny bones (the hammer, anvil, and stirrup) carry the vibrations to the inner ear, where they are picked up by the auditory nerve (see the Insight on p. 458).

Characteristics of the ear limit the perception of sound. Only sound waves with frequencies between about 20 Hz and 20 kHz (kilohertz) initiate nerve impulses that are interpreted by the human brain as sound. This frequency range is called the **audible region** of the **sound frequency spectrum** (•Fig 14.2).

Frequencies lower than 20 Hz are in the **infrasonic region**. Waves in this region, which we can't hear, are found in nature. Longitudinal waves generated by earthquakes have infrasonic frequencies, and we use these waves to study the Earth's interior (Chapter 13). Infrasonic waves, or *infrasound*, are also generated by wind and weather patterns. Elephants and cattle have hearing response in the infrasonic region and may even give early warnings of earthquakes and weather disturbances. Aircraft, automobiles, and other rapidly moving objects can also produce infrasound.

Above 20 kHz is the **ultrasonic region**. Ultrasonic waves can be generated by high-frequency vibrations in crystals. Ultrasonic waves, or *ultrasound*, cannot be detected by humans but can be by other animals. The audible region for dogs extends to about 45 kHz, so ultrasonic whistles can be used to call dogs without

- FIGURE 14.2 **Sound frequency spectrum** The audible region of sound for humans lies between about 20Hz and 20kHz. Below this is the infrasonic region, and above it is the ultrasonic region. The upper limit is about 1 GHz because of the elastic limitations of materials.

disturbing people. Cats and bats have even higher ranges, up to about 70 kHz and 100 kHz, respectively.

There are many other practical applications of ultrasound. Since ultrasound can travel for kilometers in water, it is used in sonar. Sonar is the ultrasound counterpart of radar, which uses radio waves for ranging and detection. Sound pulses generated by the sonar apparatus are reflected by underwater objects, and the resulting echoes are picked up by a detector. The time required for a sound pulse to make one round trip, together with the speed of sound in water, gives the distance of the reflecting object. In a similar manner, ultrasound is used in autofocus cameras. Distance measurement allows focus adjustments to be made. Sonar can be used not only for ranging but also to create images of the seafloor and objects resting on it, such as shipwrecks (•Fig. 14.3a).

Sonar appeared in the animal kingdom long before it was developed by human engineers. On their nocturnal hunting flights, bats use a kind of natural sonar to navigate in and out of their caves and to locate and catch flying insects. The bats emit pulses of ultrasound and track their prey by means of the reflected echoes (Fig. 14.3b). Certain species of cave-dwelling birds have evolved the same ability.

In medicine, ultrasound is used to examine internal tissues and organs that are nearly invisible to X-rays. Perhaps the best known medical application of ultrasound is its use to view a fetus without exposing it to the dangerous effects of X-rays. Ultrasonic generators (transducers) made of quartz crystals produce high-frequency waves that are used to scan a designated region of the body from several angles. Reflections from the scanned areas are monitored, and a computer constructs an image from the reflected signals. Images are recorded sev-

(a)

(b)

(c)

•FIGURE 14.3 Some uses of ultrasound (a) This computer-enhanced false-color image, constructed from sonar signals, shows the wreck of the American sailing ship *Hamilton*. The schooner, sunk in Lake Ontario during the War of 1812, lies under 91 m (300 ft) of water. (b) With the aid of their own natural sonar systems, bats hunt flying insects. They emit pulses of ultrasonic waves, which lie within their audible region, and use the echoes reflected from their prey to guide their attack. (Notice the size of the animal's ears.) (c) Ultrasound generated by crystal transducers, which convert electrical oscillations into mechanical vibrations, is transmitted through tissue and is reflected from internal structures. The reflected waves are detected by the transducers, and the signals are used to construct an image, or echogram, as shown here for a well-developed fetus.

eral times each second. The series of images provides a "moving picture" of an internal structure, such as the heart of a fetus. A still shot, or echogram, is shown in Fig. 14.3c.

In industrial and home applications, ultrasonic baths are used to clean metal machine parts, dentures, and jewelry. The high-frequency (short-wavelength) ultrasound vibrations loosen particles in otherwise inaccessible places.

Ultrasonic frequencies extend into the megahertz (MHz) range, but the sound frequency spectrum does not continue indefinitely. There is an upper limit of about 10^9 Hz, or 1 GHz (gigahertz), which is determined by the upper limit of the elasticity of the materials through which the sound propagates.

14.2 The Speed of Sound

Objectives: To (a) tell how the speed of sound differs in different media and
(b) describe the temperature dependence of the speed of sound in air.

In general, the speed at which a disturbance moves through a medium depends on the elasticity and density of the medium. For example, as you learned in Chapter 13, the wave speed in a stretched string is given by $v = \sqrt{F_T/\mu}$, where F_T is the tension in the string and μ is the linear mass density.

Similar expressions describe wave speeds in solids and liquids, for which the elasticity is expressed in terms of moduli (Chapter 9). In general, the speeds of sound in solids and liquids are given by $v = \sqrt{Y/\rho}$ and $v = \sqrt{B/\rho}$, respectively, where Y is Young's modulus, B is the bulk modulus, and ρ is the density. The speed of sound in a gas is inversely proportional to the square root of the molecular mass, but the equation is more complicated and will not be presented here.

Note: Review Section 9.1.

Solids are generally more elastic than liquids, which in turn are more elastic than gases. In a highly elastic material, the restoring forces between the atoms or molecules cause a disturbance to propagate faster. Thus, the speed of sound is generally about 2 to 4 times faster in solids than in liquids and about 10 to 15 times faster in solids than in gases such as air (Table 14.1).

Although not expressed explicitly in the preceding equations, the speed of sound also depends on the temperature of the medium. In dry air, for example, the speed of sound is 331 m/s (about 740 mi/h) at 0°C. As the temperature increases, so do the speeds of the gas molecules. As a result, the molecules collide with each other more frequently, and a disturbance is transmitted more quickly. Thus the speed of sound in air increases with increasing temperature. For *normal environmental temperatures*, the speed of sound in air increases by about 0.6 m/s for each degree Celsius above 0°C. Thus, a good approximation of the speed of sound in air for a particular (environmental) temperature is given by

$$v = (331 + 0.6 T_C) \text{ m/s} \quad \text{speed of sound in air} \quad (14.1)$$

where T_C is the air temperature in degrees Celsius.* Although not written explicitly, the units associated with the factor 0.6 are meters per second per Celsius degree [m/(s·C°)].

Let's take a comparative look at the speeds of sound in different media.

TABLE 14.1 The Speed of Sound in Various Media (typical values)

Medium	Speed (m/s)
Solids	
Aluminum	5100
Copper	3500
Iron	4500
Glass	5200
Polystyrene	1850
Zinc	3200
Liquids	
Alcohol, ethyl	1125
Mercury	1400
Water	1500
Gases	
Air (0°C)	331
Air (100°C)	387
Helium (0°C)	965
Hydrogen (0°C)	1284
Oxygen (0°C)	316

---

*A better approximation of these and higher temperatures is given by the expression

$$v = \left(331\sqrt{1 + \frac{T_C}{273}}\right) \text{ m/s}$$

In Table 14.1, see v for air at 100°C, which is outside the normal environmental temperature range.

Example 14.1 ■ Solid, Liquid, Gas: Speed of Sound in Different Media

From their material properties, find the values of the speeds of sound in (a) a solid copper rod, (b) liquid water, and (c) air at room temperature (20°C).

Thinking It Through. We know that the speeds of sound in solids and liquids depend on elastic moduli and densities. These values are available in Tables 9.1 and 9.2. The speed of sound in air is given by Eq. 14.1.

Solution.

Given: $Y_{Cu} = 11 \times 10^{10}$ N/m^2 *Find:* (a) v_{Cu} (speed in copper)
$B_{H_2O} = 2.2 \times 10^9$ N/m^2 (b) v_{H_2O} (speed in water)
$\rho_{Cu} = 8.9 \times 10^3$ kg/m^3 (c) v_{air} (speed in air)
$\rho_{H_2O} = 1.0 \times 10^3$ kg/m^3
(all values from Tables 9.1 and 9.2)
$T_C = 20°C$ (for air)

(a) To find the speed of sound in a copper rod, we use the expression $v = \sqrt{Y/\rho}$:

$$v_{Cu} = \sqrt{Y/\rho} = \sqrt{(11 \times 10^{10} \text{ N/m}^2)/(8.9 \times 10^3 \text{ kg/m}^3)} = 3.5 \times 10^3 \text{ m/s}$$

(b) For water, $v = \sqrt{B/\rho}$:

$$v_{H_2O} = \sqrt{B/\rho} = \sqrt{(2.2 \times 10^9 \text{ N/m}^2)/(1.0 \times 10^3 \text{ kg/m}^3)} = 1.5 \times 10^3 \text{ m/s}$$

(c) For air at 20°C, by Eq. 14.1, we have

$$v_{air} = (331 + 0.6T_C) \text{ m/s} = [331 + 0.6(20)] \text{ m/s} = 343 \text{ m/s}$$

Follow-up Exercise. In this Example, how many times faster is the speed of sound in copper (a) than in water and (b) than in air (at room temperature)?

A generally useful value for the speed of sound in air is $\frac{1}{3}$ km/s (or $\frac{1}{5}$ mi/s). Using this value, you can, for example, estimate how far away lightning has struck by counting the number of seconds between the time you observe the flash and the time you hear the associated thunder. Because the speed of light is so fast, you see the lightning flash almost instantaneously. The sound waves of the thunder travel relatively slowly, at about $\frac{1}{3}$ km/s. For example, if the interval between the two events is measured to be 6 s (often by counting "one thousand one, one thousand two, . . ."), the lightning stroke was about 2 km away ($\frac{1}{3}$ km/s × 6 s).

You may also have noticed the delay of sound relative to light at a baseball game. If you're sitting in the outfield stands, you see the batter hit the ball before you hear the crack of the bat.

Example 14.2 ■ Safe or Out? You Make the Call

On a cool October afternoon (air temperature = 15°C), from your seat in the centerfield stands 113 m from first base, you witness the play that will decide the World Series. You see the runner's foot touch the bag; half a second later, straining your ears, you hear the faint thud of the ball in the first baseman's glove. The umpire signals safe; half the fans boo loudly. As a student of physics, you make the call—did the ump blow it?

Thinking It Through. This Example has to do with the delay of hearing the sound, and hence the speed of sound. You can see the action almost instantaneously yet the sound takes an appreciable time to reach you, because the speed of light is much faster than that of sound. If the sound travel time is less than $\frac{1}{2}$ s, then the runner had already touched the base when the first baseman caught the ball, and the runner is safe. If that time is greater that $\frac{1}{2}$ s, he's out. (Why?)

Solution. Listing the data, we have

Given: $d = 113$ m *Find:* t (sound travel time)
$T = 15°C$
$t = \frac{1}{2}$ s

With a temperature of 15°C, the speed of sound is, from Eq. 14.1,

$$v = (331 + 0.6 T_C) \text{ m/s} = [331 + 0.6(15°C)] \text{ m/s} = 340 \text{ m/s}$$

For a constant speed, the general distance–time relation is $d = vt$, so the time for the sound to travel the distance to your seat is

$$t = \frac{d}{v} = \frac{113 \text{ m}}{340 \text{ m/s}} = 0.332 \text{ s}$$

or just under one-third of a second. The runner reached the base about $\frac{1}{2}$ s $-\frac{1}{3}$ s $= \frac{1}{6}$ s before the ball arrived. The ump was right!

To show why it is justified to say that the visual observation takes place almost instantaneously, let's see how long it takes for light to travel from first base to your seat. For practical purposes, the speed of light in air is the same as that in a vacuum, $c = 3.00 \times 10^8$ m/s, which is on the order of 10^6 (a million) times greater than the speed of sound. Then, the time for light to travel 113 m is

$$t = \frac{d}{c} = \frac{113 \text{ m}}{3.00 \times 10^8 \text{ m/s}} = 3.77 \times 10^{-7} \text{ s}$$

or 0.377 μs—not a very long time.

Follow-up Exercise. On a day when the air temperature is 20°C, a hiker shouts and hears the echo from the vertical stone cliff face 5.00 s later. (a) How far away is the cliff? (b) If the air temperature were 15°C, what would be the difference in the times of hearing the echoes?

Conceptual Example 14.3

Speed of Sound: Sound Traveling Far and Wide

Note that the speed of *dry* air for a given temperature is given to a good approximation by Eq. 14.1. However, the moisture content of the air (humidity) varies, and this variation affects the speed of sound. At the same temperature, would sound travel faster in (a) dry air or (b) moist air?

Reasoning and Answer. According to an old folklore saying, "Sound traveling far and wide, a stormy day will betide" (happen or predict). This saying implies that sound travels faster on a highly humid day when a storm or precipitation is likely. But is this saying valid?

Near the beginning of this section, we learned that the speed of sound in a gas is inversely proportional to the square root of the gas's molecular mass. So, at constant pressure, is moist air more or less dense than dry air?

In a volume of moist air, a large number of water (H_2O) molecules occupy the space normally occupied by either nitrogen (N_2) or oxygen (O_2) molecules, which make up 98% of the air. Water molecules are less massive than both nitrogen and oxygen molecules. [From Section 10.3, the molecular (formula) masses are H_2O, 18 g; N_2, 28 g; and O_2, 32 g.] Thus, the average molecular mass of moist air is less than that of dry air, and the speed of sound is greater in moist air.

We can look at this situation another way: Since water molecules are less massive, they have less inertia and respond to the sound wave faster than nitrogen or oxygen molecules do. The water molecules therefore propagate the disturbance faster.

Follow-up Exercise. Considering only molecular masses, would you expect the speed of sound to be greatest in nitrogen, oxygen, or helium (at the same temperature and pressure)? Explain.

Humidity was included here as an interesting consideration for the speed of sound in air. However, henceforth in computing the speed of sound in air at a certain temperature, we will consider only *dry* air (Eq. 14.1) unless otherwise stated.

Sound is one of our major means of communicating. For more on our speech and hearing, see the Insight on p. 458.

Insight

Speech and Hearing

Sound is one of our most important means of interpersonal communication, and our most important sound source is the human voice. Let's look at the anatomy and physics of the voice and of the ear, which serves as our receiver for sound.

The Human Voice

The energy for sounds associated with the human voice originates in the muscle action of the diaphragm, which forces air up from the lungs. To produce variations in sound, this steady stream of air must be periodically disturbed, or "modulated." The fundamental modulating organ is the larynx (the "voice box"), across which are stretched membrane-like folds called the vocal cords. The opening and closing of the vocal cords modulates the air stream to produce sounds (Fig. 1).

As illustrated in the figure, there are two sets of vocal cords: (1) *false vocal cords*, which do not produce sound but help close the larynx during swallowing, and (2) *true vocal cords*, which are elastic and are responsible for vocal sounds. When air is forced between the true vocal cords, they vibrate from side to side, producing sound. During normal breathing, these vocal cords remain relaxed and open. The opening between them is called the *glottis*.

Changing the tension of the vocal cords controls the pitch (perceived frequency) of the vocal sounds (as we shall see in Section 14.6). The loudness of the sound is related to the pressure of the air passing over the vocal cords. A strong blast of air results in greater amplitude of the vibrations and louder sound.

The sound waves from the larynx are further modulated in the numerous resonance cavities in the throat, mouth, and nose, where standing waves are set up. Some of the vocal cavities can be altered by means of controllable structures, such as the tongue and lips, so as to produce a wide variety

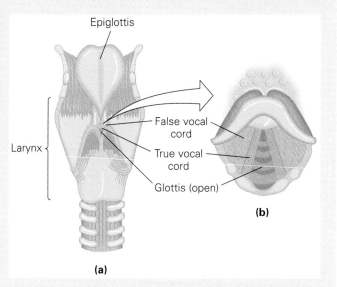

FIGURE 1 Anatomy of the "voice box" (a) A cross section of the larynx. (b) A down-the-throat view of the vocal cords.

of sounds and words. (Sound out the vowel letters *a, e, i, o,* and *u*, and notice the positions of your tongue and lips.)

The waveforms of voice sounds are quite specific for individuals and provide a "voice print" that can be used for identification, just as fingerprints are. The validity of voice prints for legal identification, however, is highly controversial.

Hearing

The anatomy of the human ear is illustrated in Fig. 2. Sound enters the outer ear and travels through the ear (or auditory) canal to the *eardrum* (the tympanum), which separates the

14.3 Sound Intensity and Sound Intensity Level

Objectives: To (a) define sound intensity and explain how it varies with distance from a point source and (b) calculate sound intensity levels on the decibel scale.

Wave motion involves the propagation of energy. The rate of the energy transfer is expressed in terms of **intensity**, which is the energy transported per unit time across a unit area. Since energy/time is power, intensity is power/area:

$$\text{intensity} = \frac{\text{energy/time}}{\text{area}} = \frac{\text{power}}{\text{area}}$$

The standard units of intensity (power/area) are watts per square meter (W/m^2).

Consider a point source that sends out spherical sound waves, as shown in •Fig. 14.4. If there are no losses, the sound intensity at a distance R from the source is

$$I = \frac{P}{A} = \frac{P}{4\pi R^2} \tag{14.2}$$

where P is the power of the source and $4\pi R^2$ is the area of a sphere of radius R, through which the sound energy passes perpendicularly.

Insight

FIGURE 2 Anatomy of the human ear The ear converts pressure in the air into electrical nerve impulses that are interpreted by the brain as sounds.

outer ear from the middle ear. The eardrum is a membrane that vibrates in response to the impinging sound waves. The vibrations are transmitted through the middle ear, which contains an intricate set of connected bones, commonly called the *hammer* (malleus), *anvil* (incus), and *stirrup* (stapes), because of their shapes.

The bones of the middle ear form a delicate set of levers with a force multiplication factor (mechanical advantage) of about 2. However, the amplification of the pressure in a wave is much greater because the area of the eardrum is about 20 times that of the *oval window*, the membrane-covered opening to the inner ear. (The stirrup is in contact with the membrane.) Thus, pressure is amplified by a factor of about 40.

The inner ear includes the semicircular canals, which are important in controlling balance, and the *cochlea*. It is in the cochlea that sound waves are translated into nerve impulses and that pitch or frequency discrimination is made. The cochlea consists of a series of liquid-filled tubes, coiled into a spiral shape that resembles a snail shell. Supported by a membrane, called the *basilar membrane*, within the tubes are thousands of special receptor cells called *hair cells*. The hair cells are specialized nerve receptors. When a region of the basilar membrane is set into vibration by sound of a particular frequency, hair cells in that region are stimulated and nerve impulses are sent to the brain, where they are interpreted as sound. The brain translates this positional information (impulses from particular fibers originating in specific regions of the basilar membrane) back into frequency information (the subjective physiological sensation of pitch).

The sensation of loudness is determined by the amplitude of vibration of the basilar membrane. The greater the amplitude, the more strongly the hair cells are stimulated and the more impulses they send to the brain in a given period of time. The brain translates this information into degrees of loudness.

Incidentally, the middle ear is connected to the throat by the Eustachian tube, the end of which is normally closed. It opens during swallowing and yawning to permit air to enter and leave, so internal and external pressures are equalized. You have probably experienced a "stopping up" of your ears with a sudden change in atmospheric pressure (for example, during rapid ascents or descents in elevators or airplanes). Swallowing opens the Eustachian tubes and relieves the excess pressure difference on the middle ear.

The intensity for a point source is therefore *inversely proportional to the square of the distance from the source* (an inverse-square relationship). Two intensities at different distances from a source of constant power can be compared as a ratio:

$$\frac{I_2}{I_1} = \frac{P/4\pi R_2^2}{P/4\pi R_1^2} = \frac{R_1^2}{R_2^2}$$

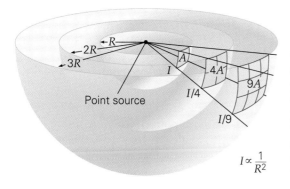

•**FIGURE 14.4 Intensity of a point source** The energy emitted from a point source spreads out equally in all directions. Since intensity is power/area, $I = P/A = P/4\pi R^2$, where the area is that of a spherical surface. The intensity then decreases with the distance from the source as $1/R^2$.

or

$$\frac{I_2}{I_1} = \left(\frac{R_1}{R_2}\right)^2 \qquad (14.3)$$

Suppose that the distance from a point source is doubled; that is, $R_2 = 2R_1$, or $R_1/R_2 = \frac{1}{2}$. Then

$$\frac{I_2}{I_1} = \left(\frac{R_1}{R_2}\right)^2 = \left(\frac{1}{2}\right)^2 = \frac{1}{4}$$

and

$$I_2 = \frac{I_1}{4}$$

Since the intensity decreases by a factor of $1/R^2$, doubling the distance decreases the intensity to a quarter of its original value.

A good way to understand this inverse-square relationship intuitively is to look at the geometry of the situation. As Fig. 14.4 shows, the greater the distance from the source, the larger the area over which a given amount of sound energy is spread, and thus the lower its intensity. (Imagine having to paint two walls of different areas. If you had the same amount of paint to use on each, you'd have to spread it more thinly over the larger wall.) Since this area increases as the square of the radius R, the intensity decreases accordingly—that is, as $1/R^2$.

Sound intensity is perceived by the ear as **loudness**. On the average, the human ear can detect sound waves (at 1 kHz) with an intensity as low as 10^{-12} W/m². This intensity (I_o) is referred to as the *threshold of hearing*. Thus, for us to hear a sound, it must not only have a frequency in the audible range, but also be of sufficient intensity. As the intensity is increased, the perceived sound becomes louder. At an intensity of 1.0 W/m², the sound is uncomfortably loud and may be painful to the ear. This intensity (I_p) is called the *threshold of pain*.

Threshold of hearing:
$I_o = 10^{-12}$ W/m²

Threshold of pain:
$I_p = 1.0$ W/m²

Note that the thresholds of pain and hearing differ by a factor of 10^{12}:

$$\frac{I_p}{I_o} = \frac{1.0 \text{ W/m}^2}{10^{-12} \text{ W/m}^2} = 10^{12}$$

That is, the intensity at the threshold of pain is a *trillion* times greater than that at the threshold of hearing. Within this enormous range, the perceived loudness is not directly proportional to the intensity. That is, if the intensity is doubled, the perceived loudness does not double. In fact, doubling of perceived loudness corresponds approximately to an increase in intensity by a factor of 10. For example, a sound with an intensity of 10^{-5} W/m² would be perceived to be twice as loud as one with an intensity of 10^{-6} W/m² (the smaller the negative exponent, the larger the number).

Sound Intensity Level—The Bel and the Decibel

It is convenient to compress the large range of sound intensities by using a logarithmic scale (base-10) to express intensity levels. The intensity level of a sound must be referenced to a standard intensity, which is taken to be that of the threshold of hearing, $I_o = 10^{-12}$ W/m². Then, for any intensity I, the intensity level is the log of the ratio of I to I_o: $\log I/I_o$. For example, if a sound has an intensity of $I = 10^{-6}$ W/m²,

$$\log \frac{I}{I_o} = \log \frac{10^{-6} \text{ W/m}^2}{10^{-12} \text{ W/m}^2} = \log 10^6 = 6 \text{ B}$$

(Recall that $\log_{10} 10^x = x$.) The exponent of the power of ten in the final log term is taken to have a unit called the **bel** (B). Thus, a sound with an intensity of 10^{-6} W/m² has an intensity level of 6 B on this scale. In this way, the intensity range from 10^{-12} W/m² to 1.0 W/m² is compressed into a scale of intensity levels ranging from 0 B to 12 B.

Note: The bel was named in honor of Alexander Graham Bell, the inventor of the telephone.

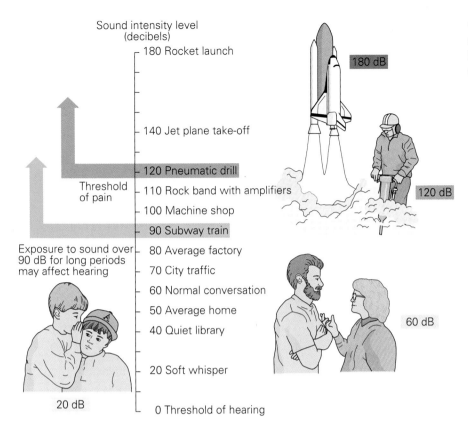

•FIGURE 14.5 **Sound intensity levels and the decibel scale** The intensity levels of some common sounds on the decibel (dB) scale.

A finer intensity scale is obtained by using a smaller unit, the **decibel** (dB), which is a tenth of a bel. The 0 to 12 B range corresponds to 0 to 120 dB. In this case, the equation for the relative **sound intensity level**, or **decibel level** (β), is

$$\beta = 10 \log \frac{I}{I_o} \quad \text{where } I_o = 10^{-12} \text{ W/m}^2 \quad (14.4)$$

Sound intensity level in decibels

Note that the sound intensity level (in decibels) is *not* the same as the sound intensity (in watts per square meter).

The decibel intensity scale and familiar sounds at some intensity levels are shown in •Fig. 14.5. Sound intensities can have detrimental effects on hearing, and because of this the U.S. government has set occupational noise-exposure limits.

Example 14.4 ■ Sound Intensity Levels: Using Logarithms

What are the intensity levels of sounds with intensities of (a) 10^{-12} W/m² and (b) 5.0×10^{-6} W/m²?

Thinking It Through. We can find the sound intensity levels directly by using Eq. 14.4.

Solution.

Given: (a) $I = 10^{-12}$ W/m²   Find: (a) β (sound intensity level)
(b) $I = 5.0 \times 10^{-6}$ W/m²   (b) β

(a) Using Eq. 14.4 we have,

$$\beta = 10 \log \frac{I}{I_o} = 10 \log \left(\frac{10^{-12} \text{ W/m}^2}{10^{-12} \text{ W/m}^2} \right) = 10 \log 1 = 0 \text{ dB}$$

The intensity is the same as that at the threshold of hearing. (Recall that $\log 1 = 0$, since $1 = 10^0$ and $\log 10^0 = 0$.)

(b) $\beta = 10 \log \dfrac{I}{I_o} = 10 \log \left(\dfrac{5.0 \times 10^{-6} \text{ W/m}^2}{10^{-12} \text{ W/m}^2} \right)$

$= 10 \log(5.0 \times 10^6) = 10(\log 5.0 + \log 10^6) = 10(0.70 + 6.0) = 67 \text{ dB}$

Follow-up Exercise. Note in this Example that the intensity of 5.0×10^{-6} W/m² is halfway between 10^{-6} and 10^{-5} (or 60 and 70 dB), yet this intensity does not correspond to a midway value of 65 dB. (a) Why? (b) What intensity *does* correspond to 65 dB? (Compute to 3 significant figures.)

Example 14.5 ■ Intensity Level Factors: Using Ratios

(a) What is the difference in the intensity levels if the intensity of a sound is doubled? (b) By what factors does the intensity increase for intensity level *differences* of 10 dB and 20 dB?

Thinking It Through. (a) If the intensity is doubled, we know the increase ratio is 2. That is, $I_2 = 2I_1$, and $I_2/I_1 = 2$. We can then use Eq. 14.4 to find the intensity difference. Recall that $\log a - \log b = \log a/b$. (b) Here it is important to note that these values are intensity level *differences*, $\Delta \beta = \beta_2 - \beta_1$, *not* intensity *levels*. The equation developed in (a) will work. (Why?)

Solution. Listing the data, we have

Given: (a) $I_2 = 2I_1$ **Find:** (a) $\Delta \beta$ (intensity level difference)
(b) $\Delta \beta = 10$ dB (b) I_2/I_1 (factors of increase)
$\Delta \beta = 20$ dB

(a) With $\log a - \log b = \log a/b$, using Eq. 14.4 and this relationship, we have for the difference $\Delta \beta = \beta_2 - \beta_1 = 10[\log(I_2/I_o) - \log(I_1/I_o)] = 10 \log (I_2/I_o)/(I_1/I_o) = 10 \log I_2/I_1$. Then,

$$\Delta \beta = 10 \log \dfrac{I_2}{I_1} = 10 \log 2 = 3 \text{ dB}$$

Thus, doubling the intensity increases the intensity level by 3 dB (for example, an increase from 55 dB to 58 dB).

(b) For a 10-decibel difference,

$$\Delta \beta = 10 \text{ dB} = 10 \log \dfrac{I_2}{I_1} \quad \text{and} \quad \log \dfrac{I_2}{I_1} = 1.0$$

Since $\log 10^1 = 1$,

$$\dfrac{I_2}{I_1} = 10^1 \quad \text{and} \quad I_2 = 10 I_1$$

Similarly, for a 20-decibel difference,

$$\Delta \beta = 20 \text{ dB} = 10 \log \dfrac{I_2}{I_1} \quad \text{and} \quad \log \dfrac{I_2}{I_1} = 2.0$$

Since $\log 10^2 = 2$,

$$\dfrac{I_2}{I_1} = 10^2 \quad \text{and} \quad I_2 = 100 I_1$$

Thus, an intensity level difference of 10 dB corresponds to increasing (or decreasing) the intensity by a factor of 10. An intensity level difference of 20 dB corresponds to increasing (or decreasing) the intensity by a factor of 100.

You should be able to guess the factor that corresponds to an intensity level difference of 30 dB. In general, the factor of the intensity change is $10^{\Delta B}$, where ΔB is the level difference in bels. Since 30 dB = 3 B and $10^3 = 1000$, the intensity changes by a factor of 1000 for an intensity level difference of 30 dB.

Follow-up Exercise. A $\Delta \beta$ of 20 dB and a $\Delta \beta$ of 30 dB correspond to factors of 100 and 1000, respectively, in intensity changes. Does a $\Delta \beta$ of 25 dB correspond to an intensity change factor of 500? Explain.

Example 14.6 ■ Combined Sound Levels: Adding Intensities

Sitting at a sidewalk restaurant table, a friend talks to you in normal conversation (60 dB). At the same time, the intensity level of the street traffic reaching you is also 60 dB. What is the total intensity level of the combined sounds?

Thinking It Through. It is tempting simply to add the two sound intensity levels together and say the total is 120 dB. But intensity levels in decibels are logarithmic, so you can't add them in the normal way. However, intensities (I) can be added arithmetically, since energy and power are scalar quantities. Then the combined intensity level can be found from the sum of the intensities.

Solution. We have

Given: $\beta_1 = 60$ dB *Find:* Total β
$\beta_2 = 60$ dB

Let's find the intensities associated with the intensity levels:

$$\beta_1 = 60 \text{ dB} = 10 \log \frac{I_1}{I_o} = 10 \log \left(\frac{I_1}{10^{-12} \text{ W/m}^2} \right)$$

By inspection, we have

$$I_1 = 10^{-6} \text{ W/m}^2$$

Similarly, $I_2 = 10^{-6}$ W/m², since both intensity levels are 60 dB. So, the total intensity is

$$I_{\text{total}} = I_1 + I_2 = 1.0 \times 10^{-6} \text{ W/m}^2 + 1.0 \times 10^{-6} \text{ W/m}^2 = 2.0 \times 10^{-6} \text{ W/m}^2$$

Then, converting back to intensity level, we get

$$\beta = 10 \log \frac{I_{\text{total}}}{I_o} = 10 \log \left(\frac{2.0 \times 10^{-6} \text{ W/m}^2}{10^{-12} \text{ W/m}^2} \right) = 10 \log (2.0 \times 10^6)$$
$$= 10(\log 2.0 + \log 10^6) = 10(0.30 + 6.0) = 63 \text{ dB}$$

This value is a long way from 120 dB! Notice that the combined intensities doubled the value and the intensity level increased by 3 dB, in agreement with our finding in part (a) of Example 14.5.

Follow-up Exercise. In this Example, suppose the added noise gave a total that *tripled* the value of the conversation intensity. What would be the total combined intensity level in this case?

14.4 Sound Phenomena

Objectives: To (a) explain sound reflection, refraction, and diffraction and (b) distinguish between constructive and destructive interference.

Reflection, Refraction, and Diffraction

A sound wave can be reflected; an echo is a familiar example of sound *reflection*. Sound *refraction* is less common than reflection, but you may have experienced this effect on a calm summer evening. Then it is sometimes possible to hear distant voices or other sounds that ordinarily would not be audible. This effect is due to the refraction, or bending, of the sound waves as they pass from one region into a region where the air density is different. The effect is similar to what would happen if the sound passed into another medium.

The required conditions for sound refraction are a layer of cooler air near the ground or water and a layer of warmer air above it. These conditions occur frequently over bodies of water because of cooling after sunset (•Fig. 14.6). As a result, the waves are refracted in an arc that may allow a distant person to receive an increased intensity of sound.

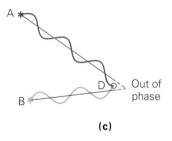

•**FIGURE 14.7 Interference**
(a) Sound waves from two point sources spread out and interfere. (b) At points where the waves arrive in phase (zero phase difference), such as point C, constructive interference occurs. (c) At points where the waves arrive completely out of phase (phase difference of 180°), such as point D, destructive interference occurs. The phase difference at a particular point depends on the path lengths the waves travel to reach that point.

•**FIGURE 14.6 Sound refraction** Sound travels slower in the cool air near the water surface than in the upper, warmer air. As a result, the waves are refracted, or bent. This bending increases the intensity of the sound at a distance where it otherwise might not be heard.

Another bending phenomenon is *diffraction,* a bending of sound around corners or around an object. We usually think of waves as traveling in straight lines. However, you can hear someone who is standing around a corner but whom you cannot see. This bending is different from that of refraction, in which no obstacle causes the bending.

Reflection, refraction, and diffraction are described in a general sense here for sound. These phenomena are important considerations for light waves and will be discussed more fully in Chapters 22 and 24.

Interference

Like waves of any kind, sound waves *interfere* when they meet. Suppose that two loudspeakers separated by some distance emit sound waves in phase at the same frequency. If we consider the speakers to be point sources, then the waves will spread out spherically and interfere (•Fig. 14.7a). The lines from a particular speaker represent wave crests (or condensations), and the troughs (or rarefactions) lie in the intervening white areas.

In particular regions of space, there will be constructive or destructive interferences. For example, if the waves meet in a region where they are exactly in phase (where two crests or two troughs coincide), there will be total **constructive interference** (Fig. 14.7b). Notice that the waves have the same motion at point C in the figure. Conversely, if the waves meet such that the crest of one coincides with the trough of the other (at point D), the two waves will cancel each other out (Fig. 14.7c). The result will be total **destructive interference**.

It is convenient to describe the path lengths traveled by the waves in terms of wavelength (λ) to determine whether they arrive in phase. In such analysis we work with points of zero displacement rather than crests or troughs. Consider the waves arriving at point C in Fig. 14.7b. The path lengths in this case are AC = 4λ and BC = 3λ. The **phase difference** ($\Delta\theta$) is related to the **path-length difference** (ΔL) by the simple relationship

$$\Delta\theta = \frac{2\pi}{\lambda}(\Delta L) \qquad \text{phase difference and path-length difference} \qquad (14.5)$$

Since 2π rad is equivalent, in angular terms, to a full wave cycle or wavelength, multiplying the path-length difference by $2\pi/\lambda$ gives the phase difference in radians. For the example illustrated in Fig. 14.7b, we have

$$\Delta\theta = \frac{2\pi}{\lambda}(AC - BC) = \frac{2\pi}{\lambda}(4\lambda - 3\lambda) = 2\pi \text{ rad}$$

When $\Delta\theta = 2\pi$ rad, the waves are shifted by one wavelength. This is the same as $\Delta\theta = 0°$, so the waves are actually in phase. Thus, the waves interfere constructively in the point C region, increasing the intensity, or loudness, of the sound detected there.

From Eq. 14.5, we see that the sound waves are in phase at any point where the path-length difference is zero or an integral multiple of the wavelength. That is,

$$\Delta L = n\lambda \quad (n = 0, 1, 2, 3, \ldots) \quad \begin{array}{c}\textit{condition for}\\ \textit{constructive interference}\end{array} \quad (14.6)$$

A similar analysis for the situation in •Fig. 14.7c, where $AD = 2\frac{3}{4}\lambda$ and $BD = 2\frac{1}{4}\lambda$, gives

$$\Delta\theta = \frac{2\pi}{\lambda}(2\tfrac{3}{4}\lambda - 2\tfrac{1}{4}\lambda) = \pi \text{ rad}$$

or $\Delta\theta = 180°$. At point D, the waves are completely out of phase, and destructive interference occurs in this region.

Sound waves will be out of phase at any point where the path-length difference is an odd number of half-wavelengths ($\lambda/2$), or

$$\Delta L = m\left(\frac{\lambda}{2}\right) \quad (m = 1, 3, 5, \ldots) \quad \begin{array}{c}\textit{condition for}\\ \textit{destructive interference}\end{array} \quad (14.7)$$

At these points, a softer or less intense sound will be heard or detected. If the amplitudes of the waves are exactly equal, the destructive interference is total and no sound is heard!

An application of destructive interference for sound waves is illustrated in •Fig. 14.8. Engine noise in commercial jet aircraft cabins can be distracting. In a new technique, microphones throughout the cabin pick up the various sounds; a computer analyzes them and generates sound waves that are 180° out of phase with the undesirable noise. Destructive interference then reduces the cabin noise.

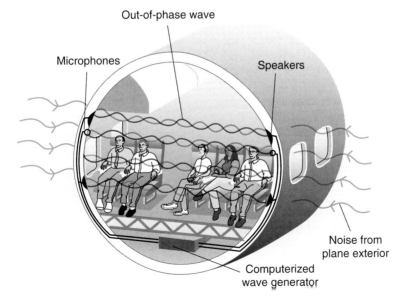

•FIGURE 14.8 **Destructive interference** Destructive interference is used to reduce the cabin noise in commercial jet aircraft. Microphones route noise to a computer, which then drives speakers to produce sound waves 180° out of phase with the noise. The resulting destructive interference makes the cabin quieter.

Example 14.7 ■ Pump up the Volume: Sound Interference

At an open-air concert on a hot day (air temperature of 25°C), you sit 7.00 m and 9.10 m, respectively, from a pair of speakers, one at each side of the stage. A musician, warming up, plays a single 494-hertz tone. What do you hear? (Consider the speakers to be point sources.)

Thinking It Through. The sound waves from the speakers interfere. Is it constructive or destructive interference or something in between? This depends on the path-length difference, which we can compute from the given distances.

Solution.

Given: $d_1 = 7.00$ m and $d_2 = 9.10$ m *Find:* ΔL (path-length difference
$f = 494$ Hz in wavelength units)
$T = 25°C$

The path-length difference (2.10 m) between the waves arriving at your location must be expressed in terms of the wavelength of the sound. To do this, we first need to know the wavelength. Given the frequency, we can find the wavelength from the relationship $\lambda = v/f$, provided we know the value of the speed of sound v at the given temperature. We calculate v from Eq. 14.1:

$$v = 331 + 0.6T_C = 331 + 0.6(25) = 346 \text{ m/s}$$

The wavelength of the sound waves is then

$$\lambda = \frac{v}{f} = \frac{346 \text{ m/s}}{494 \text{ Hz}} = 0.700 \text{ m}$$

Thus, the distances in terms of wavelength are

$$d_1 = (7.00 \text{ m})\left(\frac{\lambda}{0.700 \text{ m}}\right) = 10.0\lambda \quad \text{and} \quad d_2 = (9.10 \text{ m})\left(\frac{\lambda}{0.700 \text{ m}}\right) = 13.0\lambda$$

The path-length difference in terms of wavelengths is

$$\Delta L = d_2 - d_1 = 13.0\lambda - 10.0\lambda = 3.0\lambda$$

This is an integral number of wavelengths ($n = 3$), so constructive interference occurs. The sounds of the two speakers reinforce each other, and you hear an intense tone at 494 Hz.

Follow-up Exercise. Suppose that in this Example the tone traveled to a person sitting 7.00 m and 8.75 m, respectively, from the two speakers. What would be the situation in this case?

Another interesting interference effect occurs when two tones of nearly the same frequency ($f_1 \approx f_2$) are sounded simultaneously. The ear senses pulsations in loudness known as **beats**. The human ear can detect as many as 7 beats per second before they sound "smooth" or unpulsating.

Suppose that two sinusoidal waves with the same amplitude but slightly different frequencies interfere (•Fig. 14.9a). Figure 14.9b represents the resulting sound wave. The amplitude of the combined wave varies sinusoidally, as shown by the black curves (known as *envelopes*) that outline the wave.

What does this mean in terms of listener perception? A listener will hear a pulsating sound (beats) as determined by the envelopes. The maximum amplitude is 2A (at the point where the maxima of the two original waves interfere constructively). Detailed mathematics shows that a listener will hear these beats at a frequency called the **beat frequency** (f_b), given by

$$f_b = |f_1 - f_2| \tag{14.8}$$

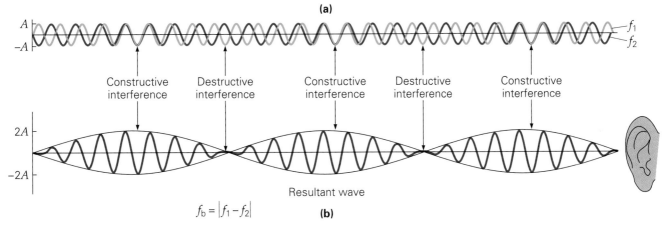

•**FIGURE 14.9 Beats** Two traveling waves of equal amplitude and slightly different frequencies interfere and give rise to pulsating tones called beats. The beat frequency is given by $f_b = |f_1 - f_2|$.

The absolute value is taken since the frequency f_b cannot be negative, even if $f_2 > f_1$. A negative beat frequency would be meaningless.

Beats can be produced when tuning forks of nearly the same frequency are vibrating at the same time. For example, using forks with frequencies of 516 Hz and 513 Hz, the beat frequency is $f_b = 516$ Hz $- 513$ Hz $= 3$ Hz, and three beats are heard each second. Musicians tune two stringed instruments to the same note by adjusting the tensions in the strings until the beats disappear ($f_1 = f_2$).

14.5 The Doppler Effect

Objectives: To (a) describe and explain the Doppler effect and (b) give some examples of its occurrences and applications.

If you stand along a highway and a car or truck approaches you with its horn blowing, the **pitch** (the perceived frequency) of the sound is higher as the vehicle approaches and lower as it recedes. You can also hear variations in the frequency of the motor noise when you watch a race car going around a track. A variation in the perceived sound frequency due to the motion of the sound source is an example of the **Doppler effect**.

As •Fig. 14.10 shows, the sound waves emitted by a moving source tend to bunch up in front of the source and spread out in back. The Doppler shift in frequency can be found by assuming that the air is at rest in a reference frame such as that depicted in •Fig. 14.11. The speed of sound in air is v, and the speed of the moving source is v_s. The frequency of the sound produced by the source is f_s. In one period, $T = 1/f_s$, a wave crest moves a distance $d = vT = \lambda$. (The sound wave would travel this distance in still air in any case, whether or not the source is moving.) But, in one period, the source travels a distance $d_s = v_s T$ before emitting another wave crest. The distance between the successive wave crests is thus shortened to a wavelength λ':

$$\lambda' = d - d_s = vT - v_s T = (v - v_s)T = \frac{v - v_s}{f_s}$$

The frequency heard by the observer (f_o) is related to the shortened wavelength by $f_o = v/\lambda'$, and substituting v/λ' gives

$$f_o = \frac{v}{\lambda'} = \left(\frac{v}{v - v_s}\right) f_s$$

The Austrian physicist Christian Doppler (1803–1853) first described what we now call the Doppler effect.

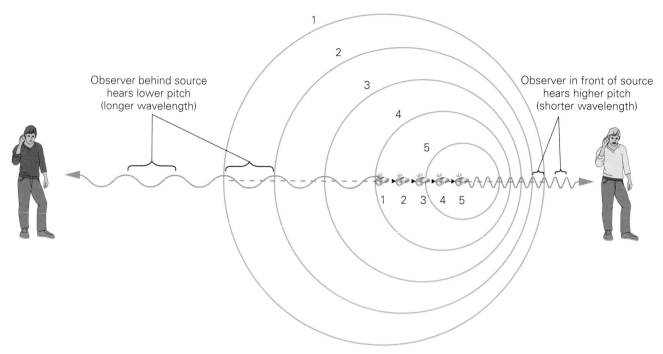

•**FIGURE 14.10 The Doppler effect** The sound waves bunch up in front of a moving source—the whistle—giving a higher frequency there. They trail out behind the source, giving a lower frequency there. (Not drawn to scale. Why?)

or

$$f_o = \left(\frac{1}{1 - \frac{v_s}{v}}\right) f_s \qquad \begin{array}{l}\text{source moving}\\\text{toward a stationary observer}\end{array} \qquad (14.9)$$

where v_s = speed of source
and v = speed of sound

Since $1 - (v_s/v)$ is less than 1, f_o is greater than f_s in this situation. For example, suppose that the speed of the source is a tenth of the speed of sound: $v_s = v/10$, or $v_s/v = \frac{1}{10}$. Then, by Eq. 14.9, $f_o = \frac{10}{9} f_s$.

Similarly, when the source is moving away from the observer ($\lambda' = d + d_s$), the observed frequency is given by

$$f_o = \left(\frac{v}{v + v_s}\right) f_s = \left(\frac{1}{1 + \frac{v_s}{v}}\right) f_s \qquad \begin{array}{l}\text{source moving}\\\text{away from a stationary observer}\end{array} \qquad (14.10)$$

Here f_o is less than f_s. (Why?)

•**FIGURE 14.11 The Doppler effect and wavelength** Sound from a moving car's horn travels a distance d in a time T. During this time, the car (the source) travels a distance d_s before putting out a second pulse, thereby shortening the observed wavelength of the sound in the approaching direction.

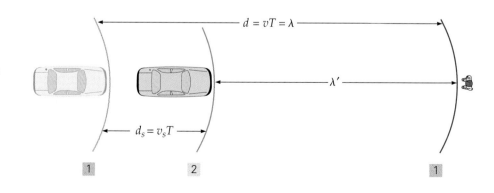

Combining Eqs. 14.9 and 14.10 yields a general equation for the observed frequency with a moving source and a stationary observer:

$$f_o = \left(\frac{v}{v \pm v_s}\right)f_s = \left(\frac{1}{1 \pm \frac{v_s}{v}}\right)f_s \quad \begin{cases} - \text{ for source moving} \\ \text{ toward stationary observer} \\ + \text{ for source moving} \\ \text{ away from stationary observer} \end{cases} \quad (14.11)$$

As you might expect, the Doppler effect also occurs with a moving observer and a stationary source. This situation is a bit different. As the observer moves toward the source, the distance between successive wave crests is the normal wavelength (or $\lambda = v/f_s$), but the measured wave speed is different. Relative to the approaching observer, the sound from the stationary source has a wave speed of $v' = v + v_o$, where v_o is the speed of the observer and v is the speed of sound in still air. (The observer moving toward the source is moving in a direction opposite that of the propagating waves and thus meets more wave crests in a given time.)

The observed frequency (f_o) is then (with $\lambda = v/f_s$):

$$f_o = \frac{v'}{\lambda} = \left(\frac{v + v_o}{v}\right)f_s$$

or

$$f_o = \left(1 + \frac{v_o}{v}\right)f_s \quad \begin{array}{l} \text{observer moving} \\ \text{toward a stationary source} \end{array} \quad (14.12)$$

where v_o = speed of observer
and v = speed of sound

Similarly, for an observer moving away from a stationary source, the perceived wave speed is $v' = v - v_o$, and

$$f_o = \frac{v'}{\lambda} = \left(\frac{v - v_o}{v}\right)f_s$$

or

$$f_o = \left(1 - \frac{v_o}{v}\right)f_s \quad \begin{array}{l} \text{observer moving} \\ \text{away from a stationary source} \end{array} \quad (14.13)$$

Equations 14.12 and 14.13 can be combined into a general equation for a moving observer and a stationary source:

$$f_o = \left(\frac{v \pm v_o}{v}\right)f_s = \left(1 \pm \frac{v_o}{v}\right)f_s \quad \begin{cases} + \text{ for observer moving} \\ \text{ toward stationary source} \\ - \text{ for observer moving} \\ \text{ away from stationary source} \end{cases} \quad (14.14)$$

There are also the cases when both the source and the observer are moving, either toward one another or away from one another. We will not consider them mathematically (but will do so conceptually in Example 14.9).

Example 14.8 ■ On the Road Again: The Doppler Effect

As a truck traveling at 96 km/h approaches and passes a person standing along the highway, the driver sounds the horn. If the horn has a frequency of 400 Hz, what are the frequencies of the sound waves heard by the person (a) as the truck approaches and (b) after it has passed? (Assume that the speed of sound is 346 m/s.)

Thinking It Through. This is an application of the Doppler effect, Eq. 14.11, with a moving source and a stationary observer. In such problems it is important to identify the data correctly.

Solution.

Given: $v_s = 96$ km/h $= 27$ m/s
$\qquad\quad f_s = 400$ Hz
$\qquad\quad v = 346$ m/s

Find: (a) f_o (observed frequency approaching)
$\qquad\quad$ (b) f_o (observed frequency moving away)

(a) From Eq. 14.11 with a minus sign (source approaching stationary observer),

$$f_o = \left(\frac{v}{v - v_s}\right) f_s = \left(\frac{346 \text{ m/s}}{346 \text{ m/s} - 27 \text{ m/s}}\right)(400 \text{ Hz}) = 434 \text{ Hz}$$

(b) A plus sign is used in Eq. 14.11 when the source is moving away:

$$f_o = \left(\frac{v}{v + v_s}\right) f_s = \left(\frac{346 \text{ m/s}}{346 \text{ m/s} + 27 \text{ m/s}}\right)(400 \text{ Hz}) = 371 \text{ Hz}$$

Follow-up Exercise. Suppose that the observer in this Example was initially moving toward and then past a stationary 400-hertz source at a speed of 96 km/h. What would be the observed frequencies? (Will they differ from those for the moving source?)

Conceptual Example 14.9

It's All Relative: Moving Source and Moving Observer

Suppose a sound source and an observer are moving away from one another in opposite directions, each at half the speed of sound in air. Then, the observer would (a) receive sound with a frequency higher than the source frequency, (b) receive sound with a frequency lower than the source frequency, (c) receive sound with the same frequency as the source frequency, (d) receive no sound from the source.

Reasoning and Answer. As we know, when a source moves away from a stationary observer, the observed frequency is lower (Eq. 14.10). Similarly, when an observer moves away from a stationary source, the observed frequency is also lower (Eq. 14.13). With both source and observer moving away from each other in opposite directions, the combined effect would make the observed frequency even less, so the answer is not (a) or (c).

It would appear that (b) is the correct answer, but we must logically eliminate (d) for completeness. Remember that the speed of sound relative to the air is constant. Therefore, (d) would be correct *only if the observer is moving faster than the speed of sound* relative to the air. Since the observer is moving at only half the speed of sound, (b) is the correct answer.

Think about it this way. Regardless of how fast the source is moving, the sound from the source is moving at the speed of sound through the air toward the observer. The observer is moving at only half the speed of sound through the air, so the sound from the source can easily reach and pass the observer.

Follow-up Exercise. What would be the situation if both the source and the observer were traveling in the same direction with the same subsonic speed? (Subsonic, as opposed to supersonic, refers to a speed that is less than the speed of sound in air.)

Problem-Solving Hint

You may find it difficult to remember whether a plus or minus sign is used in the general equations for the Doppler effect. Let your experience help you. For the commonly experienced case of being a stationary observer, the sound frequency increases when the source approaches, so the denominator in Eq. 14.11 must be smaller than the numerator. For this case, you use the minus sign. When the source is receding, the frequency is lower. The denominator in Eq. 14.11 must then be larger than the numerator, and you use the plus sign for this case. Similar reasoning will help you choose a plus or minus sign for the numerator in Eq. 14.14.

The Doppler effect also occurs for light waves, although the formulas describing it are different from those given above. When a distant light source such as a star moves away from us, the frequency of the light we receive from it is lowered. That is, the light is shifted toward the red (long-wavelength) end of the spectrum, an effect known as a *Doppler red shift*. Similarly, the frequency of light from an object approaching us is increased—the light is shifted toward the blue (short-wavelength) end of the spectrum, producing a *Doppler blue shift*. The magnitude of the frequency shift is related to the speed of the source.

The Doppler shift of light from astronomical objects is very useful to astronomers. The rotation of a planet, star, or other body can be established by looking at the Doppler shifts of light from opposite sides of the object—because of the rotation, one side is receding (and hence red-shifted), and the other approaching (blue-shifted). Similarly, the Doppler shifts of light from stars in different regions of our galaxy, the Milky Way, indicates that the galaxy is rotating.

You have been subjected to a practical application of the Doppler effect if you have ever been caught speeding in your car by police radar, which uses reflected radio waves. (Radar stands for *ra*dio *d*etecting *a*nd *r*anging and is similar to underwater sonar, which uses ultrasound.) If the radio waves are reflected from a parked car, the reflected waves return to the source with the same frequency. But for a car that is moving toward a patrol car, the reflected waves have a higher frequency, or are Doppler-shifted. Actually, there is a double Doppler shift. The moving car acts like a moving observer in receiving the wave (first Doppler shift), and in reflecting it the car acts like a moving source emitting a wave (second Doppler shift). The magnitudes of the shifts depend on the speed of the car. A computer quickly calculates this speed and displays it for the police officer.

For another important application of the Doppler effect in weather and airplane safety, see the Insight on p. 472.

Sonic Booms

Consider a jet plane that can travel at supersonic speeds. As the speed of a moving source of sound approaches the speed of sound, the waves ahead of the source come close together (•Fig. 14.12). When a plane is traveling at the speed of sound, the waves can't outrun it, and they pile up in front. At supersonic speeds, the waves overlap. This overlapping of a large number of waves produces many points of constructive interference, forming a large pressure ridge, or *shock wave*. This is sometimes called a *bow wave* because it is analogous to the wave produced by the bow of a boat moving through water at a speed greater than the speed of the water waves (Fig. 14.12c).

From aircraft traveling at supersonic speed, the shock wave trails out to the sides and downward. When this pressure ridge passes over an observer on the ground, the large concentration of energy produces what is known as a **sonic boom.** There is really a double boom because shock waves are formed at both ends of the aircraft. Under certain conditions, the shock waves can break windows and cause other damage. (Sonic booms are no longer heard as frequently as in the past. Pilots are now instructed to fly supersonically only at high altitudes and away from populated areas.)

On a smaller scale, you have probably heard a "mini" sonic boom. The "crack" of a whip is actually a sonic boom created by the transonic speed of the whip's tip—that is, the speed of the tip changes from subsonic to supersonic (and back).

A common misconception is that a sonic boom occurs only when a plane breaks the sound barrier. As an aircraft approaches the speed of sound, the pressure ridge in front of it is essentially a barrier that must be overcome with extra power. However, once supersonic speed is reached, this barrier is no longer there, and the shock wave, continuously created, trails behind the plane, producing booms along its ground path.

Insight

Doppler Radar

Radar has been used since the early 1940s to provide information about rainstorms and other forms of precipitation. This information is obtained from the intensity of the reflected signal. Such conventional radars can also detect the hooked (rotational) "signature" of a tornado, but only after the storm is well developed.

A major improvement in weather forecasting came about with the development of a radar system that could measure the Doppler frequency shift in addition to the magnitude of the echo signal reflected from precipitation. The Doppler shift is related to the *radial* velocity of the precipitation blown by the wind (the component of the wind velocity moving either toward or away from the radar installation).

A Doppler-based radar system (Fig. 1a) can penetrate a storm and monitor its wind speeds. The direction of a storm's wind-driven rain gives a wind "field" map of the storm region. Such maps provide strong clues of developing tornadoes, so meteorologists can detect them much earlier than was ever before possible (Fig. 1b). With Doppler radar, forecasters have been able to predict tornadoes as much as 20 min before they touch down, as compared to just over 2 min for conventional radar. Doppler radar has saved many lives with this increased warning time. The National Weather Service has a network of Doppler radars around the United States, and Doppler radar scans are now common on TV weather forecasts and on the Internet.

Doppler radars have also been installed at major airports for another use. Several airplane crashes and near crashes have been attributed to downward wind bursts (also known as microbursts and down bursts). Such strong down drafts cause wind shears capable of forcing an aircraft to the ground. The Doppler radar used at airports is designed to detect dangerous wind bursts in the vicinity of the landing approaches.

Wind bursts generally result from high-speed down drafts in the turbulence of thunderstorms, but they can also occur in clear air when rain evaporates high above ground. A down draft spreads out when it hits the ground and then circulates upward. An airplane entering this pattern experiences an unexpected upward headwind that lifts the plane. The pilot often cuts speed and lowers the plane's nose to compensate. Further into the circular pattern, the wind

(b)

(c)

•**FIGURE 14.12 Bow waves and sonic booms** (a) When an aircraft exceeds the speed of sound in air, the sound waves form a pressure ridge, or shock wave. As the trailing shock wave passes over the ground, observers hear a sonic boom—actually two booms, because shock waves are formed at the front and tail of the plane. (b) A bullet traveling at a speed of 500 m/s. Note the shock waves produced (and the turbulence behind the bullet). The image was made by using interferometry with polarized light and a pulsed laser, with an exposure time of 20 ns. (c) A shock wave in water. Bow waves like these are produced whenever a boat travels faster than the speed of the water waves that it creates.

Insight

FIGURE 1 Doppler radar (a) A Doppler radar installation. (b) Doppler radar depicts the precipitation inside a thunderstorm. A hook echo is a signature of a possible tornado.

quickly turns downward. The airplane can suddenly lose altitude and, when near the ground, as on landing, may crash. Since Doppler radar can detect the wind speed and the direction of raindrops in clouds, as well as dust and other objects floating in the air, it can provide an early warning against dangerous wind shear conditions.

Ideally, the sound waves produced by a supersonic aircraft form a cone-shaped shock wave (•Fig. 14.13). The waves travel outward with a speed v, and the speed of the plane (the source) is v_s. Note from the figure that the angle between a line tangent to the spherical waves and the line along which the plane is moving is given by

$$\sin \theta = \frac{vt}{v_s t} = \frac{v}{v_s} \quad (14.15)$$

The inverse ratio of the speeds is called the **Mach number** (M), named after Ernst Mach (1838–1916), an Austrian physicist, who used it in studying supersonics:

$$M = \frac{v_s}{v} \quad (14.16)$$

If v equals v_s, the plane is flying at the speed of sound, and the Mach number is 1 (that is, $v_s/v = 1$). Therefore, a Mach number less than 1 indicates a subsonic speed, and a Mach number greater than 1 indicates a supersonic speed. In the latter case, the Mach number tells the speed of the aircraft in terms of a multiple of the speed of sound. A Mach number of 2, for instance, indicates a speed twice the speed of sound.

14.6 Musical Instruments and Sound Characteristics

Objective: To explain some of the sound characteristics of musical instruments in physical terms.

Musical instruments provide good examples of standing waves and boundary conditions. On some stringed instruments, different notes are produced by using finger pressure to vary the lengths of the strings (•Fig. 14.14). As you learned in Chapter 13, the natural frequencies of a stretched string (fixed at each end, as is the case for the strings on an instrument), are $f_n = n(v/2L)$, from Eq. 13.21, where

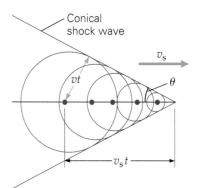

•**FIGURE 14.13 Shock wave cone and Mach number** When the speed of the source (v_s) is greater than the speed of sound in air (v), the interfering spherical sound waves form a conical shock wave that appears as a V-shaped pressure ridge when viewed in 2 dimensions. The angle θ is given by $\sin \theta = v/v_s$, and the inverse ratio v_s/v is called the Mach number.

•**FIGURE 14.14 A shorter vibrating string, a higher frequency** Different notes are produced on stringed instruments such as guitars, violins, or cellos by placing a finger on a string to change its effective, or vibrating, length.

$v = \sqrt{F_T/\mu}$. Initially adjusting the tension in a string tunes it to a particular (fundamental) frequency. Then the effective length of the string is varied by finger location and pressure.

Standing waves are also set up in wind instruments. For example, consider a pipe organ with fixed pipe lengths, which may be open or closed (•Fig. 14.15). An open pipe is open at both ends, and a closed pipe is closed at one end and open at the other (the antinode end). Analysis similar to that done in Chapter 13 for a stretched string with the proper boundary conditions shows that the natural frequencies for the pipes (where v is the speed of sound in air) are

$$f_n = \frac{v}{\lambda_n} = n\left(\frac{v}{2L}\right) = nf_1 \quad n = 1, 2, 3, \ldots \quad \text{natural frequencies for a pipe open on both ends} \quad (14.17)$$

and

$$f_m = \frac{v}{\lambda_m} = m\left(\frac{v}{4L}\right) = mf_1 \quad m = 1, 3, 5, \ldots \quad \text{natural frequencies for a pipe closed on one end} \quad (14.18)$$

•**FIGURE 14.15 Organ pipes** Longitudinal standing waves are formed in vibrating air columns in pipes (illustrated here as sinusoidal curves). **(a)** An open pipe has antinodes at both ends. **(b)** A closed pipe has a closed (node) end and an open (antinode) end. **(c)** A modern pipe organ. The pipes can be open or closed.

Note that the natural frequencies depend on the length of a pipe. This is an important consideration in a pipe organ (Fig. 14.15c), particularly in selecting the dominant or fundamental frequency. (Pipe diameter is also a factor but is not considered in this simple analysis.)

The same physical principles apply to wind and brass instruments. In all of these, the human breath is used to create standing waves in an open tube. Most such instruments allow the player to vary the effective length of the tube, and thus the pitch produced—either by opening and closing holes in the tube, as in woodwinds, or with the help of slides or valves that vary the actual length of tubing in which the air can resonate, as in most brasses (•Fig. 14.16).

Recall from Chapter 13 that a musical note or tone is referenced to the fundamental vibrational frequency. In musical terms, the first overtone is the second harmonic, the second overtone is the third harmonic, and so on. Note that for a closed organ pipe (Eq. 14.18), the even harmonics are missing.

Example 14.10 ■ Pipe Dreams: Fundamental Frequency

A particular open organ pipe has length of 0.653 m. Taking the speed of sound in air to be 345 m/s, what is the fundamental frequency of this pipe?

Thinking It Through. The fundamental frequency ($n = 1$) for an open pipe is given directly by Eq. 14.17. Physically, there is a half wavelength ($\lambda/2$) in the length of the pipe, so $\lambda = 2L$.

Solution.

Given: $L = 0.653$ m *Find:* f_1 (fundamental frequency)
$v = 345$ m/s (speed of sound)

With $n = 1$,

$$f_1 = \frac{v}{2L} = \frac{345 \text{ m/s}}{2(0.653 \text{ m})} = 264 \text{ Hz}$$

This frequency is middle C (C_4).

Follow-up Exercise. A closed organ pipe has a fundamental frequency of 256 Hz. What would be the frequency of its first overtone? Is this audible?

Perceived sounds are described by terms whose meanings are similar to those used to describe the physical properties of sound waves. Physically, a wave is generally characterized by intensity, frequency, and waveform (harmonics). The corresponding terms used to describe the sensations of the ear are loudness, pitch, and quality (or timbre). These general correlations are shown in Table 14.2. However, the correspondence is not perfect. The physical properties are objective and can be measured directly. The sensory effects are subjective and vary from person to person. (Think of temperature as measured by a thermometer and by the sense of touch.)

Sound intensity and its measurement on the decibel scale were covered in Section 14.3. Loudness is related to intensity, but the human ear responds differently to sounds of different frequencies. For example, two tones of the same intensity (in watts per square meter) but different frequencies might be judged by the ear to have different loudnesses.

Frequency and pitch are often used synonymously, but again there is an objective–subjective difference. If the same low-frequency tone is sounded at two intensity levels, most people will say that the more intense sound has a lower pitch, or perceived frequency.

(a)

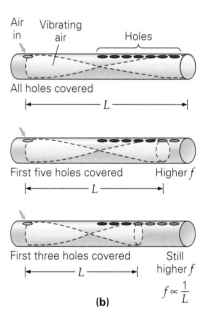

(b)

•FIGURE 14.16 **Wind instruments** (a) The saxophone, like all wind instruments, is essentially an open tube. (b) The effective length of the air column, and hence the pitch of the sound, is varied by opening and closing holes along the tube. The frequency f is inversely proportional to the effective length L of the air column.

TABLE 14.2 General Correlation between Perception and Physical Characteristics of Sound	
Sensory Effect	Physical Wave Property
Loudness	Intensity
Pitch	Frequency
Quality (timbre)	Waveform (harmonics)

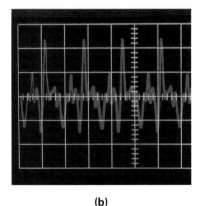

•**FIGURE 14.18 Waveform and quality** (a) The superposition of sounds of different frequencies and amplitudes gives a complex waveform. The harmonics, or overtones, determine the quality of the sound. (b) The waveform of a violin tone is displayed on an oscilloscope.

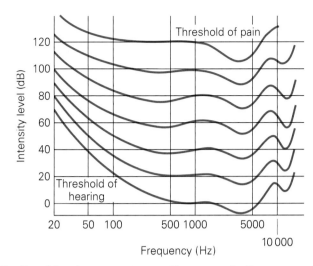

•**FIGURE 14.17 Equal loudness contours** The curves indicate tones that are judged to be equally loud, though they have different frequencies and intensity levels. For example, on the lowest contour, a 1000-hertz tone at 0 dB sounds as loud as a 50-hertz tone at 40 dB. Note that the frequency scale is logarithmic to compress the large frequency range.

The curves in the graph of intensity level versus frequency shown in •Fig. 14.17 are called equal *loudness contours* (or Fletcher-Munson curves, after the researchers who generated them). They join points representing intensity–frequency combinations that a person with average hearing judges to be equally loud. The top curve shows that the decibel level of the threshold of pain does not vary much from 120 dB, regardless of the frequency of the sound. However, the threshold of hearing, represented by the lowest contour, varies widely with frequency. For a tone with a frequency of 2000 Hz, the threshold of hearing is 0 dB. But a 20-hertz tone would have to have an intensity level of over 70 dB just to be heard (the extrapolated y-intercept of the lowest curve).

It is interesting to note the dips (or minima) in the curves. These indicate that the ear is most sensitive to sounds with frequencies around 4000 Hz and 12 000 Hz. Note that a tone with a frequency of 4000 Hz can be heard at intensity levels *below* 0 dB. These minima occur as a result of resonance in a closed cavity in the auditory canal (similar to a closed pipe). The length of the cavity is such that it has a fundamental resonance frequency of about 4000 Hz, resulting in extra sensitivity. As in a closed cavity, the next natural frequency is the third harmonic (see Eq. 14.18), which is three times the fundamental frequency, or about 12 000 Hz.

The **quality** of a tone is the characteristic that enables it to be distinguished from another of basically the same intensity and frequency. Tone quality depends on the waveform—specifically, on the number of harmonics (overtones) present and their relative intensities (•Fig. 14.18). The tone of a voice depends in large part on the vocal resonance cavities. One person can sing a tone with the same basic frequency and intensity as another, but different combinations of overtones give the two voices different qualities.

The notes of a musical scale correspond to certain frequencies; as we saw in Example 14.10, middle C (C_4) has a frequency of 264 Hz. When a note is played on an instrument, its assigned frequency is that of the first harmonic, which is the fundamental frequency. (The second harmonic is the first overtone; the third harmonic is the second overtone, and so on.) This frequency is dominant over the accompanying overtones that determine the sound quality of the instrument. Recall from Chapter 13 that the overtones produced depend on how an instrument is played. Whether a violin string is plucked or bowed, for example, can be discerned from the quality of identical notes.

Chapter Review

Important Terms

sound waves 453
audible region 453
sound frequency spectrum 453
infrasonic region 453
ultrasonic region 453
intensity 458
loudness 460
bel 460

decibel 461
sound intensity level (decibel level) 461
constructive interference 464
destructive interference 464
phase difference 464
path-length difference 464

beats 466
beat frequency 466
pitch 467
Doppler effect 467
sonic boom 471
Mach number 473
quality 476

Important Concepts

- The sound frequency spectrum is divided into infrasonic ($f < 20$ Hz), audible (20 Hz $< f <$ 20 kHz), and ultrasonic ($f > 20$ kHz) frequency regions.
- The speed of sound in a medium depends on the elasticity of the medium and its density. In general, $v_{\text{solids}} > v_{\text{liquids}} > v_{\text{gases}}$.
- The intensity of a point source is inversely proportional to the square of the distance from the source.
- The sound intensity level is a logarithmic function of the sound intensity and is expressed in decibels (dB).
- Sound wave interference of two point sources depends on phase difference as related to path-length difference. Sound waves that arrive at a point in phase reinforce each other (constructive interference); sound waves that arrive at a point out of phase cancel each other (destructive interference).
- The Doppler effect depends on the velocities of the sound source and observer relative to still air. When the relative motion of the source and observer is toward each other, the observed pitch increases; when the relative motion of source and observer is away from each other, the observed pitch decreases.

Important Equations

Speed of Sound (in meters per second):
$$v = (331 + 0.6 T_C) \text{ m/s} \tag{14.1}$$

Intensity of a Point Source:
$$I = \frac{P}{4\pi R^2} \quad \text{and} \quad \frac{I_2}{I_1} = \left(\frac{R_1}{R_2}\right)^2 \tag{14.2, 14.3}$$

Intensity Level (in decibels):
$$\beta = 10 \log \frac{I}{I_o} \quad \text{where } I_o = 10^{-12} \text{ W/m}^2 \tag{14.4}$$

Phase Difference (where ΔL is path-length difference):
$$\Delta \theta = \frac{2\pi}{\lambda}(\Delta L) \tag{14.5}$$

Condition for Constructive Interference:
$$\Delta L = n\lambda \quad (n = 0, 1, 2, 3, \ldots) \tag{14.6}$$

Condition for Destructive Interference:
$$\Delta L = m\left(\frac{\lambda}{2}\right) \quad (m = 1, 3, 5, \ldots) \tag{14.7}$$

Beat Frequency:
$$f_b = |f_1 - f_2| \tag{14.8}$$

Doppler Effect:
$$f_o = \left(\frac{v}{v \pm v_s}\right) f_s = \left(\frac{1}{1 \pm \frac{v_s}{v}}\right) f_s \tag{14.11}$$

where v_s = speed of source
and v = speed of sound
$\begin{cases} - \text{ for source moving toward stationary observer} \\ + \text{ for source moving away from stationary observer} \end{cases}$

$$f_o = \left(\frac{v \pm v_o}{v}\right) f_s = \left(1 \pm \frac{v_o}{v}\right) f_s \tag{14.14}$$

where v_o = speed of observer
and v = speed of sound
$\begin{cases} + \text{ for observer moving toward stationary source} \\ - \text{ for observer moving away from stationary source} \end{cases}$

Mach Number:
$$M = \frac{v_s}{v} \tag{14.16}$$

Natural Frequencies of Organ Pipe Open on Both Ends:
$$f_n = n\left(\frac{v}{2L}\right) = nf_1 \quad (n = 1, 2, 3, \ldots) \tag{14.17}$$

Natural Frequencies of Organ Pipe Closed on One End:
$$f_m = m\left(\frac{v}{4L}\right) = mf_1 \quad (m = 1, 3, 5, \ldots) \tag{14.18}$$

Exercises

14.1 Sound Waves
and
14.2 The Speed of Sound

1. A sound wave with a frequency of 15 Hz is in what region of the sound spectrum: (a) audible, (b) infrasonic, (c) ultrasonic, (d) supersonic?

2. A sound wave in air (a) is longitudinal, (b) is transverse, (c) has longitudinal and transverse components, (d) travels faster than a sound wave through a liquid.

3. The speed of sound is generally greatest in (a) solids, (b) liquids, (c) gases, (d) a vacuum.

4. The speed of sound in air (a) is about 1/3 km/s, (b) is about 1/5 mi/s, (c) depends on temperature, (d) all of these.

5. Suggest a possible explanation of why some flying insects produce buzzing sounds and some do not.

6. Explain why sound travels faster in warmer air than in colder air.

7. Two different-frequency sounds are emitted from a single loudspeaker. Which one will reach your ear first?

8. The speed of sound in air is temperature dependent. What effect should humidity have, if any?

9. The wave speed in a liquid is given by $v = \sqrt{B/\rho}$, where B is the bulk modulus and ρ is the density. Show that this equation is dimensionally correct. What about $v = \sqrt{Y/\rho}$ for a solid (Y is Young's modulus)?

10. ■ What is the speed of sound in air at (a) 10°C and (b) 20°C?

11. ■ The speed of sound in air on a summer day is 350 m/s. What is the air temperature?

12. ■ The thunder from a lightning flash is heard by an observer 3.0 s after she sees the flash. What is the approximate distance to the lightning strike in (a) kilometers and (b) miles?

13. ■ Sonar is used to map the ocean floor. If an ultrasonic signal is received 2.0 s after it is emitted, how deep is the ocean floor?

14. ■■ Particles approximately 3.0×10^{-2} cm in diameter are to be scrubbed loose in an aqueous ultrasonic cleaning bath. Above what frequency should the bath be operated to produce wavelengths of this size and smaller?

15. ■■ A tuning fork vibrates at a frequency of 256 Hz. What is the wavelength of the sound coming from the fork when the air temperature is (a) 0°C and (b) 20°C?

16. ■■ Brass is an alloy of copper and zinc. Does the addition of zinc to copper cause an increase or a decrease in the speed of sound in brass rods as compared to copper rods?

17. ■■ The speed of sound in steel is about 4.5 km/s. A steel rail is struck with a hammer, and there is an observer 0.30 km away with one ear to the rail. (a) How much time will elapse from the time the sound is heard through the rail until the time it is heard through the air? Assume that the air temperature is 20°C and that no wind is blowing. (b) How much time would elapse if the wind were blowing toward the observer at 36 km/h from where the rail was struck?

18. ■■ At a baseball game on a cool day (air temperature of 16°C), a fan hears the crack of the bat 0.25 s after observing the batter hit the ball. How far is the fan from home plate?

19. ■■ A 2000-hertz tone is sounded when the air temperature is 20°C and then again at 10°C. What is the percentage change in the wavelength of the sound between the higher and the lower temperature?

20. ■■ A person holds a rifle horizontally and fires at a target. The bullet has a muzzle speed of 200 m/s, and the person hears the bullet strike the target 1.00 s after firing it. The air temperature is 72°F. What is the distance to the target?

21. ■■ A hiker shouts and hears the echo reflected from a cliff wall 3.40 s later. If the air temperature is 20°C, how far is the hiker from the cliff? (Assume no wind.)

22. ■■ The driver of a truck parked on a mountain road gives the air horn a quick blast toward a rock cliff 0.760 km away. If he hears the echo 4.50 s later, what is the approximate air temperature at this mountain elevation?

23. ■■■ Sound propagating through air at 30°C passes through a vertical cold front into air that is 4.0°C. If the sound has a frequency of 2400 Hz, by what percentage does its wavelength change in crossing the boundary?

14.3 Sound Intensity and Sound Intensity Level

24. If the air temperature increases, would the sound intensity from a constant output point source (a) increase, (b) decrease, or (c) remain unchanged?

25. The decibel scale is referenced to a standard intensity of (a) 1.0 W/m², (b) 10^{-12} W/m², (c) normal conversation, (d) the threshold of pain.

26. The Richter scale, used to measure the intensity level of earthquakes, is a logarithmic scale, as is the decibel scale. Why are such scales used?

27. Can there be negative decibel levels, such as −10 dB? If so, what would these mean?

28. ■ Calculate the intensity generated by a 1.0-watt point source of sound at a location (a) 3.0 m and (b) 6.0 m from it.

29. ■ What happens to the intensity of sound at a location if the distance from the point source of sound is halved?

30. ■ By how many times must the distance from a point source be increased to reduce the sound intensity by half?

31. ■ Calculate the intensity level for (a) the threshold of hearing and (b) the threshold of pain.

32. ■ Find the intensity levels in decibels for sounds with intensities of (a) 10^{-2} W/m², (b) 10^{-6} W/m², and (c) 10^{-15} W/m².

33. ■ If the intensity of one sound is 10^{-4} W/m² and the intensity of another is 10^{-2} W/m², what is the difference in their intensity levels?

34. ■ Is a 100-watt speaker twice as loud as a 50-watt one at the same distance?

35. ■■ What is the intensity of a sound that has an intensity level of (a) 50 dB and (b) 90 dB?

36. ■■ Noise levels for some commercial jet aircraft are given in Table 14.3. Which are the lowest and highest intensities for (a) takeoff and (b) landing?

TABLE 14.3 Takeoff and Landing Noise Levels for Some Common Commercial Jet Aircraft *

Aircraft	Takeoff noise (dB)	Landing noise (dB)
737	85.7–97.7	99.8–105.3
747	89.5–110.0	103.8–107.8
DC-10	98.4–103.0	103.8–106.6
L-1011	95.9–99.3	101.4–102.8

*Noise level readings taken from 198 m (650 ft). Range depends on the aircraft model and the type of engine used.

37. ■■ What would be the intensity level of a 23-decibel sound after being amplified (a) 10 thousand times, (b) a million times, (c) a billion times?

38. ■■ A tape player has a signal-to-noise ratio of 53 dB. How many times larger is the intensity of the signal than that of the background noise?

39. ■■ The sound intensity levels for a machine shop and a quiet library are 90 dB and 40 dB, respectively. (a) What is the intensity of each? (b) How many times greater is the intensity of the sound in the machine shop than that in the library?

40. ■■ A dog's barking has a sound intensity level of 40 dB. What would the intensity level be if two identical dogs were barking?

41. ■■ At a rock concert, the average sound intensity level for a person in a front-row seat is 110 dB for a single band. If all the bands scheduled to play produce sound of that same intensity, how many of them would have to play simultaneously for the sound level to be at or above the threshold of pain?

42. ■■ At a distance of 10.0 m from a point source, the intensity level is measured to be 70 dB. At what distance from the source will the intensity level be 40 dB?

43. ■■ At a 4th of July celebration, a firecracker explodes (•Fig. 14.19). Considering the firecracker to be a point source, what are the intensities heard by observers at points B, C, and D relative to that for the observer at A?

•FIGURE 14.19 **A big bang** See Exercise 43.

44. ■■ A person standing 4.0 m from a wall shouts such that the sound strikes the wall with an intensity of 2.5×10^{-4} W/m². Assuming that the wall absorbs 20% of the incident energy and reflects the rest, what is the sound intensity level just before and after the sound is reflected?

45. ■■ A gas-powered lawnmower is rated at 95 dB. (a) What is the sound intensity for this mower? (b) How many times more intense is the sound of this mower than that of an electric-powered mower rated at 83 dB?

46. ■■■ A 1000-hertz tone issuing from a loudspeaker has an intensity level of 100 dB at a distance of 2.5 m. If the speaker is assumed to be a point source, how far from the speaker will the sound have undiminished intensity levels (a) of 60 dB and (b) barely enough to be heard?

47. ■■■ A bee produces a buzzing sound that is barely audible to a person 3.0 m away. How many bees would

have to be buzzing at that distance to produce a sound with an intensity level of 50 dB?

14.4 Sound Phenomena
and
14.5 The Doppler Effect

48. Beats are the direct result of (a) interference, (b) refraction, (c) diffraction, (d) the Doppler effect.

49. Traffic radar is based on (a) beats, (b) the Doppler effect, (c) interference, (d) the sonic boom.

50. Do interference beats have anything to do with the "beat" of music? Explain.

51. (a) Is there a Doppler effect if a sound source and an observer are moving with the same velocity? (b) What would be the effect if a moving source accelerated toward a stationary observer?

52. How fast would a "jet fish" have to swim to create an aquatic sonic boom?

53. As a person walks *between* a pair of loudspeakers that produce tones of the same amplitude and frequency, he hears a varying sound intensity. Explain.

54. Two sound waves with the same wavelength, 0.50 m, arrive at a point after having traveled (a) 2.50 m and 3.75 m and (b) 3.25 m and 8.25 m, respectively. What type of interference occurs in each case?

55. ■ Two adjacent point sources, A and B, are directly in front of an observer and emit identical 1000-hertz tones. To what closest distance behind source B would source A have to be moved for the observer to hear no sound? (Assume that the air temperature is 20°C and ignore intensity fall-off with distance.)

56. ■ A violinist and a pianist simultaneously sound notes with frequencies of 436 Hz and 440 Hz, respectively. What beat frequency will the musicians hear?

57. ■ A violinist tuning an instrument to a piano note of 264 Hz detects three beats per second. What are the possible frequencies of the violin tone?

58. ■ On a 20°C day with no wind blowing, the frequency heard by a moving person from a 500-hertz stationary siren is 520 Hz. (a) Is the person moving toward the siren or away from the siren? (b) What is the person's speed?

59. ■ What is the frequency heard by a person driving 50 km/h directly towards a factory whistle ($f = 800$ Hz) if the air temperature is 0°C?

60. ■■ While standing near a railroad crossing, you hear a train horn. The frequency emitted by the horn is 400 Hz. If the train is traveling at 90.0 km/h and the air temperature is 25°C, what is the frequency you hear (a) when the train is approaching and (b) after it has passed?

61. ■■ Two identical strings on different cellos are tuned to the 440-hertz A note. The peg holding one of the strings slips, so its tension is decreased by 1.5%. What is the beat frequency heard when the strings are then played together?

62. ■■ How fast, in kilometers per hour, must a sound source be moving toward you to make the observed frequency 5.0% greater than the true frequency? (Assume that the speed of sound is 340 m/s.)

63. ■■ What is the half-angle of the shock wave of a jet aircraft just as it breaks the sound barrier?

64. ■■ On transatlantic flights, the supersonic transport (SST) *Concorde* flies at a speed of Mach 1.5. What is the half-angle of the conical shock wave formed by the *Concorde* at this speed?

65. ■■ The half-angle of the conical shock wave formed by a supersonic jet is 35°. What are (a) the Mach number of the aircraft and (b) the actual speed of the aircraft if the air temperature is −20°C?

66. ■■■ Two point-source loudspeakers are a certain distance apart and a person stands 12.0 m in front of one of them on a line perpendicular to the baseline of the speakers. If the speakers emit identical 1000-hertz tones, what is their minimum nonzero separation so the observer hears little or no sound? (Take the speed of sound as exactly 340 m/s.)

67. ■■■ A bystander hears a siren vary in frequency from 476 Hz to 404 Hz as a fire truck approaches, passes by, and moves away on a straight street (•Fig. 14.20). What is the speed of the truck? (Take the speed of sound in air to be 343 m/s.)

•**FIGURE 14.20 The siren's wail**   See Exercise 67.

14.6 Musical Instruments and Sound Characteristics

68. The human ear can hear tones best at (a) 1000 Hz, (b) 4000 Hz, (c) 6000 Hz, (d) all frequencies.

69. The quality of sound depends on its (a) waveform, (b) frequency, (c) speed, (d) intensity.

70. (a) After a snowfall, why does it seem particularly quiet? (b) Why do empty rooms sound hollow? (c) Why do people's voices sound fuller or richer when they sing in the shower?

71. Why aren't the frets on a guitar evenly spaced?

72. Is it possible for an open and a closed organ pipe of the same length to produce notes of the same frequency? Justify your answer.

73. When you blow across the top of a bottle with water in it, why does the frequency of the sound increase with increasing levels of water?

74. ■ The first three natural frequencies of an organ pipe are 126 Hz, 378 Hz, and 630 Hz. (a) Is the pipe an open or closed one? (b) Taking the speed of sound in air to be 340 m/s, find the length of the pipe.

75. ■ A closed organ pipe has a fundamental frequency of 528 Hz (a C note) at 20°C. What is the fundamental frequency of the pipe when the temperature is 0°C?

76. ■ The human ear canal is about 2.5 cm long. It is open at one end and closed at the other (see Fig. 1 in the Insight on p. 458). What is the fundamental frequency of the ear canal (at 20°C)? To what frequency is the ear most sensitive?

77. ■■ A closed organ pipe has a length of 0.800 m. At 20°C, what are the frequencies of (a) the second harmonic and (b) the third harmonic?

78. ■■ Both an open organ pipe and a closed organ pipe have lengths of 0.52 m at 20°C. What is the fundamental frequency of each pipe?

79. ■■ Using the sign convention of Eqs. 14.11 and 14.14, show that the general equation of the Doppler effect for a moving source and a moving observer is given by

$$f_o = f_s \left(\frac{v \pm v_o}{v \mp v_s} \right)$$

80. ■■ A fire truck travels at a speed of 90 km/h with its siren emitting sound at a frequency of 500 Hz. What is the frequency heard by a passenger in a car traveling at 65 km/h in the opposite lane to the fire truck, (a) approaching it and (b) moving away from it? [Hint: See Exercise 79, and take the speed of sound to be 354 m/s (a hot day).]

81. ■■ A tuning fork with a frequency of 440 Hz is held above a resonance tube partially filled with water. Assuming that the speed of sound in air is 342 m/s, for what three smallest heights of the air column will resonance occur? [Hint: For resonance to occur, the frequency of the tuning fork must match that of the tube.]

82. ■■■ A closed organ pipe has a fundamental frequency of 660 Hz in air at 0°C. What is the fundamental frequency when the pipe is filled with the helium?

Additional Exercises

83. Two identical sources producing 440-hertz tones are located 6.97 m and 8.90 m from an observation point. If the air temperature that day is 15°C, how do the waves interfere at the point?

84. How long does it take sound to travel 3.5 km in air if the temperature is 30°C?

85. If a person standing 30.0 m from a 550-hertz point source moves 5.0 m closer to the source, by what factor does the sound intensity change?

86. An office of a mail-order company has 50 computers, which generate a sound intensity level of 40 dB (from the keyboards). The office manager tries to cut the noise in half by removing 25 computers. Does he achieve his goal? What is the intensity level generated by 25 computers?

87. How fast must an observer be moving toward a 500-hertz siren so she hears double the frequency?

88. The intensity levels of two people's speech are 60.0 dB and 65.0 dB, respectively. What is the intensity level of the combined sounds?

89. The intensity level of a sound is 90 dB at a certain distance from its source. How much energy falls on an area of 1.5 m^2 in 5.0 s?

90. A jet flies at a speed of Mach 2.0. What is the half-angle of the conical shock wave formed by the aircraft? Can you tell its speed from this information?

91. The note A (440 Hz) is produced by a stringed musical instrument. The air temperature is 20°C. Approximately how many vibrations does the string make before the sound reaches a person 30 m away?

92. A person drops a stone into a deep well and hears the splash 3.16 s later. How deep is the well? (Assume the air temperature in the well to be 10°C.)

93. The frequency of an ambulance siren is 700 Hz. What are the frequencies heard by a stationary pedestrian as the ambulance approaches and then passes her at a speed of 90.0 km/h? (Assume the air temperature is 20°C.)

94. One hunter sees another who is 300 m away fire his rifle (smoke comes out of the barrel). If the air temperature is 5°C, how long will it be until the first hunter hears the shot? (Assume no wind.)

95. An open organ pipe has a length of 0.75 m. What would be the length of a closed organ pipe whose third harmonic ($m = 3$) is the same as the fundamental frequency of the open pipe?

Appendices

APPENDIX I   Mathematical Relationships
     Algebraic Relationships
     Geometric and Trigonometric Relationships

APPENDIX II   Kinetic Theory of Gases

APPENDIX III   Planetary Data

APPENDIX IV   Alphabetical Listing of the Chemical Elements

APPENDIX V   Properties of Selected Isotopes

Answers to Follow-up Exercises

Answers to Odd-Numbered Exercises

Appendix I   Mathematical Relationships

Algebraic Relationships

$(a + b)^2 = a^2 + 2ab + b^2$
$(a - b)^2 = a^2 - 2ab + b^2$
$(a^2 - b^2) = (a + b)(a - b)$

Quadratic Formula

If $ax^2 + bx + c = 0$, then $x = \dfrac{-b \pm \sqrt{b^2 - 4ac}}{2a}$

Powers and Exponents

$x^0 = 1$

$x^1 = x$ $x^{-1} = \dfrac{1}{x}$

$x^2 = x \cdot x$ $x^{-2} = \dfrac{1}{x^2}$ $x^{\frac{1}{2}} = \sqrt{x}$

$x^3 = x \cdot x \cdot x$ $x^{-3} = \dfrac{1}{x^3}$ $x^{\frac{1}{3}} = \sqrt[3]{x}$

etc.   etc.   etc.

$x^a \cdot x^b = x^{(a+b)}$
$x^a / x^b = x^{(a-b)}$
$(x^a)^b = x^{ab}$

Logarithms

If $x = a^n$, then $n = \log_a x$.

common logarithms: base 10
(assumed when abbreviation "log" is used, unless another base is specified)

$\log 10^x = x$

$\log xy = \log x + \log y$
$\log x/y = \log x - \log y$
$\log x^y = y \log x$

natural logarithms: base $e = 2.71828\ldots$ (abbreviated "ln")

$\ln e^x = x$
$\log x = 0.43429 \ln x$
$\ln x = 2.3026 \log x$

Geometric and Trigonometric Relationships

Areas and Volumes of Some Common Shapes

Circle: $A = \pi r^2 = \dfrac{\pi d^2}{4}$ (area)
             $c = 2\pi r = \pi d$ (circumference)

Triangle: $A = \tfrac{1}{2}ab$

Sphere: $A = 4\pi r^2$
           $V = \tfrac{4}{3}\pi r^3$

Cylinder: $A = \pi r^2$ (end)
             $A = 2\pi rh$ (body)
             $A = 2(\pi r^2) + 2\pi rh$ (total)
             $V = \pi r^2 h$

Definitions of Trigonometric Functions

$\sin \theta = \dfrac{y}{r}$ $\cos \theta = \dfrac{x}{r}$ $\tan \theta = \dfrac{\sin \theta}{\cos \theta} = \dfrac{y}{x}$

$\theta°$ (rad)	$\sin \theta$	$\cos \theta$	$\tan \theta$
0° (0)	0	1	0
30° ($\pi/6$)	0.500	0.866	0.577
45° ($\pi/4$)	0.707	0.707	1.00
60° ($\pi/3$)	0.866	0.500	1.73
90° ($\pi/2$)	1	0	$\to \infty$

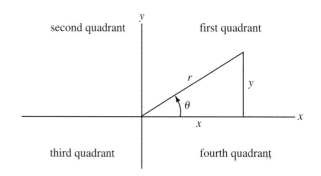

Appendix I   **A-1**

For very small angles:

$\cos\theta \approx 1 \qquad \sin\theta \approx \theta$ (radians)

$\tan\theta = \dfrac{\sin\theta}{\cos\theta} \approx \theta$ (radians)

The sign of a trigonometric function depends on the quadrant, or the signs of x and y; for example, in the second quadrant $(-x, y)$, $-x/r = \cos\theta$, and $y/r = \sin\theta$. The sign can also be assigned by using the reduction formulas.

Reduction Formulas

	(θ in second quadrant)	(θ in third quadrant)	(θ in fourth quadrant)
$\sin\theta =$	$\cos(\theta - 90°)$	$-\sin(\theta - 180°)$	$-\cos(\theta - 270°)$
$\cos\theta =$	$-\sin(\theta - 90°)$	$-\cos(\theta - 180°)$	$\sin(\theta - 270°)$

Fundamental Identities

$\sin^2\theta + \cos^2\theta = 1$

$\sin 2\theta = 2\sin\theta\cos\theta$

$\cos 2\theta = \cos^2\theta - \sin^2\theta = 2\cos^2\theta - 1 = 1 - 2\sin^2\theta$

$\sin^2\theta = \tfrac{1}{2}(1 - \cos 2\theta)$

$\cos^2\theta = \tfrac{1}{2}(1 + \cos 2\theta)$

For half-angle ($\theta/2$) identities, replace θ with $\theta/2$; for example:

$\sin^2\theta/2 = \tfrac{1}{2}(1 - \cos\theta) \qquad \cos^2\theta/2 = \tfrac{1}{2}(1 + \cos\theta)$

$\sin(\alpha \pm \beta) = \sin\alpha\cos\beta \pm \cos\alpha\sin\beta$

$\cos(\alpha \pm \beta) = \cos\alpha\cos\beta \mp \sin\alpha\sin\beta$

$\tan(\alpha \pm \beta) = \dfrac{\tan\alpha \pm \tan\beta}{1 \mp \tan\alpha\tan\beta}$

Law of Cosines

For a triangle with angles A, B, and C with opposite sides a, b, and c, respectively:

$a^2 = b^2 + c^2 - 2bc\cos A$

(with similar results for $b^2 =$ and for $c^2 =$).

If $A = 90°$, this equation reduces to the Pythagorean theorem:

$a^2 = b^2 + c^2$ (of the form $r^2 = x^2 + y^2$)

Law of Sines

For a triangle with angles A, B, and C with opposite sides a, b, and c, respectively:

$\dfrac{a}{\sin A} = \dfrac{b}{\sin B} = \dfrac{c}{\sin C}$

Appendix II Kinetic Theory of Gases

The basic assumptions are as follows:

1. All the molecules of a pure gas have the same mass (m) and are in continuous and completely random motion. (The mass of each molecule is so small that the effect of gravity on it is negligible.)
2. The gas molecules are separated by large distances and occupy a volume that is negligible compared to these distances.
3. The molecules exert no forces on each other except when they collide.
4. Collisions of the molecules with one another and with the walls of the container are perfectly elastic.

The magnitude of the force exerted on the wall of the container by a gas molecule colliding with it is $F = \Delta p/\Delta t$. Assuming that the direction of the velocity (v_x) is normal to the wall, the magnitude of the average force is

$$F = \dfrac{\Delta(mv)}{\Delta t} = \dfrac{mv_x - (-mv_x)}{\Delta t} = \dfrac{2mv_x}{\Delta t} \qquad (1)$$

After striking one wall of the container, which for convenience is assumed to be a cube with sides of dimensions L, the molecule recoils in a straight line. Suppose that the molecule reaches the opposite wall without colliding with any other molecules along the way. The molecule then travels the distance L in a time equal to L/v_x. After the collision with that wall, again assuming no collisions on the return trip, the round trip will take $\Delta t = 2L/v_x$. Thus, the number of collisions per unit time a molecule makes with a particular wall is $v_x/2L$, and the average force of the wall from successive collisions is

$$F = \dfrac{2mv_x}{\Delta t} = \dfrac{2mv_x}{2L/v_x} = \dfrac{mv_x^2}{L} \qquad (2)$$

The random motions of the many molecules produce a relatively constant force on the walls, and the pressure (p) is the total force on a wall divided by its area:

$$p = \dfrac{\Sigma F_i}{L^2} = \dfrac{m(v_{x_1}^2 + v_{x_2}^2 + v_{x_3}^2 + \cdots)}{L^3} \qquad (3)$$

The subscripts refer to individual molecules.

The average of the squares of the speeds is given by

$$\overline{v_x^2} = \dfrac{v_{x_1}^2 + v_{x_2}^2 + v_{x_3}^2 + \cdots}{N}$$

where N is the number of molecules in the container. In terms of this average, Eq. 3 can be written as

$$p = \dfrac{Nm\overline{v_x^2}}{L^3} \qquad (4)$$

However, the molecules' motions occur with equal frequency along any one of the three axes, so $\overline{v_x^2} = \overline{v_y^2} = \overline{v_z^2}$, and $\overline{v^2} = \overline{v_x^2} + \overline{v_y^2} + \overline{v_z^2} = 3\overline{v_x^2}$. Then,

$$\sqrt{\overline{v^2}} = v_{\text{rms}}$$

where v_{rms} is called the root-mean-square (rms) speed. Substituting this result into Eq. 4 along with V for L^3 (since L^3 is the volume of the cubical container) gives

$$pV = \tfrac{1}{3}Nmv_{\text{rms}}^2 \qquad (5)$$

This result is true even though collisions between molecules were ignored. Statistically, these collisions average out, so the number of collisions with each wall is as described. This result is also independent of the shape of the container. A cube merely simplifies the derivation.

We combine this result with the empirical perfect gas law:

$$pV = Nk_BT = \tfrac{1}{3}Nmv_{rms}^2$$

The average kinetic energy per gas molecule is thus proportional to the absolute temperature of the gas:

$$\overline{K} = \tfrac{1}{2}mv_{rms}^2 = \tfrac{3}{2}k_BT \tag{6}$$

The collision time is negligible compared with the time between collisions. Some kinetic energy will be momentarily converted to potential energy during a collision; however, this potential energy can be ignored because each molecule spends a negligible amount of time in collisions. Therefore, by this approximation, the total kinetic energy is the internal energy of the gas, and the internal energy of a perfect gas is directly proportional to its absolute temperature.

Appendix III   Planetary Data

Name	Equatorial Radius (km)	Mass (Compared to Earth's)*	Mean Density ($\times 10^3$ kg/m^3)	Surface Gravity (Compared to Earth's)	Semimajor Axis $\times 10^6$ km	Semimajor Axis AU†	Period Years	Period Days	Eccentricity	Inclination to Ecliptic
Mercury	2439	0.0553	5.43	0.378	57.9	0.3871	0.24084	87.96	0.2056	7°00'26"
Venus	6052	0.8150	5.24	0.894	108.2	0.7233	0.61515	224.68	0.0068	3°23'40"
Earth	6378.140	1	5.515	1	149.6	1	1.00004	365.25	0.0167	0°00'14"
Mars	3397.2	0.1074	3.93	0.379	227.9	1.5237	1.8808	686.95	0.0934	1°51'09"
Jupiter	71398	317.89	1.36	2.54	778.3	5.2028	11.862	4337	0.0483	1°18'29"
Saturn	60000	95.17	0.71	1.07	1427.0	9.5388	29.456	10760	0.0560	2°29'17"
Uranus	26145	14.56	1.30	0.8	2871.0	19.1914	84.07	30700	0.0461	0°48'26"
Neptune	24300	17.24	1.8	1.2	4497.1	30.0611	164.81	60200	0.0100	1°46'27"
Pluto	1500–1800	0.02	0.5–0.8	~0.03	5913.5	39.5294	248.53	90780	0.2484	17°09'03"

*Planet's mass/Earth's mass, where $M_E = 6.0 \times 10^{24}$ kg
†Astronomical unit, 1 AU = 1.5×10^8 km, the average distance between the Earth and the Sun.

Appendix IV   Alphabetical Listing of the Chemical Elements (perodic table on p. 881.)

Element	Symbol	Atomic Number (Proton Number)	Atomic Mass	Element	Symbol	Atomic Number (Proton Number)	Atomic Mass	Element	Symbol	Atomic Number (Proton Number)	Atomic Mass
Actinium	Ac	89	227.0278	Chromium	Cr	24	51.996	Hydrogen	H	1	1.00794
Aluminum	Al	13	26.98154	Cobalt	Co	27	58.9332	Indium	In	49	114.82
Americium	Am	95	(243)	Copper	Cu	29	63.546	Iodine	I	53	126.9045
Antimony	Sb	51	121.757	Curium	Cm	96	(247)	Iridium	Ir	77	192.22
Argon	Ar	18	39.948	Dubnium	Db	105	(262)	Iron	Fe	26	55.847
Arsenic	As	33	74.9216	Dysprosium	Dy	66	162.50	Krypton	Kr	36	83.80
Astatine	At	85	(210)	Einsteinium	Es	99	(252)	Lanthanum	La	57	138.9055
Barium	Ba	56	137.33	Erbium	Er	68	167.26	Lawrencium	Lr	103	(260)
Berkelium	Bk	97	(247)	Europium	Eu	63	151.96	Lead	Pb	82	207.2
Beryllium	Be	4	9.01218	Fermium	Fm	100	(257)	Lithium	Li	3	6.941
Bismuth	Bi	83	208.9804	Fluorine	F	9	18.998403	Lutetium	Lu	71	174.967
Bohrium	Bh	107	(264)	Francium	Fr	87	(223)	Magnesium	Mg	12	24.305
Boron	B	5	10.81	Gadolinium	Gd	64	157.25	Manganese	Mn	25	54.9380
Bromine	Br	35	79.904	Gallium	Ga	31	69.72	Meitnerium	Mt	109	(268)
Cadmium	Cd	48	112.41	Germanium	Ge	32	72.561	Mendelevium	Md	101	(258)
Calcium	Ca	20	40.078	Gold	Au	79	196.9665	Mercury	Hg	80	200.59
Californium	Cf	98	(251)	Hafnium	Hf	72	178.49	Molybdenum	Mo	42	95.94
Carbon	C	6	12.011	Hahnium	Ha	105	(262)	Neodymium	Nd	60	144.24
Cerium	Ce	58	140.12	Hassium	Hs	108	(265)	Neon	Ne	10	20.1797
Cesium	Cs	55	132.9054	Helium	He	2	4.00260	Neptunium	Np	93	237.048
Chlorine	Cl	17	35.453	Holmium	Ho	67	164.9304	Nickel	Ni	28	58.69

Element	Symbol	Atomic Number (Proton Number)	Atomic Mass	Element	Symbol	Atomic Number (Proton Number)	Atomic Mass	Element	Symbol	Atomic Number (Proton Number)	Atomic Mass
Niobium	Nb	41	92.9064	Rhodium	Rh	45	102.9055	Thallium	Tl	81	204.383
Nitrogen	N	7	14.0067	Rubidium	Rb	37	85.4678	Thorium	Th	90	232.0381
Nobelium	No	102	(259)	Ruthenium	Ru	44	101.07	Thulium	Tm	69	168.9342
Osmium	Os	76	190.2	Rutherfordium	Rf	104	(261)	Tin	Sn	50	118.710
Oxygen	O	8	15.9994	Samarium	Sm	62	150.36	Titanium	Ti	22	47.88
Palladium	Pd	46	106.42	Scandium	Sc	21	44.9559	Tungsten	W	74	183.85
Phosphorus	P	15	30.97376	Seaborgium	Sg	106	(263)	Uranium	U	92	238.0289
Platinum	Pt	78	195.08	Selenium	Se	34	78.96	Vanadium	V	23	50.9415
Plutonium	Pu	94	(244)	Silicon	Si	14	28.0855	Xenon	Xe	54	131.29
Polonium	Po	84	(209)	Silver	Ag	47	107.8682	Ytterbium	Yb	70	173.04
Potassium	K	19	39.0983	Sodium	Na	11	22.98977	Yttrium	Y	39	88.9059
Praseodymium	Pr	59	140.9077	Strontium	Sr	38	87.62	Zinc	Zn	30	65.39
Promethium	Pm	61	(145)	Sulfur	S	16	32.066	Zirconium	Zr	40	91.22
Protactinium	Pa	91	231.0359	Tantalum	Ta	73	180.9479			110	(269)
Radium	Ra	88	226.0254	Technetium	Tc	43	(98)			111	(272)
Radon	Rn	86	(222)	Tellurium	Te	52	127.60			112	(277)
Rhenium	Re	75	186.207	Terbium	Tb	65	158.9254				

Appendix V Properties of Selected Isotopes

Atomic Number (Z)	Element	Symbol	Mass Number (A)	Atomic Mass*	Abundance (%) or Decay Mode† (if radioactive)	Half-Life (if radioactive)
0	(Neutron)	n	1	1.008665	β^-	10.6 min
1	Hydrogen	H	1	1.007825	99.985	
	Deuterium	D	2	2.014102	0.015	
	Tritium	T	3	3.016049	β^-	12.33 y
2	Helium	He	3	3.016029	0.00014	
			4	4.002603	≈100	
3	Lithium	Li	6	6.015123	7.5	
			7	7.016005	92.5	
4	Beryllium	Be	7	7.016930	EC, γ	53.3 d
			8	8.005305	2α	6.7×10^{-17} s
			9	9.012183	100	
5	Boron	B	10	10.012938	19.8	
			11	11.009305	80.2	
			12	12.014353	β^-	20.4 ms
6	Carbon	C	11	11.011433	β^+, EC	20.4 ms
			12	12.000000	98.89	
			13	13.003355	1.11	
			14	14.003242	β^-	5730 y
7	Nitrogen	N	13	13.005739	β^-	9.96 min
			14	14.003074	99.63	
			15	15.000109	0.37	
8	Oxygen	O	15	15.003065	β^+, EC	122 s
			16	15.994915	99.76	
			18	17.999159	0.204	
9	Fluorine	F	19	18.998403	100	
10	Neon	Ne	20	19.992439	90.51	
			22	21.991384	9.22	
11	Sodium	Na	22	21.994435	β^+, EC, γ	2.602 y
			23	22.989770	100	

Atomic Number (Z)	Element	Symbol	Mass Number (A)	Atomic Mass*	Abundance (%) or Decay Mode (if radioactive)	Half-Life (if radioactive)
			24	23.990 964	β^-, γ	15.0 h
12	Magnesium	Mg	24	23.985 045	78.99	
13	Aluminum	Al	27	26.981 541	100	
14	Silicon	Si	28	27.976 928	92.23	
			31	30.975 364	β^-, γ	2.62 h
15	Phosphorus	P	31	30.973 763	100	
			32	31.973 908	β^-	14.28 d
16	Sulfur	S	32	31.972 072	95.0	
			35	34.969 033	β^-	87.4 d
17	Chlorine	Cl	35	34.968 853	75.77	
			37	36.965 903	24.23	
18	Argon	Ar	40	39.962 383	99.60	
19	Potassium	K	39	38.963 708	93.26	
			40	39.964 000	β^-, EC, γ, β^+	1.28×10^9 y
20	Calcium	Ca	30	39.962 591	96.94	
24	Chromium	Cr	52	51.940 510	83.79	
25	Manganese	Mn	55	54.938 046	100	
26	Iron	Fe	56	55.934 939	91.8	
27	Cobalt	Co	59	58.933 198	100	
			60	59.933 820	β^-, γ	5.271 y
28	Nickel	Ni	58	57.935 347	68.3	
			60	59.930 789	26.1	
			64	63.927 968	0.91	
29	Copper	Cu	63	62.929 599	69.2	
			64	63.929 766	β^-, β^+	12.7 h
			65	64.927 792	30.8	
30	Zinc	Zn	64	63.929 145	48.6	
			66	65.926 035	27.9	
33	Arsenic	As	75	74.921 596	100	
35	Bromine	Br	79	78.918 336	50.69	
36	Krypton	Kr	84	83.911 506	57.0	
			89	88.917 563	β^-	3.2 min
38	Strontium	Sr	86	85.909 273	9.8	
			88	87.905 625	82.6	
			90	89.907 746	β^-	28.8 y
39	Yttrium	Y	89	89.905 856	100	
43	Technetium	Tc	98	97.907 210	β^-, γ	4.2×10^6 y
47	Silver	Ag	107	106.905 095	51.83	
			109	108.904 754	48.17	
48	Cadmium	Cd	114	113.903 361	28.7	
49	Indium	In	115	114.903 88	95.7; β^-	5.1×10^{14} y
50	Tin	Sn	120	119.902 199	32.4	
53	Iodine	I	127	126.904 477	100	
			131	130.906 118	β^-, γ	8.04 d
54	Xenon	Xe	132	131.904 15	26.9	
			136	135.907 22	8.9	
55	Cesium	Cs	133	132.905 43	100	
56	Barium	Ba	137	136.905 82	11.2	
			138	137.905 24	71.7	
			144	143.922 73	β^-	11.9 s

Atomic Number (Z)	Element	Symbol	Mass Number (A)	Atomic Mass*	Abundance (%) or Decay Mode (if radioactive)	Half-Life (if radioactive)
61	Promethium	Pm	145	144.91275	EC, α, γ	17.7 y
74	Tungsten (Wolfram)	W	184	183.95095	30.7	
76	Osmium	Os	191	190.96094	β^-, γ	15.4 d
			192	191.96149	41.0	
78	Platinum	Pt	195	194.96479	33.8	
79	Gold	Au	197	196.96656	100	
80	Mercury	Hg	202	201.97063	29.8	
81	Thallium	Tl	205	204.97441	70.5	
			210	209.990069	β^-	1.3 min
82	Lead	Pb	204	203.973044	β^-, 1.48	1.4×10^{17} y
			206	205.97446	24.1	
			207	206.97589	22.1	
			208	207.97664	52.3	
			210	209.98418	α, β^-, γ	22.3 y
			211	210.98874	β^-, γ	36.1 min
			212	211.99188	β^-, γ	10.64 h
			214	213.99980	β^-, γ	26.8 min
83	Bismuth	Bi	209	208.98039	100	
			211	210.98726	α, β^-, γ	2.15 min
84	Polonium	Po	210	209.98286	α, γ	138.38 d
			214	213.99519	α, γ	164 μs
86	Radon	Rn	222	222.017574	α, β	3.8235 d
87	Francium	Fr	223	223.019734	α, β^-, γ	21.8 min
88	Radium	Ra	226	226.025406	α, γ	1.60×10^3 y
			228	228.031069	β^-	5.76 y
89	Actinium	Ac	227	227.027751	α, β^-, γ	21.773 y
90	Thorium	Th	228	228.02873	α, γ	1.9131 y
			232	232.038054	100; α, γ	1.41×10^{10} y
92	Uranium	U	232	232.03714	α, γ	72 y
			233	233.039629	α, γ	1.592×10^5 y
			235	235.043925	0.72; α, γ	7.038×10^8 y
			236	236.045563	α, γ	2.342×10^7 y
			238	238.050786	99.275; α, γ	4.468×10^9 y
			239	239.054291	β^-, γ	23.5 min
93	Neptunium	Np	239	239.052932	β^-, γ	2.35 d
94	Plutonium	Pu	239	239.052158	α, γ	2.41×10^4 y
95	Americium	Am	243	243.061374	α, γ	7.37×10^3 y
96	Curium	Cm	245	245.065487	α, γ	8.5×10^3 y
97	Berkelium	Bk	247	247.07003	α, γ	1.4×10^3 y
98	Californium	Cf	249	249.074849	α, γ	351 y
99	Einsteinium	Es	254	254.08802	α, γ, β^-	276 d
100	Fermium	Fm	253	253.08518	EC, α, γ	3.0 d
101	Mendelevium	Md	255	255.0911	EC, α	27 min
102	Nobelium	No	255	255.0933	EC, α	3.1 min
103	Lawrencium	Lr	257	257.0998	α	$\approx$35 s
104	Rutherfordium	Rf	261	261.1087	α	1.1 min
105	Hahnium	Ha	262	262.1138	α	0.7 min

*The masses given throughout this table are those for the neutral atom, including the Z electrons.
†EC stands for electron capture.

Answers to Follow-up Exercises

Chapter 1

1.1 A mass of 1000 kg has a weight of 2200 lb. The weight of a metric ton is equivalent to the British *long* ton (2200 lb), or 200 lb greater than the British *short* ton (2000 lb).
1.2 Yes, m = m.
1.3 (a) 50 mi/h [(0.447 m/s)/(mi/h)] = 22 m/s. (b) Since (1609 m/mi) = 1 and (1 h/3600 s) = 1, then multiplying the two factors gives 1. Then, trading denominators, we get (1609 m/3600 s)(h/mi) = 1. Thus, 0.447 m/s = 1 mi/h.
1.4 13.3 times
1.5 $1 \text{ m}^3 = 10^6 \text{ cm}^3$
1.6 (a) To get the smallest numerical value, we must make the number in the numerator as small as possible and the number in the denominator as large as possible. It is clear from Example 1.6 that, since 1 mi = 1.61 km, the distance in the numerator should be in miles. Similarly, since there are about 4 L in 1 gal, the volume in the denominator should be in liters. Expressing fuel economy in miles per liter will give the smallest numerical value. (b) To demonstrate the comparative numerical values, we convert one set of units to the other three. Arbitrarily selecting 1 mi/gal, on converting, we obtain 1 mi/gal = 1.61 km/gal = 0.264 mi/L = 0.425 km/L.
1.7 (a) $7.0 \times 10^5 \text{ kg}^2$ (b) 3.02×10^2 (no units)
1.8 (a) 23.70 (b) 22.09
1.9 $2.3 \times 10^{-3} \text{ m}^3$, or $2.3 \times 10^3 \text{ cm}^3$
1.10 16.9 m
1.11 $1.03 \times 10^3 \text{ cm}^2$
1.12 1.8×10^{11} metric tons (180 000 million metric tons, or 3000 times the annual U.S. output)

Chapter 2

2.1 (a) No. Using meters per minute, the answer would be in minutes. (b) 0.200 m/min (or 0.003 33 m/s)
2.2 (a) Not if there is motion; $d \neq 0$, so $avg.sp. = d/t \neq 0$. A zero speed would simply imply that $d \neq 0$, or the object is at rest. (b) 4.4×10^4 mi/h (44 000 mi/h—pretty fast)
2.3 33 km/h
2.4 No. The + and − signs indicate the vector directions with respect to the reference axis. If the velocity and acceleration are in *opposite* directions, a moving object will slow down. For example, suppose that a car traveling in the negative x direction ($-v_o$) experiences an acceleration in the positive x direction ($+a$). The car would slow down, so a positive acceleration ($+a$) can produce a deceleration. Similarly, if both the velocity and acceleration are in the *same* direction, the car will speed up. For example, if the car is initially traveling in the negative x direction ($-v_o$), a negative acceleration ($-a$) will speed it up in that direction.
2.5 9.0 m/s in the direction of the original motion
2.6 96 m (A lot quicker, isn't it?)
2.7 0.904 s (The positive is root taken—why?)
2.8 The speeds of the cars as a function of time are given by Eq. 2.7, $v = v_o + at$. With $v_o = 0$, $v = at$. Hence, speed is proportional to t (that is, $v \propto t$); if car B accelerates twice as long, it will have twice the speed. (No t^2 dependence here.)
2.9 20 ft (Speed in a school zone *can* make a difference.)
2.10 1.16 s longer
2.11 Time for bill to fall its length = 0.179 s. This is less than the average reaction time (0.192 s) computed in the Example, so most people cannot catch bill.
2.12 $y_u = y_d = 5.12$ m, as measured from reference $y = 0$ at release point.

Chapter 3

3.1 $v_x = -0.40$ m/s (in $-x$ direction), $v_y = +0.30$ m/s
3.2 $x = 9.00$ m, $y = 12.6$ m (same)
3.3 Relative to the ground, or "fixed" reference frame
3.4 $\mathbf{v} = 0\,\hat{\mathbf{x}} + (3.7 \text{ m/s})\,\hat{\mathbf{y}}$
3.5 No, the boat would curve back and forth with both components acting.
3.6 $d = 524.976$ m = 525 m (rounding differences.)
3.7 14.5° W of N
3.8 $\mathbf{v} = (8.25 \text{ m/s})\,\hat{\mathbf{x}} + (-22.1 \text{ m/s})\,\hat{\mathbf{y}}$
3.9 $(x, y) = (73.8 \text{ m}, 7.50 \text{ m})$
3.10 Yes, the ball is in the air 1.77 s before reaching the hoop.
3.11 Consider the velocity components of the player's motion. Note that for the vertical motion near the maximum height (similar to Fig. 3.17b), the upward speed (v_y) is quite small, going to zero at the maximum height and then increasing from zero in the descent. During this time, the combination of the slow vertical motions and the constant horizontal motion gives the illusion that the player hangs in the air.

Chapter 4

4.1 6.0 m/s in the direction of the net force
4.2 (a) 11 lb (b) weight in pounds ÷ 2.2 lb/kg
4.3 8.3 N
4.4 (a) 50° above the $+x$ axis
(b) $\mathbf{v} = (9.8 \text{ m/s})\,\hat{\mathbf{x}} + (4.5 \text{ m/s})\,\hat{\mathbf{y}}$
4.5 2000 N
4.6 $a = \dfrac{(m_2 - m_1) - f}{m_1 + m_2}$
4.7 (a) 7.35 N (b) neglecting air resistance, 7.35 N, downward
4.8 (a) $m_2 > 1.7$ kg (b) $\theta < 17°$
4.9 Move supports closer together so angle (and $\sin\theta$) are greater, and $T (\propto 1/\sin\theta)$ is reduced.
4.10 $\mu_s = 1.41\mu_k$ (for three cases in Table 4.1)
4.11 F varies with angle, with the angle for minimum applied force being around 33° in this case (greater forces are required for 20° and 50°). In general, the optimum angle depends on the coefficient of friction.
4.12 $mg \sin\theta - f_k = ma$, block accelerates down the plane. No, would have to measure acceleration.
4.13 Air resistance depends not only on speed, but also on size and shape. If the heavier ball were larger, it would have more exposed area to collide with air molecules, and the retarding force would build up faster. Depending on the size difference, the heavier ball might reach terminal velocity first and the lighter ball strike the ground first. Or the balls might reach terminal velocity together.

Chapter 5

5.1 -20 J
5.2 Work is done initially in lifting the handles and load. But when you move the wheelbarrow forward at a constant height,

no work is done by the vertical component because that component is perpendicular to the motion.

5.3 No, speed would decrease (and if it stopped, the block would then move up the plane).

5.4 $W_{x_1} = 0.034$ J, $W_x = 0.64$ J (measured from x_o)

5.5 No, $W_2/W_1 = 4$, or 4 times as much

5.6 Here we have $m_s = m_g/2$ as before. However, $v_s/v_g = (6.0 \text{ m/s})/(4.0 \text{ m/s}) = \frac{3}{2}$. Using a ratio, $K_s/K_g = \frac{9}{8}$, and the safety still has more kinetic energy than the guard. (Answer could also be obtained from direct calculations of kinetic energies, but for a relative comparison, a ratio is usually quicker.)

5.7 $W_3/W_2 = 1.4$, or 40% larger. More work, but a smaller percentage increase.

5.8 $\Delta K_{total} = 0$, $\Delta U_{total} = 0$

5.9 9.9 m/s

5.10 To determine whether the speed depends on the mass of a ball *in principle*, look at the physical relationships that apply to the situation. Here, energy is the major consideration, and as always, we should keep in mind the versatility of the conservation of energy in analyzing phenomena. The mechanical energy is conserved while the balls are in flight, so let's consider the initial energy (E_o) of a ball and its energy (E) just before striking the ground. By the conservation of energy,
$$E_o = E$$
$$\tfrac{1}{2}mv_o^2 + mgh = \tfrac{1}{2}mv^2$$
and
$$v = \sqrt{v_o^2 + 2gh}$$
The mass cancels and doesn't appear in the equation. Thus, speed is independent of mass. (Recall that in free fall, all objects or projectiles fall with the same vertical acceleration g—see Section 2.5.)

5.11 0.025 m

5.12 (a) 59% (b) $E_{loss}/t = mg(y/t) = mgv = (60mg)$ J/s

5.13 52%

5.14 (a) The replacement motor does the same work in twice the time, or half the work in the same time. (b) This motor does the same work in half the time, or twice the work in the same time.

5.15 (a) No. This would mean that all the energy input could be converted into useful work with no losses. Practically, friction and losses are always present. (b) The work output would be greater than the energy input—you'd get more out than you put in. That is, energy would be created. (That would be nice, but it would violate the conservation of energy, one of the cornerstones of physics.)

Chapter 6

6.1 5.0 m/s. Yes, this is 18 km/h or 11 mi/h, a speed at which humans can run.

6.3 $(-3.0 \text{ kg·m/s})\hat{x} + (4.0 \text{ kg·m/s})\hat{y}$

6.4 It would increase to 60 m/s: greater speed, longer drive, ideally. (There is also a directional consideration.)

6.5 (a) Yes, for the m_1/m_2/Earth system. But with m_2 attached to the Earth, the mass of this part of the system would be vastly greater than that of m_2, so its change in velocity would be negligible. (b) Assuming the ball is tossed in the + direction: for the tosser, $v_t = -0.50$ m/s; for the catcher, $v_c = 0.48$ m/s. For the ball: $p = 0, +25$ kg·m/s, $+1.2$ kg·m/s.

6.6 yes: $x_1 = \dfrac{m_1 v_{1_o}^2}{f}$, and $x_2 = \dfrac{m_2 v_{2_o}^2}{2f}$, then $\dfrac{x_1}{x_2} = 1$, and $x_1 = x_2$.

6.7 (a) Yes, $K_o = K_1 + K_2$. (b) No, the y component of momentum is zero before the collision, so it must be zero afterward; thus one puck must be below the $+x$ axis and one above.

6.8 No; all of the kinetic energy cannot be lost to make the dent. The momentum after the collision cannot be zero, since it was not zero initially. Thus, the balls must be moving and have kinetic energy. This can also be seen from Eq. 6.11: $K_f/K_i = m_1/(m_1 + m_2)$, and K_f cannot be zero (unless m_1 is zero, which is not possible).

6.9 5.0 m

6.10 No. If $v_1 = v_2$, then $m_1 - m_2 = 2m_1$, which requires that $m_2 = -m_1$. Negative mass is not possible, so $m_1 - m_2$ is always less than $2m_1$ and the velocities cannot be equal.

6.11 All of the balls swing out, but to different degrees. With $m_1 > m_2$, the stationary ball (m_2) moves off with a greater speed after collision than the incoming, heavier ball (m_1), and the heavier ball's speed is reduced after collision, in accordance with Eq. 6.16 (see Fig. 6.14b). Hence, a "shot" of momentum is passed along the row of balls with equal mass (see Fig. 6.14a), and the end ball swings out with the same speed as was imparted to m_2. Then, the process is repeated: m_1, *now moving more slowly*, collides again with the initial ball in the row (m_2), and another, but smaller, shot of momentum is passed down the row. The new end ball in the row receives less kinetic energy than the one that swung out just a moment previously, and so doesn't swing as high. This process repeats itself instantaneously for each ball, with the observed result that all of the balls swing out to different degrees.

6.12 $M = 11$ kg, initially at origin

6.13 $(X_{CM}, Y_{CM}) = (0.47 \text{ m}, 0.10 \text{ m})$; same location as in Example, two-thirds of the length of the bar from m_1. Note: The location of the CM does not depend on the frame of reference.

6.14 Yes, the CM does not move.

Chapter 7

7.1 1.61×10^3 m = 1.61 km (about a mile)

7.2 1% (still a pretty good approximation)

7.3 (a) 4.7 rad/s, 0.38 m/s; 4.7 rad/s, 0.24 m/s (b) To equalize the running distances, because the curved sections of the track have different radii and thus different lengths.

7.4 120 rpm

7.5 (a) 86% (b) No; the astronaut still has mass and thus, with gravity acting, has weight by definition. (See the end of Section 7.6.)

7.6 106 rpm

7.7 The string cannot be exactly horizontal; it must make some small angle to the horizontal so that there will be an upward component of the tension force to balance the ball's weight.

7.8 No; it depends on mass: $F_c = \mu_s mg$.

7.9 No. Both masses have the same angular frequency or speed ω, and $a_c = r\omega^2$, so actually $a_c \propto r$. Remember, $v = 2\pi r/T$, and note that $v_2 > v_1$, with $a_c = v^2/r$.

7.10 The direction of both $\boldsymbol{\omega}$ and $\boldsymbol{\alpha}$ for the CD would be downward, perpendicular to the plane of the CD.

7.11 -0.031 rad/s^2

7.12 0.0028 m/s^2 (a large force, but a small acceleration)

7.13 (a) No, they do not vary linearly; $\Delta U = 2.4 \times 10^9$ J, only a 9.1% increase. (b) A smaller negative value means a greater energy in the Earth's negative potential well (higher up in the well).

7.14 This is the amount of *negative* work done by an external force or agent when the masses are brought together. The masses attract each other gravitationally, so without an external force they would accelerate toward each other and collide. An external agent must apply forces in opposite directions to prevent this and get the masses in the configuration, hence the work is negative. To separate the masses by infinite distances, an equal amount of positive work (against gravity) would have to be done.

7.15 4.7 times

7.16 Because of air resistance, the spacecraft would lose more energy and continue its downward spiral. With this frictional effect, the spacecraft would heat up. (Recall that meteors burn up in the atmosphere.) To keep a spacecraft from overheating, a material is used that evaporates from the outer surface tiles. This requires work, which is supplied by heat energy. Enough heat is carried off for the spacecraft to remain relatively cool. (See latent heat, Chapter 11.)

7.17 You wouldn't want to apply a forward thrust as in this Example, since the radius of the orbit of your spacecraft would increase and the spacecraft traveled more slowly, falling farther behind. By using a reverse thrust (Example 7.16), you would move to a lower orbit and speed up. Once you had overtaken the station, you could apply an appropriate forward thrust to put the spacecraft back into the proper orbit for docking.

Chapter 8

8.1 $s = r\theta = (0.12 \text{ m})(1.7) = 0.20$ m; $s = v_{CM}t = (0.10 \text{ m/s})(2.00 \text{ s}) = 0.20$ m

8.2 The weights of the balls and the forearm produce torques that tend to cause rotation in the direction opposite that of the applied torque.

8.3 $T \propto 1/\sin\theta$, and as θ gets smaller, so does $\sin\theta$ and T increases. In the limit, $\sin\theta \to 0$ and $T \to \infty$.

8.4 $\Sigma\tau$: $Nx - m_1gx_1 - m_2gx_2 - m_3gx_3 = (200 \text{ g})g(50 \text{ cm}) - (25 \text{ g})g(0 \text{ cm}) - (75 \text{ g})g(20 \text{ cm}) - (100 \text{ g})g(85 \text{ cm}) = 0$, where $N = Mg$.

8.5 No. With f_{s_2}, the reaction force N would not generally be the same (f_{s_2} and N are perpendicular components of the force exerted on the ladder by the wall). In this case, we still have $N = f_{s_1}$, but $Ny - (m_1g)x_1 - (m_mg)x_m - f_{s_2}x_3 = 0$.

8.6 5 bricks

8.7 (d) no (equal masses) (e) Yes; with larger mass farther from axis of rotation, $I = 360$ kg·m².

8.8 The long pole (or your extended arms) increases the moment of inertia by placing more mass farther from the axis of rotation (the tightrope or rail). When the walker leans to the side, a gravitational torque tends to produce a rotation about the axis of rotation, causing a fall. However, with a greater rotational inertia (greater I), the walker has time to shift his or her body so that the center of gravity is again over the rope or rail and thus again in (unstable) equilibrium. With very flexible poles, the CG may be below the wire, thus ensuring stability.

8.9 $t = 0.63$ s

8.10 $\alpha = \dfrac{2mg - (2\tau_f/R)}{(2m + M)R}$; $\dfrac{N}{\text{kg·m}}$, and $\dfrac{N}{\text{kg·m}} = \dfrac{\text{kg·m/s}^2}{\text{kg·m}} = \dfrac{1}{\text{s}^2}$

8.11 No; in general, they have different moments of inertia.

8.12 The yo-yo would roll back and forth, oscillating about the critical angle.

8.13 (a) 0.24 m (b) The force of *static* friction, f_s, acts at the point of contact, which is always instantaneously at rest and so does no work. Some frictional work may be done due to rolling friction, but this is considered negligible for hard objects and surfaces.

8.14 $v_{CM} = 2.2$ m/s; using a ratio, 1.4 times greater; no rotational energy

8.15 You already know the answer: 5.6 m/s. (It doesn't depend on the mass of the ball.)

Chapter 9

9.1 (a) +0.10% (b) 39 kg

9.2 2.3×10^{-4} L, or 2.3×10^{-7} m³

9.3 3.03×10^4 N (or 6.82×10^3 lb—about 3.4 tons!) This is roughly the force on your back right now. Our bodies don't collapse under atmospheric pressure because cells are filled with incompressible fluids (mostly water!), bone, and muscle, which react with an equal outward pressure (equal and opposite forces). As with forces, it is a pressure *difference* that gives rise to dynamic effects.

9.4 $d_o = \sqrt{\dfrac{F_o}{F_i}} d_i = \sqrt{\dfrac{1}{10}}(8.0 \text{ cm}) = 2.5$ cm

9.5 10.3 m (about 34 ft). You can see why we don't use water barometers.

9.6 Pressure in veins is lower than that in arteries (120/80).

9.7 greater, by 7.75×10^2

9.8 The object would sink, so the buoyant force is less than the object's weight. Hence, the scale would have a reading greater than 40 N. Note that with a greater density, the object would not be as large and less water would be displaced.

9.9 11%

9.10 The water level falls. When in the boat, the block is supported by buoyant forces; that is, it is essentially floating and displaces its own weight in water. When the block is in the water, however, it displaces only its own *volume* of water. Since the block is denser than water, less water is displaced in the second case than in the first.

9.11 -18%

9.12 69%

9.13 As the water falls, speed (v) increases and area (A) must decrease to have $Av = $ a constant.

9.14 0.38 m

Chapter 10

10.1 (a) $-40°$C (b) You should immediately know the answer—this is the temperature at which the Fahrenheit and Celsius temperatures are numerically equal.

10.2 (a) $T_R = T_F + 460$ (b) $T_R = \frac{9}{5}T_C + 492$ (c) $T_R = \frac{9}{5}T_K$

10.3 96°C

10.4 273°C; no, not on Earth

10.5 50 C°

10.6 It depends on the metal of the bar. If the thermal expansion coefficient (α) of the bar is less than that of iron, it will not expand as much and not be as long as the diameter of the circular ring after heating. However, if the bar's α is greater than that of iron, the bar will expand more than the ring and the ring will be distorted.

10.7 Basically, the situations would be reversed. Faster cooling would be achieved by submerging the ice in Example 10.7—the cooler water would be less dense and would rise, promoting mixing. For a lake with cooling at the surface, cooler, less-dense water would remain at the surface until minimum density was achieved. With further cooling, the denser water would sink and freezing would occur from the bottom up.

10.8 v_{rms}, 1.69%; K, 3.41%

10.9 The rotational kinetic energy for oxygen is the difference between the total energies, 2.44×10^3 J. The oxygen is less massive and so has the higher v_{rms}.

Chapter 11

11.1 21.4 m (about the height of a five-story building)
11.2 -2.5×10^4 J
11.3 40°C
11.4 Mercury has the greater temperature change—29.9 times greater than that of water.
11.5 The final temperature (T_f) would be less, so the computed value of c would be lower.
11.6 2.28×10^6 J
11.7 24°C
11.8 42 J/s
11.9 No; the air spaces provide good insulation because air is a poor conductor. The many small pockets of air between the body and the outer garment form an insulating layer that minimizes conduction and so retards the loss of body heat. (There is little convection in the small spaces.) Similarly, for a diver in a wet suit, a thin film of water acts as an insulating layer.
11.10 2.1×10^4 s (350 min, or 5.8 h)
11.11 Drapes reduce heat loss by limiting radiation through the window and by keeping convection currents away from the glass.

Chapter 12

12.1 Pumping quickly amounts to an adiabatic compression of the gas in the pump cylinder (why?). Work done *on* a system is negative ($-W$), and since $Q = 0$, the first law of thermodynamics tell us that the change in internal energy is positive since $\Delta U = Q - W = 0 - (-W) = +W$. Thus all the work the bicyclist does goes into increasing the internal energy U. Since the internal energy of a gas is proportional to its temperature, there will be a corresponding increase in temperature: $\Delta T \propto \Delta U > 0$.
12.2 122°C
12.3 -1.22×10^3 J/K
12.4 Overall zero entropy change requires $|\Delta S_w| = |\Delta S_m|$ or $|Q_w/\overline{T}_w| = |Q_m/\overline{T}_m|$. Because the system is isolated, the magnitudes of the two heat flows *must* be the same $|Q_w| = |Q_m|$. Thus no overall entropy change requires the water and the metal to have the same average temperature $\overline{T}_w = \overline{T}_m$. This is not possible, unless they are initially at the *same* temperature. This represents a very uninteresting situation, involving no heat flow! So if there is any heat flow (due to initial temperature differences), zero overall entropy change is impossible.

12.5 As the blades of grass break down, they become more disordered. Thus they experience an entropy increase. The overall universe entropy must also increase, since this degradation is a naturally occurring process.
12.6 An increase in humidity would inhibit evaporation from the bird's head, thus decreasing the dunking rate. A humidity decrease would increase the dunking rate. A change in atmospheric pressure would have no significant effect on the liquid in the bird's body and therefore no noticeable effect on its operation.
12.7 150 J/cycle
12.8 (a) $COP_{ref} = 3.3$, $COP_{hp} = 4.3$ (b) COP_{ref}
12.9 an increase of 7.5%

Chapter 13

13.1 No, its maximum speed is $(\sqrt{k/m})A = 4.0$ m/s. Thus it is traveling at 75% of the maximum speed.
13.2 0.49 J
13.3 (a) 0.63 m/s (b) $x = (0.20 \text{ mm})\cos(1000\pi t)$, $x = -0.19$ mm at $t = 1.1 \times 10^{-3}$ s
13.4 9.76 m/s²; no. Since this is less than the accepted value at sea level, the park is probably at an altitude above sea level.
13.5 (a) 0.50 m (b) 0.10 Hz
13.6 440 Hz
13.7 increase the tension

Chapter 14

14.1 (a) 2.3 (b) 10.2
14.2 (a) 858 m (b) $\Delta t = 0.05$ s
14.3 It would be greatest in He, because it has the smallest molecular mass. (It would be lowest in oxygen, which has the largest molecular mass.)
14.4 (a) The dB scale is logarithmic, not linear.
(b) 3.16×10^{-6} W/m²
14.5 No, $I_2 = 316 I_1$.
14.6 65 dB
14.7 destructive interference: $\Delta L = 2.5\lambda = 5(\lambda/2)$, and $m = 5$ No sound would be heard if the waves from the speakers had equal amplitudes. Of course, during a concert the sound would not be single-frequency tones but would have a variety of frequencies and amplitudes. Listeners at certain locations might not hear certain parts of the audible spectrum, but this probably wouldn't be noticed.
14.8 toward, 431 Hz; past, 369 Hz
14.9 With the source and the observer traveling in the same direction at the same speed, their relative velocity would be zero. That is, the observer would consider the source to be stationary. Since the speed of source and observer is subsonic, the sound from the source would overtake the observer without a shift in frequency. Generally, for motions involved in a Doppler shift, the word *toward* is associated with an *increase* in frequency and *away* with a *decrease* in frequency. Here, the source and observer remain a constant distance apart. (What would be the case if the speeds were supersonic?)
14.10 768 Hz; yes

Chapter 15

15.1 (a) positive charge (b) $+2.0$ μC (c) deficiency (d) It loses 1.3×10^{13} electrons.

15.2 No; if the comb were positive, it would polarize the paper in the reverse direction and still attract it.

15.3 $(0.17\text{ N})\hat{x}$ As the x distance increases, the angle θ decreases. Thus the x components approach a value equal to the magnitude of each force. Their total approaches a value equal to twice each force's magnitude.

15.4 0.12 m

15.5 $F_e/F_g = 4.16 \times 10^{42}$; F_g is much *less* than F_e because the electron's mass is considerably smaller than a proton's, and F_e (magnitude) doesn't change.

15.6 The field is zero to the left of q_1 at $x = -0.60$ m. The fields *oppose* one another both to the left of q_1 and the right of q_2. Both point to the right between the charges and cannot cancel there. Since q_1 is the smaller of the two in magnitude, the location of zero net field must be closer to it than to q_2.

15.7 $\mathbf{E} = (-65.8\text{ N/C})\hat{x} + (360\text{ N/C})\hat{y}$, or 366 N/C at 79.6° above the $-x$ axis.

15.8 By symmetry, the y components cancel and the net electric field is in the $+x$ direction and given by $E_{net} = 2E\cos\theta$. For large distances, $\theta \to 0°$ and $\cos\theta \to 1$. Similarly, $x \to r$ and E from each charge becomes $E = kq/r^2 \to kq/x^2$. Thus $E_{net} = 2E\cos\theta \to 2(kq/x^2)(1) = k(2q)/x^2$. At large values of x, the separation distance d between the charges appears to be zero, so it is not in the answer. The charges appear to be superimposed and located at the origin. Thus the field approaches that of a single point charge with a charge of $Q = 2q$.

15.9 (a) $\mathbf{E}$ points upward, from ground to cloud.
(b) 5.6×10^{11} electrons/m^2

15.10 Positive charge would move to the outside surface. The leaves of the electroscope attached to the outside would deflect. The leaves of the electroscope attached to the inside surface would not deflect.

15.11 Their sign is negative since electric field lines points toward negative charges—inward at our Gaussian surface. In general, the net number of lines passing through a Gaussian surface is proportional to the net charge enclosed by it. In this case, the net charge is negative and so is the net count on the field lines.

Chapter 16

16.1 (a) $\Delta U_e = +7.20 \times 10^{-19}$ J (b) ΔV is unchanged (why?). (c) ΔV_{mid} is unchanged. (d) $v = 4.65 \times 10^4$ m/s

16.2 $\mathbf{E}$ is in the $-x$ direction, so the proton gains electric potential energy and lose kinetic energy as it travels in the $+x$ direction. By equating the electric potential energy gain to the loss in kinetic energy (the proton stops, so $\Delta K = K - K_o = -K_o$ here), the electric field strength is $E = 9.39 \times 10^4$ V/m.

16.3 (a) $\Delta U_e = +4.35 \times 10^{-18}$ J (b) The electron moves to a region of lower electric potential since it moves farther from the positively charged proton.

16.4 $U_{12} = -3.27 \times 10^{-19}$ J, $W = +3.27 \times 10^{-19}$ J

16.5 (a) No, you know only the potential difference, not the absolute value of the potentials. There is no mention of where the value of zero electric potential has been assigned. (b) Under storm conditions, $\mathbf{E}$ is upward. Thus the potential decreases upward and the higher potential surface is closer to the Earth's surface.

16.6 $d = 2.7 \times 10^{-11}$ m; this value is much smaller than an atomic diameter, so such a capacitor is not practical.

16.7 $V = 7.91 \times 10^3$ V

16.8 If the spacing increased, C would decrease. Since $Q = CV$, then $Q \propto C$ when the voltage across the capacitor is kept constant. Hence, Q would decrease and charge would flow away from the capacitor: $\Delta Q = -3.30 \times 10^{-12}$ C.

16.9 $V_1 = 8.00$ V; $V_2 = 4.00$ V; $U_{C_1} = 8.00 \times 10^{-5}$ J; $U_{C_2} = 4.00 \times 10^{-5}$ J

16.10 (a) $Q_1 = 8.0 \times 10^{-7}$ C; $Q_2 = 1.6 \times 10^{-6}$ C; $Q_3 = 2.4 \times 10^{-6}$ C (b) $U_{C_1} = 3.2 \times 10^{-6}$ J; $U_{C_2} = 6.4 \times 10^{-6}$ J; $U_{C_3} = 4.8 \times 10^{-6}$ J

Chapter 17

17.1 1.0×10^9 s, or about 32 y!

17.2 (a) 12 Ω (b) 4.8×10^2 Ω (c) 1.2×10^3 Ω

17.3 0.82 m

17.4 $R = 0.67$ Ω; for a sensitive electrical thermometer, use material with a high temperature coefficient of resistivity to get a large resistance change for a given temperature change.

17.5 print mode: $R = 82.3$ Ω; standby mode: $R = 1.8$ kΩ

17.6 First form: $P_o = I_o V_o$. Since the resistor is ohmic, when $V_o \to 2V_o$, then $I_o \to 2I_o$ and the power becomes $P \to (2I_o)(2V_o) = 4I_o V_o = 4P_o$.
Second form: $P_o = V_o^2/R_o$. Since the resistor is ohmic, when $V_o \to 2V_o$, the power becomes $P_o \to (2V_o)^2/R_o = 4V_o^2/R_o = 4P_o$.
Third form: $P_o = I_o^2 R_o$. Since the resistor is ohmic, when $V_o \to 2V_o$, then $I_o \to 2I_o$ and the power becomes $P \to (2I_o)^2 R_o = 4I_o^2 R_o = 4P_o$.

17.7 The dryer's power would increase, because the wire's cross-sectional area is larger, thus reducing its resistance ($P = V^2/R$). The new power rating would increase by 0.25 kW to 1.45×10^3 W. Since the power goes up, this could potentially be dangerous, since the filament could overheat and start a fire.

17.8 8.3 h

17.9 At best, power plants produce electric energy with efficiencies of 35% (ignoring joule heat losses during transmission). Thus, in terms of primary fuels, 35% is the maximum efficiency of *any electrical appliance*. The electric hot water heater would then convert 95% of the delivered electric energy to heat. The net result is that 95% of 35%, or only 33%, of the original fuel energy content is actually used to heat water. For a natural gas system, the gas is delivered at essentially no energy cost to the water heater, where it can be burned. Thus approximately 95% of the gas's heat content goes into heating the water, and gas heating systems are overall almost three times more efficient than the electric type.

Chapter 18

18.1 (a) series: $P_1 = 4.0$ W, $P_2 = 8.0$ W, $P_3 = 12.0$ W; parallel: $P_1 = 144$ W, $P_2 = 72$ W, $P_3 = 48$ W (b) In series, the most power is dissipated in the largest resistance. In parallel, the most power is dissipated in the smallest resistance. (c) The series arrangement requires less power: It dissipates less total power, because it has a larger equivalent resistance and carries a much smaller current at the same voltage.

18.2 The voltage of each light bulb increases by a factor of 18/17. Since the power dissipated in each is proportional to the *square* of the voltage (resistance stays the same), the power increases by a factor of $(18/17)^2 = 1.12$, or a 12% increase.

18.3 $P_1 = I_1^2 R_1 = 54.0$ W, $P_2 = I_2^2 R_2 = 9.0$ W, $P_3 = I_3^2 R_3 = 0.87$ W, $P_4 = I_4^2 R_4 = 2.55$ W, $P_5 = I_5^2 R_5 = 5.63$ W; the sum of

the power dissipated in each resistor is 72.1 W, rounded to three significant figures. For the battery, we have agreement to within rounding error since $P_{batt} = I_{batt}V_{batt} = 72.0$ W.
18.4 (a) Since there are no junctions, $I_1 = I_2$, which we call I. Using the loop theorem, we have $12 - 6I - 3I = 0$, so $I = 1.33$ A. (b) $V_1 = IR_1 = 8.0$ V, $V_2 = IR_2 = 4.0$ V
18.5 At the junction, we still have $I_1 = I_2 + I_3$ (Eq. 1). Using the loop theorem around loop 3 in the clockwise direction gives $6 - 6I_1 - 9I_2 = 0$ (Eq. 2). For loop 1, we still get $6 - 6I_1 - 12 - 2I_3 = 0$ (Eq. 3). Solve Eq. 1 for I_2 and substitute into Eq. 2. Then solve Eq. 2 and Eq. 3 simultaneously for I_1 and I_3. All answers are the same as those in the Example.
18.6 4.2 s
18.7 10 A
18.8 1.99×10^{-4} A, or 199 μA

Chapter 19

19.1 south
19.2 (a) Since the proton has a positive charge, it would initially be deflected in the opposite direction, or $+z$. Use the right-hand force rule. (b) 0.10 T
19.3 1.5×10^3 A
19.4 5.0×10^3 turns/m, or 1.5×10^3 turns in the solenoid length of 0.30 m
19.5 (a) 3.3×10^{-3} N/m (b) The force is repulsive, since the currents are oppositely directed. You should be able to show this by using the right-hand source and force rules.
19.6 0.13 A·m^2, about 1.4×10^{22} times larger than that due to the orbital motion of the electron
19.7 0.50 V

Chapter 20

20.1 (a) any way that will increase the flux, such as increasing the loop area or the number of loops, changing to a lower resistance wire would also help (b) A dc current creates a steady magnetic field, thus a steady flux and no emf.
20.2 $\overline{P} = 2.2 \times 10^{-4}$ W, $E = \overline{P}t = 1.8 \times 10^{-6}$ J
20.3 clockwise
20.4 17%
20.5 (a) 6.1×10^3 J (b) 5.0×10^2 J, 12 times more energy
20.6 (a) You would use it as a step-down transformer—the primary coil would have 100 windings and the secondary only 50, because U.S. appliances are designed to work at half of the European voltage of 240 V. (b) The voltage would double to 480 V. The appliance, designed to run at 120 V, would have four times the required voltage. Since $P = V^2/R$, with ohmic resistance, the power would go up by a factor of 16 to about 19 kW. It would burn up, melt, and perhaps start a fire.
20.7 (a) Higher voltages allow lower current usage. This in turn reduces joule heat losses in the delivery wires and in the motor windings, making more energy available for doing mechanical work and therefore a more efficient use of the electric energy. (b) Since the voltage is doubled, the current is halved. The heat loss in the wire is proportional to the *square* of the current. Thus losses will be cut by a factor of 4 or be reduced to 25% of their value at 120 V.
20.8 The round-trip time for electromagnetic radiation is about 8.8 min. In that time, *Sojourner* will travel about 0.44 m and will not collide with the rock.

20.9 (a) With increasing distance, the Sun's light intensity drops, as would the force of light pressure on the sail. In turn, the ship's acceleration would be reduced. (b) by enlarging the sail area to catch more light

Chapter 21

21.1 (a) 0.25 A (b) 0.35 A (c) 9.6×10^2 Ω, larger than the 240 Ω required for a bulb of the same power in the United States. The voltage in Great Britain is larger than that in the United States; to keep the power constant, the current must be reduced by using a larger resistance.
21.2 If the motor of the razor runs off the voltage frequency, then it, and the electric clock as well, might run slower.
21.3 (a) $\sqrt{2}(120$ V$) = 170$ V (b) 30 Hz
21.4 (a) $\sqrt{2}(2.55$ A$) = 3.61$ A (b) 20 Hz
21.5 (a) would increase to 0.896 A (b) The capacitor; with a frequency increase, X_C decreases. Since resistance is independent of frequency, it remains constant and overall Z decreases.
21.6 They would be equal.
21.7 (a) 41.1 Hz (b) 576 W
21.8 (a) As the frequency driving an RL circuit increases, inductive reactance increases ($X_L = 2\pi fL$). Thus Z increases since $Z = \sqrt{R^2 + X_L^2}$, and current I decreases. Since the power loss occurs entirely across the resistor (why?) ($P = I^2R$), power decreases. (b) As the frequency in an RC circuit increases, capacitive reactance *decreases* ($X_C = 1/2\pi fC$). Thus Z also decreases. This results in an increase in both current and power.
21.9 If you have a receiver tuned to a frequency *between* the two station frequencies, you would not receive the maximum strength signal from either station but there might be enough power from each to hear them simultaneously.
21.10 The voltage across the inductor opposes the voltage across the capacitor. Since they have equal magnitudes, when we add the voltages, including their polarity, they cancel: 25.0 V $+ (1.08 \times 10^3$ V$) - (1.08 \times 10^3$ V$) = 25.0$ V. We get the source voltage, in agreement with Kirchhoff's loop theorem.

Chapter 22

22.1 Light travels in straight lines; by the law of reflection, with $\theta_i = \theta_r$, if you can see someone in a mirror, that person can see you. Conversely, if you can't see the trucker's mirror, then he or she can't see your image in that mirror and won't know that your car is behind the truck.
22.2 $n = 1.25$
22.3 By Snell's law, $n_2 = 1.25$, so $v = c/n_2 = 2.40 \times 10^8$ m/s.
22.4 It would be smaller.
22.5 (a) The frequency of the light is unchanged in the different media, so the emerging light has the same frequency as that of the source. (b) The wavelength in air is independent of the water and glass media, as can be shown by adding another step (medium) to the Example solution. By reverse analysis, $\lambda_{air} = n_{water}\lambda_{water} = (c/v_{water})\lambda_{water} = c/f$. Thus, the wavelength in air is c/f.
22.6 Because of total internal reflections, the diver would see the reflection of something on the sides and/or bottom of the pool. (Use reverse ray tracing.)
22.7 $n = 1.4574$

Chapter 23

23.1 No effect. Note that the solution to the Example does not include the distance. The geometry of the situation is the same regardless of the distance from the mirror.

23.2 $d_i = 30$ cm, real, inverted, and same size. Note from the Example results (a) and (b) that when the object is closer to the mirror than is the center of curvature, the image becomes magnified.

23.3 The image always upright and reduced (and virtual).

23.4 $d_i = -d_o$ and $M = +1$; virtual (behind mirror, same distance as object is in front), upright, and same size.

23.5 $|M| = 0.75$

23.6 $|M| = \frac{1}{3}$ (Recall that the image is always reduced with a diverging lens.)

23.7 Blocking off half of the lens would halve the *amount* of light focused at the image plane, so the resulting image would be less bright but still full size.

23.8 3.0 cm to right of L_2; real, inverted, and $|M| = \frac{3}{4}$

Chapter 24

24.1 $\Delta y = y_r - y_b = 1.20 \times 10^{-2}$ m

24.2 twice as thick, $t = 199$ nm

24.3 In brass instruments, the sound comes from a relatively large, flared opening. There is thus little diffraction, so most of the energy is radiated in the forward direction. In woodwind instruments, much of the sound comes from tone holes along the column of the instrument. These holes are small compared to the wavelength of the sound, so there is appreciable diffraction. As a result, the sound is radiated in nearly all directions, even backward.

24.4 The width would increase by a factor of 1.65.

24.5 fewer; $n \leq 1.43$, so $n = 0, 1$

24.6 $\theta_2 = 35.9°$

24.7 $\lambda = 588$ nm; yellow

24.8 It follows that the particles making up clouds scatter all wavelengths of incident (sun) light. Clouds are made up of small water droplets and ice crystals (high clouds). But, though small, they are still much larger than the wavelengths of light, so there is no preferential scattering. (Storm clouds, which have more, and larger, droplets appear dark because they block out the sunlight that is strongly scattered on the other side of the cloud.)

Chapter 25

25.1 It wouldn't work, a real image would form on person's side of lens ($d_i = +75$ cm).

25.2 For an object at $d_o = 25$ cm, the image for eye 1 would be formed at 1.0 m; this is beyond the near point for that eye, so the object could be seen clearly. The image for eye 2 would be formed at 0.77 m; this is inside the near point for that eye, so the object would not be seen clearly.

25.3 glass for near-point viewing, 2.0 cm longer

25.4 length doubles

25.5 $f_i = 8.0$ cm

25.6 The erecting lens (of focal length f_e) should go between the objective and the eyepiece, positioned a distance of $2f_e$ from the image formed by the objective, which acts as an object. The erecting lens then produces an inverted image of the same size at $2f_e$ on the opposite side of the lens, which acts as an object for the eyepiece. The use of the erecting lens lengthens the telescope by $4f_e$.

25.7 3.4×10^{-7} rad, an order of magnitude better

25.8 3.46×10^{-2} m

25.9 (a) $v = 2.26 \times 10^8$ m/s, $\lambda' = 338$ nm, $f' = 6.67 \times 10^{14}$ Hz (b) blue (unchanged) (c) Wavelength of 338 nm lies in the nonvisible ultraviolet region, so the object would not be seen or would appear black.

Chapter 26

26.1 Light waves from two simultaneous events on the y axis meet at some midpoint receptor on the y axis. Since there is no relative motion along that axis, a simultaneous recording of the two events will also be recorded along the y' axis. Hence the two observers agree on simultaneity for this situation.

26.2 $v = 0.9995c$; no, not twice as fast. Travel is limited to less than c, so this is only about a 0.15% increase.

26.3 (a) 0.667 μs (b) 0.580 μs (c) The observer watching the ship measures the proper time interval. To the person on the ship, that time interval is dilated.

26.4 $v = 0.991c$

26.5 The traveler measures the proper time interval of 20.0 y, but Earth inhabitants measure the dilated version of this (why?). The gamma factor is based on a recalculated value for traveler speed $v = 0.99504c$. Keeping five places after the decimal and then rounding to three significant figures we get

$$\gamma = 1/\sqrt{1 - (0.99504c/c)^2} = 10.04988$$

and

$$t = \gamma t_o = (10.04988)(20.0 \text{ y}) = 200.99 \text{ y} \approx 201 \text{ y}$$

26.6 (a) 1.17 MeV (b) 0.207 MeV

26.7 (a) Total momentum is zero both nonrelativistically and relativistically (both particles have the same mass but opposite velocities, so their equal and opposite momenta cancel to zero). Hence the final speed is the same using both approaches, $v' = 0$ (b) Nonrelativistically, the combined mass afterward is $m' = 2m$, exactly. Relativistically, the combined mass is $m' = 3.34m$. (c) Both approaches give a final kinetic energy of zero, since the combined particle must be at rest due to momentum conservation.

26.8 1.95×10^7 more massive

26.9 $u = +0.69c$, which, as expected, is lower in magnitude than the nonrelativistic (and wrong) result of $+0.80c$.

26.10 The velocity addition equation always yields the result c for light. Here the results are $-c$ for the observer heading in the $+x$ direction and $+c$ for the observer headed in the $-x$ direction, as expected.

Chapter 27

27.1 (a) The wavelength is 967 nm, which is infrared. With some emissions in the red region, it would appear reddish. (b) The wavelength is 290 nm, which is ultraviolet. With significant visible emissions, it would appear bluish-white.

27.2 496 nm, shorter than in the Example, indicating a larger photon energy. Thus the maximum kinetic energy of the photoelectrons is larger, requiring a higher stopping voltage.

27.3 2.00 V

27.4 (a) 3.60×10^{-2} (wavelength will increase by 3.60%) (b) This is a larger increase than that in the Example, because in a head-on collision the photon transfers the maximum possible amount of energy to the electron. The photon will lose more energy than in the "glancing" collision of the Example.

27.5 1.09×10^6 m/s

27.6 365 nm

27.7 The minimum energy required for ionization from the $n = 3$ (second excited state) in hydrogen is 1.51 eV (see Fig. 27.12). If an incoming photon carried exactly 1.51 eV, it could be absorbed and the electron would escape the atom, but essentially with no kinetic energy. The atom would "barely" be ionized. To carry 1.51 eV, the light needs a wavelength of $\lambda = hc/E = 1.24 \times 10^3$ eV·nm/1.51 eV = 821 nm, which is infrared.

Chapter 28

28.1 (a) 8.8×10^{-33} m/s (b) 2.3×10^{33} s, or 7.2×10^{25} y—about 4.8×10^{15} times longer than the age of the universe. This movement would not be noticeable.

28.2 The proton's de Broglie wavelength under the same conditions is 4.1×10^{-12} m, about twenty times smaller than atomic spacing distances. Thus protons would *not* be expected to exhibit significant diffraction effects under conditions in which electrons might.

28.3 Only one electron; a second electron would have all three quantum numbers (n, ℓ, and m_ℓ) the same as the first, which is not allowed by the Pauli exclusion principle.

28.4 (a) $1s^2 2s^2 2p^6$ (b) $-2e$, or -3.2×10^{-19} C

28.5 1.2×10^6 m/s

Chapter 29

29.1 $^{12}_{6}C + ^{4}_{2}He \rightarrow ^{16}_{8}O$; the resulting nucleus is therefore ^{16}O.

29.2 Neglecting the emitted neutrino, the decay is $^{22}_{11}Na \rightarrow ^{22}_{10}Ne + ^{0}_{+1}e$. The daughter nucleus is ^{22}Ne.

29.3 (a) 48 d (reducing the activity by a factor of 64 requires six half lives; $1/2^6 = 1/64$) (b) The biological process of excretion in the body also remove the ^{131}I from the body. Radioactive decay is not the only way to eliminate it.

29.4 The closest integer is 20, since $2^{20} \approx 1.1 \times 10^6$. (Neither 19 nor 21 works; $2^{19} \approx 0.52 \times 10^6$ and $2^{21} \approx 2.1 \times 10^6$.) Thus it takes 20 half-lives, or about 560 y.

29.5 to four ^{14}C half lives, or 2.3×10^4 y

29.6 About four half lives, or 5.1×10^9 y; this is in decent agreement with the estimate given (using other methods) but is an *overestimate* (why?).

29.7 (a) Starting with 29 protons and 29 neutrons: $^{58}_{29}Cu$, $^{59}_{29}Cu$, $^{60}_{29}Cu$, $^{61}_{29}Cu$, $^{62}_{29}Cu$, $^{63}_{29}Cu$, $^{64}_{29}Cu$, and so on. We delete the odd–odd isotopes (why?) and get the most likely (stable) isotopes: $^{59}_{29}Cu$, $^{61}_{29}Cu$, $^{63}_{29}Cu$, $^{65}_{29}Cu$, and so on. (b) We can further trim the list by deleting those with $N \approx Z$ and those with N significantly larger than Z (why?). From Fig. 29.14, for this proton number, we expect neutron numbers in the mid-30s. Hence we would guess just $^{63}_{29}Cu$ and $^{65}_{29}Cu$. According to Appendix V, these are in fact the only two stable isotopes of copper.

29.8 (a) The average binding energy per nucleon for ^3He is 2.573 MeV/nucleon, which is considerably smaller than the value of 7.075 MeV/nucle for ^4He. (b) Unlike ^3He, all protons and neutrons in ^4He *are* paired, making it the more tightly bound of the two isotopes.

29.9 (a) Removal of a single proton from ^{12}C takes only 15.96 MeV; removal of a single neutron takes a minimum of 18.72 MeV. (b) One factor is the proton repulsion due to other protons. We expect protons, in general, to be less bound to a nucleus than are neutrons.

Chapter 30

30.1 $Q = -15.63$ MeV, so it is endoergic (takes energy to happen).

30.2 The increase in mass has an energy equivalent of 1.193 MeV. The rest of the incident kinetic energy (1.534 MeV − 1.193 MeV, or 0.341 MeV) must be distributed between the kinetic energies of the proton and of the oxygen.

30.3 6.17×10^{11} kg of He produced per fusion

30.4 The neutrino and the daughter nucleus would recoil in opposite directions, conserving linear momentum (assuming the original nucleus was at rest and therefore had zero linear momentum).

30.5 For the negative pion (π^-) to have a charge of $-e$, its quark structure must be $\bar{u}d$. For the same reason, the quark structure of the positive pion (π^+) is $u\bar{d}$, and the antiparticle of the π^+ would have the composition of $\bar{u}\bar{\bar{d}}$. However, the antiquark of an antiquark is the original quark—for example, $\overline{(\bar{u})} = u$, and so on. Hence the antiparticle of the π^+ has the quark structure $\bar{u}\bar{\bar{d}} = \bar{u}d$, which is the π^-, as expected.

Answers to Odd-numbered Exercises

Chapter 1
1. (b)
3. (c)
5. A duodecimal system might have a dime worth 12¢ and a dollar worth 12 "dimes," or $1.44.
7. 1 nautical mile = 6076 ft
9. (d)
11. (d)
13. (d)
15. $m^2 = m^2$
17. no, $V = \pi d^3/6$
19. yes; m/s = m/s − m/s
21. yes; $[L]^2 = [L]^2 + [L]^2$
23. 1/s, or s^{-1}
25. (a) kg·m^2/s^2 (b) yes
27. (a)
29. (a) yd (b) dm (c) cm
31. 37 000 000 times
33. 63.7 m, 60.9 m, 6.20 m, and 19.5 m
35. 68 L
37. metric; 9.4×10^2 m^2
39. (a) 341 m/s (b) 0.268 s
41. (a) 16 km/h for each 10 mi/h (b) 105 km/h
43. (a) 59 mL (b) 3.5 oz
45. 6.1 cm
47. (a) 26 m^3 (b) 9.2×10^2 ft^3
49. (a) 2.4×10^3 kg/m^3 (b) 80 kg
51. (a)
53. 5.05 cm; 5.05×10^{-1} dm; 5.05×10^{-2} m
55. no; only one doubtful digit; best 25.48 cm
57. (a) 1.0 m (b) 8.0 cm (c) 16 kg (d) 1.5×10^{-2} μs
59. (a) 10.1 m (b) 775 km (c) 2.55×10^{-3} kg (d) 9.30×10^7 mi
61. 1.27×10^{14} m^2
63. 470 cm^2
65. (a) 2.0 kg·m/s (b) 2.1 kg·m/s, rounding difference
67. (c)
69. 1.03×10^3 g = 1.03 kg
71. (a) 52% (b) 64 g, 20 g
73. 9.47×10^{15} m
75. 0.87 m
77. same area for both, 1.3 cm^2
79. 25 min
81. 17 m
83. yes, $1.21/gal
85. 42 cm^2
87. 4.9×10^2 cm^3
89. 0.60 L
91. not reasonable (56 mi/h)
93. 4.48×10^3 cm^3

Chapter 2
1. (c)
3. (a)
5. no final position can be given; may be 0 to 500 m from start
7. no
9. 1.65 m down
11. 30 km, yes
13. (a) 0.50 m/s (b) 8.3 min
15. 45 m at 27° west of north
17. (a) 2.7 cm/s (b) 1.9 cm/s
19. (a) 7.35 s (b) no
21. (a) 1.0 m/s; 0; 1.3 m/s; 2.8 m/s; 0; 1.0 m/s (b) 1.0 m/s; 0; 1.3 m/s; −2.8 m/s; 0; 1.0 m/s (c) 1.0 m/s; 0; 0; −3.8 m/s (d) −0.89 m/s
23. 8.7×10^2 km
25. 60.0 mi/h; no
27. impossible
29. yes, velocity is changing due to direction change
31. v_o
33. 1.85 m/s^2
35. 3.7 s
37. 15.5 m/s
39. $a_{0-4} = 2.0$ m/s^2; $a_{4-10} = 0$; $a_{10-18} = -1.0$ m/s^2
41. 150 s
43. (a)
45. no, 9.9 m/s^2
47. 3.5 m/s^2
49. (a) 81.4 km/h (b) 0.794 s
51. 3.09 s
53. 23 s
55. 1.43×10^{-4} s
57. 75 m/s
59. no, x = 13.3 m
61. 96 m
63. $x - x_o = \dfrac{v_o + v}{2}(t - t_o)$;
$v = v_o + a(t - t_o)$;
$x - x_o = v_o(t - t_o) + \tfrac{1}{2}a(t - t_o)^2$;
$v^2 = v_o^2 + 2a(x - x_o)$
The x in Eq. 2.9 stands for displacement; the x here stands for position.
65. (d)
67. 9.80 m/s^2 downward
69. 78.4 m
71. (a) straight line, negative slope (b) parabola
73. 15.9 m
75. 67 m
77. slightly less than 8.0 m/s
79. (a) 48 m (b) 38 m/s downward
81. (a) same, 1.14 s (b) 11.2 m/s, −11.2 m/s (equal but opposite)
83. 3.61 m (no)
85. hits 14 cm in front of the prof.
87. (a) 2.45 (b) Moon: 99.2 m; 22.0 s, Earth: 16.5 m; 3.67 s
89. (a) −20.9 m/s (b) 2.87 s
91. (a) 21 m/s (b) 19 m
93. 45 mi/h
95. 71 m
97. 16.6 m/s
99. 1.2×10^2 m
101. (a) 27 m/s (b) 4.3 s
103. 51.5 m
105. (a) 5.0 m/s^2 (b) 13 m/s (c) and (d) see ISM

Chapter 3
1. (a)
3. (a) velocity increases or decreases
(b) parabolic path (c) circle
5. 5.0 m/s^2, 53° above +x axis
7. (a) 6.0 m/s (b) 3.6 m/s
9. (a) 70 m (b) 0.57 min, 0.43 min
11. $x = 1.75$ m; $y = -1.75$ m
13. (a) $x = 6.56$ m; $y = 3.25$ m (b) 5.41 m/s at 13.9° above +x axis
15. 1.0 m/s
17. (a) 4.3 m/s (b) 2.1×10^2 m
19. opposite direction, same direction
21. $v_x = 8.66$ m/s; $v_y = 5.00$ m/s
23. yes
25.

27. 109 mi/h
29. (a)

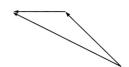

(b) 372 mi at 28.4° north of west
31. (a) $(-3.4$ cm$)\hat{x} - (2.9$ cm$)\hat{y}$ (b) 4.5 cm, 63° above −x axis (c) $(4.0$ cm$)\hat{x} - (6.9$ cm$)\hat{y}$
33. (a) $(14.4$ N$)\hat{y}$ (b) 12.7 N at 85.0° above +x axis
35. 16 m/s at 79° above the −x axis
37. 23 m/s at 3.0° above the +x axis
39. 8.5 N at 21° below −x axis
41. parallel 30 N, perpendicular 40 N
43. 27 m at 72° above −x axis
45. 102 mi/h at 61.1° north of west
47. 242 N at 48° below −x axis
49. (a) face (b) back and face (c) back
51. 13 s
53. (a) +85 km/h (b) −5.0 km/h
55. 4.0 min
57. 146 s = 2.43 min
59. (a) 68 m (b) 0.76 m/s, 11° relative to shore
61. (a)

(b) 4.7 m/s
63. (a) $v_{BA} = 39$ km/h, 40° north of east (b) $v_{AB} = 39$ km/h, 40° south of west
65. see Fig. 3.17
67. the vertical motion does not affect the horizontal motion
69. 2.7×10^{-13} m
71. 40 m
73. 26 m/s and 15 m/s
75. yes, 0.11 m

A-15

77. (a) 0.77 m (b) would not fall back in
79. (a) 1.15 km (b) 30.6 s (c) 6.13 km
81. 35° or 55°
83. (a) 26 m (b) 23 m/s at 68° below horizontal
85. 40.9 m/s, 11.9° above horizonal
87. the pass is short
89. (a) 66.0 m/s (b) too long
91. $(4.0 \text{ m/s})\hat{x} + (2.0 \text{ m/s})\hat{y}$; 4.5 m/s at 27° above the $+x$ axis
93. 1.3 m at 43° south of east
95. 6.02 m/s
97. (a) $y_{max} = 10.2$ m (at $x = 20.4$ m) (b) yes, barely
99. (a) 30.3° or 59.7° (b) no
101. (a) 21.7 m/s (b) 33.3 m/s, 49° below horizontal
103. (a) 53° above or below $+x$ axis (b) ±8.0 m/s
105. 6.1×10^2 km

Chapter 4

1. no, no net force
3. (c)
5. no, same mass, same inertia
7. balloon moves (a) forward (b) backward
9. the dishes at rest tend to remain at rest
11. $m_{Al} = 1.35 m_{water}$
13. $\mathbf{F}_3 = (-7.6 \text{ N})\hat{x}$
15. (b)
17. longer contact time, smaller force
19. 1.7 kg
21. 7.0×10^6 N
23. 75.5 kg
25. 78 N, 18 lb
27. (a) yes (b) no, 2.7 kg (1 lb)
29. (a) 1.2 m/s² (b) same
31. 2.40 m/s²
33. (a) 0.133 m/s² (b) 300 N, constant v
35. 8.9×10^4 N
37. (c)
39. yes, forces on different objects (one on ball, one on bat) cannot cancel
41. brick exerts a force of 800 N on the fist, no
43. 1.5 m/s², opposite to hers
45. yes, if it accelerates downward at 1.8 m/s²
47. (a) 735 N (b) 735 N (c) 585 N
49. 2.0 m/s², 19° north of east (b) 1.3 m/s² 30° south of east
51. (a) 0.96 m/s² (b) 2.6×10^2 N
53. 64 m
55. (a) 3.0 m/s² (b) 2.7 m/s²
57. (a) 0.711 m/s² (b) 1067 N
59. (a) 2.5 m/s² right (b) 2.0 m/s² left
61. 10 m/s²
63. 3.9×10^2 N
65. 1.2 m/s², m_1 up and m_2 down (b) 21 N
67. (a) no friction (b) opposite its direction of velocity (c) sideways (d) forward, in direction of velocity
69. kinetic friction (sliding) is less than static friction (rolling)
71. increasing normal force, therefore friction; yes, more difficult to accelerate
73. treads displace water, wide and smooth tires increase friction

75. 2.7×10^2 N
77. (a) zero (b) 3.1 m/s²
79. 0.064
81. 0.28 m
83. (a) 30° (b) 22°
85. 0.73
87. (a) 26 N (b) 21 N
89. 0.33
91. (a) between 0.72 kg and 1.7 kg (b) between 0.88 kg and 1.5 kg
93. 1.5×10^3 N
95. no
97. (a) 75 N (b) 7.3 N
99. 0.32 N
101. 5.2 m
103. no

Chapter 5

1. (d)
3. (a) no (b) yes (c) no (d) yes
5. positive on the way down, negative on the way up; not constant; maximum at points B and D ($\theta = 0°$ or 180°); minimum at points A and C ($\theta = 90°$)

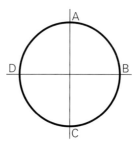

7. -98 J
9. 6.65 m
11. 3.7 J
13. 1.47×10^5 J
15. 2.4×10^3 J
17. (a) 1.48×10^5 J (b) -1.23×10^5 J (c) 2.50×10^4 J
19. (c)
21. no, more work because the force increases as the spring stretches
23. 2.5 cm
25. 1.25×10^5 N/m
27. 0.21 J
29. (a) 15 kg (b) 5.1 kg more
31. -3.0 J
33. Because the heel can lift from the blade, the blade will be in contact with the ice longer. This will make the displacement a bit greater (longer stride), so more work is done. More work translates to faster speed according to the work–energy theorem.
35. 19 J
37. small car is four times as long
39. (a) 45 J (b) 21 m/s
41. 4.2 m/s in $+x$ direction
43. 8.3×10^3 J
45. (d)
47. 4.9×10^2 J
49. 5.9 J
51. (a) 8.0 J (b) 5.6 J

53. (a) $U_b = -44$ J, $U_a = 66$ J (b) to the attic: -1.1×10^2 J; to the ground: -1.1×10^2 J; to the basement: -1.1×10^2 J
55. (d)
57. Due to the conservation of energy, the initial potential energy is equal to the final potential energy ($U_i = U_f$), so the initial height is equal to the final height ($h_i = h_f$).
59. Each time you land on the trampoline, you coil your legs on the way down, then push down on the trampoline as you land, thereby compressing it more. Storing more energy, you rebound and go higher. The limit is determined by the "spring constant" of the trampoline and how much you can compress the trampoline.
61. (a) $K = 15.0$ J, $U = 0$, $E = 15.0$ J (b) $K = 7.65$ J, $U = 7.35$ J, $E = 15.0$ J (c) $K = 0$, $U = 15.0$ J, $E = 15.0$ J
63. (a) $K = 20$ J, $U = 60$ J (b) 8.9 m/s (c) 80 J
65. (a) 1.03 m (b) 0.841 m (c) 2.32 m/s
67. (a) 11 m/s (b) no (c) 7.7 m/s
69. (a) 2.7 m/s (b) 0.38 m (c) 29°
71. 12 m/s
73. no, $E = Pt$, so you are paying for energy
75. the one that arrives first, due to a shorter time interval
77. 97 W
79. 5.7×10^{-5} W
81. 6.7×10^2 J
83. 1.8 hp
85. (a) 2.3 hp; the horse is working hard (b) 10 hp
87. $0.36
89. 13 J
91. (a) $+2.2 \times 10^5$ J (b) -2.2×10^5 J
93. (a) at least 2.9 m/s (b) yes, if starting with $v_o < 2.9$ m/s; insufficient energy
95. (a) 52 km/h = 14 m/s (b) 37 km/h = 10 m/s
97. no, because $E_{top} > E_{bottom}$
99. (a) 5.5×10^2 W (b) 0.74 hp

Chapter 6

1. (b)
3. no, mass is also a factor
5. 75 kg
7. the lineman
9. 24 J
11. 4.05 kg·m/s in the direction opposite the initial velocity v_o
13. (a) 3.7 kg·m/s down (b) 7.0 kg·m/s down
15. 3.5 m/s in the same direction or 8.2 m/s in the opposite direction
17. $\Delta \mathbf{p} = (-1.1 \text{ kg·m/s})\hat{x} + (-2.8 \text{ kg·m/s})\hat{y}$
19. 6.5×10^3 N
21. (a) 491 kg·m/s downward (b) yes; 519 kg·m/s downward
23. According to the impulse–momentum theorem, a shorter contact time will result in a greater force if all other factors (m, v_o, v) remain the same.
25. (a), (b) Increasing the contact time Δt reduces the average force F.
27. 6.0×10^3 N

29. (a) 1.2×10^3 N (b) 1.2×10^4 N
31. 1.2×10^2 N in the direction opposite initial velocity v_o
33. 1.1×10^3 N; 4.7×10^3 N
35. 8.2 N upward
37. 0.057 s
39. Air moves backward and the boat moves forward according to momentum conservation. If a sail were installed behind the fan on the boat, the boat would not go forward, because the forces between the fan and the sail are internal forces of the system.
41. Throw something or even blow a strong breath of air out of your mouth.
43. 1.0 m/s westward
45. 1.67 m/s in the original direction of the bullet
47. 7.6 m/s, 12° above the $+x$ axis
49. (a) 11 m/s to the right (b) 22 m/s to the right (c) at rest (0)
51. 0.78 m/s
55. (a)
57. Momentum is a vector, and kinetic energy is a scalar.
59. $v_1 = -0.48$ m/s; $v_2 = +0.020$ m/s
61. $v_p = -1.8 \times 10^6$ m/s; $v_\alpha = +1.2 \times 10^6$ m/s
63. 2.1×10^5 J
65. 13.9 m/s at 53.1° south of east
67. no, 1.1 kg·m/s at 291°
69. (a) $m_1/m_2 = \frac{1}{2}$ (b) $v_2 = \frac{2}{3}v_{1o}$
71. (a) $v = v_o/90$ (b) 2.5×10^2 m/s (c) no, 99%
73. (b) $e_{\text{elastic}} = 1.0$; $e_{\text{inelastic}} = 0$
75. It lowers the center of mass and increases the moment of inertia.
77. (d)
79. (a) (0, -0.45 m) (b) no, only that they are equidistant from CM
81. (a) 4.6×10^6 m from the center of the Earth (b) 1.8×10^6 m below the surface of the Earth
83. (8.0 m, 0)
85. remains at the center of the sheet
89. (a) The 65-kilogram skater travels 3.3 m, and the 45-kilogram skater travels 4.7 m. (b) same distances as in part (a)
91. (a) no, only if the masses are the same (b) in the same direction
93. (a) $(4.6d, 1.3d)$ (b) $(3.9d, 6.4d)$ (c) $(4.3d, 3.8d)$
95. (a) 2.3 kg·m/s at 45° below the $-x$ axis (b) Not necessarily, a collision does not have to occur; the momentum would be the same as in part (a).
97. (a) 6.0×10^4 kg·m/s (b) 7.5×10^3 N
99. (a) 4.0 m/s (b) 4.0 m/s
101. (a) 1.7 m/s (b) 32 J
103. (a) 28% (b) 1.3×10^7 m/s
105. (8.7 cm, 15 cm)

Chapter 7

1. (c)
3. (2.5 m, 53°)
5. (a) 0.26 rad (b) 0.79 rad (c) 1.6 rad (d) 2.1 rad
7. 4.7 cm
9. (a) 4.00 rad (b) 229°
11. (a) 1 rad = 57.3° (b) 2π rad = 360°
13. 10.7 rad
15. (a) $0.53° = 9.2 \times 10^{-3}$ rad (b) $1.9° = 3.4 \times 10^{-2}$ rad
17. No, we can cut six such pieces and one 0.28-radian piece.
19. (b)
21. yes; no
23. The point farthest from the center has the greatest tangential speed, and the point closest to the center has the smallest tangential speed.
25. 2.4 rev/day
27. 1.8 s
29. B
31. (a) 0.84 rad/s (b) 3.4 m/s, 4.2 m/s
33. (a) 7.27×10^{-5} rad/s (b) 1.99×10^{-7} rad/s
35. (a) 60 rad (b) 3.6×10^3 m
37. (d)
39. There is insufficient centripetal force (provided by friction and adhesive forces) on the water droplets, so they fly out along a tangent and the clothes get dry.
41. The inertia of our body tends to keep us moving forward along a straight line (Newton's first law), and the car makes a turn by the centripetal force between the tires and the road. So we feel as if we were being thrown outward.
43. 0.049 rad/s, or about 680 rev/day
45. 1.3 m/s
47. no; 1.33 m/s² > 1.25 m/s²
49. 13.4°
51. 9.9 m/s
53. (a) $v = \sqrt{rg}$ (b) $(5/2)r$
55. (b) The angle is independent of the mass. (c) $\tan\theta = \dfrac{v^2}{gr} - \mu_s$
57. (d)
59. No, a car in circular motion always has centripetal acceleration.
61. 1.1×10^{-3} rad/s²
63. 7.5 rev
65. (a) 1.82 rad/s² (b) 28.7 rev
67. (a) 53 s (b) $\mathbf{a} = (8.5 \text{ m/s}^2)\hat{\mathbf{r}} + (1.4 \text{ m/s}^2)\hat{\mathbf{t}}$
69. (b)
71. No. Gravity acts on the astronauts and the spacecraft, providing the necessary centripetal force for the orbit, so g is not zero and, by definition, there is weight ($w = mg$). The "floating" occurs because the spacecraft and astronauts are "falling" (accelerating toward Earth at the same rate).
73. yes, if we also know the radius of the Earth
75. 9.80 m/s²
77. (a) 2.4×10^{20} N toward the Sun (b) 6.4×10^{20} N toward the Sun
79. 8.0×10^{-10} N, toward the opposite corner
81. 3.4×10^5 m
83. The gravitational force by the Sun is greater by 1.7×10^2 times.
85. (a) -2.5×10^{-10} J (b) 0
87. (c)
89. (a) to get more velocity relative to space (b) The tangential speed of the Earth is higher in Florida.
91. (a) 3.0×10^{-19} s²/m³ (b) 3.0×10^{-19} s²/m³
93. 1.53×10^9 m
95. 2.0 min
97. (a) 0 (b) 1.2×10^3 N (c) 1.8×10^3 N
99. (a) 1.10×10^{10} m/s² (b) $(1.12 \times 10^9)g$
101. 1.1×10^4 m/s
103. 1 y²/AU³

Chapter 8

1. (a)
3. Yes; rolling motion is a good example.
5. (b)
7. 0.10 m
9. (a) 0.50 m/s (b) 1.0 m/s
11. yes
13. 0.58 rotations
15. (a)
17. Yes, the toy clown is in stable equilibrium. Its center of gravity is directly below the tightrope. If the clown leans to one side, his own weight will restore its equilibrium position. If the weights are removed, the clown will be in a unstable equilibrium and he will fall.
19. 3.3×10^2 N
21. 6 stable (faces) and 20 unstable (12 edges and 8 corners)
23. 2.3 m
27. 1.6×10^2 N
29. 3.6 N
31. $m_2 = 0.20$ kg, $m_3 = 0.50$ kg, $m_4 = 0.40$ kg
33. Yes; the center of gravity of each stick is at or to the left of the edge of the table.
35. 1.2 m from left end of board
37. (a) to satisfy $\Sigma\tau = 0$ (b) 0.19 (c) 3.5 m/s
39. (a)
41. The moment of inertia depends on how mass is distributed about an axis. Physically, it means that under a constant torque, the angular acceleration depends on the location of the axis of rotation.
43. When the paper is pulled quickly (a large force is required to accelerate the roll), the force the paper can provide is not great enough to accelerate the paper roll. However, if the paper is pulled slowly, the paper is strong enough to accelerate the roll because the force required is smaller. The amount of paper on the roll affects the results: The more paper on the roll, the easier to tear.
45. (a) 22.5 kg·m² (b) 62.5 kg·m² (c) 85.0 kg·m²
47. 0.64 m·N
49. (a) 8.0 kg·m² (b) 7.5 kg·m² (c) yes
51. 1.1 rad
53. 3.5 times
55. 1.2 m/s²
57. 2.4 N
59. (a) 1.5g (b) 67 cm position
61. 6.5 m/s²

A-17

63. $\theta = \tan^{-1}\dfrac{7\mu_s}{2}$
65. (c)
67. (a) 28 J (b) 14 W
69. 0.47 m·N
71. 0.16 m
73. cylinder goes higher by 7.1%
75. (a) 1.31×10^8 J (b) 1.46×10^6 W
77. (a) 29% (b) 40% (c) 50%
79. (a) $v = \sqrt{gR}$ (b) $h = 2.7R$
(c) weightlessness
81. Walking toward the center decreases the moment of inertia and so increases the rotational speed.
83. The arms and legs are put into these positions to decrease the moment of inertia, thereby increasing the rotational speed.
85. The cat manipulates its body to change the moment of inertia to rotate or flip over. It twists part of the body one way and then rotates the feet the other way.
87. 1.4 rad/s
89. $L_{\text{rot}} = 2.4 \times 10^{29}$ kg·m^2/s; $L_{\text{rev}} = 2.8 \times 10^{34}$ kg·m^2/s
91. $T_2 = 3.3T_1$
93. (a) 4.3 rad/s (b) $K = 1.1K_o$ (c) the work done by the skater in tucking her arms
95. (a) The lazy Susan will rotate in the opposite direction. (b) 0.56 rad/s (c) no, 2.1 rad; the Earth would have to rotate slower in the eastward direction. This would result in a longer day.
97. 52.4 J
99. (a) 4.9 m/s^2 (b) 9.8 rad/s^2
101. zero
103. (a) 2.5 cm (b) 36 cm
105. 1.3 rad/s
107. $T_1 = 21$ N; $T_2 = 15$ N
109. smaller side; 1.3 times

Chapter 9

1. (c)
3. It will have less strain for a given stress or greater stress for a given strain.
5. N/m
7. 0.020
9. 1.1×10^{11} N/m^2
11. 47 N
13. 3.92×10^{-5} m
15. 2.0×10^{-7} m
17. 6.7×10^3 N/m^2
19. (a) Ethyl alcohol has the smallest B (greatest compressibility).
(b) $p_w/p_{ea} = 2.2$
21. 5.4×10^{-5}
23. (c)
25. When the bowl is full, the atmospheric pressure on the water does not allow water out of the bottle. When the water level in the bowl drops below the neck of the bottle, air bubbles in and water flows out until the pressures are equalized again. No.
27. Bicycle tires have a much smaller contact area with the ground, so they need a higher pressure to balance the weight of the bicycle and the rider.
29. 10 m; no
31. 22 cm
33. 6.39×10^{-4} m^2
35. 1.6×10^4 N = 3600 lb
37. The air density is not a constant but decreases rapidly with altitude.
39. 1.07×10^5 Pa
41. (a) 1.1×10^8 Pa (b) 1.9×10^6 N
43. 470 N
45. 2.6 mm
47. (a)
49. There is no change. As the ice melts, the volume of the newly converted water decreases; however, the ice that was initially above the water surface is now under the water. This compensates for the decrease in volume. It does not matter whether the ice is hollow.
51. 10 000 metric tons
53. (a) water displacement (b) no
55. 33 N; same
57. no
59. 0.33 m
61. 17.7 m
63. no
65. (a)
67. The water on the inside of the curtain increases the speed of the air inside. This in turn decreases the pressure inside, according to Bernoulli's principle. The pressure difference moves the curtain inward.
69. (a) The air flow above the paper decreases the pressure there. This creates a pressure difference, and a lift force results. (b) The egg is kept aloft by the pressure of the air coming out of the end of the tube. As the eggs moves to one side, a change in the flow speed around the egg creates an inward pressure that makes the egg move back to midstream.
71. 4
73. (a) 3.5 cm^3/s (b) 0.031% (c) The slow speed allows time for the exchange of substances such as oxygen between the blood and the tissues.
75. (a) 0.99 m/s; 1.7 m/s; 2.2 m/s; 2.6 m/s (b) 0.44 m, from $y = 20$ cm
77. 2.2 Pa
79. (b)
81. (a)
83. 2.0×10^2 Pa
85. 99.99 cm
87. (a) 0.09 m (b) 8.1 kg
89. 1.32×10^5 N = 2.96×10^4 lb
91. 8.5×10^2 kg/m^3
93. 0.45 m
95. 2.2 m/s
97. (a) $v_s = 11$ cm/s; $v_n = 66$ cm/s (b) 1.2×10^3 Pa
99. (a) 3.3×10^4 N (b) 3.8×10^4 N

Chapter 10

1. (d)
3. Not necessarily; internal energy U does not depend solely on temperature. It also depends on mass.
5. (a) 538°C (b) −18°C (c) −29°C (d) −40°C
7. 77.8°C
9. 56.7°C; −62°C
11. 136°F; −128°F
13. (a) −101 F° (b) 27 C°
15. (c)
17. (a) increases (b) constant
19. The pressure of the gas is held constant. If the temperature increases, so does the volume and vice versa, according to the ideal gas law. Therefore temperature is determined by volume.
21. The balloons collapsed.
23. (a) −273°C (b) −23°C (c) 0°C (d) 52°C
25. 300 K is lower
27. (a) 2.2 mol (b) 2.5 mol (c) 3.0 mol (d) 2.5 mol
29. 1.2×10^5 Pa
31. 0.0247 m^3
33. increases by 4 times
35. 33.4 lb/in^2
37. (a) expand (b) 10.6%
39. (d)
41. (a) The ice moves upward. (b) The ice moves downward. (c) copper
43. Water expands when cooled from 4°C to 2°C as it exhibits abnormal expansion between 0°C and 4°C.
45. no; yes
47. 0.06%
49. 0.0027 cm
51. (a) 60.07 cm (b) 3.91×10^{-3} cm^2; yes, the flow speed will be affected.
53. ring; 353°C
55. 0.48 gal; morning
57. 13.4×10^3 kg/m^3
59. (a) 1.05×10^3 °C (b) no
61. (c)
63. The gases diffuse through the porous membrane, but the helium gas diffuses faster because it has a smaller mass. Eventually there will be equal concentrations of the two gases on both sides of the container.
65. (a) 6.1×10^{-21} J (b) 7.7×10^{-21} J
67. (a) 6.21×10^{-21} J (b) 1.37×10^3 m/s
69. O_2 by 1.22 times
71. 273°C
73. (b)
75. 6.1×10^3 J
77. 0.46 gal
79. 6.8×10^{-4} m
81. 7.8 cm
83. 33 C°
85. (a) 3.3×10^{-2} mol (b) 2.0×10^{22} molecules (c) 0.94 g
87. (a) 18 F° (b) 5.6 C°
89. 9.9×10^{20}
93. 3.3 cm
95. 2.026×10^5 Pa

Chapter 11

1. (d)
3. 2.54×10^5 J
5. 5.86×10^3 W
7. (a) 60 000 times (b) 83 h

9. Water has a high specific heat, so it takes longer to cool off than does the air.
11. Yes, a negative heat corresponds to removal of heat.
13. only on the temperature difference
15. (a) copper (b) 2.1×10^4 J
17. 1.7×10^6 J
19. 0.13 kg
21. 0.18 C°
23. 55 L
25. (a) 3.1×10^2 J/(kg·C°) (b) The final temperature will be higher, and the measured specific heat value will be in error and higher than the accepted value.
27. 21.5°C
29. (d)
31. Different substances have different internal energies, different molecular structures, and different intrinsic heat values. These quantities affect the temperature at which phase changes take place due to the different effects of the addition or removal of heat on different substances. Latent heats will also be different for different substances because of different molecular structures, or bonds. The latent heat energy goes into breaking these bonds.
33. The latent heat of vaporization of water is high; the condensing of steam on the skin releases 2.26×10^6 J of heat.
35. 1.84×10^6 J more
37. 8.0×10^4 J
39. 4.1×10^3 J
41. 0.81 kg
43. 11°C
45. 0.94 L or 0.94 kg
47. 4.7×10^{14} J; yes
49. 0.17 L
51. convection
53. (a) It convects the heat from the hot soup to the cooler air. (b) No; the ice blocks air flow and cooling of the air. Also, ice is a poor conductor.
55. They increase the surface area for better conduction and radiation.
57. The double-walled and partially evacuated container counteracts conduction and convection because both processes depend on a medium to transfer the heat. (The double walls are more for holding the partially evacuated region than for reducing conduction and convection.) The mirrored interior minimizes radiation.
59. 4.5 times
61. 4.54×10^6 L
63. (a) 5.5×10^5 W (b) 73 kg; not reasonable
65. 7.8 h
67. (a) 4.9 in. (b) 6.9 in.
69. 1.7
71. 2.3 cm
73. 7.8×10^5 J
75. 1.4×10^8 J
77. 65°C
79. 88.7°C
81. 1.9 g; heat loss to the skates and to the environment
83. Al by 1.8×10^4 J
85. 2.3×10^6 J

Chapter 12

1. (c)
3. (a)
5.

7. (b)
9. (d)
11. This is an adiabatic compression. When the plunger is pushed in, the work done goes into the increase of the internal energy of the air. The increase in internal energy increases the temperature of the air and causes the paper to catch fire.
13. (a) 400 J added (b) $\Delta U = 0$, so T constant
15. (a) decrease (b) zero (c) -500 J
17. (a) 3.3×10^3 J (b) yes, -2.3×10^3 J
19. (a) zero (b) $-p_1V_1$ (on the system) (c) $-p_1V_1$ (out of the system)
21. 2.09×10^3 J
23. 3.6×10^4 J
25. (a) AB: -1.66×10^3 J; BC: zero; CD: 3.31×10^3 J; DA: zero (b) $\Delta U = 0$, $Q = W = 1.65 \times 10^3$ J (c) 800 K
27. (b)
29. Energy must be created for the change in entropy to be negative.
31. 1.2×10^3 J/K
33. -2.1×10^2 J/K
35. -26 J/K
37. 126°C
39. 1.33 J/K
41. (a) 2.73×10^4 J (b) adiabatic or isentropic
43. (a) 61.0 J/K (b) -57.8 J/K (c) 3.2 J/K
45. (b)
47. Heat could be completely converted to work for a single process (not a cycle), such as an isothermal process of an ideal gas.
49. No, the warm air rises to the higher altitude, and gravity and buoyancy do work. Since it is a natural process with work input, the entropy increases and the second law is not violated.
51. 25%
53. 1.47×10^5 J
55. 40%
57. (a) 6.6×10^8 J (b) 27%
59. (a) 6.1×10^5 J (b) 1.9×10^6 J
61. 3.0 kW
63. (a) 2.4×10^5 J (b) 10 C°
65. (a) 5.6×10^5 W (b) 4.5×10^5 W
67. (a)
69. (a) No, this change does not make the cycle a Carnot cycle because a Carnot cycle consists of two adiabatic and two isothermal processes. (b) There are heat transfers for all four legs. There are heat inputs for Q_3 and Q_4 and heat outputs for Q_1 and Q_2. The transfers at legs 1 and 3 occur at constant temperatures as the processes are isothermal.
71. Diesel engines run hotter because diesel fuel burns at a higher temperature. According to Carnot efficiency, the higher the hot reservoir temperature, the higher the efficiency.
73. 0°C
75. (a) 6.7% (b) Probably not, due to the poor return; fossil fuels are much better.
77. 9.1×10^3 J
79. 100 C°
81. raising the high-temperature reservoir's temperature
83. (a) 50% (b) 39 kW
85. (a) 25% (b) $T_{hot} = (4/3)T_{cold}$
87. (a) 64% (b) ϵ_c is the upper limit of the efficiency. In reality, much more energy is lost than is ideally predicted.
89. (a) 13 (b) no, COP = 11
95. (a) isobaric expansion (b) 146 J
97. 3.2×10^3 J
99. 0.10 J/K
101. 3500 J
103. The process is an adiabatic free expansion. There is no transfer of heat, and the gas does no work. So the internal energy of the gas remains unchanged according to the first law of thermodynamics.

Chapter 13

1. (b)
3. (a) four times as large (b) twice as large
5. (b)
7. $4A$
9. 0.025 s
11. 41 N/m
13. (a) 10^{-12} s (b) 63 m/s
15. (a) $x = \pm A$ (b) $x = 0$ (c) $x = \pm A$
17. (a) 1.0 m/s (b) at the equilibrium position (c) 0.90 m/s
19. decreases by a factor of $1/\sqrt{2}$
21. (b) 1.8×10^2 times more
23. (a) 2.5 m/s (b) 2.5 m/s (c) 2.7 m/s, equilibrium position
25. (a) 0.14 m (b) 10.0 cm (original position)
27. $\sqrt{2}$ times as large
29. No; the tangent goes to ∞.
31. 1.0 kg
33. (a) 2.0 s (b) 0.50 Hz
35. (a) $x = A \sin \omega t$ (b) $x = A \cos \omega t$
37. (a) 5.0 cm (b) 10 Hz (c) 0.10 s
41. 9.787 m/s^2
43. (a) (8.0 cm) cos (6.3 rad/s)t (b) 90°, or $\pi/2$ rad (c) -8.0 cm, or at the other amplitude
45. increase by 2.4 times
47. (5.0 cm) cos ($\pi t/4$)
51. (d)
53. (a) transverse and longitudinal (b) longitudinal (c) longitudinal
55. 0.34 m
57. 1.5 m/s
59. 4.7×10^{14} Hz
61. (a) 188 m to 545 m (b) 2.70 m to 3.41 m
63. 3.0 m/s

65. (a) 3.8×10^2 s **(b)** yes **(c)** P waves take 1.6×10^3 s; S waves do not go through the liquid core.
67. (a) 0.20 s **(b)** 2.0 s
69. (b)
71. reflection
73. (d)
75. At a frequency of $f_o/2$ (or a period of $2T_o$), the swing is pushed every other oscillation. So only half the energy is going into the swing, but it is pushed smoothly.
77. (a) 200 Hz **(b)** 300 Hz
79. (a) 6.0 m **(b)** 2.0 m
81. $n = 5$
83. 10 Hz, 20 Hz, 30 Hz, and 40 Hz
85. 503 Hz
87. 0.22 kg, 0.055 kg, 0.024 kg, 0.014 kg
89. (a) 12 cm **(b)** 3.0 cm, 9.0 cm, 15 cm
91. 5.0 m/s; 2.5×10^2 m/s²
93. 210 Hz
95. (a) Since $v = \sqrt{\dfrac{F_T}{\mu}}$, doubling the tension will make the speed $\sqrt{2}$ times as large. The wavelength is unchanged, as it is determined by the length of the string and the mode of vibration. **(b)** The frequency is increased by a factor of $\sqrt{2}$ due to the speed increase.
97. 1.1 cm/s
99. 3.0 s

Chapter 14

1. (b)
3. (a)
5. Some insects produce sounds that are not in our audible range.
7. They arrive at the same time.
11. 32°C
13. 1.5 km
15. (a) 1.29 m **(b)** 1.34 m
17. (a) 0.81 s **(b)** 0.78 s
19. −1.75%
21. 583 m
23. −4.5%
25. (b)
27. yes, an intensity below the threshold intensity
29. four times as large
31. (a) 0 dB **(b)** 120 dB
33. 20 dB
35. (a) 10^{-7} W/m² **(b)** 10^{-3} W/m²
37. (a) 63 dB **(b)** 83 dB **(c)** 113 dB
39. (a) 10^{-3} W/m², 10^{-8} W/m² **(b)** $I_M/I_L = 10^5$
41. 10 bands
43. $I_B = 0.56 I_A$; $I_C = 0.25 I_A$; $I_D = 0.17 I_A$
45. (a) 3.2×10^{-3} W/m² **(b)** 16
47. 10^5 bees
49. (b)
51. (a) no **(b)** increasing sound frequency
53. The varying sound intensity is caused by the interference effect. At certain locations, interference is constructive; at other locations, it is destructive.
55. 0.172 m

57. 267 Hz and 261 Hz
59. 834 Hz
61. 3.3 Hz
63. 90°
65. (a) 1.74 **(b)** 555 m/s
67. 28 m/s
69. (a)
71. The spacing (change in length) depends not only on the frequency difference Δf, but also on the values of the frequencies themselves. For different frequencies, ΔL is therefore different.
73. As the level of water increases in the bottle, the length of the air column above the water decreases. This decrease in length of the air column decreases the wavelength and increases the frequency.
75. 510 Hz
77. (a) f_2 does not exist, only odd harmonics in this case **(b)** 322 Hz
81. 0.194 m; 0.583 m; 0.972 m
83. destructively
85. increases by 1.44 times
87. speed of sound
89. 7.5×10^{-3} J
91. 38 vibrations
93. approaching: 755 Hz; moving away: 652 Hz
95. 1.1 m

Chapter 15

1. (c)
3. (a) We know there are two types of charges because attractive and repulsive forces can be produced by different combinations of just two types of charges. **(b)** no effect
5. decrease; increase
7. 3.1×10^{14} electrons
9. (a) $+4.8 \times 10^{-9}$ C **(b)** 2.7×10^{-20} kg
11. negative
13. No, the wall is neutral but polarized.
15. The spheres can be charged through polarization by induction. If you bring a positively charged object near one of the two spheres, the sphere near that object will have a net negative charge, and the other sphere will have a net positive charge (polarization by induction). Then separate the two spheres (with the positively charged object nearby), and they will have opposite charges according to charge conservation.
17. (d)
19. Both the Sun and the planets are electrically neutral.
21. (a) same **(b)** 1/4 as large **(c)** 1/2 as large
23. (a) 5.8×10^{-11} N **(b)** zero
25. 2.24 m
27. (a) 50 cm **(b)** 50 cm
29. (a) $x = 0.25$ m **(b)** nowhere **(c)** $x = -0.94$ m for $\pm q_3$
31. 2.3×10^{39}
33. (a) 96 N at 39° below the $+x$ axis **(b)** 61 N at 84° above the $-x$ axis
35. (b)
37. (b)

39. It is determined by the relative density or spacing of the field lines. The closer the lines, the greater the magnitude.
41. Since the charge is positive, the electric field inside the sphere is not zero. Field lines run radially outward to the sphere's inside surface and stop at the induced negative charges on that surface. The lines reappear on the outside surface (positively charged) and continue radially outward as if emanating from the central point charge. If the charge were negative, the field lines would reverse direction.
43. 2.0×10^5 N/C
45. 1.2×10^{-7} m
47. 1.0×10^{-7} N/C upward; 5.6×10^{-11} N/C downward
49. $\mathbf{E} = (2.2 \times 10^5 \text{ N/C})\hat{\mathbf{x}} - (4.1 \times 10^5 \text{ N/C})\hat{\mathbf{y}}$, or 4.6×10^5 N/C at 61° below the $+x$ axis
51. 5.4×10^6 N/C toward the charge of $-4.0\ \mu$C
53. 3.8×10^7 N/C in the $+y$ direction
55. 15 μC/m²
57. $\mathbf{E} = (-4.4 \times 10^6 \text{ N/C})\hat{\mathbf{x}} + (7.3 \times 10^7 \text{ N/C})\hat{\mathbf{y}}$
59. (b)
61. no
63. (a) zero **(b)** $+Q$ **(c)** $-Q$ **(d)** $+Q$
65. (a) zero **(b)** $\dfrac{kQ}{r^2}$ **(c)** zero **(d)** $\dfrac{kQ}{r^2}$
67.

69. (a)
71. The net charges are equal and opposite in sign.
73. 10 field lines entering (negative)
75. no, just more positives than negatives: net positive
77. 3.1×10^{12} electrons
79. (a) 1.6×10^{-7} s **(b)** $(-0.25$ m, 0$)$
81. $+3.0\ \mu$C on the $-y$ axis or $-3.0\ \mu$C on the $+y$ axis
83.

85. If d is zero, the positive and negative charges overlap, so the electric field is zero.
87. It will still rotate, but it will be drawn into the region where the field is highest, negative end first.
89. 4.9 mm

Chapter 16

1. (d)
3. (a)
5. Approaching a negative charge means moving toward a region of larger negative

potential values—that is, losing potential. Positive charges tend to move toward regions of lower potential, thereby losing potential energy and gaining kinetic energy (speeding up).
7. The electron will lose potential energy.
9. (b)
11. (a) 2.7 μC (b) positive to negative
13. 1.4×10^6 m/s
15. (a) 1.3×10^6 m/s down (b) It loses potential energy.
17. (a) 0.90 m (b) -6.7×10^3 V (-6.7 kV), a decrease
19. (a) $+6.2 \times 10^{-19}$ J (b) -6.2×10^{-19} J (c) $+4.8 \times 10^{-19}$ J
21. 1.1 J
23. (a) $+0.27$ J (b) no
25. -0.72 J
27. 3.1×10^5 V
29. -1.3×10^6 V
31. (a) 5.9×10^7 m/s (b) 5.9×10^{-9} s
33. (a)
35.

37. yes
39. (a)
41. It increases to a larger positive value.
43. An electron volt is $e\,\Delta V = q\,\Delta V = \Delta U$. It is also the kinetic energy gained by an electron that goes through a potential difference of 1 V. So, it is a unit of energy. And 1 GeV is larger than 1 MeV by 1000 times.
45. Equipotential surfaces cannot cross. Since electric field lines are perpendicular to the equipotential surfaces, electric field lines cannot cross either.
47. 1.0 cm
49. $+6.0$ V
51. 12.6 m
53. $+300$ eV $= +4.8 \times 10^{-17}$ J
55. (a) 2.0×10^7 eV (b) 2.0×10^4 keV (c) 20 MeV (d) 2.0×10^{-2} GeV (e) 3.2×10^{-12} J
57. 6.2×10^7 m/s (proton); 4.4×10^7 m/s (alpha particle)
59. (a) 3.5 V; 1.1×10^6 m/s (b) 41 kV; 3.8×10^7 m/s (c) 5.0 kV; 4.2×10^7 m/s
61. $+2.8$ V
63. (b)
65. 2.4×10^{-5} C
67. 0.71 mm
69. (a) 4.2×10^{-9} C (b) 2.5×10^{-8} J
71. 2.2 V
73. (d)
75. The charged rubber rod polarizes the stream of water, and the stream is attracted to it.
77. 3.1×10^{-9} C; 3.7×10^{-8} J
79. 2.4
81. (b)
83. equal capacitance
85. 6.0 μF

87. (a) 3.0 μC (b) 9.0 μC
89. max. 6.5 μF; min. 0.67 μF
91. (a) 2.4 μC for C_1, 2.4 μC for C_2, 1.2 μC for C_3, 3.6 μC for C_4 (b) 6.0 V for all the capacitors
93. $-\dfrac{kq^2}{d}$
95. -1.1×10^2 eV
97. (a) 8.4×10^{-13} V (b) 1.5×10^{-9} V (c) In (a), the top is $+$; in (b) top is $-$.
99. 2.6×10^{-16} J

Chapter 17

1. (b)
3. (a)
5. As the internal resistance increases, the voltage across it also increases. This decreases the terminal voltage.
7. (a) 3.0 V (b) 1.5 V
9. (a) 24 V (b) the two 6.0-volt batteries in series, combined in parallel with the 12-volt battery
11. (d)
13. 8.3 min
15. 4.7×10^{20} electrons
17. 3.6×10^5 C
19. (a) 5.9×10^{16} protons (b) 1.9×10^5 J/s
21. (d)
23. (a) 11.4 V (b) 0.32 Ω
25. (a) same (b) 1/4 the current
27. Aluminum wire is thicker than copper by 29%
29. 1.0 V
31. 1.3×10^{-2} Ω
33. 3.0×10^{19} electrons/s
35. The shorter wire carries 4 times the current.
37. 0.13 Ω
39. 4.6 mΩ
41. not ohmic
43. It increases by 1.6 times.
45. 8.2°C to 32°C
47. 0.77 A
49. (d)
51. (d)
53. The wire in a 60-watt bulb would be thicker.
55. 2.0×10^3 W
59. 18 min
61. (a) 4.3×10^3 J (b) 13 Ω
63. 58 Ω
65. (a) 0.60 kWh (b) $0.09
67. (a) 0.15 A (b) 1.4×10^{-4} $\Omega \cdot$m (c) 2.3 W
69. (a) 1.1×10^2 J (b) 6.8 J
71. 21 Ω
73. $R_{120}/R_{60} = 4/3$
75. $151
77. 64 C°
79. (a) 0.33 A (b) 3.6×10^2 Ω
81. It increases to 1.78 kW.
83. copper at 117 °C or aluminum at -73°C
85. 3.5×10^6 J
87. (a) 5.0×10^{-2} A (b) 5.0 V
89. (a) 4.0×10^2 Ω (b) 4.0 W (c) 4.8×10^2 J
91. 1.1×10^{22} electrons

93. Since $P = VI$, a higher voltage lowers the current if P is kept constant. The resistance of the transmission lines is a fixed number. With lower current, the power loss (I^2R loss) on the lines is reduced: $P_{\text{loss}} = I^2R = \dfrac{P^2}{V^2} R \propto \dfrac{1}{V^2}$. So if V is raised by a factor of 10, P_{loss} will be reduced by 1/100.
95. 6.6×10^{-6} m/s

Chapter 18

1. (b)
3. No, not generally; however, if all resistances are equal, the voltages across them are also the same.
5. The 5-ohm resistor gets the most power, because it gets the most voltage and all resistors have the same current.
7. (a) series: 60 Ω (b) parallel: 5.5 Ω
9. 30 Ω
11. (a) 30 Ω (b) 0.30 A (c) 1.4 W
13. (a) 0.57 Ω (b) 6.0 V (c) 9.0 W
15. 3.0 $\mu\Omega$
17. (a) 1.0 A (b) 1.0 A (c) 2.0 W; 4.0 W; 6.0 W (d) $P_{\text{sum}} = P_{\text{total}} = 12$ W
19. $I = 1.0$ A for all; $V_{8.0} = 8.0$ V; $V_{4.0} = 4.0$ V
21. 2.7 Ω
23. (a)

(b) 31
25. 0.25 A, 2.5 V
27. (a) 1.0 A; 0.50 A; 0.50 A (b) 20 V; 10 V (c) 30 W
29. no
31. 100 s = 1.7 min
33. (a) 0.085 A (b) 7.0 W; 2.6 W; 0.24 W; 0.41 W
35. 35 W
37. $I_1 = 2.7$ A; $I_2 = 3.3$ A; $I_3 = I_4 = 0.55$ A
39. (d)
41. positive
45. $I_1 = 1.0$ A; $I_2 = I_3 = 0.50$ A
47. $I_1 = 0.33$ A left; $I_2 = 0.33$ A right
49. $I_1 = 3.75$ A up; $I_2 = 1.25$ A left; $I_3 = 1.25$ A right
51. 0.664 A left; 0.786 A right; 1.450 A up; 0.770 A down; 0.016 A down; 0.664 A down
53. (c)
55. (a) $V_R = V_o$, $V_C = 0$ (b) $V_R = 0.37V_o$, $V_C = 0.63V_o$ (c) $V_R = 0$, $V_C = V_o$
57. 2.0 MΩ
59. (a) 1.50 MΩ (b) 11.4 V
61. (a) 9.4×10^{-4} C (b) $V_C = 24$ V; $V_R = 0$
63. (a) 0 V; 2.0 μA (b) 1.7×10^{-6} C
65. (c)
67. (a) An ammeter has very low resistance, so if it were connected in parallel in a

circuit, the circuit current would be very high and the galvanometer could burn out. **(b)** A voltmeter has very high resistance, so if it were connected in series in a circuit, it would read the voltage of the source, because it has the highest resistance (most probably) and therefore the largest voltage drop among the circuit elements.
69. 7.4 kΩ
71. 50 kΩ
73. 0.20 mA
75. (b) If $R_a = 0$, $R = \dfrac{V}{I}$ (that is, the measurement is "perfect").
77. (c)
79. Since current is caused by voltage, a high voltage can produce high "harmful" current even though the resistance of a body is high.
81. It is safer to jump. If you step off the car one foot at a time, there will be a high voltage between your feet. If you jump, the voltage between your feet is zero, because both feet will be at the same potential all the time.
83. 75 Ω
85. 6.0 Ω
87. 8.1 J/s
89. (a) 1.0 A; 0.40 A; 0.20 A; 0.40 A **(b)** 100 W; 4.0 W; 2.0 W; 4.0 W
91. 0.440 A; 0.323 A; 0.117 A; 0.0878 A; 0.0293 A
95. 11.8 A
97. (a) 4.47 V **(b)** 0.447 A
99. 10 mΩ, 2.0 mΩ, and 1.0 mΩ

Chapter 19

1. (e)
3. (c)
5. Similarities: Two kinds of poles, north and south (charges, positive and negative); like poles (charges) repel, and unlike poles (charges) attract. Differences: Poles come in pairs (single charge can exist).
7. (b)
9. They are either parallel or opposite.
11. (a) The bottom half would have a magnetic field directed into the page, and the top half would have a magnetic field directed out of the page. **(b)** the same
13. 2.5×10^3 m/s
15. 2.0×10^{-14} T to the left, looking in the direction of the velocity
17. (a) 3.8×10^{-18} N **(b)** 2.7×10^{-18} N **(c)** zero
19. (a) 8.6×10^{12} m/s^2; horizontal toward right, looking in direction of velocity. **(b)** same magnitude but in opposite direction
21. (b)
23. (b)
25. 0.25 m
27. 2.4 A
29. both 2.9×10^{-6} T
31. 1.4×10^{-5} T at 38° below a horizontal line to the left

33. 1.9×10^{-5} T toward observer (up)
35. 1.0×10^{-4} T away from observer
37. 3.2×10^{-2} A
39. 4.2×10^{-5} T
41. (a) $f = \dfrac{qB}{2\pi m}$ **(b)** $T = \dfrac{2\pi m}{qB}$, independent of m and v **(c)** 5.7×10^{-3} m; $f = 2.8 \times 10^6$ Hz
43. (d)
45. away from you
47. 12 T
49. attract for currents in the same direction, repel for opposite currents
51. yes, with the plane of the loop perpendicular to the magnetic field
53. 1.2 N perpendicular to the plane of current and field
55. 5.0×10^{-3} T north to south
57. (a) zero **(b)** 4.0 N/m in $+z$ direction **(c)** 4.0 N/m in $-y$ direction **(d)** 4.0 N/m in $-z$ direction **(e)** 4.0 N/m in $+y$ direction
59. 0.40 N/m in $+z$ direction
61. 6.7×10^{-6} N/m, attractive
63. 0.75 N upward
65. 2.7×10^{-5} N/m toward wire 1
67. Left segment: 2.5 N to the left; right segment: 2.5 N to the right; top segment: 7.5 N upward
69. zero, yes
71. (d)
73. Pushing the button in both cases completes the circuit. The current through the wires activates the electromagnet. Doorbell: This causes the clapper to be attracted and ring the bell. However, this breaks the armature contact and opens the circuit. Holding the button causes this to repeat, so the bell rings continuously. Door chimes: When the circuit is completed, the electromagnet attracts the core and compresses the spring. Inertia causes it to hit one tone bar, and the spring force then sends the core in the opposite direction to strike the other bar.
75. 2.0×10^5 m/s
77. (a) 1.0×10^5 m/s **(b)** the same **v**, independent of charge
79. (a) 4.8×10^{-26} kg **(b)** 2.4×10^{-18} J **(c)** no, work equals zero
81. 2.9×10^{-2} T
83. (b)
85. north magnetic pole
87. 0.44 T
89. 1.0×10^{-14} J
91. (a) 1.0 m **(b)** 1.5×10^{-5} s
93. 5.6×10^{-7} m/s
95. 3.0×10^{-2} m·N

Chapter 20

1. (d)
3. (a) When the bar magnet enters the coil, the needle deflects to one side. When the magnet leaves the coil, the needle reverses direction.

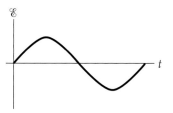

(b) no
5. counterclockwise (in head-on view).
7. Sound waves cause the resistance of the button to change as described. This changes the current, so the sound wave produces electrical pulses that travel through the phone lines and to a receiver. The receiver has a coil wrapped around a magnet. The pulses create a varying magnetic field as they pass through the coil, causing the diaphragm in the receiver to vibrate in air and producing sound waves.
9. (a) zero **(b)** 1.9×10^{-2} T·m^2 **(c)** 1.4×10^{-2} T·m^2
11. 3.3×10^{-2} T·m^2
13. 1.3×10^{-6} T·m^2
15. 1.6 V
17. 0.30 s
19. 0.35 V
21. (a) No, the emf is zero. **(b)** near the Earth's magnetic poles
23. (a) 0.60 V **(b)** zero
25. 4.0 V
27. 3.0×10^{-4} J
29. (c)
31. As it moves up and down in the coil, the magnet will induce a current in the coil, which will light the bulb. However, the magnet produces the current at the expense of the kinetic energy and potential energy. The magnet's motion will damp out rather rapidly (Lenz's law).
33. (a) 0.057 V **(b)** 0.57 V
35. ±104 V; the initial voltage direction was not specified, so there are two possible directions.
37. (a) 100 V **(b)** zero
39. 16 Hz
41. 100 V
43. (a) 216 V **(b)** 160 A **(c)** 8.1 Ω
45. (c)
47. The dc voltage must first be converted to changing voltage. This is achieved by the points in older cars and electronic devices in newer cars. The coil functions as a step-up transformer to give high-voltage pulses. The distributor distributes these pulses to different spark plugs to ignite the gas–air mixture.
49. (a) 16 **(b)** 5.0×10^2 A
51. 24:1
53. (a) 17.5 A **(b)** 15.7 V
55. (a) no **(b)** 45%
57. (a) $N_s/N_p = 1:20$ **(b)** 2.5×10^{-2} A
59. (a) 128 kWh **(b)** $1840
61. (a) $N_s/N_p = 1:2$; 1:14; 1:30 **(b)** 2.0; 14; 30 **(c)** 833
63. (a) 53 W **(b)** $N_p/N_s = 200$

65. (b)
67. Infrared (heat) radiation is absorbed by clouds (water molecules), but the sunburning ultraviolet radiation is not.
69. AM: 1.8×10^2 m to 5.7×10^2 m; FM: 2.8 m to 3.4 m; TV: 0.34 m to 5.6 m
71. 2.6 s
73. AM: 67 m; FM: 0.77 m
75. (a) 30 A (b) 2.5 A
77. 0.60 m
79. once every 0.01 s
81. (a) up (b) 25 mA
83. 0.75 T·m^2
85. 4.0 T/s
87. 0.33 A and 30 A
89. (a) 50 rotations/s (b) No, if they are ideal (without losses), the same turn ratios would work. In reality, the 50-hertz delivery would have fewer losses due to eddy currents because these currents would be generated less frequently.

Chapter 21

1. (c)
3. The voltage and current reach maximum at the same time, reach minimum at the same time, are zero at the same, and so on.
5. 7.1 A
7. 1.2 A
9. (a) 10 A (b) 14 A (c) 12 Ω
11. (a) 4.5 A; 6.3 A (b) 112 V; 158 V
13. $V = (170\ V)\sin 119\pi t$
15. 0.33 A; 0.47 A
17. 2.4×10^2 W
19. (a) 60 Hz (b) 1.4 A (c) 1.2×10^2 W
(d) $V = (120\ V)\sin 380t$
(e) $P = (240\ W)\sin^2 380t$
21. (b)
23. For a capacitor, the *lower the frequency*, the longer the charging time in each cycle. A longer charging time means more charge accumulation on the plates and therefore *more* opposition to the current. For an inductor, the *lower the frequency*, the more slowly the current in the inductor changes, the smaller the rate of change in its magnetic flux, and thus the *smaller* the self-induced emf that opposes the current.
25. In an ac circuit, a capacitor can oppose current, because as the capacitor charges, the voltage across its plates increases. Also, an inductor can oppose current, because the induced emf opposes the change in flux.
27. 2.65×10^{-5} F
29. 1.3×10^3 Ω
31. 2.3 A
33. 60%, an increase
35. 255 Hz
37. 0.70 H
39. (d)
41. (d)
43. (a) 1.7×10^2 Ω (b) 2.0×10^2 Ω
45. (a) 38 Ω; 1.1×10^2 Ω (b) 1.1 A

47. $+27°$
49. 72 Ω
51. 50 W
53. 5.3×10^{-11} F
55. ab: 1.3 A; ac: 1.2 A; bc: 4.0 A; cd: 1.8 A; bd: 1.6 A; ad: 2.9 A
57. 13 A
59. $(V_{rms})_R = 12$ V; $(V_{rms})_L = 2.7 \times 10^2$ V; $(V_{rms})_C = 2.7 \times 10^2$ V
61. (a) 362 Ω (b) no; 1.1×10^2 Hz
63. 30%
65. 0.71 A
67. 1.7×10^{-12} F
69. 37°
71. The capacitor opposes current more strongly at lower frequencies, and the inductor opposes current more strongly at high frequencies. In Fig. 21.18a, the inductor in series with R_L filters out the high-frequency current, so only the low-frequency current reaches R_L. In Fig. 21.18b, the capacitor in series with R_L filters out the low-frequency current, so only the high-frequency current reaches R_L.

Chapter 22

1. (c)
3. (d)
5. They are visible because of the diffuse reflections off the particulate matter in the air.
7. 70°
9. 47°
11. 40°
13. 27°
15. 12 m
17. $\alpha = 90°$, any θ_{i_1}
19. (d)
21. (b)
23. The angles of refractions are different for air–glass and for water–glass interfaces.
25. The laser beam has a better chance to hit the fish. Due to refraction, the fish appears to the first hunter at a location different from its true location. The laser beam obeys the same law of refraction and retraces the light by which the second hunter sees the fish.
27. 26% greater in zircon
29. 27°
31. (a) 24° (b) diamond to air
33. 47°
35. 6.5×10^{14} Hz; 2.8×10^{-7} m
37. 16/15
41. (a) 1.41 (b) 1.88
43. (a) 4.2 cm (b) 3.2 cm
45. 75%
47. seen for 40° but not for 50°
49. 43°
51. (a) no, $\theta_c = 38.7°$ (b) transmitted, $\theta_c = 48.6°$
53. 2.0 m
55. (a) 12.5° (b) 26.2°
57. (b)
59. No, the light will be further dispersed by the second prism.

61. (a) The angle of incidence is approximately zero. (b) No (no); the speeds are different.
63. Red is transmitted, and blue is internally reflected.
65. (a) 21.7° (b) 0.22° (c) 0.37°
67. (a) 58° (b) 30°
69. 1.64
71. 0°; not refracted

Chapter 23

1. (b)
3. (a) During the day, light passes both ways through the pane, so it is difficult to see the reflection due to the light coming through. At night, there is little light coming through the pane, so the reflections are much more clear. (b) The two images are due to reflections on both sides of the pane of glass, producing two similar images. (c) It is a combination of half-silvering and bright light on one side and dark on the other.
5. (a) The left–right reversal is an apparent one, caused by the front–back reversal. Right and left are directional senses (like clockwise and counterclockwise) rather than fixed directions referenced to a coordinate system. (b) no
7. When viewed by a driver through a rearview mirror, the right–left reversal property of the image formed by a plane mirror makes it read "AMBULANCE" to the driver in front.
9. (a) 0.80 m (b) 5.0 cm (c) $+1.0$
11. (a) 1.5 m behind the mirror
(b) 1.0 m/s
13. 3.0 m and 13 m behind north mirror; 5.0 m and 11 m behind south mirror
17. (d)
19. (a) The plane mirror gives a large view of the area immediately around that side of the truck. The small convex mirror gives a wide-angle perspective of the road in back of both sides of the truck. (b) These are convex mirrors that give a better field of view, but the images are smaller than the objects and hence appear closer than they actually are. (c) Yes, it can be considered a converging mirror, because it collects a large amount of radio waves and focuses them onto a small area.
21. (a) The image is smaller than the object, so it is possible to "see your full body in 10 cm" in a diverging mirror. (b) As the ball swings toward the mirror and approaches the focal point, the image enlarges. An enlarged image appears to be closer to the observer's eyes, so it appears to move toward the observer and therefore to jump out of the mirror.
23. 5.0 cm
25. $d_i = -30$ cm; $h_i = 9.0$ cm; image is virtual, upright, and magnified
27. (a) $f = 10$ cm, $R = 20$ cm (b) 3.0 cm

A-23

29. (a)

(b)

(c)

(d)

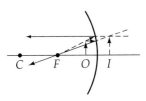

33. 120 cm
35. 2.3 cm
37. (a) $d_i = 60$ cm; $M = -3.0$, real and inverted
(b)

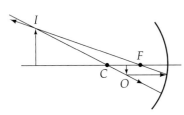

39. 10 m
41. $d_i = 60$ cm; image is real, inverted, and magnified.
43. 5.0 cm, 15 cm
45. yes: 13 cm, 27 cm
47. (c)
49. When the fish is inside the focal point, the image is upright, virtual, and magnified.
51. $d_i = 12.5$ cm; $M = -0.25$
53. 22 cm
55. (a) $d_i = -6.4$ cm; $M = 0.64$, virtual and upright

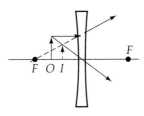

(b) $d_i = -10.5$ cm; $M = 0.42$, virtual and upright

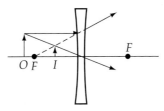

57. (a) 18 cm (b) 6.0 cm
59. 8.1 cm
61. (a) 20 cm (b) $M = -1$
63. (a) $d = 4f$ (b) approaches zero
65. (a) -18 cm (b) -90 cm
67. (a) 4.6 cm (b) 0.80 cm, inverted
69. 8.6 cm; $M = 0.86$, real and inverted
73. (a) focal length decreases by a factor of 0.34 (b) The diverging lens becomes the converging lens and vice versa, and f decreases by a factor of 0.075.
75. ∞; it has no focusing power.
77. -0.70 D
79. 0.55 mm (inverted)
81. $d_i = 37.5$ cm; $h_i = 3.0$ cm, real and inverted
83. The image formed by the converging lens is at the mirror. This image is the object for the diverging lens. If the mirror is at the focal point of the diverging lens, the rays refracted after the diverging lens will be parallel to the axis. These rays will be reflected back parallel to the axis by the mirror and will form another image at the mirror. This second image is now the object for the converging lens. By reversing the rays, a sharp image is formed at the location on the screen where the original object is. Therefore the distance from the diverging lens to the mirror is the focal length of the diverging lens.
85. 14 cm
87. (a) $d_i = \dfrac{d_o f}{d_o - f} = \dfrac{f}{1 - f/d_o}$

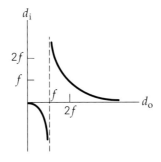

$|M| = \dfrac{d_i}{d_o} = \dfrac{f}{d_o - f}$

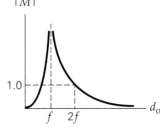

(b) $d_i = \dfrac{d_o(-f)}{d_o + f} = \dfrac{-f}{1 + f/d_o}$

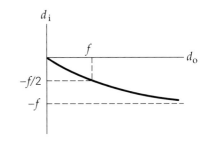

$|M| = \dfrac{d_i}{d_o} = \dfrac{-f}{d_o + f}$

89. (a)

(b)

91. 85 cm
93. (a)

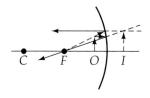

(b) 60 cm

Chapter 24

1. (b)
3. blue
5. The path-length difference will change because the airplane is moving. This change results in a change in the condition of interference—constructive is no longer constructive, and so on. Therefore the pictures flutter.
7. 3.4%
9. (a) 0.80 m (b) 3.2 m
11. 2.0 m
13. (a) 4.40×10^{-7} m (b) 4.40 cm
15. (a) 1.9 mm (b) decreases to 1.2 mm
17. (a) $\lambda = 4.00 \times 10^{-7}$ m, violet (b) 3.45 cm
19. 450 nm
21. (a)
23. all wavelengths except bluish purple
25. (a) 30λ (b) destructively
27. 5.4×10^{-8} m
29. 160 nm; 200 nm
31. (a) 158.2 nm (b) 316.4 nm
33. 1.51×10^{-6} m
35. (b)
37. a second diffraction pattern perpendicular to the first
39. (a) 5.4 cm (b) 2.7 cm
41. (a) 4.3 mm (b) microwave
43. 4.9 mm
45. blue: $x_1 = 0.50$ m, 18.4°, $x_2 = 1.2$ m, 39.2°; red: $x_1 = 0.89$ m, 30.7°, x_2 not possible
47. (a) 2.44×10^3 lines/cm (b) 11
49. 15.4°
53. (d)
55. Looking through a lens of each pair, rotate one of the lenses. If the intensity changes as the glasses are rotated, both pairs are polarized.
57. (a) twice (b) four times (c) none (d) six times
59. 54° to 60°
61. 31.7°
63. 39.3°
67. no
69. Blue light scatters more efficiently than red light. In the morning and evening, the blue component of the light from the Sun is scattered more in the denser atmosphere near the Earth, so we see red when we look in the direction of the rising or setting Sun. During the day, we see mainly the blue component from overhead scattering.
71. 1.74
73. 56°
75. (b) $\lambda/2n$
77. 6.4×10^{-4} rad
79. 560 nm
81. 4.9×10^{-7} m

Chapter 25

1. (b)
3. According to the lens maker's equation, the eye focuses by changing the shape of its lens to adjust the focal length to form a sharp image. The image distance is fairly constant, as is the distance from the lens to the retina. From the thin-lens equation, the eye must have a short focal length for looking at close objects, so the radius is small; the eye must have a long focal length for looking at distant objects, so the radius is large.
5. The pre-flash occurs before the aperture is open and the film exposed. The bright light causes the iris to reduce the area of the pupil so that when the second flash comes momentarily, you don't have a wide opening through which you get the red-eye reflection from the retina.
7. (a) +5.0 D (b) −2.0 D
9. converging
11. diverging, −5.0 D
13. (a) +3.0 D (b) take them out
15. (a) converging (b) +3.3 D
17. +3.0 D
19. 21 cm
21. right: +1.42 D, −0.46 D; left: +2.16 D, −0.46 D
23. (d)
25. inside the focal length
27. (a) 2.3× (b) 2.5×
29. (a) 2.7× (b) 1.7×
31. 1.9×
33. (a) 1.8× (b) 1.3×
35. 360×
37. (a) 340× (b) 3900%
39. 25×
41. 3.6×
43. (b)
45. No, the whole star can still be seen. The obstruction will reduce the intensity or brightness of the image.
47. (a) 4.0× (b) 75 cm
49. (a) 110× (b) 88.3 cm
51. (a) 2 (b) 1
55. (a)
57. smaller
59. 550 nm
61. 1.32×10^{-7} rad; Yerkes/Hale = 4.98
63. yes
65. (a) 1.63×10^{-6} rad (b) 9.76×10^{17} m
67. (a) 5.55×10^{-5} rad (b) blue (c) 33.3%
69. (d)
71. Since white is obtained by adding colors, it cannot be obtained by the subtractive method. That method subtracts colors, and the one we see is the one that is not absorbed. Black objects do not absorb all wavelengths of light. We see the objects because we perceive the extremely faint light as black. (Think of twilight vision.)
73. White light enters the blue filter, which allows the green, blue, and violet to pass through; when it passes through the yellow filter, only the green emerges.
75. 1.6×10^{-2} rad
77. 6.7 m
79. 1.1×10^{-2} rad
81. 2.1×
83. 83 cm
85. 500×
87. objects as large as typical houses

Chapter 26

1. (d)
3. (a)
5. (a)
7. Since the speed of light is constant in all directions through all frames of references and no ether (wind) would be present, there is no extra time difference between the two beams and no fringe shift is observed.
9. (a) 3.38 s (b) 3.58 s
11. 55 m/s; 45 m/s
13. no
17. (c)
19. (b), since your friend is traveling to the front, he sees the front mark first.
21. (a) 300 m/s
23. (c), her friend does appear the same height because the height is perpendicular to the velocity.
25. Yes, you will find yourself younger after a high-speed space trip (close to c) due to time dilation.
27. 42 beats/min
29. 66 m; 100 m
31. Earth twin is 64 y; traveling twin is 37 y.
33. length of 20.4 m and diameter 8.25 m
35. $0.87c$
37. 5.3 m; assume relative velocity parallel to length of stick
39. She is traveling at $0.40c$ in the direction of the lengths of the two sticks.
41. (d)
43. No, there are no such limits on momentum and energy as $p = \gamma m v$ and $E = \gamma m c^2$.
45. yes; no
47. (a) $0.985c$ (b) 2.50 MeV (c) 1.56×10^{-21} kg·m/s
49. 0.15 kg
51. 1.6×10^{24} J, or 150 000 times more than 1.00×10^{19} J = 3 million kWh
53. 79 keV

55. (a)

(b)

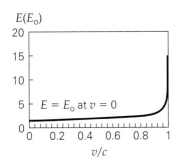

57. (a) $0.92c$ **(b)** 1.4×10^3 MeV
59. (a) 1.13 MeV **(b)** 1.64 MeV
61. 1.43×10^5 y
63. (a) $E_o = 939$ MeV > 600 MeV
(b) $0.792c$ **(c)** 6.50×10^{-19} kg·m/s
65. (d)
67. The stick would be elongated or "stretched" by the gravity gradient (difference).
69. 1.8×10^{19} kg/m³
71. Drop the cup with the pole vertical.
73. $0.27c$, same direction as spacecraft
75. (a) $0.988c$ to the left **(b)** $0.988c$ to the right
77. 4.8 h
79. length 38 m; height 2.5 m; width 2.0 m
81. 50 min
83. $0.68c$
85. 1.9 kg

Chapter 27

1. (d)
3. No, it is actually $2^4 = 16$ times since it depends on T^4.
5. (a)

(b)

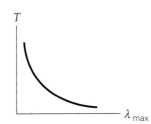

7. 1.06×10^{-5} m; 2.83×10^{13} Hz
9. 3.1×10^{13} Hz
11. 2.56×10^{-20} J
13. (b)
15. The packets of radio signals arrive in such large quantity and so quickly that our ear cannot distinguish between discrete arrivals. It is similar to a movie, in which the picture is actually single frames passing in front of us rapidly, but we see it as a smooth and continuous moving picture. Likewise, although the radio signals are single frames, we hear them continuously.
17. The energy of radiation is proportional to its intensity, according to the wave theory, and its frequency, according to the particle theory. It takes a certain amount of energy to eject a photoelectron. Since only the frequency, not the intensity, matters in this case, it favors the particle theory.
19. $\lambda = 6.0 \times 10^{-11}$ m, X-ray
21. (a) 1.32×10^{-18} J **(b)** 8.27 eV
23. 3.0 eV
25. (a) 6.7×10^{-34} J·s **(b)** 2.9×10^{-19} J
27. 6.8×10^{14} Hz
29. 2.48 eV
31. no
33. (a) sodium **(b)** $\lambda_{silver} < 262$ nm; $\lambda_{sodium} < 504$ nm
35. 6.6×10^{-34} J·s; 1.5 eV

37. (d)
39. 200 000 scatterings can shift the wavelength from X-ray to visible.
41. 4.86×10^{-3} nm
43. 0.281 nm
45. 0.127 nm
47. 18.5°
49. (d)
51. The theory applies only to atoms with a single electron, because it does not include electron–electron interactions.
53. According to momentum conservation, the atom recoils, so it carries some kinetic energy and therefore the energy of the photon is smaller than expected. Thus the wavelength of the photon is longer.
55. (a) ultraviolet **(b)** visible (red)
57. (a) 6.89×10^{14} Hz **(b)** 8.21×10^{14} Hz
59. $n \approx 100$
61. (a) -1.51 eV **(b)** -0.378 eV **(c)** -0.136 eV
63. yes; 15.4 eV
65. (a) 0.967 eV **(b)** 2.97 eV **(c)** 10.2 eV; (c)
67. (a) 54.4 eV **(b)** 122 eV
69. $(2.2 \times 10^6$ m/s$)/n$
71. (a) potential $= -27.2$ eV; kinetic $= +13.6$ eV **(b)** $|U| = 2K$, potential energy is twice as large in magnitude.
73. Through stimulated emission, the photon from a laser is caused by electron transitions between two discrete energy levels, so there are only a few colors (frequencies). A light bulb emits thermal radiation at many frequencies.
75. 1.14×10^{31} photons
79. (a) 12.1 eV **(b)** 13.1 eV
81. 4.4×10^{-16} J
83. (a) 1.5×10^{15} Hz **(b)** 1.3 V
85. four
87. 6.59×10^{15} Hz

Chapter 28

1. (c)
3. The wavelength associated with a moving car is too short for us to observe.
5. (a) 7.28×10^{-6} m **(b)** 3.97×10^{-9} m
7. $\lambda_e/\lambda_p = 43$
9. 1.5×10^4 V
11. -53% (a decrease)
13. 3.7×10^{-63} m
15. 24 V
17. (b)
19. (a)

(b)

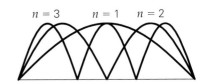

21. (c)
23. (a)
25. The periodic table groups elements according to the values of quantum numbers n and ℓ. The elements within a group have the

same or very similar electronic configurations for the outermost electrons.
27. (a) 2 (b) 14
29. (a) 2 (b) 3
31. (a) Be (b) N (c) Ne (d) S
33. (a) $1s^2 2s^2 2p^1$ (b) $1s^2 2s^2 2p^6 3s^2 3p^6 4s^2$
(c) $1s^2 2s^2 2p^6 3s^2 3p^6 3d^{10} 4s^2$
(d) $1s^2 2s^2 2p^6 3s^2 3p^6 3d^{10} 4s^2 4p^6 4d^{10} 5s^2 5p^2$
35. It would be $1s^3$ since the first state can now populate three electrons.
37. (b)
39. 0.21 m
41. 2.1×10^{-29} m/s
43. 1.1×10^{-27} J
45. 10^4 times
47. (a)
49. yes
51. 1.9 GeV
53. 83.5 V
55. (a) 4.0×10^{-21} kg·m/s (b) 1.7×10^{-13} m
59. $x = 0$
61. Yes, when $n = 1$ and $\ell = 0$.
63. no

Chapter 29

1. (c)
3. (d)
5. The isotopes of an element have the same number of protons but a different number of neutrons.
7. (a) ^{1}H; ^{2}D; ^{3}T (b) H$_2$O; D$_2$O; T$_2$O; HDO; HTO; DTO (c) T$_2$O, HTO, DTO
9. $^{41}_{19}$K
11. (a) $R_{He} = 1.9 \times 10^{-15}$ m;
$R_{Ne} = 3.3 \times 10^{-15}$ m; $R_{Ar} = 4.1 \times 10^{-15}$ m;
$R_{Kr} = 5.3 \times 10^{-15}$ m; $R_{Xe} = 6.1 \times 10^{-15}$ m;
$R_{Rn} = 7.3 \times 10^{-15}$ m (b) They are roughly ten thousand to fifty thousand times smaller.
13. No; in beta decay, a proton, which is a nucleon, is created.
15. (a) $^{237}_{93}$Np → $^{233}_{91}$Pa + $^4_2$He
(b) $^{32}_{15}$P → $^{32}_{16}$S + $^0_{-1}$e (c) $^{56}_{27}$Co → $^{56}_{26}$Fe + $^0_{+1}$e
(d) $^{56}_{27}$Co + $^0_{-1}$e → $^{56}_{26}$Fe (e) $^{42}_{19}$K → $^{42}_{19}$K + γ
17. α-decay: $^{214}_{84}$Po → $^{210}_{82}$Pb + $^4_2$He;
β-decay: $^{210}_{82}$Pb → $^{210}_{83}$Bi + $^0_{-1}$e
19. α–β: $^{213}_{83}$Bi; β–α: $^{213}_{83}$Bi
21. (a) $^4_2$He (b) $^0_{-1}$e (c) $^{102}_{39}$Y (d) $^{23}_{11}$Na*
(e) $^{22}_{10}$Ne
23. (a) α to ^{233}Pa; β to ^{233}U; α to ^{229}Th; α to ^{225}Ra; β to ^{225}Ac; α to ^{221}Fr; α to ^{217}At; α to ^{213}Bi; α to ^{209}Tl or β to ^{213}Po; β to ^{209}Pb or α to ^{209}Pb; β to ^{209}Bi (b) They decay because there are too many neutrons in the nucleus.
25. (c)
27. No, decay is exponential, not linear.
29. (a) 6.76×10^{-5} Ci (b) 2.50×10^6 Bq
31. (a) 1/8 (b) $1/2^{24}$, or about 6×10^{-6}% of original
33. 28.6 y
35. 91.7%
39. ^{104}Tc
41. 0.98 μg
43. (a) 6.75×10^{19} nuclei (b) 2.11×10^{18} nuclei
45. 9.2 mg
47. 7.6×10^6 kg

49. (d)
51. ^{1}H and ^{2}H are stable, and ^{3}H is unstable because its nucleus has too many neutrons.
53. (a) $^{17}_8$O, because it has an unpaired neutron beyond a magic number. (b) $^{42}_{20}$Ca, because the magic number difference makes it more stable (both are paired). (c) $^{10}_5$B, because it lacks pairing. (d) Approximately the same, as both have 126 neutrons (paired and magic) and the different number of protons does not affect neutrons.
55. (b) and (d); others are odd–odd
57. 2.013 553 u
59. (a) 92.2 MeV (b) 7.68 MeV/nucleon
61. deuterium
63. 104.7 MeV
65. 10.1 MeV
67. 7.59 MeV/nucleon
69. (a) two alphas (b) two alphas (c) Yes, since two alphas gives a more stable arrangement.
71. (d)
73. (a) α particles (b) X-rays: 1.0 rem; α particles: 20 rem
75. yes
77. $^{75}_{33}$As, $^{76}_{34}$Se ($^1_0$n + $^{75}_{33}$As → $^{76}_{33}$As → $^{76}_{34}$Se + $^0_{-1}$e)
79. (a) 28.3 MeV (b) 4.001 501 u
81. (a) 3.4×10^{22} nuclei (b) 1.8×10^{16} nuclei (c) 9.7×10^{19} decays/min; 5.2×10^{15} decays/min; no
83. (a) $^0_{-1}$e (b) $^{222}_{86}$Rn (c) $^{237}_{94}$Pu (d) γ (e) $^0_{+1}$e
85. (a) 1.80×10^3 MeV (b) 237.997 821 u
87. (c) and (d)
89. $^3_2$He
91. (b) 2.3×10^{17} kg/m^3
93. (a) Atomic density is about 10^{14} times less than the nuclear density. (b) The masses are the same but are averaged over a much larger volume. (c) No, the average atomic density is much higher than hydrogen gas (most gases are typically ≈ 2 kg/m^3) because the gas density includes mostly empty space between gas molecules.

Chapter 30

1. (d)
3. As the target mass M_A becomes very large, $K_{min} \to |Q|$. If $Q = 0$, the incident particle would not need kinetic energy to start the reaction.

5. (a) $^{14}_7$N (b) $^2_1$H (c) $^{30}_{15}$P (d) $^{17}_8$O (e) $^{10}_5$B
7. (a) $^{14}_7$N* (b) $^{14}_7$N* (c) $^{31}_{15}$P* (d) $^{18}_9$F* (e) $^{14}_7$N*
9. (a) $^{22}_{11}$Na; no (b) $^{222}_{86}$Rn; yes (c) $^{12}_6$C; no

11. +4.28 MeV
13. 4.36 MeV
15. 1.53 MeV
17. exoergic
19. 5.38 MeV
21. the first reaction, by 3.05 times
23. (a) 0.24 b (b) 0.66 b (c) 1.6 b (d) 1.7 b
25. (d)
27. (d)
29. The reaction products collide with the core materials and coolant and in the process produce heat, which converts water to steam that powers a turbine.
31. (a) 2.22 MeV (b) 12.9 MeV
33. 36 collisions
35. (c)
37. 4.68×10^{-13} m
39. (a) 0.86 MeV (b) 1.41×10^{-21} kg·m/s opposite that of the neutrino (c) both zero
41. 13.37 MeV
45. β^-: $M_P > M_D$; β^+: $M_P > M_D + 2m_e$
47. (b)
49. 3.5×10^{-28} kg
51. 135 MeV
53. 1.98×10^{-16} m
55. (a)
57. All hadrons contain quarks and/or antiquarks. Quarks are not believed to exist freely outside the nucleus.
59. (a) π^+ (b) K^0 (c) Σ^0 (d) Ξ^-
61. udd
63. (a)
65. (a) $^7_4$Be* (b) $^{60}_{29}$Cu* (c) $^{236}_{92}$U* (d) $^{13}_6$C* (e) $^{17}_8$O*
67. 0.156 MeV
69. (a) 0.86 MeV (b) no

Photo Credits

All Demonstration photos: Michael Freeman

Chapter 1—**CO.1** Photo Researchers, Inc. **Fig. 1.2b** U.S. Department of Commerce **Fig. 1.3c** U.S. Department of Commerce **Fig. 1.5** Pearson Education/PH College **Fig. 1.7a** Jerry Wilson/Lander College **Fig. 1.7b** Fundamental Photographs **Fig. 1.8** Visuals Unlimited **Fig. 1.9** John Smith **Fig. 1.16** Liaison Agency, Inc. **Fig. 1.18** Pearson Education/PH College **Fig. 1.19** Pearson Education/PH College

Chapter 2—**CO.2** Corbis Corporation Media **Fig. 2.2** John Smith **Fig. 2.3** Liaison Agency, Inc. **Fig. 2.7** Focus on Sports Inc. **Fig. 2.15b** James Sugar/Black Star **IN, p. 52 Fig. 1** North Wind Picture Archives **IN, p. 52 Fig. 2** J. M. Charles Rapho/Photo Researchers, Inc. **Fig. 2.16a-b** Pearson Education/PH College

Chapter 3—**CO.3** FPG International LLC **Fig. 3.12** Tony Stone Images **Fig. 3.13c** Photri-Microstock **Fig. 3.16a** Richard Megna/Educational Development Center/Fundamental Photographs **Fig. 3.18a** Photo Researchers, Inc. **Fig. 3.18b** Rainbow **Fig. 3.22** Focus on Sports Inc.

Chapter 4—**CO.4** Rainbow **Fig. 4.1** The Granger Collection **Fig. 4.4** John Smith **IN, p. 112 Fig. 1** AP/Wide World Photos **Fig. 4.17a-b** Arnold & Brown **Fig. 4.20a** Photo Researchers, Inc. **Fig 4.20b** Gabe Palmer/Mug Shots **Fig. 4.25** PhotoEdit **Fig. 4.27** Photo Researchers, Inc. **Fig. 4.29** Arnold & Brown **Fig. 4.31** Allsport Photography USA Inc. **Fig. 4.32** Arnold & Brown **Fig. 4.39** The Stock Market **Fig. 4.40L-R** The Goodyear Tire and Rubber Company **Fig. 4.41** Focus on Sports Inc.

Chapter 5—**CO.5** Asarco, Inc. **Fig. 5.9a** Guntram Gerst/Peter Arnold, Inc. **Fig. 5.9b** Michael Collier **Fig. 5.11a** Ken Straiton/Photo Researchers, Inc. **Fig. 5.11b** Vince Streano/The Image Works **Fig. 5.19** The Image Works **Fig. 5.21** SuperStock, Inc. **IN, p. 163 Fig. 1** Michael Probst/AP/Wide World Photos **Fig. 5.22** Daemmrich/Stock Boston **Fig. 5.25** Bont Skates Pty Ltd **Fig. 5.27** David Lissy/Focus on Sports, Inc.

Chapter 6—**CO.6** John Cordes/Focus on Sports, Inc. **Fig. 6.1a** Gary S. Settles/Photo Researchers, Inc. **Fig. 6.1b** David G. Curran/Rainbow **Fig. 6.1c** Stephen J. Krasemann/Photo Researchers, Inc. **Fig. 6.5a** Omikron/Science Source/Photo Researchers, Inc. **IN, p. 180 Fig. 1** Llewellyn/Uniphoto Picture Agency **Fig. 6.8a** M. Hans/Vandystadt/Photo Researchers, Inc. **Fig. 6.8b** Globus Brothers/The Stock Market **Fig. 6.11a** Ann Purcell/Photo Researchers, Inc. **Fig. 6.11b** H.P. Merten/The Stock Market **Fig. 6.15** Richard Megna/Fundamental Photographs **Fig. 6.16a** Richard Megna/Fundamental Photographs **Fig. 6.16b** R.D. Rubic/Precision Chromes, Inc. **Fig. 6.20** Paul Silverman/ Fundamental Photographs **Fig. 6.22** Focus on Sports, Inc. **Fig. 6.23b-c** NASA Headquarters **Fig. 6.27** Jonathan Watts/Science Photo Library/Photo Researchers, Inc. **Fig. 6.29** Runk/Schoenberger/Grant Heilman Photography, Inc. **Fig. 6.33** Addison Geary/Stock Boston

Chapter 7—**CO.7** Tony Savino/The Image Works **Fig. 7.6b** Tom Tracy/The Stock Market **Fig. 7.7** SuperStock, Inc. **Fig. 7.11** Chris Priest/Science Photo Library/Photo Researchers, Inc. **Fig. 7.13c** Michael Livenston/The Stock Market **Fig. 7.14** Focus on Sports, Inc. **Fig. 7.24a** NASA/Science Source/Photo Researchers, Inc. **Fig. 7.24c** Johan Elbers **Fig. 7.24d** C. Seghers/Photo Researchers, Inc. **IN, p. 242 Fig. 1** Photo Researchers, Inc. **Fig. 7.33** NASA Headquarters **Fig. 7.34** Frank Labua/Pearson Education/PH College **Fig. 7.36** Alexander Lowry/Photo Reseachers, Inc.

Chapter 8—**CO.8** Jerry Wachter/Focus on Sports, Inc. **Fig. 8.11a** Richard Hutchings/Photo Researchers, Inc. **Fig. 8.11b** Ed. Degginger/Color-Pic, Inc. **Fig. 8.11c** Jean-Marc Loubat/Agence Vandystadt/Photo Researchers, Inc. **IN, p. 264 Fig. 1** (*FPG* International LLC) **Fig. 8.12b** Beaura Katherine Ringrose **IN, p. 266 Fig. 1** Albert Einstein (TM) represented by The Roger Richman Agency, Inc. Beverly Hills, CA 90012 www.hollywoodlegands.com. Photo courtesy of the Archives California Institute of Technology **Fig. 8.25a** Frank Labua/Pearson Education/PH College **Fig. 8.25b** Frank Labua/Pearson Education/PH College **Fig. 8.25c** NASA Headquarters **Fig. 8.28a** Vince Streano/The Image Works **Fig. 8.28b** NASA Headquarters **Fig. 8.29** Jerry Wilson **Fig. 8.43b** Tony Savino/The Image Works **Fig. 8.46** Gerard Lacz/Natural History Photographic Agency

Chapter 9—**CO.9** Focus on Sports, Inc. **IN, p. 300 Fig. 1** Jonathan Watts/Science Photo Library/Photo Researchers, Inc. **Fig. 9.11** James Holmes/Reed Nurse/Science Photo Library/Photo Researchers, Inc. **IN, p. 309 Fig. 2** Blair Seitz/Photo Researchers, Inc. **Fig. 9.12** Jan Halaska/Photo Researchers, Inc. **Fig. 9.14** Compliments of Clearly Canadian Beverage Corporation **Fig. 9.15** Bill Curtsinger/Photo Researchers, Inc. **Fig. 9.17b** Diane Schiumo/ Fundamental Photographs **Fig. 9.23b** Hermann Eisenbeiss/Photo Researchers, Inc. **Fig. 9.24a** David Spears/Science Photo Library/Photo Researchers, Inc. **Fig. 9.24b** Richard Steedman/The Stock Market **Fig. 9.26** Patrick Watson/Medichrome/The Photo Shop **Fig. 9.28** Underwood & Underwood/Corbis **Fig. 9.29** Frank Labua/Pearson Education/PH College **Fig. 9.30a-b** Courtesy of CENCO **Fig. 9.34** Michael J. Howell/Stock Boston **Fig. 9.35a-b** Stephen T. Thornton **Fig. 9.35c** John Smith **Fig. 9.36** Stephen T. Thornton **Fig. 9.39** Tom Cogill

Chapter 10—**CO.10** Jim Corwin/Photo Researchers, Inc. **Fig. 10.2c** Frank Labua/Pearson Education/PH College **Fig. 10.3a** Richard Megna/Fundamental Photographs **Fig. 10.3b** Leonard Lessin/Peter Arnold, Inc. **Fig. 10.5b** John Smith **IN, p. 337 Fig. 1** Maximilian Stock/Science Photo Library/Photo Researchers, Inc. **Fig. 10.6** Sinclair Stammers/Science Photo Library/Photo Researchers, Inc. **Fig. 10.11a** Richard Choy/Peter Arnold, Inc. **Fig. 10.11b** Joe Sohm/The Image Works **Fig. 10.15** Paul Silverman/Fundamental Photographs **Fig. 10.20** Stephen T. Thornton

Chapter 11—**CO.11** Tom Cogill **Fig. 11.2** Jerry Wilson **Fig. 11.4** John Smith **Fig. 11.10** Frank Labau/Pearson Education/PH College **Fig. 11.11c** Richard Lowenberg/Science Source/Photo Researchers, Inc. **Fig. 11.15** Dr. Ray Clark and Mervyn Goff/Science Photo Library/Photo Researchers, Inc. **Fig. 11.16** John Smith **Fig. 11.17** Carl Glassman/The Image Works **Fig. 11.19** John Smith **Fig. 11.20** Leonard Lessin/Peter Arnold, Inc. **Fig. 11.23** Joe Sohm/Photo Researchers, Inc.

Chapter 12—**CO.12** Helen Marcus/Photo Researchers, Inc. **IN, p. 401 Fig. 2** Garry Ladd/Photo Researchers Inc. **Fig. 12.8** Lowell Georgia/Photo Researchers, Inc. **Fig. 12.10** Leonard Lessin/Peter Arnold, Inc.

Chapter 13—**CO.13** J.R. Berintenstein/Photo Researchers, Inc. **Fig. 13.9** John Matchett/Picture Perfect USA, Inc. **IN, p. 436 Fig. 1** Peter Menzel/Peter Arnold, Inc. **IN, p. 436 Fig. 2** H. Yamaguchi/Sygma **IN, p. 437 Fig. 4b** Russell D. Curtis/Photo Researchers, Inc. **IN, p. 437 Fig. 4c** Russell D. Curtis/Photo Researchers, Inc. **Fig. 13.14** Education Development Center, Inc. **Fig. 13.18** Fundamental Photographs **Fig. 13.19a** Richard Megna/Fundamental Photographs **Fig. 13.20** Richard Megna/Fundamental Photographs **Fig. 13.21** SuperStock, Inc. **Fig. 13.22** Lawrence Migdale/Stock Boston **IN, p. 445 Fig. 1** University of Washington Libraries, Special Collections Farquharson 12 **Fig. 13.26** Courtesy of CENCO **Fig. 13.28** Courtesy of Central Scientific Company

Chapter 14—**CO.14** Michael A. Keller/The Stock Market **Fig. 14.1b** Leonard Lessin/Peter Arnold, Inc. **Fig. 14.3a** NASA Headquarters **Fig. 14.3b** Merlin D. Tuttle/Photo Researchers, Inc. **Fig. 14.3c** Howard Sochurek/Woodfin Camp & Associates **Fig. 14.12b** Philippe Plaily/Science Photo Library/Photo Researchers, Inc. **Fig. 14.12c** Jim Kahnwieler/Positive Images **IN, p. 473 Fig. 1** University of Illinois, Department of Atmospheric Sciences **IN, p. 473 Fig. 2** David R. Frazier/Photo Researchers, Inc. **Fig. 14.14** Jeff Greenberg/Visuals Unlimited **Fig. 14.16** Michael Furman Photographer Ltd./The Stock Market **Fig. 14.18b** Richard Megna/Fundamental Photographs

Chapter 15—**CO.15** Zefa/London/The Stock Market **Fig. 15.7a** Jerry Wilson **Fig. 15.7c** Charles D. Winters/Photo Researchers, Inc. **IN, p. 500, Fig. 1b** Keith Kent/Peter Arnold, Inc. **IN, p. 495, Fig. 1c** Philippe Wojazer/Reuters/Corbis-Bettmann

Chapter 16—**CO.16** Richard Megna/Fundamental Photographs **Fig. 16.13** Spencer Grant/Photo Researchers, Inc. **Fig. 16.16a** Larry Mulvehill/Photo Researchers, Inc.

Chapter 17—**CO.17** Phil Jude/Photo Researchers, Inc. **IN, p. 543 Fig. 1** Courtesy of International Business Machines Corporation. Unauthorized use not permitted **Fig. 17.7** Alfred Pasieka/Science Photo Library/Photo Researchers, Inc. **Fig. 17.11a** Ronald Brown/Arnold & Brown **Fig. 17.11b** Frank Labua/Pearson Education/PH College **IN, p. 555, Fig. 1 and Fig. 2** Frank Labua/Pearson Education/PH College **Fig. 17.12** NASA/Mark Marten/Photo Researchers, Inc. **Fig. 17.13** Frank Labua/Pearson Education/PH College

Chapter 18—**CO.18** Richard Megna/Fundamental Photographs **Fig. 18.14a** Courtesy of Central Scientific Company IN, p. 584, **Fig. 2** Zircon **Fig. 18.17c** Richard Megna/Fundamental Photographs **Fig. 18.19b** Frank Labua/Pearson Education/PH College **Fig 18.19c** Paul Silverman/ Fundamental Photographs **Fig. 18.20b** Frank Labua/Pearson Education/PH College **Fig. 18.20c** U.S. Department of Energy/Science Photo Library/Photo Researchers, Inc. **Fig. 18.23a-b** Frank Labua/Pearson Education/PH College **Fig. 18.41** M. Antman/The Image Works

Chapter 19—**CO.19** Peter Menzel/Stock Boston **Fig. 19.1** Courtesy of Central Scientific Company **Fig. 19.3** Richard Megna/Fundamental Photographs **Fig. 19.4b** Richard Megna/Fundamental Photographs **Fig. 19.8a** Richard Megna/Fundamental Photographs **Fig. 19.9a** Richard Megna/Fundamental Photographs **Fig. 19.10a** Richard Megna/ Fundamental Photographs **Fig. 19.13c** Grant Heilman Photography **Fig. 19.23b** Jerry Wilson **Fig. 19.26** James Holmes/Oxford Centre for Molecular Sciences/Science Photo/Photo Researchers, Inc. **Fig. 19.28** Courtesy of Central Scientific Company IN, p. 622 **Fig 1** and **Fig. 2** Richard Blakemore, University of New Hampshire **Fig. 19.32** Pekka Parviainen/Science Photo Library/Photo Researchers, Inc.

Chapter 20—**CO.20** Tim Barnwell/Stock Boston **Fig. 20.1d** Richard Megna/Fundamental Photographs IN, p. 638 **Fig. 2b** Richard Megna/Fundamental Photographs **Fig. 20.6a** Beaura Katherine Ringrose **Fig. 20.10a** U.S. Department of the Interior **Fig. 20.10b** Jim Steinberg/Photo Researchers, Inc. **Fig. 20.12d** Westinghouse Electric Corp. **Fig. 20.15a** Malcolm Fife PhotoDisc, Inc. **Fig. 20.15b** U.S. Department of Energy/Mark Marten/Photo Researchers, Inc. **Fig. 20.19** Munshi Ahmed Photography **Fig. 20.20** AP/Wide World Photos IN, p. 648 Richard Megna/Fundamental Photographs **Fig. 20.22a** Larry Mulvehill/Science Source/Photo Researchers, Inc. **Fig. 20.22b** GCA/CNRI/Phototake NYC IN, p. 657 **Fig. 1** National Oceanic and Atmospheric Administration **Fig. 20.32** Beaura Katherine Ringrose

Chapter 21—**CO.21** Jeffry Myers/The Stock Market **Fig. 21.5** Frank Labua/Pearson Education/PH College **Fig. 21.14** Frank Labua/ Pearson Education/PH College IN, p. 670, **Fig. 1** Rip Griffith/Photo Researchers, Inc. **Fig. 21.15** Tom Lyle/Medichrome/The Stock Shop, Inc.

Chapter 22—**CO.22** Alise and Mort Pechter/The Stock Market **Fig. 22.5** Peter M. Fisher/The Stock Market **Fig. 22.6** Photri/The Stock Market **Fig. 22.7** Richard Megna/Fundamental Photographs **Fig. 22.9** Richard Megna/Fundamental Photographs **Fig. 22.11a** Kent Wood/Photo Researchers, Inc. **Fig. 22.11c** DiMaggio/Kalish/Peter Arnold, Inc. IN, p. 698 **Fig 1a-e** © Edward Pascuzzi **Fig. 22.13** Ken Kay/Fundamental Photographs **Fig. 22.15** Stuart Westmoreland/Corbis **Fig. 22.16a** © Gemological Institute of America **Fig. 22.17a** Courtesy of Central Scientific Company **Fig. 22.18** Richard Megna/ Fundamental Photographs **Fig. 22.19a** Nick Koudis/PhotoDisc, Inc. **Fig. 22.19b** Hank Morgan/Photo Researchers, Inc. **Fig. 22.20a** David Parker/Photo Researchers, Inc. IN, p. 703 **Fig. 1** P. Gontier/The Image Works IN, p. 703 **Fig. 2** SIU/Photo Researchers Inc. IN, p. 704 **Fig. 1** Doug Johnson/Photo Researchers, Inc. **Fig. 22.21** Martin Pond/Science Photo Library/Photo Researchers, Inc. **Fig. 22.23** SIU/Photo Researchers, Inc. **Fig. 22.24** Frank Labua/Pearson Education/PH College

Chapter 23—**CO.23** Peticolas/Megna/Photo Researchers, Inc. **Fig. 23.6** Paul Silverman/Fundamental Photographs **Fig. 23.13** Richard Hutchings/Photo Researchers, Inc. **Fig. 23.16a** John Smith **Fig. 23.16b** Richard Megna/Fundamental Photographs IN, p. 730 **Fig. 1b** John Smith IN, p.730 **Fig. 1c** Bohdan Hrynewych/Stock Boston **Fig. 23.22** Frank Labua/Pearson Education/PH College **Fig. 23.23b** Michael Freeman **Fig. 23.24** Tom Tracy/The Stock Market **Fig. 23.25** John Smith

Chapter 24—**CO.24** Adrienne Hart Davis/Science Photo Library/Photo Researchers, Inc. **Fig. 24.1** Richard Megna/Fundamental Photographs **Fig. 24.2b** From the "Atlas of Optical Phenomena," Michel Cagnet, Maurice Francon, Jean Claude Thrierr. (c) by Springer-Verlag OHG, Berlin, 1962. **Fig. 24.6c** Peter Aprahamian/Photo Researchers, Inc. **Fig. 24.7a** David Parker/Science Photo Library/Photo Researchers **Fig. 24.7b** Gregory G. Dimijian, M.D./Photo Researchers, Inc. **Fig. 24.8b** H.R. Bramaz/Peter Arnold, Inc. IN, p. 749, **Fig. 2** Kristen Brochmann/Fundamental Photographs **Fig. 24.9b** Ken Kay/Fundamental Photographs **Fig. 24.10a** Courtesy of Jay Boleman **Fig. 24.10b** NASA/Science Photo Library/Photo Researchers, Inc. **Fig. 24.12** Ken Kay/Fundamental Photographs **Fig. 24.13b** Eduquip-Malacalaster Corporation **Fig. 24.15** Dan McCoy/Rainbow **Fig. 24.16b** A.J. Stosick **Fig. 24.16c** Courtesy of Dr. M.F. Perutzle/Medical Research Council **Fig. 24.19b** Ed Degginger **Fig. 24.21b** Diane Schiumo/Fundamental Photographs **Fig. 24.22b** Nina Barnett Photography **Fig. 24.23b** Peter Aprahamian/Sharples Stress Engineers Ltd./Science Photo Library/Photo Researchers, Inc. IN, p. 765, **Fig. 2** Leonard Lessin/Peter Arnold, Inc. **Fig. 24.25** Jim Baron/The Image Finders **Fig. 24.26** Roger Ressmeyer/© 1994 CORBIS **Fig. 24.27** Photri-Microstock

Chapter 25—**CO.25** Patrick Bennett/Corbis IN, p. 776 **Fig. 1** James King-Holmes/Science Photo Library/Photo Researchers, Inc. **Fig. 25.5b** Frank Labua/Pearson Education/PH College **Fig. 25.9b** Rocher/Jerrican/Photo Researchers, Inc. **Fig. 25.14a-b** California Institute of Technology/Palomar Observatory **Fig. 25.15** NASA Headquarters IN, p. 792 **Fig. 1** Tony Craddock/Science Photo Library/Photo Researchers, Inc. **Fig. 25.18** The Image Finders **Fig. 25.19** Bill Bachmann/Photo Researchers, Inc. **Fig. 25.21a** Fritz Goro/Life Magazine, Time Warner, Inc.

Chapter 26—**CO.26** Space Design International **Fig. 26.3b** PASCO Scientific **Fig. 26.5** Science Photo Library/Photo Researchers, Inc. **Fig. 26.17b** Space Telescope Science Institute IN, p. 831, **Fig. 1** National Radio Astronomy Laboratory **Fig. 2** David Hardy/Science Photo Library/Photo Researchers, Inc.

Chapter 27—**CO.27** Stephane Husain/Liaison Agency, Inc. **Fig. 27.7a** Larry Albright **Fig. 27.7b** Ann Purcell/Photo Researchers, Inc. **Fig. 27.8** Wabash Instrument Corp./Fundamental Photographs **Fig. 27.13** Gary Retherford/Photo Researchers, Inc. **Fig. 27.14** Dan McCoy/Rainbow **Fig. 27.18a** Will & Deni McIntyre/Photo Researchers, Inc. **Fig. 27.18b** Michal Rizza/Stock Boston **Fig. 27.20a** Hank Morgan/Rainbow **Fig. 27.20b** Chuck O'Rear/Westlight/Corbis

Chapter 28—**CO.28** Courtesy of International Business Machines Corporation. Unauthorized use not permitted. **Fig. 28.3a** Hank Morgan/Rainbow **Fig. 28.3b** Will & Deni McIntyre/Photo Researchers, Inc. IN, p. 873, **Fig. 2** Lawrence Migdale/Stock Boston; **Fig. 3a** Manfred Kage/Peter Arnold, Inc.; **Fig. 3b** Dr. R. Kessel/Peter Arnold, Inc.; **Fig. 3c** David Scharf/Peter Arnold, Inc. **Fig. 28.5** Courtesy of International Business Machines Corporation. Unauthorized use not permitted. Almaden Research Center IN, p. 876, **Fig. 1a** Omikron/Science Source/Photo Researchers, Inc.; **Fig. 1b** Mehau Kulyk/Science Photo Library/Photo Researchers, Inc.; **Fig. 3b** Will & Deni McIntyre/Photo Researchers, Inc. **Fig. 28.13** Lawrence Berkeley/Science Photo Library/Photo Researchers, Inc.

Chapter 29—**CO.29** Roger Tully/Tony Stone Images **Fig. 29.5** Mary Evans Picture Library/Photo Researchers, Inc. Fig. 29.13 Tom McHugh/Photo Researchers, Inc. **Fig. 29.17** AP/Wide World Photos **Fig. 29.20** Lawrence Berkeley Lab/Photo Researchers, Inc. **Fig. 29.21** Larry Mulvehill/Photo Researchers, Inc. **Fig. 29.22** Larry Mulvehill/Photo Researchers, Inc. **Fig. 29.23a** Dan McCoy/Rainbow **Fig. 29.23b** Monte S. Buchsbaum, M.D. Mount Sinai School of Medicine, New York, NY **Fig. 29.24a** S. Wanke/Photo Disk, Inc.

Chapter 30—**CO.30** Alexander Tsiaras/Stock Boston **Fig. 30.1b** Fermilab Visual Media Services **Fig. 30.5a** Tom Tracy/The Stock Market **Fig. 30.7a** Igor Kostin/Imago/Sygma **Fig. 30.7b** U.S. Dept. of Energy/Science Photo Library/Photo Researchers, Inc. **Fig. 30.8b** Plasma Physics Laboratory, Princeton University **Fig. 30.9c** Gary Stone/Lawrence Livermore National Laboratory Jerry Wilson author photo, Jeremey Kwasney

Index

Aberration(s)
 chromatic, 732
 lens, 732–33
 spherical, 723–24, 732
 of starlight, 807
Absolute pressure, 305–6
Absolute reference frame, 804–5, 808
Absolute temperature
 gas laws and, 338–43
 molecular speed and, 348–49
Absolute zero, 340–43, 349
Absorption
 photo, 856
 resonance, 378
Absorption spectrum, 850
Accelerated rolling without slipping, 253
Acceleration, 42–46
 angular, 224–27
 apparent, 243
 average, 42–44
 centripetal, 217–22, 226
 constant, 44–45, 46–50
 due to gravity, 50–51
 electron, 515–16, 823–25
 force and, 100–101
 instantaneous, 42, 111
 Newton's second law on, 105
 in simple harmonic motion, 430–31
 tangential, 225–26
 torque and, 271–73
ac circuits, 664–86
 capacitive reactance and, 668–70
 impedance and, 671–77
 inductive reactance, 670–72
 RC series, 580–83, 584, 672
 resistance in, 665–68
 resonance in, 677–81
 RL series, 673–74, 678
 RLC series, 674–75
Accommodation, 775
Accretion disk, 831
ac current (alternating current), 555, 641, 642–43, 665–66
Achromatic doublet, 732
Actinide series, 883
Action-at-a-distance forces, 101, 944
Action potential, 529
Action-reaction forces, 109, 483
Activity, of a radioactive isotope, 903, 905
ac generators, 641
Acuity, visual, 780
ac voltage, 666–68
Adapters, 667, 668
Addition
 with significant figures, 19
 vector, 69–78. *See also* Vector addition
Additive method of color production, 795

Additive primary colors, 795
Adiabat, 393, 408–9
Adiabatic process, 393–94, 408–9
Air bag, automobile, 180–81
Air conditioners, 405–8
Air foils, 128
Airplane flight dynamics, 82–83, 319
Air, refraction of light in, 695
Air resistance, 50, 51, 52
 defined, 128
 range and, 89, 90
 terminal velocity and, 128–29
Air speed, 82
Air wedges, 770
Alpha decay, 897–98, 909
Alpha particles, 893–94, 896, 897, 901, 916
Alternating current (ac), 555, 641, 642–43, 665–66
Alternators. *See* Generators
Altitude, boiling point and, 371
Ammeters, 583–85, 586
Amorphous substances, 367
ampere (A), 6, 546, 552
Ampére, Andre, 546
Amplitude, 419, 420, 425, 433
AM (amplitude-modulated) radio band, 653, 679
Analytical component method (vectors), 74–75
Analyzers, 761
Anderson, C. D., 886
Aneroid barometer, 307
Angle(s)
 critical, 697–99
 of incidence, 690–91, 697–98
 phase, 674, 675
 polarizing (Brewster), 757–58
 projection at, 86–88
 of reflection, 688
 of refraction, 690–91, 693, 697
 of repose, 127
 shear, 298
Angular acceleration, 224–27
Angular displacement, 211
Angular frequency, 425
Angular magnification, 781–83
Angular measure, 211–13
Angular momentum, 277–83
 conservation of, 278–80
 real-life, 280–83
Angular speed, 214–17
Angular velocity, 214–17
Anisotropic material, 759
Anodes, 544, 545
Antennas, broadcast, 680
Antilock brakes, 278
Antimatter, 887–88
Antineutrino, 943
Antinodes, 441

Antiparticles, 886–88
Anti-quarks, 948
Anti-Sun, 698
Anvil (incus), 459
Aphelion, 232, 236
Apparent acceleration, 243
Apparent weightlessness, 241–43
Appliances, power ratings of, 587
Applied force, 150
Aqueous humor, 775
Archimedes, 310
Archimedes' principle, 310–11, 313
Arc length, 211–13
Arcs, parabolic, 86
Area expansion, 344, 345, 346
Areas, Kepler's law of, 236, 280
Area vectors, 635
Aristotle, 33, 51, 52, 101
Armature, 617
Artificial gravity, 242
Asperities, 121
Astigmatism, 733, 779–80
Astronomical telescopes, 785–86, 787
atmosphere (atm), 302, 307
 standard, 306
Atmospheric lapse rate, 359
Atmospheric pressure, 306–7
Atmospheric scattering of light, 762–67
Atom(s). *See also* specific elements
 combining, 883
 multielectron, 876–78
 neutral, 911
 Rutherford–Bohr model of, 893
Atomic bomb, 934, 940
Atomic clock, 6
Atomic energy levels, 852–55, 856
atomic mass unit (u), 911
Atomic number, 894
Atomic orbitals, 878
Atomic physics. *See* Quantum mechanics
Atomic quantum numbers, 874–80
 hydrogen atom, 874–75
 multielectron atoms, 876–78
 Pauli exclusion principle, 878–80
Attractive forces, 483, 517
Atwood, George, 115
Atwood machine, 114–15, 118–19
Audible region, 453
Audio tape recorder, 638
Aurora australis, 624
Aurora borealis, 624
Autocollimation, 740
Average acceleration, 42–44

Average angular acceleration, 224–25
Average angular speed, 214
Average angular velocity, 214
Average speed, 33–35
 average velocity vs., 37–39
 for round trip, 41
Average velocity, 37–39
Avogadro's number, 340
Axes, Cartesian coordinate, 35
Axis of rotation, instantaneous, 253
Axis, transmission, 760

Back emf, 643–45
Back-scattering, 893–94
Bacteria, magnetotactic, 622–23
Balance, electronic, 619
Balanced forces, 100, 256
Ballistic pendulum, 206
Balloons, hot-air, 333
Ballpark calculations, 175–176
Balmer, J. J., 851
Balmer series, 851, 854, 855
Banked roads, 223, 246
Bardeen, John, 553
Barometer(s), 306, 523
 aneroid, 307
Barometric pressure, 306
Barrier penetration (tunneling), 898
Baryons, 948, 949
Base units, 3, 6
Basilar membrane, 459
Basketball, projection at an angle, 87–88
Batteries
 car, 548–49
 dielectrics and, 528–30
 direct current and, 543–45
 joule heat and testing of, 555
BCS theory, 553
Beat(s), 466–67
Beat frequency, 466–67
becquerel (Bq), 906
Becquerel, Henri, 896
bel (B), 460–63
Bernoulli, Daniel, 316
Bernoulli's equation, 316–20
Beta decay, 898–901, 909, 941–43
 β^- decay, 899
 β^+ decay, 899
Beta particles, 896, 897, 901, 916
Biconcave lenses, 725, 727, 777–78
Biconvex lenses, 724–25, 727–28, 777, 779
Bicycle stability, 266
Bifocals, 778
Big Bang, 951
Bimetallic coils, 335
Binding energy, 853, 911–14

I-1

Biological radiation hazards, 917
Biometrics, 776
Bird, drinking, 402–3
Birefringence, 759–62
Blackbody, 380, 840
Blackbody radiation, 380, 840–41, 842
Black holes, 828–30, 831
Blinker circuits, 582–83
Blood components, separating, 220–21
Blood flow, 316
Blood pressure, 308–9
Blood transfusion, 324
Blue shift, Doppler, 471
Body heat, 381
Body temperature, human, 337
Body waves, 436
Bohr, Niels, 851
Bohr orbit, 867
Bohr theory of the hydrogen atom, 850–55, 866, 867, 871
Boiling point, 367, 371
Boltzmann's constant, 339
Bone scan, 892
Bottom (beauty) quarks, 949
Boundary conditions, 441
Bow (shock) waves, 471, 472
Boyle, Robert, 338
Boyle's law, 338, 339
Bragg, Lawrence, 757
Bragg's law, 757
Brahe, Tycho, 234–35
Braking car, 113–14
  antilock brakes, 278
Braking (reverse) thrust, 201–2
Branch, 576
Branch currents, 579–80
Brass instruments, 475
Breeder reactors, 936–37
Brewster, David, 757
Brewster (polarizing) angle, 757
British thermal unit (Btu), 362
British units, 2, 5, 6–7, 10
Broadcast antennas, 680
Broadening, natural (line), 886
Brownouts, 564
Bubble chamber, 917
Bulk modulus, 296, 298–99, 455
Buoyancy (buoyant force), 308–14

Californium-252, 920
calorie (cal), 361
Calorimeter, 366
Calorimetry, 366–67
Cancer, 631, 655, 656, 917–18
Camera
  electronic flashes, 540
  operation of, 774, 801
  resolution of, 789
candela (cd), 6
Capacitance, 524–27
  equivalent parallel, 534
  equivalent series, 534
Capacitive circuits, 674
Capacitive reactance, 668–70

Capacitor(s), 499, 524. See also Parallel plates
  in cardiac defibrillators, 527
  changing through a resistor, 581
  dielectric material in, 527–32
  discharging through a resistor, 582–83
  energy storage in, 526
  in parallel, 532–33, 534–35
  parallel-plate, 524–26
  in series, 532–36
  in series–parallel combination, 535–36
  as transducers, 531–32
Capillary system, 15–16
Car battery, 548–49
Carbon-14 dating, 907–8
Cardiac defibrillators, 527
Carnot, Sadi, 408
Carnot cycle, 408–11
Carnot (ideal) efficiency, 409–10
Cartesian coordinates, 35
Cassegrain focus, 788
Cassini, 234
Cathodes, 543–44, 545
Cathode ray tube (CRT), 617–18
Cavendish, Henry, 228, 230, 489
Celsius temperature scale, 335–38, 363
Center of curvature, 715, 725, 727
Center of gravity (CG)
  center of mass and, 199–200
  locating, 267–68
  stable equilibrium and, 261–65
Center of mass (CM), 194–200
centipoise (cP), 322
Central maximum
  double slit, 743
  single slit, 751–53
Centrifuge, 220–21
Centripetal acceleration, 217–22, 226
Centripetal force, 222–24
Ceramic alloys, of rare earth elements, 552
CFCs (chlorofluorocarbons), 654, 656–57
cgs system, 6, 314
Chain reaction, 933–34
Chamber
  cloud, 917
  spark, 917
Characteristic frequencies, 441
Charge(s), 483–85. See also Electric charge
  and current, 546
  force per unit, 495
  point, 516
  quantized, 484, 485
  test, 495
Charge–force law (law of charges), 483
Charging
  by conduction, 487

by contact, 487
electrostatic. *See* Electrostatic charging
  by friction, 487
  by induction, 488
Charles, Jacques, 338
Charles' law, 338
Charm quark, 949
Chernobyl reactor, 936, 938
Chief (radial) ray, 716, 717, 726
China syndrome, 938
Chlorofluorocarbons (CFCs), 654, 656–57
Cholesterol, 316
Christmas tree lights, 572–73
Chromatic aberration, 732
Circle of least confusion, 732, 779
Circuit(s), 566–99. *See also* ac circuits; Kirchhoff's rules; Resistors
  blinker, 582–83
  capacitive, 674
  complete, 546
  completely capacitive, 676
  completely inductive, 676
  completely resistive, 676
  defined, 543
  household, 587–91
  inductive, 674
  multiloop, 575–80
  open, 587
  oscillator, 680–81
  resonance in, 677–81
  safety and, 586–91
Circuit breakers, 588–90
Circuit diagrams, 544
Circuit elements (components), 575
Circular motion, 210–50. *See also* Gravity; Kepler's laws; Uniform circular motion
  angular acceleration, 224–27
  angular measure, 211–13
  angular speed and velocity, 214–17
  earth satellites, 237–43
  relationship of radius, speed, and energy, 239
Circumference, of Earth, 2
Clap skates, 163
Classical (Newtonian) relativity, 803–5
Clausius, Rudolf, 397
Closed systems, 153, 183, 188, 399
Closed-tube manometer, 306
Cloud chamber, 917
Cochlea, 459
Coefficient(s)
  of friction, 122–27
  of kinetic friction, 122
  of performance (COP), 407
  of static friction, 122
  temperature, of resistivity, 551–52
  thermal. *See* Thermal coefficient(s)
  of viscosity, 322

Coherent sources, 742
Collision(s), 179, 181, 185–86
  conservation of momentum and, 187–94
  elastic, 187, 190–94
  inelastic, 187–90
Color(s), 793–96
Color blindness, 794
Color charge, 950
Color force, 950
Color vision, 793–95
Compact disc (CD), 225, 859
Compass, 622
Complementary colors, 795
Complete circuits, 546
Completely capacitive circuits, 676
Completely inductive circuits, 676
Completely inelastic collisions, 189–90
Completely resistive circuits, 676
Component form, 73
Component method of vector addition, 72–77
  analytical, 74–75
Components of force, 108–9, 116–17
Components of motion, 65–69
  kinematic equations for, 67–68
  Newton's second law on, 108–9
Compound lens systems, 730
Compound microscopes, 783–85
Compound nucleus, 928
Compressibility, 299
Compressional stress, 295, 296
Compressional waves, 435, 436, 437, 453
Compton, Arthur H., 848, 850
Compton effect, 848–50
Compton scattering, 916
Compton wavelength, 849
Computerized tomography (CT), 655, 656, 920
Concave (converging) mirrors, 715, 716, 717, 718, 721–22, 723
Concurrent forces, 256
Condensation(s), 453, 464
Condensation point, 367
Condition for rolling without slipping, 253
Condition for translational equilibrium, 120
Conduction
  electrostatic charging by, 487–88
  heat transfer by, 373–76
Conductivity
  resistivity and, 549–52
  thermal, 373–76
Conductor(s), 485–86
  electric fields and, 502–4
  Gauss's law and, 505
  thermal, 373
Cones, of eye, 775

I-2

Configuration(s), 233
  of charges, 495
  electric potential energy of various, 517–19
Confinement
  inertial, 941
  magnetic, 941
  quark, 950
Conservation
  of angular momentum, 278–80
  of (electric) charge, 484–85, 897
  of energy, 153–60, 240
    conservative and nonconservative forces in, 153–54
    first law of thermodynamics on, 390–95
    in Lenz's law, 636
    Q value and, 929–32
  of linear momentum, 183
  of mass, 929–32
  of momentum, 182–87
    collisions and, 187–94
  of nucleons, 897
  of relativistic momentum and energy, 824–25
  of total mechanical energy, 154–58
Conservative forces, 153–54, 157–58
Conservative systems, 154–55
Constancy of the speed of light, 808
Constant(s)
  Boltzmann's, 339
  decay, 903, 905
  dielectric, 528
  phase, 430
  Planck's, 842
  Rydberg, 851
  spring. *See* Spring (force) constant
  Stefan–Boltzmann, 378
  universal gas, 339
  universal gravitational, 227, 228
Constant acceleration, 44–45, 46–50. *See also* Free fall; Kinematic equations
  components of motion and, 67–68
  Newton's laws of motion and, 111–13
Constant forces
  Newton's laws of motion and, 111–13
  work done by, 138–42
Constant-volume gas thermometer, 340, 341
Constructive interference, 438, 464, 743
Contact, electrostatic charging by, 487–88
Contact forces, 101
Continuity, equations of, 315–16
Continuous spectrum, 850
Contraction, thermal, 335

Control rods, 936
Convection, 376–77
Conventional current, 546
Converging (biconvex) lenses, 724–25, 727–28, 774, 777, 779
Converging (concave) mirrors, 715, 716, 717, 718, 721–22, 723
Conversion factors, 13–17
Conversions, unit, 13–17
Converters, 667, 668
Convex (diverging) mirrors, 715, 716, 717, 722–23
Coolants, 935–36
Cooper, Leon, 553
Coordinates
  Cartesian, 35
  polar, 211
Copiers, electrostatic, 490–91
Copper-bottomed pots, 374
Core, of nuclear reactor, 935
Core material, 610
Cornea, 774
Cost
  of electric power, 558
  of portable (battery) electricity, 560
coulomb (C), 484
Coulomb, Charles A., 484, 489
Coulomb force, 851, 894
Coulomb's law, 489–94, 495
Counter torque, 645
Couple, 256
Crane hoist, 161–62
Critical angle, 697–99
Critical mass, 933–34
Crookes, William, 940
Crossed polarizers, 760
Cross section, of reactions, 932
Crystal (X-ray) diffraction, 756
Crystalline lens, 775
Crystalline solids, 367
CT (computerized tomography), 655, 656, 920
cubic meter (m^3), 8–9
curie (Ci), 906
Curie, Marie, 896, 906
Curie, Pierre, 896, 906
Curie temperature, 610, 621
Current(s), 486
  alternating (ac), 555, 641, 642–43, 665–66
  branch, 579–80
  conventional, 546
  defined, 546
  direct (dc), 546–47, 580–81
  drift velocity and, 547
  eddy, 647–48
  in generators, 640
  induced, 633, 636
  magnetic fields and, 605, 606–8
  magnetic forces and, 610–15
  photo, 843
  rms (effective), 666
Current-carrying
  solenoid, 607–8
  wire, 605–6
  wire loop, 606

Curveball, 320
Curvilinear motion, 64, 67–68
Cutoff frequency, 844
Cygnus X-1, 831

d'Alibard, 501
Damped harmonic motion, 432
Dark matter, 828
Data, implicit, 54–56
Dating, radioactive, 619, 906–9
Daughter nucleus, 897–98, 899, 900
Day–night atmospheric convection cycles, 376
Day–night rearview mirror, 736
Davisson, C. J., 869, 870
dc current, 546–47, 580–81
dc motor, 616–17
D–D reaction, 938
de Broglie, Louis, 866
de Broglie equation, 867
de Broglie hypothesis, 866–70
de Broglie (matter) wave, 866
de Broglie wavelength, 867–70
Decay
  alpha, 897–98, 909
  beta, 898–901, 909, 941–43
  gamma, 900–901
  muon, 814–15, 819
  positron, 899
Decay constant, 903, 905
Decay rate, 903–9
Deceleration, 43
decibel (dB), 460–63
decibel level (β), 461
Decimal system, 7
Declination, magnetic, 622–24
Dedicated grounding, 589–91
Defibrillators, cardiac, 527
Deformation, 101
Degree of freedom, 352
Degrees, and radians, 211–12
Δ (delta), 35
Dendrites, 528
Density
  buoyancy and, 311–14
  determination of, 12–13
  of fluids, 301
  linear turn, 607
  optical, 692
  probability, 871
  surface charge, 503
  temperature and, 347
Deposition, 367
Depth, 301–3
Depth perception, 697
Derived units, 3
Destructive interference, 438, 464, 465, 743, 747, 751
Deuterium, 896
Diamonds, 699–700, 702
Diastole, 308
Diastolic pressure, 308–9
Diatomic gases, 348
  internal energy of, 353
  kinetic theory of, 351–54
  monatomic gases vs., 353–54

Dichroism, 759–62
Dielectric(s), 527–32
Dielectric constant, 528
Dielectric permittivity, 530
Differential transformer, 638
Diffraction, 440, 750–57
  de Broglie wavelength and, 867–68
  defined, 750
  in optical instruments, 789–93
  radio reception and, 752
  single-slit, 751–53
  of sound, 464, 750
  X-ray, 756–57
Diffraction gratings, 753–56
Diffuse (irregular) reflection, 689
Diffusion, 349–51
Dimensional analysis, 9–12
Diopters, 734
Dipoles
  electric, 498–99
  magnetic, 600
Dirac, Paul A. M., 886
Direct current (dc), 546–47, 580–81
  back emf and, 644
  batteries and, 543–45
Direction, 35–36
Dish (parabolic collector), 792
Disintegration energy, 931
Dispersion, 440, 701–2
Displacement, 35–36, 419, 420
  angular, 211
  work and, 138–39
Displacement reference, 144–45
Distance, 33
  acceleration and, 42
  focal, 719
  image, 712, 719, 720
  object, 712
Diverging (biconcave) lenses, 725, 727, 777–78
Diverging (convex) mirrors, 715, 716, 717, 722–23
Division, and significant figures, 18, 19
Doppler, Christian, 467
Doppler blue shift, 471
Doppler effect, 467–73, 654
  for light waves, 470–71
  in sonic booms, 471–73
Doppler radar, 472
Doppler red shift, 471
Double refraction, 759–62
Double-slit experiment, Young's, 742–45, 747
Down quark, 948
Drift velocity, 547
Drinking bird, 402–3
D–T reaction, 938
Dual nature of light, 850
Dynamics, 33

Ear, human, 458–59
Eardrum (tympanum), 458–59
Earth
  circumference of, 5

I-3

gravitational attraction between Moon and, 101, 228
magnetic field of, 601, 621–25
rotational motion of, 282–83
Earthquakes, 432, 436–37, 453
Earth satellites, 237–43
Echograms, 454–55
Echolocation, 450
Eclipse, solar, 826
Eddy currents, 647–48
Effective (rms)
current, 666
voltage, 666
Efficiency, 162–64
Carnot (ideal), 409–10
electrical, 559–60
mechanical, 162–64
relative, 409–10
thermal, 403–4, 405
Effusion, 350
Einstein, Albert, 266, 807, 808, 821, 826, 832, 842, 844, 846, 847, 848, 850, 856, 887, 950
Elastic collisions, 187, 190–94
Elastic limit, 143, 295, 296, 297
Elastic moduli, 295–99
Electrical efficiency, 559–60
Electrical motion detectors, 531–32
Electrical plugs, 589, 591
Electrical resistance, 544, 545, 547–52. See also Resistor(s)
in an ac circuit, 665–68
in a car battery, 548–49
defined, 548
equivalent parallel, 570
equivalent series, 568
factors influencing, 549
impedance without, 671–72
joule heat and, 556–57
temperature and, 551–52
of a wire, 550
Electrical safety, 587–91
Electrical shock, 589
Electrical systems, British vs. American, 667–68
Electrical thermometer, 551–52
Electric charge, 483–85
conservation of, 484–85, 897
quantized, 484, 485
storm clouds and, 499–500
Electric circuits. See Circuit(s)
Electric current, 546. See also Current(s)
Electric dipoles, 498–99
Electric energy transmission, 547
Electric eye, 847
Electric field(s), 494–501
in computer monitor, 510
conductors and, 502–4
equipotential surfaces and, 519–24
Gauss's law on, 504–5
Maxwell's equation on, 650

in one dimension, 496–97
in storm clouds, 499–500
time-varying, 650
in two dimensions, 497–98
Electric force, 489–94
gravitational force vs., 483, 494
per unit charge, 495, 512
Electric heaters, 556–57
Electricity, 482–510, 511–42, 542–65. See also Circuit(s); Current; Electrical resistance; Electric charge; Electric field(s); Electric force; Electrostatic charging
capacitance, 524–27
dielectrics, 527–32
electric potential, 513, 528–29
electric potential difference, 512–16
electric potential energy, 512, 517–19
electric power, 552–60
safety and, 586–91
Electric lines of force (electric field lines), 498
Electric potential, 513, 528–29
Electric potential difference, 512–16
Electric potential energy, 512, 517–19
Electric power, 552–60
Electrodes, 543
Electrolytes, 543
Electromagnetic force, 944
Electromagnetic induction, 632–64. See also Electromagnetic waves (radiation)
in action, 638
back emf, 643–45
as computer equipment hazard, 639
defined, 634
Faraday's law on, 636–40, 641, 645, 670
generators and, 640, 641–45
induced emfs, 633–35
Lenz's law on, 636–40, 647, 670
power transmission and, 648–49
transformers and, 645–49
Electromagnetic inertia, 636
Electromagnetic spectrum, 653
Electromagnetic waves (radiation), 649–57
gamma rays, 655, 896, 897, 901, 916, 917
infrared radiation, 377–78, 654
microwaves, 654
oscillator circuits and, 680–81
power, 652–53
radiation pressure, 651, 652
radio and television, 653–54
types of, 652–55
ultraviolet radiation, 654–55

visible light, 654, 854
X-rays, 655
Electromagnetism, 601–2, 605–8
applications of, 615–20
Electromagnets, 609–10
Electromotive force (emf), 544
back, 643–45
induced, 633–35
motional, 640
Electron(s), 947
Compton wavelength of, 849
current and charge, 546–47
de Broglie wavelength of, 868–70
electric charge of, 483, 484
energy level of, 852–54
Heisenberg uncertainty principle of, 885
magnetic moment in hydrogen atom, 615
pair production and, 886–87
photoelectrons, 843
speed of, 845–46
valence, 485–86, 883
X-rays and, 655
Electron acceleration, 515–16, 823–25
Electron capture, 899–900
Electron clouds, 872
Electron configuration, 880, 882
Electron flow, 547
Electronic balance, 619
Electronic camera flashes, 540
Electron microscope, 872–83
Electron neutrino, 946, 947
Electron period, 880
Electron spin, 608–9, 874–75
electron volt (eV), 523–24
Electroscope, 486, 487
Electrostatic charging, 485–89
by contact or conduction, 487–88
by friction, 487
by induction, 488
by polarization, 488, 489
Electrostatic copiers, 490–91
Electrostatic force, 489–94
Electrostatic potential energy, 518–19
Electrostatics, 483
Electroweak force, 950–51
Elementary particles, 947–48
Elliptical orbits, 235–36
emf (electromotive force), 544
Emission
photo, 844, 856
stimulated, 856–57, 858
Emission spectrum, 850
Emissivity, 378–80
Endoergic (endothermic) reaction, 930
Endoscopy, 703
Energy, 137–38
binding, 853, 911–14
conservation of, 153–60, 240
disintegration, 931
electric, 547
forward thrust and, 240

fusion as a source of, 940–41
internal, 334–35
kinetic. See Kinetic energy
mechanical, 159
potential. See Potential energy
quantum of, 842
relationship to radius, speed, and circular motion, 239
relativistic total, 822–25
rest, 822–25
of a spring–mass system in simple harmonic motion, 420–23
storage in capacitor, 526
threshold, 887, 931–32
total, 159–60
total mechanical, 154–58
in work–energy theorem, 445–49, 247–77
zero-point, 349
Energy levels
of hydrogen atom, 852–55
of multielectron atom, 876–78
Engines, heat. See Heat engines
Engines, internal combustion, 402
Entropy, 397–400, 401
change in as isothermal process, 397–98
decrease in, 400
Envelope, 466
Equation(s)
Bernoulli's, 316–20
of continuity, 315–16
de Broglie, 867
flow rate, 316, 317
kinematic, 46–50, 67–68
len maker's, 733–34
Maxwell's, 649–50, 804, 805
of motion, 424–32
Schrödinger's wave, 870–74, 876
spherical mirror, 718–19, 720, 722
of state, 389
thin-lens, 727
translational and rotational, 275
Equilibrium, 256–61
mechanical, 256
rotational, 256
stable, 261–65
static. See Static equilibrium
thermal, 371
translational, 119–21, 256
unstable, 261, 262
Equipartition theorem, 352
Equipotential surfaces, 519–24
gravitational, 520–21
Equivalence principle, 825–26
Equivalence statements, 13
Equivalent length, 13
Equivalent parallel capacitance, 534
Equivalent parallel resistance, 534, 570
Equivalent series capacitance, 534

I-4

Equivalent series resistance, 568
Eratosthenes, 3
Erecting lens, 786
Escape speed, 237–38, 828–29
Ether, 804–7
Evaporation, 372–73
Event, 811
Event horizon, 829
Exact numbers, 17
Exchange particles, 944–47
Excited state, 852
 lifetime of, 853
Exhaling, 395
Exoergic (exothermic) reaction, 930
Expansion, thermal, 335, 343–47
Expansion gaps, 346
Extension cords, 550
External forces, 185
Extraordinary ray, 759
Eye, human, 774–80
 structure of, 774–75
 vision defects, 775–80
Eyepiece (ocular), 783, 785

Fahrenheit temperature scale, 335–38, 342
False vocal cords, 458
farad (F), 525, 526
Faraday, Michael, 634
Faraday's law of induction, 636–40, 641, 645, 670
Far point, 775
Farsightedness (hyperopia), 777, 779
Fermi, Enrico, 946
Ferromagnetic materials, 609, 610, 621
Feynman, Richard, 944
Feynman diagram, 944, 945
Fiber optics, 699–701, 703
Fibroscopy, 703
Field(s). *See also* Electric field(s); Magnetic field(s)
 force, 495
 vector, 495
Filters, photographic, 796
Fine structure, 874
First law of thermodynamics, 390
Fission reaction, 932–38
 induced, 932
 in reactors, 934–38
 spontaneous, 932
FitzGerald, George F., 807
Fletcher–Munson curves, 476
Flow
 blood, 316
 incompressible, 314–15
 irrotational, 314
 laminar, 322
 nonviscous, 314
 steady, 314
 turbulent, 322
Flow rate equation, 316, 317
Fluid(s), 294, 295, 300–308
 buoyancy and, 308–14
 dynamics, 314–20

ideal, 314–15
surface tension and, 321
viscosity of, 314, 321–24
Fluid volume expansion, 346
Fluorescence, 855
FM (frequency-modulated) radio band, 653, 679
Foam insulation, 377
Focal distance, 719
Focal length, 715, 723, 725
Focal point, 715, 718, 725
Focal rays, 716, 717, 726–27
Foci, 235–36
Following through, 180
Football spiral, 280
foot-pound (ft·lb), 139
Force(s), 99
 action-at-a-distance, 101, 944
 action-reaction, 109, 483
 applied, 150
 attractive, 483, 517
 balanced, 100, 256
 buoyant, 308–14
 centripetal, 222–24
 color, 950
 components of, 108–9, 116–17
 concept of, 100–101
 concurrent, 256
 conservative, 153–54, 157–58
 constant. *See* Constant forces
 contact, 101
 Coulomb, 851, 894
 defined, 100
 electric, 489–94
 electromagnetic, 944
 electromotive (emf), 544
 electroweak, 950–51
 external, 185
 of friction, 121, 122–25
 as a function of position, 143
 fundamental, 944–47
 gravitational, 153–54
 inertia and, 102–4
 instantaneous, 111
 lift, 112
 magnetic, 610–15
 momentum and, 177–78
 net (unbalanced). *See* Net (unbalanced) force
 nonconservative, 153–54, 159–60
 normal, 110, 122, 125–26
 nuclear, 895
 strong nuclear, 944–46
 van der Waals, 321
 variable, 142–45
 weak nuclear, 943, 944, 946
Force constant, 142–45
Forced convection, 377
Force field, 495
Force per unit area (pressure), 300–1
Force per unit charge, 495
Forward thrust, 240
fps system, 7
Frame of reference. *See* Reference frame
Franklin, Ben, 487, 501, 778

Free-body diagrams, 115, 116–19
Free fall, 50–56, 104
Free space, permittivity of, 526
Freezing point, 367, 371, 372
Freon, 405, 654
Frequency(ies), 216–17
 angular, 425
 beat, 466–67
 color of light and, 794
 cutoff, 844
 fundamental, 441, 443, 444, 474, 475, 476
 for musical instruments, 475–76
 of pendulum, 427
 resonance, 677
 resonant (natural), 441–43, 474
 threshold, 846–47
 wave, 420, 425, 434
Fresnel, Augustin, 730
Fresnel lenses, 730
Friction, 121–29, 153
 centripetal force and, 223
 coefficients of, 122–27
 electrostatic charging by, 487
 forces of, 121, 122–25
 kinetic (sliding), 121, 122–25, 126–27
 rolling, 121
 static, 121, 122–25
 total energy and, 159–60
 and walking, 121
Fringe shift, 807
f-stops, 801
Fuel rods, 935
Fundamental forces, 944–47. *See also* Electromagnetic force; Gravitational force; Strong nuclear force; Weak nuclear force
Fundamental frequency, 441, 443, 444, 474, 475, 476
Fuses, 587–90
Fusion, as energy source, 940, 941
Fusion, latent heat of, 368–69
Fusion reaction, 938–41

Galilean telescope, 786
Galileo Galilei, 33, 51, 52, 101–2, 306, 773
Galloping Gertie, 445
Galvanometer, 583–86, 616
Gamma decay, 900–901
Gamma radiation, and poultry, 921
Gamma rays, 655, 896, 897, 901, 916, 917
Gas(es), 294, 295. *See also* Diatomic gases; Fluid(s); Kinetic theory of gases
 monatomic, 348, 349, 351–52, 353–54
 noble (inert), 351–52
 speed of sound in, 455–56
 as thermal conductors, 373
Gaseous (vapor) phase, 367

Gas laws, 338–43
Gas leaks, 358
Gauge pressure, 305–6
gauss (G), 602
Gauss, Karl Friedrich, 504
Gaussian surfaces, 504
Gauss's law, 504–5
Gedanken (thought) experiments, 809, 817, 820
Geiger, Hans, 916
Geiger counter, 916
Gell-Mann, Murray, 948
General theory of relativity, 809, 825–30
 equivalence principle, 825–26
 light and gravity in, 826–30
Generators, 640, 641–45
Geomagnetic poles, 621
Geomagnetism, 621
Geometrical optics, 687–710
 defined, 688
 dispersion, 440, 701–2
 rays, 688
 reflection. *See* Reflection
 refraction. *See* Refraction
 wave fronts, 688
Geometric methods of vector addition, 70–72
Geothermal energy, 402
Germer, L. H., 869, 870
gigaelectron volt (GeV), 524
Gilbert, William, 621
Glare, reducing, 761
Glashow, Sheldon, 950, 951
Glazer, D. A., 917
Global warming, 379
Glottis, 458
Gluons, 950
Goeppert-Mayer, Maria, 914
Golfing, projection at an angle, 86–87
Grand unified theory (GUT), 951
Graphical analysis, 39–41
Gravitational attraction, between Earth and Moon, 228
Gravitational constant, universal, 227, 228
Gravitational equipotential surfaces, 520–21
Gravitational force, 153–54, 944
 in attraction of Earth and Moon, 101, 228
 electric force vs., 483, 494
 gravitons and, 946
Gravitational lensing, 827–28
Gravitational potential energy, 150–52, 231–33, 391
Gravitational red shift, 830
Gravitons, 946
Gravity. *See also* Center of gravity (CG); Circular motion
 acceleration due to, 50–51
 artificial, 242
 in general theory of relativity, 826–30

I-5

Newton's law of, 211, 227–34, 238
specific, 313–14
zero, 242
Gravity assists (swingbys), 234–35
gray (Gy), 918
Great Wall of China, view from space, 791–93
Greenhouse effect, 378, 379, 654
Grounded plugs, 589–91
Ground fault circuit interrupter (GFCI), 638
Grounding, 589–90
Ground state, 852
Ground (neutral) wires, 587
Group, of elements, 882
Gyroscope, 280–82

Hadrons, 947, 948, 949
Hair cells (ear), 459
Hale Observatories, 788–89
Half-life, 903–9
declining source strength, 906
defined, 903
and medicine, 905
radioactive dating, 906–9
Hammer (malleus), 459
Hard iron, 610
Harmonic series, 441, 443, 476
Harrier "jump" jet, 112
Hearing, 458–59
threshold of, 460, 475–76
Heart, human, as pump, 308
Heat, 360–87
body, 381
joule, 554–59, 587
latent, 367–72
mechanical equivalent of, 362
phase changes and, 367–73
specific, 363–67, 370
temperature distinguished from, 334–35
units of, 361–62
Heat capacity, 365
Heat engines, 400–405
cyclic, 402, 408
defined, 406
ideal, 408–11
Heaters, electric, 556–57
Heat pumps, 407, 408
Heat reservoirs, 389
Heat transfer, 373–82
by conduction, 373–76
by convection, 376–77
by radiation, 377–81
Heisenberg, Werner, 884
Heisenberg uncertainty principle, 883–86
Helicopters, 283
Helium–neon (He–Ne) gas laser, 857–58
henry (H), 670
Henry, Joseph, 634
Herapath, W., 760
Herapathite, 760
hertz (Hz), 216, 420

High-potential wires, 586–87, 588
Hoist, crane, 161–62
Hologram, 858
Holography, 858–59, 860
Hooke, Robert, 143, 297
Hooke's law, 143, 297, 419, 425
Horizontal projections, 83–85
horsepower (hp), 161
Hot-air balloons, 333
Household circuits, 587–91
Hubble Space Telescope, 789, 828
Human body temperature, 337
Human ear, 458–59
Human eye, 774–80
Human heart, as pump, 308
Human voice, 458
Humidity, 457
Hurricanes, 280
Huygens, Christian, 234, 690
Huygens' principle, 690
Hydraulic brakes, 303
Hydraulic lift, 304–5
Hydroelectric power, 633
Hydrogen atom, 874–75
Bohr theory of, 850–55, 866, 867, 871
electric potential difference and, 516–17
magnetic moment of electron, 615
Hydrogen (H) bomb, 940
Hydrogen isotope, 896
Hyperopia (farsightedness), 777, 779

Ice, 371
Ice pail experiment, 503–4
Ice point, 336
Ice skating, 372
Ideal fluid, 314–15
Ideal (perfect) gas law, 338–40, 389–90
absolute temperature and, 341, 342–43
isobaric process for, 391
macroscopic form of, 339–40
Ideal heat engines, 408–11
Image(s)
holographic, 858–60
real, 711, 716, 718
virtual, 711, 716, 718, 727
Image distance, 712, 719, 720
Image side of lenses, 725–26, 727
Impedance, 672–77
without resistance, 671–72
Implicit data, 54–56
Impulse, 178–82
Impulse–momentum theorem, 178–79, 181
Incidence,
angle of, 690–91, 697–98
plane of, 689
Inclined plane, 118–19, 271–72
Incompressible flow, 314–15
Incus (anvil), 459
Index of refraction, 691, 692–93, 701

Induced currents, 633, 636, 639
Induced emfs, 633–35
Induced fission, 932
Inductance, 670
Induction. *See* Electromagnetic induction
Inductive circuits, 674
Inductive reactance, 670–72
Inductor(s), 670
Inelastic collisions, 187–90
Inertia
electromagnetic, 636
force and, 102–4
moment of, 265–68, 281
in Newton's first law of motion, 101–2, 103
pulley, 270–71
rotational, 266, 267, 283
Inertial confinement, 941
Inertial reference frame, 803–4, 808–9
relativity of simultaneity and, 809–12
Infrared radiation, 377–78, 654
Infrasonic region, 453
Infrasonic waves (infrasound), 453
Inner transition elements, 882–83
Input (primary) coil, 645
Instantaneous
acceleration, 42, 111
angular speed, 214
angular velocity, 214
axis of rotation, 253
force, 111
speed, 34
velocity, 37, 39
Insulation, 374–76
Insulator(s), 485–86
foam, 377
thermal, 373
Intensity, of sound, 458–63, 475–76
Interference, 437–39
constructive, 438, 743
destructive, 438, 464, 465, 743, 747, 751
sound, 464–67
thin-film, 745–50
total constructive, 438, 464
total destructive, 439
Young's double-slit experiment, 742–43
Interference maxima, 754
Interference minima, 751
Interferometer, 805–7, 815
Internal combustion engines, 402
Internal energy, 334–35
of diatomic gases, 353
of monatomic gases, 349
in thermodynamics, 390–95
Internal motion, 198
Internal reflection, and fiber optics, 700
Internal resistance, 544
International System of Units (SI), 2–6

Intravenous (IV) blood transfusions, 324
Inverse-square law, 227, 228
Inversion, population, 857
Iodine-123, 919
Ion(s), 484, 618, 619
Ion channels, 528
Ionization, 487
Iris, 775
Irregular (diffuse) reflection, 689
Irreversible process, 390
Irrotational flow, 314
Isentropic process, 399
Isobaric expansion, 394–95
Isobaric process, 391–92
Isochoric process, 392
Isolated (closed) systems, 153, 183, 188, 399
Isomet, 392
Isometric process, 392
Isotherm(s), 393, 396, 408–9
Isothermal process, 393, 397–98
Isotope(s), 895–96. *See also* Half-life
Isotropic expansion, 344
Isotropic material, 759
I^2R **losses,** 554

Jansky, Carl, 792
Jet propulsion, 111, 200–2
joule (J), 139, 361, 362, 524
Joule, James, 362
Joule heat, 554–59, 587
battery testing and, 555
electrical resistance and, 556–57
loss, 544
Jumper, 572–73
Junctions (nodes), electrical, 576
Junction theorem, 576, 583

Karate chops, 300
K-capture, 900
Keck telescope, 789
kelvin (K), 6, 341
Kelvin, Lord, 341, 404, 408
Kelvin temperature scale, 339, 340–43
Kepler, Johannes, 228, 234–35
Kepler's laws of planetary motion, 228, 234–37
first law (law of orbits), 235–36
second law (law of areas), 236, 280
third law (law of periods), 236–37
Kidneys, 351
kilocalorie (kcal), 361
kiloelectron volt (keV), 900
kilogram (kg), 5
kilowatt (kW), 161
kilowatt-hour (kWh), 557
Kinematic equations, 46–50, 67–68. *See also* Constant acceleration; Free fall
Kinematics, 32–63. *See also* Acceleration; Displacement;

Distance; Free fall;
Speed; Velocity
rotational, 226–27
Kinetic energy
in beta decay, 943
collisions and, 187–94
defined, 146
gravitational potential energy and, 151
momentum vs., 174–75, 179, 181–82
Q value and, 929–32
relativistic, 821–22
rotational work and, 274–77
translational, 275–76, 334, 348, 352
work–energy theorem and, 145–49, 274–77
Kinetic (sliding) friction, 121
coefficient of, 122–25, 126–27
Kinetic theory of gases, 348–54, 372
absolute temperature in, 348–49
diffusion in, 349–51
equipartition theorem in, 352
Kirchhoff plots, 578–79
Kirchhoff's rules, 575–80
application of, 578–79
first rule (junction theorem), 576, 583
second rule (loop theorem), 576–77, 578–79, 585, 672, 673
K shell, 900

Laminar flow, 322
Land, of compact disc, 859
Land, Edwin H., 760, 795
Lanthanide series, 883
Lapse rate, atmospheric, 359
Large Electron–Positron (LEP) Collider, 929
Larynx (voice box), 458
Laser(s), 839, 855–60
for correction of nearsightedness, 778
in eye surgery, 778, 858
helium–neon, 857–58
in holography, 858–59, 860
Laser printers, 490–91
Latent heat, 367–72
of fusion, 368–69
of sublimation, 367, 368
of vaporization, 368–69, 372
Lateral magnification, 711
Law
of areas (Kepler's second law), 236, 280
of charges (charge–force law), 483
of conservation of linear momentum, 183
of conservation of mechanical energy, 155
of conservation of total energy, 153
of inertia. See Newton's laws of motion, first

of orbits (Kepler's first law), 235–36
of periods (Kepler's third law), 236–37
of poles (pole–force law), 600
of reflection, 688
LCDs (liquid crystal displays), 367, 764–65
Lead-206 dating, 908
Leaning Tower of Pisa, 51, 52, 264
LEDs (light-emitting diodes), 765
Length
arc, 211–13
equivalent, 13
focal, 715, 723, 725
proper, 816, 817
SI unit, 4
Young's modulus and, 296–97
Length contraction, 815–18, 822
Lens(es), 710, 724–34
aberrations in, 732–33
blocking off half, 729
combinations of, 730–72
converging (biconvex), 724–25, 727–28, 774, 777, 779
crystalline, 775
diopters, 734
diverging (biconcave), 725, 727, 777–78
erecting, 786
Fresnel, 730
image side, 725–26, 727
meniscus, 725
nonreflecting, 749
object side, 725–26
oil immersion, 793
optical flats, 748–50
transition point, 728–29
Lens maker's equation, 733–34
Lens power, 734
Lenz's law, 636–40, 647, 670
LEP (Large Electron–Positron) Collider, 929
Leptons, 947–48, 949
Lever (moment) arm, 254
Life, and second law of thermodynamics, 401
Lifetime, of excited state, 853
Lift, airplane, 319
Lift, hydraulic, 304–5
Lift force, 112
Light
atmospheric scattering of, 762–67
determination of color, 794
dual nature of, 850
in general theory of relativity, 826–30
monochromatic, 701
quantization of, 842–47
relativistic velocity addition for, 833
speed of. See Speed, of light
visible, 654, 854

white, 701
Light bulbs, 548, 554–55
Light-emitting diodes (LEDs), 765
Light microscopes, 872
Lightning, 500–501, 527
Lightning rods, 500–501
Light meters, photographic, 847
Light polarization. See Polarization
Light rays. See Ray(s)
Light waves, 470–71. See also Physical optics
light-year (ly), 820
Linear expansion, 344
Linear momentum, 173. See also Momentum
Linear motion, 39–41
Linear turn density, 607
Liquid(s), 294, 295. See also Fluid(s); Thermal expansion
speed of sound in, 455–56
as thermal conductors, 373
volume stress and bulk modulus and, 299
Liquid crystal(s), 367
Liquid crystal displays (LCDs), 367, 764–65
Liquid drop model, 932–33
Liquid-in-glass thermometers, 335, 338
Liquid phase, 367
liter (L), 9
LOCA (loss-of-coolant accident), 937–38
Longitudinal (compressional) waves, 435, 436, 437, 453
Long jump event, 90
Loop theorem, 576–77, 578–79, 585, 672, 673
Lorentz, H. A., 807
Lorentz–FitzGerald contraction, 807, 815
Loudness, 460, 475
Loudness contours, 475
Lyman series, 853
Lymphocytes (white blood cells), 873

Mach, Ernst, 472
Mach number, 472, 473
Magic numbers, 914, 915
Magnetically levitated trains, 599
Magnetic confinement, 941
Magnetic declination, 622–24
Magnetic dipoles, 600
Magnetic domains, 609
Magnetic equator, 621
Magnetic field(s), 634. See also Electromagnetism
current-carrying solenoid and, 606–7
current-carrying wires and, 605, 606, 607–8
defining direction, 601
of Earth, 601, 621–25
Maxwell's equation on, 650

strength of, 601–4
time-varying, 637, 650
Magnetic flux, 602, 634–35
Magnetic forces, 610–15
Magnetic materials, 608–10
Magnetic moment, 614, 615
spin, 876
Magnetic moment vector, 614
Magnetic monopoles, 600
Magnetic permeability, 609–10
Magnetic poles, 600
Magnetic quantum number, 874
Magnetic resonance imaging (MRI), 876–77
Magnetic vectors, 602
Magnetism, 599–632. See also Electromagnetism; Magnetic fields
magnetic materials, 608–10
magnetic permeability, 609–10
magnetic poles, 600
pole–force law (law of poles), 600
right-hand force rule, 603, 604, 611–12
right-hand source rule, 605, 606, 607
Magnetotactic bacteria, 622–23
Magnets
electro, 609–10
permanent, 610
superconducting, 610
Magnification
angular, 781–83
lateral, 712
Magnification factor, 719–20
Magnifying glass, 780–83
Magnitude, 35–36
Magnitude-angle form, 72
Magnus, Gustav, 320
Magnus effect, 320
Malleus (hammer), 459
Malus, E. L., 761
Malus's law, 761
Manhattan Project, 936
Manometers, 305, 306
Maps, topographic, 520
Mars, 767
Mass
center of, 194–200
conservation of, 929–32
critical, 933–34
as fundamental property, 104
inertia and, 102
momentum and, 173–74
Newton's second law on, 104, 106
SI unit, 4–5
speed vs., 148–49
weight vs., 5, 104
Mass defect, 912
Mass distribution, 267
Mass–energy equivalence, 822–25, 911
Mass number, 895
Mass spectrometer, 618–20
Matter
and antimatter, 840

dark, 828, 829
phases of, 294
Matter (de Broglie) waves, 866–70
Maxwell, James Clerk, 649, 805
Maxwell's equations, 649–50, 804, 805
Measured numbers, 17
Measurement, 1–20. *See also* British unit(s); Metric system; SI unit(s)
 dimensional analysis, 9–12
 of pressure, 305–7
 significant figures, 17–20
 standard units, 2
 unit analysis, 11–14
 units of quantities, 12–13
Mechanical
 efficiency, 162–64
 energy, 159
 equilibrium, 256
 equivalent of heat, 362
 load, 644
 work, 138, 139–40
Mechanics, 32–33. *See also* Dynamics; Kinematics
Medical ultrasound, 454
Medicine, nuclear, 918
megaelectron volt (MeV), 524, 900
Melanin, 656
Melanoma, 656
Meltdown, 938
Melting, surface, 372
Melting point, 367
Mendeleev, Dmitri, 880
Meniscus lenses, 725
Mercury, as superconductor, 553
Meson(s), 944–46, 948, 949
Metals, 373
Metastable state, 856
meter (m), 4
Meter of the Archives, 4
Method of mixtures, 366–67
Metric system, 2, 3, 6–9. *See also* SI units
metric ton, 9
meV (megaelectron volt), 524, 900
Michelson, Albert A., 805
Michelson–Morley experiment, 805–7, 809, 815
microampere (μA), 546
microcoulomb (μC), 484
microfarad (μF), 524
Microscopes, 780–85
 compound, 783–85
 electron, 872–73
 magnifying glass, 780–83
 scanning electron, 873
 scanning tunneling, 865, 874
 transmission electron, 872–73
Microwave(s), 654
Microwave ovens, 226–27, 380
Mile, number of feet in, 8
milliampere (mA), 546
millihenry (mH), 670
milliliter (mL), 9

Mirage, 696
Mirror images, 713
Mirrors, 710–24
 parabolic, 723–24, 788–89
 plane, 711–14
 spherical. *See* Spherical mirrors
Mixed units, 12
Mixtures, method of, 366–67
mks system, 6
Moderator, 936
mole (mol), 6, 339–40
Molecular speed, 348–49
Moment (lever) arm, 254
Moment of inertia, 265–68, 281
Momentum, 172–209. *See also* Angular momentum; Collision(s); Conservation, of momentum
 change, and projectiles, 178
 defined, 173
 force and, 177–78
 impulse and, 178–82
 jet propulsion and, 200–202
 kinetic energy vs., 174–75, 179, 181–82
 relativistic, 822, 824–25
 total, 173, 176–77
Momentum selector, 619
Monatomic gases, 348, 349, 351–52, 353–54
Monochromatic light, 701
Montreal Protocol, 657
Moon
 angular momentum of, 282, 283
 centripetal acceleration and, 221
 gravitational attraction between Earth and, 101, 228
 lack of air resistance, 51
 mass of, 5
 weight on, 104
Morley, E. W., 807
Motion, 33, 64–98. *See also* Circular motion; Components of motion; Friction; Kinematics; Newton's laws of motion; Projectile motion; Rotational motion; Simple harmonic motion; Translational motion
 curvilinear, 64, 67–68
 damped harmonic, 432
 defined, 33
 equations of, 424–32
 force and, 100
 internal, 198
 nonuniform linear, 39–41
 periodic, 419
 relative velocity, 78–83
 uniform linear, 39
Motional emf, 640
Motion detectors, 531–32
Motor oils, and viscosity, 323
Motors
 back emf of, 643–45
 dc, 616–17

MRI (magnetic resonance imaging), 876–77
Multielectron atoms, 876–78
Multiloop circuits, 575–80
Multimeters, 586
Multiplication, and significant figures, 18, 19
Multiplier resistors, 585, 586
Muon(s), 945, 947
Muon decay, 814–15, 819
Muon neutrino, 946, 947
Muscles, and torque, 255–56
Musical instruments, 473–76
Mutual inductance, 634, 670
Myopia (nearsightedness), 776–78

National Institute of Standards and Technology (NIST), 6, 613
Natural (line) broadening, 886
Natural convection, 376–77
Natural (resonant) frequencies, 441–43, 474
Nautical mile, 8
Near point, 775, 777
Nearsightedness (myopia), 776–78
Negative lens, 727
Neon lights, 664
Neon-tube relaxation oscillator (blinker circuit), 582–83
Nerve signal transmission, 528–29
Net charge, 484
Net (unbalanced) force, 100–101, 256
 inertia and, 102–4
 weight and, 104–5
Net (total) work, 140–42
Neurons, 528
Neutral atoms, 911
Neutrino(s), 899, 943, 947
 anti, 943
 beta decay and, 941–43
 electron, 946, 947
 muon, 946, 947
 tau, 947
Neutron(s), 483, 894, 899, 909–10
Neutron activation analysis, 920–21
Neutron capture reactions, 932
Neutron number, 895
newton (N), 103, 104
Newton, Isaac, 33, 99, 100, 173, 701, 750
Newtonian relativity. *See* Classical (Newtonian) relativity
newton-meter (N·m), 361
Newton's law of gravitation, 211, 227–34, 238
Newton's laws of motion, 101–21
 first, 101–2, 103, 182, 256
 second, 102–9, 111, 146
  center of mass and, 195
  centripetal force and, 222, 223–24

  in component form, 108–9
  free-body diagrams and, 116–19
  momentum in, 177–78, 182–83
  rotational form of, 266, 268
  weight in, 104–5
 third, 109–11, 112, 183, 200
Newton's rings, 748, 749
Noble (inert) gases, 351–52
Nodes
 electrical (junctions), 576
 in standing waves, 441
Noise reduction, 465
Noise, white, 445
Nonconservative forces, 153–54, 159–60
Nonconservative systems, 155
Nondispersive waves, 439–40
Noninertial reference frame, 803–4
Nonisobaric process, 392
Nonmetals, 373
Nonreflecting lenses, 749
Nonreflective coatings, 747–48, 749
Nonuniform linear motion, 39–41
Nonviscous flow, 314
Nonvisual radiation, and telescopes, 792
Normal force, 110, 122, 125–26
Northern lights, 624
Nuclear chain reactions, 937
Nuclear force, 895
Nuclear medicine, 918
Nuclear notation, 895–96
Nuclear reactions, 927–56
 cross sections in, 932
 elementary particles in, 947–48
 endoergic (endothermic), 930
 exoergic (exothermic), 930
 fission, 932–38
 fundamental forces and exchange particles in, 944–47
 fusion. *See* Fusion reaction
 quark model, 948–50
Nuclear reactors, 934–38
 breeder, 936–37
 power, 934–36
 safety measures, 937–38
Nuclear stability, 909–15
Nuclear structure, 893–96
Nucleon(s), 894, 897
Nucleon populations, 909
Nucleus, 892–926. *See also* Decay; Half-life; Radioactivity
 compound, 928
 daughter, 897–98, 899, 900
 force, 895
 notation, 895–96
 stability, 909–15
 structure, 893–96
Nuclide(s), 896

Object distance (d), 712
Objective, 783, 785

Objective coil, 872
Object side, of lenses, 725–26
Ocean, as heat engine, 416
Ocean tides, 249
Ocular (eyepiece), 783, 785
Oersted, Hans Christian, 605
ohm (Ω), 548, 672
Ohm, Georg, 548
Ohm's law, 548, 643, 672
Oil immersion lens, 793
Onnes, Heike Kamerlingh, 553
Open circuits, 587
Open-tube manometer, 305
Optic(s). See Fiber optics; Geometrical optics; Physical optics
Optical activity, 762
Optical density, 692
Optical flats, 748–50
Optical instruments, 773–801. See also Eye, human; Microscopes; Telescopes
 diffraction and resolution in, 789–93
Orbital(s), 878
Orbital motion, 219–20, 235–36
 gravitational potential energy and, 230–32
Orbital quantum number, 874, 875
Order-of-magnitude calculations, 175–76
Ordinary ray, 759
Organ, pipe, 474–75
Oscillation(s), 419, 425–28
 equations of motion for, 426–27
 in a parabolic potential wall, 422
Oscillator circuits, 680–81
Osmosis, 351
Otto, Nickolas, 404
Otto cycle, 404
Output (secondary) coil, 645
Oval window, of ear, 459
Overheating the human body, 372
Ozone layer depletion, 654, 656–57

Pain, threshold of, 460, 47
Paint, and pigment mixing, 795
Pair annihilation, 887
Pairing effect, 909–10
Pair production, threshold energy for, 887
Parabola, 86
Parabolic arcs, 86
Parabolic collector (dish), 792
Parabolic mirrors, 723–24, 788–801
Parabolic potential well, 422
Parallel, arrangement of capacitors, 532–33, 534–35
Parallel axis theorem, 268
Parallelogram method, 71
Parallel-plate capacitor, 524–26

Parallel plates, 512, 525–26
 for determining charge on storm clouds, 499–500
 electric potential difference and, 513, 514–15
 equipotential surfaces and, 519–20
Parallel rays, 716, 717, 726–27
Partially polarized light, 757
Particle(s)
 alpha, 893–94, 896, 897, 901, 916
 antiparticles and, 886–88
 beta, 896, 897, 901, 916
 elementary, 947–48
 exchange, 944–47
 quantum, 848–50
 size of in scattering, 765–66
 system of, 194–96
 virtual, 944
 W, 946, 951
Particle accelerators, 928, 929
pascal (Pa), 301, 306–7
Pascal, Blaise, 303
Pascal's principle, 303–5
Paschen series, 853
Path independence, 152
Path-length difference, 464–65
Pauli, Wolfgang, 878, 943
Pauli exclusion principle, 878–80
PDLC (polymer dispersed liquid crystal), 765
Peak current, 665
Peak-to-peak value, 667
Peak voltage, 665
Pendulum, 427
Penetration, of nuclear radiations, 901
Perfect gas law, 339
Performance, coefficient of, 407
Perihelion, 232, 236
Period(s), 216–17, 419–20, 425, 434
 electron, 880
 of elements, 880
 of pendulum, 427
Periodic motion, 419
Periodic table of elements, 880–83
Periodic waves, 433, 434
Permanent magnets, 610
Permeability, magnetic, 609–10
Permittivity of free space, 526
Perpetual motion machines, 397, 400
PET, 920
Phase angle, 674, 675
Phase changes, 367–73
 latent heat and, 367–72
 specific heat and, 370
Phase constant, 430
Phase diagrams, 672
Phase differences, 464–65
Phasors, 672
Phosphorescent materials, 856
Photocells, 843, 847
Photocurrents, 843
Photoelectric effect, 842–47

 characteristics of, 844
 electron speed and, 845–46
 threshold frequency and wavelength, 846–47
Photoelectrons, 843
Photoemission, 844, 856
Photographic
 filters, 796
 light meters, 847
Photomultiplier tube, 916
Photon(s), 842–47, 944
 Compton effect and, 848–49
 defined, 843
 electromagnetic force and, 944
 lasers and, 856–57
Photonuclear effect, 914, 915
Photosensitive materials, 843
Physical (wave) optics, 741–72
 atmospheric scattering of light, 762–67
 diffraction. See Diffraction
 Newton's rings, 749, 750
 optical flats, 748–50
 polarization. See Polarization
 thin-film interference, 745–50
 Young's double-slit experiment, 742–45, 747
Physiological diffusion, 351
picocoulomb (pC), 484
picofarad (pF), 524
Pigments, 795
Pion(s), 945
Pipe organ, 474–75
Pit, of compact disc, 859
Pitch, 467, 475
Planck, Max, 842, 844
Planck's constant, 842
Planck's hypothesis, 840–42
Plane, inclined, 118–19, 271–72
Plane mirrors, 711–14
Plane of incidence, 689
Plane (linearly) polarized light, 757
Planetary motion, Kepler's laws of, 228, 234–37
Plane wave fronts, 688
Plaque, and cholesterol, 316
Plasma, 940–41
Plimsoll mark, 328
Plugs, electrical, 589, 591
Plum pudding model, 893
Point charges, 516
poise (P), 322
poiseuille (Pl), 322
Poiseuille, Jean, 322, 323
Poiseuille's law, 323–24
Polar coordinates, 211
Polarization, 757–62
 by double refraction, 759–62
 electrostatic charge separation by, 488, 489
 by reflection, 757–59
 by refraction, 757
Polarization direction, 760
Polarized plugs, 591
Polarizers, crossed, 760
Polarizing (Brewster) angle, 757–58
Polaroid, 760

Pole–force law (law of poles), 600
Poles, magnetic, 600
Polygon method, 71
Polymer dispersed liquid crystal (PDLC), 765
Polymer foam insulation, 377
Population inversion, 857
Positive lens, 727
Positron(s), 886–87
Positron decay, 899
Positron emission tomography (PET), 920
Positronium atom, 887
Potential difference, electric, 512–516
Potential energy, 145, 149–52
 conservative forces and, 153, 154
 electric, 512, 517–19
 electrostatic, 518–19
 gravitational, 150–52, 231–33, 391
 of a spring–mass system in simple harmonic motion, 420–23
 zero reference point and, 151–52
Potential energy barrier, 873
Potentiometers, 598
Power, 160–62
 ac circuits and, 665–66
 defined, 160
 electric, 552–60
 horse, 161
 resolving, 791
 rotational, 274–77
Power factor, 675–77
Power rating, 587
Power reactors, 934–36
Power supply, 545
Power transmission, 648–49
Power waves, 652–53
Precession, 282
Presbyopia, 777
Pressure, 300–308
 absolute, 305–6
 atmospheric, 306–7
 defined, 300
 depth and, 301–3
 gauge, 305–6
 measurement of, 305–7
 Pascal's principle and, 303–5
 temperature vs., 341
Pressure cookers, 371
Primary (input) coil, 645
Primary colors, 795
Principal quantum number, 851
Principle of classical (Newtonian) relativity, 804–5
Principle of equivalence, 825
Principle of relativity, 808
Principle of superposition, 437
Printers, laser, 490–91
Prism
 diffraction grating vs., 754–55
 dispersion by, 701–2
 internal reflection in, 698, 699

Prism binoculars, 786
Probability density, 871
Problem solving, 21–25
Problem-solving hints and strategies
   on body heat, 381
   on Celsius–Fahrenheit conversions, 337
   on components of motion, 67
   on "correct" answer, 20
   on Coulomb's law in vector addition, 493
   on de Broglie wavelength, 869
   on Doppler effect, 470
   on free-body diagrams, 116–17
   on kinematic equations, 47–48
   on latent heat, 370
   on order-of-magnitude calculations, 175–76
   on photon calculations, 847
   on Q values, 931
   on ratios for working with gravity, 230
   on relative velocity, 80
   on resistors in parallel, 570–71
   on spherical mirror equation, 722
   on spring constant, 144–45
   on static equilibrium, 261
   on time dilation, 815
   on trigonometric functions, 213
   on vertical projectile motion, 56
   on work–energy theorem, 148, 149
Process, 390
   irreversible, 390
   reversible, 390
Projectile motion, 83–90
   at arbitrary angles, 85–90
   horizontal, 83–85
   and momentum change, 178
   vertical, 54–56
Projectile range, 86, 89–90
Projector coil, 872
Propagation, of waves, 419
Proper length, 816, 817
Proper time, 813
Proportional limit, 296
Propulsion, jet, 111, 200–2
Proton(s), 894
   electric charge of, 483, 484
   pairing effect, 909–10
   potential differences and, 514–15, 516–17
Proton number, 895
Proton–proton cycle, 939
Pulleys, and inertia, 270–71
Pulse, of human body, 308
Pulse, wave, 419
Pump(s)
   heat, 407, 408
   human heart as, 308
Pumping, laser, 857

Pupil, of eyes, 775
p-V diagram, 390
P (primary) waves, 436, 437

Quality, 476
Quality factor (QF), 918
Quantities, units of, 12–13
Quantization
   of light, 842–47
   Planck's hypothesis and, 840–42
Quantized charge, 484, 485
Quantum, 842
Quantum chromodynamics (QCD), 950
Quantum electrodynamics (QED), 944, 950
Quantum mechanics, 865–91
   atomic quantum numbers. See Atomic quantum numbers
   de Broglie hypothesis, 866–70
   Heisenberg uncertainty principle, 883–86
   particles and antiparticles, 886–88
   Schrödinger's wave equation, 870–74, 876
Quantum number, 851
Quantum particles, 848–50
Quantum physics, 839–65
   Bohr theory, 850–55, 866, 867, 871
   Compton effect, 848–50
   lasers, 855–60
   Planck's hypothesis, 840–42
Quark confinement, 950
Quark model, 949
Quarks, 484, 948–50
Quarter-wavelength antennas, 663
Quasars, 827–28
Quinine sulfide periodide (herapathite), 760
Q value, 929–32

rad, 918
Radar, 471, 472
   Doppler, 472
Radial (chief) ray, 716, 717, 726
radian (rad), 211–13, 918
Radiation
   biological effects and medical application, 917–20
   blackbody, 380, 840–41, 842
   damage, 901
   detection of, 916–17
   domestic and industrial applications, 920–21
   dosage, 918
   electromagnetic. See Electromagnetic waves (radiation)
   hazards, 917
   heat transfer by, 377–81
   nonvisible, for telescopes, 792
   thermal, 840
   ultraviolet, 654–55

Radiation detectors, 916–17
Radiation dosage, 918
Radiation penetration, 901
Radiation pressure, 651, 652
Radioactive dating, 619, 906–9
Radioactive half-lives, and medicine, 905
Radioactive tracers, 919–20
Radioactive waste, 938
Radioactivity, 896–902
   alpha decay, 897–98, 909
   beta decay, 898–901, 909, 941–43
   defined, 896
   gamma decay, 900–901
Radioisotopes, 919
Radio reception, 679, 752
Radio telescopes, 792
Radio waves, 653–54
Radius of curvature, 715, 725
Radon, 902
Rainbows, 702, 704
Range, projectile, 86, 89–90
Rapid-transit rail cars, and braking, 648
Rarefactions, 435, 453, 464
Ray(s), 688
   chief, 716, 717, 726
   extraordinary, 759
   focal, 716, 717, 726–27
   gamma, 655, 896, 897, 901, 916, 917
   ordinary, 759
   parallel, 716, 717, 726–27
Ray diagrams, 716–24
Rayleigh, Lord, 763, 790
Rayleigh criterion, 790–91
Rayleigh scattering, 763, 764–65
RC circuits, 580–83, 584, 672
Reactance
   capacitive, 668–70
   inductive, 670–72
Reaction time, 53–54
Reactors, nuclear. See Nuclear reactors
Real image, 711, 716, 718
Recoil, of rifle, 200
Rectangular vector components, 72–73
Red shift
   Doppler, 471
   gravitational, 830
Reference circle, 425
Reference frame
   absolute, 804–5, 808
   inertial, 803–4, 808–9
Reflecting telescopes, 788–89
Reflection, 688–90
   angle of, 688
   irregular (diffuse), 689
   law of, 688
   polarization by, 757–59
   regular (specular), 689
   sound, 463–64
   total internal, 697–701
   wave, 439
Reflection gratings, 753
Refracting telescopes, 785–88
Refraction, 690–97

   angle of, 690–91, 693, 697
   defined, 690
   index of, 691, 692–93, 701
   marching analogy, 692
   polarization by, 757
   polarization by double, 759–62
   sound, 463–64
   wave, 439–40
Refrigeration, 405–7
Regular (specular) reflection, 689
Relative biological effectiveness (RBE), 918
Relative efficiency, 410
Relative permeability, 610
Relative velocity, 78–83, 803
   in one dimension, 78–80
   in two dimensions, 81–83
Relativisitc kinetic energy, 821–22
Relativistic momentum, 822
   conservation of, 824–25
Relativistic total energy, 822–25
Relativistic velocity addition, 830–33
Relativity, 802–39
   classical (Newtonian), 803–5
   general theory of, 809, 825–30
   length contraction, 815–18
   Michelson–Morley experiment, 805–7, 809, 815
   relativistic kinetic energy, 821–22
   relativistic momentum, 822, 824–25
   relativisitc total energy, 822–25
   relativistic velocity addition, 830–33
   rest energy, 822–25
   of simultaneity, 809–12
   space travel and, 820–21
   special theory of, 807, 808–9
   time dilation, 812–15, 818, 819–20, 821–22
rem, 918
Repose, angle of, 127
Representative elements, 882
Repulsive forces, 483, 486, 493–94, 517
Resistance. See Air resistance; Electrical resistance
Resistivity
   conductivity and, 549–52
   temperature coefficient of, 551–52
Resistor(s), 567–575
   changing a capacitor through, 581
   discharging a capacitor through, 582–83
   multiplier, 585, 586
   in parallel, 569–73
   RLC circuits and, 675–76
   in series, 567–69
   in series–parallel combinations, 571–72, 573–75
   shunt, 583, 584–85

Resolution, 789–93
Resolving power, 791
Resonance, of circuits
  circuit, 677–81
  applications of, 678–79
  physical explanation of, 677–78
Resonance(s), of nuclear reaction, 932
Resonance, of waves, 444–45
Resonance absorption, 378
Resonance frequency, 677
Resonant (natural) frequencies, 441–43, 474
Rest energy, 822–25
Resting membrane potential, 529
Resultant, 70
Retina, 775
Retrorockets, 240
Return stroke, of lightning, 501
Reverse (braking) thrust, 201–2
Reversible process, 390
revolutions per minute (rpm), 214
Richter scale, 478
Right-hand force rule, 603, 604
  for current-carrying wire, 611–612
Right-hand rule, 214, 215, 255
Right-hand source rule, 605, 606, 607
Rigid body
  angular momentum of, 278–79
  defined, 252
  in equilibrium, 256–57, 261, 262
  moment of inertia and, 265–68
  motions of, 252–54
RLC circuits, 674–75
  power factor for, 675–77
  resonance in, 677–81
RL circuits, 673–74, 678
rms (effective) current, 666
rms (effective) voltage, 666
rms speed, 348
Rockets, 200–202
Rods, of eye, 775
roentgen (R), 918
Roentgen, Wilhelm, 655
Rolling, and components of motion, 66–67
Rolling friction, 121
Rolling motion, 253–54, 271–73
Root-mean-square (rms)
  current, 666
  speed, 348
  velocity, 666
Rotation, instantaneous axis of, 253
Rotational dynamics, 265–74
  applications of, 268–74
  moment of inertia and, 265–68

parallel axis theorem and, 268
Rotational equilibrium, 256
Rotational inertia, 266, 267, 283
Rotational kinematics, 226–27
Rotational kinetic energy, 274
Rotational motion, 251–93. *See also* Rotational dynamics
  angular momentum and, 277–83
  equations for, 275
  of a rigid body, 252–54
  translational motion vs., 252–54
Rotational power, 274–77
Rotational static equilibrium, 258–59
Rotational work, 274–77
Rounding, 18–19, 20
rpm (revolutions per minute), 214
Rubbia, Carlo, 951
Rumford, Count (Benjamin Thompson), 362
Russell traction, 286
Rutherford, Ernest, 893, 928, 930
Rutherford, Lord, 916
Rutherford–Bohr model of the atom, 893
R-values, 386
Rydberg constant, 851

Safety
  airplane, and Doppler radar, 472
  electrical, 587–91
  and impulse, 180–81
  nuclear reactor, 937
Sailing, and vector components, 77–78
Salam, Abdus, 950, 951
San Andreas fault, 436
Satellites
  centripetal acceleration in, 219–20
  Earth, 237–43
Scalar quantities, 33–35, 139. *See also* Distance; Speed
Scanning electron microscope (SEM), 873
Scanning tunneling microscope (STM), 865, 874
Scattering
  of atmospheric light, 762–67
  back-, 893–94
  Compton, 916
  Rayleigh, 763, 764–65
  X-ray, 849
Schrieffer, Robert, 553
Schrödinger, Erwin, 870
Schrödinger's wave equation, 870–74, 876
Schwarzschild, Karl, 829
Schwarzschild radius, 829–30, 831
Scientific method, 52
Scintillation counter, 916–17
Scott, David, 51
second (s), 5, 6

Secondary (output) coil, 645
Second law of thermodynamics, 396
Sedimentation, 220
Seismic waves, 436–37
Seismology, 436–37
Selectors
  momentum, 619
  velocity, 618
Self-induction, 647, 670
Semiconductor (solid state) detector, 917
Semiconductors, 486, 552
Series, arrangement of capacitors, 532–36
Series–parallel, arrangement of capacitors, 535–36
Series RC circuits, 672–73
Series RLC circuits, 674–76
Series RL circuits, 673–74
Shape, and shear modulus, 298
Shear angle, 298
Shear modulus, 296, 298
Shear strain, 298
Shear stress, 298
Shear waves. *See* Transverse (shear) waves
Shell(s), electron, 876–78, 879, 882, 883, 914
Shock (bow) waves, 471, 472
Short circuits, 589–90
Short-wave bands, 653
Shunt, 572–73
Shunt resistor, 583, 584–85
sievert (Sv), 918
Significant figures, 17–20
Simple harmonic motion (SHM), 419–24
  energy and speed of spring–mass system in, 420–23
  equations of motion for, 424–32
  initial conditions and phase, 429–30
  velocity and acceleration in, 430–31
Simple microscope (magnifying glass), 780–83
Simultaneity, relativity of, 809–12
Single-photon emission tomography (SPET), 920
Single-slit diffraction, 751–53
Sinusoidal functions, 428, 433
SI unit(s), 2–6. *See also* Metric system
  base units, 3, 6
  British units compared with, 10
  derived units, 3
  length, 4
  mass, 4–5
  time, 5, 6
  volume, 8–9
Sky, color of, 763–66
Skydiving, 128, 129, 160
Sliding friction, 121–27

Sliding motion, 276–77
Slipping, 253–54
slug, 6
Smoke detector, 920
Smudge pots, and heat transfer, 381
Snell's law, 691, 693, 697, 701
Snell, Willebord, 691
Soft iron, 609–10
Solar clock, 5
Solar collectors, 387, 401
Solar eclipse, 826
Solar energy, 847
Solar flares, 624–25
Solar panels, 381–82
Solenoids, 606–7, 608
Solid(s), 294–99
  crystalline, 367
  elastic moduli and, 295–99
  speed of sound in, 455–56
  thermal expansion of, 335, 343–47
Solid phase, 367
Solid state (semiconductor) detector, 917
Sonar, 454
Sonic booms, 471–73
Sound, 452–81
  diffraction of, 464, 750
  Doppler effect. *See* Doppler effect
  interference in, 464–67
  of musical instruments, 473–76
  phenomena, 463–67
  reflection in, 463–64
  refraction in, 463–64
  speed of, 455–57
Sound frequency spectrum, 453
Sound intensity, 458–63, 475–76
Sound intensity level, 460–63
Sound waves, 435, 453–55
Space colonies, 242
Space diagrams, 116
Space travel, 820–21
Spark chamber, 917
Spark plugs, 661
Special theory of relativity, 807, 808–9
Specific gravity, 313–14
Specific heat, 363–67, 370
Spectrometers, 755
Spectroscopy, 754
Spectrum
  absorption, 850
  continuous, 850
  electromagnetic, 653
  emission, 850
  sound frequency, 453
Specular (regular) reflection, 689
Speech, 458
Speed, 33–35
  acceleration and, 42
  angular, 214–17
  average, 33–35
  conservation of energy and, 156–57

of electrons, 845–46
escape, 237–38, 828–29
instantaneous, 34
kinetic energy and, 149
of light, 4, 651–52
    ether and, 804–7
    refraction and, 691–93, 701
    relativistic velocity addition vs., 832–33
    in special relativity, 808–9
mass vs., 148–49
molecular, 348–49
relationship to radius, energy, and circular motion, 239
root-mean-square (rms), 348
of sound, 455–57
of a spring-mass system in simple harmonic motion, 420–23
tangential, 215–16, 238
warp, 818
wave, 434
SPET, 920
Spherical aberration, 723–24, 732
Spherical mirror(s), 711–12, 714–24
  concave (converging), 715, 716, 717, 718, 721–22, 723
  convex (diverging), 715, 716, 717, 722–23
  ray diagrams and, 716–24
  spherical aberration in, 723–24
Spherical mirror equation, 718–19, 720, 722
Sphygmomanometer, 309
Spin magnetic moment, 876
Spin quantum number, 874–75
Split-ring commutator, 616–17
Spontaneous fission, 932
Spring (force) constant, 142–45
  conservative forces and, 157–58
  determining, 144
  experimental determination, 423–24
  simple harmonic motion and, 419, 423–24, 425–26
Spring force, 142–44, 153, 157–58
Spring–mass system, 420–23
Stability, 266. *See also* Equilibrium
Stable equilibrium, 261–65
stadium (unit), 3
Standard atmosphere, 306–7
Standard temperature and pressure (STP), 340
Standard units, 2
Standing waves, 440–45, 866–67
Stapes (stirrup), 459
Starlight, aberration of, 807
State, equations of, 389
State of a system, 389–90
State variables, 389

Static charges, 490
Static equilibrium, 256, 257
  rotational, 258–59
  translational, 257–58
  without translation or rotation, 259–60
Static friction, 121, 122–25
Static translational equilibrium, 119, 120–21
Statute mile, 8
Steady flow, 314
Steam engines, 402
Steam point, 336
Stefan–Boltzmann constant, 378
Stefan's law, 378
Step-down transformers, 646–47
Stepped leader, and lightning, 501
Step-up transformers, 646–47
Stimulated emission, 856–57, 858
Stirrup (stapes), 459
Stopping distance of a vehicle, 49–50
Stopping potential, 844, 845–46
Storm clouds, and electric charge, 499–500
STP (standard temperature and pressure), 340
Strain, 295–96
  shear, 298
  tensile, 295
  volume, 298
Strange quarks, 948
Streamer, lightning, 501
Streamlines, 314
Stress, 295–96
  compressional, 295, 296
  on femur, 297–98
  shear, 298
  tensile, 295, 296
  thermal expansion and, 345–46
  volume, 298, 299
Stringed instruments, 442, 473
Strong nuclear force, 895, 944–46
Studsensor, 584
Sublimation, latent heat of, 367, 368
Subshells, 876–78, 880, 882–83
Subtend, 211
Subtraction
  with significant figures, 19
  vector, 71–72
Subtractive method of color production, 795–96
Subtractive primary pigments, 795
Sun
  as a blackbody, 841
  flattened and anti-, 698
  fusion power of, 939–40
  gravitational potential energy and, 232–33
Sunlight, 758–59
Sunrises, 766–67

Sunsets, 766–67
Superconducting magnets, 610
Superconductivity, 552, 553
Superforce, 951–52
Supernova, 946
Superposition, 437–39, 495–98
Surface charge density, 503
Surface melting, 372
Surface tension, 321
Surface waves, 436
S (secondary) waves, 436, 437
Swingbys (gravity assists), 234–235
Symmetry reversal, 713
Synaptic terminals, 528
System(s)
  conservative, 154–55
  defined, 153
  isolated (closed), 153, 183, 188, 399
  nonconservative, 155
  of particles, 194–96
  thermally isolated, 389
  thermodynamic, 389–90
  of units, 2
Systole, 308
Systolic pressure, 308–9

Tangential acceleration, 225–26
Tangential speed, 215–16, 238
Tangential velocity, 215
Tape recorders, 638
Tau neutrino, 947
Tauon(s), 947
Telescopes, 785–89
  astronomical, 785–86, 787
  constructing, 787–88
  Galilean, 786
  Hubble Space, 789, 828
  Keck, 789
  nonvisible radiation used for, 792
  radio, 792
  reflecting, 788–89
  refracting, 785–88
  resolution and, 790–91
  terrestrial, 786
Temperature, 333–59
  absolute. *See* Absolute temperature
  Celsius scale, 335–38, 363
  Curie, 610, 621
  density and, 347
  doubling, 343
  electrical resistance and, 551–52
  Fahrenheit scale, 335–38, 342
  heat distinguished from, 334–35
  human body, 337
  Kelvin scale, 339, 340–43
  specific heat and, 363
  speed of sound and, 455
Temperature coefficient of resistivity, 551–52
Tensile strain, 295
Tensile strength, 120–21

Tensile stress, 295, 296
Tension, 106–7
  surface, 321
Terminal velocity, 128–29
Terminal voltage, 544
Terrestrial telescopes, 786
tesla (T), 602
Tesla, Nikola, 602
Test charge, 495
Thermal coefficients
  of area expansion, 344, 345
  of linear expansion, 344
  of volume expansion, 344, 345
Thermal conductivity, 373–76
Thermal conductors, 373
Thermal contact, 335
Thermal contraction, 335
Thermal cycles, 395, 402
Thermal efficiency, 403–4, 405
Thermal equilibrium, 371
Thermal expansion, 335, 343–47
Thermal gradient, 373
Thermal insulators, 373
Thermally isolated systems, 389
Thermal pumps, 405–8
Thermal radiation, 840
Thermionic emission, 893
Thermodynamics, 388–418
  first law of, 390–95
  of heat engines. *See* Heat engines
  processes, 390
  second law of, 395–97, 399, 401, 404, 408. *See also* Entropy
  systems, 389–90
  third law of, 410–11
Thermograms, 378
Thermography, 378
Thermometer(s), 335
  constant-volume gas, 340, 341
  electrical, 551–52
  liquid-in-glass, 335, 338
Thermonuclear fusion reactions, 940
Thermos bottle, 385
Thermostats, 335
Thin-film interference, 745–50
Thin-lens equation, 727
Third law of thermodynamics, 410
Thompson, Benjamin (Count Rumford), 362
Thomson, G. P., 870
Thomson, J. J., 893
Thomson, William (Lord Kelvin), 341
Thought (gedanken) experiments, 809, 817, 820
Three finger rule, 603
Three Mile Island (TMI) reactor, 927, 938
Three-wire system, 587
Threshold energy, 931–32
  for pair production, 887
Threshold frequency, 846–47

Threshold of hearing, 460, 475–76
Threshold of pain, 460, 475
Time
 proper, 813
 reaction, 53–54
 SI unit, 5–6
 work and, 162
Time constant, 581, 582
Time dilation, 812–15, 818, 821–22
 twin paradox, 819–20
Time-varying magnetic field, 637, 650
Tire gauges, 305
Tip-to-tail (triangle) method, 70
TMI (Three Mile Island) reactor, 927, 938
Tokamak, 941
ton, metric, 9
Tone, 453, 476
Topographic maps, 520
Top (truth) quarks, 949
Tornadoes, 280
Torque, 254–56
 acceleration and, 271–73
 angular momentum and, 278–80, 283
 counter, 645
 on a current-carrying loop, 613–15
 defined, 254
 door opening and, 269–70
 moment of inertia and, 265–67
 yo-yo roll and, 273–74
torr, 307
Torricelli, Evangelista, 306, 307
Total binding energy, 912, 914
Total constructive interference, 438, 464
Total destructive interference, 439
Total energy, 159–60
Total internal reflection, 697–701
Total linear momentum, 173
Total mechanical energy, 154–58
Total momentum, 173, 176–77
Total (net) work, 140–42
Tracers, radioactive, 919–20
Transducers, capacitors as, 531–32
 differential, 638
Transformers, 645–49
 differential, 638
Transition elements, 882
Translational equations, 275
Translational equilibrium, 119–21, 256
 condition for, 120
 static, 119, 120–21, 257–58
Translational kinetic energy, 275–76, 334, 348, 352
Translational motion
 equations for, 275
 of a rigid body, 252–54

static equilibrium and, 259–60
translational static equilibrium and, 257–58
Translational static equilibrium, 119, 120–21, 257–58
Transmission axis (polarization direction), 760
Transmission electron microscope (TEM), 872–83
Transmission gratings, 753
Transmutation, 897
Transverse (shear) waves, 434–35, 436
Triangle, finding length of one side, 23–24
Triangle (tip-to-tail) method, 70–71
Triple point of water, 342
Tritium, 896
True vocal cords, 458
Tuning
 a radio, 445, 678
 stringed instruments, 443–44
Tuning forks, 467
Tunneling, 865, 873–74, 898
Turbines, 643
Turbulent flow, 322
TV waves, 653–54
Twin (clock) paradox, 819–20
Twisted-nematic display, 764
Tympanum (eardrum), 458–59
Tyndall, John, 700
Type-S fuse, 588

UHF (ultrahigh frequency) band, 653
Ultrasonic region, 453
Ultrasonic waves (ultrasound), 453–55
Ultrasound, medical, 454
Ultraviolet catastrophe, 841–42
Ultraviolet radiation, 654–55
Unbalanced force. *See* Net (unbalanced) force
Uncertainty principle, Heisenberg, 883–86
Unification theories, 950–52
Uniform circular motion, 217–24
 centripetal acceleration and, 217–22
 centripetal force and, 222–24
 simple harmonic motion and, 424
Uniform motion, 39
Unit(s)
 of heat, 361–62
 of quantities, 12–13
 system of, 2
Unit analysis, 11–14
Unit conversions, 13–17
Unit vectors, 73
Universal gas constant, 339
Universal gravitational constant (big *G*), 227, 228
Universal law of gravitation, 211, 227–34, 238

Universe, evolution of, 951–52
Unpolarized light, 757
Unstable equilibrium, 261, 262
Up quarks, 948
Uranium, 897–98, 934
Uranium-238, 901, 902, 908

Valence electrons, 485–86, 883
Van Allen belts, 624
Van der Meer, Simon, 951
Van der Waals forces, 321
Vapor (gaseous) phase, 367
Vaporization, latent heat of, 368–69, 372
Variable forces, 142–45
Variables, state, 389
Vector addition, 69–78
 component method, 72–77
 Coulomb's law on, 491–93
 geometric method, 70–72
Vector components, 68–69, 77–78
 rectangular, 72–73
Vector field, 495
Vector quantities, 33, 35–41. *See also* Displacement; Velocity
Vector subtraction, 71–72
Vector sum(s), 183
 total momentum as, 176–77
Velocity, 36–39
 angular, 214–17
 average, 37–39
 drift, 547
 graphical analysis of, 39–41
 instantaneous, 37, 39
 momentum and, 173–74
 relative, 78–83, 803
 in simple harmonic motion, 430–31
 tangential, 215
 terminal, 128–29
Velocity addition, relativistic, 830–33
Velocity selector, 618
Venturi tunnel, 329
Vertical projectile motion, 54–56
Very Large Array (VLA), 792
Vibration(s), 441. *See also* Oscillation(s)
Virtual image, 711, 716, 718, 727
Virtual particle, 944
Viscosity, 314, 321–24
 coefficient of, 322
Visible light, 654, 854
Visible spectrum, 701–2
Vision
 color, 793–95
 defects of, 775–80
Visual acuity, 780
Vitreous humor, 774
Viviani, Vincenzo, 52
VLA (Very Large Array), 792
Vocal cords, 458
Voice, human, 458
volt (V), 513, 552
 electron, 523–24
 gigaelectron, 524

 kiloelectron, 900
 megaelectron, 524, 900
Volta, Allesandro, 543
Voltage, 513, 543
 ac, 666–68
 peak, 665
 rms (effective), 666
 terminal, 544
Voltmeters, 583, 585–86
Volume
 bulk modulus and, 298–99
 SI unit, 8–9
Volume expansion, 344
Volume strain, 298
Volume stress, 298, 299
Von Laue, Max, 756

Walking, friction and, 121
Waste, radioactive, 938
Water
 electrostatic potential energy of molecule, 518–19
 ice point, 336
 steam point, 336
 thermal equilibrium, 371
 triple point, 342
watt (W), 161, 552, 554
Watt, James, 161
Wave(s), 418–51
 body, 436
 characteristics of, 433–34
 defined, 433
 diffraction. *See* Diffraction
 dispersion, 440
 electromagnetic. *See* Electromagnetic waves (radiation)
 infrasonic (infrasound), 453
 light, 470–71
 longitudinal (compressional), 435, 436, 437, 453
 matter (de Broglie), 866–70
 nondispersive, 439–40
 P (primary), 436, 437
 periodic, 433, 434
 properties of, 435–40
 reflection and, 439
 refraction and, 439–40
 S (secondary), 436, 437
 seismic, 436–37
 shock (bow), 471
 simple harmonic motion. *See* Simple harmonic motion (SHM)
 sound, 435, 453–55
 standing, 440–45, 866–67
 superposition and, 437–39
 surface, 436
 transverse (shear), 434–35, 436
 types of, 434–35
 ultrasonic (ultrasound), 453–55
 water, 435, 750
Wave fronts, 688
Wave function, 870
Wave interference. *See* Interference
Wavelength
 color of light and, 794

I-13

Compton, 849
de Broglie, 867–70
defined, 434
photoelectric effect and, 846–47
refraction and, 695–96
Young's double-slit experiment, 742–45
Wave motion, 432–35
Wave nature of light, 843
Wave optics. *See* Physical optics
Wave propagation, 419
Wave pulse, 419, 433
Waves, water, 435, 750
Wave speed, 434
Weak charge, 951
Weak nuclear force, 943, 944, 946
Weather, and Doppler radar, 472

Weight
buoyant force and, 310–11
mass vs., 5, 104
in Newton's second law of motion, 104–5
Weightlessness, apparent, 241–43
Weinberg, Steven, 950, 951
Wheatstone bridge, 594
White blood cells (lymphocytes), 873
White light, 701
White noise, 445
Wien's displacement law, 841
Wilson, C. T. R., 917
Wind, flying into, 82–83
Wind instruments, 475
Window pane, as mirror, 735
Wires
electrical resistance of, 550
ground (neutral), 587

high-potential, 586–87, 588
magnetic field in current-carrying, 605, 606, 607–8
magnetic forces on current-carrying, 610–15
solenoid vs., 608
Work, 137–49
by a constant force, 138–42
mechanical, 138, 139–40
power and, 160
rotational, 274–77
in thermodynamics, 390–91
time and, 162
total (net), 140–42
by a variable force, 142–45
Work–energy theorem, 145–49, 274–77
impulse–momentum theorem and, 179, 181
Work function, 845
W particle, 946, 951

Xerography, 490–91
X-ray(s), 655
X-ray diffraction, 756–57
X-ray scattering, 849

Yerkes Observatory, 788
Young, Thomas, 742
Young's double-slit experiment, 742–45, 747
Young's modulus, 296–97, 455
Yo-yo, and rolling, 273
Yukawa, Hideki, 944–45

Zeppelins, 329
Zero, absolute, 340–43, 349
Zero gravity, 241
Zero-point energy, 349
Zero reference point, 151–52
Zweig, George, 948

Mathematical Symbols

$=$	is equal to
$\neq$	is not equal to
$\approx$	is approximately equal to
$\sim$	about
$\propto$	is proportional to
$>$	is greater than
$\geq$	is greater than or equal to
$\gg$	is much greater than
$<$	is less than
$\leq$	is less than or equal to
$\ll$	is much less than
$\pm$	plus or minus
$\mp$	minus or plus
$\bar{x}$	average value of x
Δx	change in x
$|x|$	absolute value of x
Σ	sum of
∞	infinity

The Greek Alphabet

Alpha	A	α	Nu	N	ν
Beta	B	β	Xi	Ξ	ξ
Gamma	Γ	γ	Omicron	O	o
Delta	Δ	δ	Pi	Π	π
Epsilon	E	ε	Rho	P	ρ
Zeta	Z	ζ	Sigma	Σ	σ
Eta	H	η	Tau	T	τ
Theta	Θ	θ	Upsilon	Y	υ
Iota	I	ι	Phi	Φ	ϕ
Kappa	K	κ	Chi	X	χ
Lambda	Λ	λ	Psi	Ψ	ψ
Mu	M	μ	Omega	Ω	ω

Quadratic Formula

If $ax^2 + bx + c = 0$, then

$$x = \frac{-b \pm \sqrt{b^2 - 4ac}}{2a}$$

Values of Some Useful Numbers

$\pi = 3.141\,59\ldots$ $\sqrt{2} = 1.414\,21$

$e = 2.718\,28\ldots$ $\sqrt{3} = 1.732\,05$

Trigonometric Relationships

Definitions of Trigonometric Functions

$$\sin \theta = \frac{y}{r} \qquad \cos \theta = \frac{x}{r} \qquad \tan \theta = \frac{\sin \theta}{\cos \theta} = \frac{y}{x}$$

$\theta°$ (rad)	$\sin \theta$	$\cos \theta$	$\tan \theta$
0° (0)	0	1	0
30° ($\pi/6$)	0.500	$\sqrt{3}/2 \approx 0.866$	$\sqrt{3}/3 \approx 0.577$
45° ($\pi/4$)	$\sqrt{2}/2 \approx 0.707$	$\sqrt{2}/2 \approx 0.707$	1.00
60° ($\pi/3$)	$\sqrt{3}/2 \approx 0.866$	0.500	$\sqrt{3} \approx 1.73$
90° ($\pi/2$)	1	0	∞

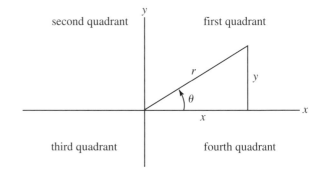

Physical Data

Quantity	Symbol	Approximate Value
Universal gravitational constant	G	6.67×10^{-11} N·m²/kg²
Acceleration due to gravity (generally accepted value on surface of Earth)	g	9.80 m/s² = 980 cm/s² = 32.2 ft/s²
Speed of light	c	3.00×10^8 m/s = 3.00×10^{10} cm/s = 1.86×10^5 mi/s
Boltzmann's constant	k_B	1.38×10^{-23} J/K
Avogadro's number	N_A	6.02×10^{23} mol^{-1}
Gas constant	$R = N_A k$	8.31 J/(mol·K) = 1.99 cal/(mol·K)
Coulomb's law constant	$k = 1/4\pi\epsilon_o$	9.00×10^9 N·m²/C²
Electron charge	e	1.60×10^{-19} C
Permittivity of free space	ϵ_o	8.85×10^{-12} C²/(N·m²)
Permeability of free space	μ_o	$4\pi \times 10^7$ T·m/A = 1.26×10^{-6} T·m/A
Atomic mass unit	u	1.66×10^{-27} kg $\leftrightarrow$ 931 MeV
Planck's constant	h	6.63×10^{-34} J·s
	$\hbar = h/2\pi$	1.05×10^{-34} J·s
Electron mass	m_e	9.11×10^{-31} kg = 5.49×10^{-4} u $\leftrightarrow$ 0.511 MeV
Proton mass	m_p	1.67265×10^{-27} kg = 1.007 267 u $\leftrightarrow$ 938.28 MeV
Neutron mass	m_n	1.67500×10^{-27} kg $\times$ 1.008 665 u $\leftrightarrow$ 939.57 MeV
Bohr radius of hydrogen atom	r_1	0.053 nm

Solar System Data*

Equatorial radius of Earth	6.378×10^3 km = 3963 mi
Polar radius of Earth	6.357×10^3 km = 3950 mi
Mass of Earth	5.98×10^{24} kg
Diameter of Moon	3500 km $\approx$ 2160 mi
Mass of Moon	7.4×10^{22} kg $\approx \frac{1}{81}$ mass of Earth
Average distance of Moon from Earth	3.8×10^5 km = 2.4×10^5 mi
Diameter of Sun	1.4×10^6 km $\approx$ 864 000 mi
Mass of Sun	2.0×10^{30} kg
Average distance of Earth from Sun	1.5×10^8 km = 93×10^6 mi

*See Appendix III for additional planetary data.